DATE DUE

			PRINTED IN U.S.A.

NOBEL LECTURES IN CHEMISTRY
1981–1990

NOBEL LECTURES

Including Presentation Speeches
And Laureates' Biographies

PHYSICS

CHEMISTRY

PHYSIOLOGY OR MEDICINE

LITERATURE

PEACE

ECONOMIC SCIENCES

NOBEL LECTURES
INCLUDING PRESENTATION SPEECHES
AND LAUREATES' BIOGRAPHIES

CHEMISTRY

1981–1990

EDITOR-IN-CHARGE

TORE FRÄNGSMYR
Uppsala University, Sweden

EDITOR

BO G MALMSTRÖM
Department of Biochemistry and Biophysics
Chalmers University of Technology and
Goteborg University
Sweden

World Scientific
Singapore • New Jersey • London • Hong Kong

lge, NJ 07661

UK office: 57 Shelton Street, Covent Garden, London WC2H 9HE

First published 1992
Reprinted 1997, 1999

NOBEL LECTURES IN CHEMISTRY (1981–1990)

ISBN 981-02-0788-3
ISBN 981-02-0789-1 (pbk)

Printed in Singapore.

FOREWORD

Since 1901 the Nobel Foundation has published annually "Les Prix Nobel" with reports from the Nobel Award Ceremonies in Stockholm and Oslo as well as the biographies and Nobel lectures of the laureates. In order to make the lectures available to people with special interests in the different prize fields the Foundation gave Elsevier Publishing Company the right to publish in English the lectures for 1901–1970, which were published in 1964–1972 through the following volumes:

Physics 1901–1970	4 volumes
Chemistry 1901–1970	4 volumes
Physiology or Medicine 1901–1970	4 volumes
Literature 1901–1967	1 volume
Peace 1901–1970	3 volumes

Elsevier decided later not to continue the Nobel project. It is therefore with great satisfaction that the Nobel Foundation has given World Scientific Publishing Company the right to bring the series up to date.

The Nobel Foundation is very pleased that the intellectual and spiritual message to the world laid down in the laureates' lectures will, thanks to the efforts of World Scientific, reach new readers all over the world.

Lars Gyllensten
Chairman of the Board

Stig Ramel
Executive Director

Stockholm, June 1991

PREFACE

This volume reproduces the Nobel Lectures in Chemistry for the period 1981–1990, as originally published in "Les Prix Nobel". These can be expected to continue to be a source of education and inspiration for chemists as well as for other persons interested in chemistry. In the lectures some of the foremost chemical investigators of this century describe their breakthrough work. They also present short autobiographical sketches before their lectures, and these add a human touch to the volume.

The Nobel Committee for Chemistry, of which I was a member (1972–1988) and its chairman (1977–1988), takes a broad view of the field. It follows closely the development of research in classical areas such as inorganic, organic and physical chemistry, and also in structural and theoretical chemistry as well as in biochemistry. Some chemists in the more classical branches have occasionally criticized the inclusion of biochemistry, but already in 1907 the prize was awarded to Eduard Buchner "for his biochemical researches and his discovery of cell-free fermentation".

The decade covered by this book illustrates beautifully that revolutionary advances are being made in many of the chemical disciplines. The award to Henry Taube in 1983 and to Elias Corey in 1990 were made for contributions in hard inorganic and organic chemistry, respectively. The prize to Donald Cram, Jean-Marie Lehn and Charles Pedersen in 1987, on the other hand, can be said to belong to both of these areas. Bruce Merrifield, chemistry laureate in 1984, has made contributions on the border between organic chemistry and biochemistry. Physical chemists, Dudley Herschbach, Yuan Lee and John Polanyi, were selected by the committee in 1986 and theoretical chemists, Kenichi Fukui and Roald Hoffmann, in 1981. In 1985 the prize went to Herbert Hauptman and Jerome Karle for methodological developments in structural chemistry, but two awards in this field border on biochemistry. Thus, Aaron Klug got the prize in 1982 for his structural characterization of nucleic acid-protein complexes, and in 1988 Johann Deisenhofer, Hartmut Michel and Robert Huber were awarded for their determination of the structure of a photosynthetic reaction center, which also represented the first structure of an integral membrane protein. Finally, true biochemical discoveries, although with an implication for chemical catalysis, was the basis of the prize to Sidney Altman and Thomas Cech in 1989.

The text of the lectures is taken directly from the earlier publication in "Les Prix Nobel", but some errors have been corrected in consultation with the laureates. In addition, the laureates have been given the opportunity to update their biographical sketches, but not all of them have availed themselves of this possibility.

Bo G Malmström
Editor

CONTENTS

Chemistry 1981

KENICHI FUKUI and ROALD HOFFMANN

*for their theories, developed independently, concerning the course of chemical
reactions*

THE NOBEL PRIZE FOR CHEMISTRY

Speech by Professor INGA FISCHER-HJALMARS of the Royal Academy of Sciences.
Translation from the Swedish text

Your Majesties, Your Royal Highnesses, Ladies and Gentlemen,
The laureates in chemistry of this year have studied the theory of chemical reactions. Chemical reactions is something that fills our daily life. All of us are constantly starting chemical reactions, by turning the key of the car or by cooking at the stove, to say nothing about the endless number of reactions in our own body that follows every breath.

In chemical reactions new compounds are created. It is possible to make designs for their preparation. But these designs are not realiable until the events at the micro level have been understood and the laws have been found that are governing the transformations of the molecules.

A molecule is composed of atoms that are tied together by aid of the electrons. Atomic nuclei and electrons are not at rest but are constantly moving. The paths of the electrons are usually called orbitals. The forms of these orbitals are determining the bonds between the atoms.

In a reaction molecules are impinging against each other. During the collision the electrons are influenced by new atomic nuclei and the orbitals are changed. Some of the bonds are broken and others are created. Afterwards, new molecules have been formed.

What is it then that decides the sequence of events during the collision? One governing factor is the energy. The new molecules are found at a lower energy level than the original ones. Sometimes, the change will take place without difficulty. The reaction complex is simply sliding down an energy slope. But in general some hindrance must be overcome. It is necessary first to go upwards before starting the downhill ride. Then, the problem is to find the lowest passage over the height barrier. Frequently, rather much is known about the starting material and the final product, about the energy valley of the starting point and the valley of the final destination. But about the character of the ridge in between very little has been known. This year's laureates in chemistry have helped us to foresee the obstacles and tb find the best way to the final goal.

The barriers depend on the fact that the electronic orbitals must be transformed. Now, there is a large number of electrons in every molecule, each with its orbital. A drastic simplification of the barrier problem was achieved in the beginning of the 1950's when Kenichi Fukui discovered that only a few orbitals, those with highest energy, are dominating the frontier of the reaction. He therefore called them frontier orbitals. By use of his theory Fukui found the laws for many groups of organic-chemical reactions. An example: Naphthalene is an important initial material in dye-stuff industry, among others. For a long time the puzzling fact had been known that hydrogen atoms at different positions in the naphthalene molecule are reacting most unequally. The explanation came first through Fukui's theory.

Many molecules have no stereo-symmetry. Then, the molecule and its mirror image have very different effects, as the right-hand and left-hand are functioning differently. The finger-tips of a right-hand can easily match those of a left-hand, but not those of another right-hand. Usually, chemical reactions will give rise to a mixture of right-and left-molecules, but our bodies are only producing one kind. For the preparation of vitamins and drugs that are to react in the body it is therefore often necessary to use methods giving only right- or only left-molecules. Such methods have been found through an exquisite combination of theory and experiment. The problem to prepare vitamin B_{12} was attacked among others by the brilliant molecule builder Woodward at Harvard. There was also the theoretician Roald Hoffmann. Together, Hoffmann and Woodward discovered that not only the energy of the orbital but also its symmetry is decisive for the reactivity. Thus, the Woodward−Eschenmoser synthesis of vitamin B_{12} could be accomplished.

Hoffmann continued to develop the theory of orbital symmetry to an exceedingly practical instrument for synthetic work of widely different character. At the same time, Fukui showed that the frontier orbital theory leads to another powerful method of solving the intricate problems of stereochemistry. In this way, the theoreticians Fukui and Hoffmann have radically changed the conditions for the design of chemical experiments.

Professor Fukui, Professor Hoffmann. Each of you have independently developed important theories of chemical reactivity. The concepts of frontier orbitals and conservation of orbital symmetry have revealed completely new aspects of the interaction between molecules in collision. Through drastic simplifications you have been able to make beautiful generalizations. From your theoretical work new tools have emerged of the greatest importance for the design of chemical experiments. In recognition of your outstanding work the Royal Academy of Sciences has decided to award this year's Nobel Prize for Chemistry to you.

On behalf of the Royal Academy of Sciences I wish to convey to you our warmest congratulations, and now I ask you to receive your Prizes from the hands of His Majesty the King.

KENICHI FUKUI

I was born the eldest of three sons of Ryokichi Fukui, a foreign trade merchant and factory manager, and Chie Fukui, in Nara, Japan, on October 4, 1918. In my high school years, chemistry was not my favourite subject, but the most decisive occurrence in my educational career came when my father asked the advice of Professor Gen-itsu Kita of Kyoto Imperial University concerning the course I should take. Prof. Kita suggested that Ryokichi, one of his juniors from the same native province, should send me to the Department of Industrial Chemistry with which he was then affiliated.

For a few years after my graduation from Kyoto Imperial University in 1941, I was engaged in experimental research on synthetic fuel chemistry in the Army Fuel Laboratory. The result brought me a prize in 1944. I became lecturer in the Fuel Chemistry Department of Kyoto Imperial University in 1943, assistant professor in 1945, and professor in 1951. In 1947 I married Tomoe Horie. I have two children, Tetsuya (son) and Miyako (daughter).

While I started originally as an experimentalist, I had built up a subgroup of theoreticians in my group before 1956. My work on experimental organic chemistry continued along with this, and the results were mostly published in Japanese papers, the number of which amounted to 137 during the period 1944—1972, together with my papers on reaction engineering and catalytic engineering.

But the nature of my main work in chemistry can be better represented by more than 280 English publications, of which roughly 200 concern the theory of chemical reactions and related subjects. Other English papers relate to statistical theory of gellation, organic synthesis by inorganic salts, and polymerization kinetics and catalysts.

My first scientific delight came in 1952 when I found a correlation between the frontier electron density and the chemical reactivity in aromatic hydrocarbons. This success led my theoretical group to the chemical reactivity theory, extending more and more widely the range of compound and reactions that were discussed.

The year in which my 1952 paper was published was the same as that of Professor Mulliken's publication of the important paper on the charge-transfer force in donor-acceptor complexes. Influenced by this paper, I gave a theoretical foundation for the findings mentioned above. The basic idea was essentially the consideration of the importance of the electron delocalization between the frontier orbitals of reactant species. The frontier orbital approach was further developed in various directions by my own group and many other scientists, both theoretical and experimental.

I was also interested in formulating the path of chemical reactions. The first paper appeared in 1970. This simple idea served to provide information on the geometrical shape of reacting molecules, and I was able to make the role of the frontier orbitals in chemical reactions more distinct through visualization, by drawing their diagrams.

I must confess that, when I was writing the 1952 paper, I never imagined I would be coming to Stockholm to receive the Nobel Prize 30 years later. But I have to add that already at that time Professor Gen-itsu Kita encouraged me by suggesting the possibility of the growth of my theory leading me one day to this supreme prize. The possibility became a reality through the good circumstances in which I found myself: with my teachers, my colleagues and students, and, of course, my parents and family.

Some other details which perhaps should be mentioned are:

The Japan Academy Medal, May 1962.

National Science Foundation Senior Foreign Scientist Fellow, Feb.−July, 1970.

Member, International Academy of Quantum Molecular Science (France), from July 1970 to the present.

Councillor, Kyoto University, Nov. 1970−March 1973.

Dean, Faculty of Engineering, Kyoto University, April 1971−March 1973.

US−Japan Eminent Scientist Exchange Program Chemist, Sept. 1973.

Counsellor, Institute for Molecular Science, from Jan. 1976 to the present.

Vice-President, Chemical Society of Japan, March 1978−Feb. 1979.

Chairman, Executive Committee, 3rd International Congress of Quantum Chemistry, Kyoto, Oct.−Nov. 1979.

Foreign Associate, National Academy of Sciences, April 1981.

Order of Culture, Nov. 1981.

Person of Cultural Merits, Nov. 1981.

Member, European Academy of Arts, Sciences and Humanities, Dec. 1981.

Professor Emeritus, Kyoto University, April 1982.

President, Kyoto Institute of Technology, June 1982–May 1988.

President, Chemical Society of Japan, March 1983–Feb. 1984.

Foreign Honorary Member, American Academy of Arts and Sciences, May 1983.

Member, the Japan Academy, Dec. 1983.

Member, Pontifical Academy of Sciences, Dec. 1985.

Professor Emeritus, Kyoto Institute of Technology, June 1988.

Director, Institute for Fundamental Chemistry, June 1988 to the present.

Imperial Honour of the Grand Cordon of the Order of the Rising Sun, Nov. 1988.

Honorary Member, the Royal Institution of Great Britain, March 1989.

Foreign Member, the Royal Society, June 1989.

THE ROLE OF FRONTIER ORBITALS
IN CHEMICAL REACTIONS

Nobel lecture, 8 December, 1981

by

KENICHI FUKUI

Department of Hydrocarbon Chemistry, Kyoto University, Sakyo-ku,
Kyoto 606, Japan

Since the 3rd century for more than a thousand years chemistry has been
thought of as a complicated, hard-to-predict science. Efforts to improve even a
part of its unpredictable character are said to have born fruit first of all in the
success of the "electronic theory". This was founded mainly by organic chemists, such as Fry, Stieglitz, Lucas, Lapworth and Sidgwick, brought to a
completed form by Robinson and Ingold, and developed later by many other
chemists.[1] In the electronic theory, the mode of migration of electrons in
molecules is noted and is considered under various judgements. For that
purpose, a criterion is necessary with respect to the number of electrons which
should originally exist in an atom or a bond in a molecule. Therefore, it can be
said to be the concept by Lewis of the sharing of electrons that has given a firm
basis to the electronic theory.[2]

In the organic electronic theory, the chemical concepts such as acid and
base, oxidation and reduction and so on, have been conveniently utilized from
a long time ago. Furthermore, there are terms centring closer around the
electron concept, such as electrophilicity and nucleophilicity, and electron
donor and acceptor both being pairs of relative concepts.

One may be aware that these concepts can be connected qualitatively to the
scale of electron density or electric charge. In the electronic theory, the static
and dynamic behaviours of molecules are explained by the electronic effects
which are based on nothing but the distribution of electrons in a molecule.

The mode of charge distribution in a molecule can be sketched to some
extent by the use of the electronegativity concept of atoms through organic
chemical experience. At the same time, it is given foundation, made quantitative, and supported by physical measurements of electron distribution and
theoretical calculations based on quantum theory.

The distribution of electrons or electric charge—with either use the result is
unchanged—in a molecule is usually represented by the total numbers (generally not integer) of electrons in each atom and each bond, and it was a concept
easily acceptable even to empirical chemists as having a tolerably realistic
meaning. Therefore, chemists employed the electron density as a fundamental
concept to explain or to comprehend various phenomena. In particular, for the
purpose of promoting chemical investigations, researchers usually rely upon
the analogy through experience, and the electron density was very effectively
and widely used as the basic concept in that analogy.

When the magnitude of electron density is adopted as the criterion the electrostatic attraction and repulsion caused by the electron density are taken into account. Therefore, it is reasonable to infer that an electrophilic reagent will attack the position of large electron density in a molecule while a nucleophilic reaction will occur at the site of small electron density. In fact, Wheland and Pauling[3] explained the orientation of aromatic substitutions in substituted benzenes along these lines, and theoretical interpretations of the mode of many other chemical reactions followed in the same fashion.

However, the question why one of the simple reactions known from long before, the electrophilic substitution in naphthalene, for instance, such as nitration, yields α-substituted derivatives predominantly was not so easy to answer. That was because, in many of such unsubstituted aromatic hydrocarbons, both the electrophile and the nucleophile react at the same location. This point threw some doubt on the theory of organic reactivity, where the electron density was thought to do everything.

THE CONCEPT OF FRONTIER ORBITAL INTERACTIONS

The interpretation of this problem was attempted by many people from various different angles. Above all, Coulson and Longuet-Higgins[4] took up the change of electron density distribution under the influence of approaching reagent. The explanation by Wheland[5] was based on the calculation of the energy required to localize electrons forcibly to the site of reaction. But I myself tried to attack this problem in a way which was at that time slightly unusual. Taking notice of the principal role played by the valence electrons in the case of the molecule formation from atoms, only the distribution of the electrons occupying the highest energy π orbital of aromatic hydrocarbons was calculated. The attempt resulted in a better success than expected, obtaining an almost perfect agreement between the actual position of electrophilic attack and the site of large density of these specified electrons as exemplified in Fig. 1.[6]

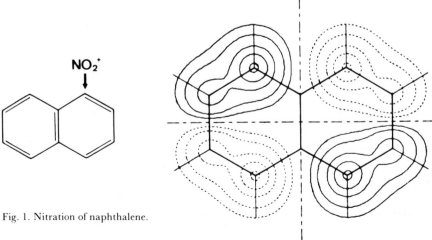

Fig. 1. Nitration of naphthalene.

The "orbital" concept, which was established and developed by many scientists, such as Pauling, Slater, Mulliken, Roothaan, Löwdin, Hückel, Parr and so on, had till then been employed to construct the wave function of a molecule, through which molecular properties were usually interpreted.[7] It seemed that the electron distribution in an orbital was directly connected to chemical observations and this fact was certainly felt to be interesting by many chemists.

But the results of such a rather "extravagant" attempt was by no means smoothly accepted by the general public of chemists. That paper received a number of controversial comments. This was in a sense understandable, because, for lack of my experiential ability, the theoretical foundation for this conspicuous result was obscure or rather improperly given. However, it was fortunate for me that the paper on the charge-transfer complex of Mulliken[8] was published in the same year as ours.

The model of Mulliken et al. for protonated benzene was a good help.[9] Our work in collaboration with Yonezawa, Nagata and Kato provided a simple and pointed picture of theoretical interpretation of reactions,[10] as well as the "overlap and orientation" principle proposed by Mulliken with regard to the orientation in molecular complexes.[11] Subsequent to the electrophilic substitution, the nucleophilic substitution was discussed and it was found that in this case the lowest energy vacant orbital played this particular part.[12] In reactions with radicals, both of the two orbitals mentioned above, became the particular orbitals.

There was no essential reason to limit these particular orbitals to π orbitals, so that this method was properly applied not only to unsaturated compounds but also to saturated compounds. The applicability to saturated compounds was a substantial advantage in comparison with many theories of reactivity which were then available only for π electron compounds. The method displayed its particular usefulness in the hydrogen abstraction by radicals from paraffinic hydrocarbons, the S_N2 and E2 reactions in halogenated hydrocarbons, the nucleophilic abstraction of α-hydrogen of olefins, and so forth.[13]

These two particular orbitals, which act as the essential part in a wide range of chemical reactions of various compounds, saturated or unsaturated, were referred to under the general term of "frontier orbitals", and abbreviated frequently by HOMO (highest occupied molecular orbital) and LUMO (lowest unoccupied molecular orbital).

In this way, the validity of the theory became gradually clearer. The vein of ore discovered by chance was found to be hopefully more extensive than expected. But it was attributed to the role of the symmetry of particular orbitals pointed out in 1964 with regard to Diels-Alder reactions[14] that the utility of our studies was further broadened. It was remarked that as is seen in Fig. 2, the symmetries of HOMO and LUMO of dienes and those of LUMO and HOMO of dienophiles, respectively, were found to be in a situation extremely favourable for a concerted cyclic interaction between them.

This signified the following important aspects: First, it pointed out a possible correlation between the orbital symmetry and the rule determining the sub-

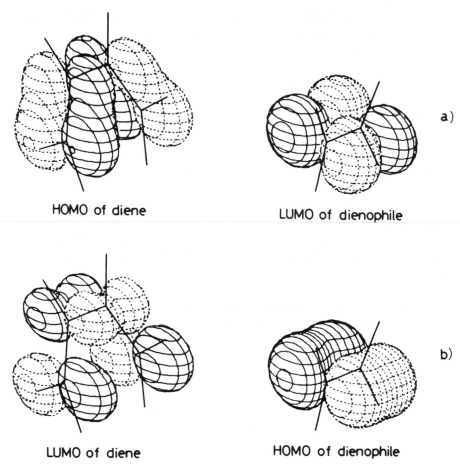

HOMO of diene   LUMO of dienophile   a)

LUMO of diene   HOMO of dienophile   b)

Fig. 2. The significance of orbital symmetry in the HOMO-LUMO overlapping in Diels-Alder reactions.

stantial occurrence or non-occurrence of a chemical reaction, which may be called the "selection rule", in common with the selection rule in molecular spectroscopy. Second, it provided a clue to discuss the question concerning what was the "concertedness" in a reaction which forms a cycle of electrons in conjugation along the way.

In 1965 Woodward and Hoffmann proposed the stereoselection rules which are established today as the "Woodward-Hoffmann" rules.[15, 16] An experimental result developed in Havinga's important paper[17] was extended immensely. It is only after the remarkable appearance of the brilliant work by Woodward and Hoffmann that I have become fully aware that not only the density distribution but also the nodal property of the particular orbitals have significance in such a wide variety of chemical reactions. In fact, we studied previously the noted $(4n+2)$ rule proposed by Hückel[18] and noticed that the sign of the bond order in the highest energy electron orbital of an open-chain conjugation should be closely related to the stabilization of the corresponding conjugated rings.[19] We did not imagine, however, on that occasion that the discussion might be extended to the so-called Möbius-type ring-closure![20]

By considering the HOMO-LUMO interactions between the fragments of a conjugated chain divided into parts,[21] the frontier orbital theory can yield selection rules which are absolutely equivalent to those obtained from the principle called "the conservation of orbital symmetry" by Woodward and Hoffmann. One point that I may stress here is, as was pointed out by Fujimoto, Inagaki and myself,[22] that the electron delocalization between the particular orbitals interprets definitely in terms of orbital symmetries the formation and breaking of chemical bonds which, I believe, should be a key for perceiving chemical reaction processes.

In the cycloaddition of butadiene and ethylene shown in Fig. 2, both the interaction between the HOMO of diene and the LUMO of dienophile and that between the LUMO of diene and the HOMO of dienophile stabilize the interacting system. If one is interested in the local property of interaction, however, one may recognize the clear distinction between the roles of the two types of orbital interactions. The HOMO of ethylene and the LUMO of butadiene are both symmetric with regards to the symmetry plane retained throughout the course of cycloaddition. This signifies that each of the carbon atoms of ethylene are bound to both of the terminal carbons of butadiene. The chemical bonding between the diene and dienophile thus generated may be something like the one in a loosely bound complex, e.g., protonation to an olefinic double bond. On the contrary, the HOMO of butadiene and the LUMO of ethylene are antisymmetric. The interaction between these orbitals leads, therefore, to two separated chemical bonds, each of which combines a carbon atom of ethylene and a terminal carbon atom of butadiene. Needless to say it is the interaction between the HOMO of diene and the LUMO of dienophile that is of importance for the occurrence of concerted cycloaddition.[22]

In this way, it turned out in the course of time that the electron delocalization between HOMO and LUMO generally became the principal factor determining the easiness of a chemical reaction and the stereoselective path, irrespective of intra- and intermolecular processes, as illustrated in Fig. 3. Besides our own school, a number of other chemists made contributions. I want to refer to several names which are worthy of special mention.

First of all, the general perturbation theory of the HOMO-LUMO interaction between two molecules was built up by Salem.[23-25] One of Salem's papers[25] was in line with the important theory of Bader,[26] which specified the mode of decomposition of a molecule or a transition complex by means of the symmetry of the normal vibration. Furthermore Pearson[27] investigated the relation between the symmetry of reaction coordinates in general and that of HOMO and LUMO.

The discussion so far may seem to be an overestimation of these selected orbitals, HOMO and LUMO. This point was ingeniously modified by Klopman.[28] He carefully took into account the factors to be considered in the perturbation theory of reacting systems and classified reactions into two cases: the one was "frontier-controlled" case in which the reaction was controlled by the particular orbital interaction, and the other was the "charge-controlled"

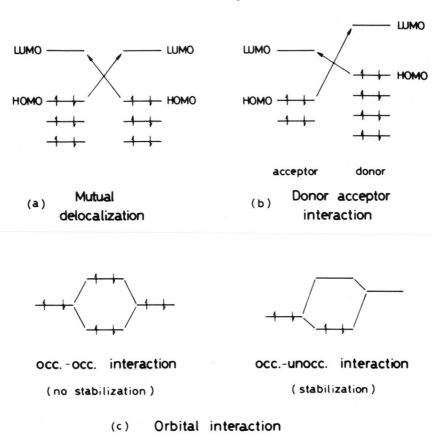

Fig. 3. The mode of interaction between orbitals of two molecules.

case, where it was controlled by the electrostatic interaction of charges. This classification was conveniently used by many people. In this context the review articles of Herndon[29] and of Hudson[30] appeared to be very useful. The names of Coulson[4] and Dewar[31] should also be noted here as those who contributed to the development of reactivity theories.

Returning to the subject again, let us assume that two molecules approach each other and orbital overlapping takes place. The perturbation theory[32] of this sort of interaction indicates that, the larger the orbital overlapping is and the smaller the level separation of two overlapping orbitals is, the larger is the contribution of the orbital pair to the stabilization of an interacting system. Accordingly, at least at the beginning, a reaction will proceed with a mutual nuclear configuration which is most favourable for the HOMO-LUMO overlapping.

Now let us suppose an electron flow from the HOMO of molecule I to the LUMO of molecule II. In each molecule the bonds between the reaction centre—the place at which the orbitals overlap with those of the other molecule—and the remaining part of the molecule are weakened. On this occasion, in molecule I the bonds which are bonding in HOMO are weakened and those antibonding in HOMO are strengthened, while in molecule II the bonds which are antibonding in LUMO are weakened and those bonding in LUMO are strengthened. Consequently, the HOMO of molecule I particularly destabilizes as compared with the other occupied orbitals, and the LUMO of molecule II discriminatively stabilizes among unoccupied orbitals, so that the HOMO-LUMO level separation between the two molecules is decreased. Such a circumstance is clearly understandable in Fig. 3.

The following tendency is further stressed. When the bond weakenings specified above have arisen, the HOMO and the LUMO tend to become more localized at these weakened bonds in each molecule. Besides, the weakening of the bonds between the reaction centre and the remaining part causes an increase of the amplitudes of HOMO and LUMO at the reaction centres, resulting in a larger overlapping of HOMO and LUMO.[33] Such a trend of the characteristic change in the orbital pattern is made numerically certain by actual calculations. The role of interaction between HOMO and LUMO turns out in this way to become more and more important as the reaction proceeds.

A series of studies on chemical interactions were attempted in which the interaction of reactants was divided into the Coulomb, the exchange, the polarization, and the delocalization interactions, and their magnitude of contribution to the interaction energy was quantitatively discussed.[32, 34] The interactions discussed by this method were the dimerization[35] and the addition to ethylene[36] of methylene and the dimerization of BH_3,[37] and also several donor-acceptor interactions—BH_3-NH_3,[38] BH_3-CO,[39] NH_3-HF,[40] etc. The method was applied also to reactions of radicals, such as the abstraction of a methane hydrogen by methyl radical, the addition of methyl radical to ethylene,[41] and recombinations, disproportionations, and self-reactions of two radicals.[42] In these calculations, the configuration analysis proposed originally by Baba[43] was also utilized conveniently. We could show numerically the mode of increase of the electron delocalization from HOMO to LUMO along with the proceeding of reaction, the increasing weight of contribution of such a delocalization to the stabilization of the reacting system, the driving force of the reaction in terms of orbital interactions, and so on.

The question "Why HOMO and LUMO solely determine the reaction path?" was one which I very frequently received from the audiences in my lectures given in the past in different places. The discussion so far made here is thought to correspond, at least partly, to that answer. But one may not adhere so strictly to the HOMO and LUMO. In one-centre reactions like substitutions, which the orbital symmetry has nothing to do with, any occupied orbitals which are very close to HOMO should properly be taken into account.[12] In large paraffin molecules a number of HOMO's (high-lying occupied MO's), and furthermore as will shortly be referred to later, in metal crystals, even "HOMO-band" must be taken along the line of reactivity argument. If HOMO or LUMO happens to be inadequate owing to its extension, the symmetry, or the nodal property, the next orbital should be sought for. One of the simplest examples of such an instance is the protonation of pyridine. In this case, the nitrogen lone-pair orbital is not HOMO, but the addition of proton to the nitrogen lone-pair so as not to disturb the π conjugation will evidently be more advantageous than the addition to higher occupied π orbitals which may intercept the π conjugation. Thus, the reason why proton dare not add to the positions of large amplitude of π HOMO in this case will easily be understood. It is not completely satisfactory to dispose of a disagreement between the HOMO-LUMO argument and the experimental fact formally as an *exception* to the theory. A so-called exception does possess its own reason. To investigate what the reason is will possibly yield a novel finding.

The HOMO-LUMO interaction argument was recently pointed out[44] to be in an auxiliary sense useful for the interpretation of the sign of a reaction constant and the scale of a substituent constant in the Hammett rule[45] which has made an immeasurable contribution to the study of the substituent effect in chemical reactivity. In the cyclic addition, like Diels-Alder reactions and 1,3-dipolar additions, the relative easiness of occurrence of reactions, various subsidiary effects, and interesting phenomena like regioselection and periselection were interpreted with considerable success simply by the knowledge of the height of the energy level of HOMO and LUMO, the mode of their extension, their nodal structure, etc.[46]—I defined these in a mass: the *"orbital pattern"*.

Other topics that have been discussed in terms of HOMO-LUMO interactions are thermal formation of excited states,[47] singlet-triplet selectivity,[48] the chemical property of biradicals and excited molecules,[49] the interaction of the central atom and ligands in transition metal complexes,[50] the interaction of three or more orbitals[51] and so forth. Inagaki et al. included in the theory the polarization effect in HOMO and LUMO due to the mixing in of other orbitals and gave an elucidation for a number of organic chemical problems which were not always easy to explain. The unique stereoselection in the transannular cross-bond formation, the lone-pair effect, the d orbital effect, and the orbital polarization effect due to substituents were the cases.[52]

As was partly discussed above, the method of orbital interaction was applied not only to the ground electronic state but to the excited states, giving an explanation of the path of even complicated photochemical isomerizations.[13,21] In a majority of cases the HOMO and the LUMO of the ground-state molecule were also found to be the essential orbitals. Even the ground-state reaction of a strong electron acceptor (or donor) causes a mixing in of an ionized electron configuration or an excited electron configuration in another molecule. In consequence, a partial HOMO-HOMO or LUMO-LUMO interaction, which would be trivial if there were no influence of the acceptor (or donor), becomes important in stabilizing the interacting system.[47]

The problems so far discussed have been limited to chemical reactions. However, the HOMO-LUMO interaction must come into relation also with other chemical phenomena in almost the same mechanism—with the exception of one different point that they usually do not bring about so remarkable a change in the nuclear configuration as in the case of chemical reactions. Now let us examine the possibility of applying the theory to so-called "aromaticity"—one of the simplest, but the hardest-to-interpret problems. There seem to be few problems so annoying to theoreticians as the explanation of this chemically classical concept. I greatly appreciate the contribution of Dewar's theory[53, 54] based on a quantitative energy values argument. Here, however, I want to give a quality comment through a totally different way of consideration.

It is easily ascertained[55] in Fig. 4 that in benzene, naphthalene, phenanthrene, etc., any virtual division of the molecule into two always produces the parts in which their HOMO and LUMO overlap in-phase at the two junctions.

Benzene

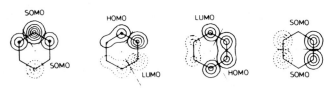

Naphthalene

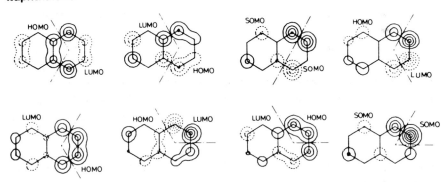

Phenanthrene

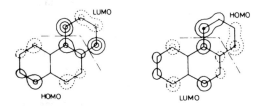

Anthracene

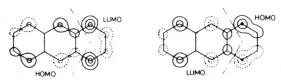

Fig. 4. The HOMO-LUMO phase relationship in virtual division of aromatic hydrocarbons. (SOMO: a singly occupied MO of a radical)

But these circumstances are not seen in anthracene which is usually looked upon as one of the typical representatives of aromatic compounds. Hosoya[55] pointed out from the comparison with phenanthrene indicated in Fig. 4, that the ring growth of type (II) was less stable than that of (I).

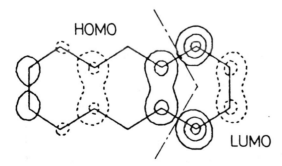

(I)   (II)

It is well known that anthracene occasionally exhibits a reactivity of olefin-like additions.

In view of so-called Hückel's $(4n+2)$-rule mentioned above, an anthracene molecule has 14π electrons and fulfils the stability condition for "aromaticity." Actually, if one considers a molecule of anthracene with the two inside bonds deleted,

it is really seen that the HOMO and the LUMO of the two parts overlap in an in-phase manner at both of the junctions:

HOMO

LUMO

In this way, it is understood that the two bonds which were deleted above exerted a certain unfavourable influence for aromaticity. Such an influence bears a close resemblance to that of impurity scattering in the wave of a free electron moving in a metal crystal.

This discussion seems to be a digression but, as a matter of fact, it relates to the essential question as to how an electron in a molecule can delocalize. As will be mentioned later, Anderson[56] solved the question how an electron in a random system can localize. In a molecule, there are potential barriers between atoms which should be got over by the aid of a certain condition to be satisfied, in order for an electron to move around it freely. Although the question how valence electrons can delocalize in a molecule have not yet been solved satisfactorily under the condition of unfixed nuclear configuration, the in-phase relation of HOMO and LUMO at the junctions of the two parts of the molecule seems to be at least one of the conditions of intramolecular delocalization of electrons.

Generally speaking, the electron delocalization gives rise to a stabilization due to "conjugation" which is one of the old chemical concepts. If so, similar stabilization mechanisms must be chemically detected in other systems than aromatic compounds. The discussion of this delocalization stabilization at the transition state or on the reaction path was nothing but the reactivity theory hitherto mentioned. The term "delocalizability" was attached to the reactivity indices we derived,[10] and our reactivity theory itself was sometimes called "delocalization approach."[14] The "hyperconjugation" of various sorts is explained in the same manner. The stabilization due to homoaromaticity or bicycloaromaticity of Goldstein,[57] the stability in spirocycles, pericycles,[58] "laticycles" and "longicycles" of Hoffmann and Goldstein,[59] that of spirarenes of Hoffmann and Imamura,[60] and so on, are all comprehended as examples of the stabilization due to the delocalization between HOMO and LUMO, although other explanations may also be possible.

You may be doubtful to what extent such a qualitative consideration is reliable. In many cases, however, a considerably accurate nonempirical determination of the stable conformation of hydrocarbon molecules[61, 62] results in a conclusion qualitatively not much different from the expectation based on the simple orbital interaction argument mentioned above.

CHEMICAL REACTION PATHWAYS

It has already been pointed out that the detailed mechanism of a chemical reaction was discussed along the reaction path on the basis of the orbital interaction argument. For that purpose, however, it is required that the problem as to how the chemical reaction path is determined should have been solved. The method in which the route of a chemical reaction was supposed on the potential energy surface and the rate of the reaction was evaluated by the aid of a statistical-mechanical formulation was established by Eyring.[63] Many people wrote papers where the rate expression was derived wave-mechanically with the use of the potential energy function. Besides, the problem of obtaining the trajectory of a given chemical reaction with a given initial condition was treated by Karplus.[64]

The centre line of the reaction path, so to speak, the idealized reaction coordinate—which I called "intrinsic reaction coordinate" (IRC)[65]—seemed to have been, rather strangely, not distinctly defined till then. For that reason, I began with the general equation which determines the line of force mathematically. [34, 66, 67] Although my papers themselves were possibly not very original, they turned out later to develop in a very interesting direction.[68-74] These papers opened the route to calculate the quasistatic change of nuclear configuration of the reacting system which starts from the transition state proceeding to a stable equilibrium point.[66] I termed the method of automatic determination of the molecular deformation accompanying a chemical reaction as "reaction ergodography." [34] [67] This method was applied to a few definite examples by Kato and myself[67] and by Morokuma.[72, 73] Those examples were: abstraction and substitution of methane hydrogen by hydrogen atom,[67] nucleophilic replacement in methane by hydride anion,[72] and isomerization of methylcarbylamine to acetonitrile.[73] All of these reactions thus far treated are limited to the simplest cases, but there seems to be no principal difficulty in extending the applicability to larger systems. Once IRC was determined in this way, the driving force of a chemical reaction was analyzed on the basis of the orbital interaction argument.[66]

In the reacting system with no angular momentum it is possible to obtain the IRC by the use of the space-fixed Cartesian coordinate system. All of the calculated examples mentioned above belong to this case. However, in the reaction in which rotational motion exists, it is required to discuss the IRC after separating the nuclear configuration space from the rotational motion.[74-77] For that purpose, it is essential to derive the general classical Hamiltonian of the reacting systems and then to separate the internal motion which is determined only by the internal coordinates. The nuclear configuration space thus separated out is in general a Riemannian space. The classical Lagrangian form to be obtained in that process of constructing the Hamiltonian is used to derive the IRC equation in the presence of rotational motion. It is thus understood that the rotational motion of the reacting system generally causes a deviation of IRC.[74]

Once the method of determining unique reaction pathways is obtained, the next problem we are concerned with is to see if the calculated pathways are interpreted in terms of the frontier orbital interactions. A method referred to as the "interaction frontier orbitals" or "hybrid molecular orbitals" has been developed very recently by Fujimoto and myself in order to furnish a lucid scheme of frontier orbital interactions with the accuracy of nonempirical calculations now and in the future.[78-80] By including properly contributions of other MOs than the HOMO and the LUMO, we realized in terms of orbital diagrams how ingenious the empirically established chemical concepts—"reaction sites" and "functional groups"—and the empirical concept of reaction pathways could be. Fig. 5 compares the HOMO of styrene and its interaction frontier orbital for protonation to the olefinic double bond. The latter is seen to be localized very efficiently in the frontier of chemical interaction. The double bond is evidently the functional unit in this case. Innovation of the frontier orbital concept will hopefully be continued by young people to make it useful for one of our ultimate targets: theroretical design of molecules and chemical reactions.

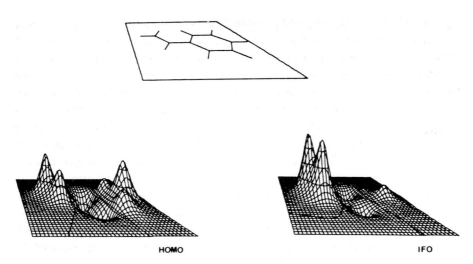

Fig. 5. A comparison of HOMO and the interaction frontier orbital for protonation in styrene.

FRONTIER ORBITALS IN RELATED FIELDS

Theoretical treatments of the property of solid crystals, or chemisorption on a solid surface, appear to have hitherto been almost monopolistically treated by the methodology of physics. But the orbital pattern technique has also advanced gradually in this field.

The "cluster approach," [81, 82] in which a portion of the metal crystal is drawn out as the form of a cluster of atoms and its catalytic actions or other properties are investigated, has contributed to the development of the orbital pattern approach, because the physical methods mentioned above can hardly be applied to such sizable systems. It is expected that, if clusters of various sizes and various shapes are studied to look into the characteristic feature of their HOMO's (high-lying occupied MO's) and LUMO's (low-lying unoccupied MO's), the nature of chemisorption and catalytic action, the mode of surface chemical reaction, and several related subjects of interest can be investigated theoretically.

As is the case of molecular interactions in usual chemical reactions, only the HOMO- and LUMO-bands lying in the range of several electron volts near the Fermi level can participate in the adsorption of molecules and surface reactions on solid crystals. You may recollect here that, in the BCS (Bardeen-Cooper-Schrieffer) theory of superconductivity, too, only the HOMO's and LUMO's in close proximity to the Fermi surface can be concerned in forming electron pairs as the result of interaction with lattice vibration. In the case of solid catalysts mentioned above, the discrimination of particular orbitals and electrons from the others have made the situation much easier.

Consider a system composed of a regular repetition of a molecular unit, for instance, a one-dimensional high polymer chain or a one-dimensional lattice, in which a certain perturbation is imposed at a definite location. Sometimes it is convenient to discuss the influence of this perturbation by transforming the orbitals belonging to the HOMO-band to construct the orbitals localized at that place. One such technique was proposed by Tanaka, Yamabe and myself.[83] This method is expected to be in principle applied to a local discussion of such problems as the adsorption of a molecule on the two-dimensional surface of catalysts, surface reactions, and related matters. This approach may be called a little more chemical than the method using the function of local density of states[84] or similar ones, in that the former can be used for the argument of the reactivity of molecules on a catalyst surface in terms of the phase relationship of localized orbitals.

What is called low-dimensional semiconductors and some superconductors have also been the objects of application of the orbital argument. In these studies, the dimerization of S_2N_2 to S_4N_4[85] and the high-polymerization to $(SN)_x$[86] were discussed, and the energy band structure of $(SN)_x$ polymer chain was analyzed to investigate the stable nuclear arrangement and the mode of inter-chain interactions.[87]

The modern technique employed in solid state physics to interpret the interesting characteristic behaviours of noncrystalline materials, in particular of amorphous materials, in which the nuclear arrangements were not regular, was certainly striking. Anderson showed generally that in a system of random lattice the electron localization should take place.[56] Mott, stating in his 1977 Nobel lecture that he thought is the first prize awarded for the study of amorphous materials, answered the question, "How can a localized electron be conducted?" with the use of the idea of hopping. Here, too, the HOMO-LUMO interaction—in this case the consideration of spin is essential—would play an important part.

Here in a few words, I want to refer to the meaning and the role of virtual orbitals. The LUMO, which has been one of the stars in orbital arguments hitherto discussed, is the virtual orbital which an external electron is considered to occupy to be captured by a molecule to form an anion. Virtual orbitals always play an essential part in producing metastable states of molecules by electron capture.[88] To discuss such problems generally, Tachibana et al.[89] systematized the theory of resonant states from the standpoint of complex eigenvalue problem. The idea of resonant states will take a principal part in chemical reactions, particularly in high-energy reactions which will be developed more in the future.

PROSPECT

In introducing above a series of recent results of the studies carried out mainly by our group, I have ventured to make those things the object of my talk which are no more than my prospective insight and are not yet completely established. This is just to stimulate, by specifying what are the fields I believe promising in the future, the intentional efforts of many younger chemists in order to develop them further.

In my opinion, quantum mechanics has two different ways of making participations in chemistry. One is the contribution to the nonempirical comprehension of empirical chemical results just mentioned. However, we should not overlook another important aspect of quantum mechanics in chemistry. That is the promotion of empirical chemistry from the theoretical side. But, also for this second purpose, as a matter of course, reliable theoretical foundations and computational methods are required. The conclusions of theories should be little affected by the degree of sophistication in approximations adopted.

On the other hand, for theoreticians to make the second contribution, the cases where predictions surpassing the experimental accuracy are possible by very accurate calculations are for the present limited to those of a very few, extremely simple molecules. In order to accomplish this object in regard to ordinary chemical problems, it becomes sometimes necessary to provide qualitative theories which can be used even by experimental chemists. If one can contribute nothing to chemistry without carrying out accurate calculations with respect to each problem, one can not be said to be making the most of quantum mechanics for the development of chemistry. It is certainly best that the underlying concepts are as close to experience as possible, but the sphere of chemical experience is steadily expanding. Quantum chemistry has then to perform its duty by furnishing those concepts with the theoretical basis in order to make them chemically available and serviceable for the aim of promoting empirical chemistry.

Even the same atoms of the same element, when they exist in different molecules, exhibit different behaviours. The chemical symbol H even seems to signify atoms of a completely different nature. In chemistry, this terrible individuality should never be avoided by "averaging," and, moreover, innumerable combinations of such atoms form the subject of chemical research, where it is not the "whole assembly of compounds of different kinds but each individual kind of compounds" that is of chemical interest. On account of this formidable complexity, chemistry possesses inevitably one aspect of depending on the analogy through experience. This is in a sense said to be the fate alloted to chemistry, and the source of a great difference in character from physics. Quantum chemistry, too, so far as it is chemistry, is required to be useful in promoting empirical chemistry as mentioned before.

ACKNOWLEDGEMENTS

Lastly, I want to mention at this opportunity out of a sense of gratefulness the names of many people in our group who have been walking on the same road as mine since my first paper (1952) on quantum chemistry, particularly Drs. T. Yonezawa, C. Nagata, H. Kato, A. Imamura, K. Morokuma, T. Yamabe and H. Fujimoto, and also I can not forget the names of younger doctors mentioned in the text who made a contribution in opening new circumstances in each field. Among them, Prof. T. Yonezawa was helpful in performing calculations in our 1952 paper, and also, it is to be mentioned with appreciation that the attractive title "frontier orbitals" of my lecture originated from the terminology I adopted in that paper by the suggestion of Prof. H. Shingu, who kindly participated in that paper as an organic chemist to classify the relevant experimental results. Furthermore, many other collaborators are now distinguishing themselves in other important fields of chemistry, which, however, have not been the object of the present lecture.

It was the late Prof. Yoshio Tanaka of the University of Tokyo and Prof. Masao Horio of Kyoto University who recognized the existence and significance of my early work in advance to others. I owe such a theoretical work, which I was able to carry out in the Faculty of Engineering, Kyoto University, and moreover in the Department of *Fuel Chemistry,* to the encouragement and kind regard of Prof. Shinjiro Kodama, who fostered the Department. What is more, it was the late Prof. Gen-itsu Kita, my life-teacher, and the founder of the Department, who made me enter into chemistry, one of the most attractive and promising fields of science, and led me to devote my whole life to it. For all of these people no words of gratitude can by any means be sufficient.

REFERENCES

1. For instance, see Ingold, C. K., Structure and Mechanism in Organic Chemistry, Cornell University Press (1953), Ithaca, N.Y.
2. For instance, see Lewis, G. N., Valence and the Structure of Atoms and Molecules, Chemical Catalog Co. (1923), New York, N.Y.
3. Wheland, G. W. and Pauling, L., J. Am. Chem. Soc. *57*, 2086 (1935).
4. Coulson, C. A. and Longuet-Higgins, H. C., Proc. Roy. Soc. (London) *A191*, 39; *A192*, 16 (1947).
5. Wheland, G. W., J. Am. Chem. Soc. *64*, 900 (1942).
6. Fukui, K., Yonezawa, T. and Shingu, H., J. Chem. Phys. *20*, 722 (1952).
7. For instance, see Parr, R. G., The Quantum Theory of Molecular Electronic Structure, Benjamin (1963), New York, N.Y. and references cited therein.
8. Mulliken, R. S., J. Am. Chem. Soc. *74*, 811 (1952).
9. Picket, L. W., Muller, N. and Mulliken, R. S., J. Chem. Phys. *21*, 1400 (1953).
10. Fukui, K., Yonezawa, T. and Nagata, C., Bull. Chem. Soc. Jpn. *27*, 423 (1954); Fukui, K., Kato, H. and Yonezawa, T., ibid. *34*, 1112 (1961).
11. Mulliken, R. S., Rec. Trav. Chim. *75*, 845 (1956).
12. Fukui, K., Yonezawa, T., Nagata, C. and Shingu, H., J. Chem. Phys. *22*, 1433 (1954).
13. For instance, see Fukui, K., Theory of Orientation and Stereoselection, Springer (1970), Berlin.
14. Fukui, K. In: Molecular Orbitals in Chemistry, Physics and Biology, Löwdin, P.-O. and Pullman, B. eds., Academic (1964), New York, N.Y., p. 513.
15. Woodward, R. B. and Hoffmann, R., Angew. Chem. *81*, 797 (1969); The Conservation of Orbital Symmetry, Academic Press (1969), New York, N.Y.; see papers In: Orbital Symmetry Papers, Simmons, H. E. and Bunnett, J. F., eds., ACS (1974), Washington D.C.
16. Woodward, R. B. and Hoffmann, R., J. Am. Chem. Soc. *87*, 395 (1965).
17. Havinga, E., de Kock, R. J. and Rappoldt, M. P., Tetrahedron *11*, 276 (1960); Havinga, E. and Schlatmann, J. L. M. A., ibid. *15*, 146 (1961).
18. Hückel, E., Z. Phys. *70*, 204 (1931); *76*, 628 (1932).
19. Fukui, K., Imamura, A., Yonezawa, T. and Nagata, C., Bull. Chem. Soc. Japan *33*, 1501 (1960).
20. Heilbronner, E., Tetrahedron Lett., *1964*, 1923.
21. Fukui, K., Acc. Chem. Res. *4*, 57, (1971).
22. Fujimoto, H., Inagaki, S. and Fukui, K., J. Am. Chem. Soc. *98*, 2670 (1976).
23. Salem, L., J. Am. Chem. Soc. *90*, 543, 553 (1968).
24. Devaquet, A. and Salem, L., J. Am. Chem. Soc. *91*, 3743 (1969).

25. Salem, L., Chem. Brit. *5*, 449 (1969).
26. Bader, R. F. W., Can., J. Chem. *40*, 1164 (1962).
27. Pearson, R. G., Symmetry Rules for Chemical Reactions, John Wiley (1976), New York, N.Y. and references cited therein.
28. Klopman, G., J. Am. Chem. Soc. *90*, 223 (1968). Also see Klopman, G., Chemical Reactivity and Reaction Paths, John Wiley (1974), New York, N.Y.
29. Herndon, W. C., Chem. Rev. *72*, 157 (1972).
30. Hudson, R. F., Angew. Chem. Int. Ed. Eng. *12*, 36 (1973).
31. For instance, see Dewar, M. J. S. and Dougherty, R. C., The PMO Theory of Organic Chemistry, Plenum (1975), New York, N.Y. and many papers cited therein; see also Dewar, M. J. S., Thetrahedron *S8*, Part I, p. 85 (1966).
32. Fukui, K. and Fujimoto, H., Bull. Chem. Soc. Japan *41*, 1989 (1968).
33. Fukui, K. and Fujimoto, H., Bull. Chem. Soc. Japan *42*, 3392 (1969).
34. Fukui, K. In: The World of Quantum Chemistry, Daudel, R. and Pullman, B. eds., Reidel (1974), Dordrecht, p. 113.
35. Fujimoto, H., Yamabe, S. and Fukui, K., Bull. Chem. Soc. Japan *45*, 1566 (1972).
36. Fujimoto, H., Yamabe, S. and Fukui, K., Bull. Chem. Soc. Japan *45*, 2424 (1972).
37. Yamabe, S., Minato, T., Fujimoto, H. and Fukui, K., Theoret. Chim. Acta (Berl.) *32*, 187 (1974).
38. Fujimoto, H., Kato, S., Yamabe, S. and Fukui, K., J. Chem. Phys. *60*, 572 (1974).
39. Kato, S., Fujimoto, H., Yamabe, S., Fukui, K., J. Am. Chem. Soc. *96*, 2024 (1974).
40. Yamabe, S., Kato, S., Fujimoto, H. and Fukui, K., Bull. Chem. Soc. Japan *46*, 3619 (1973); Yamabe, S., Kato, S., Fujimoto, H. and Fukui, K., Theoret. Chim. Acta (Berl.) *30*, 327 (1973).
41. Fujimoto, H., Yamabe, S., Minato, T. and Fukui, K., J. Am. Chem. Soc. *94*, 9205 (1972).
42. Minato, T., Yamabe, S., Fujimoto, H. and Fukui, K., Bull. Chem. Soc. Japan *51*, 1 (1978); Minato, T., Yamabe, S., Fujimoto, H. and Fukui, K., Bull. Chem. Soc. Japan *51*, 682 (1978).
43. Baba, H., Suzuki, S. and Takemura, T., J. Chem. Phys. *50*, 2078 (1969).
44. Henri-Rousseau, O. and Texier, F., J. Chem. Education *55*, 437 (1978).
45. For instance, see Hammett, L. P., Physical Organic Chemistry, McGraw-Hill (1940), New York, N.Y.
46. Houk, K. N., Acc. Chem. Res. *8*, 361 (1975) and references cited therein.
47. Inagaki, S., Fujimoto, H. and Fukui, K., J. Am. Chem. Soc. *97*, 6108 (1975).
48. See Fleming, I., Frontier Orbitals and Organic Chemical Reactions, John Wiley (1976), New York, N.Y.; Gilchrist, T. L. and Storr, R. C., Organic Reactions and Orbital Symmetry, Cambridge Univ. Press, 2nd ed. (1979), London.
49. Fukui, K. and Tanaka, K., Bull. Chem. Soc. Japan *50*, 1391 (1977).
50. Fukui, K. and Inagaki, S., J. Am. Chem. Soc. *97*, 4445 (1975) and many papers cited therein.
51. Inagaki, S., Fujimoto, H. and Fukui, K., J. Am. Chem. Soc. *98*, 4693 (1976).
52. Inagaki, S. and Fukui, K., Chem. Lett. *1974*, 509; Inagaki, S., Fujimoto, H. and Fukui, K., J. Am. Chem. Soc. *98*, 4054 (1976).
53. Dewar, M. J. S., The Molecular Orbital Theory of Organic Chemistry, McGraw-Hill (1969), New York, N.Y.
54. Dewar, M. J. S., Angew. Chem. *83*, 859 (1971).
55. Fukui, K., Kagaku to Kogyo *29*, 556 (1976); Hosoya, H., Symposium on Electron Correlation in Molecules, Res. Inst. for Fund. Phys., Dec. 18, 1976.
56. Anderson, P. W., Phys. Rev. *109*, 1492 (1958); cf. 1977 Nobel Lectures in Physics by Anderson, P. W., and by Mott, N. F.
57. Goldstein, M. J., J. Am. Chem. Soc. *89*, 6357 (1967).
58. Simmons, H. E. and Fukunaga, T., J. Am. Chem. Soc. *89*, 5208 (1967).
59. Goldstein, M. J. and Hoffmann, R., J. Am. Chem. Soc. *93*, 6193 (1971).
60. Hoffmann, R., Imamura, A. and Zeiss, G., J. Am. Chem. Soc. *89*, 5215 (1967).
61. For instance, see Hehre, W. J. and Pople, J. A., J. Am. Chem. Soc. *97*, 6941 (1975).
62. Hehre, W. J., Acc. Chem. Res. *8*, 369 (1975).
63. For instance, see Glasstone, S., Laider, K. J. and Eyring, H., The Theory of Rate Processes, McGraw-Hill (1941), New York, N.Y. and references cited therein.

64. Wang, I. S. Y. and Karplus, M., J. Am. Chem. Soc. *95*, 8060 (1973) and references cited therein.
65. Fukui, K., J. Phys. Chem. *74*, 4161 (1970).
66. Fukui, K., Kato, S. and Fujimoto, H., J. Am. Chem. Soc. *97*, 1 (1975).
67. Kato, S. and Fukui, K., J. Am. Chem. Soc. *98*, 6395 (1976).
68. Tachibana, A. and Fukui, K., Theoret. Chim. Acta (Berl.) *49*, 321 (1978).
69. Tachibana, A. and Fukui, K., Theoret. Chim. Acta (Berl.) *51*, 189, 275 (1979).
70. Tachibana, A. and Fukui, K., Theoret. Chim. Acta (Berl.) *5*, (1979).
71. Fukui, K., Rec. Trav. Chim., Pays-Bas *98*, 75 (1979).
72. Joshi, B. D. and Morokuma, K., J. Chem. Phys. *67*, 4880 (1977).
73. Ishida, K., Morokuma, K. and Komornicki, A., J. Chem. Phys. *66*, 2153 (1977).
74. Fukui, K., Tachibana, A. and Yamashita, K., Intern. J. Quant. Chem., Quant. Chem. Symp. *15*, 621 (1981).
75. Miller, W. H., Handy, N. C. and Adams, J. E., J. Chem. Phys. *72*, 99 (1980); Gray, S. K., Miller, W. H., Yamaguchi, Y. and Schaefer, H. F., III., ibid. *73*, 2732 (1980).
76. Fukui, K., Intern. J. Quant. Chem., Quant. Chem. Symp. *15*, 633 (1981).
77. Fukui, K., Acc. Chem. Res. *14*, 363 (1981).
78. Fukui, K., Koga, N. and Fujimoto, H., J. Am. Chem. Soc. *103*, 196 (1981).
79. Fujimoto, H., Koga, N., Endo, M. and Fukui, K., Tetrahedron Letters *22*, 1263; 3427 (1981).
80. Fujimoto, H., Koga, N. and Fukui, K., J. Am. Chem. Soc. *103*, 7452 (1981).
81. For instance, see Johnson, K. H. In: The New World of Quantum Chemistry, Pullman, B. and Parr, R. G. eds., Reidel (1976), Dordrecht, p. 317 and references cited therein.
82. Kobayashi, H., Kato, H., Tarama, K. and Fukui, K., J. Catalysis *49*, 294 (1977); Kobayashi, H., Yoshida, S., Kato, H., Fukui, K. and Tarama, K., Surface Science *79*, 189 (1979).
83. Tanaka, K., Yamabe, T. and Fukui, K., Chem. Phys. Letters *48*, 141 (1977).
84. For instance, Schrieffer, J. R. In: The New World of Quantum Chemistry, Pullman, B. and Parr, R. G. eds., Reidel, D., Dordrecht (1976), p. 305; Danese, J. B. and Schrieffer, J. R., Intern. J. Quant. Chem., Quant. Chem. Symp. *10*, 289 (1976).
85. Tanaka, K., Yamabe, T., Noda, A., Fukui, K. and Kato, H., J. Phys. Chem. *82*, 1453 (1978).
86. Yamabe, T., Tanaka, K., Fukui, K. and Kato, H., J. Phys. Chem. *81*, 727 (1977).
87. Tanaka, K., Yamabe, T. and Fukui, K., Chem. Phys. Letters *53*, 453 (1978).
88. For instance, see Ishimaru, S., Yamabe, T., Fukui, K. and Kato, H., J. Phys. Chem. *78*, 148 (1974); Ishimaru, S., Fukui, K. and Kato, H., Theoret. Chim. Acta (Berl.) *39*, 103 (1975).
89. Tachibana, A., Yamabe, T. and Fukui, K., J. Phys. B. Atom. Molec. Phys. *10*, 3175 (1977); Adv. Quant. Chem. *11*, 195 (1978).

Roald Hoffmann

ROALD HOFFMANN

I came to a happy Jewish family in dark days in Europe. On July 18, 1937 I was born to Clara (née Rosen) and Hillel Safran in Zloczow, Poland. This town, typical of the Pole of the Settlement, was part of Austria-Hungary when my parents were born. It was Poland in my time and is part of the Soviet Union now. I was named after Roald Amundsen, my first Scandinavian connection. My father was a civil engineer, educated at the Lvov (Lemberg) Polytechnic, my mother by training a school teacher.

In 1939 the war began. Our part of Poland was under Russian occupation from 1939—1941. Then in 1941 darkness descended, and the annihilation of Polish Jewry began. We went to a ghetto, then a labor camp. My father smuggled my mother and me out of the camp in early 1943, and for the remainder of the war we were hidden by a good Ukrainian in the attic of a school house in a nearby village. My father remained behind in the camp. He organized a breakout attempt which was discovered. Hillel Safran was killed by the Nazis and their helpers in June 1943. Most of the rest of my family suffered a similar fate. My mother and I, and a handful of relatives, survived. We were freed by the Red Army in June 1944. At the end of 1944 we moved to Przemysl and then to Krakow, where I finally went to school. My mother remarried, and Paul Hoffmann was a kind and gentle father to me until his death, two months prior to the Nobel Prize announcement.

In 1946 we left Poland for Czechoslovakia. From there we moved to a displaced persons' camp, Bindermichl, near Linz, in Austria. In 1947 we went on to another camp in Wasseralfingen bei Aalen in Germany, then to München. On Washington's Birthday 1949 we came to the United States.

I learned English, my sixth language at this point, quite quickly. After P.S. 93 and P.S. 16, Brooklyn, I went on to the great Stuyvesant High School, one of New York's selective science schools. Among my classmates were not only future scientists but lawyers, historians, writers — a remarkable group of boys. In the summers I went to Camp Juvenile in the Catskills, a formative experience. Elinor, my younger sister, was born in 1954.

In 1955 I began Columbia College as a premedical student. That summer and the next I worked at the National Bureau of Standards in Washington with E.S. Newman and R.E. Ferguson. The summer after I worked at Brookhaven National Laboratory, with J.P. Cumming. These summers were important because they introduced me to the joys of research, and kept me going through some routine courses at Columbia. I did have some good chemistry teachers, G.K. Fraenkel and R.S. Halford, and a superb teaching assistant, R. Schneider. But I must say that the world that opened up before me in my non-science courses is what I remember best from my Columbia days. I almost switched to art history.

In 1958 I began graduate work at Harvard. I intended to work with W.E. Moffitt, a remarkable young theoretician, but he died in my first year there. A young instructor, M.P. Gouterman, was one of the few faculty members at Harvard who at that time was interested in doing theoretical work, and I began research with him. In the summer of 1959 I got a scholarship from P-.O. Löwdin's Quantum Chemistry Group at Uppsala to attend a Summer School. The school was held on Lidingö, an island outside of Stockholm. I met Eva Börjesson who had a summer job as a receptionist at the school, and we were married the following year.

I came back to Harvard, began some abortive (and explosive) experimental work, and Eva and I took off for a year to the Soviet Union. It was the second year of the U.S.–U.S.S.R graduate student exchange. I worked for 9 months at Moscow University with A.S. Davydov on exciton theory. Eva and I lived in one of the wings, Zona E, of that great central building of Moscow University. My proficiency in Russian and interest in Russian culture date from that time.

On returning to the U.S. I switched research advisors and started to work with W.N. Lipscomb, who had just come to Harvard. Computers were just coming into use. With Lipscomb's encouragement and ebullient guidance, L.L. Lohr and I programmed what was eventually called the extended Hückel method. I applied it to boron hydrides and polyhedral molecules in general. One day I discovered that one could get the barrier to internal rotation in ethane approximately right using this method. This was the beginning of my work on organic molecules.

In 1962 I received my doctorate, as the first Harvard Ph.D. of both Lipscomb and Gouterman. Several academic jobs were available, and I was also offered a Junior Fellowship in the Society of Fellows at Harvard. I chose the Junior Fellowship. The three ensuing years in the Society (1962–65), gave me the time to switch my interests from theory to applied theory, specifically to organic chemistry. It was E.J. Corey who taught me, by example, what was exciting in organic chemistry. I began to look at all kinds of organic transformations, and so I was prepared when in the Spring of 1964 R.B. Woodward asked me some questions about what subsequently came to be called electrocyclic reactions. That last year at Harvard was exciting. I was learning organic chemistry at a great pace, and I had gained access to a superior mind. R.B. Woodward possessed clarity of thought, powers of concentration, encyclopedic knowledge of chemistry, and an aesthetic sense unparalleled in modern chemistry. He taught me, and I have taught others.

The 1962–65 period was creative in other ways as well: Our two children, Hillel Jan and Ingrid Helena, were born to Eva and me.

In 1965 I came to Cornell where I have been ever since. A collegial department, a great university and a lovely community have kept me happy. I am now the John A. Newman Professor of Physical Science. I have received many of the honors of my profession. I am especially proud that in addition to the American Chemical Society's A.C. Cope Award in Organic Chemistry, which I received jointly with R.B. Woodward in 1973, I have just been selected for the Society's Award in Inorganic Chemistry in 1982, the only person to receive these two awards in different subfields of our science.

I have been asked to summarize my contributions to science.

My research interests are in the electronic structure of stable and unstable molecules, and of transition states in reactions. I apply a variety of computational methods, semiempirical and nonempirical, as well as qualitative arguments, to problems of structure and reactivity of both organic and inorganic molecules of medium size. My first major contribution was the development of the extended Hückel method, a molecular orbital scheme which allowed the calculation of the approximate σ and π electronic structure of molecules, and which gave reasonable predictions of molecular conformations and simple potential surfaces. These calculations were instrumental in a renaissance of interest in σ electrons and their properties. My second major contribution was a two-pronged exploration of the electronic structure of transition states and intermediates in organic reactions. In a fruitful collaboration R.B. Woodward and I applied simple but powerful arguments of symmetry and bonding to the analysis of concerted reactions. These considerations have been of remarkable predictive value and have stimulated much productive experimental work. In the second approach I have analyzed, with the aid of various semiempirical methods, the molecular orbitals of most types of reactive intermediates in organic chemistry — carbonium ions, diradicals, methylenes, benzynes, etc.

Recently I and my collaborators have been exploring the structure and reactivity of inorganic and organometallic molecules. Approximate molecular orbital calculations and symmetry-based arguments have been applied by my research group to explore the basic structural features of every kind of inorganic molecule, from complexes of small diatomics to clusters containing several transition metal atoms. A particularly useful theoretical device, the conceptual construction of complex molecules from ML_n fragments, has been used by my research group to analyze cluster bonding and the equilibrium geometries and conformational preferences of olefin and polyene metal carbonyl complexes. A satisfactory understanding of the mode of binding of essentially every ligand to a metal is now available, and a beginning has been made toward understanding organometallic reactivity with the exploration of potential energy surfaces for ethylene insertion, reductive elimination and alkyl migrative insertion reactions. Several new structural types, such as the triple-decker and porphyrin sandwiches, have been predicted, and recently synthesized by others. On the more inorganic side, we have systematically explored the geometrics, polytopal rearrangement and substitution site preferences of five, six, seven and eight coordination, the factors that influence whether certain ligands will bridge or not, the constraints of metal-metal bonding, and the geometry of uranyl and other actinide complexes. I and my coworkers are beginning work on extended solid state structures and the design of novel conducting systems.

The technical description above does not communicate what I think is my major contribution. I am a teacher, and I am proud of it. At Cornell University I have taught primarily undergraduates, and indeed almost every year since

1966 have taught first-year general chemistry. I have also taught chemistry
courses to non-scientists and graduate courses in bonding theory and quantum
mechanics. To the chemistry community at large, to my fellow scientists, I
have tried to teach "applied theoretical chemistry": a special blend of
computations stimulated by experiment and coupled to the construction of
general models – frameworks for understanding.

Added in 1992:

In the last decade I and my coworkers have begun to look at the
electronic structure of extended systems in one-, two-, and three dimensions.
Frontier orbital arguments find an analogue in this work, in densities of
states and their partitioning. We have introduced an especially useful tool,
the COOP curve. This is the solid state analogue of an overlap population,
showing the way the bond strength depends on electron count. My group
has studied molecules as diverse as the platinocyanides, Chevrel phases,
transition metal carbides, displacive transitions in NiAs, MnP and NiP, new
metallic forms of carbon, the making and breaking of bonds in the solid
state and many other systems. One focus of the solid state work has been on
surfaces, especially on the interaction of CH_4, acetylene and CO with specific metal faces. The group has been able to carry through unique
comparisons of inorganic and surface reactions. And in a book "Solids and
Surfaces. A Chemist's View of Bonding in Extended Structures," I've tried
to teach the chemical community just how simple the concepts of solid state
physics are. And, a much harder task, to convince physicists that there is
value in chemical ways of thinking.

In 1986-88 I participated in the production of a television course in
introductory chemistry. "The World of Chemistry" is a series of 26 half-hour
episodes developed at the University of Maryland and produced by Richard
Thomas. The project has been funded by Annenberg/the Corporation for
Public Broadcasting. I am the Presenter for the series which began to be
aired on PBS in 1990, and will also be seen in many other countries.

My first real introduction to poetry came at Columbia from Mark Van
Doren, the great teacher and critic whose influence was at its height in the
1950's. Through the years I maintained an interest in literature, particularly
German and Russian literature. I began to write poetry in the mid-seventies,
but it was only in 1984 that a poem was first published. I own much to a
poetry group at Cornell that includes A. R. Ammons, Phyllis Janowitz and
David Burak, as well as to Maxine Kumin. My poems have appeared in many
magazines and have been translated into French, Portuguese, Russian and
Swedish. My first collection, "The Metamict State", was published by the
University of Central Florida Press in 1987, and is now in a second printing.
A second collection, "Gaps and Verges", was also published by the University
of Central Florida Press, in 1990. Articles on my poetry have appeared in
Literaturnaya Gazeta and *Studies in American Jewish Literature*. I received the

1988 Pergamon Press Fellowship in Literature at the Djerassi Foundation, Woodside, California, where I was in residence for three years.

It seems obvious to me to use words as best as I can in teaching myself and my coworkers. Some call that research. Or to instruct others in what I've learned myself, in ever-widening circles of audience. Some call that teaching. The words are important in science, as much as we might deny it, as much as we might claim that they just represent some underlying material reality.

It seems equally obvious to me that I should marshal words to try to write poetry. I write poetry to penetrate the world around me, and to comprehend my reactions to it.

Some of the poems are about science, some not. I don't stress the science poems over the others because science is only one part of my life. Yet there are several reasons to welcome more poetry that deals with science.

Around the time of the Industrial Revolution – perhaps in reaction to it, perhaps for other reasons – science and its language left poetry. Nature and the personal became the main playground of the poet. That's too bad for both scientists and poets, but it leaves lots of open ground for those of us who can move between the two. If one can write poetry about being a lumberjack, why not about being a scientist? It's experience, a way of life. It's exciting.

The language of science is a language under stress. Words are being made to describe things that seem indescribable in words – equations, chemical structures and so forth. Words do not, cannot mean all that they stand for, yet they are all we have to describe experience. By being a natural language under tension, the language of science is inherently poetic. There is metaphor aplenty in science. Emotions emerge shaped as states of matter and more interestingly, matter acts out what goes on in the soul.

One thing is certainly not true: that scientists have some greater insight into the workings of nature than poets. Interestingly, I find that many humanists deep down feel that scientists have such inner knowledge that is barred to them. Perhaps we scientists do, but in such carefully circumscribed pieces of the universe! Poetry soars, all around the tangible, in deep dark, through a world we reveal and make.

It should be said that building a career in poetry is much harder than in science. In the *best* chemical journal in the world the acceptance rate for full articles is 65%, for communications 35%. In a *routine* literary journal, far from the best, the acceptance rate for poems is below 5%.

Writing, "the message that abandons", has become increasingly important to me. I expect to publish four books for a general or literary audience in the next few years. Science will figure in these, but only as a part, a vital part, of the risky enterprise of being human.

BUILDING BRIDGES BETWEEN INORGANIC AND ORGANIC CHEMISTRY

Nobel lecture, 8 December 1981

by

ROALD HOFFMANN

Department of Chemistry, Cornell University,
Ithaca, N.Y. 14853

R. B. Woodward, a supreme patterner of chaos, was one of my teachers. I dedicate this lecture to him, for it is our collaboration on orbital symmetry conservation, the electronic factors which govern the course of chemical reactions, which is recognized by half of the 1981 Nobel Prize in Chemistry. From Woodward I learned much: the significance of the experimental stimulus to theory, the craft of constructing explanations, the importance of aesthetics in science. I will try to show you how these characteristics of *chemical* theory may be applied to the construction of conceptual bridges between inorganic and organic chemistry.

FRAGMENTS

Chains, rings, substituents—those are the building blocks of the marvelous edifice of modern organic chemistry. Any hydrocarbon may be constructed on paper from methyl groups, CH_3, methylenes, CH_2, methynes, CH, and carbon atoms, C. By substitution and the introduction of heteroatoms all of the skeletons and functional groupings imaginable, from ethane to tetrodotoxin, may be obtained.

The last thirty years have witnessed a remarkable renaissance of inorganic chemistry, and the particular flowering of the field of transition metal organometallic chemistry. Scheme 1 shows a selection of some of the simpler creations of the laboratory in this rich and ever-growing field.

Structures 1—3 illustrate at a glance one remarkable feature of transition metal fragments. Here are three iron tricarbonyl complexes of organic moieties—cyclobutadiene, trimethylenemethane, an enol, hydroxybutadiene—which on their own would have little kinetic or thermodynamic stability. Yet complexed to $Fe(CO)_3$ these molecules are relatively stable, they exist in a bottle. The inorganic fragment is not merely a weakly attached innocent bystander. It transforms essentially and strongly the bonding relationships in the molecule.

Structures 4—6 contain the ubiquitous cyclopentadienyl (Cp) ligand, two of them in the archetypical ferrocene, one in $CpMn(CO)_3$, two bent back in $Cp_2Ti(CO)_2$. Structures 7—9 introduce us to the simplest representatives of the burgeoning class of clusters—assemblages of two or more metal atoms embellished with external ligands.

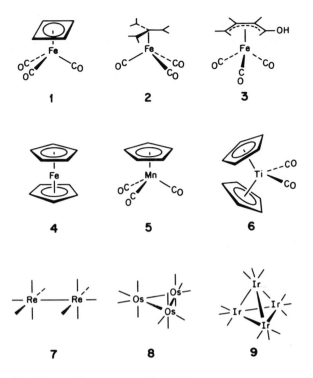

If we seek order, unity, a way of thinking about these complexes it is not difficult to perceive that the molecules contain as building blocks transition metal-ligand fragments, ML_n, such as $M(CO)_5$, $M(CO)_4$, $M(CO)_3$, MCp. It must be said immediately that there is nothing special about the carbonyl ligand. It is merely a representative and common component of organometallic complexes. Phosphines, olefins, alkyls will do as well.

To reconstruct the complexes 1–9, we need to know the electronic structure of the fragments. For the simple qualitative picture of the bonding in these molecules that we seek, we do not need to know every last detail of the electronic structure of each molecule. It will suffice that we know the frontier orbitals of the fragments—the higher occupied and the lower unoccupied levels—in other words the valence active orbitals of each fragment. It is K. Fukui who taught us the importance of the frontier orbitals. We shall soon see that it is the resemblance of the frontier orbitals of inorganic and organic moieties that will provide the bridge that we seek between the subfields of our science.

Over the last eight years my coworkers and I have built up a library of the orbitals of ML_n fragments. (1–3) We have done so using entirely qualitative, approximate molecular orbital calculations of the extended Hückel type (a procedure for its time, developed with another of my teachers, W. N. Lipscomb) and symmetry arguments (the value of which I first learned from still another of my teachers, M. P. Gouterman). Molecular orbital theory, R. S. Mulliken's great contribution to chemistry, is fundamental to our approach, be it in the construction of the very orbitals of the fragments, their changes on molecular deformations, or the interaction of several such fragments to restore

the composite molecule. Yet when I seek the simplest of all possible ways to tell
you of the orbitals of these fragments, I am led back to the valence bond picture
introduced into chemistry by L. Pauling. (4)

Let us go back to the building blocks. The common fragments ML_n, 10–13,
may be viewed in many ways. One convenient approach is to see them as pieces
of an octahedron. This is quite analogous to perceiving CH_3, CH_2, and CH in
a tetrahedron. If not a unique viewpoint, it is a useful one. Given that we have
an octahedron, or pieces thereof, let us prepare the metal atom for octahedral
bonding, and then bring in the appropriate number of ligands.

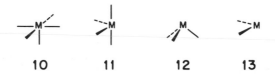

| 10 | 11 | 12 | 13 |

The valence orbitals of the transition metals are nd, $(n+1)$ s and $(n+1)$ p,
with n=3, 4, 5. To prepare the metal atom for bonding we must form six
equivalent octahedral hybrids. This is accomplished, 14, by using all of the s
and p functions and two of the d's. Three d functions, d_{xz}, d_{xy}, and d_{yz}, are left
unhybridized. They may be described, and we will do so often, as the t_{2g} set of
the crystal field, ligand field, or molecular orbital theories of an octahedral
complex. (5)

To form an octahedral complex we would bring in six ligands to make use of
the six octahedral hybrids. Perhaps it is appropriate to digress here and make
clear our ligand convention, which is to consider the ligand always as an even
electron Lewis base. While acceptor character or Lewis acidity is a desirable
feature in a ligand, Lewis basicity or donation is essential. We see the basicity
in the lone pairs of CO, PH_3, and CH_3^- 15–17, in bidentate four-electron
ligands, be they ethylenediamine 18, or butadiene 19, or in the electronically
tridentate $C_5H_5^-$, Cp^- 20, the equivalent of three two-electron bases. (6)

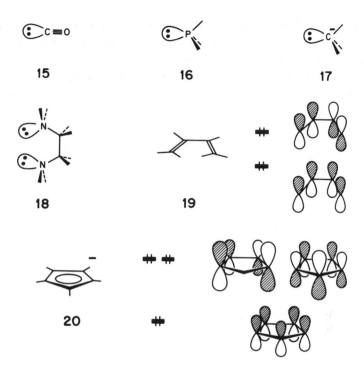

Let six two-electron ligands approach the metal atom prepared for octahedral bonding, 21. Sizable σ overlaps lead to formation of strongly metal-ligand σ bonding combinations, and their strongly metal-ligand σ*antibonding counterparts. The six electron pairs of the ligands enter the six bonding combinations. Any electrons the metal contributes enter the t_{2g} orbital left behind.

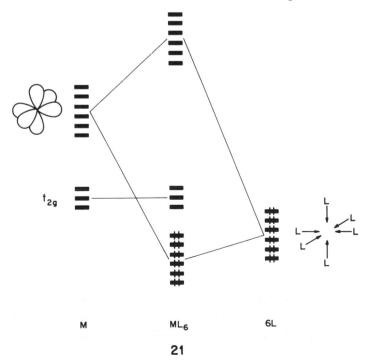

Indeed, for $Cr(CO)_6$ with its 6 metal electrons we attain a nice closed shell configuration, a situation we have learned to associate with relative kinetic and thermodynamic stability in organic chemistry.

What if we have not six ligands coming in, but only five? This situation is depicted in 22. Five hybrids interact strongly, are removed from the frontier orbital region, just as all six were in 21. One hybrid, the one pointing toward

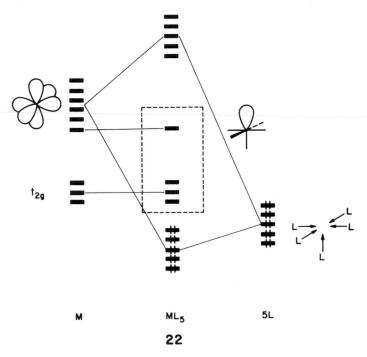

M   **ML₅**   **5L**

22

where the ligands are not, remains roughly untouched, relatively low-lying in energy. The frontier orbitals, enclosed in a dashed box in 22, now contain the t_{2g} set plus one hybrid.

What if we have four ligands, ML_4, or three, ML_3? Much the same things happens. In ML_4 two hybrids are left behind, in ML_3, three. We have thus reached the simplest of all possible pictures of the electronic structure of the ML_n fragments with $N = 5, 4, 3$, namely that given in 23–25. The ML_n

"t_{2g}"   "t_{2g}"

23   **24**   **25**

fragment's frontier orbitals consist of the descendants of an octahedral t_{2g} set at relatively low energy, and above them $6-n$ (one for $n = 5$, two for $n = 4$, three for $n = 3$) hybrids pointing toward the missing octahedral vertices.

What remains is to decide how many electrons to place into these frontier orbitals, and here the ever-useful Mendeleev Table, modified in 26 for electron counting purposes, tells us that Fe in oxidation state zero will have eight electrons in $Fe(CO)_4$ or $Fe(CO)_3$, and so will Co(I), or Ni(II).

4	5	6	7	8	9	10
Ti	V	Cr	Mn	Fe	Co	Ni
Zr	Nb	Mo	Tc	Ru	Rh	Pd
Hf	Ta	W	Re	Os	Ir	Pt

26

The reader had best beware. The account given here is simplified, as much as I dare simplify it. In that process, perforce, is lost the beautiful detail and complexity that makes $Fe(CO)_3$ different from $FeCl_3^{3-}$. There is a time for detail and there is a time for generality. The reader of my papers will know that I and my coworkers do not stint on detail, whether it is in explication or in perusal of the literature. But the time now, here, is for building conceptual frameworks and so similarity and unity take temporary precedence over difference and diversity.

Recall that the reason for building up the frontier orbitals of inorganic fragments is that we wish to use these orbitals in the construction of organometallic and inorganic complexes. We are now ready for that task. For instance, if we want trimethylenemethane iron tricarbonyl we construct a molecular orbital interaction diagram, 27. On one side are the newly learned orbitals of $Fe(CO)_3$, on the other side the older, better known frontier orbitals of $C(CH_2)_3$. We interact the two, using the full armament of group theory and perturbation theory (7) to follow what happens.

I will not trace this argument any further, for the primary purpose of this lecture is not the description of the electronic structure of organometallic complexes. My coworkers and I have done this comprehensively elsewhere. (1, 8) Instead, I wish to describe a bridge between organic and inorganic chemistry that becomes possible the moment we gain knowledge of the orbitals of the ML_n fragments.

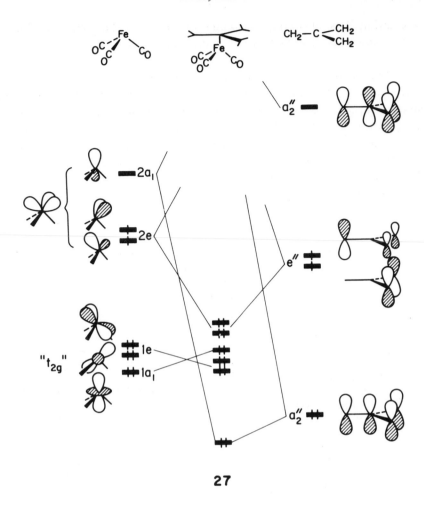

27

THE ISOLOBAL ANALOGY

Consider the d^7 fragment, $Mn(CO)_5$ (or $Co(CN)_5^{3-}$). Above three lone pairs in the t_{2g} set this doublet molecule has a single electron in a hybrid pointing away from the ML_5. The similarity to CH_3, the methyl radical, is obvious, **28**.

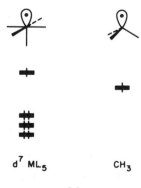

d^7 ML_5 CH_3

28

The drawing 28 is, of course, schematic. In Figure 1 I show the a_1 orbitals of MnH_5^{5-} and CH_3, so as to provide a more realistic comparison.

If d^7 ML_5 is like CH_3 then they should both behave similarly. Let us think about what a methyl radical does. It dimerizes to ethane and starts radical chains. $Mn(CO)_5$ or $Co(CN)_5^{3-}$ do similar things. They dimerize to $Mn_2(CO)_{10}$ or $Co_2(CN)_{10}^{6-}$, 29, and there is a rich radical-type chemistry of each. (9) One can even codimerize the organic and inorganic fragments to give $(CO)_5MnCH_3$. That may not be the preferred way to make this quite normal organometallic alkyl complex in the laboratory, but the construction on paper is quite permissible.

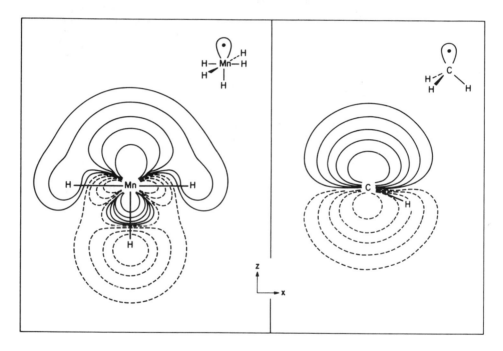

Figure 1. Contour diagram of the isolobal a_1 orbitals of MnH_5^{5-} (left) and CH_3 (right), as computed by the extended Hückel method. The contours of ψ, plotted in a plane passing through Mn and three H's (left) and C and one H (right), are ±0.2, ±0.1, ±0.55, ±0.25, ±0.1.

CH_3 and d^7 ML_5 resemble each other. Another way we can see that resemblance, traceable to their singly occupied a_1 orbitals, is to compare the overlap of both orbitals with a probe ligand, let us say a hydrogen. This is done in Figure 2. Note the remarkable parallelism of the two overlaps. The H-CH_3 overlap is everywhere smaller than the H-MnL_5 overlap, but the dependence of both on the distance is quite similar.

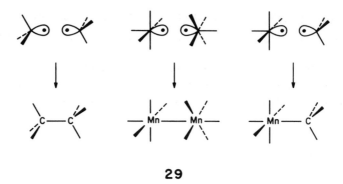

29

A word is needed to describe the resemblance of the two fragments, CH_3 and d^7 ML_5. They are certainly not isostructural, nor are they isoelectronic. However, both possess a frontier orbital which looks approximately the same for the two fragments. We will call two fragments *isolobal* if the number, symmetry properties, approximate energy and shape of the frontier orbitals and the number of electrons in them are similar—not identical, but similar.[10] Thus CH_3 is isolobal with $Mn(CO)_5$. We will introduce a symbol for the isolobal relationsship: a "two-headed" arrow with half an orbital below. Thus,

$$CH_3 \longleftrightarrow Mn(CO)_5$$

Let's extend the definition a little.

(1) If $Mn(CO)_5$ is isolobal with CH_3, so are $Tc(CO)_5$ and $Re(CO)_5$, as well as $Fe(CO)_5^+$. The shape of the a_1 hybrid will vary slightly with different principal quantum number, but essentially it is only the d-electron count that matters.

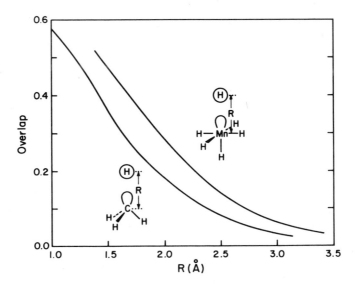

Figure 2. Overlap integrals between the a_1 frontier orbital of MnH_5^{5-} and CH_3 and a H 1s orbital at a distance R from the Mn or C.

(2) If $Mn(CO)_5$ is isolobal with CH_3, then $Cr(CO)_5$, $Mo(CO)_5$ or $W(CO)_5$ are isolobal with CH_3^+, and $Fe(CO)_5$ (square pyramidal!) is isolobal with CH_3^-.

(3) If $Mn(CO)_5$ is isolobal with CH_3, so are $Mn(PR_3)_5$ or $MnCl_5^{5-}$ or any d^7 ML_5 species. And so is $CpFe(CO)_2$, a ubiquitous fragment. The procedure here is to write $CpFe(CO)_2$ as $Cp^-Fe^+(CO)_2$ and to replace the Cp^- by its isolobal equivalent of three carbonyls, reaching $Fe(CO)_5^+$, isoelectronic with $Mn(CO)_5$.

Let us go on the ML_4 fragment. It is clear that a d^8 ML_4, e.g. $Fe(CO)_4$, is isolobal with a methylene or carbene, CH_2.

$$Fe(CO)_4 \longleftrightarrow{}_{\underset{O}{}} CH_2$$

As 30 reveals, both fragments have two electrons in delocalized a_1 and

b_2 orbitals which are the equivalent of two localized hybrids. There are explicable differences in the ordering of the two combinations. (11) The different ordering has, however, no grave consequences— recall that we are not so much interested in the fragments themselves as in their bonding capabilities. The moment we interact $Fe(CO)_4$ or CH_2 with another ligand, the initial ordering of a_1 and b_2 becomes relatively unimportant since both are typically strongly involved in the bonding.

Dimerize, conceptually, the isolobal fragments $Fe(CO)_4$ and CH_2. One gets the known ethylene, 31, the carbene-iron tetracarbonyl complex, 32, derivatives of which are known, (12) and $Fe_2(CO)_8$, 33. The last is an unstable molecule, so far observed only in a matrix. (13) We come here to a cautionary note on the isolobal analogy. The isolobal analogy carries one between organic and inorganic molecules of similar electronic structure. But there is no guarantee that the result of such an isolobal mapping $31 \longleftrightarrow{}_{\underset{O}{}} 33$ leads one to a molecule of great kinetic stablity. It might, and it might not.

Lest the reader be concerned about this limitation of the analogy let me remind him or her of what happens as one proceeds from ethylene down Group IV. Si, Ge, Sn, Pb substitution leads to olefin analogues, but they are kinetically and thermodynamically so unstable that it has taken great effort to provide evidence for their fleeting existence.

$Fe_2(CO)_8$, 33, has π and π^* levels similar to those of ethylene. But the low energy of its π^* makes this molecule coordinatively unsaturated. It can, for instance, add another CO to reach the stable diiron enneacarbonyl. More interesting, as we will soon see, it is the strategy of stabilizing the unstable $Fe_2(CO)_8$ by making a complex of it, just as is routinely done for unstable organic molecules 1–3.

$Fe(CO)_4$, $Ru(CO)_4$, or $Os(CO)_4$ may be trimerized in various combinations with methylene, in 34–37. These cyclopropanes, ranging from all organic to all inorganic, are known. But note that when I show the "all-metallic" three-

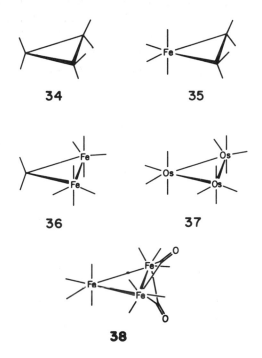

membered ring I have to go to Os. It is well known that the ground state structure of $Fe_3(CO)_{12}$ is 38, with two bridging carbonyls. (14) Another limitation of the isolobal analogy is exposed here: the unbridged Fe structure analogous to 37 is certainly not much higher in energy than 38, but nevertheless the lowest energy structure is bridged. Movement of some ligands (e.g. carbonyls, but not phosphines) in and out of bridging sites is an experimental reality, a facile process, for transition metal complexes, especially of the first transition series. Such easy terminal to bridging interconversions are rare in organic chemistry, with the exception of carbonium ions. Bridging in inorganic structures, when it does occur, does not cause a major perturbation in the nature of the frontier orbitals.

$$d^9 \; ML_3 \quad \longleftrightarrow \quad CR$$

39

Consider next the $d^9 \; ML_3$ fragment, e.g. $Co(CO)_3$. This is isolobal with a carbyne CH, as 39 shows. Once again there are differences, though of no great significance, in the a_1 versus e energy ordering between the two fragments. Their similarity is revealed most strikingly by the existence of the entire series of mixed organic and inorganic tetrahedranes, 40–44. To be sure, 41 can be called a cyclopropenium complex and 42 a binuclear acetylene complex, but I do believe that something is gained in seeing the entire series as a progression of isolobal substitution.

40   **41**   **42**

43   **44**

The fundamentals of the isolobal analogy have now been exposed. Just how far reaching the relationships written down here

$$d^7 \; ML_5 \quad \longleftrightarrow \quad CH_3$$

$$d^8 \; ML_4 \quad \longleftrightarrow \quad CH_2$$

$$d^9 \; ML_3 \quad \longleftrightarrow \quad CH$$

are will become clearer in a while. For the moment it is important to note that the isolobal analogy is not solely the creation of my research group. In his fruitful explorations of the reactivity of d^8 square planar complexes J. Halpern often made use of the similarity of such an ML_4 entity to a carbene. (9a) He did the same for d^7 ML_5 and organic free radicals. (9b) L. F. Dahl, in a beautiful series of structural studies of transition metal clusters, saw clearly the relationship of the orbitals of an ML_n fragment to a chalcogen or pnicogen atom, which of course are easily related to CR. (15, 16) And most importantly, K. Wade (17) and D. M. P. Mingos (18) independently developed a comprehensive and elegant picture of the electronic structure of transition metal clusters by relating them to the polyhedral boron hydrides (which W. N. Lipscomb and I studied – the circle closes!) It is a trivial step from BH to CH^+. All of these workers saw the essence of the isolobal analogy.

STRUCTURAL IMPLICATIONS OF THE ISOLOBAL ANALOGY

How quickly do the hands and mind of man provide us with the problem of choice! The molecules I would need to illustrate the isolobal analogy at work did not exist thirty years ago. Now they are around us, in superabundance. I have made a selection, based in part on the ease with which these lovely molecules illustrate the principles, in part on the ambiguous and ephemeral basis of recent (1981) appearance in the literature.

One obvious use of the isolobal analogy is in the structural sense. The analogy allows us to see the simple essence of seemingly complex structures. I should like to show you some examples centering on the ML_4 fragment.

Last summer there appeared a structure of the cluster $HRe_3(CO)_{12}$ $Sn(CH_3)_2$ from the work of H. D. Kaesz and collaborators. (19) The unique hydrogen was not located; presumably it bridges one Re-Re bond. If we remove the hydrogen as a proton, a convention we have found useful, [8n] we reach Re_3 $(CO)_{12}$ $SnMe_2^-$.45. Not a usual molecule, but the isolobal chain

$$Re(CO)_4^- \longleftrightarrow Fe(CO)_4 \longleftrightarrow CR_2 \longleftrightarrow SnR_2$$

allows us immediately to see the very close similarity of this structure to the previously known 46 (20) and 47 (21).

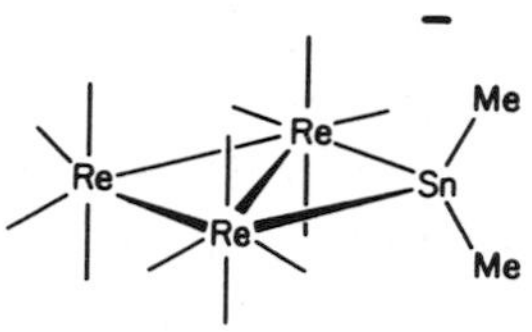

45

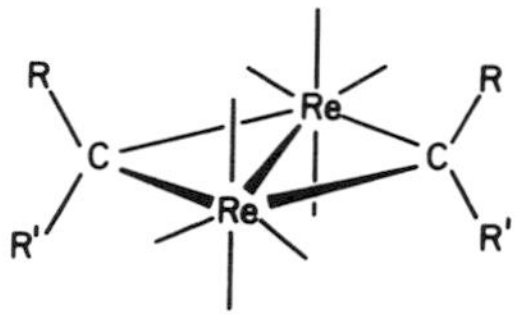

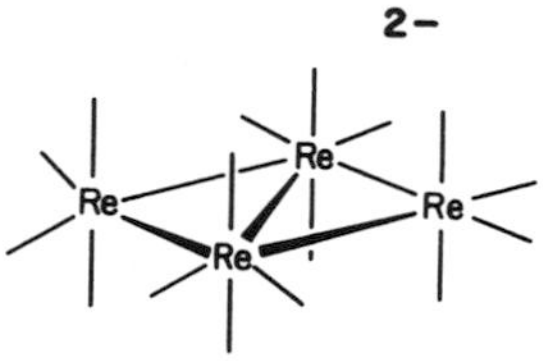

46

47

It is interesting to speculate when we might see the missing members of the series on the organic side, $(CO)_4 Re(CH_2)_3^+$ and $(CH_2)_4^{2+}$.

Two ML_4 fragments united yield the forementioned unstable $Fe_2(CO)_8$ system, **48**. It was hinted before that one should think of stabilizing this species

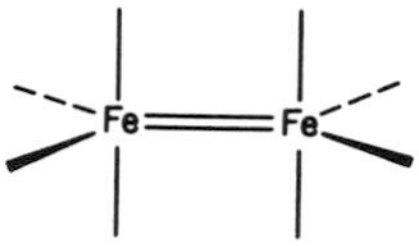

48

by complexation. A pretty example is at hand, **49**. (22) Two $Fe_2(CO)_8$ units are complexed by a tin atom! Note the pinning back of the equatorial carbonyls, analogous to the bending back of hydrogens in a transition metal complexed olefin. Alternatively, and interestingly, this is spiropentane.

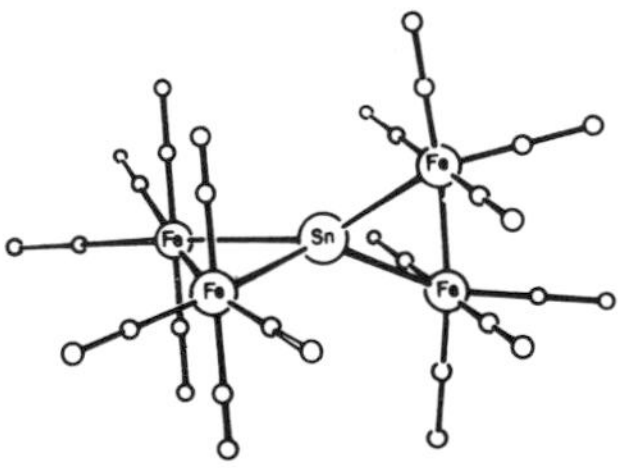

49

Earlier in 1981, J. Lewis, B. F. G. Johnson and their collaborators published a synthesis and structure of $Os_5(CO)_{19}$. (23) The structure appears terribly complicated, **50**, until one realizes it is really **51**, a typical trigonal bipyramidal $Os(CO)_5$ derivative, with two of the equatorial carbonyls substituted by olefins, or rather by the $Os_2(CO)_8$ olefin analoques. The "olefin" orientation is just as it should be. (1k, o)

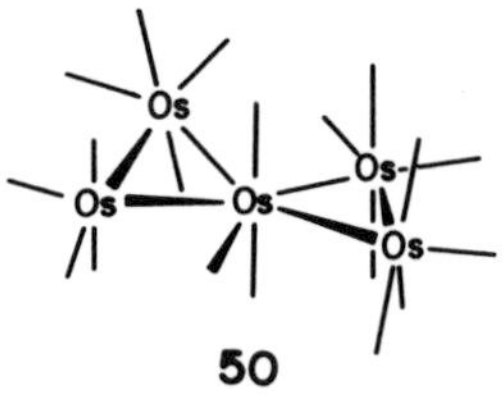

50 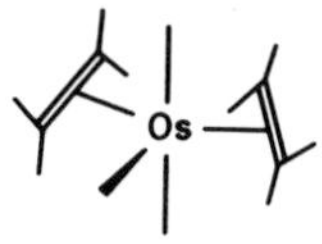   **51**

Another system isolobal to ethylene and $Fe_2(CO)_8$ is the "mixed dimer" $(CO)_4FeCH_2$. Upon formally shifting an electron from the metal to the carbon, a bit of alchemy, one gets to a phosphido complex, 52:

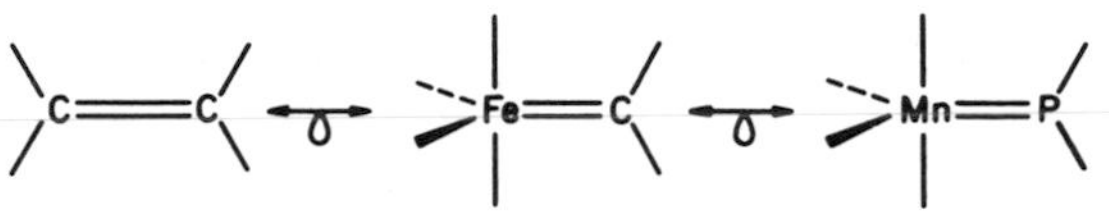

52

The reason for this transformation sequence is that complexes of $Mn(CO)_4$ PR_2 have been made. P. Braunstein, D. Grandjean and coworkers have reported a remarkable set of structures, among them the three shown with their isolobal analogues in 53–55. (24) In each structure we can see the obvious ethylene-like $(CO)_4 MnPR_2$ entity.

53

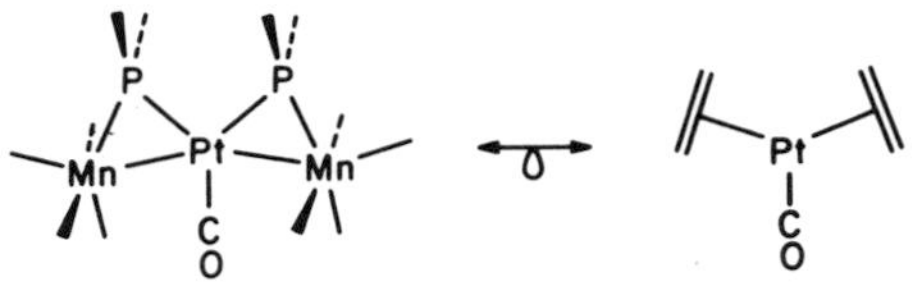

54

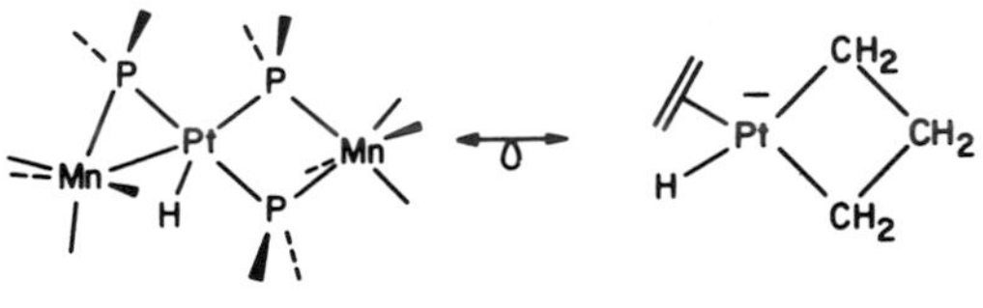

55

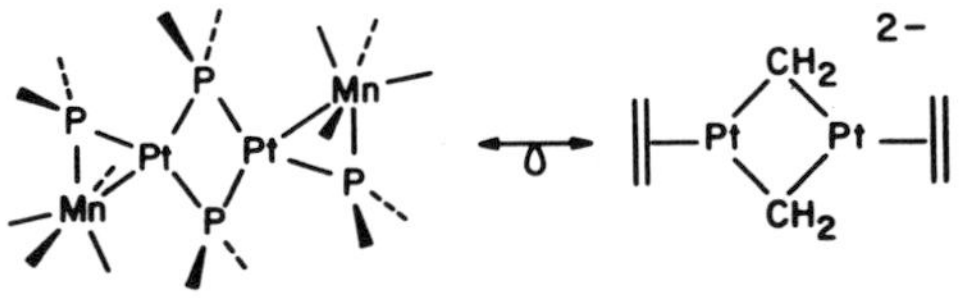

At the same time that these structures were published there appeared a structure of 56, synthesized in an entirely different way by R. J. Haines, N. D. C. T. Steen and R. B. English. (25) Unbridge the two semibridged carbonyls, do a bit more of electronic alchemy relating Mn-Pt to Fe-Rh, and the relationship to 55 becomes crystal clear.

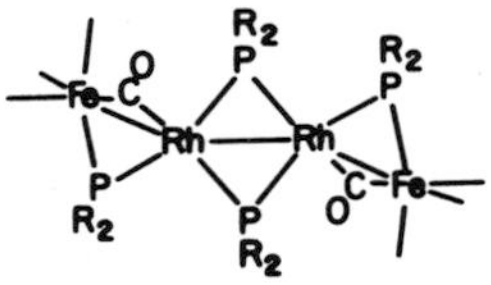

56

Realizing that $Fe(CO)_4$ ⟷ $CpFe(CO)^-$ ⟷ $CpRh(CO)$ we see immediately that 57 is still another $Fe_2 (CO)_8$ analogue.

57

Thus, the W. A. Herrmann methylene complex, 58, is a two-thirds inorganic cyclopropane. (26)

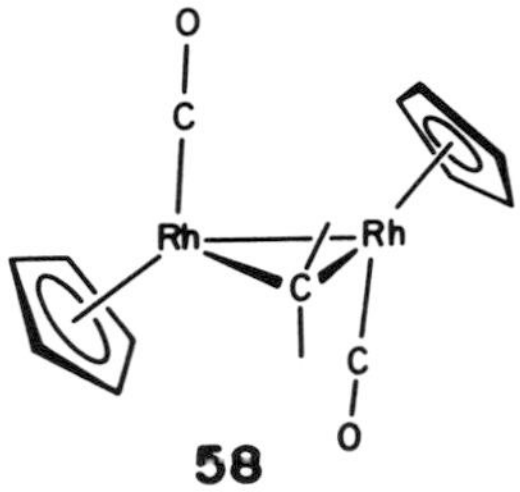

58

There are a few more fascinating $Cp_2 Rh_2 (CO)_2$ structures to be shown, but first we need to examine one extension of the isolobal concept.

THE RELATIONSHIP BETWEEN ML_n and ML_{n-2} FRAGMENTS

Earlier in the discussion we looked at two octahedral fragments, ML_5 and ML_4, in which a pair of axial ligands remained. If we remove these ligands, 59, an interesting extension of the isolobal analogy emerges.

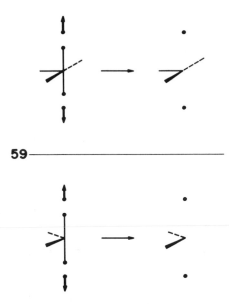

If the z axis is oriented along the direction of the vanishing ligands, then it is clear that the main result of this perturbation is that the metal d_{z^2} is lowered in energy. It returns from the metal-ligand σ antibonding manifold to become a non-bonding orbital, 60, 61. (1o, 3c)

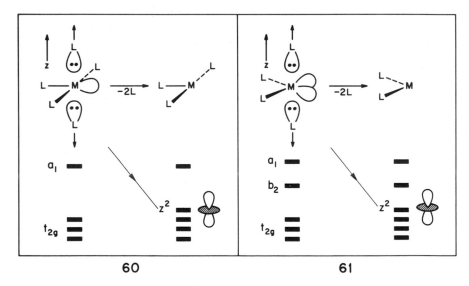

60   61

The high-lying orbitals (one in ML_5, two in ML_4) remain. (27) The obvious relationships that emerge then are those between a d^n ML_5 and a d^{n+2} C_{2v} or T-shaped ML_3; and between d^n $C_{2v}ML_4$ and d^{n+2} ML_2. Or to put it explicitly in terms of the most common fragments, 62.

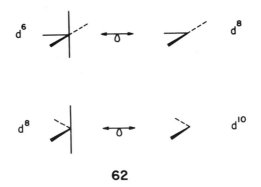

62

So we now add some further diversity to the non-isomorphic mapping which is the isolobal relationship:

$$CH_3^+ \quad \longleftrightarrow \quad Cr(CO)_5 \quad \longleftrightarrow \quad PtCl_3^-$$

$$CH_2 \quad \longleftrightarrow \quad Fe(CO)_4 \quad \longleftrightarrow \quad Ni(PR_3)_2$$

An obvious application is to olefin complexes; the similarity of $(CO)_4Fe(C_2H_4)$ and $(PR_3)_2 Ni(C_2H_4)$, and that of $(CO)_5 Cr(C_2H_4)$ and Zeise's salt emerges directly. (1*o, x*)

Returning to the $[CpRh(CO)]_2$ analogue of ethylene, we can now think about some other structures. First, it turns out that moving the carbonyls into the bridge does little to change the ethylene-like nature of the dirhodium fragment, 63. (1*u*)

$$\underset{/}{\overset{\backslash}{\text{C}}}=\underset{\backslash}{\overset{/}{\text{C}}} \quad \longleftrightarrow \quad \text{Fe}=\text{Fe} \quad \longleftrightarrow \quad \text{Rh}=\text{Rh} \quad \longleftrightarrow \quad \text{Cp}-\text{Rh}=\text{Rh}-\text{Cp}$$

63

Since $CH_2 \longleftrightarrow Fe(CO)_4 \longleftrightarrow Pt(CO)_2 \longleftrightarrow Rh(CO)_2^-$ it is possible to see in the compound of R. G. Bergman and coworkers, 64, (28) an analogue of W. A. Herrmann's 58.

64

Essentially the same fragment, 63, reappears in the fantastic $[Cp'Rh(CO)]_4$ Pt structure of F. G. A. Stone and coworkers, 65, (29) $(Cp'=\eta^5-C_5Me_5)$ and can be related to the $(RC{\equiv}CR)_2Pt$ structure earlier synthesized by the same group, 66. (30)

65

66

INTO THE t_{2g} SHELL

It turns out that not only is d^9 $Co(CO)_3$ isolobal with CH, but so is d^5 $CpW(CO)_2$. To see how this comes about let us first relate the Cp complex to a simple ML_n.

$$CpW(CO)_2 \quad \overleftrightarrow{\quad 0 \quad} \quad CpCr(CO)_2 \quad \overleftrightarrow{\quad 0 \quad} \quad Cr(CO)_5^+$$

As was shown earlier, $Cr(CO)_5^+$ is isolobal with CH_3^{2+}. That is not a very productive analogy. So let us examine $Cr(CO)_5^+$ in more detail. The electronic structure of an ML_5 fragment was given earlier. It is repeated in more detail at left in 67. The ML_5-CH_3 analogy concentrates on the hybrid of σ (a_1) symmetry. But the t_{2g} set, even if it is less "directional" than the hybrids, has extent in space and well-defined symmetry properties. In particular, two of the t_{2g} orbitals are of π pseudosymmetry, one of δ. If, as we are forced to do by the electron deficiency, we extend our view at least to the π component of the t_{2g} set (dotted lines at right in 67), we see a clear relationship between d^5 ML_5 and CH, just as there is between d^7 ML_5 and CH_3.

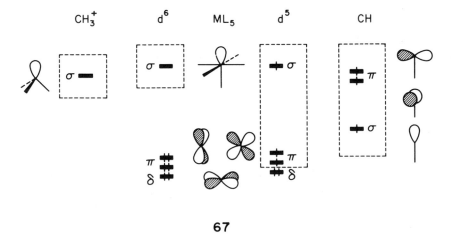

67

A little further reflection will show that by using one half of the π set of the t_{2g} we can get a relationship between d^6 ML_5 and CH_2.

To summarize:

$$d^7 \ ML_5 \ \rightleftharpoons \ CH_3$$

$$d^6 \ ML_5 \ \rightleftharpoons \ CH_2$$

$$d^5 \ ML_5 \ \rightleftharpoons \ CH$$

or to put it another way

$$d^6 \ ML_5 \ \rightleftharpoons \ CH_3^+$$

$$d^6 \ ML_5 \ \rightleftharpoons \ CH_2$$

$$d^6 \ ML_5 \ \rightleftharpoons \ CH^-$$

This gives us another way of looking at things, a *deprototonation analogy*. In what way is CH_3^+ like CH_2 or like CH^-? Let us draw out their orbitals schematically, including CH_4 for good measure, in 68.

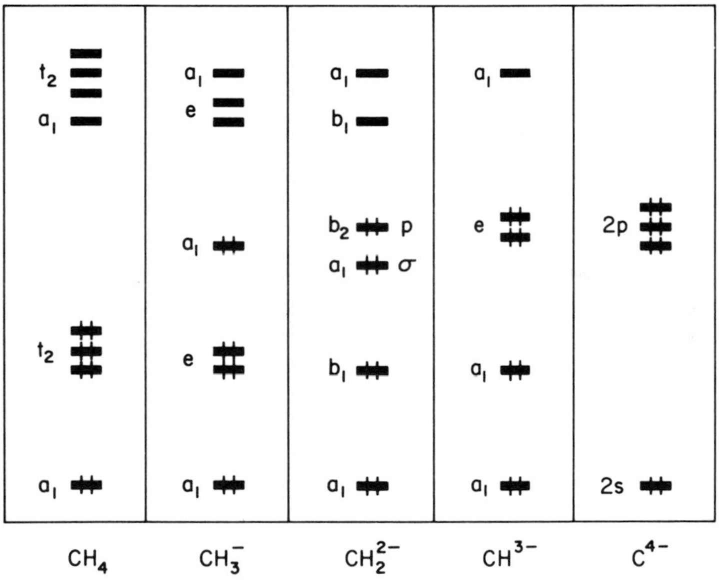

68

Taking away a proton from each molecular fragment does not change its ability to function as a donor (though its quality or donor strength will be *very* different). Each fragment, from CH_4 to C^{4-}, is in principle an eight electron donor.

To recapitulate: the isolobal analogy is not a one-to-one mapping. A $d^6 \ ML_5$ fragment is isolobal with CH_3^+ and CH_2 and CH^-. This is why the d^5 $CpW(CO)_2$ is isolobal with CH.

The isolobal analogy for low d-electron count metals has been exploited most notably in the work of F.G.A. Stone's group at Bristol. Just four compounds from their many beautiful examples are shown in 69–72. (31)

Since d^6 $Cr(CO)_5$ ←→ CH_2 ←→ d^{10} $Pt(PR_3)_2$ 69 is cyclopropane. Since $CpW(CO)_2$ is isolobal with CR, 70 is cyclopropene. Both isomers 71 and 72 are related to $(CO)_3Fe(cyclobutadiene)$, 1, or for that matter to the organic square pyramidal $C_5H_5^+$. (32)

FROM INORGANIC TO ORGANIC CHEMISTRY

The psychological direction of the isolobal analogy in general has been to make one feel more comfortable about the structures of complex inorganic molecules by relating them to known, presumably simpler, organic molecules. It is interesting to reverse this process and think about as yet unsynthesized organic structures related to known inorganic ones. The mapping from one realm of chemistry to the other must be accompanied by the warning already given: there is no guarantee that the "product" of an isolobal transformation is as stable, kinetically or thermodynamically, as the "reactant". (33)

$Fe(CO)_3$ is isolobal with CH^+. Thus, 1 is related to $C_5H_5^+$, 73, (32) and the ubiquitous ferroles, 74, (34) are related to $C_6R_6^{2+}$, 75. 35) Another product of

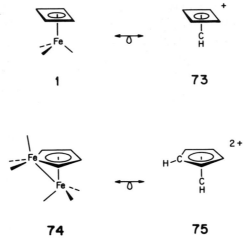

1   **73**

74   **75**

the interaction of acetylenes with iron carbonyls is the flyoverbridge, 76, a binuclear ring-opened fulvene complex. (34) The isolobal replacement carries over to 77. This is an unusual $C_8H_8^{2+}$ of C_2 symmetry, a hypothetical doubly homoallylic cation. It is not a geometry one would normally have thought of for

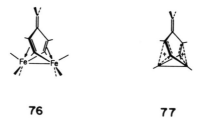

76   **77**

a heptafulvene dication, yet once reached by the isolobal mapping it appears to be geometrically reasonable. (36) More such mappings await exploitation.

FROM ORGANIC REACTION MECHANISMS TO INORGANIC ONES

R. J. Puddephatt, C. F. H. Tipper and co-workers have discovered a remarkable rearrangement of a platinacyclobutane, 78, in which a carbon adjacent to

78

the metal, with its substituents, exchanges in a very specific way with the carbon opposite the metal. (37) The labeling experiment of C. P. Casey that shows this most directly is given in 79. (38) How does this happen?

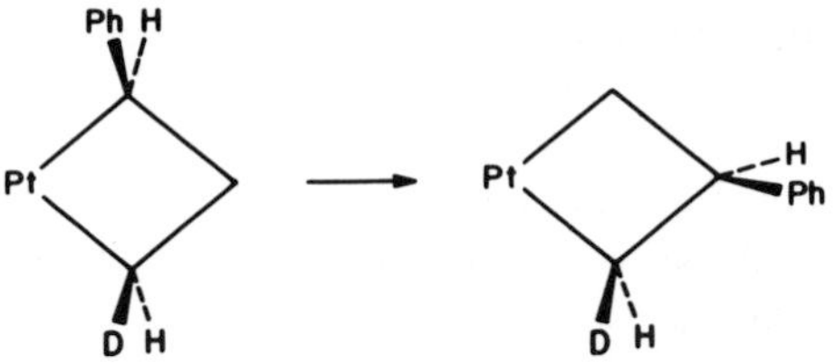

79

Kinetic evidence for a primary dissociative step to $Cl_2(py)Pt(CH_2)_3$ exists. (37) Suppose the ML_3 fragment can distort from its original T shape to a C_{3v} geometry. Since $d^8PtCl_2(py)$ is isolobal with CH^+ we can see a relationship to the cyclobutyl cation, 80.

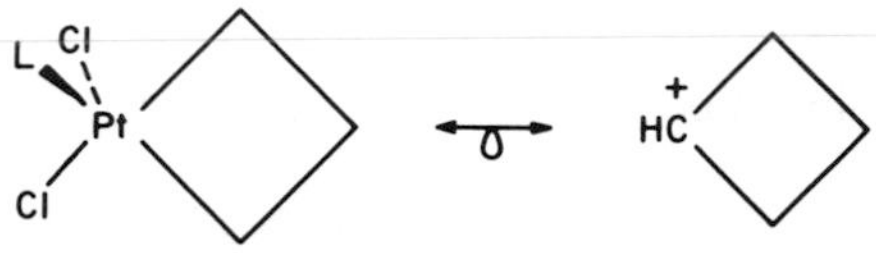

80

This association immediately brings to mind the entire complex of speculations and facts surrounding the facile rearrangement of cyclobutyl cations through bicyclobutonium waypoints. (39) The motions likely to occur are shown in 81. Ligand loss is followed by geometric reorganization at the metal, approach to a "bicyclobutonium structure", an itinerary around the periphery of a Jahn-Teller wheel through "cyclopropyl carbinyl" waypoints and exit through an isomeric "bicyclobutonium" structure. This is but one instance among many where the isolobal analogy is useful in moving between organic and inorganic reaction mechanisms.

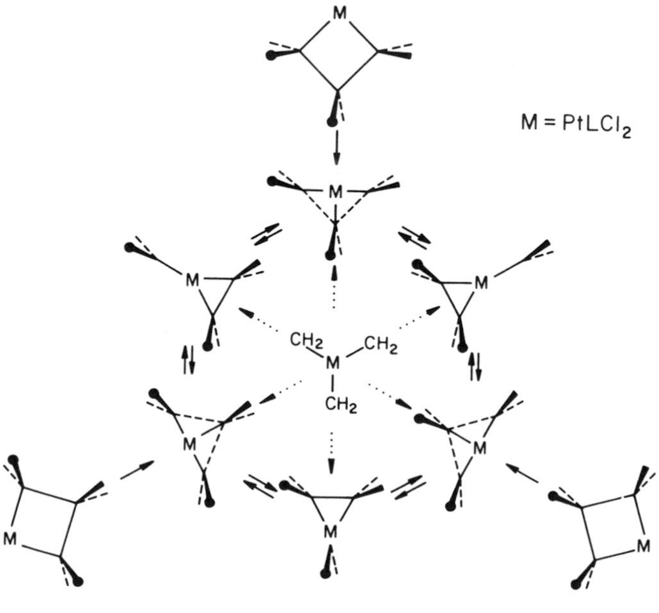

$M = PtLCl_2$

81

BEYOND THE OCTAHEDRON

The octahedron was a most useful starting point for generating fragment frontier orbitals, thereby engendering the isolobal analogy. But the octahedral polytype is not unique for six-coordinate complexes, and higher coordination numbers are feasible. We seek another more far-ranging derivation and find one based on the eighteen electron rule.

An (unoriginal) justification of this rule goes as follows: Consider n ligands, n ≤ 9, coming up to a metal with its 9 valence orbitals, **82**. A little group theory shows that for the octahedron and most, but not quite all, coordination geometries the n ligand orbitals will find a match in number, symmetry properties and extent in space among the hybrid sets that can be formed from the nine metal orbitals. The exceptions are very well understood. (40) Given this match, n M-L σ bonding combinations will go down in energy, n M-L σ * antibonding

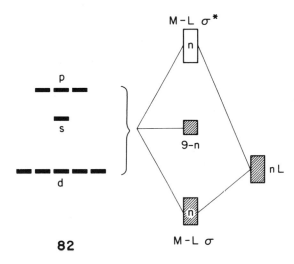

combinations will go up, and 9-n metal orbitals will remain relatively untouched, approximately non-bonding. The eighteen electron rule then is the statement: "Thou shalt not fill antibonding orbitals". Filling bonding (n) + nonbonding (9-n) orbitals leads to 9 electron pairs or eighteen electrons.

This "proof" is trivial but not silly. Upon a little reflection it will lay bare the limitations of the eighteen electron rule on the left and right side of the transition series, for special symmetry cases, and for weak-field ligands.

Next remove a ligand, a base, from the 18 electron complex. A localized hole on the metal, a directional hybrid, is created. The electron pair leaves with the ligand, **83**. To put it in another way, in some localized description of the

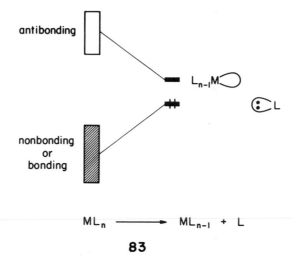

$$ML_n \longrightarrow ML_{n-1} + L$$

83

bonding, one M-L σ bond was formed by interaction of a ligand pair of electrons with a metal based hybrid. Reversing the process, breaking the bond, frees that hybrid.

A parallel analysis for main group elements leads to the octet rule, since only s and p are considered as valence orbitals. Hybrids are freed by removing ligands, so that CH_3^+ has one vacant directional orbital, CH_2^{2+} has two such.

The parallel between ML_n and EL_n fragments (M = transition metal, E = main group element) derives from the generation of similar hybrid patterns on removal of ligands from 18 or 8 electron configurations. For instance, if the octahedral polytope is used as a starting point, the eighteen electron rule is satisfied for a d^6 ML_6. The d^6 ML_5 will have one hybrid and no electrons in the gap between antibonding and bonding or nonbonding levels, just like CH_3^+, 84. d^6 ML_4 will have two empty hybrids, so will CH_2^{2+}. The common form of the isolobal analogy follows.

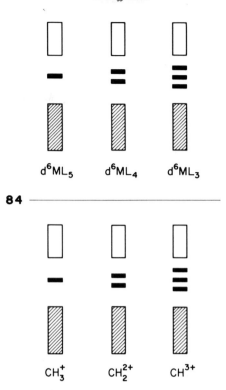

$d^6ML_5 \qquad d^6ML_4 \qquad d^6ML_3$

84

$CH_3^+ \qquad CH_2^{2+} \qquad CH^{3+}$

The advantage of this alternative derivation is that it is easily extended to higher coordination numbers. For instance in any of the multitude of seven coordinate geometries (8g) the 18 electron configuration is d^4. It follows immediately that for fragments derived from these seven-coordinate structures

$$d^5 \ ML_6 \ \longleftrightarrow \ CH_3$$

$$d^6 \ ML_5 \ \longleftrightarrow \ CH_2$$

$$d^7 \ ML_4 \ \longleftrightarrow \ CH$$

From an eight-coordinate starting point, (8p) where the 18 electron configuration is d^2:

$$d^3 \ ML_7 \ \longleftrightarrow \ CH_3$$

$$d^4 \ ML_6 \ \longleftrightarrow \ CH_2$$

$$d^5 \ ML_5 \ \longleftrightarrow \ CH$$

The conclusions may be summarized in Table 1. Note once again the non-isomorphic, many-to-one nature of the isolobal analogy. Also, the results of the previous section on "Into the t_{2g} Shell" are contained in the present discussion.

Table 1. Isolobal Analogies

| Organic Fragment | Transition Metal Coordination Number on which Analogy is Based | | | | |
	9	8	7	6	5
CH_3	d^1ML_8	d^3ML_7	d^5ML_6	d^7ML_5	d^9ML_4
CH_2	d^2ML_7	d^4ML_6	d^6ML_5	d^8ML_4	$d^{10}ML_3$
CH	d^3ML_6	d^5ML_5	d^7ML_4	d^9ML_3	

L = neutral two electron ligand

MISCELLANEA

The general rules in hand, the cautions understood, one can proceed to apply the isolobal analogy. Indeed my strategy has been to show the applications as I introduced extensions of the model. Here are some additional examples:

In a recent study of binuclear acetylene complexes the discussion focused on four structural types, 85–88. (41)

85   86   87   88

That these were isolobal with tetrahedrane, an olefin, bicyclobutane and cyclobutene was not only a curiosity, but actually made easier for us the complicated analysis of the interconversions of these molecules. And the isolobal analogy points to the synthesis of the as yet unknown "isomers" in the series, dimetallacyclobutadienes and butadienes, already known in complexed form.

Finally for amusement, consider the chain assembled by H. Vahrenkamp and coworkers, 89. (42) (No implication is made in the simplified drawing of the chain stereochemistry.) There had better a relationship to n-heptane, and so there is. We proceed using Table 1 as needed. $CpCr(CO)_3$ appears in a seven-coordinate guise here, so does $CpCr(CO)_2$; while $Co(CO)_3$ and $Fe(CO)_4$ are in a five-coordinate environment. It follows that:

$$CpCr(CO)_3 \rightleftharpoons d^5 ML_6 \rightleftharpoons CH_3$$

$$AsMe_2 \rightleftharpoons CH_2^-$$

$$CpCr(CO)_2 \rightleftharpoons d^5 ML_5 \rightleftharpoons CH_2^+$$

$$Co(CO)_3 \rightleftharpoons d^9 ML_3 \rightleftharpoons CH_2^+$$

$$Fe(CO)_4 \rightleftharpoons d^8 ML_4 \rightleftharpoons CH_3^+$$

Now it is simple—the inorganic chain **89** *is* really n-heptane.

89

ONE OF MANY BRIDGES

The isolobal analogy is a model. It is the duty of our scientific craft to push it to its extremes, and being only a model it is certain to fail somewhere. For any model, as ingenious a construction as it might be, is bound to abstract only a piece of reality. The reader has seen just how far the model can be pushed and he or she has seen where it breaks down.

The pleasing aspect of this particular model is that it brings together different subfields of our central science. We are separated, split asunder—organic, inorganic, physical, biological, analytical chemists—by the very largesse of our creation. The variety of molecules we create, and the methods we use to study them breed jargon and specialization. Yet underneath the seeming complexity there must be a deep unity. I think this approach would have pleased R. B. W.

ACKNOWLEDGMENT

It should be obvious to the reader that the spirit of this new line of work of my group owes much to what went on before. In particular I owe a direct debt of gratitude to my teachers M. P. Gouterman, W. N. Lipscomb, Jr., E. J. Corey, R. B. Woodward and my younger collaborators in the "organic days". E. L.

Muetterties helped me learn inorganic chemistry. But it is my coworkers—graduate students, postdoctoral fellows, senior visitors—who in long group meetings, patiently yet with inspiration, helped me shape this view of a piece of chemical experience. In rough chronological order of passing through Baker Lab in the "inorganic days" they are: M. Elian, N. Rösch, A. R. Rossi, J. M. Howell, K. Haraki, M. M.-L. Chen, D. M. P. Mingos, A. B. Anderson, P. D. Mollère, P. J. Hay, J. C. Thibeault, P. Hofmann, J. W. Lauher, R. H. Summerville, T. A. Albright, D. L. Thorn, D. L. DuBois, Nguyen Trong Anh, A. Dedieu, E. Shustorovich, P. K. Mehrotra, M.-H. Whangbo, B. E. R. Schilling, K. Tatsumi, J. K. Burdett, H. Berke, A. R. Pinhas, S. Shaik, E. D. Jemmis, D. Cox, A. Stockis, R. D. Harcourt, R. Bach, O. Eisenstein, R. J. Goddard, H. H. Dunken, P. Kubáček, D. M. Hoffman, C. Mealli, Z. Havlas, C. N. Wilker, T. Hughbanks, S.-Y. Chu, S. Wijeyesekera, C. Minot, S. Cain, S. Sung, M. Kertesz, C. Zheng. In the "organic days" they were preceded by C.-C. Wan, D. Hayes, J. R. Swenson, P. Clark, G. W. Van Dine, A. Imamura, G. D. Zeiss, W. J. Hehre, R. Gleiter, M. Gheorghiu, R. Bissell, R. R. Gould, D. B. Boyd, S. Z. Goldberg, B. G. Odell, S. Swaminathan, C. A. Zeiss, R. B. Davidson, R. C. Dobson, W.-D. Stohrer, J. E. Williams, A. Devaquet, C. C. Levin, H. Fujimoto, C. S. Kim, and L. Libit.

I am in the business of communicating ideas to people. The graphical aspect of this enterprise, be it lecture slides or published articles, is critical. Throughout these years nearly all of my drawings, containing countless "lined orbitals", have been expertly and beautifully executed by Jane S. Jorgensen and Elisabeth Fields, to whom I'm most grateful. The typing of my manuscripts and the associated details of production are the outcome of hard work by Eleanor R. Stolz and Eda J. Kronman, and I thank them for their help.

Throughout this period my research has been generously supported almost entirely by a research grant from the National Science Foundations' Quantum Chemistry Program. Other support has come from the Materials Science Center at Cornell, and smaller unrestricted research grants from Eli Lilly, Allied and Exxon Foundation.

REFERENCES

(1)   Some selected papers from my goup on this subject are the following:
    (a)   Elian, M. and Hoffmann, R., Inorg. Chem., 14, 1058 (1975).
    (b)   Hoffmann, R. and Hofmann, P., J. Amer. Chem. Soc., 98, 598 (1976).
    (c)   Lauher, J. W. and Hoffmann, R., ibid, 98, 1729 (1976).
    (d)   Lauher, J. W., Elian, M., Summerville, R. H. and Hoffmann, R., ibid, 98, 3219 (1976).
    (e)   Elian, M., Chen, M. M.-L., Mingos, D. M. P. and Hoffmann, R., Inorg. Chem., 15, 1148 (1976).
    (f)   Summerville, R. H. and Hoffmann, R., J. Amer. Chem. Soc., 98, 7240 (1976); 101, 3821 (1979).
    (g)   Hoffmann, R., Chen, M. M.-L. and Thorn, D. L., Inorg. Chem., 16, 503 (1977).
    (h)   Albright, T. A., Hofmann, P. and Hoffmann, R., J. Amer. Chem. Soc., 99, 7546 (1977).
    (i)   Thorn, D. L. and Hoffmann, R., Inorg. Chem., 17, 126 (1978).

(j)   Nguyen Trong Ahn, Elian, M. and Hoffmann, R., J. Amer. Chem. Soc., 100, 110 (1978).

(k)   Hoffmann, R., Albright, T. A. and Thorn, D. L., Pure Appl. Chem., 50, 1 (1978).

(l)   Dedieu, A. and Hoffmann, R., J. Amer. chem. Soc., 100, 2074 (1978).

(m)   Schilling, B. E. R., Hoffmann, R. and Lichtenberger, D. L., *ibid*, 101, 585 (1979).

(n)   Schilling, B. E. R. and Hoffmann, R., *ibid*, 101, 3456 (1979).

(o)   Albright, T. A., Hoffmann, R., Thibeault, J. C. and Thorn, D. L., *ibid*, 101, 3801 (1979).

(p)   Albright, T. A., Hoffmann, R., Tse, Y.-C. and D'Ottavio, T., *ibid*, 101, 3812 (1979).

(q)   Dedieu, A., Albright, T. A. and Hoffmann, R., *ibid*, 101, 3141 (1979).

(r)   Shaik, S., Hoffmann, R., Fisel, C. R. and Summerville, R. H., *ibid*, 102, 7667 (1980).

(s)   Goddard, R. J., Hoffmann, R. and Jemmis, E. D., *ibid*, 102, 7667 (1980).

(t)   Jemmis, E. D., Pinhas, A. R. and Hoffmann, R., *ibid*, 102, 2576 (1980).

(u)   Pinhas, A. R., Albright, T. A., Hofmann, P. and Hoffmann, R., Helv. Chim. Acta, 63, 29 (1980).

(v)   Hoffmann, R., Science, 211, 995 (1981).

(w)   Tatsumi, K. and Hoffmann, R., J. Amer. Chem. Soc., 103, 3328 (1981).

(x)   Eisenstein, O. and Hoffmann, R., *ibid*, 103, 4308 (1981); 102, 6148 (1980).

(y)   Beautiful contour plots of the frontier orbitals of the most important fragments have been made in a rare unpublished pamphlet: T. A. Albright "A Holiday Coloring Book of Fragment Molecular Orbitals", Cornell University, 1977. Some orbitals are reproduced in Reference 3a.

(2) Others have contributed much to the development of the fragment formalism in inorganic chemistry. For some early references see:

(a)   Burdett, J. K., J. Chem. Soc., Faraday Trans., 2, 70, 1599 (1974).

(b)   Whitesides, T. H., Lichtenberger, D. L. and Budnik, R. A., Inorg. Chem., 14, 68 (1975).

(c)   Kettle, S. F. A., J. Chem. Soc., A, 420 (1966); Inorg. Chem., 4, 1661 (1965).

(d)   Green, M. L. H., "Organometallic Compounds", Vol. 2, Methuen, London, 1968, p. 115.

(e)   Wade, K., Chem. Commun., 792 (1971); Inorg. Nucl. Chem. Lett., 8, 559, 563 (1972); "Electron Deficient Compounds", Nelson, London, 1971.

(f)   Mingos, D. M. P., Nature (London), Phys. Sci., 236, 99 (1972).

(g)   Braterman, P. S., Struct. Bonding (Berlin), 10, 57 (1972).

(h)   Cotton, F. A., Edwards, W. T., Rauch, F. C., Graham, M. A., Perutz, R. N. and Turner, J. J., J. Coord. Chem., 2, 247 (1973).

(i)   Korolkov, D. V. and Miessner, H., Z. Phys. Chem. (Leipzig), 253, 25 (1973).

(3) For three recent overviews of the fragment formalism by some of the most active workers in the field, see:

(a)   Burdett, J. K. "Molecular Shapes", Wiley Interscience, New York 1980.

(b)   Mingos, D. M. P. in "Comprehensive Organometallic Chemistry" ed. Wilkinson, G., Stone, F. G. A. and Abel, E. W., Pergamon, Oxford 1982.

(c)   Albright, T. A., Tetrahedron, in press.

(4) (a)   Pauling, L. "The Nature of the Chemical Bond", 3rd Ed., Cornell Univ. Press, Ithaca, New York 1960.

(b)   In recent times L. Pauling has returned to the problem of the electronic structure of transition metal complexes. See Pauling, L., Proc. Nat. Acad. Sci. USA, 72, 3799, 4200 (1975); 73, 274, 1403, 4290 (1976); 74, 2614, 5235 (1977); 75, 12, 569 (1978); Acta Crystallogr., Sect. B, 34, 746 (1978).

(5) For a leading reference see C. J. Ballhausen "Introduction to Ligand Field Theory", McGraw-Hill, New York, 1962.

(6) An excellent introduction to the conventions and achievements of organometallic chemistry is J. P. Collman and L. S. Hegedus "Principles and Applications of Organotransition Metal Chemistry", University Science Books, Mill Valley, 1980.

(7)   For an introduction to the methodology see E. Heilbronner and H. Bock "Das HMO Modell und seine Anwendung", Verlag Chemie, Weinheim, 1968; W. L. Jorgensen and L. Salem "The Organic Chemist's Book of Orbitals", Academic Press, New York, 1973; R. Hoffmann, Accts. Chem. Res., 4, 1 (1971).

(8)   (a)   Rösch, N. and Hoffmann, R., Inorg. Chem., 13, 2656 (1974).

      (b)   Hoffmann, R., Chen, M. M.-L., Elian, M., Rossi, A. R. and Mingos, D. M. P., *ibid*, 13, 2666 (1974).

      (c)   Rossi, A. R. and Hoffmann, R., *ibid*, 14, 365 (1975).

      (d)   Hay, P. J., Thibeault, J. C. and Hoffmann, R., J. Amer. Chem. Soc., 97, 4884 (1975).

      (e)   Hoffmann, R., Howell, J. M. and Rossi, A. R., *ibid*, 98, 2484 (1976).

      (f)   Komiya, S., Albright, T. A., Hoffmann, R. and Kochi, J., ibid, 98, 7255 (1976); 99, 8440 (1977).

      (g)   Hoffmann, R., Beier, B. F., Muetterties, E. L. and Rossi, A. R., Inorg. Chem., 16, 511 (1977).

      (h)   Barnett, B. L., Krüger, C., Tsay, Y.-H., Summerville, R. H. and Hoffmann, R., Chem. Ber., 110, 3900 (1977).

      (i)   Hoffmann, R., Thorn, D. L. and Shilov, A. E., Koordinats. Khim. 3, 1260 (1977).

      (j)   DuBois, D. L. and Hoffmann, R., Nouv. J. Chim., 1, 479 (1977).

      (k)   Thorn, D. L. and Hoffmann, R., J. Amer. Chem. Soc., 100, 2079 (1978); Nouv. J. Chim., 3, 39 (1979).

      (l)   Albright, T. A. and Hoffmann, R., Chem. Ber., 111, 1578 (1978); J. Amer. Chem. Soc., 100, 7736 (1978).

      (m)   Albright, T. A., Hoffmann, R. and Hofmann, P., Chem. Ber., 111, 1591 (1978).

      (n)   Hoffmann, R., Schilling, B. E. R., Bau, R., Kaesz, H. D. and Mingos, D. M. P., J. Amer. Chem. Soc., 100. 6088 (1978).

      (o)   Mehrotra, P. K. and Hoffmann, R., Inorg. Chem., 17, 2187 (1978).

      (p)   Burdett, J. K., Hoffmann, R. and Fay, R. C., *ibid*, 17, 253 (1978).

      (q)   Schilling, B. E. R. and Hoffmann, R., J. Amer. Chem. Soc., 100, 7224 (1978); Acta Chem. Scand. B33, 231 (1979).

      (r)   Berke, H. and Hoffmann, R., J. Amer. Chem. Soc., 100, 7224 (1978).

      (s)   Schilling, B. E. R., Hoffmann, R. and Faller, J. W., *ibid*, 101, 592 (1979).

      (t)   Pinhas, A. R. and Hoffmann, R., Inorg. Chem., 18, 654 (1979).

      (u)   Krüger, C., Sekutowski, J. C., Berke, H. and Hoffmann, R., Z. Naturf., 33b, 1110 (1978).

      (v)   Jemmis, E. D. and Hoffmann, R., J. Amer. Chem. Soc., 102, 2570 (1980).

      (w)   Stockis, A. and Hoffmann, R., *ibid*, 102, 2952 (1980).

      (x)   Tatsumi, K., Hoffmann, R. and Whangbo, M.-H., J. Chem. Soc., Chem. Commun., 509, (1980).

      (y)   Tatsumi, K. and Hoffmann, R., Inorg. Chem., 19, 2656 (1980); J. Am. Chem. Soc., 103, 3328 (1981); Inorg. Chem., 20, 3781 (1981).

      (z)   McKinney, R. J., Thorn, D. L., Hoffmann, R. and Stockis, A., J. Amer. Chem. Soc., 103, 2595 (1981).

      (aa) Tatsumi, K., Hoffmann, R., Yamamoto, A., Stille, J. K., Bull. Chem. Soc., Jpn., 54 1857 (1981).

      (ab) Kubáček, P. and Hoffmann, R., J. Amer. Chem. Soc., 103, 4320 (1981).

      (ac) Eisenstein, O., Hoffmann, R. and Rossi, A. R., *ibid*, 103, 5582 (1981).

      (ad) Kamata, M., Hirotsu, K., Higuchi, T., Tatsumi, K., Hoffmann, R. and Otsuka, S., *ibid*, 103, 5772 (1981).

      (ae) Cox, D. N., Mingos, D. M. P. and Hoffmann, R., J. Chem. Soc., Dalton Trans., 1788 (1981).

      (af) Hoffman, D. M. and Hoffmann, R., Inorg. Chem., 20, 3543 (1981).

(9)   (a)   Halpern, J., Advances in Chemistry; "Homogeneous Catalysis", Ser. 70, 1 (1968); Disc. Far. Soc., 46, 7 (1968).

      (b)   Kwiatek, J. and Seyler, J. K., *ibid*, 70, 207 (1968).

(10) The term "isolobal" was introduced in Reference le, but the concept, as we will see below, is older.

(11) In methylene a_1 is below b_2 because the latter is a pure p orbital, while the former has some s character. In $Fe(CO)_4$ the a_1 and b_2 can be thought of as being derived from the e_g orbital (z^2, x^2-y^2) of the octahedron. Removal of two *cis* ligands stabilizes x^2-y^2 (b_2) more than it does z^2 (a_1). Were the ligands removed *trans*, to reach the square planar ML_4 fragment, the situation would be reversed. See also Reference 3.

(12) Fischer, E. O., Beck, H.-J., Kreiter, C. G., Lynch, J., Müller, J. and Winkler, E., Chem. Ber., 105, 162 (1975); Pfiz, R. and Daub, J., J. Organometal. Chem., 152, C32−C34 (1978); LeBozec, H., Gorgues, A. and Dixneuf, P. H., J. Amer. Chem. Soc., 100, 3946 (1978); Lappert, M. F. and Pye, P. L., J. Chem. Soc., Dalton Trans., 2172 (1977).

(13) Poliakoff, M. and Turner, J. J., J. Chem. Soc., A, 2403 (1971).

(14) Wei, C. H. and Dahl, L. F., J. Amer. Chem. Soc., 91, 1351 (1969).

(15) Foust, A. S., Foster, M. S. and Dahl, L. F., J. Amer. Chem. Soc., 91, 5631−5633 (1969) and references therein.

(16) See also Ellis, J. E., J. Chem. Educ., 53, 2 (1976) for an extensive exposition of this analogy.

(17) Wade, K., Adv. Inorg. Chem. Radiochem., 18, 1 (1976); Chemistry in Britain, 11, 177 (1975) and Reference 2 e.

(18) Mingos, D. M. P., Adv. Organometal. Chem., 15, 1 (1977); Mason, R. and Mingos, D. M. P., MTP Int. Rev. Sci. Phys. Sci. Ser. II, 11, 121 (1975); Mingos, D. M. P., Trans. Amer. Cryst. Assoc., 16, 17 (1980); Reference 2f and 3b.

(19) Huie, B. T., Kirtley, S. W., Knobler, C. B. and Kaesz, H. D., J. Organometal. Chem., 213, 45 (1981).

(20) Fischer, E. O., Lindner, T. L., Fischer, H., Huttner, G., Friedrich, P. and Kreissl, F. R., Z. Naturforsch., 32b, 648 (1977).

(21) Bau, R., Fontal, B., Kaesz, H. D. and Churchill, M. R., J. Amer. Chem. Soc., 89, 6374 (1967).

(22) Cotton, J. D., Duckworth, J., Knox, S. A. R., Lindley, P. F., Paul, I., Stone, F. G. A. and Woodward, P., Chem. Comm. 253 (1966). Cotton, J. D., Knox, S. A. R., Paul, I. and Stone, F. G. A., J. Chem. Soc. (A), 264 (1967); Lindley, P. F. and Woodward, P., *ibid*, 382 (1967).

(23) Farrar, D. H., Johnson, B. F. G., Lewis, J., Nicholls, J. N., Raithby, P. R. and Rosales, M. J., J. Chem. Soc., Chem. Commun., 273 (1981).

(24) Braunstein, P., Matt, D., Bars, O., Louër, M., Grandjean, D., Fischer, J. and Mitschler, A., J. Organometal. Chem., 213, 79 (1981).

(25) Haines, R. J., Steen, N. D. C. T. and English, R. B., J. Chem. Soc., Chem. Commun., 587 (1981).

(26) Herrmann, W. A., Krüger, C., Goddard, R. and Bernal, I., Angew. Chem., 89, 342 (1977).

(27) There are some further details, not discussed here but treated elsewhere (Reference 1o, 1x, 3c). In particular the ML_2 will have another relatively low-lying orbital when L is a π acceptor.

(28) Jones, W. D., White, M. A. and Bergman, R. G., J. Am. Chem. Soc., 100, 6770 (1978).

(29) Green, M., Howard, J. A. K., Mills, R. N., Pain, G. N., Stone, F. G. A. and Woodward, P., J. Chem. Soc., Chem. Commun., 869 (1981).

(30) Boag, N. M., Green, M., Grove, D. M., Howard, J. A. K., Spencer, J. L. and Stone, F. G. A., J. Chem. Soc., Dalton Trans., 2170 (1980).

(31) (a) Stone, F. G. A., Accounts Chem. Res., 14, 318 (1981) and references therein; Busetto, L., Green, M., Howard, J. A. K., Hessner, B., Jeffery, J. C., Mills, R. M., Stone, F. G. A. and Woodward, P., J. Chem. Soc., Chem. Commun., 1101 (1981). Ashworth, T. V., Chetcuti, M. J., Farrugia, L. J., Howard, J. A. K., Jeffery, J. C., Mills, R., Pain, G. N., Stone, F. G. A. and Woodward, P., A. C. S. Symposium Series 155 "Reactivity of Metal-Metal Bonds", ed. M. H. Chisholm, Washington, 1981, p. 299−313.

(b) For further molecules of type **71** and **72** see Sappa, E., Manotti Lanfredi, A. M. and Tiripicchio, A., J. Organometal. Chem., 221, 93 (1981); Shapley, J. T., Park, J. T., Churchill, M. R., Bueno, C. and Wasserman, H. J., J. Am. Chem. Soc., 103, 7385

(1981); Jaouen, G., Marinetti, A., Mentzen, B., Mutin, R., Saillard, J.-Y., Sayer, B. G. and McGlinchey, M. J., Organometallics, 1, 753 (1982).

(32) (a)   Stohrer, W.-D. and Hoffmann, R., J. Amer. Chem. Soc., 90, 1661 (1972).

    (b)   Williams, R. E., Inorg. Chem., 10, 210 (1971); Adv. Inorg. Chem. Radiochem., 18, 67 (1976).

    (c)   Masamune, S., Sakai, M. and Ona, H., J. Amer. Chem. Soc., 94, 8955 (1972); Masamune, S., Sakai, M., Ona, H. and Jones, A. L., *ibid*, 94, 8956 (1972).

    (d)   Hart, H. and Kuzuya, M., ibid, 94, 8958 (1972).

(33) See in this context Chandrasekhar, J., Schleyer P. v. R., and Schlegel, H. B., Tetrahedron Lett., 3393 (1978).

(34) See Reference li for the relevant litterature citations.

(35) Hogeveen, H. and Kwant, P. W., Accounts Chem. Res., 8, 413 (1976). For some related main group structures see: Jutzi, P., Kohl, F., Hofmann, P., Krüger, C. and Tsay, Y.-H., Chem. Ber., 113, 757 (1980).

(36) But apparently not a stable structure after all: Sevin, A. and Devaquet, A., Nouv. J. Chim., 1, 357 (1977); Clark, T. and Schleyer, P. v. R., *ibid*, 2, 665 (1978).

(37) Al-Essa, R. J., Puddephatt, R. J., Thompson, P. J. and Tipper, C. F. H., J. Amer. Chem. Soc., 102, 7546 (1980) and references therein.

(38) Casey, C. P., Scheck, D. M. and Shusterman, A. J., J. Amer. Chem. Soc., 101, 4233 (1979).

(39) See Saunders, M. and Siehl, H.-U., *ibid*, 102, 6868 (1980) and references therein.

(40) The exceptions include $Zr(BH_4)_4$, $W(RCCR)_3$ (CO), Cp_3M and Cp_3MR, Cp_4U, UO_2L_6 among others. For the relevant references see Chu, S.-Y. and Hoffmann, R., J. Phys. Chem., 86, 1289 (1982), and Ref. 3a.

(41) Hoffman, D. M. and Hoffmann, R., to be published.

(42) Langenbach, H. J., Keller, E. and Vahrenkamp, H., J. Organometal. Chem., 191, 95 (1980).

Chemistry 1982

AARON KLUG

for his development of crystallographic electron microscopy and his structural elucidation of biologically important nuclei acid-protein complexes

THE NOBEL PRIZE FOR CHEMISTRY

Speech by Professor BO G. MALMSTRÖM of the Royal Academy of Sciences.
Translation from the Swedish text

Your Majesties, Your Royal Highnesses, Ladies and Gentlemen,

Life is order, death is disorder. A fundamental law of Nature states that spontaneous chemical changes in the universe tend toward chaos. But life has, during milliards (*American English* billions) of years of evolution, seemingly contradicted this law. With the aid of energy derived from the sun it has built up the most complicated systems to be found in the universe — living organisms. Living matter is characterized by a high degree of chemical organisation on all levels, from the organs of large organisms to the smallest constituents of the cell. The beauty we experience when we enjoy the exquisite form of a flower or a bird is a reflection of a microscopic beauty in the architecture of molecules.

The chemical order of life is not maintained by some mysterious vital force. The secret of life is instead to be found in chemical properties which are the consequence of the very structural organisation of atoms and molecules. Modern structural chemistry has here provided tools for opening the door to one of the greatest mysteries of science.

The structure of chemical substances, i.e. the exact position in space of all the atoms in a molecule, can generally be determined if the substance can be obtained in the form of crystals. When X-rays are scattered from the periodic arrangement of the atoms in a crystal, a specific pattern is formed, which can be recorded photographically and translated to the original structure with the aid of a complicated mathematical analysis. The principle of this method, called X-ray diffraction, has been known since the beginning of this century, and its discovery was awarded with a Nobel Prize in physics in 1915. Almost half a century elapsed, however, before the technique had developed to a point allowing the determination of the structure of the giant molecules which are the building blocks of life. In 1962 the Nobel Prizes in chemistry as well as in medicine were awarded to scientists who had studied the molecular structure of the key substances of life, proteins and nucleic acids.

A nucleic acid, DNA, is carrier of the traits of heredity in the cell. Consequently it possesses all information necessary to direct the entire chemical machinery of the cell, and it does so by determining which proteins the cell shall manufacture. Proteins in turn determine the chemical pattern of the cell by their ability to speed up certain chemical reactions. Life can thus be regarded as the result of an interplay between nucleic acids and proteins.

Giant molecules have a tendency to aggregate, and biological function is often associated with complicated molecular aggregates. In the chromosomes of the cell nucleus, for example, the hereditary material is present in the form of chromatin, a giant aggregate between DNA and thousands of protein molecules. In viruses, which represent the border between living and dead matter, there are simpler aggregates between nucleic acids and proteins. A virus can be said to be genetic material without a cell of its own, and the structure of viruses can provide clues to the more complicated organisation of the hereditary material in higher organisms.

Large molecular aggregates can seldom be obtained in a form which allows structural determination by X-ray diffraction. The investigator who has been awarded with this year's Nobel Prize in chemistry, *Aaron Klug*, has developed a method to study the structure of molecular aggregates from biological systems. His technique is based on an ingenious combination of electron microscopy with principles taken from diffraction methods. Electron microscopy has long been used to depict the structural components of the cell, but its power of resolution is often limited by a lack of contrast in the picture. Klug has shown that even pictures seemingly lacking in contrast may contain a large amount of structural information, which can be made available by a mathematical manipulation of the picture.

With this technique, in combination with other methods of structural chemistry, Klug has *inter alia* investigated viruses and chromatin of the cell nucleus. His virus studies have illuminated an important biochemical principle, according to which the complicated molecular aggregates in the cell are formed spontaneously from their components. The chromatin investigations have provided clues to the structural control of the reading of the genetic message in DNA. In a long-term perspective they will undoubtedly be of crucial importance for our understanding of the nature of cancer, in which the control of the growth and division of cells by the genetic material no longer functions.

Dr. Klug,

I have tried to say — in Swedish — that with your ingenious development of crystallographic electron microscopy you have given science an important tool for determining the chemical structure of complicated components in the most refined chemical systems found in the universe — living organisms. You have applied your methods to investigations of viruses and of chromatin, the complex molecular aggregate between DNA and proteins in cell nuclei, and your structural results have clarified important biochemical principles. It is for these fundamental contributions that the Royal Academy of Sciences has decided to award this year's Nobel Prize in Chemistry to you.

On behalf of the Academy I wish to convey to you our warmest congratulations, and I now ask you to receive your prize from the hands of His Majesty the King.

Aaron Klug

AARON KLUG

I was born in 1926 to Lazar and Bella (nee Silin) Klug in Zelvas, Lithuania, but remember nothing of the place, because I was brought to South Africa as a child of two and grew up there. My father was trained as a saddler, but in fact as a young man worked in his father's business of rearing and selling cattle, so he grew up in the countryside. He had a traditional Jewish education and secular schooling, and though not a conventionally well educated man, he had some gift for writing, and had a number of articles published in the newspapers of the capital, for which he acted as what would now be called a stringer. Shortly after I was born he emigrated to Durban, where members of my mother's family had settled at the turn of the century, and the rest of the family followed soon thereafter.

Durban was then a relatively sleepy town in subtropical surroundings. It was a fine place for a boy—there was the beach and the bush and school was not too taxing. I went to a good school, Durban High School, which was run on traditional English lines, with a curriculum somewhat adapted to South African circumstances. We had some good masters particularly in History and English. However, by the standards of to-day, there were few challenges other than Advanced Latin Prose Composition in the 6th Form. The philosophy of the school was quite simple—the bright boys specialised in Latin, the not so bright in science and the rest managed with geography or the like. There was a good library but it was the playing fields that kept one out of mischief. I did not feel a particularly strong call to any one subject, but read voraciously and widely and began to find science interesting. It was the book called *Microbe Hunters* by Paul de Kruif, well known in its time, which influenced me to begin medicine at university as a way into microbiology.

At the University of Witwatersrand in Johannesburg, I took the pre-medical course and, in my second year, I took, among other subjects, biochemistry, or physiological chemistry as it was then called, which stood me in good stead in later years when I came to face biological material. However, I felt the lack of a deeper foundation, and moved to chemistry and this in turn led me to physics and mathematics. So finally I took a science degree.

I had by then decided that I wanted to do research in physics and I went to the University of Cape Town which was then offering scholarships which enabled one to do an M.Sc. degree, in return for demonstrating in laboratory classes. The University lay in a beautiful site on the slopes of Table Mountain, which one climbed at week-ends. I was lucky to find as Professor there, R.W. James, the X-ray crystallographer, who had brought to Cape Town the traditions of the Bragg school at Manchester. He was an excellent teacher and I used to attend his undergraduate lectures as well as those in the M.Sc. course. From him I acquired a feeling for optics, and a knowledge of Fourier theory, and I remember particularly certain optical experiments on rather abstruse phenomena such as external and internal conical refraction which fascinated me. After taking my M.Sc. degree, I stayed on and worked on the X-ray analysis of some small organic compounds, in the course of which I developed a method of using molecular structure factors for solving crystal structures, and taught myself some quantum chemistry to calculate bond lengths and so on. During this time, I developed a strong interest, broadly speaking, in the structure of matter, and how it was organised. I had now acquired a good knowledge of X-ray diffraction, not only through my own work, but through having helped James check the proofs of his fine book—*The Optical Principles of the Diffraction of X-rays*—still a standard work. James wrote beautifully and fully and took great pains to make everything clear.

Supported by an 1851 Exhibition Scholarship and also by a research studentship to Trinity College, I went to Cambridge in 1949. Cambridge was the place for someone from the Colonies or the Dominions to go on to, and it was to the Cavendish Laboratory that one went to do physics. I wanted to work on some form of "unorthodox" X-ray crystallography, for example protein structure, but the MRC Unit where Perutz and Kendrew were working was full, and Bragg, then the Cavendish Professor, had closed down a project on order-disorder phenomena in alloys, which interested me. I finally found myself a research student of D.R. Hartree, who had been a colleague of both Bragg and James at Manchester. He suggested to me a theoretical problem left over from his work during the war on the cooling of steel through the austenite-pearlite transition, and I learned a fair amount of metallurgy in order to understand the physical basis of the phenomenon. It turned out however in the end that it was not special crystallographic insight that was called for—the course of the transition was in practice governed by the diffusion of the latent heat and I ended up using numerical methods to solve the partial differential equations for heat flow in the presence of a phase transition. I learned a good deal during this time, particularly in computing and solid state physics, and the idea of nucleation and growth in a phase change had its echo when I came later to think about the assembly of tobacco mosaic virus.

After taking my Ph.D., I spent a year in the Colloid Science department in Cambridge, working with F.J.W. Roughton, who had asked Hartree for someone to help him tackle the problem of simultaneous diffusion and chemical reaction, such as occurs when oxygen enters a red blood cell. The methods I had developed for the problem in steel were applicable here, and I was glad to put them to use on an interesting new problem. The quantitative data came from experiments in which thin layers of blood were exposed to oxygen or carbon monoxide. In the course of my stay there, I also showed how one could analyse the experimental kinetic curves for the reaction of haemoglobin with carbon dioxide or oxygen by simulations in the computer, and so fit the rate constants.

This work made me more and more interested in biological matter, and I decided that I really wanted to work on the X-ray analysis of biological molecules. I obtained a Nuffield Fellowship to work in J.D. Bernal's department in Birkbeck College in London and I moved there at the end of 1953. I joined a project on the protein ribonuclease, but shortly afterwards met Rosalind Franklin, who had moved to Birkbeck earlier and had begun working on tobacco mosaic virus. Her beautiful X-ray photographs fascinated me and I was also able to interpret some pictures which had apparently anomalous curved layer lines in terms of the splitting which occurs when the helical parameters are non-rational. From then on my fate was sealed. I took up the study of tobacco mosaic virus, and in four short years, together with Kenneth Holmes and John Finch, who had joined us as research students, we were able to map out the general outline of the structure of tobacco mosaic virus. This work was done partly in parallel with that of Donald Caspar, then at Yale, but he spent 1955–56 in Cambridge, and I formed an association with him which continued across the Atlantic for many years. It was during this time that I met Francis Crick and we published a paper together on diffraction by helical structures. I was fortunate to work with him again later, and so be able to learn, as he once wrote of Bragg, from watching the way he went about a problem.

Rosalind Franklin died in 1958 and, supported by an N.I.H. grant, Finch, Holmes and I continued the work on viruses, now extended to spherical viruses. We were joined soon after by Reuben Leberman, a biochemist. In 1962 we moved to the newly built MRC Laboratory of Molecular Biology in Cambridge which, under the leadership of Perutz, was to house the original unit from the Cavendish Laboratory (Perutz, Kendrew, Crick and, later, Brenner), enlarged by Sanger's group from the Biochemistry Department and Hugh Huxley from University College London. I was thus privileged to join the laboratory at this stage in its expansion and so be able to take advantage of, and to help build up, its unique environment of intellectual and technological sophistication. The rest of my scientific career is largely a matter of record and much of this is dealt with in the lecture that follows.

However, I should perhaps add that during the 20 years I have been back in Cambridge, I have been actively involved in the teaching of undergraduates, as well as of course supervising research students. I am still a Director of Studies in Natural Science at my College, Peterhouse, and under the tutorial or—as it is called in Cambridge—supervision, system, I teach undergraduates myself. I like teaching and the contact with young minds keeps one on one's toes, but increasing responsibilities have forced me to shed much of it in recent years.

Before I came to Cambridge I married Liebe Bobrow whom I had met in Cape Town. She trained in modern dance at the Jooss-Leeder School in London and later became a choreographer and coordinator for the Cambridge Contemporary Dance Group. More recently she has directed and acted in the theatre. We have two sons, Adam and David, born in 1954 and 1963. Adam, after studying History and Economics at Oxford and the London School of Economics, is now doing research in Econometrics. David is a second year student of Physics.

CURRICULUM VITAE — A. KLUG

Date of birth:	11th August 1926

Education:

1937—1941	Durban High School
1942—1945	University of Witwatersrand, Johannesburg: B.Sc. 1945
1946—1948	University of Capetown: M.Sc. 1946
1948—1949	Junior Lecturer in Department of Physics, University of Capetown
1949—1952	University of Cambridge: Ph.D. 1952 1851 Exhibition Overseas Scholarship Rouse Ball Student of Trinity College, Cambridge
1953	Senior Assistant in Research in Colloid Science Department, Cambridge
1954—1961	Crystallography Laboratory, Department of Physics, Birkbeck College, London. Nuffield Fellow (1954—1957); Head, Virus Resarch Project (1958—1961)
1962—present	MRC Laboratory of Molecular Biology, Cambridge; since 1978, Joint Head (with Dr. H.E. Huxley) of the Division of Structural Studies
1962—present	Fellow of Peterhouse, Cambridge and Director of Studies in Natural Sciences

Honours & Lectures:

1969	Fellow of the Royal Society, London
1969	Honorary Foreign Member, American Academy of Arts and Sciences
1972	Carter-Wallace Lecturer, Princeton University
1973	Leeuwenhoek Lecturer, Royal Society
1974	Molecular Biology Institute Distinguished Lecture Series, U.C.L.A.
1975	Dunham Lecturer, Harvard Medical School
1978	Hon. D.Sc. Chicago University
1978	Hon. D.Sc. Columbia University, New York
1978	Dr. Honoris Causa, Strasbourg University
1979	Heineken Prize of the Royal Netherlands Academy of Arts and Sciences
1979	Harvey Lecture, New York
1980	Hon. Dr. Fil., University of Stockholm
1981	Louisa Gross Horwitz Prize, Columbia University

FROM MACROMOLECULES TO BIOLOGICAL ASSEMBLIES

Nobel lecture, 8 December, 1982

by

AARON KLUG

MRC Laboratory of Molecular Biology,
Cambridge CB2 2QH, U.K.

Within a living cell there go on a large number and variety of biochemical processes, almost all of which involve, or are controlled by, large molecules, the main examples of which are proteins and nucleic acids. These macromolecules do not of course function in isolation but they often interact to form ordered aggregates or macromolecular complexes, sometimes so distinctive in form and function as to deserve the name of organelle. It is in such biological assemblies that the properties of individual macromolecules are often expressed in a cell. It is on some of these assemblies on which I have worked for over 25 years and which form the subject of my lecture today.

The aim of our field of structural molecular biology is to describe the biological machinery, in molecular, i.e. chemical, detail. The beginnings of this field were marked just over 20 years ago in 1962 when Max Perutz and John Kendrew received the Nobel prize for the first solution of the structure of proteins. In the same year Francis Crick, James Watson, and Maurice Wilkins were likewise honoured for elucidating the structure of the double helix of DNA. In his Nobel lecture Perutz recalled how 40 years earlier, in 1922, Sir Lawrence Bragg, whose pupil he had been, came here to thank the Academy for the Nobel prize awarded to himself and his father, Sir William, for having founded the new science of X-ray crystallography, by which the atomic structure of simple compounds and small molecules could be unravelled. These men have not only been my predecessors, but some of them have been something like scientific elder brothers to me, and I feel very proud that it should now be my turn to have this supreme honour bestowed upon me. For the main subjects of my work have been both nucleic acids and proteins, the interactions between them, and the development of methods necessary to study the large macromolecular complexes arising from these interactions.

In seeking to understand how proteins and nucleic acids interact, one has to begin with a particular problem, and I can claim no credit for the choice of my first subject, tobacco mosaic virus. It was the late Rosalind Franklin who introduced me to the study of viruses and whom I was lucky to meet when I joined J.D. Bernal's department in London in 1954. She had just switched from studying DNA to tobacco mosaic virus, X-ray studies of which had been begun

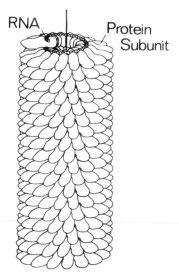

Fig. 1. Diagram summarizing the results of the first stage of structure analysis of tobacco mosaic virus (71). There are three nucleotides per protein subunit and $16^{1}/_3$ subunits per turn of the helix. Only about one-sixth of the length of a complete particle is shown.

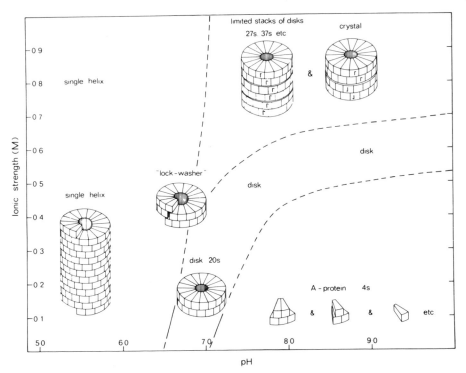

Fig. 2. Diagram showing the ranges over which particular forms of TMV protein participate significantly in the equilibrium (17). This is not a conventional phase diagram: a boundary is drawn where a larger species becomes detectable and does not imply that the smaller species disappears sharply. The "lock washer" indicated on the boundary between the 20 S disk and the helix is not well defined and represents a metastable transitory state observed when disks are converted to helices by abrupt lowering of the pH.

by Bernal in 1936. It was Rosalind Franklin who set me the example of tackling large and difficult problems. Had her life not been cut tragically short, she might well have stood in this place on an earlier occasion.

TOBACCO MOSAIC VIRUS

Tobacco mosaic virus (TMV) is a simple virus consisting only of a single type of protein molecule and of RNA, the carrier of the genetic information. Its simple rod shape results from its design, namely a regular helical array of these protein molecules, or subunits, in which is embedded a single molecule of RNA. This general picture was already complete by 1958 when Rosalind Franklin died (Fig. 1). It is clear that the protein ultimately determines the architecture of the virus, an arrangement of $16^1/_3$ subunits per turn of a rather flat helix with adjacent turns in contact. The RNA is intercalated between these turns with 3 nucleotide residues per protein subunit and is situated at a radial distance of 40 Å from the central axis and is therefore isolated from the outside world by the coat protein. The geometry of the protein arrangement forces the RNA backbone into a moderately extended single-strand configuration. Running up the central axis of the virus particle is a cylindrical hole of diameter 40 Å, which we then thought to be a trivial consequence of the protein packing, but which later turned out to figure prominently in the story of the assembly.

At first sight, the growth of a helical structure like that of TMV presents no problem of comprehension. Each protein subunit makes identical contacts with its neighbours so that the bonding between them repeats over and over again. Subunits can have a precise built in geometry so that they can assemble themselves like steps in a spiral staircase in a unique way. Subunits would simply add one or a few at a time onto the step at the end of a growing helix, entrapping the RNA that would protrude there and generating a new step, and so on. It was in retrospect thus not too surprising when the classic experiments of Fraenkel-Conrat and Williams in 1955 (6) demonstrated that TMV could be reassembled from its isolated protein and nucleic acid components. They showed that, upon simple remixing, infectious virus particles were formed that were structurally indistinguishable from the original virus. Thus all the information necessary to assemble the particle must be contained in its components, that is, the virus "self assembles". Later experiments (7) showed that the reassembly was fairly specific for the viral RNA, occuring most readily with the RNA homologous to the coat protein.

All this was very satisfactory but there were yet some features which gave cause for doubt. First, other experiments (8) showed that foreign RNAs could be incorporated into virus-like rods and these cast doubt on the belief that specificity *in vivo* was actually achieved during the assembly itself. Another feature about the reassembly that suggested that there were still missing elements in the story was its slow rate. Times of 8 to 24 hours were required to give maximum yields of assembled particles. This seemed to us rather slow for the assembly of a virus *in vivo*, since the nucleic acid is fully protected only on completion. These doubts, however, lay in the future and before we come to their resolution, I return to the structural analysis of the virus and the virus protein.

X-ray analysis of TMV: the protein disc

After Franklin's death, Holmes and I continued the X-ray analysis of the virus. Specimens for X-ray work can be prepared in the form of gels in which the particles are oriented parallel to each other, but randomly rotated about their own axes. These gels give good X-ray diffraction patterns but because of their nature the three-dimensional X-ray information is scrambled into two dimensions. Unscrambling these data to reconstruct the 3-dimensional structure has proved to be major undertaking, and it was only in 1965 that Holmes and I obtained the first 3-dimensional Fourier maps to a resolution of about 12 Å. In fact, only recently has the analysis by Holmes and his colleagues in Heidelberg (where he moved in 1968) reached a resolution approaching 4 Å in the best regions of the electron density map, but falling off significantly in other parts (9). At this resolution it is not possible to identify individual amino-acid residues with any certainty and ambiguities are too great to build unique atomic models. However, the map, taken together with the detailed map of the subunit we obtained in Cambridge (see below) yields a considerable amount of information about the nature of the contacts with RNA (10).

These difficulties in the X-ray analysis of the virus were foreseen, and by the early 1960's I came to realize that the way around this difficulty was to try to crystallize the isolated protein subunit of the virus, solve its structure by X-ray diffraction and then try to relate this to the virus structure solved to low resolution. We therefore began to try to crystallize the protein monomer. In order to frustrate the natural tendency of the protein to aggregate into a helix, Leberman introduced various chemical modifications in the hope of blocking the normal contact sites, but none of these modified proteins crystallised. The second approach was to try to crystallize small aggregates of the unmodified protein subunits. It had been known for some time, particularly from the work of Schramm and Zillig (11), that the protein on its own, free of RNA, can aggregate into a number of distinct forms, besides that of the helix. I chose conditions under which the protein appeared to be mainly aggregated in a form with a sedimentation constant of about 4S, identified by Caspar as a trimer (12). We obtained crystals almost immediately but we found (13) them to contain not the small aggregate hoped for, but a large one, corresponding to an aggregate with a sedimentation constant of 20S. The X-ray analysis showed that this was built from two juxtaposed layers, or rings, of 17 subunits each and we named this form the two-layer disc (Fig. 3 and 4). Our inital dismay in being faced with such a large structure, of molecular weight 600,000, was tempered by the fact that the geometry of the disc was clearly related to that of the virus particle. The cylindrical rings contained 17 subunits each compared with $16^1/_3$ units per turn of the virus helix, so that the lateral bonding within the discs was therefore likely to be closely related to that in the virus. We also

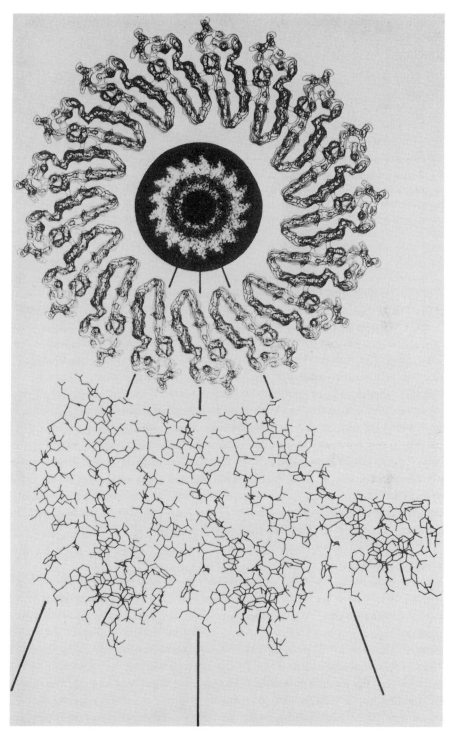

Fig. 3. The disk viewed from above at successive stages of resolution. From the centre outward there follow (i) a rotationally filtered electron microscope image at about 25 Å resolution (72); (ii) a slice through the 5 Å electron density map of the disk obtained by X-ray analysis, showing rod-like α-helices (26) and (iii) part of the atomic model built from the 2.8 Å map (Bloomer *et al.*, ref. 15).

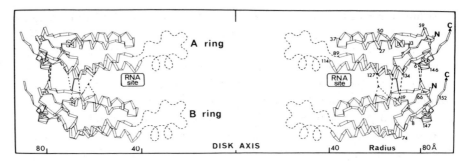

Fig. 4. Section through a disk along its axis reconstructed from the results of X-ray analysis to a resolution of 2.8 Å (15). The ribbons show the path of the polypeptide chain of the protein subunits. Subunits of the two rings can be seen touching over a small area toward the outside of the disk but opening up into the "jaws" toward the centre. The dashed lines at low radius indicate schematically the mobile portion of the protein in the disk, extending in from near the RNA binding site to the edge of the central hole.

showed, by analysing electron micrographs, that the disc was polar, i.e. that its two rings faced in the same direction as do successive turns of the virus helix.

This was the first very large structure ever to be tackled in detail by X-ray analysis and it took about a dozen years to carry through the analysis to high resolution. The formidable technical problems were overcome only after the development in our laboratory of more powerful X-ray tubes and of special apparatus (cameras, computer-linked densitometers) for data collection from a structure of this magnitude. (In fact we had begun building better X-ray tubes in London to use on weakly diffracting objects like viruses). The 17-fold rotational symmetry of the disc also gives rise to redundant information in the X-ray data, which was exploited in the final analysis (14), to improve and extend the resolution of a map based originally on only one heavy atom derivative. The map at 2.8 Å resolution (15) has been interpreted in terms of a detailed atomic model for the protein (Figs. 3 and 4), although the individual interactions upon RNA binding have yet to be deduced.

Protein polymorphism
These results on the structure of the disc which showed that it was fairly closely related to the virus helix made me wonder whether the disc aggregate might not be fulfilling some vital biological role. It had been easy to dismiss it as perhaps an adventitious aggregate of a sticky protein or a storage form. The polymorphism of TMV protein was first considered in some detail by Caspar in 1963 (12) who foresaw that some of the aggregation states might give insight into the way the protein functions. Quantitative studies of aggregation started by Lauffer in the 1950's (16) concentrated upon a rather narrow range of conditions, the main interest being in understanding the forces driving the aggregation (these are largely entropic). Because of the scattered nature of the earlier observations, Durham, Finch and I began a systematic survey of the aggregation states as a result of which the broad outline became clear (17, 18). The results can be summarised as a phase diagram (Fig. 2).

At low or acid pH, the protein alone will form helices of indefinite lengths that are structurally very similar to the virus except for the lack of the RNA. Above neutrality the protein tends to exist as a mixture of smaller aggregates from about trimer upwards, in rapid equilibrium with each other, commonly referred to as A-protein. Near pH 7 and at about room temperature the dominant form present is the disc which is in a relatively slow equilibrium with the A-form in the ratio of about 4:1. The dominant factor controlling the state of aggregation of the coat protein is thus the pH. The control is mediated through groups, probably carboxylic acid residues, as identified by Caspar (12), that bind protons abnormally in the helical state, but not in the disc or A-form. Thus the helical structure can be stabilised either in the virus by the interaction of the RNA with the protein, or, in the case of the free protein, by protonating the acid groups. These groups thus act as a "negative switch", ensuring that under physiological conditions the helix is not formed, and thus that enough protein in the form of discs or A-protein is available to interact with the RNA during virus assembly.

A role for the disc

The disc aggregate of the protein therefore has a number of significant properties. It is not only closely related to the virus helix, but also is the dominant form of the protein under "physiological" conditions; moreover, disc forms had also been observed for other helical viruses. These strengthened my conviction that the disc form was not adventitious but might play a significant role in the assembly of the virus. What could this role be?

Assembly of any large aggregate of identical units such as a crystal can be considered from the physical point of view in two stages: first nucleation and then the subsequent growth, or, in more biochemical language, as initiation and subsequent elongation. The process of nucleation—or, crudely, getting started—is frequently more difficult than the growth. Thus, a simple mode of initiation in which the free RNA interacts with individual protein subunits does pose problems in getting started. At least 17 separate subunits would have to bind to the flexible RNA molecule before the assembling linear structure could close round on itself to form the first turn of the virus helix. This difficulty could be avoided if a preformed disc were to serve as a jig upon which the first few turns of the viral helix could assemble to reach sufficient size to be stable. This mode of nucleation of helix assembly could also furnish a mechanism for the recognition by the protein of its homologous RNA. The surface of the disc presents a set of 51 (= 17×3) nucleotide binding sites which could interact with a special long run of bases, resulting in an amplified discrimination that might not be possible with a few nucleotides. It thus seemed that the disc could solve both the physical and biological requirements for initiating virus growth and conferring specificity on the interaction. This hypothesis is illustrated in Fig. 5. It turned out that all the details in this diagram are wrong, but yet the spirit is correct. As A.N. Whitehead once observed, it is more important that an idea should be fruitful than it should be true.

This proposed mechanism of nucleation required that the disc be able to

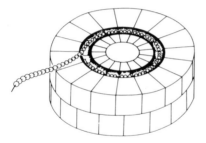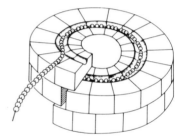

Fig. 5. The role of the disk as originally conceived: the specific recognition of a special (terminal) sequence of TMV-RNA initiates conversion of the disc form of the protein into two turns of helix. (See Fig. 7, for the mechanism finally established.)

dislocate into a two-turn helix to form the beginning of the growing nucleoprotein rod. To test this, we carried out a very simple experiment, the pH drop experiment (19). This showed that an abrupt lowering of the pH would convert discs directly, within seconds, into short helices—or lockwashers (Fig. 2), which stack on each other to give longer nicked helices, which in due course anneal to give more perfect helices. This conversion is an *in situ* one, not requiring dissociation and then reassociation into a different form. The success of this experiment encouraged us to proceed to experiments with RNA itself, the natural "substrate" of the virus protein.

The first reconstitution experiments carried out by Butler and myself proved to be dramatic (20). When a mixture was made at pH 7 of the viral RNA and a disc preparation, complete virus particles were formed within 10 to 15 minutes, rather than over a period of hours, as was the case in the early reassembly experiments in which protein had been used in the disaggregated form (6).

The notion that discs are involved in the natural biological process of initation was strengthened by companion experiments (20) in which assembly was carried out with RNAs from different sources. These showed a preference, by several orders of magnitude, of discs for the viral RNA over foreign RNAs or synthetic polynucleotides of simple sequence. It is thus the disc state of the protein that is needed to achieve specificity in the interaction with the RNA. In the experiments cited earlier, in which virus-like rods were made containing TMV A-protein and foreign RNA (8), reactions were carried out at an acid pH, and under these artificial conditions the protein alone would tend to form helical rods and so could entrap any RNA present.

Besides this effect of discs on the rate of initiation, which had been predicted, we also found to our surprise that the discs appeared to enhance the rate of elongation, and we concluded that they must be therefore actively involved in growth. This result has been questioned by some other workers in the field and is still the subject of argument (21, 22), but recent discoveries on the configuration of RNA during incorporation into a growing particle, discussed below, have made the involvement of discs in the elongation, as well as in nucleation, much more intelligible.

The disc form of the protein therefore provided the elements which were missing from the simple reconstitution experiments using disaggregated pro-

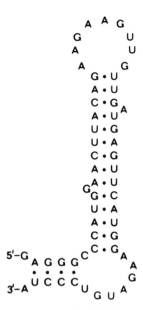

Fig. 6. Postulated secondary structure of the RNA in the nucleation region (24). This gives a weakly bonded double-helical stem and a look at the top probably the actual origin of assembly. The sequence at and near the top contains a repeating motif of three bases having G in the middle position and A, or U in the outer positions.

tein, namely speed and specificity. We now knew what the disc did, the next question was how did it do it?

The interaction of the protein disc with the initiation sequence on the RNA

Specificity in initiation ensures that only the viral RNA is picked out for coating by the viral protein. This must be brought about by the presence of a unique sequence on the viral RNA for interaction with the protein disc. Zimmern and Butler isolated the nucleation region containing this site by supplying limited quantities of disc protein, sufficient to allow nucleation to proceed, but not subsequent growth, then digesting away the uncoated RNA with nuclease (23, 24). With the varying protein: RNA ratios and different digestion conditions, they found they could isolate a series of RNA fragments, all of which contained a unique common core sequence with variable extents of elongation at either end. These fragments could be rebound to the coat protein when it was in the form of discs. Among this population of fragments was a fragment only about 60 nucleotides long—just over the length necessary to bind round a single disc —and it appeared to represent the minimum protected core. Because of the strong rebinding of this fragment back to the disc, it seemed likely that it constituted the "origin of assembly", where the normal nucleation reaction began.

However, the work on the RNA produced, in turn, another puzzle: the obvious expectation that the nucleation region would be near one end of the RNA turned out to be wrong. The nucleation occurs about one sixth of the way along the RNA from the 3' end (25), so that over 5000 nucleotides have to be

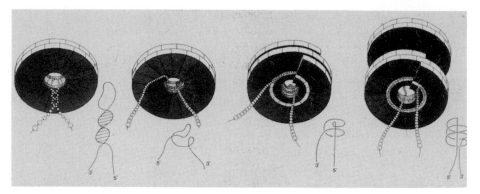

Fig. 7. Nucleation of virus assembly occurs by the insertion of a hairpin of RNA (Fig. 6) into the central hole of the protein disk and between the two layers of subunits. The loop at the top of the hairpin binds to form part of the first turn, opening up the base-paired stem as it does so, and causes the disk to dislocate into a short helix. This presumably "closes the jaws", entrapping the RNA between the turns of protein subunits, and gives a start to the nucleoprotein helix (which can then elongate rapidly to some minimum stable size).

coated in the major direction of elongation $(3'-5')$ and 1 000 have to be coated in the opposite direction. Yet growing nucleoprotein rods observed in the electron microscope (20) were always found to have all the uncoated RNA only at one end: why were rods never seen with a tail at each end? The resolution of this conundrum came from considering the structure of the protein disc, to which I now turn.

Although the structure of the disc was solved in detail only in 1977, an earlier stage in the X-ray analysis gave the clue as to how it might interact with the RNA. At 5 Å resolution (26) the course of the polypeptide chain could be traced and the basic design of the disc established (cf Fig. 4). The subunits of the upper ring of the disc lie in a plane perpendicular to the disc axis while those of the lower ring are tilted downward towards the centre, so that the two rings touch only towards the outside of the disc. In the neighbourhood of the central hole they are thus far apart, like an open pair of jaws which could, as it were, "bite" a stretch of RNA entering through the central hole. Moreover, entry through the centre would be facilitated because the inner region of the protein, from around the RNA binding site inward, was found to be disordered and not packed into a regular structure.

It therefore looked very much as though the disc were designed to permit the RNA to enter through the central hole, effectively enlarged by the flexibility of the inner loop of protein, and intercalate between its two layers. The RNA which would enter thus would of course be the nucleation sequence which lies rather far from an end of the RNA molecule. This could, however, be achieved if the RNA doubled back on itself at a point near the origin of assembly and so entered as a hairpin loop. Indeed, the smallest RNA fragment that is protected during nucleation has a base sequence which can fold into a weakly paired double-helical stem with a loop at the top, that is a hairpin (Fig. 6). This was proposed by Zimmern (24). The loop and top of the stem have an unusual sequence, containing a repeating motif of three nucleotides, with guanine G in one specific position, and usually A or some times U in the other two. Since there are three nucleotide binding sites per protein subunit, such a triplet repeat pattern will place a specific base in a particular site on the protein molecule and could well lead to the recognition of the exposed RNA loop by the disc during the nucleation process.

Nucleation and growth

The hypothesis for nucleation (27) then is that the special RNA hairpin would insert through the central hole of the disc into the jaws formed by the two layers of protein subunits (Fig. 7). The dimensions are quite suitable for this to occur and the open loop could then bind to the RNA binding sites on the protein. More of the rather unstable double helical stem would melt out and be opened as more of the RNA was bound within the jaws of the nucleating disc. Some, as yet unknown, feature of this interaction would cause the disc to dislocate into a short helical segment, entrapping the RNA and, after the rapid addition of a few more discs (23), would provide the first stable nucleoprotein particle.

The subsequent events after nucleation can be called growth and as stated above there is a controversy about the particular way in which this proceeds. Our view is that elongation in the major direction of growth very likely takes place through the addition of further discs, as indeed our first reconstitution experiments drove us to conclude. The special configuration generated during the insertion of the loop into the centre of the disc must be perpetuated as the rod grows, by pulling further RNA up through the central hole. Thus, elongation could occur by a substantially similar mechanism to nucleation, only now, rather than requiring the specific nucleation loop of the RNA, it occurs by means of a "travelling loop" which can be inserted into the centre of the next incoming disc. This mechanism therefore overcomes the main difficulty in envisaging how a whole disc of protein subunits could interact with the RNA in the growing helix. There is now more evidence for growth by incorporation of blocks of subunits of roughly disc size (22), but the subject is still controversial and I will therefore not proceed further with it.

On the other hand, there is now clear experimental confirmation of our hypothesis for the mechanism of nucleation. This predicts (1) that two tails of the RNA will be left at one end of the growing nucleoprotein rod formed, and (2) that one of these tails would project directly from one end but the other would be doubled back all the way from the active growing point at the far end of the rod down the central hole of the growing rod. Both of these predictions have now been confirmed. Hirth's group in Strasbourg has obtained electron micrographs of growing rods in which the RNA is spread by partial denaturation, and many particles show two tails protruding from the same end (28). In Cambridge my colleagues have used high resolution electron microscopy, in which the two ends of the rods can be identified by their shapes to show that it is indeed the longer tail that is doubled back through the growing rod (29). Other experiments show that the RNA configuration has a substantial effect on the rate of assembly (29).

Design and construction: physical and biological requirements

We have seen that the formation of the protein disk is the key to the mechanism of the assembly of TMV. The protein subunit is designed not to form an endless helix, but a closed two-layer variant of it, the disc, which is stable and which can be readily converted to the lockwasher or helix-going form. The disc therefore represents an intermediate sub-assembly by means of which the entropically difficult problem of nucleating helical growth is overcome. At the same time the nucleation by the disc sub-assembly furnishes a mechanism for recognition of the homologous viral RNA (and rejection of foreign RNAs) by providing a long stretch of nucleotide binding sites for interaction with the special sequence of bases on the RNA. The disc is thus an obligatory intermediate in the assembly of the virus, which simultaneously fulfils the physical requirement for nucleating the growth of the helical particle and the biological requirement for specific recognition of the viral RNA. TMV is self-assembling, self-nucleating and self-checking.

There are a number of morals to be derived from the story of TMV assembly (1). The first is that one must distinguish between the design of a structure and the construction process used to achieve it. That is, while TMV looks like a helical crystal and its design lends itself to a process of simple addition of subunits, its construction actually follows a more complex path that is highly controlled. It illustrates the point that function is inextricably linked with structure and how much can be done by one single protein. A most intricate structural mechanism has been evolved to give the assembly an efficiency and purposefulness whose basis we now understand. The general moral of all this is that not merely does nature once again confound our obvious preconceptions, but it has left enough clues for us to be able to puzzle out finally what is happening. As Einstein once put it, "Raffiniert ist der Herr Gott, aber bösartig ist er nicht: The Lord is subtle, but he is not malicious".

CRYSTALLOGRAPHIC OR FOURIER ELECTRON MICROSCOPY

In 1955, Finch and I in London, and Caspar, then in Cambridge, took up the X-ray analysis of crystals of spherical viruses. These had first been investigated by Bernal and his colleagues just before and after the war, using "powder" and "still" photography. Finch and I worked on Turnip Yellow Mosaic virus and its associated empty shell, and Caspar on Tomato Bushy Stunt virus. Crick and Watson had predicted that spherical viruses ought to have one of the forms of cubic symmetry, and we showed that both viruses had icosahedral symmetry. Later, when Finch and I showed that poliovirus also had the same symmetry, we realised that there was some underlying principle at work, and this eventually led Caspar and me to formulate our theory of virus shell structure (30).

When my research group moved to Cambridge in 1962, we turned to electron microscopy for the speed with which it enables one to tackle new subjects, and also because it produces a direct image, or so we thought. Armed with a theory of virus design and some X-ray data, we had some notion of how spherical shells of viruses might be constructed and thought we would be able to see the fine detail in electron **micrographs**. Thus, we knew what we were looking *for*, but we soon found that we did not understand what we were looking *at*: the micrographs did not present simple direct images of the specimens. We soon discovered the limitations of electron microscopy. First, there were preparation artefacts and also radiation damage during observation. Secondly, artificial means of contrast enhancement had to be used as the majority of atoms in biological specimens have an atomic number too low to give sufficient contrast on their own. Thirdly, the image formed depends on the operating conditions of the microscope and on the focussing conditions and aberrations present. Above all, because of the large depth of focus of the conventional microscope, all features along the direction of view are superimposed in the image. Finally, in the case of strongly scattering or thick specimens, there is multiple scattering within the specimen, which can destroy even this relation between object and image.

For these reasons, the detail one sees in a raw image is often unreliable and not easily interpretable without methods which correct for the operating conditions of the microscope and which can separate contributions to the image from different levels of the specimen. It is also important to be able to assess the degree of specimen preservation in each particular case. These procedures for image processing of electron micrographs were developed by myself and my colleagues over a period of about 10 years. Their aim is to extract from the information recorded in electron micrographs the maximum amount of reliable information about the 2- or 3-dimensional structures which are being examined. Some applications of these methods to various problems studied in the MRC laboratory over the first 15 years are given in Table 1. Electron microscopy combined with image reconstruction, supplemented wherever possible

Table 1. Some applications of electron microscope image reconstruction in the MRC Laboratory of Molecular Biology, Cambridge, 1964–1979.

Viruses	Organelles	Enzymes, etc.
Helical 　TMV, TMV protein disc, 　Paramyxoviruses	Microtubules from flagellar doublets and brain; tubulin sheets	Haemocyanin Glutamate dehydrogenase
Icosahedral 　Polyoma, wart, TBSV, 　TYMV, R17, Nudaurelia, 　CPMV	Muscle filaments: actin; actin + tropomyosin; actin + myosin + tropomyosin (inhibited and relaxed)	Catalase; crystals and tubes Sickle cell haemoglobin fibres
Adenovirus hexon Aberrant hex. & pent tubes of polyoma	Bacterial flagella; Bacterial cell walls	Purple membrane (Bacteriohodopsin)
Phage T2 and T4 　Head and its tubular 　variants (polyheads)	Ribosome crystals Chromatin: crystals of nucleosome cores; tubes of histone octamers	Cytochrome oxidase
Tail: sheath + core Baseplate	Gap junctions	

by X-ray studies on wet, intact material, has provided what are now generally accepted models of the structural organisation of a large number of biological systems such as those listed in the table. In this lecture I will describe a limited number of examples which serve to demonstrate the power of various techniques and the nature of the results they can give. Fuller accounts of the methods and the theory are given elsewhere (2, 3), but I would like to emphasize here that these methods arose out of practical concerns and grew in the course of tackling concrete problems; nervertheless they have proved to be of wide application.

Two dimensional reconstruction: digital computer processing
We began our studies on viruses, both spherical and helical, using the method of negative staining which had been recently introduced by Huxley, and by Brenner and Horne (31). In this method the specimen is embedded in a thin amorphous layer of a heavy metal salt which simultaneously preserves and maps out the shape of the regions from which it is excluded. Much fine detail was to be seen, but one could not easily make sense of it in most cases. People simply thought that the specimens were being disordered, because it was assumed that the negative stain gave, as it were, a footprint of the particle. We gradually came to realise that the confusion arose, not so much because of the disorder that the stain produced, but because there was a superposition of detail from the front and back of the particle; i. e., the stain was enveloping the whole particle, so forming a cast rather than a footprint. This interpretation was proved in two different ways which proceeded in parallel. First, in the case of the spherical viruses, one could build a model and compute or otherwise display it in projection and we found that this could account for many if not all of the previously uninterpretable images (32). The uniqueness of the model could be proved by tilting experiments in which the specimens on the grid and the model were tilted in the same manner through large angles (cf. Fig. 10, ref. 73). The second approach was applied to helical structures, which are translationally periodic and therefore lend themselves to a direct image analysis, which I shall now illustrate.

Figure 8a shows an electron micrograph of a negatively stained specimen of a "polyhead", which is a variant of the head of T4 bacteriophage, consisting mainly of the major head protein. The particle has been flattened and so its original tubular form lost. The image clearly shows some structural periodicities, but these are difficult to discern and such interpretations used to be left to subjective judgement. I realised that the optical (Fraunhofer) diffraction pattern produced from such an image would allow an objective analysis of all the periodicities present to be made (33). This is shown in Figure 8b. Here clear diffraction maxima can be seen: these fall into two sets which can be accounted for as arising respectively from the near and far sides of the specimen. In this way it was established that the negative stain was producing a complete cast of the particle rather than a one sided footprint of it (33). Since this is a helically periodic structure, the diffraction maxima tend to lie on a lattice and so they pick out genuine repeating features within the structure. In

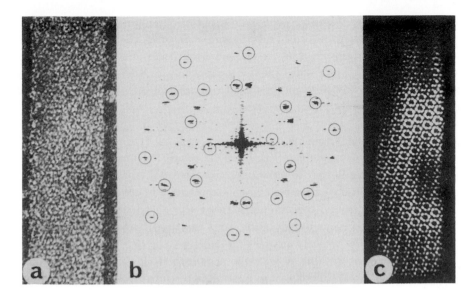

Fig. 8. Optical diffraction and image filtering of the tubular structures known as "polyheads", consisting of the major head protein of T4 bacteriophage (35). (a) Electron micrograph of negatively stained flattened particle × 200,000.

(b) Optical diffraction pattern of (a), with circles drawn around one set of diffraction peaks corresponding to one layer of the structure.

(c) Filtered image of one layer in (a) using the diffraction mask shown in (b). The aperatures in the mask ars chosen so that the averaging here extends locally only over a few unit cells. Individual molecules arranged in hexamers can be seen.

this case the regular diffraction maxima extended to a spacing of about 20 Å which demonstrated that the long range order in the specimen was preserved to this resolution, which is indeed sufficient to resolve individual protein molecules.

The confusion in the image is largely due to the superposition of the near and far sides of the particle, and any one such side can be filtered out in an optical system by a suitably positioned mask which transmits only the desired diffracted rays (34). The filtered image, Figure 8c, is immediately interpretable in terms of a particular arrangement of protein molecules (35).

The clarity of the processed image derives also from the fact that the background noise in the diffraction pattern has been filtered out. This noise arises because of the individual variations between molecules in the specimen, i.e. the disorder, and these contribute randomly in all parts of the diffraction pattern. Indeed, what has been done is that the signal to noise ratio in the image has been enhanced by averaging over the copies of the molecules present in the arrangement. This idea of averaging over many copies of a repeated motif is central to the most powerful techniques developed so far for producing reliable images of biological specimens, and the three dimensional procedures which I will describe later can also use this technique.

The essence of image processing of this type is that it is a two-step procedure after the first image has been obtained. First the Fourier transform of the raw image is produced. Fourier coefficients are then manipulated or otherwise corrected and then transformed back again to reproduce the reconstructed image. These operations can be carried out most easily on a digital computer, and digital imaging processing as first introduced by DeRosier and myself (36) allows a much greater flexibility than our original optical method and makes three dimensional procedures possible.

Three-dimensional image reconstruction

The first example I have given (Fig. 8) is of a relatively simple case where the problem is essentially that of separating contributions from two overlapping crystalline layers and we have seen how the method of Fourier analysis resolves the superposition in real space into separated sets of contributions in Fourier space. It was, however, already clear from the simple analysis of spherical viruses that in order to get a unique or reliable picture of a three dimensional structure one must be able to view the specimen from very many different directions (32). These different views were often provided by specimens lying in different orientations but they can also be realised by tilting the specimen in the microscope, as mentioned above. Originally, as described above, the different views were interpreted by the building of models, but eventually I saw that a set of transmission images taken in different views could be combined objectively to give a reconstruction of a three-dimensional object.

This happened when DeRosier and I were studying the tail of bacteriophage T4 and our analysis showed that there were contributions to the image from the internal structure as well as from the front and back surfaces (36). To work in three dimensions a generalised form of the two-dimensional filtering process had to be found, and—by making a connection with X-ray analysis—I realised that what is required is a three-dimensional Fourier synthesis. In the analysis of the X-ray diffraction patterns of TMV, I had used the idea that a helical structure could be built up mathematically out of a set of cylindrical harmonic functions; there is a relation between the number of functions that could be obtained and the number of different views available. Each new view would give additional harmonics of higher spatial frequency, and so, if one had enough views, one could build up the complete structure. Later we came to see (36) that this synthesis was only a special case of a general theorem known to crystallographers as the projection theorem.

The general method of reconstruction which we developed (Fig. 9) is based on the projection theorem, which states that the two-dimensional Fourier transform of a plane projection of a three-dimensional density distribution is identical to the corresponding central section of the three-dimensional transform normal to the direction of view. The three-dimensional transform can therefore be built up section by section using transforms of different views of the object, and the three-dimensional reconstruction then produced by Fourier inversion. The important feature of the method is that it tells one how many different views are needed for a required resolution and how these are to be recombined into a three-dimensional map of the object (36, 37). The process is both quantitative and free from arbitrary assumptions. The approach is similar

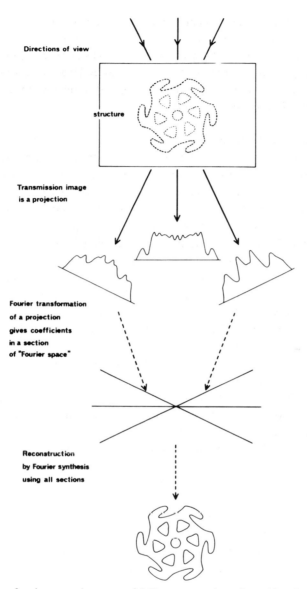

Fig. 9. Scheme for the general process of 3-D reconstruction of an object from a set of 2-D projections (36).

to conventional X-ray crystallography, except that the phases of the X-ray diffraction pattern cannot be measured directly, whereas here they can be computed from a digitised image. Were it not for radiation damage, the different views could be collected from a single particle by using a tilting stage in the microscope, but more realistically one must use several particles in different but identifiable orientations. In general, it is desirable to combine data from different particles so that imperfections can be averaged out.

The Fourier method is only one way out of several for solving the sets of mathematical equations which relate the unknown three dimensional density

direction of tilt axis

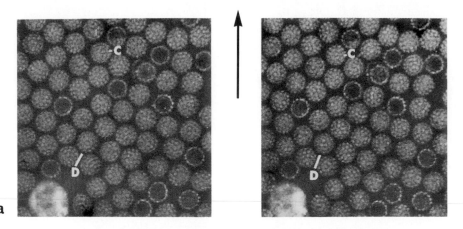

a

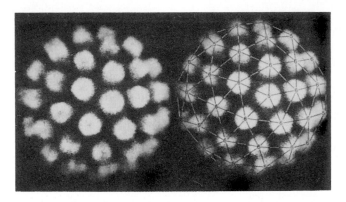

b

Fig. 10. (a) Electron micrographs of the same field of negatively-stained close-packed particles of human wart virus (HWV) before (i) and after (ii) tilting the specimen grid through an angle close to 18° (73). × 100,000.

(b) A three-dimensional reconstructed image of human wart virus (38, 39). Alongside is shown the underlying icosahedral surface lattice (30) with the 5-fold and 6-fold vertices marked.

distribution with known projections in different directions (37), but in fact no other reliable method has been shown to be superior and it is used in the CAT scanner. Moreover, the Fourier method has the advantage that because it is carried out in steps, i.e. formation of the two-dimensional transforms, and then recombination in three dimensions, it is possible as described above, to assess, select, and correct the data going into the final reconstruction.

Many applications have been made. The first application was in fact to the phage tail of T4, the problem in which it had arisen. Particles with helical symmetry are the most straightforward to reconstruct, because a reconstruction can be made from a single view of the whole particle, to a limited resolution, set by the helix symmetry. In physical terms, this is because a single image of a helical particle presents many different views of the repeating subunit, and it was this simplification that led us to use the phage tail as a first specimen for 3-D image reconstruction. Generally, more than one view is necessary, but any symmetry present will reduce the number required. Typically for small icosahedral viruses, three or four views are sufficient, but many more specimens must be investigated before the appropriate number can be found and averaging carried out (38). An example, from Crowther and Amos (39), is given in Fig. 10.

Phase contrast microscopy

Electron microscopy, combined with some method of image analysis, when applied to negatively stained specimens, has proved ideal for determining the arrangement and shape of small protein subunits within natural or artificial arrays, including two dimensional crystals and macromolecular assemblies such as viruses and microtubules (2). The structural information obtainable has proved to be highly reliable with respect to detail down to about the 20 Å or 15 Å level. It became clear, however, that the degree of detail revealed was limited by the granularity of the negative stain and the fidelity with which it follows the surface of the specimen (40). To obtain much higher resolution information, better than about 10 Å, one should dispense with the stain and view the protein itself. At high resolution, there is a second problem: irradiation damage. This can be reduced by cutting down the illuminating beam, but the statistical noise is then increased, and the raw image becomes less and less reliable. However, this difficulty can be overcome satisfactorily by imaging ordered arrays of molecules, so that the information from the different molecules can be averaged, as described above, to give a statistically significant picture. The first problem of replacing the negative stain, yet avoiding dehydration, can be solved in two ways. One, now being intensively studied, is to use frozen hydrated specimens (41). The second, tried method is that of Unwin and Henderson, who, in their radical approach to determining the structure of unstained biological specimens by electron microscopy (42, 43) used a dried-down solution of glucose to preserve the material.

The question then arose as to how this unstained specimen, effectively transparent to electrons, is to be visualized. In the light microscopy of transparent specimens the well-known Zernike phase contrast method is used. Here the phase of the scattered beams relative to the unscattered beam are shifted by means of a phase plate and then the scattered and unscattered beams are allowed to interfere in the image plane to produce an image. A successful electrostatic phase contrast device for electron microscopy, quite analogous to the phase plate used in light microscopy, was constructed by Unwin (44), but it is not easy to make or use. A practical way of producing phase contrast in the electron microscope is simply to record the image, with the objective lens underfocussed, and this was the method used by Unwin and Henderson.

The defocussing phase contrast method arose out of an academic study by Erickson and myself of image formation in the electron microscope (45). This was undertaken because of a controversy that had developed concerning the nature of the raw image itself. When three-dimensional image reconstruction was introduced and applied to biological particles embedded in negative stain, objections were raised by various workers in the field of materials science, accustomed to dynamical effects in strongly scattering materials, to the premise that the image essentially represented the simple projection of the distribution of stain. It was asked whether multiple or dynamical scattering might not vitiate this assumption. To investigate this question, Erickson and I undertook an experimental study of negatively stained thin crystals of catalase as a function of the depth of focussing (45). We found that a linear or first order theory of image formation would explain almost entirely the changes in the Fourier transform of the image. We concluded that the direct image, using a suitable value of underfocus dependent on the frequency range of interest, is a valid picture of the projection of the object density. When greater values of underfocus were used to enhance the contrast, the image could be corrected to give a valid picture.

This study, although confined to the medium resolution range, included a practical demonstration that *a-posteriori* digital image processing could be used to measure and compensate for the effects of defocussing, and we suggested that this approach could be directly extended to high resolution to compensate for the effects of spherical aberration as well as defocussing. It also provided a convenient way of producing phase contrast in the electron microscope in the case of unstained specimens. The image is recorded with the objective lens underfocussed, so changing the phases of the scattered beams relative to the unscattered (or zero order) beam. Defocussing does not however act as a perfect phase plate analogous to that of Zernike, since the phases are not all changed by the same amount, and successive bands of spatial frequencies contribute to the image with alternately positive and negative contrast. In order to produce a "true" image, the electron image must be processed to correct for the phase contrast transfer of the microscope so that all spatial frequencies contribution with the same sign of contrast.

To produce their spectacular three-dimensional reconstructed image of the purple membrane of Halobacterium to a resolution of about 7 Å (44), Henderson and Unwin took a series of very low-dose images of different pieces of membrane tilted at different angles. The final map represented an average over some 100,000 molecules. The small amount of contrast present in the individual micrographs was produced by underfocussing which was then compensated for in the computer reconstruction by the method described above. For the first time the internal structure of a protein molecule was "seen" by electron microscopy.

THE STRUCTURE OF CHROMATIN

The work on viruses has given results not only of intrinsic interest, but as I indicated above, the difficulties in tackling large molecular aggregates led to the development of methods and techniques which could be applied to other systems. A recent example of this approach, and one which I think would not have gone so fast without our earlier experience, is that of chromatin. Chromatin is the name given to the chromosomal material when extracted. It consists mainly of DNA, tightly associated with an equal weight of a small set of rather basic proteins called histones. We took up the study of chromatin in Cambridge about ten years ago when the protein chemists had shown that there were only five main types of histones, the apparent proliferation of species being due to post-synthetic modifications, so that the structural problem appeared tractable.

The DNA of the eukaryotic chromosome is probably a single molecule, amounting to several centimeters in length if laid out straight, and it must be highly folded to make the compact structure one can see in a chromosome. At the same time it is organised into separate genetic or functional units, and the manner in which this folding is achieved, genes organised and their expression controlled, is the subject of intense study throughout the world. The aim of our research group has been to try to understand the structural organisation of chromatin at various levels and to see what connections could be made with functional controls.

The large amounts in which histones occur suggested that their role was structural, and it was shown over the years 1972−1975 that the four histones H2A, H2B, H3 and H4 are responsible for the first level of structural organisation in chromatin. They fold successive segments of the DNA about 200 base pairs long into compact bodies of about 100 Å in diameter, called nucleosomes. A string of nucleosomes or repeating units is thus created and when these are closely packed they form a filament about 100 Å in diameter. The role of the fifth histone H1 was at first not clear. It is much more variable in sequence than the other four, being species and tissue specific. In the years 1975−1976 we showed that H1 is concerned with the folding of the nucleosome filament into the next higher level of organisation, and later how it performed this role.

This is not the place to tell in detail how this picture of the basic organisation of chromatin emerged (4), but the idea of a nucleosome arose from the convergence of several different lines of work. The first indications for a regular structure came from X-ray diffraction studies on chromatin which showed that there must be some sort of repeating unit, albeit not well ordered, on the scale of about 100 Å (46, 47). The first biochemical evidence for regularity came from the work of Hewish and Burgoyne (48) who showed that an endogenous nuclease in rat liver could cut the DNA into multiples of a unit size, which was later shown by Noll, using a different enzyme, micrococcal nuclease, to be about 200 base pairs (49). The fact that the nuclease cuts the DNA of chromatin at regularly spaced sites, quite unlike its action on free DNA, is attributed to the fact that the DNA is folded in such a way as to make only short stretches of free DNA between these folded units available to the enzyme. The third piece of evidence which led to the idea of a nucleosome was the observation by Kornberg and Thomas (50) that the two highly conserved histones, H3 and H4, existed in solution as a specific oligomer, the tetramer $(H3)_2(H4)_2$, which behaved rather like an ordinary multi-subunit globular protein. On the basis of these different lines of evidence, Kornberg in 1974 (51) proposed a definite model for the basic unit of chromatin as a bead of about 100 Å diameter, containing a stretch of DNA 200 base pairs long condensed around the protein core made out of 8 histone molecules, namely the $(H3)_2(H4)_2$ tetramer and 2 each of H2A and H2B. The fifth histone, H1, was somehow associated with the outside of each nucleosome. A quite unexpected feature of the model was that it was the DNA which "coated" the histones, rather than the reverse.

However, in 1972, when Kornberg came to Cambridge, all this lay in the future. We began using X-ray diffraction to follow the reconstitution of histones and DNA, because the X-ray pattern given by nuclei, or by chromatin isolated from them, limited as it was, was the only assay then available to follow the ordered packaging of the DNA. These X-ray studies showed that almost 90 % reconstitution could be achieved when the DNA was simply mixed with an unfractionated total histone preparation, but all attempts to reconstitute chromatin by mixing DNA with a set of all four purified single species of histone failed, as if the process whereby the histones were being separated was denaturing them. We therefore looked for milder methods of histone extraction and found that the native structure could be reformed readily if the four histones were kept together in two pairs, H3 and H4 together, and H2A and H2B together, but not once they had been taken apart. It was this work which led Kornberg to investigate further the physicochemical properties of the histones and to the discovery (50) of the histone tetramer $(H3)_2(H4)_2$, which in turn led him to the model of the nucleosome as described above.

The structure of the nucleosome

Approaches such as nuclease digestion and X-ray scattering on unoriented specimens of chromatin or nucleosomes in solution could reveal certain features of the nucleosome, but a full description of the structure can only come from crystallographic analysis, which gives complete three-dimensional structural information. In the summer of 1975 my colleagues and I therefore set about trying to prepare nucleosomes in forms suitable for crystallisation. Nucleosomes purified from the products of micrococcal nuclease digestion contain an average of about 200 nucleotide pairs of DNA, but there is a rather wide distribution about the average, and such preparations are not homogeneous enough to crystallize. However, this variability in size can be eliminated by further digestion with micrococcal nuclease. While the action of micrococcal nuclease on chromatin is first to cleave between nucleosomes, it subsequently acts as an exonuclease on the excised nucleosome, shortening the DNA first to about 166 base pairs, where there is a brief pause in the digestion (52), and then about 146 base pairs, where there is a clear plateau in the course of digestion, before more degradation occurs. During this last stage the histone H1 is released (52), leaving as a major metastable intermediate a particle containing 146 base pairs of DNA complexed with a set of 8 histone molecules. This enzymatically reduced form of the nucleosome is called the core particle and its DNA content was found to be constant over many different species. The DNA removed by the prolonged digestion, which had previously joined one nucleosome to the next, is called the linker DNA.

A core particle therefore contains a well-defined length of DNA and is homogeneous in its protein composition. We naturally tried to crystallize preparations of core particles, but we were not at first successful probably because of small traces of the fifth histone H1. Eventually my colleague Leonard Lutter found a way to produce exceptionally homogeneous preparations of nucleosome core particles, and these formed good single crystals (53). The conditions for growing the crystals were based on our previous experience in crystallising transfer-RNA, because we reasoned that a good part of the nucleosome core surface would consist of DNA. These experiments perhaps surprised biologists in showing dramatically that almost all the DNA in the nucleus is organised in a highly regular manner.

The derivation of a three-dimensional structure from a crystal of a large molecular complex is, as for the TMV disk, a process that can take many years. We have therefore concentrated on obtaining a picture of the nucleosome core particle at low resolution by a combination of X-ray diffraction and electron microscopy, supplemented where possible by biochemical and physicochemical studies. We first solved the packing in the crystals by analysing electron micrographs of thin crystals and then obtained projections of the electron density along the three principal axes of the crystals, using X-ray diffraction amplitudes and electron microscope phases (53, 54). The nucleosome core particle turned out to be a flat disc-shaped object, about 110 Å by 110 Å by 57 Å, somewhat wedge-shaped, and strongly divided into two layers. We proposed a model in which the DNA was wound into about $1^3/_4$ turns of a shallow superhelix of pitch about 27 Å around the histone octamer. There are thus about 80 nucleotides in each turn of the superhelix. This model for the organisation of DNA in a nucleosome core also provided an explanation for the results of certain enzyme digestion studies on chromatin (53, 55) thus showing that what we had crystallised was essentially the native structure.

The first crystals we obtained were found to have the histone proteins within them partly proteolysed, but their physicochemical properties remained very similar to those of the intact particle. We have since grown crystals from intact nucleosome cores which diffract to a resolution of about 5 Å and a detailed analysis is in progress (56). Over the years Daniela Rhodes, Ray Brown and Barbara Rushton have grown crystals of core particles prepared from seven different organisms: all give essentially identical X-ray patterns testifying to the universality of nucleosomes. There is a dyad axis of symmetry within the particle, which is not surprising since the 8 histones occur in pairs and DNA is studded with local dyad axes. High angle diffuse X-ray scattering from the crystals shows that the DNA of the core particle is in the B-form.

An electron density map of one of the principal projections of the crystal is shown in Fig. 11a. This map gives the total density in the nucleosome, the density of the DNA not being distinguished from that of the protein. The contributions of protein and DNA can be distinguished by using neutron scattering combined with the method of contrast variation and such a study was therefore begun by John Finch and a group at the Institue Laue Langevin, Grenoble, when sufficiently large crystals were available (57). They obtained maps of the DNA and protein along the three principal projections (see Figs.

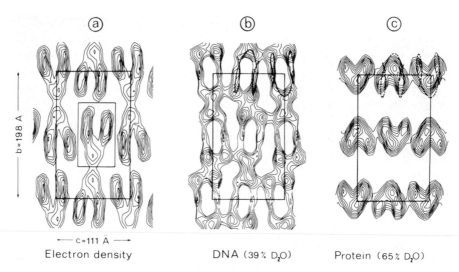

Fig. 11. Fourier projection maps of the nucleosome core particle. (a) Map from X-ray data (56); (b) and (c) from neutron scattering data using contrast variation (57): (b) the DNA component with the path of the superhelix drawn superimposed on the density; (c) the protein core component.

11b and c). The map of the DNA is consistent with the projection of about $1^3/_4$ superhelical turns as proposed earlier, and the map of the protein shows that the histone octamer itself is consistent with a wedge shape.

Three dimensional image reconstruction of the histone octamer and the spatial arrangement of the inner histones

An alternative to separating the contributions of the DNA and the protein by neutron diffraction is to study the histone octamer directly. The histone octamer which forms the protein core of the nucleosome can exist in that form free in solution in high salt, which displaces the DNA (58). In the course of attempts to crystallize it, we obtained ordered aggregates − hollow tubular structures − which were investigated by electron microscopy (59). The image reconstruction method described above was used to produce a low resolution three-dimensional map and model of the octamer (fig. 12a). As a check that the removal of DNA had not led to a change in the structure of the histone octamer, projections of this model were calculated and compared with the projections of the protein core of the nucleosome obtained from the neutron scattering study mentioned above. There was a good agreement between the three maps showing that the gross structure was not altered.

To the resolution of the anlysis (20 Å) it was shown that the histone octamer possesses a two-fold axis of symmetry, just as does the nucleosome core particle itself. Like the nucleosome core, the histone octamer is a wedgeshaped particle of bipartite character. Its periphery shows a system of ridges which form a more or less continuous helical ramp of external diameter 70 Å and pitch about 27 Å, exactly suitable for it to act as a spool on which could be wound about $1^3/_4$ turns of superhelix of DNA in the appropriate dimensions (Fig. 12b).

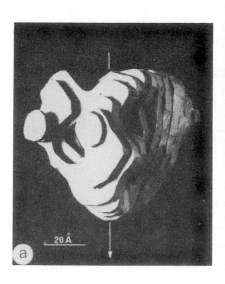
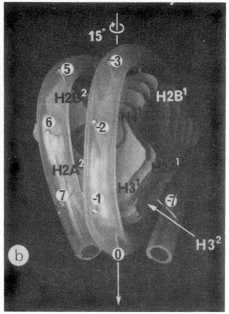

Fig. 12. (a) Model of the histone octamer obtained by three-dimensional image reconstruction from electron micrographs (59). The dyad axis is marked. The ridges on the periphery of the model form a left-handed helical ramp on which 1 3/4 to 2 turns of a superhelix of DNA could be wound.

(b) The histone octamer structure (a) with two turns of a DNA superhelix wound around it. (Note that for clarity, the diameter of the plastic tube has been chosen smaller than the true scale for DNA.) Distances along the DNA are indicated by the numbers -7 to $+7$, taking the dyad axis as origin, to mark the 14 repeats of the double helix contained in the 146 base pairs of the nucleosome core. The assignment of the individual histones to various locations on the model is described in the text.

The resolution of the octamer map is too low to define individual histone molecules, but we have exploited the relation of the octamer to the superhelix of DNA to interpret them in terms of individual histones (59). This interpretation uses the results of Mirzabekov and his colleagues (60) on the chemical cross-linking of histones to nucleosomal DNA, and also information on histone/histone proximities given by protein crosslinking. This data cannot be inter-preted reliably without a three-dimensional model because a knowledge of the points of contact of histones along a strand of the DNA is not sufficient to fix a spatial arrangement of the histones in the nucleosome core. Furthermore, because the two superhelical turns of DNA are close together the pattern of histone/DNA crosslinks need not directly reflect the linear order of histones along the DNA. The three-dimensional density map restricts the number of possiblities and enables choices to be made.

In the spatial arrangement proposed, the helical ramp of density in the octamer map is composed of a particular sequence of the eight histones, in the order H2A−H2B−H4−H3−H3−H4−H2B−H2A, with a dyad in the middle. The $(H3)_2(H4)_2$ tetramer has the shape of a dislocated disc or single turn of a helicoid, which defines the central turn of a DNA superhelix. The structure for the histone tetramer explains the findings of many workers, expanding on the original observations of Felsenfeld (61), that H3 and H4 alone, in the absence of H2A and H2B, can confer nucleosome-like properties on DNA, in particular supercoiling and resistance to micrococcal nuclease digestion, whereas H2A and H2B alone cannot. It also explains the asymmetric dissociation of the histone octamer when the salt concentration is lowered: the octamer dissociates, through a hexameric intermediate, into a $(H3)_2(H4)_2$ tetramer and two H2A.H2B dimers (58, 62).

The role of H1 and higher order structures

These studies have given a fairly detailed picture of the internal structure of the nucleosome, but until 1975 there was still no clear idea of the relation of one nucleosome to another along the nucleosome chain or basic chromatin filament, nor of the next higher level of organisation. It had been known for some time that the thickness of fibres observed in electron microscopical studies of whole mount chromosome specimens varied from about 100 to 250 Å in diameter, depending on whether chelating agents had been used or not in the preparation. Taking this as a clue, Finch and I carried out some experiments *in vitro* on short lengths of chromatin prepared by brief micrococcal digestion of nuclei (63). In the presence of chelating agents this native chromatin appeared as fairly uniform filaments of 100 Å diameter. When Mg^{++} ions were added, these coiled up into thicker, knobbly fibres about 250−300 Å diameter, which are transversely striated at intervals of about 120−150 Å, corresponding apparently to the turns of an ordered, but not perfectly regular helix or supercoil. Since the term "supercoil" had already been used up in a different context, we called it a solenoid, because the turns were spaced close together. On the basis of these micrographs and companion X-ray studies (64), we suggested that the second level of folding of chromatin was achieved by the winding of the nucleosome filament into a helical fibre with about 6 nucleosomes per turn. Moreover, we found that when the same experiments were carried out on H1-depleted chromatin, only irregular clumps were formed, showing that the fifth histone H1 is needed for the formation or stabilisation of the ordered fibre structure.

   These experiments told us the level at which H1 performs its function of condensing chromatin, but the way in which the H1 molecule mediates the coiling of the 100 Å filament into the 300 Å fibre only became clear later by putting together evidence from the biochemistry, from the crystallographic analysis, and from more refined electron microscope observations.

   From observations on the course of nuclease digestion, taken in conjunction with the known X-ray structure of the nucleosome core, one can deduce where the H1 might be on the complete nucleosome. I have mentioned that there is an intermediate in the digestion of chromatin by micrococcal nuclease at about 166 base pairs of DNA and it is during this step from 166 to 146 base pairs that H1 is released (52). Since the 146 base pairs of the particle correspond to $1^3/_4$ superhelical turns, we therefore suggested that the 166 base pair particle

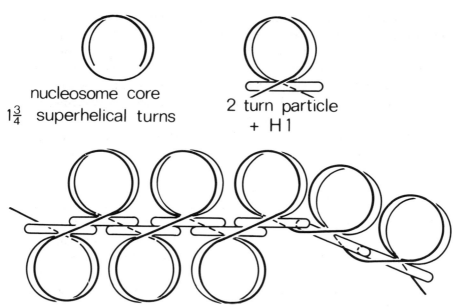

nucleosome core
$1\frac{3}{4}$ superhelical turns

2 turn particle
+ H1

Fig. 13. (Top) If the 146 base pairs of DNA in the nucleosome core correspond to 1 3/4 superhelical turns, then the 166 base particle corresponds to about 2 full superhelical turns. Since the 166 base pair particle is the limit point for the retention of Hl (52), it must be located as shown. (Bottom) Schematic diagram of the nucleosome filament at low ionic strength, showing origin of the zigzag structure (Fig. 14). At the right of the drawing is shown a variant of the zigzag structure which is often observed: this is formed by flipping a nucleosome by 180° about the filament axis.

contains two full turns of DNA (53). This brings the two ends of the DNA on the nucleosome close together so that both can be associated with the same single molecule of H1 (Fig. 13). A particle consisting of the histone octamer and 166 base pairs has been called the chromatosome (65) and has been suggested by us and others to constitute the basic structural element of chromatin. In this particle, the H1 would therefore be on the side of the nucleosome in the region of the entry and exit of the DNA superhelix.

This location follows in logic: but was histone H1 really there? Although H1 is too small a molecule to be seen directly by electron microscopy, its position in the nucleosome can be inferred from its effect on the appearance of chromatin, in the intermediate range of folding between the 100 Å nucleosome filament and the 300 Å solenoidal fibre. These intermediate stages were revealed in the course of a systematic study by Thoma and Koller (66), of the folding of chromatin with increasing ionic strength. By employing monovalent salts rather than divalent ones, they exposed a range of structures showing increasing degrees of compaction as the ionic strength was raised. Thus, from the filament of nucleosomes around 1 mM, the extent of structure increased through a family of intermediate helical structures until, by 60 mM, the compact 300 Å fibre structure was formed, in all respects identical to that originally observed by Finch and myself.

The location of H1 can be deduced by considering the difference between the structures observed in the range of ionic strength 1−5 mM in the presence or

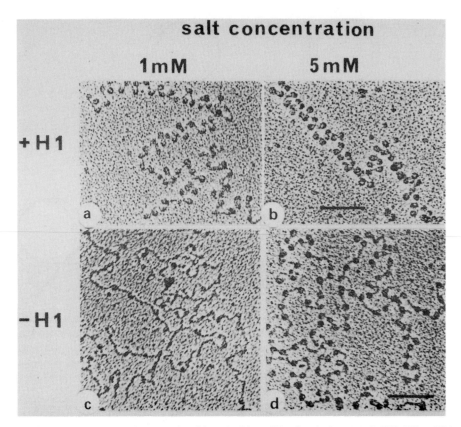

Fig. 14. The appearance of chromatin with and without Hl at low ionic strength (66). When Hl is present the first recognizable ordered structure is (a) a loose zigzag in which the DNA enters and leaves the nucleosome at sites close together; at a somewhat higher salt concentration (b) the zigzag is tighter. In the absence of Hl, there is no order in the sense of a defined filament direction; (c) at the lower salt concentration, nucleosome beads are no longer visible, the structure having opened to produce a fibre of DNA coated with histones; (d) at a higher ionic strength, beads are again visible but the DNA enters and leaves the nucleosome more or less at random. The bar represents 100 nm.

absence of H1 (Fig. 14). In chromatin containing H1, an ordered structure is seen in which the nucleosomes are arranged in a regular zigzag with their flat faces down on the supporting grid. The zigzag form arises because the DNA enters and leaves the nucleosome at sites close together, as one would expect from the combination of X-ray and biochemical evidence mentioned in the last paragraph (Fig. 13). In chromatin depleted of H1, entrance and exit points are more or less on opposite sides and in any case randomly located. Indeed, at very low ionic strength, the nucleosomal structure unravels into a linearised form in which individual beads are no longer seen. When H1 is present this is prevented from happening. We therefore concluded that H1, or strictly part of it, must be located at, and stabilises, the region where DNA enters and leaves the nucleosome, as was predicted.

    In the zigzag intermediates the H1 regions on adjacent nucleosomes appear

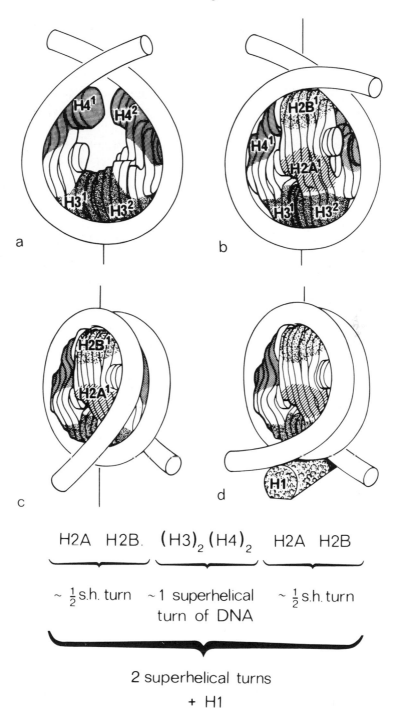

Fig. 15. "Exploded" views of the nucleosome, showing the roles of the constituent histones. The patches on the histone core indicate locations of individual histone molecules, but the boundaries between them are not known and are thus left unmarked. (a) the $(H3)_2(H4)_2$ tetramer has the shape of a lock washer and can act as a spool for 70–80 b.p. of DNA, forming about one superhelical turn. (b) an H2A.H2B dimer associates with one face of the tetramer. (c) H2A.H2B dimers on opposite faces each bind 30–40 b.p. DNA, or one-half a superhelical turn, to give a complete 2-turn particle. (d) histone Hl interacts with the unique configuration of DNA at the entry and exit points to seal off the nucleosome.

to be close together or touching. We therefore suggested that, with increasing ionic strength, more of the H1 regions interact with one another, eventually aggregating into a helical polymer along the centre of the solenoid and thus accounting for its geometrical form. Polymers of H1 have indeed been shown to exist by chemical crosslinking experiments at both low and high ionic strength (61), but it remains to be shown that they are located in the centre of the fibre. The important point, however, is that it appears to be the aggregation of H1 which accompanies, and indeed may control, the formation of the 300 Å fibre.

The roles of the histones

From the spatial arrangements of molecules proposed for the histone octamer and from the location deduced for histone H1, one can see (59) the roles of the individual histones in folding the DNA on the nucleosome (Fig. 15). The $(H3)_2(H4)_2$ tetramer has the shape of roughly a single turn of a helicoid and this defines the central turn of the DNA superhelix. H2A and H2B add as two heterodimers, H2A.H2B, one on each face of the H3—H4 tetramer, each binding one extra half-turn of the DNA, thereby completing the two-turn superhelix. Finally, H1 then binds to the unique region at the side of the two-turn particle where three segments of DNA come together, stabilizing and "sealing off" the nucleosome, and also mediating the folding to the next level of organisation. Such a sequence of events in time would provide a structural rationale for the temporal order of assembly of histones on to newly replicated DNA (68, 69, 70).

We now have arrived at a moderately detailed model of the nucleosome and a description for the next higher level of folding. There is thus a firm structural and chemical framework in which to consider the dynamic processes which take place in chromatin in the cell, that is, transcription, replication and mitosis.

CONCLUDING REMARKS

I particularly wanted to outline the chromatin work because it may serve as a contemporary paradigm for structural studies which try to connect the cellular and the molecular. One studies a complex system by dissecting it out physically, chemically, or in this case enzymatically, and then tries to obtain a detailed picture of its parts by X-ray analysis and chemical studies, and an overall picture of the intact assembly by electron microscopy. There is, however, a sense in which viruses and chromatin, which I have described in this lecture, are still relatively simple systems. Much more complex systems, ribosomes, the mitotic apparatus, lie before us and future generations will recognise that their study is a formidable task, in some respects only just begun. I am glad to have had a hand in the beginnings of the foundation of structural molecular biology.

Acknowledgements

It will be obvious that I could not have accomplished all that has been summarised here without the help of many highly able and valued colleagues and collaborators. After Rosalind Franklin's death, I was able to continue and extend the virus work with John Finch and Kenneth Holmes, who were then students, and who became colleagues. Over the years I have had a transatlantic association with Donald Caspar and have benefitted from his advice, criticism and insights. I can mention here only some of the names of my other collaborators in the several branches in which I have been involved: in the study of virus chemistry and assembly, Reuben Leberman, Tony Durham, Jo Butler and David Zimmern; in virus crystallography, William Longley, Peter Gilbert, John Champness, Gerard Bricogne and Anne Bloomer; in electron microscopy and image reconstruction, David DeRosier, Harold Erickson, Tony Crowther, Linda Amos, Jan Mellema, Nigel Unwin and, throughout, John Finch; in the structural studies on transfer RNA, Brian Clark, who provided the biochemical background without which the work could not have begun, Jon Robertus, Jane Ladner and Tony Jack; in chromatin, Roger Kornberg, whose skill and insight transformed a "messy" project into a clear problem, Markus Noll, Len Lutter, and also Daniela Rhodes and Ray Brown, who fruitfully transferred their experience from tRNA to nucleosomes, and finally Tim Richmond and John Finch who are engaged in the higher resolution X-ray studies now in progress.

REFERENCES

Reviews

(1)   Klug, A. The assembly of tobacco mosaic virus: structure and specificity. The Harvey Lectures, *74*, 141–172, (1979).

(2)   Crowther, R. A. and Klug, A. Structural analysis of macromolecular assemblies by image reconstruction from electron micrographs. Ann. Rev. Biochem. *44*, 161–182, (1975).

(3)   Klug, A. Image analysis and reconstruction in the electron microscopy of biological macromolecules. Chemica Scripta, *14*, 245–256 (1979).

(4) *a*. Kornberg, R. D. Structure of chromatin. Ann Rev. Biochem. *46*, 931–954 (1977).

     b. Kornberg, R. D. and Klug, A. The nucleosome. Scientific American, *244*, 52–64 (1981).

(5)   Butler, P. J. G. and Klug, A. The structure of nucleosomes and chromatin. in "Horizons in biochemistry and Biophysics". (ed. F. Palmieri), Wiley, 1983 in press.

Others

(6)   Fraenkel-Conrat, H. and Williams, R. C. Proc. Natl. Acad. Sci. U.S.A. *41*, 690–698 (1955).

(7)   Fraenkel-Conrat, H. and Singer, B. Biochim. Biophys. Acta. *33*, 359–370 (1959).

(8)   Matthews, R. E. F. (1966) Virology *30*, 82–96 (1966).

(9)   Strbbs, G., Warren, S., and Holmes, K. Nature (London) 267, 216–221 (1977).

(10)   Holmes, K. C. J. Supramolec. Structure *12*, 305–320 (1979).

(11)   Schramm, G., and Zillig, W. Z. Naturforsch. Ser. b. *10*, 493–498 (1955).

(12)   Caspar, D. L. D. Adv. Protein Chem. *18*, 37–121 (1963).

(13)   Finch, J. T., Leberman, R., Chang, Y.-S., and Klug, A. Nature (London), *212*, 349–350 (1966).

(14)   Bricogne, G. Acta Crystallogr. Sect. A *32*, 832–847 (1976).

(15)   Bloomer, A. C., Champness, J. N., Bricogne, G., Staden, R., and Klug, A. Nature (London) *276*, 362–368 (1978).

(16)   Lauffer, M. A., and Stevens, C. L. Adv. Virus Res. *13*, 1–6 (1968).

(17)   Durham, A. C. H., Finch, J. T., and Klug. A. Nature (London) New Biol. *299*, 37–42 (1971).

(18) Durham, A. C. H. and Klug, A. Nature (London), New Biol. *299*, 42–46 (1971).

(19) Klug, A., and Durham, A. C. H. Cold Spring Harbor Symp. Quant. Biol. *36*, 449–460 (1971).

(20) Butler, P. J. G., and Klug, A. Nature (London), New Biol. *229*, 47–50 (1971).

(21) Fukuda, M., Ohno, T., Okada, Y., Otsuki, Y., and Takebe, I. Proc. Natl. Acad, Sci. U.S.A. *75*, 1727–1730 (1978).

(22) Butler, P. J. G., and Lomonossoff, G. P. Biophys. J. *32*, 295–312 (1980).

(23) Zimmern, D., and Butler, P. J. G. Cell *11*, 455–462 (1977).

(24) Zimmern, D. Cell *11*, 463–482.

(25) Zimmern, D., and Wilson, T. M. A. FEBS Lett. *71*, 294–298 (1976).

(26) Champness, J. N., Bloomer, A. C., Bricogne, G., Butler, P. J. G., and Klug, A. Nature (London) *259*, 20–24 (1976).

(27) Butler, P. J. G., Bloomer, A. C., Bricogne, G., Champness, J. N., Graham, J., Guiley, H., Klug, A., and Zimmern, D. In "Structure-Function Relationships of Proteins" (R. Markham and R. W. Horne, eds.), 3rd John Innes Symp. pp. 101–110. North-Holland/Elsevier, Amsterdam. (1976).

(28) Lebeurier, G. (1976), Nicolaieff, A., and Richards, K. E. Proc. Natl. Acad. Sci. U.S.A. *74*, 149–153 (1977).

(29) Butler, P. J. G., Finch, J. T., and Zimmern, D. Nature (London) *265*, 217–219 (1977).

(30) Caspar, D. L. D., and Klug, A. Cold Spring Harbor Symp. Quant. Biol. *27*, 1–24 (1962).

(31) Brenner, S., and Horne R. W. Biochim. Biphys. Acta. *34*, 103–110 (1959).

(32) Klug, A., and Finch, J. T. J. Mol. Biol. *11*, 403–423 (1965).

(33) Klug, A., and Berger, J. E. J. Mol. Biol. *10*, 565–569 (1964).

(34) Klug, A. and DeRosier, D. J. Nature (London), *212*, 29–32 (1966).

(35) DeRosier, D. J. and Klug, A. J. Mol. Biol. *65*, 469–488 (1972).

(36) DeRosier, D. J., and Klug, A. Nature (London), *217*, 130–134 (1968).

(37) Crowther, R. A., DeRosier, D. J., and Klug, A. Proc. Roy. Soc. Lond. A. *317*, 319–340 (1970).

(38) Crowther, R. A., Amos, L. A., Finch, J. T., DeRosier, D. J., and Klug, A. Nature (London) *226*, 421–425 (1970).

(39) Crowther, R. A., and Amos, L. A. Cold Spring Harbor Symp. Quant. Biol. *36*, 489–494 (1971).

(40) Unwin, P. N. T. J. Mol. Biol. *87*, 657–670 (1974).

(41) Taylor, K. A. and Glaeser, R. M. J. Ultrastructure Res. *55*, 448–456 (1976).

(42) Unwin, P. N. T., and Henderson, R. J. Mol. Biol. *94*, 425–440 (1975).

(43) Henderson, R., and Unwin, P. N. T. Nature *257*, 28–32 (1975).

(44) Unwin, P. N. T. Proc. Roy. Soc. Lond. A. *329*, 327–359 (1972).

(45) Erickson, H. P., and Klug, A. Ber. Bunsenges, Phys. Chem. *74*, 1129–1137 (1970); Phil. Trans. Roy. Soc. ser B. *261*, 105–118 (1971).

(46) Wilkins, M. H. F., Zubay, G., and Wilson, H. R. J. Mol. Biol. *1*, 179–185 (1959).

(47) Luzzati, V., and Nicolaieff, A. J. Mol. Biol. *1*, 127–133 (1959).

(48) Hewish, D. R., and Burgoyne, L. A. Biochem. Biophys. Res. Commun. *52*, 504–510 (1973).

(49) Noll, M. Nature (London) *251*, 249–251 (1974)

(50) Kornberg, R. D., and Thomas, J. O. Science *184*, 865–868 (1974).

(51) Kornberg, R. D. Science *184*, 868–871 (1974).

(52) Noll, M., and Kornberg, R. D. J. Mol. Biol. *109*, 393–404 (1977).

(53) Finch, J. T., Lutter, L. C., Rhodes, D., Brown, R. S., Rushton, B., Levitt, M., and Klug, A. Nature (London) *269*, 29–36 (1977).

(54) Finch, J. T., and Klug, A. Cold Spring Harbor Symp. Quant. Biol. *42*, 1–15 (1978).

(55) Lutter, L. C. J. Mol. Biol. *124*, 391–420 (1978).

(56) Finch, J. T., Brown, R. S., Rhodes, D., Richmond, T., Rushton, B., Lutter, L. C., and Klug, A. J. Mol. Biol. *145*, 757–769 (1981).

(57) Finch, J. T., Lewit-Bentley, A., Bentley, G. A., Roth, M., and Timmins, P. A. Phil. Trans. Roy. Soc. Lond. B. 2901, 635–638 (1980).

(58) Thomas, J. O., and Kornberg, R. D. Proc. Natl. Acad. Sci. U.S.A. *72*, 2626–2630 (1975).

(59) Klug, A., Rhodes, D., Smith, J., Finch, J. T., and Thomas, J. O. Nature (London) *287*, 509–516 (1980).

(60) Mirzabekov, A. D., Shick, V. V., Belyavsky, A. V., and Bavykin, S. G. Proc. Natl. Acad. Sci. U.S.A. *75*, 4184-4188 (1978).

(61) Camerini-Otero, R. D., and Felsenfeld, G. Nucleic Acids Res. *4*, 1159–1181 (1977a).

(62) Thomas, J. O., and Kornberg, R. D. FEBS Lett. *58*, 353–358 (1975).

(63) Finch, J. T., and Klug, A. Proc. Natl. Acad. Sci. U.S.A. *73*, 1897–1901 (1976).

(64) Sperling, L., and Klug, A. J. Mol. Biol. *112*, 253–263 (1977).

(65) Simpson, R. T. Biochemistry *17*, 5524–5531 (1978).

(66) Thoma, F., Koller, Th., and Klug, A. J. Cell Biol. *83*, 403–427 (1979).

(67) Thomas, J. O., and Khabaza, A. J. A. Eur. J. Biochem. *112*, 501–511 (1980).

(68) Worcel, A., Han, S., and Wong, M. L. Cell *15*, 969–977 (1978).

(69) Senshu, T., Fukuda, M., and Ohashi, M. J. Biochem. (Japan) *84*, 985–988 (1978).

(70) Cremisi, C., and Yaniv, M. Biochem. Biophys. Res. Commun. *92*, 1117–1123 (1980).

(71) Klug, A., and Caspar, D. L. D. Adv. Virus Res. *7*, 225–325 (1960).

(72) Crowther, R. C., and Amos, L. A. J. Mol. Biol. *60*, 123–130 (1971).

(73) Klug, A., and Finch, J. T. J. Mol. Biol. *31*, 1–12 (1968).

Chemistry 1983

HENRY TAUBE

for his work on the mechanisms of electron transfer reactions, especially in metal complexes

THE NOBEL PRIZE FOR CHEMISTRY

Speech by Professor INGVAR LINDQVIST of the Royal Academy of Sciences
Translation from the Swedish text

You Majesties, Your Royal Highnesses, Ladies and Gentlemen,

Henry Taube has been awarded the 1983 Nobel prize in chemistry for his studies of the mechanisms of electron transfer-reactions, particularly of metal complexes. I will not, during these few minutes, try to give a survey of the rich scientific production of Taube. Instead, I will choose one chemical reaction and attempt to show how our way of looking at this and similar reactions has changed drastically thanks to Taube.

Chemistry is a science which is rapidly aging. Most of the chemical investigations, basic or applied, which were published fifty years ago are to-day forgotten, not because they were wrong, but because our outlook of the science chemistry has changed so much during this time. Some few contributions are, however, still alive because they have determined in a decisive way our ideas about the fundamental relations in chemistry.

It must be in the spirit of Alfred Nobel that works of this type should be awarded the prize, each in its time. It can then be remembered that these new ways of thinking are in the long run also of utmost importance for applied chemistry, although it might be difficult to point directly to their usefulness at the occasion when the prize is awarded. Starting with Svante Arrhenius, who got the Nobel prize 1903, I will show how Taube fits in with this line of "new thinkers". Arrhenius got the prize because he convinced his contemporary chemists that salts in aqueous solutions exist as positive and negative ions and not as neutral molecules. The reaction which I will discuss could, according to Arrhenius, be described as follows: Trivalent cobolt ions oxidize bivalent chromium ions and thereby form bivalent cobolt ions and trivalent chromium ions. The net result is an electron transfer between two positively charged ions. Arrhenius did not speculate further in molecular terms about the nature of the ions in the solution.

It was another Nobel prize winner (1913), Alfred Werner, who carried the development further in a decisive way when he proved that metal ions in solution are in many cases surrounded by a fixed number of neighbouring negative ions or neutral molecules. He also suggested that these neighbours are arranged in a certain way, e.g., in the corners of an octahedron if they are six in number. If Werner had known what we know to-day thanks to Taube, about the reaction I am discussing, he would have expressed the conditions in the following way: A trivalent cobolt ion surrounded by five ammonia molecules and one chloride ion is reacting with a bivalent chromium ion which probably is surrounded by six water molecules. During the reaction there are formed a bivalent cobolt ion surrounded by five ammonia molecules and one water

Arrhenius:

$$Co^{3+} + Cr^{2+} \rightarrow Co^{2+} + Cr^{3+}$$

Werner:

$$Co^{III}(NH_3)_5Cl^{2+} + Cr^{II}(H_2O)_6^{2+} \rightarrow$$
$$\rightarrow Co^{II}(NH_3)_5H_2O^{2+} + Cr^{III}Cl(H_2O)_5^{2+}$$

Taube:

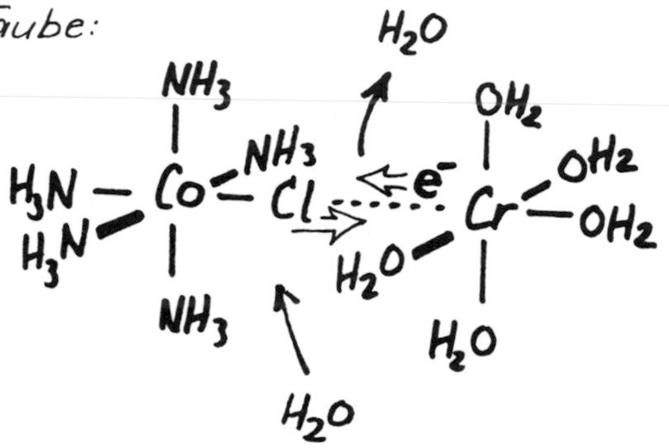

molecule and a trivalent chromium ion surrounded by five water molecules and one chloride ion. The electron transfer is thus connected with a chloride ion transfer from cobolt to chromium. This description is certainly much more complete and gives a clear picture of the conditions before and after the reaction. It does not tell anything, however, about *how* the reaction has taken place.

Taube has now proceeded one step further and has shown us exactly how the reaction occurs. The first step is the formation of a larger complex where the chromium ion is attached to the cobolt ion via the chloride ion which thus functions as a bridge between the two metal ions. This requires that a water molecule leaves the chromium ion to give place for the bridging chloride ion. The chromium ion will thus be surrounded by five water molecules and the bridging chloride ion. It is only when this bridge has formed that the electron transfer can take place making the cobolt ion bivalent and the chromium ion trivalent. Finally the bridge is opened and the chloride ion follows the trivalent chromium ion while the cobolt ion must take up a molecule of water to replace the chloride ion. In an impressive series of investigations Taube has developed and refined this idea and it will in the future be a natural part of the set of paradigms of chemistry, which thus in a decisive way has been enriched by the contributions of Henry Taube.

Professor Henry Taube,

I have not tried, during these few minutes, to give a comprehensive presentation of all your outstanding contributions to inorganic chemistry, but have rather preferred to try to show the conceptual importance of one of your achievements, which is an essential part of that work for which you have been awarded the Nobel prize. It is an honour and a pleasure for me to extend to you the congratulations of the Royal Academy of Sciences and to ask you to receive your prize from the hands of His Majesty the King.

HENRY TAUBE

Born: Neudorf, Saskatchewan, Canada, 1915; naturalized U.S. citizen, 1942.

Education:
B.S.: 1935, University of Saskatchewan
M.S.: 1937, University of Saskatchewan (Research Supervisor, Prof. J.W.T. Spinks)
Ph.D.: 1940, University of California, Berkeley (Research Supervisor, Prof. W.C. Bray)

Professional Experience:
Instructor, University of California, Berkeley, 1940−41
Instructor and Assistant Professor, Cornell University, 1941−46
Assistant Professor, Associate Professor, Professor, University of Chicago, 1946−61
Professor, Stanford University, 1962–86
Professor Emeritus, 1986
Chairman, Department of Chemistry, University of Chicago, 1956−59
Chairman, Department of Chemistry, Stanford University, 1972−74 & 1978−79

Honors and Awards:
Guggenheim Fellow, 1949 and 1955
American Chemical Society Award for Nuclear Applications in Chemistry, 1955
Harrison Howe Award, Rochester Section, ACS, 1960
Chandler Medal, Columbia University, 1964
John Gamble Kirkwood Award, New Haven Section, ACS, 1966
ACS Award for Distinguished Service in the Advancement of Inorganic Chemistry, 1967
Nichols Medal, New York, ACS, 1971
Willard Gibbs Medal, Chicago Section, ACS, 1971
F. P. Dwyer Medal, University of New South Wales, Australia, 1973
Honorary Doctorate (L.L.D.) University of Saskatchewan, 1973
Marguerite Blake Wilbur Endowed Professorship, 1976
National Medal of Science, Washington, D.C., 1977
Allied Chemical Award for Excellence in Graduate Teaching & Innovative Science, 1979
Degree of Ph. D. Honoris Causa of the Hebrew University of Jerusalem, 1979
T. W. Richards Medal of the Northeastern Section, ACS, 1980
ACS Award in Inorganic Chemistry of the Monsanto Company, 1981
The Linus Pauling Award, Puget Sound Section, ACS, 1981
National Academy of Sciences Award in Chemical Sciences, 1983
Bailar Medal, University of Illinois, 1983
Doctor of Science, University of Chicago, 1983
Robert A. Welch Foundation Award in Chemistry, 1983

Nobel Prize in Chemistry, 1983
Doctor of Science, Polytechnic Institute, New York, 1984
Honorary Member, College of Chemists of Catalonia and Beleares, 1984
Priestley Medal, ACS, 1985
Doctor of Science, State University of New York, 1985
Corresponding Member, Academy of Arts and Science of Puerto Rico, 1985
Honorary Member, Canadian Society for Chemistry, 1986
Distinguished Achievement Award, International Precious Metals Institute, 1986
The Oesper Award, The Cincinnati Section of the American Chemical Society, 1986
Doctor of Science, University of Guelph, 1987
Honorary Member, Hungarian Academy of Sciences, 1988
Doctor of Science, *honoris causa*, Seton Hall University, 1988
Doctor of Science, Lajos Kossuth University of Debrecen, Hungary, 1988
Honorary Fellowship, Royal Society of Chemistry, 1989
Honorary Fellowship, Indian Chemical Society 1989
G. M. Kosolapoff Award, Auburn Section, ACS, 1990
Doctor of Science, Northwestern University, 1990

Membership in Societies:
American Chemical Society
National Academy of Sciences
American Academy of Arts & Sciences
Phi Beta Kappa
Sigma Xi
Phi Lamda Upsilon (honorary member)
Royal Physiographical Society of Lund
American Philosophical Society
Royal Danish Academy of Sciences & Letters
Foreign Member of the Finnish Academy of Science & Letters
Foreign Member, Royal Society
Corresponding Member, Brazilian Academy of Sciences
Foreign Associate, Engineering Academy of Japan
Corresponding Member, Australian Academy of Science

Consultantship:
Catalytica Associates, Inc., Mountain View, California

Research Interests:
Current research interests include: charge transfer as affecting properties
including the reactivity of ligands; mixed valence molecules; mechanisms of
"atom" and electron transfer reactions; basic chemistry of osmium and
ruthenium, effects arising from back-bonding.

Publications:

Over 350 scientific articles and a book have been published as a result of this research.

ELECTRON TRANSFER BETWEEN METAL COMPLEXES — RETROSPECTIVE

Nobel lecture, 8 December, 1983

by
HENRY TAUBE

Department of Chemistry, Stanford University,
Stanford, CA 94305

This will be an account in historical perspective of the development of part of the field of chemistry that I have been active in for most of my professional life, the field that is loosely described by the phrase "electron transfer in chemical reactions". In the short time available to me for the preparation of this paper, I can't hope to provide anything significant in the way of original thought. But I can add some detail to the historical record, especially on just how some of the contributions which my co-workers and I have made came about. This kind of information may have some human interest and may even have scientific interest of a kind which cannot easily be gathered from the scientific journals. For publication there, the course of discovery as it actually took place may be rewritten to invest it with a logic that it did not fully acquire until after the event.

Simple electron transfer is realized only in systems such as $Ne + Ne^+$. The physics already becomes more complicated when we move to $N_2 + N_2^+$ for example, and with the metal ion complexes which I shall deal with, where a typical reagent is $Ru(NH_3)_6^{2+}$, and where charge trapping by the solvent, as well as within the molecule, must be taken into account, the complexity is much greater. Still, a great deal of progress has been made by a productive interplay of experiment, qualitative ideas, and more sophisticated theory, involving many workers. Because of space limitations, I will be unable to trace all the ramifications of the field today, and will emphasize the earlier history of the subject, when some of the ideas basic to the field were being formulated. This choice of emphasis is justified because, by an accident of history, I was a graduate student at the University of California, Berkeley, about the time the first natal stirrings of the subject of this article occurred, and at a place where these stirrings were most active. As a result, I may be in a unique position to deal knowledgeably and fairly with the early history of the subject. The emphasis on the early history is all the more justified because most of the topics touched on in this article, and also closely related topics, are brought up to date in a very recent volume of the series, Progress in Inorganic Chemistry (1).

Chemical reactions are commonly classified into two categories: substitution or oxidation-reduction. The latter can always be viewed as involving electron transfer, though it is agreed that when we consider the mechanisms in solution, electron transfer is not as simple as it is in the $Ne + Ne^+$ case. Rearrangement of atoms always attend the changes in electron count at each center, and these must be allowed for. I will, however, simplify the subject by considering only processes of simple chemistry: those in which electron transfer leaves each of the reaction partners in a stable oxidation state. While substitution reactions can be discussed without concern for oxidation-reduction reactions, the reverse is not true. The changes that take place at each center when the electron count is changed is an essential part of the "electron transfer" process, and may be the dominating influence in fixing the rate of the reaction. Moreover, most of the early definitive experiments have depended on exploiting the substitution characteristics of the reactants, and of the products. Thus, the attention which will be devoted to the substitution properties of the metal ions is not a digression but is an integral part of the subject.

An appropriate place to begin this account is with the advent of artificial radioactivity. This enormously increased the scope of isotopic tracer methods applied to chemistry, and made it possible to measure the rates of a large number of oxidation reduction reactions such as:

$$*Fe^{2+}(aq) + Fe^{3+}(aq) = *Fe^{3+}(aq) + Fe^{2+}(aq) \tag{1}$$

(The first demonstration of a redox electron exchange was made by von Hevesy and co-workers (2), who used naturally occurring isotopes to follow $Pb(IV)/Pb(II)$ exchange in acetic acid.) Because chemists were there involved in the discovery of many of the new isotopes (3), an early interest in this kind of possibility developed in the chemistry community at the University of California, Berkeley, and was already evident when I was a graduate student there (1937–40). Mention is made in a review article by Seaborg (3) devoted to artificial radioactivity, of an attempt (4) to measure the rate of the $Fe^{3+/2+}$ exchange in aqueous chloride media, the result of this early attempt being that the exchange was found to be complete by the time the separation of $Fe(III)$ from $Fe(II)$ was made. It was appreciated by many that the separation procedure, in this case extraction of the $Fe(III)$-Cl^- complex into ether, might have induced exchange. It was also appreciated that Cl^- might have affected the reaction rate, possibly increasing it, and that quite different results might be obtained were the experiment done with Cl^- being replaced by an indifferent anion.

There were several reasons for the interest, among many physical-inorganic chemists, in a reaction such as (1). That the interest in chemical applications of the new isotopes was keen in Berkeley may be traced in part to the involvement of much of the research body in teaching in the introductory chemistry course. We all had a background of qualitative observations on oxidation-reduction reactions of simple chemistry—as an example, on the reaction of $Ce(IV)(aq)$ with $Fe^{2+}(aq)$,—from experience in qualitative or quantitative analysis. Still, to my knowledge, at the time I was a graduate student, not a single measurement had been made of the rate of this kind of reaction. That a field of research, which has since grown enormously, was started by studying "self-exchange reactions" (5) such as (1), rather than net chemical changes (descriptor, "cross-reaction") (5), may reflect the intervention of a human factor. Measuring the rate of a virtual process such as (1), today made commonplace for many systems by the introduction of new spectroscopic methods, seemed more glamorous than measuring the rate of oxidation of V^{2+}(aq) by Fe^{2+}(aq), for example. But I also recall from informal discussions that it was felt that driving force would affect the rate of reaction, and thus there would be special interest in determining the rates for reactions for which ΔG^0 (except for the entropic contributions to the driving force) is zero.

The interest in the measurement of the rates of self-exchange reactions which I witnessed as a graduate student, is not reflected in the literature of the years immediately following. Many of those who might have had plans to do the experiments were engaged in war related activities. Post war, at least five different studies on the rate of reaction (1), all carried out in non-complexing media, were reported, with conflicting results, some indicating a half-life for exchange on the order of days at concentration levels of 10^{-2}M. The discrepancies led to considerable controversy, and in informal discussions, strong opinions were expressed on just what the true rate of self-exchange might be. The basis for this kind of judgment, exercized in the absence of any body of quantitative measurements, is worth thinking about. I believe it reflected an intuitive feeling that there would be a relation between the rates of self-exchange and of the related cross-reactions, and of course each of us had at least some qualitative information on redox rates for the $Fe^{3+/2+}$ couple. The definitive measurements on the rate of reaction (1) in non-complexing media were made by Dodson (6). These measurements were soon extended (7) to reveal the effect of $[H^+]$ and of complexing anions on the rate and yielded rate functions such as $[Fe^{3+}][X^-][Fe^{2+}]$ (because substitution is rapid compared to electron transfer, this is kinetically equivalent to $[FeX^{2+}][Fe^{2+}]$ and to $[Fe^{3+}][FeX^+]$), in addition to $[Fe^{3+}][Fe^{2+}]$, none specifying a unique structure for an activated complex. Particularly the terms involving the anions provided scope for speculation about mechanism. The coefficient for the simple second order function was found (7) to be 4 $M^{-1}s^{-1}$ at 25^0, $\mu=0.5$, and those who had argued for "fast" exchange won out.

Another important experimental advance during the same period, important for several reasons, was the measurement (8) of the rate of self-exchange for $Coen_3^{3+/2+}$ (k = 5.2 x $10^{-5}M^{-1}s^{-1}$ at 25^0, $\mu = 0.98$). This is, I believe, the first quantitative measurement of a rate for a self-exchange reaction and it may also be the first time that the oxidizing capability of a cobalt (III) ammine complex was deliberately exploited. In the article of ref. 8, the rate of the reaction of $Co(NH_3)_6^{3+}$ with $Coen_3^{2+}$ is also reported; this is, I believe, the first deliberate measurement of the rate of an electron transfer cross reaction. In contrast to the $Fe^{3+/2+}$ system, where both reactants are labile to substitution, $Coen_3^{3+}$ is very slow to undergo substitution and thus an important feature of the structure of the activated complex for the $Coen^{3+/2+}$ self exchange appeared to be settled. In considering the observations, the tacit assumption was made that the coordination sphere of $Coen_3^{3+}$ does not open up on the time scale of electron transfer, so that it was concluded that the activated complex for the reaction does not involve interpenetration of the coordination spheres of the two reactants. We were thus obliged to think about a mechanism for electron transfer through two separate coordination spheres (descriptor "outer-sphere" mechanism) (9).

In 1951, an important symposium on Electron Transfer Processes was held at the University of Notre Dame, and the proceedings are reported in J. Phys. Chem., Vol. 56, (1952). Though the meeting was organized mainly for the benefit of the chemists, the organizers had the perspicacity to include physicists in the program. Thus, the gamut of interests was covered, ranging from electron transfer in the gas phase in the simplest kind of systems, for example $Ne + Ne^+$, to the kind of system that the chemist ordinarily deals with. Much of the program was devoted to experimental work, the chemistry segment of which included reports on the rates of self-exchange reactions as well as of reactions involving net chemical change—but none on cross reactions. Two papers devoted to theory merit special mention: that by Holstein (10) whose contributions to the basic physics are now being applied in the chemistry community, and the paper by Libby, (11) in which he stressed the relevance of the Franck-Condon restriction (12) to the electron transfer process, and applied the principle in a qualitative way to some observations. It is clear from the discussion which several of the papers evoked that many of the participants had already appreciated the point which Libby made in this talk. Thus, in the course of the discussion, the slowness of the self-exchange in the cobaltammines was attributed (13) to the large change in the Co-N distances with change in oxidation state, then believed to be much larger than it actually is. During the meeting too, the distinction between outer- and inner-sphere activated complexes was drawn, and the suggestion was made that the role of Cl^- in affecting the rate of the $Fe^{3+/2+}$ self-exchange might be that it bridges the two metal centers. (14) But of course, because of the lability of the high spin Fe(II) and Fe(III) complexes, no unique specification of the geometry of the activated complex can be made on the basis of the rate laws alone. During the discussion of Libby's paper, a third kind of mechanism was proposed (15), involving "hydrogen atom" transfer from reductant to oxidant.

PREPARATION

My own interest in basic aspects of electron transfer between metal complexes became active only after I came to the University of Chicago in 1946. During my time at Cornell University (1941–1946), I had been engaged in the study of oxidation-reduction reactions, and I was attempting to develop criteria to distinguish between $1e^-$ and $2e^-$ redox changes, and as an outgrowth of this interest, using $1e^-$ reducing agents to generate atomic halogen, X, and studying the ensuing chain reactions of X_2 with organic molecules. My eventual involvement in research on electron transfer between metal complexes owes much to the fact that I knew many of the protagonists personally (A. C. Wahl, C. N. Rice, C. D. Coryell, C. S. Garner; the first two were fellow graduate students at Berkeley), and to the fact that I had W. F. Libby, with whom I had many provocative discussions, and J. Franck as well as F. H. Westheimer as colleagues at the University of Chicago. By the time of the meeting, at the University of Notre Dame, a meeting I did not participate in, I appreciated in a general way the special advantages which the $Cr^{3+/2+}$ couple offered in the investigation of the mechanism of electron transfer, and outlined my ideas to N. Davidson when he visited me en route to the meeting, but the experimental work which led to the first two papers (16, 17) was not done until 1953. In the interim, I had failed to interest any of my graduate students in the work, because, of course, no one foresaw what it might lead to, and because it seemed less exciting than other work in progress in my laboratories, much of it concerned with isotopic effects, tracer and kinetic. The bulk of the work reported in the two papers just cited was done by my own hands; I shall now outline the background for those early experiments.

My interest in coordination chemistry did not develop until I elected it as a topic for an advanced course given soon after coming to the University of Chicago. Instead of using the standard textbook material, I used as major source the relevant volume of the reference series by Gmelin in which the chemistry of the cobaltammines is described. At this time I already had a good background in the literature devoted to substitution at carbon and understood the issues raised in that context, and I soon became interested in raising the same issues for substitution at metal centers. Furthermore, it was evident that the complexes based on metal ions which undergo substitution slowly were readily amenable to experimental study. I became curious as well about the reasons underlying the enormous difference in rates of substitution for metal ions of the same charge and (approximately) the same radii. The ideas that resulted were presented in my course the next time it was given, but the extensive literature study that led to the paper published in February, 1952 (18) was not done until 1949 when I was on leave from the University of Chicago as a Guggenheim Fellow.

In this paper, a correlation with electronic structure was made of observations, mainly qualitative ("labile" complexes arbitrarily defined as those whose reactions appear to be complete on mixing and "inert" as those for which continuing reaction can be observed), for complexes of coordination no. 6. To make the correlation, it was necessary to break away from the practice which was common in the USA of classifying complexes as "ionic" and "covalent" according to electronic structure. Thus, for example, the comparison of the affinities of Cr(III) and Fe(III), the latter high spin and thus "ionic", convinced me that in the Fe(III) complexes, the bonds to the ligands might actually be somewhat more covalent than in the Cr(III) complexes. Furthermore, it appeared to me that in earlier discussions of relative rates of substitution, where attempts were made to understand the observations in terms of the existing classification, there was a failure to distinguish between thermodynamic stability, and inertia, the latter being understood as referring solely to rate. The affinity of Cl^- for Cr^{3+}(aq) is considerably less than it is for Fe^{3+}(aq), yet the aquation rate of $CrCl^{2+}$(aq) is much less than it is for $FeCl^{2+}$(aq). Rates, of course, cannot be accounted for by considering ground state properties alone, but the stability of the activated complex relative to the ground state must be taken into account. When the effect of electronic structure on the relative stabilities is allowed for, a general correlation of rates with electronic structure emerges. (In the language of ligand field theory, for complexes of coordination number 6, substitution tends to be slow when the metal ions have each of the πd(non-bonding) orbitals, but none of the σd (anti-bonding) orbitals occupied. The specific rates of exchange of water between solvent and the hexaaquo ions of $V^{3+}(\pi d^2)$, $Cr^{3+}(\pi d^3)$ and Fe^{3+} $(\pi d^3 \sigma d^2)$ are $1 \times 10^3 s^{-1}$, $5 \times 10^{-6} s^{-1}$ and $1 \times 10^3 s^{-1}$ respectively (19)).

A shortcoming of this early effort is that such rationalizations of the correlation as were offered were given in the language of the valence bond approach to chemical binding, because, at the time I wrote the paper, I did not understand the principles of ligand field theory even in a qualitative way. The valence bond approach provides no simple rationale of the difference in rates of substitution for labile complexes, and these have been found to cover a very wide range, thanks to a pioneering study by Bjerrum and Poulsen (20), in which they used methanol as a solvent to make possible measurements at low temperature, and those of Eigen (21), in which he introduced relaxation methods to determine the rates of complex formation for labile systems.

Activated complexes have compositions and structures, and it is necessary to know what these are if rates are to be understood. It is hardly likely that these features can be established for the activated complexes if they are not even known for the reactants, and in 1950 we were not certain of the formula for any aquo cation in water. It seemed to me important, therefore, to try to determine the hydration numbers for aquo cations. Hydration number as I use it here does not mean the average number of water molecules affected by a metal ion as this is manifested in some property such as mobility, but has a structural connotation: how many water molecules occupy the first sphere of coordination? Because the rates of substitution for Cr^{3+}(aq) were known to be slow (22), J. P. Hunt and I undertook to determine the formula for Cr^{3+}(aq) in water (23), with some confidence that we would be successful even with the slow method we applied, isotopic dilution using ^{18}O enriched water. That the formula turned out to be $Cr(H_2O)_6^{3+}$ was no great surprise—although I must admit that at one point in our studies, before we had taken proper account of isotopic fractionation effects, $Cr(H_2O)_7^{3+}$ was indicated and, faced with the apparent necessity, I was quite prepared to give up my preconceived notions. It was also no great surprise that the exchange is slow ($t_{1/2}$ at 25°, ca. 40 hr). Even so, the experiments were worth doing. They were the first of their kind, and they attracted the attention also of physical chemists, many of whom were astonished that aquo complexes could be as kinetically stable as our measurements indicated, and were impressed by the enormous difference in the residence time of a water molecule in contact with a cation, compared with water molecules just outside. That we dealt with hydration in terms of detailed structure rather than in terms of averaged effects may have encouraged the introduction into the field of other methods, such as nmr, to make the distinction between cation-bound and free water (19). As I will now detail, it also led directly to the experiments described in the papers of references 16 and 17.

THE INNER SPHERE ACTIVATED COMPLEX

R. A. Plane undertook to measure the rate of self-exchange for $Cr(H_2O)_6^{3+}-$
Cr^{2+}(aq) by using Cr^{2+}(aq) as a catalyst for the exchange of water between
$Cr(H_2O)_6^{3+}$ and solvent. The expected catalysis was found, but owing to our
inexperience in handling the air sensitive catalyst, the data were too irreproducible to lead to a value for the self-exchange rate. Catalysis on electron transfer
was expected because the aquo complex of Cr^{2+}(aq) was known to be much
more labile than $Cr(H_2O)_6^{3+}$—the lability is now known (24) to decrease by a
factor of at least 10^{14} when Cr^{2+}(aq) is oxidized to $Cr(aq)^{3+}$ (note that Cr^{2+},
but not Cr^{3+}, has an anti-bonding electron). It occurred to me in the course of
Plane's work that it would be worthwhile to test the potential of the $Cr(III)/$
$Cr(II)$ couple for diagnosis of mechanism using a non-metal oxidant. Following
up on the idea, I did a simple test tube experiment, adding solid I_2 to a solution
of Cr^{2+}(aq) which Plane had prepared for his own experiments. I observed that
reaction occurs on mixing, that the product solution is green, and that the
green color fades slowly, to produce a color characteristic of $Cr(H_2O)_6^{3+}$. The
fading is important because it demonstrates that $(H_2O)_5CrI^{2+}$, which is responsible for the green color, is unstable with respect to $Cr(H_2O)_6^{3+} + I^-$, and
thus we could conclude that the $Cr(II)$-I bond is established before $Cr(II)$ is
oxidized.

The principle having been demonstrated with a non-metal oxidant, I turned
to the problem of finding a suitable metal complex as oxidant. What was
needed was a reducible robust metal complex, having as ligand a potential
bridging group, and the idea of using $(NH_3)_5CoCl^{2+}$ surfaced during a discussion of possibilities with another of my then graduate students, R. L. Rich.
Because at that time virtually nothing was known about the rates of reduction
of $Co(III)$ ammines, and because they were not thought about as useful
oxidants, I was by no means sanguine about the outcome of the first experiment, which again was done in a test tube. I was delighted by the outcome.
Reaction was observed to be rapid (the specific rate has since been measured
(25) as $6 \times 10^5 M^{-1}s^{-1}$ at 25°) and the green color of the product solution
indicated that $(H_2O)_5CrCl^{2+}$ is formed. Further work (16, 17) showed that this
species is formed quantitatively, and that in being formed it picks up almost no
radioactivity when labelled Cl^- is present in the reaction solution, thus demonstrating that transfer is direct, i. e., Cl^- bridges the two metal centers, and this
occurs before Cr^{2+} is oxidized.

$$(NH_3)_5CoCl^{2+} + Cr(H_2O)_6^{2+} \rightleftharpoons (NH_3)_5Co^{III}Cl \ldots Cr^{II}(H_2O)_5$$
$$\text{Precursor complex}$$
$$\rightarrow [(NH_3)_5Co \ldots Cl \ldots Cr(H_2O)_5]^{4+} \rightarrow (NH_3)_5Co^{II} \ldots ClCr^{III}(H_2O)_5$$
$$\text{activated complex} \qquad \text{successor complex}$$
$$\rightarrow Co(H_2O)_6^{2+} + NH_4^+ + (H_2O)_5CrCl^{2+}$$

These early results were presented at a Gordon Conference on inorganic
chemistry, which was held only a short time after they had been obtained, and
they were received with much interest. E. L. King was present at the meeting,

and together we planned the experiment in which self-exchange by an "atom" transfer mechanism was first demonstrated (26).

$$(H_2O)_5{}^*CrCl^{2+} + Cr^{2+}(aq) = {}^*Cr^{2+}(aq) + (H_2O)_5CrCl^{2+} \tag{2}$$

Note that in a process of this kind, the bridging group will remain bound to a Cr(III) center, but the auxiliary ligands about the original Cr(III) center will be exchanged or replaced because of the high lability of Cr(II).

(Several years later, King and co-workers (27) extended the self-exchange work to include other halides as bridging ligands, and still later (28) encountered the first example of "double bridging."

$$cis\text{-}[(H_2O)_4{}^*Cr(N_3)_2]^+ + Cr^{2+}(aq) \approx {}^*Cr^{2+}(aq) + cis\text{-}[(H_2O)_4Cr(N_3)_2]^+$$

A good case can be made for this also being the first demonstration of "remote attack," that is, inner-sphere electron transfer where more than one atom separates the metal atoms in the activated complex.)

There is a brief hiatus in my work after the early experiments on inner-sphere mechanisms, caused by my taking leave from the University of Chicago in 1956. But before leaving, I hurriedly did some experiments (29), which, though semi-quantitative at best, showed that not only atoms but groups such as N_3^-, NCS^-, and carboxylates transfer to chromium when the corresponding pentaamminecobalt(III) complexes are reduced by $Cr(aq)^{2+}$, and that there are large differences in rate for different dicarboxylate complexes (maleate much more rapid than succinate). Moreover, it was shown that Cr(II) in being oxidized can incorporate other ligands such as $H_2P_2O_7^{2-}$ which are present in solution. In this paper the possibility of electron transfer through an extended bond system of a bridging group was raised, but was by no means demonstrated by the results. While I was away, Ogard (30) began his studies on the rates of aquation of $(NH_3)_5CrX^{2+}$ (X = Cl, Br, I) catalyzed by Cr^{2+}. By following the arguments made in connection with reaction (2), it can be seen that if an inner sphere mechanism operates, $(H_2O)_5CrX^{2+}$ and NH_4^+ (acid solution) will be products. The contrast with uncatalyzed aquation is worth noting, where $(NH_3)_5CrH_2O^{3+}$ and X^- are the products.

GENERAL PROGRESS

The hiatus referred to above provides me with a suitable opportunity to outline some of the advances that were being or were soon to be made on other fronts. An important one is that the rates of numerous self-exchange reactions were being measured. Here I want especially to acknowledge the contributions from the laboratories of A. C. Wahl and C. S. Garner. Some of the experiments by Wahl and co-workers made use of rapid mixing techniques—see for example, the measurement (31) of the rate of self-exchange for $MnO_4^{1-/2-}$. An important experimental contribution was also made by N. Sutin who introduced the rapid flow method (32), and later other rapid reaction techniques, into this field of study, and who helped others, including myself, to get started with the rapid flow technique. A spate of activity on the measurements of the rates of cross reactions followed, motivated in large part by the desire to test the validity of the cross reaction correlation (5).

At the quantitative theoretical level, attempts were being made to account for the barrier to reaction attending encounter and separation and that contributed by charge trapping within the metal complex and by the surrounding medium. The papers which most influenced the experimentalists, at least during the formative period in question, are those by Marcus (33) and Hush (34), dealing with adiabatic (35) electron transfer. Other theoretical approaches were being advanced during this period, and in some, attempts were made to account for non-adiabaticity, and the various theories are compared and evaluated in reference (33). (The current state of theory can be gathered from a recent article by Sutin (36)). Because this very important aspect of electron transfer reactions is not dealt with in this paper, it is essential to mention that the processes as they occur at electrodes were not being overlooked.

The correlation of the rates of cross reactions with the rates of the component self-exchange reactions (5), has been widely applied, especially to outer-sphere reactions. The limits of its validity were clearly set down by the author: allowance must be made for the work of bringing the reaction partners together and separating the products, electron delocalization must be great enough to ensure adiabatic behavior, but not so great as materially to reduce the activation energy. (The last condition, it should be noted, does not necessarily limit the applicability of the Marcus equation to outer sphere processes.) Hush's treatment (34) also leads to a correlation of rates of self-exchange and cross reactions. It also takes into account the contribution by driving force, and in fact the first calculation of the rate of a cross reaction, in this instance:

$$Pu^{3+}(aq) + PuO_2^{2+}(aq) = Pu^{4+}(aq) + PuO_2^{+}(aq)$$

from those of the self exchange processes and the equilibrium constant appears in reference 34.

Theory of another kind has profoundly influenced the development of the field, even though it is qualitative. It responds to the question of how the choice of mechanism, and relative rates, are to be understood in terms of the electronic structures of the reactants. As is true also of rates of substitution, the observations are so sensitive to electronic structure that even qualitative ideas are useful in correlating observations, and in pointing the way to new experiments. Orgel (37) early applied qualitative ligand field theory in discussing the inner-sphere mechanisms for the reduction of Cr(III) and low spin Co(III) complexes. When electron transfer takes place through a bridging group, it is important to distinguish a chemical or "hopping" mechanism—here a low lying orbital of the ligand is populated by the reductant, or a hole is generated by the oxidant in an occupied orbital—from resonance transfer, that is, electron tunneling through the barrier separating the two metal centers. This distinction was drawn rather early by George and Griffith (38) who moreover proposed alternative mechanisms for resonance transfer. Shortly thereafter, Halpern and Orgel (39) gave a more formal treatment of resonance transfer through bridging ligands. Concerns about the relation between electronic structure and the observations on electron transfer strongly influenced my own work, but before tracing this theme, I want to report on the progress made, mainly by others, in extending the descriptive chemistry of electron transfer reactions.

The unambiguous demonstration of an inner sphere mechanism in a sense introduced a second dimension to the field of electron transfer mechanisms. That in certain systems reaction perforce took place by an outer sphere mechanism had long been known, but until the experiments of references 16 and 17 were done, the inner sphere mechanism was only conjecture, and, as frequently happens in research in chemistry, only after conjecture, however reasonable, is upgraded by proof, is it accepted as a base for further development. That the distinction between the two reaction classes is meaningful, not only in terms of chemistry but also in rates, will be illustrated by a single comparison: the specific rates of reaction of $Cr^{2+}(aq)$ with $(NH_3)_5CoCl^{2+}$ is ca. 10^8 greater than it is with $Co(NH_3)_6^{3+}$ (40) (the latter can only react by an outer sphere mechanism). The classification of reaction paths as inner sphere or outer sphere, on the basis of rate comparisons, involving effects (such as those exerted by non-bridging ligands) established with reactions of known mechanism, became the focus of experiment and discussion when direct proof based on product or intermediate identification was lacking.

Early in the 1960s, new metal centers were added to the roster of those proven to react by inner sphere mechanisms. For $Co(CN)_5^{3-}$ as reducing agent (41), the demonstration of an inner sphere path again depended on the characterization of product complexes by orthodox means. Sutin and co-workers have been particularly resourceful in using flow techniques to provide direct proof of mechanism for oxidizing centers ordinarily considered as labile to substitution: thus note the proof of inner sphere mechanism for $FeCl^{2+}(aq) + Cr^{2+}(aq)$ (42), $CoCl^{2+}(aq) + Fe^{2+}(aq)$ (43), even the much studied $Fe^{3+/2+}$ exchange (44). It was early appreciated (17) that atom or group transfer is not a necessary concomitant of an inner sphere process. Whether the bridging group transfers to reductant, remains with the oxidant, or transfers from reductant to oxidant depends on the substitution labilities of reactants and products. Early qualitative observations (17) on the $Cr^{2+}(aq) + IrCl_6^{2-}$ system had apparently exposed an example of reaction by an inner sphere mechanism, but leading to no net transfer of the bridging atom. Here $(H_2O)_5CrClIrCl_5$ is formed as an intermediate, but this then aquates to $Cr(H_2O)_6^{3+} + IrCl_6^{3-}$ (later work (45) has shown the inner sphere path to be minor compared to the outer sphere, and that $Cr(H_2O)_6^{3+}$ is the lesser product of the former path). Experiments (46, 47) with $V(H_2O)_6^{2+}$ as reducing agent provided numerous examples of systems in which substitution on the reducing complex is rate determining for the net redox process (note that $V(H_2O)_6^{2+}$ because of its electronic structure (πd^3) is expected (18) to undergo substitution relatively slowly). Unstable forms of linkage isomers were prepared by taking advantage of the chemistry of the inner sphere mechanism: $(H_2O)_5CrSCN^{2+}$ (48) by the reaction of $Cr^{2+}(aq)$ with $FeNCS^{2+}$; (49) $(H_2O)_5CrNC^{2+}$ (50) by the reaction of $Cr^{2+}(aq)$ with $(NH_3)_5CoCN^{2+}$. Oxygen atom transfer was shown to be complete (51) in the reaction of $(NH_3)_5CoOH_2^{3+}$ with $Cr^{2+}(aq)$. (The path involving direct attack on the aquo complex was later (52) shown to be unobservable compared to attack on the hydroxo. Bridging by H_2O has to date not been demonstrated.) The inner sphere path was demonstrated (53) also for net $2e^-$

processes in an elegant series of studies involving the reactions of Pt(IV) complexes with those of Pt(II). In an important departure, Anet (54) showed that the "capture" property of Cr^{2+}(aq) in being oxidized can be exploited to produce complexes in which an organic radical is ligated to Cr(III). An entire chapter of the volume of reference 1 is devoted to the chemistry of similar organochromium complexes(55).

ELECTRONIC STRUCTURE AND MECHANISM

A major theme in my own research, on returning from leave, was to try to understand the large differences in rate, noted qualitatively in my early work (29), and later made more quantitative, for the reactions Cr^{2+}(aq) with carboxylate complexes of $(NH_3)_5Co(III)$. Many of the ligands were dicarboxylic acids, and to explain the observation that when a conjugated bond system connects the two carboxylates, reaction is usually more rapid than it is for the saturated analog, it seemed reasonable then to assume that in the case of conjugated ligands, the reducing agent attacks the exo carboxyl (remote attack), the conjugated bond system serving as a "conduit" for electron transfer. In adopting this view the tacit assumption was made that the reactions are non-adiabatic so that the extent of electronic coupling would be reflected in the rate. In retrospect, this assumption is naive, because the effect of conjugation would be exerted even if the reducing agent attacked at the endo carboxyl.

A false start was made in demonstrating remote attack defined as above. Activation effects accompanying electron transfer were reported (56), which if true, would have constituted proof of remote attack for these systems. These effects could not be reproduced (57) in later work (I had by now moved from University of Chicago to Stanford University). Remote attack for the large organic ligands was finally demonstrated (58) in the reaction of

$$\left[(NH_3)_5CoN\!\!\left\langle\bigcirc\right\rangle\!\!-\underset{\underset{NH_2}{|}}{C}=O\right]^{3+} \text{ with } Cr^{2+}(aq).$$ This work, which also yielded

a measurement of the rate of reaction of Cr^{2+}(aq) with the analogous Cr(III) species, provided the clue to understanding, at least in a qualitative way, the rate differences observed for different conjugated ligands. The astonishing result was that the rate of reduction of the Co(III) complex is only about 10-fold greater than that of the Cr(III) complex. When the bridging group is a non-reducible species such as acetate, the ratio is $> 10^4$. The insensitivity of rate to the nature of the oxidant suggested that the electron does not transfer directly from Cr(II) to the oxidizing center but that the mechanism rather involves the $1e^-$ reduction (59) of the ligand by the strong reducing agent Cr^{2+}, followed by reduction of the oxidizing center by the organic radical—i.e., a "hopping" mechanism obtains. This view provided satisfactory rationalizations of most of the observations of rates made with organic ligands. For example, the fact that the rate is considerably greater for $HO_2C-CH=CH-CO_2^-$ (fumarate) than for $CH_3CO_2^-$ as ligand on Co(III), may have little to do with the opportunity that Cr(II) has to attack the remote carboxyl in the former case, and only reflects the fact that fumarate can be reduced by Cr^{2+}(aq). Moreover, the otherwise puzzling observation that the rate for the fumarate complex is increased (60) by H^+ is now easily understood; positive charge added to the ligand makes it easier to reduce. That many reactions of the class under consideration proceed by a stepwise mechanism has been convincingly demonstrated and amply illustrated in subsequent work, most of it done by Gould and co-workers (61).

The rationalization offered for the operation of the stepwise mechanism in the Co(III)-Cr(II) systems is that the carrier orbital on the ligand has π symmetry, while the donor and acceptor orbitals have σ symmetry. This renders as highly improbable an event in which the four conditions: Franck-Condon restrictions at each center, and the symmetry restrictions at each center, are simultaneously met. Whether or not this is the correct explanation, it led me to search for an oxidizing center of the ammine class in which the acceptor orbital has π symmetry.

When the important condition that the complexes undergo substitution slowly was added, only one couple within the entire periodic table, $Ru^{3+/2+}$ $(\pi d^5/\pi d^6)$ then qualified (62). In principle, the $Os^{3+/2+}$ couple is also a candidate, but unless some strong π acid ligands are present, the couple is too strongly reducing to be useful. The ruthenium species had the added advantage that much more in the way of preparative work was known (63), and further that the redox potentials are close to those of the much studied cobalt couples. Since the π orbital on Ru(III) can overlap with the π^* orbital on the ligand, we expected that the "hopping" mechanism would no longer obtain. Reaction with Cr^{2+}(aq) is in fact much more rapid (2×10^4) than it is in the case of the Co(III) isonicotinamide complex, and moreover, the rates are now quite sensitive to changes in the redox potential of the oxidizing center (64). The chemistry also differs in an interesting way from the Co(III)-Cr(II) case. The bond between Ru and the ligand is not severed when Ru(III) is reduced to Ru(II), and a kinetically stable binuclear intermediate is formed, as is expected (18) from the electronic structures of the products, πd^6 for Ru(II) and πd^3 for Cr(III).

Though the main intent of this paper is to provide historical background rather than the develop the subject itself in detail, because the reaction properties are so sensitive to electronic structure, it may be appropriate in concluding this section to illustrate the connection with a few examples. Effects arising from differences in electronic structure are manifested in several different ways: by affecting the rates of substitution, they can affect the choice of mechanism, and, for an inner sphere reaction path, can determine whether binuclear intermediates are easily observable, and whether there is net transfer of a group from one center to another; even after the precursor complex is assembled, orbital symmetry can affect the mechanism itself, as in the example offered in an earlier paragraph, and can profoundly affect the rate of conversion of the precursor to the successor complex.

The Cr(III)/Cr(II) $(\pi d^3/\pi d^3 \sigma d^1)$ and Ru(III)/Ru(II) $(\pi d^5/\pi d^6)$ couples offer perhaps the greatest contrasts in behavior. It should be noted that the σ

electron in Cr(II) is antibonding, thus accounting for the tetragonal distortion in Cr(II) complexes, and their enormous lability compared to those of Cr(III). By contrast, the complexes of both oxidation states of ruthenium undergo substitution slowly, with the useful exception of water as a ligand, where the residence time on Ru(II) is a fraction of a second.

The reducing agent Cr(II) shows preference for an inner sphere mechanism, and this is especially marked if the acceptor orbital has σ symmetry. There is a great economy of motion for electron transfer by an inner sphere path for the σ donor$-\sigma$ acceptor cases which arises from the reciprocity, at the two centers, of the events which are required in overcoming the inner sphere Franck-Condon barrier. This point is illustrated for the $(H_2O)_5{}^*CrCl^{2+} + Cr(H_2O)_6{}^{2+}$ self exchange reaction in Fig. 1 where the electronic levels are shown for the precursor complex, for the activated complex, and for the successor complex. Motion of the bridging Cl^- from Cr(III) to Cr(II) lowers the energy of a acceptor orbital on Cr(III), and at the same time raises that of the donor orbital on Cr(II), and although other nuclear motions are also required, there is some correlation of the events required for activation to electron transfer, a correlation which is absent in the case of an outer sphere mechanism. The high substitution lability of Cr(II) of course means that the precursor complex can be formed rapidly.

The comparison of the rates of self exchange for $Cr(H_2O)_6{}^{3+}/Cr(H_2O)_6{}^{2+}$ vs. $(NH_3)_5RuH_2O^{3+}/(NH_3)_5RuH_2O^{2+}$ and how these respond when the higher oxidation state for each couple is converted to the hydroxo complex is quite instructive. The upper limit for the specific rate of self-exchange for the Cr(III)/Cr(II) couple is $2 \times 10^{-5}M^{-1}s^{-1}$ (65,66); although it has not been directly measured for the ruthenium couple, the specific rate can reasonably be taken to be close to that (67) for $Ru(NH_3)_6{}^{3+/2+}$, namely $\sim$

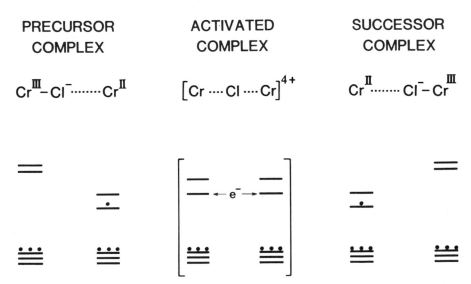

| PRECURSOR | ACTIVATED | SUCCESSOR |
| COMPLEX | COMPLEX | COMPLEX |

$$Cr^{III}-Cl^-\cdots\cdots Cr^{II} \qquad \left[Cr\cdots Cl\cdots Cr\right]^{4+} \qquad Cr^{II}\cdots\cdots Cl^--Cr^{III}$$

Fig. 1. Electronic structure and "atom" transfer in the self-exchange reaction: $(H_2O)_5CrCl^{2+} + Cr(H_2O)_6{}^{2+}$.

$1 \times 10^3 M^{-1}s^{-1}$. Because the **redox** change for the ruthenium couple involves a π electron, it causes only a small change in the dimensions of the complex (68). Thus the Franck-Condon barrier to electron transfer arising from inner sphere reorganization is small, and facile transfer by an outer sphere mechanism is observed. By contrast, because Cr(III) has no anti-bonding electrons, and the electron added on reduction is anti-bonding, there is a large change in dimensions and shape attending the reduction, and the slowness of the self-exchange can be attributed in part at least to the inner sphere barrier (69). Why the water molecule, which as a ligand on the oxidizing agent still has available an electron pair for sharing with the reducing agent, is such a poor bridging group, remains to be understood.

On deprotonation of a water molecule in each oxidant, the inner sphere path for the Cr(III)/Cr(II) system opens up, and a marked increase in rate is observed (65) ($k = 0.66$ $M^{-1}s^{-1}$ at 25°) – the increase may be as large as a factor (66) of 10^9. The self-exchange rate for $(NH_3)_5RuOH^{2+} + (NH_3)_5$-$RuOH_2^{2+}$ has not been measured, but it can be asserted with confidence that the rate by either an inner sphere path or an orthodox outer sphere path will be much less than it is for the aquo couple. The rate by the inner sphere path will be limited by the rate of bridge formation, and thus will be no greater than $1 \times 10^{-2} M^{-1}s^{-1}$ (neutral ligands in substituting on $(NH_3)_5RuOH_2^{2+}$ show specific rates (70) of the order of 0.1 $M^{-1}s^{-1}$). The orthodox outer sphere path now has a composition barrier, as well as a Franck-Condon barrier to overcome (K_{eq} for the production of $(NH_3)_5RuOH^+$ and $(NH_3)_5RuOH_2^{3+}$ from $(NH_3)_5RuOH^{2+}$ and $(NH_3)_5RuOH_2^{2+}$ is ca. 10^{-9}). Reaction by "hydrogen atom" transfer (15) is a reasonable possibility, that is, electron transfer concomitant with proton transfer from the Ru(II) complex to the Ru(III), and some evidence in support of this kind of mechanism has been advanced to explain observations (71) made in the oxidation by Fe^{3+} and $FeOH^{2+}$ of $(NH_3)_5RuOH_2^{2+}$. Reaction by such a path might be quite facile and a specific rate in excess of 1 $M^{-1}s^{-1}$ would be strong evidence in its favor.

APPLICATIONS OF THE RU(III)/RU(II) COUPLE

Some unexpected benefits have accrued from introducing Ru(III)-Ru(II) ammine couples into this field of research. Ruthenium(II) engages in back bonding interactions to a degree unprecedented among the dipositive ions of the first transition series. A discovery (72) which forcefully brought this message home, is that $(NH_3)_5RuN_2^{2+}$ is formed in aqueous solution by the direct reaction of $(NH_3)_3RuOH_2^{2+}$ with N_2. When the heteroligand in $(NH_3)_5RuL^{2+}$ is pyridine or a derivative, the complexes of both oxidation states are slow to undergo substitution, and by changing the number of π acid ligands, a versatile series of outer-sphere redox couples is made available, spanning a range in redox potential of over 1 volt. When derivatives of the Os(III)-Os(II) ammines are included, the useful range in aqueous solution is extended by approximately 0.5 V. These reagents are finding wide application in research on redox processes.

1. *Intramolecular Electron Transfer*

It has occurred to many that in trying to arrive at a basic understanding of electron transfer processes, it would be a great advantage if the reactions could be studied in an intramolecular mode rather than, as is commonly done, in the bimolecular or (intermolecular) mode, particularly if the geometrical relation between the two metal centers were unambiguosly defined. Such systems had been encountered in studies of "induced" electron transfer: (73) when a powerful

1 e$^-$ oxidant acts on $[(NH_3)_5CoN\langle O \rangle - \underset{H}{C} = O]^{3+}$ for example, the result-

ing coordinated organic radical can undergo intramolecular electron transfer, the oxidation of the ligand to the carboxylic acid being completed by Co(III). In some cases, intramolecular electron transfer can be intercepted by reaction of the radical with the external oxidant, but at best, only relative rates were obtained for these systems. In an elaboration of this kind of approach, in which pulse radiolysis is used to convert the organic ligand to a radical — usually by reduction — intramolecular transfer rates can be measured (74). These results are important in their own right, but they do not substitute for experiments in which metal-to-metal transfer is studied.

A strategy for dealing with the metal-to-metal case was devised (75a), which depends on the special properties of the Co(III)/Co(II) and Ru(III)/Ru(II) couples. The principle is the following: when a molecule (76) such as

$$(NH_3)_5Co^{III}N\langle O \rangle \langle O \rangle NRu^{III}(NH_3)_5$$

which has both metal centers in the oxidized state, is treated with an external reducing agent, Ru(III) is reduced more rapidly than Co(III). This is a direct result of the differences in electronic structure, πd^6 and πd^5 for Co(III) and Ru(III) respectively, the former requiring much more in the way of reorganization energy because the incoming electron is anti-bonding. In a subsequent step, Ru(II) reduces Co(II) by an intramolecular process, at least if the solution of the binuclear complex is sufficiently dilute.

The first method (75a) devised to produce the $[III,III]$ molecule is rather ingenious, but it involves many steps, and it has the disadvantage that SO_4^{2-} rather than NH_3 is trans to the pyridine on Ru(III). Schäffer (77) has greatly simplified the preparative procedure by taking advantage of chemistry developed by Sargeson and co-workers (78) and has studied intramolecular electron

transfer for $(NH_3)_5Co^{III}NC\langle O \rangle CNRu^{II}(NH_3)_5$ and for the related mole-

cules with the ortho and meta isomers as the bridging ligands. Quite independently of our work, Haim et al. (79) have done experiments similar to those described, but with $Fe(CN)_5H_2O^{3-}$ as the reducing agent. Substitution on $Fe(CN)_5H_2O^{3-}$ takes place readily so that the simple mixing procedure attempted by Roberts (75b) often can be used with this particular kind of reducing agent.

A point of interest in all of these studies is to learn how the rate of reaction responds to changes in the structure of the bridging group. In the bipyridine case, the coupling between the pyridine rings has been variously modified (76); since the immediate environment about each metal is left unaltered, the driving force for the reaction is but little affected, and changes in rate can then be attributed to changes in electronic coupling. The results obtained in studies of this kind are outlined in a recent review article by Haim (80). Here I will only mention an extension of this kind of strategy to a system of biochemical interest. Gray et al (81) and Isied et al (82) have succeeded in placing $(NH_3)_5Ru$ on cytochrome C at a position remote from the porphyrin (Ru(III)-Fe(III) separation 15Å). Different pulse methods were used by the two groups to reduce Ru(III) preferentially over Fe(III), and though the results of the two studies differ somewhat ($k=20\pm5s^{-1}$, $82\pm20s^{-1}$ respectively), it is clear that the general strategy is effective.

2. Robust Mixed Valence Molecules

The resurgence of interest (83) in the properties of mixed valence compounds can be traced to review articles (84, 85, 86) which appeared in 1967, and to the first deliberate synthesis of a robust mixed valence molecule, the species shown below, which is commonly referred to as the Creutz-Taube (87) ion

$$[(NH_3)_5RuN\bigcirc NRu(NH_3)_5]^{5+}\ I$$

(Quite independently of our work, Cowan and Kaufman (88) prepared a molecule based on the ferricinium/ferrocene couple.) Peter Ford and I first produced the Creutz-Taube ion in 1967. In undertaking its preparation, we were motivated by simple curiosity rather than by questions which might arise from a deep understanding of the issues raised by the properties of the mixed valence compounds. The fully reduced ([II,II]) state is readily prepared by direct substitution using pyrazine and $(NH_3)_5RuOH_2^{2+}$, and in undertaking the project, we were taking advantage of our knowledge of affinities and rates of substitution for both oxidation states of ruthenium. Complexes with Ru(II) attached to heterocyclic nitrogen show very strong absorption in the visible region of the spectrum ($\pi^*\leftarrow\pi d$) (89) and on observing that the quality of the color was not significantly altered when the [II,II] species is half oxidized, we did not pursue the matter further. Fortunately Carol Creutz took up the subject again. The electrochemical results which she obtained about June 1968 showed that the mixed valence state is very stable relative to the isovalent, and this suggested to us that electronic coupling in the mixed valence species is strong. By now, the review papers by Hush (85) and Robin & Day (86) had appeared, and taking their content to heart, we felt certain that an intervalence band must exist, which Carol Creutz then located in the near infra red region ($\lambda_{max} = 1570$ nm) where it does not affect the color (heretofore this region of the spectrum had been little investigated by chemists). Intervalence absorption corresponds to using the energy of a photon to transfer an electron from the reduced to the oxidized metal center, subject to the Franck-Condon restriction. Intervalence

absorption confers on Prussian Blue its characteristic blue color. The intervalence absorption is at longer wave length for species I than for Prussian Blue
because the two iron sites in the latter are not substitutionally equivalent. This
leads to a ground state energy difference which is then added to that associated
with the Franck-Condon barrier when the process is induced by a quantum of
light.

One of my main interests in the field of mixed valence molecules has been to
explore and to try to understand the energetics of the systems. I will illustrate
by a single example the kind of conclusion which we have reached in pursuing
these interests and where we have relied on theory introduced into the field by
Hush (85), by Mulliken (90), and for the correlation of extent of electron
delocalization with electronic structure, Mayoh and Day (91), and choose for
illustration the localized mixed valence molecule (92)

$$[(NH_3)_5RuN\bigcirc\!\!-\!\!\bigcirc NRu(NH_3)_5]^{5+}$$

The stability conferred on the ground state of the molecule by charge delocalization is only of the order of 50 cal, far below the upper limit of 5×10^2 cal set by
the electrochemical results, which measure the total stabilization of the mixed
valent compared to the isovalent state. When the nuclear coordinates about
each center are adjusted so that the Franck-Condon condition is met, the
energy separating the bonding and anti-bonding states which arise from electron delocalization is calculated as 2.2 kcal (93). This is taken to be sufficient to
ensure adiabatic transfer (94), in agreement with the conclusion reached in the
course of studying intramolecular electron transfer in Co(III)-Ru(II) systems
with 4,4'-bipyridine and related molecules as bridging ligands (76). If electron
transfer is assumed to be adiabatic, the specific rate for intramolecular electron
transfer is calculated as 3×10^8 s^{-1}, in reasonable agreement with an estimate
$(1.6\times10^8$ s$^{-1})$ reached from measurements of intermolecular electron transfer
rates for pyrinedinepentaammineruthenium species (95).

Mixed valence molecules have been prepared (96) which are delocalized
even though the bridging group is so large that direct metal-to-metal orbital
overlap cannot be responsible for the delocalization. These have remarkably
interesting properties in their own right, and are the subject of current studies
(97).

CONCLUDING REMARKS

·In this paper I have focussed rather narrowly on electron transfer reactions
between metal complexes. The separation of this subclass from other possible
ones which can be assembled from the reactant categories: metal complexes,
organic molecules (98), molecules derived from other non-metallic elements,
any of the above in excited states (99, 100), electrodes, is not totally arbitrary as
it might be were it dictated solely by limitations of space. Admittedly, all the
possible electron transfer processes are governed by the same principles, at
least when these are stated in a general enough way. But as these principles

manifest themselves in the different subclasses, the descriptive chemistry can be quite different, and these differences are the fabric of chemistry. The subclass which has been treated has perhaps been the most thoroughly studied, yet, as the article by Sutin shows (36), our understanding at a basic level is far from complete. Even for the much studied Fe^{3+}/Fe^{2+} self exchange reaction, which served to introduce this subject, the important question of whether the reaction is adiabatic or not has not been settled to everyone's satisfaction. Still a great deal of progress has been made. The descriptive matter has increased enormously since 1940, and our understanding of it, both in scope and depth, has more than kept pace with observations. A great deal of progress has also been made in many of the other subclasses—for example in the study of electrode processes, and in "atom transfer" reactions (as a specific case the use of transition metal species to carry the oxidizing capacity of O_2 to a substrate such as an organic molecule).

Both are of the greatest importance in industrial applications, and the latter also in reaching an understanding of the chemistry of living cells. Because the subclasses are interrelated, progress in one leads to progress in another.

ACKNOWLEDGEMENT

Many of the co-workers who have contributed to progress in the subject of this article are cited in the references, and this is an implicit acknowledgment of their contributions. Because this account is incomplete, others who have contributed directly to this work have not been cited, nor still others who have had interests that are not reflected in the account I have given. I am grateful to them all for the help they have given, and for what I have learned in the course of working with them. The nature of the contributions made by my co-workers is not evident either from the acknowledgment I have made, nor from my expression of gratitude. I need to add that I have always relied on the independence of my co-workers and to a large extent my contribution to the effort has been that of maintaining continuity.

I also wish to acknowledge financial support of my research by the agencies of the U. S. Government, beginning with the Office of Naval Research in about 1950. Later, I derived partial support from the U. S. Atomic Energy Commission, and still later from the National Science Foundation and the National Institutes of Health (General Medical Sciences). The Petroleum Research Fund of the American Chemical Society has also been a source of research support.

REFERENCES

(1) Progress in Inorganic Chemistry, Vol. 30. (Editor, S. J. Lippard), Wiley & Sons, N. Y., 1983.

(2) von Hevesy, G.; Zechmeister, L., Ber. (1920) 53, 410.

(3) Seaborg, G. T. Chem. Rev. (1940) 27, 199.

(4) Kennedy, J.; Ruben S.; Seaborg, G. T.: see Ref. 3, p 256.

(5) This descriptor and that following were introduced by R. A. Marcus (Faraday Society Discussions (1960) 29, 21. J. Phys. Chem. (1963) 67, 853) in proposing an equation which correlates the specific rate for a "cross reaction" such as

$$Fe^{3+} + V^{2+} = Fe^{2+} + V^{3+}$$

with the specific rates of the component "self exchange" reactions $Fe^{3+/2+}$ and $V^{3+/2+}$ Allowance is made in the equation for the effect of driving force on the rate.

(6) Dodson, R. W., J. Am Chem. Soc. (1950) 72, 3315.

(7) Silverman, J.; Dodson, R. W., J. Phys, Chem. (1952) 56, 846.

(8) Lewis, W. B.; Coryell, C. D.; Irvine, S. W., Jr. J. Chem. Soc. (1949) S386.

(9) The descriptor "inner sphere" is used for a reaction in which oxidant and reductant metal centers are linked through primary bonds to a bridging group. In the early work it was assumed that the bridging group would play a special role in the electron transfer reaction. In more recent work on intramolecular transfer (vida infra) examples have been found of systems in which the bridging group seems to serve only to hold the two partners together. "Inner sphere" or "outer sphere"?

(10) Holstein, T., J. Phys, Chem. (1956) 56, 832.

(11) Libby, W. F., J. Phys. Chem. (1956) 56, 863.

(12) As a historical note, I wish to add that James Franck was very much interested in electron transfer in chemical reactions and fully appreciated the importance of keeping the Franck-Condon restriction in mind in trying to understand the observations.

(13) Brown, H. C., discussion of the paper of ref. 11, p 868.

(14) Libby, W. F., discussion following the paper of ref. 11, p 866. See also Brown, H. C., discussion following paper of ref. 7, p 852.

(15) Dodson, R. W.; Davidson, N., discussion following the paper of ref. 11, p 866.

(16) Taube, H.; Myers, H.; Rich, R. L., J. Am. Chem. Soc. (1953) 75, 4118.

(17) Taube, H.; Myers, H., J. Am. Chem. Soc. (1954) 26, 2103.

(18) Taube, H., Chem. Rev. (1952) 50, 69.

(19) The subject of ionic hydration is brought up to date in an article by J. P. Hunt and H. L. Friedman, p 359, ref 1.

(20) Bjerrum, J.; Poulsen, K., Nature (1952) 169, 463.

(21) Eigen, M., Faraday Soc. Discussions, (1954) No. 17, p 1.

(22) Bjerrum, J., Z. Physik. Chem. (1907) 59, 336, 581; Z. Anorg. Allgem. Chem. (1921) 119, 179.

(23) Hunt, J. P.; Taube, H., J. Chem, Phys. (1950) 18, 757; ibid (1951) 19, 602.

(24) Meredith, C. W., U. S. Atomic Energy Rept. UCRL-11704 (1965) Berkeley. Work done under supervision of R. E. Connick.

(25) Candlin, J. P.; Halpern, J., Inorg. Chem. (1965) 4, 766

(26) Taube, H.; King, E. L., J. Am. Chem. Soc. (1954) 76, 4053.

(27) Ball, D. L.; King E. L., J. Am. Chem. Soc. (1958) 80, 1091.

(28) Snellgrove, R.; King, E. L., J. Am. Chem. Soc. (1962) 84, 8609

(29) Taube, H., J. Am. Soc. (1955) 77, 4481.

(30) Ogard, A. E.; Taube, H. J., Am. Chem. Soc. (1958) 80, 1084.

(31) Gjertsen, L.; Wahl, A. C., J. Am. Chem. Soc. (1959) 81, 1572; Sheppard, J. C.; Wahl, A. C., J. Am. Chem. Soc. (1957) 79, 1020.

(32) Sutin, N.; Gordon, B. M., J. Am. Chem. Soc. (1961) 83, 70.

(33) Marcus, R. A., Ann. Rev. Phys. Chem. (1964) 15, 155; also earlier papers.

(34) Hush, N. S., Trans. Faraday Soc. (1961) 57, 557.

(35) Consider the process: $A \cdot B^- \rightarrow [A \cdot B]^- \rightarrow A^- \cdot B$

$\qquad\qquad\qquad\quad$ I $\qquad\quad$ II

State I is the precursor complex; in state II the energy is independent of whether the electron is on atom A or B, i. e., the Franck-Condon condition has been met. In adiabatic transfer, electron delocalozation is great enough so that whenever state II is reached, electron transfer can take place, and the rate of the chemical reaction is determined solely by the rate at which state II is reached. In non adiabatic transfer, the system passes through state II a number of times before electron transfer occurs, and both the Franck-Condon barrier, and the rate of electron transfer in state II are rate determining.

(36) Sutin, N., p 441 of ref. 1.
(37) Orgel, L. E., Rept Xe Conseil Inst. Intern, Chem. Solvay (1956) 289.
(38) George, P.; Griffith, J., Enzymes (1959) 1, 347.
(39) Halpern, J.; Orgel, L. E., Discussions Faraday Soc. (1960) 29, 32.
(40) Zwickel, A. M.; Taube. H., J. Am. Chem. Soc. (1959) 81, 2915.
(41) Halpern, J.; Nakamura, S., J. Am. Chem. Soc. (1965) 87, 3002.
(42) Dulz, G.; Sutin, N., J. Am. Chem. Soc. (1964) 86, 829.
(43) Connochioli T. J.; Nancolles, G. H.; Sutin, N., J. Am. Chem. Soc. (1969) 86, 1453.
(44) Connochioli T. J.; Nancolles, G. H.; Sutin, N., J. Am. Chem. Soc. (1964) 86, 459.
(45) Sykes, A. G.; Thorneley, R. N. V., J. Chem. Soc. (1970) A232.
(46) Espenson, J. H., J. Am. Chem. Soc. (1967) 89, 1276.
(47) Price, H. J.; Taube, H., Inorg. Chem. (1968) 7, 1
(48) Haim, A.; Sutin, N., J. Am. Chem. Soc. (1965) 87, 4210.
(49) Fronaeus, S.; Larsson, R., Acta Chem. Scand. (1962) 16, 1447.
(50) Espenson, J. H.; Birk. J. P., J. Am. Chem. Soc. (1965) 87, 3280; ibid, (1968) 90, 1153.
(51) Kruse, W.; Taube, H., J. Am. Chem. Soc. (1960) 82, 526
(52) Toppen, D. L.; Linck, R. G., Inorg. Chem. (1971) 10, 2635.
(53) Basolo, F.; Morris, M. L.; Pearson, R. G., Disc. Faraday Soc. (1960) 29, 80.
(54) Anet, F. A. L.; Leblanc, E., J. Am. Chem. Soc. (1957) 79, 2649.
(55) Espenson, J. H.; p. 189 of the volume of ref. 1.
(56) Last in the series. Fraser, R. T. M. and Taube, H., J. Am. Chem. Soc. (1961) 83, 2239.
(57) I owe my associates during my first years at Stanford an enormous debt of gratitude for helping to set the record straight. Special thanks are due to E. S. Gould, who first uncovered discrepancies, and to J. K. Hurst who repeated much of the dubious earlier work.
(58) Nordmeyer, F. R.; Taube, H., J. Am. Chem. Soc. (1966) 88, 4295; ibid (1968) 90, 1162.
(59) The relation between reducibility of the ligands, and their effectiveness in mediating electron transfer was developed in an earlier paper. Gould, E. S.; Taube, H., J. Am. Chem. Soc. (1964) 86, 1318.
(60) Sebera, D. K.; Taube, H., J. Am. Chem. Soc. (1961) 83, 1785.
(61) See, for example: Gould E. S., J. Am. Chem. Soc. (1972) 94, 4360.
(62) Endicott, J. F.; Taube, H., J. Am. Chem. Soc. (1962) 84, 4989; ibid (1964) 86, 1686; Inorg. Chem. (1965) 4, 437.
(63) Gleu, K.; Breuel, K., Z. Anorg. Allg. Chem. (1938) 237, 335.
(64) Gaunder, R.; Taube, H., Inorg. Chem. (1970) 9, 2627.
(65) Anderson, A.; Bonner, N. A., J. Am. Chem. Soc. (1954) 76, 3826.
(66) Indirect evidence, which is quite convincing, suggests the outer sphere self exchange rate for $Cr(H_2O)^{3+}/Cr(H_2O)_6^{2+}$ to be $\sim 5 \times 10^{-10}$ M^{-1}s^{-1}. Melvin, W. S.; Haim. A., Inorg. Chem. (1977) 16, 2016.
(67) Meyer, T. J.; Taube, H., Inorg. Chem. (1968) 7, 2369.
(68) Stynes, H. D.; Ibers, J. A., Inorg. Chem. (1971) 10, 2304.
(69) Calculations of the barrier associated with inner sphere electron reorganization leave room for a non-adiabaticity factor of a few orders of magnitude. See Endicott, J. F.; Krishan, K.; Ramasami, T; Rotzinger, F. P., p 141 of ref. 1.
(70) Isied, S. S.; Taube, H., Inorg. Chem. (1976) 15, 3070.
(71) Meyer, T. J.; Taube, H., Inorg. Chem. (1968) 7, 2361.
(72) Harrrison, D. E.; Taube, H., J. Am. Chem. Soc. (1967) 89, 5706.
(73) Robson, R.; Taube, H., J. Am. Chem. Soc. (1967) 89, 6487; French, J.; Taube, H., J. Am. Chem. Soc. (1969) 91, 6951, earliest example: Saffir, P., J. Am. Chem. Soc. (1960) 82, 13.

(74) Hoffman, M. Z.; Simic, M., J. Am. Chem. Soc. (1972) 94, 1957.
(75) (a) Isied, S. S., J. Am. Chem. Soc. (1973) 95, 8198.
(b) In an earlier effort, Kirk Roberts tried the simple procedure of mixing the Co(III) complex with $(NH_3)_5RuOH_2^{2+}$. Substitution is too slow relative to intramolecular transfer for the method to work in these systems.
(76) Fischer, H.; Tom, G. M.; Taube, H., J. Am. Chem. Soc. (1976) 98, 5512.
(77) Schäffer, L. – work in progress.
(78) Dixon, N. E.; Lawrance, G. A.; Lay, P. A.; Sargeson, A. M., Inorg. Chem. (1983) 22, 846.
(79) Gaswick, D. G.; Haim. A., J. Am. Chem. Soc. (1974) 96, 7845.
(80) Haim, A. see p 273 of ref. 1.
(81) Winkler, J. R.; Nocera, D. G.; Yocom, K. M.; Bordignon, E.; Gray, H. B., J. Am. Chem. Soc. (1982) 104, 5798.
(82) Isied, S. S.; Worosila, G.; Atherton, S. J., J. Am. Chem. Soc. (1982) 104, 7659.
(83) The level of current activity in the field can be gauged by the recent review of the subject of mixed valence molecules based on $\pi d^5/\pi d^6$ couples by C. Creutz: see ref. 1, p 1.
(84) Allen, G. C.; Hush, N. S., Prog. Inorg. Chem. (1967) 8, 357.
(85) Hush, N. S., Prog. Inorg. Chem. (1967) 8, 391.
(86) Robin, M. B.; Day, P., Ad. Inorg. Chem. Radiochem (1967) 10, 247.
(87) Creutz, C.; Taube, H., J. Am. Chem. Soc. (1969) 91, 3988; ibid (1973) 95, 1086.
(88) Cowan, D. O.; Kaufman, F., J. Am. Chem. Soc. (1970) 92, 219.
(89) Ford, P.; Rudd, de F. P.; Gaunder, R.; Taube, H., J. Am. Chem. Soc. (1968) 90, 1187.
(90) Mulliken, R. S.; Person, W. B., Molecular Complexes; Wiley, New York (1969) Chapter 2.
(91) Mayoh, B.; Day, P., J. Am. Chem. Soc. (1972) 94, 2885; Inorg. Chem. (1974) 13, 2273.
(92) Tom, G. M.; Creutz, C.; Taube, H., J. Am. Chem. Soc. (1974) 96, 7828.
(93) Sutton, J. E.; Sutton, P. M.; Taube, H., Inorg. Chem. (1979) 18, 1017. Sutton, J. E.; Taube, H., Inorg. Chem. (1981) 20, 3125.
(94) Sutin, N., Inorganic Diochemistry, Vol. 2, G. L. Eichhorn, Ed., American Elsevier, N. Y. (1973), p 611.
(95) Brown, G. M.; Krentzien, H. J.: Abe, M.; Taube, H., Inorg. Chem. (1979) 18, 3374.
(96) Lay, P. A.; Magnuson, R. H.; Taube, H., J. Am. Chem. Soc. (1983) 105, 2507.
(97) Spectroscopic studies by J. Ferguson and co-workers (Australia National University) are in progress.
(98) Sheldon, R. A.; Kochi. J., Metal Catalyzed Oxidations of Organic Compounds. Academic Press, New York, 1981.
(99) Ford, P.; Wink, D.; Dibenedetto, J., p 213 of ref. 1.
(100) Meyer, T. J., p 389 of ref. 1.

Chemistry 1984

BRUCE MERRIFIELD

for his development of methodology for chemical synthesis on a solid matrix

THE NOBEL PRIZE FOR CHEMISTRY

Speech by Professor BENGT LINDBERG of the Royal Academy of Sciences.
Translation from the Swedish text

Your Majesties, Your Royal Highnesses, Ladies and Gentlemen,

The chemical reactions which take place in living organisms are not spontaneous, but require the involvement of catalysts. These catalysts are called proteins and are composed of chains of amino acids called peptides. A number of hormones and other substances which regulate different life processes are also peptides. There are about 20 naturally occurring amino acids which are found in such peptides and since the chains can be very long, the number of possible variations is virtually unlimited.

Today we know the structures of a very large number of proteins and peptides. Important contributions to this area of knowledge were made by Fredrick Sanger, who received the Nobel prize in 1958, and Stanford Moore and William H. Stein, Nobel prizewinners in 1972. A very important contribution was also made by the Swedish researcher Per Edman, who unfortunately died relatively young and whose method for the controlled degradation of peptides is now generally used.

The chemical synthesis of peptides is an important task. The principle used in such synthesis is simple and was developed a relatively long time ago by Emil Fischer, who received a Nobel prize in 1902, although for completely different discoveries. Expressed simply, this principle involves the binding together of two amino acids which have been appropriately modified to give a dipeptide. This dipeptide is then combined with a third modified amino acid to give a tripeptide and so on.

Even if the principle is simple, in practice it is difficult to synthesize peptides, since a large number of individual steps is involved. After each step the desired product must be separated from by-products and unreacted starting material and this takes time and involves loss of the product. When Vincent du Vigneaud synthesized a peptide hormone, oxytocin, which is a nonapeptide, for the first time, this represented a great step forward which was rewarded with the Nobel prize for 1955. To use a similar approach for synthesizing a peptide containing 100 or more amino acid residues is truly a heroic task, requiring a very large amount of work and chemicals. This task can be compared to climbing a high mountain peak in the Himalayas, which begins with a large expedition carrying much equipment and ends, if all goes well, with a few lightly equipped alpinists reaching the top.

Therefore, Merrifield's development during the 1960's of a method for carrying out peptide synthesis on a solid matrix revolutionized the field. He attached the first amino acid to an insoluble polymer, a plastic material in the form of small spheres. Subsequently, the other amino acids were added one after one and only after the entire peptide chain had been synthesized was it released from the polymer. The advantages of this method are considerable. The complicated purification of the product after each synthetic step is replaced by simply washing the polymer to which the peptide is attached, so that loss of product is avoided completely. At the same time, the yield for each individual step is increased to 99.5% or better, a goal which cannot be achieved with conventional methods, but which is extremely important in syntheses involving a large number of steps. Finally, this method can be automated and automatic peptide synthesizers are now commercially available.

Thousands of different peptides of different sizes, as well as proteins, peptide hormones and analogues of these compounds have now been synthesized using this method. One milestone in this respect was the synthesis of an active enzyme, ribonuclease, containing 124 amino acid residues, by Merrifield and his coworkers.

The approach of performing a multistep synthesis with a compound attached to a solid matrix as the starting material has also been used in other areas. The most important of these is undoubtedly the synthesis of oligonucleotides, which are needed in hybrid DNA research. In this case as well an automated apparatus which can be programmed to synthesize desired products has been constructed. Although Merrifield has not worked in this area himself, it is clearly his ideas which have found a new application here.

Professor Merrifield,

Your methodology for chemical synthesis on a solid matrix is a completely new approach to organic synthesis. It has created new possibilities in the fields of peptide-protein and nucleic acid chemistry. It has greatly stimulated progress in biochemistry, molecular biology, medicine and pharmacology. It is also of great practical importance, both for the development of new drugs and for gene technology.

On behalf of the Royal Swedish Academy of Sciences I wish to convey our warmest congratulations and ask you to receive your prize from the hands of His Majesty the King.

Bruce Merrifield

BRUCE MERRIFIELD

Bruce Merrifield was born in Fort Worth, Texas, July 15, 1921, the only son of George E. and Lorene (Lucas) Merrifield. In the spring of 1923 they drove across the southwest desert to settle in California where they lived in several cities throughout the state. He attended nine grade schools and two high schools before graduating from Montebello High School in 1939. His interest in chemistry began there and he also enjoyed the astronomy club where he ground a mirror and built a small reflecting telescope. As a senior he managed to be runner up in the annual science contest and in the process learned a valuable lesson in the scientific method.

College began at Pasadena Junior College and at the end of two years he transferred to the University of California at Los Angeles (UCLA). After graduation in chemistry he worked for a year at the Philip R. Park Research Foundation taking care of an animal colony and assisting with growth experiments on synthetic amino acid diets. One of these was the experiment by Geiger that first demonstrated that the essential amino acids must be present simultaneously for growth to occur.

It soon became clear that more education was necessary and he returned to graduate school at the UCLA chemistry department with professor of biochemistry M. S. Dunn to develop microbiological methods for the quantitation of the pyrimidines. Graduation was on June 19, 1949, on June 20 he and Elizabeth Furlong were married, and on June 21 they left California for New York City and the Rockefeller Institute for Medical Research.

At the Institute, later Rockefeller University, he worked as an Assistant for Dr. D. W. Woolley who was to have a profound influence on his career. They worked on a dinucleotide growth factor he discovered in graduate school and on peptide growth factors that Woolley had discovered earlier. These studies led to the need for peptide synthesis and, eventually, to the idea for solid phase peptide synthesis in 1959. The development and application of the technique have occupied him and his laboratory up to the present date. He is very proud of the fact that his office was once occupied by the great pioneer peptide chemist, Max Bergmann, and has been inspired by the knowledge that his laboratories were once filled with names like Leonidas Zervas, Joseph Fruton, Klaus Hofmann, Emil Smith, William Stein, and Stanford Moore.

In the meantime his wife, Libby, who was a biologist by training, stayed home in Cresskill, New Jersey, and raised their six children who now range in age from 19 to 32 years. They have been the great joy in the life of their parents; and now Jim has a daughter, Kelly, who is the pride of the whole family. Three years ago Libby joined the Merrifield laboratory at Rockefeller University.

He was a Nobel Guest Professor at Uppsala University in 1968 and was elected a member of the U.S. National Academy of Sciences in 1972. He has received several awards for his work on peptide chemistry including the Lasker Award for Basic Medical Research (1969), the Gairdner Award (1970), the Intra-Science Award (1970), the American Chemical Society Award for Creative Work in Synthetic Organic Chemistry (1972), the Nichols Medal (1973), the Instrument Specialties Company Award of the University of Nebraska (1977), and the 2nd Alan E. Pierce Award of the American Peptide Symposium (1979).

He has received honorary degrees from the University of Colorado (1969), Uppsala University (1970), Yale University (1971), Newark College of Engineering (1972), the Medical College of Ohio (1972), Colgate University (1977), and Boston College (1984). In 1984 he was appointed the John D. Rockefeller Jr. Professor of the Rockefeller University.

SOLID PHASE SYNTHESIS

Nobel lecture, 8 December, 1984

by

BRUCE MERRIFIELD

The Rockefeller University, 1230 York Avenue,
New York, N.Y. 10021–6399

The proteins, as the Greek root of their name implies, are of first rank in living systems, and their smaller relatives, the peptides, have now also been discovered to have important roles in biology. Among their members are many of the hormones, releasing factors, growth factors, ion carriers, antibiotics, toxins, and neuropeptides. My purpose today is to describe the chemical synthesis of peptides and proteins and to discuss the use of the synthetic approach to answer various biological questions.

The story begins with Emil Fischer (1) at the turn of this century when he synthesized the first peptide and coined the name. The general chemical requirements were to block the carboxyl group of one amino acid and the amino group of the second amino acid. Then, by activation of the free carboxyl group the peptide bond could be formed, and selective removal of the two protecting groups would lead to the free dipeptide. Fischer himself was never able to find a suitable reversible blocking group for the amine function, but his former student Max Bergmann, with Zervas, was successful (2). Their design of the carbobenzoxy group ushered in a new era. When I began working on the synthesis of peptides many years later this same general scheme was universally in use and was very effective, having led, for example, to the first synthesis of a peptide hormone by Du Vigneaud in 1953 (3). It soon became clear to me, however, that such syntheses were difficult and time consuming and that a new approach was needed if large numbers of peptides were required or if larger and more complex peptides were to be made.

SYNTHESIS ON A SOLID MATRIX

One day I had an idea about how the goal of a more efficient synthesis might be achieved. The plan (4) was to assemble a peptide chain in a stepwise manner while it was attached at one end to a solid support. With the growing chain covalently anchored to an insoluble matrix at all stages of the synthesis the peptide would also be completely insoluble and, furthermore, would be in a suitable physical form to permit rapid filtration and washing. Therefore, after completion of each of the synthetic reactions the mixture could be filtered and thoroughly washed to remove excess reactants and by-products. The intermediate peptides in the synthesis would thus be purified by a very simple, rapid procedure rather than by the usual tedious crystallization methods. When a multistep process, such as the preparation of a long polypeptide or protein is contemplated, the saving in time and effort and materials could be very large. The fact that all of the steps just described are heterogeneous reactions between a soluble reagent in the liquid phase and the growing peptide chain in the insoluble solid phase led to the introduction of the name "solid phase peptide synthesis".

The general scheme for solid phase synthesis is outlined in Fig. 1. It begins

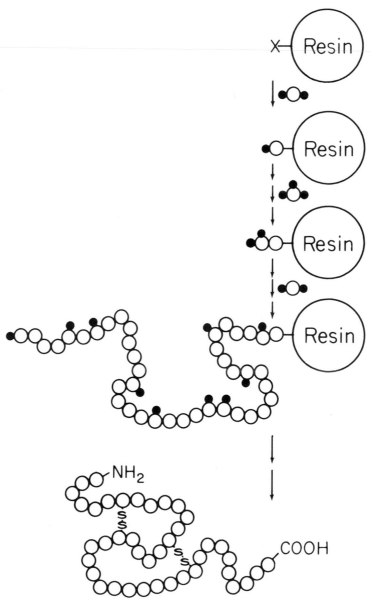

Fig. 1. The general scheme for solid phase synthesis.

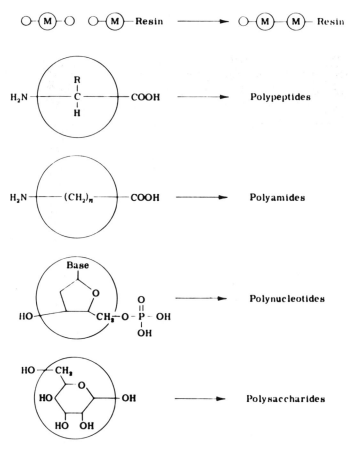

Fig. 2. Monomer units for solid phase synthesis.

with an insoluble particle, indicated by the large circles, which is functionalized with a group, X. The first monomer unit, small circles, is blocked at one end and at the reactive side chain groups (black dots) and anchored to the support by a stable covalent bond. The α protecting group is then removed and the second monomer unit is added to the first by a suitable reaction. In a similar way the subsequent units are combined in a stepwise manner until the entire polymeric sequence has been assembled. Finally, the bond holding the chain to the solid support is selectively cleaved, together with the side chain protecting groups, and the product is liberated into solution. Such a system offers four main advantages: it simplifies and accelerates the multistep synthesis because it is possible to carry out all the reactions in a single reaction vessel and thereby avoid the manipulations and attendant losses involved in the repeated transfer of materials; it also avoids the large losses which normally are encountered during the isolation and purification of intermediates; it can result in high yields of final products through the use of excess reactants to force the individual reactions to completion; and it increases solvation and decreases aggregation of the intermediate products. It only remained to translate the general idea into a workable set of reactions.

Although the plan was originally conceived as a way to synthesize peptides, the general scheme does not specify the nature of the monomer units. It soon became apparent that the technique should be applicable to units other than amino acids, such as those shown in Fig. 2. We extended it to the synthesis of depsipeptides (5) and other laboratories succeeded in synthesizing polyamides (6), polynucleotides (7) and polysaccharides (8). In principle the monomer may be any bifunctional compound that can be selectively blocked at one end and activated at the other. In addition, the solid support idea can be applied to a variety of conventional reactions in organic chemistry to aid in directing the course of the reaction or in the separation of the products from reagents and by-products. It also led to the solid phase sequencing technique.

SOLID PHASE PEPTIDE SYNTHESIS

A detailed scheme for the synthesis of peptides is shown in Fig. 3. Each of the steps has been modified in many ways, but the chemistry shown here has served well and has been applied to the synthesis of large numbers of peptides (9). The carboxyterminal amino acid is blocked at the amino end by a tert-butyloxycarbonyl (Boc) group and is covalently attached to the resin support as a benzyl ester via the chloromethyl group. Side chain functional groups must

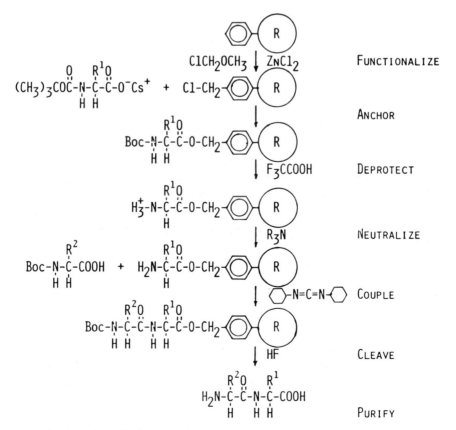

Fig. 3. A scheme for solid phase peptide synthesis.

also be blocked, usually with benzyl-based derivatives. The synthesis depends on the differential sensitivity of these two classes of protecting groups to acid, which is greater than 1000:1. The Boc group is completely removed with 50 % trifluoroacetic acid in dichloromethane, with minimal loss of the anchoring bond or of the other protecting groups. The resulting α amine salt is neutralized with a tertiary amine such as diisopropylethylamine and the free amine of the resin-bound amino acid is then ready to couple with a second Boc-amino acid. The latter must be activated for the reaction to occur. The simplest and most often used procedure is activation with dicyclohexylcarbodiimide (10) as shown, but active esters (11), anhydrides (12), and many other activated derivatives have been successfully applied. All of these reactions are carried out under non-aqueous conditions in organic solvents that swell the resin and accelerate the rates. Dichloromethane and dimethylformamide are the solvents of choice.

To extend the peptide chain the deprotection, neutralization and coupling steps are repeated for each of the succeeding amino acids until the desired sequence has been assembled. Finally, the completed peptide is deprotected and cleaved from the solid support. With the chemistry described here, this is accomplished by treatment with a strong anhydrous acid such as HF (13). The free peptide is then purified by suitable procedures.

It is very important that the repetitive steps proceed rapidly, in high yields, and with minimal side reactions in order to prevent the accumulation of excessive amounts of by-products. Much of our effort has been directed toward developing and evaluating these requirements.

SOLID PHASE NUCLEOTIDE SYNTHESIS

Similar schemes for the solid phase synthesis of oligonucleotides have now been developed which are rapid and give relatively high yields (14, 15). They employ protected nucleotides as monomer units and make use of either phosphotriester or phosphite triester chemistry. One such procedure is outlined in Figure 4. The resin is first functionalized with an aminomethyl group and the nucleotide derivative is coupled to it, through a spacer, by a stable amide bond. In this example the 5'hydroxyl is esterified to the spacer and the 3'hydroxyl is temporarily blocked with a dimethoxytrityl (DMT) group. The latter is removed with acid or $ZnBr_2$ and the chain is extended at the 3' end by coupling with the next protected nucleotide by activation with 1-(mesitylene-2-sulfonyl)-3-nitro-1, 2, 4-triazolide (MSNT). The completed oligonucleotide is finally cleaved from the solid support and deprotected by treatment with a base such as NH_4OH or tetramethylguanidine and then with hot acetic acid. The products are readily purified by ion exchange chromatography or by electrophoresis where the desired product always has the greatest negative charge. I will not deal further with polynucleotides and their use in site directed mutagenesis or with synthetic genes in this presentation but will concentrate instead on peptides and proteins.

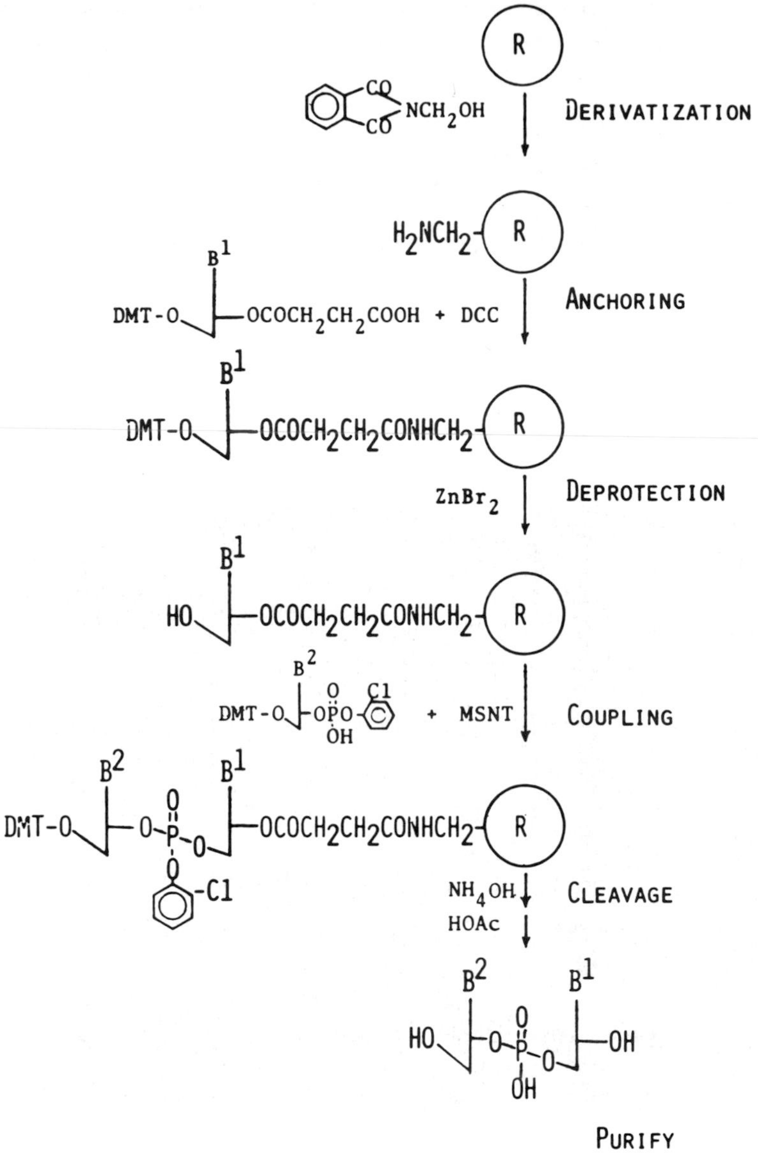

Fig. 4. A scheme for solid phase nucleotide synthesis.

THE SUPPORT

The first requirement for the development of solid phase synthesis was a suitable support. After examination of many potential supports it was found that the most satisfactory one was a gel prepared by suspension copolymerization of styrene and 1% of divinylbenzene as crosslinking agent (4). The resulting spherical beads (Fig. 5) are about 50 μm in diameter when dry, but in organic solvents such as dichloromethane they swell to 5 or 6 times their original volume. Furthermore, as peptide chains grow the dry volume increases to accomodate the added mass and, most importantly, the swollen volume

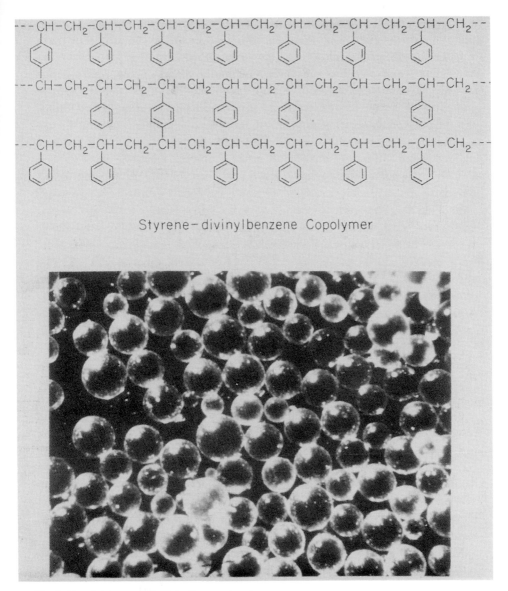

Fig. 5. Copoly(styrene-1%-divinylbenzene) resin.

continues to increase. Values up to 25 fold have been measured and calculations indicate that the maximum expansion should be about 200 fold (16). This means that the polystyrene matrix and the pendant peptide are highly solvated during the chemical reactions and are freely accessible to diffusing reagents. The reactions occur not only at the surface of the bead but, in major part, within the interior of the crosslinked polymeric matrix. This could be demonstrated by autoradiography of a cross section of a bead containing a synthetic tritium-labeled peptide (17). At this resolution the silver grains were located uniformly throughout the bead, although the distribution is not known at the molecular level. Because of the solvation and swelling of the beads, the reactions are very fast, with half times in the order of seconds for both the coupling and the deprotection steps. Current efforts to evaluate the effects of mass transfer and diffusion indicate that they are not rate limiting. We believe the solid matrix not only does not have detrimental effects on the synthesis but actually has beneficial effects in certain instances. One of the well recognized difficulties with the classical synthesis in homogeneous solution is insolubility of some intermediates. This problem can be overcome in many cases by the use of solid supports, where the peptide chain and the lightly crosslinked polymer chain become intimately mixed and exert a mutual solvating effect on one another. It becomes thermodynamically less favorable for the peptide to self aggregate and it remains available for reaction. For this to occur the solvated state of the bound peptide needs only to be favorable relative to the amorphous unsolvated state within the peptide-resin matrix (16). Similar solubilizing properties of linear polymers for covalently attached components are known, but the effect will be greater for a lightly cross-linked polymer network. The phenomenon can be illustrated by the synthesis of oligoisoleucines (18). The standard solution synthesis failed after the tetrapeptide stage because of aggregation and insolubility, whereas the chain could be extended up to 8 residues on linear polyethyleneglycol. A solid phase synthesis proceeded smoothly at least as far as the dodecamer, where the experiment was stopped. There is very significant polymer chain motion in these crosslinked polystyrene resins. Both ^{1}H and ^{13}C NMR measurements (19) have shown that the motional rates for the aromatic groups and the aliphatic backbone atoms in CH_2Cl_2 are high and equivalent to linear soluble polystyrene (τ_c 10^{-8} sec). The α carbon ^{13}C resonances of model resin-supported peptides were as sharp as the solvent peak in CH_2Cl_2 or dimethylformamide and similar to small molecules in solution (τ_c 10^{-10} sec). A variety of chemical experiments also have shown polymer flexibility. For example short resin-bound peptides that were too far apart on average to reach one another if the resin were rigid could be shown to react to the extent of 99.5 % indicating considerable motion of the polystyrene segments within the matrix (20).

Many other solid supports have also been examined and several have been satisfactory for peptide synthesis. These have included polymethylmethacrylate, polysaccharides, phenolic resins, silica, porous glass, and polyacrylamides, but only the latter have seen widespread use (21). Comparative studies with polystyrene and polyacrylamide have shown that they can be equally effective, even with difficult peptides.

AUTOMATION

The ability to purify after each reaction by simple filtration and washing and the fact that all reactions could be conducted within a single reaction vessel appeared to lend themselves ideally to a mechanized and automated process. Initially, a simple manually operated apparatus was constructed (Fig. 6). This system was first used to work out the methodology and to synthesize bradykinin (22), angiotensin (23), oxytocin (24), and many other small peptides. In order to accelerate the process we undertook the design and construction (25)

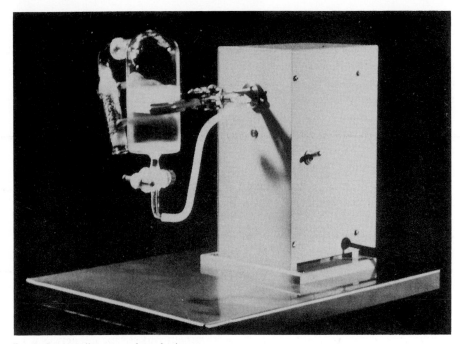

Fig. 6. A manually operated synthesis apparatus.

of the automated instrument shown in Fig. 7. The essential features were the reaction vessel, containing the resin with its growing peptide chain, and the necessary plumbing to enable the appropriate solvents and reagents to be pumped in, mixed, and removed in the proper sequence. These mechanical events were under the control of a simple stepping drum programmer and a set of timers. In the past few years many commercial instruments have been constructed in several countries. They differ considerably in detail, particularly in the sophistication of the electronic program mechanisms but are designed to carry out the same chemistry.

THE SYNTHESIS OF RIBONUCLEASE A

The idea of chemically synthesizing an enzyme must have occurred to many people over the years. There was a time when such a thought would have been unacceptable even on philosophical grounds, but from the period when enzymes were shown to be proteins and proteins were shown to be discrete organic molecules it was a goal that chemists could begin to think about. If an enzyme could be made in the laboratory, then it should become possible to learn new things about how these large and very complex molecules function. Specific changes could be made in their structures that could not be made readily by altering the native protein and data should be forthcoming that would supplement the information already obtained from the natural enzymes themselves. In this regard, a quotation from Fischer in 1906 (26) is pertinent: "Whereas cautious professional colleagues fear that a rational study of this class of compounds [proteins], because of their complicated structure and their

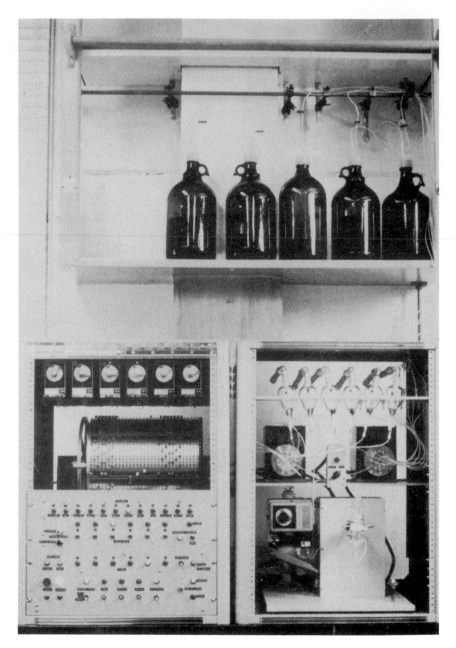

Fig. 7. An automated peptide synthesizer.

highly inconvenient physical characteristics, would today still uncover insur-
mountable difficulties, other optimistically endowed observers, among which I
will count myself, are inclined to the view that an attempt should at least be
made to besiege this virgin fortress with all the expedients of the present;
because only through this hazardous affair can the limitations of the ability of
our methods be ascertained.''

With the development of solid phase peptide synthesis and its automation the time seemed right to attempt the total synthesis of an enzyme. Dr. Bernd Gutte and I selected bovine pancreatic ribonuclease A because it was a small stable protein of known amino acid sequence (27), and the three dimensional structure was known from X-ray diffraction studies (28). Much of the detailed mechanism by which this enzyme hydrolyzes and depolymerizes ribonucleic acid was also known. The purpose of a chemical synthesis of this 124-residue molecule was, first, simply to demonstrate that a protein with the high catalytic activity and specificity of a naturally occurring enzyme could be synthesized in the laboratory. For the long range the more important purpose was to provide a new approach to the study of enzymes. We believed it should be possible to modify the structure and to alter the activity and the substrate specificity of the enzyme.

The synthesis (29) was carried out on a copoly(styrene-1 %-divinylbenzene)-resin support using the general methods described above. The C-terminal Boc-Val was anchored to the solid matrix by a benzyl ester bond, the usual benzyl-based side chain protecting groups were used and the Boc group provided the reversible N^α protection. The deprotection steps were with trifluoroacetic acid and the coupling reactions were with dicyclohexylcarbodiimide activation. Fig. 8 shows the final protected derivative of ribonuclease. It contained a total of 67 side chain protecting groups and had a molecular weight of 19,791. The synthesis is summarized in Table I. The overall yield after several purification procedures was about 3 % based on the original amount of valine attached to the resin. There was a large (83 %) loss of chains during the assembly of the peptide chain due to partial instability of the anchoring bond, and the accumulated losses during HF cleavage from the resin and the purification steps were another 80 %. The crude cleaved product was air oxidized to form the four disulfide bonds and the monomer fraction was isolated by gel filtration. The monomers with incorrect disulfide pairing or incorrect folding were digested by trypsin and the small fragments were separated from the stable protein with the correct structure. An ammonium sulfate fractionation gave the final purified enzyme possessing approximately 80 % specific activity compared with native ribonuclease A. We could not claim that our product was completely pure or that the synthesis constituted a structure proof for RNase, only that the molecule showed a close chemical and physical resemblance to the native protein and that it was a true enzyme. The chemical and physical comparisons were based on amino acid analysis, enzyme digestions, peptide maps, paper electrophoresis, gel filtration, ion-exchange chromatography, and antibody neutralization. At that time we did not have HPLC or an affinity chromatography system.

Table II summarizes the activity data at various stages of purification of the synthetic enzyme. Both the specific activity and the total number of units of RNase increased as the purification proceeded, indicating either that inhibitory impurities were being removed or that the molecule was gradually refolding into a conformation that more closely resembled the native structure. The substrate specificity of the synthetic enzyme was consistent with that to be

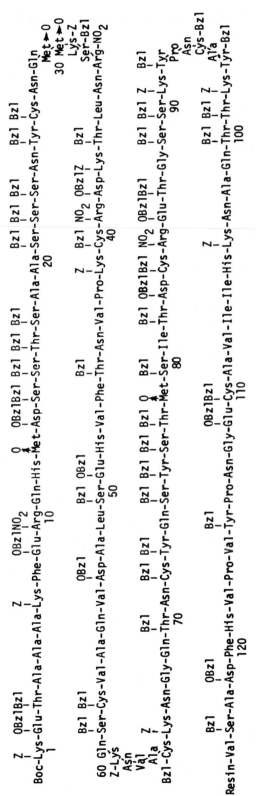

Fig. 8. Protected-ribonuclease-resin.

Summary of **Ribonuclease** A Synthesis

Stage of Synthesis	Overall Yield	
	mg	%
Boc-Val-Resin	2000	100

Deprotect
    Neutralize } Repeat 123 times
    Couple
17%		

Protected RNase-Resin	3430	17
Cleave and Deprotect HF 71%		
Crude RNase $(SH)_8$	697	12
Sephadex G-75 53%		
RNase A (monomer fraction)	373	6.4
69% Trypsin digestion Sephadex G-50		
RNase A (Trypsin resistant fraction)	256	4.4
66% $(NH_4)_2SO_4$ fractionation		
RNase A	169	2.9

Table I

Ribonuclease A Activity

Purification stage	Specific activity (%)	Total activity $\frac{mg\ RNase}{2\ g\ resin}$
HF cleavage	2	14
Sephadex G-75	9	33
IRC-50	13	53
Trypsin	61	156
$(NH_4)_2SO_4$	78	132

Table II

expected for RNase A: it was able to cleave both large (RNA) and small (C>p) substrates and therefore to catalyze both the transphosphorylation and the hydrolysis steps; it was specific for D-ribose instead of D-deoxyribose and for a pyrimidine instead of a purine at the 3' position of the phosphodiester substrate (Table III). The Km values toward RNA were also the same for the natural and synthetic enzymes.

The purified RNase A was compared on a CM-cellulose column with natural RNase A and with reduced and reoxidized natural RNase A. They were identical by this criterion, which was the one first used by White (30) to show that RNase A after reduction and reoxidation of the disulfide bonds was indistinguishable from the native enzyme. His was the demonstration that led to the hypothesis that the primary structure of the protein determined its tertiary structure (31). Our synthesis provided a new kind of evidence for this hypothesis. The fact that the only information put into the synthesis was the linear sequence means that the primary structure must be sufficient to direct the final folding of the molecule into its active tertiary structure. The synthesis of an active enzyme containing no substituents except amino acids also provided a new proof for the now well established belief that enzymatic activity can be attributed to a simple protein containing no other components.

Table III. Substrate Specificity of Synthetic Ribonuclease A

Substrate		Activity
Ribonucleic acid	(RNA)	78
Deoxyribonucleic acid	(DNA)	0
2',3'-Cyclic-cytidine phosphate	(C>p)	65
2',3'-Cyclic-guanosine phosphate	(G>p)	0
5'-(3'-guanylyl)-cytidylic acid	(GpCp)	0
5'-(3'-adenylyl)-adenylic acid	(ApAp)	0

STRUCTURE-FUNCTION STUDIES ON RIBONUCLEASE

The synthesis of ribonuclease provided answers to several fundamental questions and laid the foundation for new studies on the relation of structure to function in the enzyme. The classic S-peptide/S-protein system discovered by Richards (32) provided an ideal way to study such relations because a small peptide (residues 1−20) and a large protein component (residues 21−124) could be combined noncovalently with regeneration of nearly full enzymatic activity. The extensive work from the Hofmann (33) and Scoffone (34) laboratories on the synthesis of the S-peptide and its combination with the natural S-protein had already provided a great amount of information about the role of individual residues in the N-terminal region of the enzyme. We undertook to study this region of RNase by total synthesis (Fig. 9). During the initial synthesis we removed samples after coupling Cys^{26} and again after Ser^{21} and in that way prepared synthetic S-protein (21−124) and S-protein (26−124). The

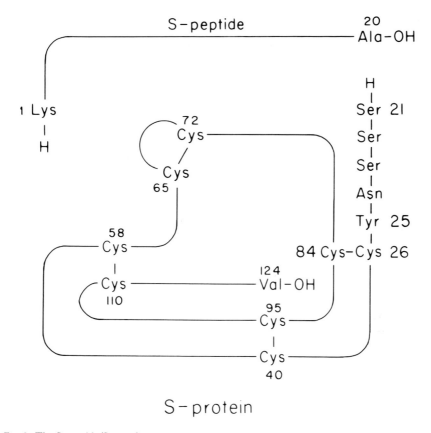

Fig. 9. The S-peptide/S-protein system.

partially purified proteins were reduced at their 4 disulfide bonds, each mixed with synthetic S-peptide, reoxidized and assayed for enzymatic activity. Each of the crude mixtures was found to have as much activity as the product derived from native S-protein by the same treatment. From these data it was concluded first, that S-protein had been synthesized and second, that the five residues 21–25 were definitely not necessary for the binding and reactivation to occur. Earlier X-ray data (35) had predicted that the serines at positions 21, 22 and 23 would probably not be necessary, but Asn^{24} and Tyr^{25} appeared to be involved in a total of 5 hydrogen bonds in RNase S and it was expected that they might be necessary for the formation of an active complex. The synthetic studies showed that they were not.

Several years earlier I had been interested in the question of whether or not a peptide from the carboxyl end of RNase might function in a manner similar to that of the S-peptide at the amino end. Consequently, the RNase 111–124 tetradecapeptide was synthesized and purified. RNase was inactivated by carboxymethylation of the imidazole ring of His^{119}. Attempts to reactivate the enzyme by addition of the synthetic peptide were uniformly unsuccessful. Somewhat later in a separate study Lin et al. (36) succeeded in preparing a series of shortened RNases. They made RNase 1–120, RNase 1–119 and

RNase 1—118 by enzymatic digestion. When the synthetic peptide 111—124 was assayed in the presence of these inactive proteins, high enzymatic activity was generated (37) and it became clear that a system existed at the C-terminus of RNase that was similar to the one at the N-terminus.

We then made the interesting discovery that the C-terminal peptide 111—124 containing His[119], the N-terminal peptide 1—20 containing His[12], and the central protein component 21—118 containing Lys[41] could be mixed together non-covalently and ribonuclease activity would be generated. Therefore, three components each containing one of the known residues required for enzymatic activity could bind together and form the specific well ordered structure necessary for substrate binding and catalytic activity.

A series of synthetic studies was then undertaken to define the roles of some of the individual residues in the C-terminal region. These can be summarized and discussed by referring to Fig. 10. When peptides shorter than 111—124 were prepared and combined with RNase 1—118, both the binding constant and the activity were progressively decreased and peptide 117–124 was inactive, indicating that each residue contributed to the binding energy (38). However, even in the complex 1–118 + 116–124 there were 3 overlapping residues. It was then found that the complex 1–115 + 116–124, in which there were no overlapping residues, had a binding constant 100 times larger (39). In these experiments it could also be shown that Tyr[115] was not necessary for enzymatic activity.

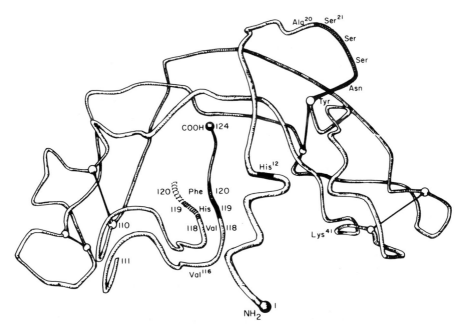

Fig. 10. A 3-dimensional representation of ribonuclease fragments 1—20, 21—118 and 111—124 summarizing the synthetic structure-function studies.

Phenylalanine-120 was shown to be important in stabilizing the ribonuclease structure by transition temperature studies and to interact with the pyrimidine substrate by X-ray and NMR studies. Our synthetic analog work on the $1-118$ $+$ $111-124$ system showed that replacement of Phe^{120} by Leu^{120} or Ile^{120} reduced the binding by 5 and 17 fold and reduced the maximum enzymatic activity to 10 % and 5 % respectively, indicating that the aromatic side chain of phenylalanine was of considerable importance in binding the peptide to the protein (40). It could only partially be replaced by a hydrophobic aliphatic chain, indicating an inexact alignment of the catalytic site. The small residue Ala^{120} and the bulky aromatic residue Trp^{120} were inactive. Replacement of Phe^{120} by an aromatic residue of similar size, Tyr^{120}, in the $111-124$ peptide gave a complex with $1-118$ that was fully active toward C>p as substrate and 190 % as active toward U>p (41). A semisynthetic enzyme with enhanced activity was a novel finding. Km and Ki data led to the conclusion that Phe^{120} does not have a unique role in the binding of substrate but is important for stabilizing the peptide-protein complex and the native enzyme itself. Nevertheless, the presence of substrate increased the binding constant between $1-118$ and $111-124$ by a factor of 50.

Similar experiments with the aspartic acid residue at position 121 have shown that it can be replaced partially ($\sim$20 %) by glutamic acid, but the Asn^{121} and Ala^{121} analogs did not show measurable binding. Removal of Val^{124} from RNase A does not affect the enzymatic activity and removal of Val^{124} from S-protein does not reduce the activity of the complex with S-peptide. In contrast, omission of Val^{124} from the C-terminal tetradecapeptide produced an essentially inactive complex with RNase $1-118$, indicating an important hydrophobic interaction necessary for peptide-protein binding. The smaller aliphatic residue Ala^{124} could only restore half of the binding energy (41).

X-ray data (42) indicate that the uracil and cytosine residues of RNA and the cyclic nucleotides probably bind to ribonuclease through the series of hydrogen bonds shown in Fig. 11. For uracil the hydroxyl of Thr^{45} is a hydrogen acceptor and for cytosine it is a hydrogen donor. Conversely the hydroxyl of Ser^{123} is a donor for uracil and an acceptor for cytosine. We reasoned that if these two hydroxyls were blocked as methyl ethers they could only be hydrogen acceptors and if replaced by Ala they could be neither donor nor acceptor. A suitable combination of these residues in replacement analogs might, therefore, lead to a synthetic ribonuclease with altered substrate specificity. Such analogs have been made for Ser^{123} (43).

The tetradecapeptide containing Ala^{123} gave a complex with RNase $1-118$ that showed appreciable selectivity for substrates containing cytosine relative to those containing uracil (either the 2',3' cyclic nucleotides or polynucleotides) (Table IV). Replacement with 0-methylserine did not result in differential substrate specificity. It was concluded that a hydrogen bond between the hydroxyl of Ser^{123} and the C^4 amino group of cytosine is not important for substrate binding and catalytic activity, but that the hydrogen bond between the hydroxyl of Ser^{123} and the C^4 carbonyl of uracil contributes significantly to the binding and activity; when Ser is replaced by Ala the H-bond is absent and

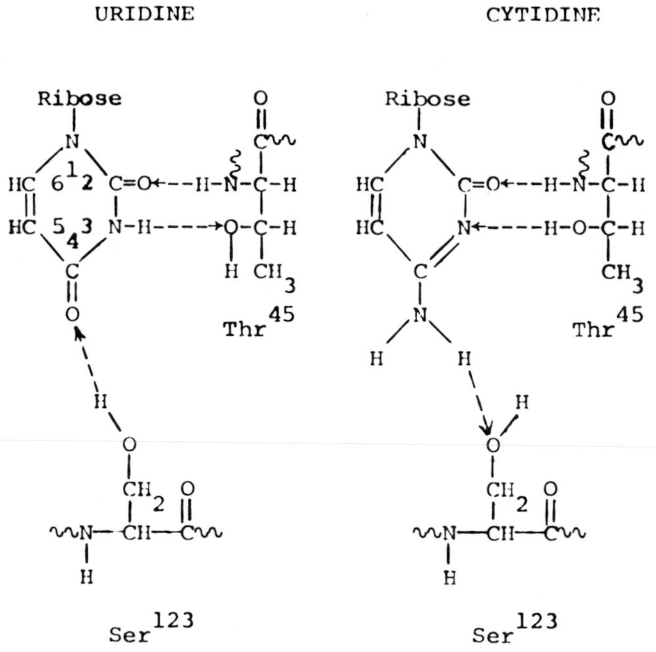

URIDINE

CYTIDINE

Fig. 11. Proposed hydrogen bonding of uracil and cytosine substrates to ribonuclease.

the activity is reduced. The corresponding studies with replacement of Thr[45] by Ala[45] and Ser(Me)[45] involve total synthesis of the enzyme and these much more difficult experiments have not yet been completed. We believe that the substrate binding at Thr[45] is much tighter than at Ser[123] and that changes at this residue will lead to much greater substrate selectivity.

Table IV. Substrate Selectivity of [Ala[123]]-RNase Complex

Enzyme	Selectivity (ks/Km) $\dfrac{C>p}{U>p}$
RNase A (natural)	4.6
[Ser[123]]-RNase 111−124 + RNase 1−118	5.0
[Ala[123]]-RNase 111−124 + RNase 1−118	19

RECENT IMPROVEMENTS IN SOLID PHASE PEPTIDE SYNTHESIS

Although the earlier solid phase chemistry was very useful for these studies on ribonuclease, it was clear that there was a need for improvement in several areas. One was the mode of attachment of the peptide to the resin. If the strategy of differential stability toward acid for the N^α and C^α groups was to be

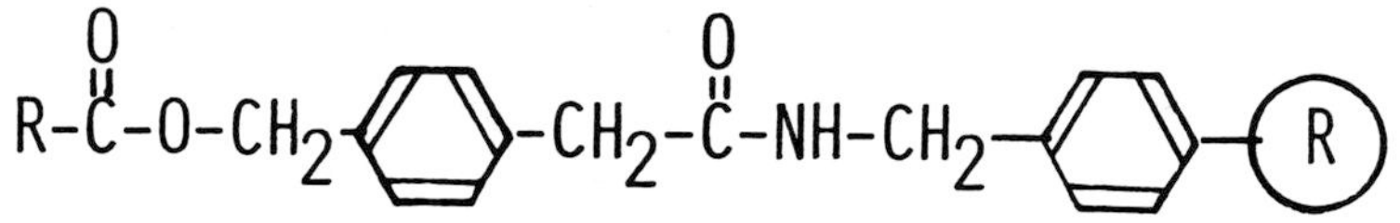

Fig. 12. Acyloxymethyl-Pam-resin.

continued, a more acid stable anchoring bond was needed. We predicted that the insertion of an acetamidomethyl group between the benzyl ester and the polystyrene matrix would increase the stability of the benzyl ester to trifluoroacetic acid by a factor of approximately 25 to 400 times. When such a linkage was finally constructed it was found to be 100 times more stable (44). A new synthesis of aminomethyl-resin was first developed in which N-hydroxymethyl-phthalimide and polystyrene resin were reacted under acid catalysis with F_3CSO_3H, HF, or S_nCl_4 (45). This product was then coupled with a derivative of the C-terminal amino acid. Thus, N^α-Boc-aminoacyloxymethylphenylacetic acid was prepared and activated with dicyclohexylcarbodiimide for the reaction. The product was the acyloxymethylphenylacetamidomethylcopoly(styrene-1%-divinylbenzene) resin (acyloxymethyl-Pam-resin) (Fig. 12). This new preparation has the advantages that it is more acid stable, and it is made from purified, well characterized intermediates, which give a cleaner product with fewer side reactions. It is free of chloromethyl groups that can give rise to quaternization and ion exchange reactions and is free of hydroxyl groups that can lead to peptide chain terminations via trifluoroacetylation (46).

An alternative protecting group strategy is to make use of an orthogonal system (47) in which the N^α, C^α, and side chain groups represent three different classes of compounds that are cleavable by three different kinds of reactions. In that way any one of the functional groups can be selectively removed in the presence of the other two. Figure 13 illustrates such a system in which the anchoring o-nitrobenzyl ester is photolabile but stable to acid or nucleophiles, the side chain groups are based on tert-butyl derivatives that are very acid labile but stable to light or nucleophiles, and the N^α protecting group is the dithiasuccinoyl group which is removed by nucleophilic thiols but is

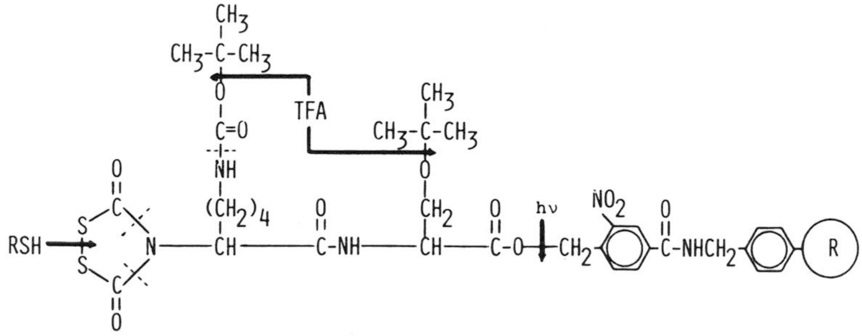

Fig. 13. An orthogonal protecting group scheme.

stable to acid and photolysis. This scheme has recently been put to the test and found to give excellent results (48).

Anhydrous hydrogen fluoride, the usual cleavage reagent for solid phase peptide synthesis, is a very strong acid (H_0-10.8) and is known to promote a number of side reactions. In particular it leads to the formation of carbonium ions which then can alkylate tyrosine, tryptophan, methionine and cysteine residues of the peptide. In addition, HF can protonate and dehydrate the side chain carboxyl of glutamic acid residues with formation of the very reactive acylium ion, which has been shown to acylate the aromatic rings of anisole and other scavengers present in the mixture. Activated glutamic residues can also form pyrrolidone (pyroglutamic)containing products. Aspartyl residues can close in HF to the aspartimide derivative and subsequently open to produce β-aspartyl residues. All of these undesired reactions result from the S_N1 mechanism of the cleavage reaction under the usual conditions (90 % HF + 10 % anisole, 0° C, 1 hr). We reasoned that if conditions could be found that would change the reactions to an S_N2 mechanism in which the acidolysis is aided by a nucleophile and carbocation is never formed (Fig. 14) it should be possible to minimize or avoid these problems. Dr. James Tam and W.F. Heath, a graduate student, have succeeded in developing such conditions and in demonstrating marked improvements in solid phase peptide synthesis (49).

The problem was to find a suitable weak base which would reduce the acidity function of the HF but which would remain largely unprotonated and nucleophilic under the resulting acidic conditions. It should be a weaker base than the groups to be cleaved so that they would be largely protonated under the same conditions. Dimethylsulfide (DMS) was found to be an ideal base for this purpose. It has a pKa of −6.8 compared with values of −2 to −5 for the benzyl ethers, esters and carbamates to be cleaved. It is a good solvent for HF and it is volatile and easily removed from the reaction mixture. A 1:1 molar

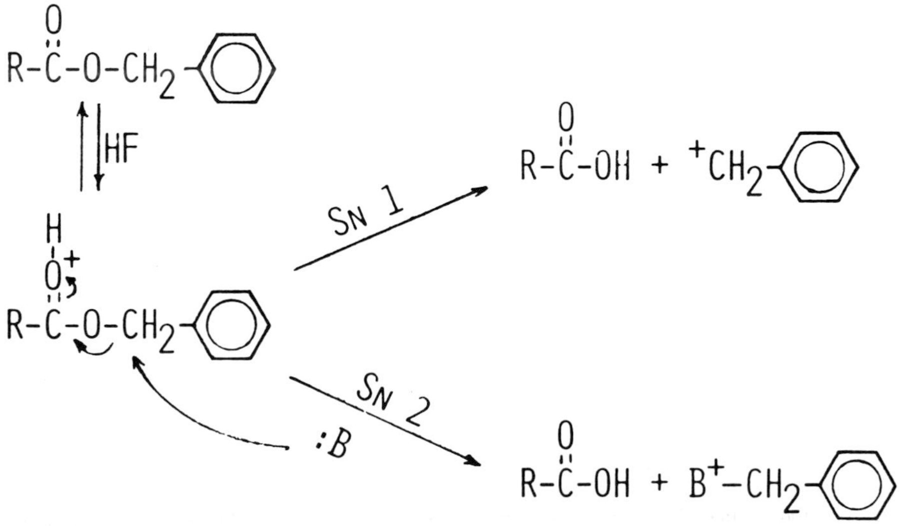

Fig. 14. The S_N1 and S_N2 acidolysis mechanisms.

mixture of HF and DMS (1:3 by volume) was determined by Hammett indicators to have an H_0 between −4.6 and −5.2. The mechanisms of removal of various benzyl-based protecting groups by HF/DMS mixtures were tested by kinetic and product analysis experiments. Based on earlier work with H_2SO_4 hydrolysis of alkyl acetates (50), a sharp upward break in the rate constant was expected when the acid concentration was increased. At the break point the mechanism changed from S_N2 to S_N1. A similar change was found in the cleavage of O-benzyl serine by HF/DMS mixtures; above 50% HF by volume the rate increased rapidly, indicating the change in mechanism. Product analysis for the deprotection of tyrosine benzyl ether as a function of HF concentration is shown in Fig. 15. Above 15% HF in DMS the yield (after 1 hr, 0° C) of tyrosine was quantitative and the other product was the benzyl-dimethylsulfonium salt. In the range of 40−50% HF the amount of sulfonium salt began to decrease and the level of the undesirable byproduct, 3-benzyltyrosine, increased. Again, there was a change from the S_N2 to the S_N1 mechanism around 40−50% HF. The reactions were accelerated in the presence of 5−10% of cresol. We selected 25% HF/65% DMS/10% cresol as the best reagent and refer to it as "low HF".

This reagent was also effective in preventing acylium ion formation in glutamyl and aspartyl peptides and avoided the acylation and imide side reactions. It was also found to be very effective in converting methionine

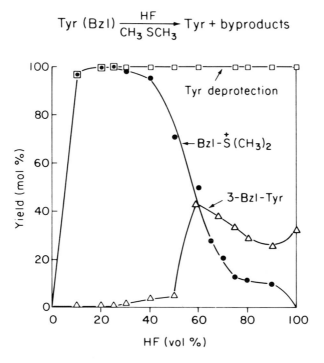

Tyr (Bzl) $\xrightarrow[\text{CH}_3\text{SCH}_3]{\text{HF}}$ Tyr + byproducts

Fig. 15. Product analysis for the deprotection of tyrosine benzyl ether in mixtures of HF and dimethylsulfide.

column at exactly the same time as natural EGF (Fig. 17). In the sensitive and discriminating Leydig cell growth assay the synthetic and natural EGF had identical activity.

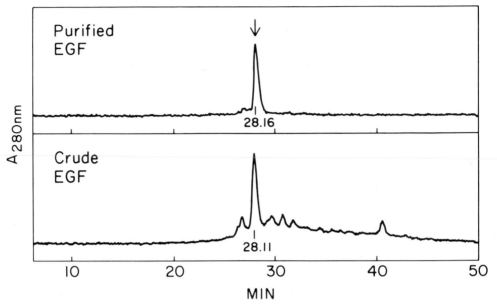

Fig. 17. HPLC analysis of synthetic EGF. The arrow indicates the position of natural EGF.

From the accumulated data presented, we conclude that the solid phase synthesis of peptides up to 50 or somewhat more residues can be readily achieved in good yield and purity; and this is a far better situation than I could have expected when this technique was first proposed.

As an example of a synthesis of a protein I have selected our recent studies on interferon. The sequence of human leucocyte interferon α_1 was first deduced from the DNA sequence of the cloned gene (67). It contains 166 amino acids with 5 cysteine residues (Fig. 18). The amino acid sequence of the isolated protein of human leucocyte interferon α_2 was also determined (68) and found to have only 155 residues. There is a high degree of homology between the two, but the latter has one deletion at Asp^{44} and is missing the last 10 residues predicted from the DNA sequence (Fig. 18). We have synthesized these two proteins and also their Ser^1 analogs and purified them by reduction, gel filtration, reoxidation, gel filtration, and affinity purification on a column of supported polyclonal antibodies to human leucocyte interferon (69). The synthetic proteins and the natural and recombinant interferon all had 10^8 to 10^9 units/mg in antiviral assays against a broad spectrum of cell lines. The development and duration of the antiviral state were also similar. Synthetic $[Ser^1]IFN-\alpha_2$ and natural Hu-Le-IFN-α showed similar growth inhibition of K 562 cells. $[Cys^1]IFN-\alpha_2$ and natural Hu-Le-IFN-α caused a similar increase of natural killer cell activity whereas synthetic $[Ser^1]IFN-\alpha_2$ caused a decrease. All four synthetic interferons bind to and are eluted from polyclonal anti-Hu-Le-IFN-α antibodies under similar conditions.

α1 1 Cys-Asp-Leu-Pro $\binom{Glu}{Gln}$ Thr-His-Ser-Leu $\binom{Asp-Asn}{Gly-Ser}$ Arg-Arg-Thr-Leu-
α2

16 Met-Leu-Leu-Ala-Gln-Met $\binom{Ser-Arg}{Arg-Lys}$ Ile-Ser $\binom{Pro-Ser}{Leu-Phe}$ Ser-Cys-Leu-

31 $\binom{Met}{Lys}$ Asp-Arg-His-Asp-Phe-Gly-Phe-Pro-Gln-Glu-Glu-Phe $\binom{Asp}{-}$ Gly-

46 Asn-Gln-Phe-Gln-Lys-Ala $\binom{Pro-Ala}{Glu-Thr}$ Ile $\binom{Ser}{Pro}$ Val-Leu-His-Glu $\binom{Leu}{Met}$

61 Ile-Gln-Gln-Ile-Phe-Asn-Leu-Phe $\binom{Thr}{Ser}$ Thr-Lys-Asp-Ser-Ser-Ala-

76 Ala-Trp-Asp-Glu $\binom{Asp}{Thr}$ Leu-Leu-Asp-Lys-Phe $\binom{Cys}{Tyr}$ Thr-Glu-Leu-Tyr-

91 Gln-Gln-Leu-Asn-Asp-Leu-Glu-Ala-Cys-Val $\binom{Met}{Ile}$ Gln $\binom{Glu-Glu-Arg}{Gly-Val-Gly}$

106 Val $\binom{Gly}{Thr}$ Glu-Thr-Pro-Leu-Met $\binom{Asn-Ala}{Lys-Glu}$ Asp-Ser-Ile-Leu-Ala-Val-

121 $\binom{Lys}{Arg}$ Lys-Tyr-Phe $\binom{Arg}{Gln}$ Arg-Ile-Thr-Leu-Tyr-Leu $\binom{Thr}{Lys}$ Glu-Lys-Lys-

136 Tyr-Ser-Pro-Cys-Ala-Trp-Glu-Val-Val-Arg-Ala-Glu-Ile-Met-Arg-

166
151 Ser $\binom{Leu}{Phe}$ Ser-Leu-Ser-Thr Asn-Leu-Gln-Glu-Arg-Leu-Arg-Arg-Lys-Glu

Fig. 18. Sequences of leucocyte interferons α$_1$ and α$_2$.

These results are encouraging, but much more needs to be done to assure that even small proteins can be synthesized readily in high yield and purity. I think we can be optimistic about the future.

ACKNOWLEDGEMENTS

I owe a very special debt of gratitude to my teachers, Dr. M. S. Dunn of U.C.L.A. and Dr. D. W. Woolley of The Rockefeller University. Several of the past and present members of my laboratory have been referred to here, but to the many others who have not been specifically mentioned I am equally grateful because they all have contributed to the progress of our work. Finally, I wish to acknowledge the continuing support of The Rockefeller University and of the National Institutes of Health of the United States.

REFERENCES

1. Fischer, E. and Fourneau, E., *Ber. 34*, 2868 (1901).
2. Bergmann, M. and Zervas, L., *Ber, 65*, 1192 (1932).
3. Du Vigneaud, V., Ressler, C., Swan J. M., Roberts C. W., Katsoyannis, P. G. and Gordon, S., *J. Am. Chem. Soc., 75*, 4879 (1953).
4. Merrifield, R. B., *J. Am. Chem. Soc., 85*, 2149 (1963).
5. Gisin, B. F., Merrifield, R. B. and Tosteson, D. C., *J. Am. Chem. Soc. 91*, 2691 (1969).
6. Kusch, P., *Angew. Chem., 78*, 611 (1966).
7. Letsinger, R. L. and Mahadevan, V., *J. Am. Chem. Soc., 87*, 3526 (1965).
8. Fréchet, J. M. and Schuerch, C., *J. Am. Chem. Soc., 93*, 492 (1971).
9. Merrifield, R. B., *Advan. Enzymol., 32*, 221 (1969).
10. Sheehan, J. C. and Hess, G. P., *J. Am. Chem. Soc., 77*, 1067 (1955).
11. Bodanszky, M. and Sheehan, J. T., *Chem. Ind.* (London) 1423 (1964).
12. Wieland, T., Birr, C. and Flor, F., *Angew. Chem. Int. Ed. Engl. 10*, 336 (1971).
13. Sakakibara, S. and Shimonishi, Y., *Bull. Chem. Soc. Jap. 38*, 1412 (1965).
14. Narang, S. A., *Tetrahedron, 39*, 3 (1983).
15. Itakura, K., Rossi, J. J. and Wallace, R. B., *Ann. Rev. Biochem. 53*, 323 (1984).
16. Sarin, V. K., Kent, S. B. H. and Merrifield, R. B., *J. Am. Chem. Soc., 102*, 5463 (1980).
17. Merrifield R. B. and Littau V., in "Peptides 1968", E. Bricas, ed., North-Holland Publ., Amsterdam, 1968, pp. 179–182.
18. Kent, S. B. H. and Merrifield, R. B., in "Peptides 1980", K. Brunfeldt, ed., Scriptor, Copenhagen, 1981, pp. 328–333.
19. Live, D. H. and Kent, S. B. H. in "Peptides: Structure and Function", V. Hruby and D. Rich, eds., Pierce Chem. Co., Rockford, Ill., 1983, pp. 65–68.
20. Bhargava, K. K., Sarin, V. K., Trang, N. L., Cerami, A. and Merrifield, R. B., *J. Am. Chem. Soc., 105*, 3247 (1983).
21. Atherton, E., Clive, D. L. J. and Sheppard, R. C., *J. Am. Chem. Soc., 97*, 6584 (1975).
22. Merrifield, R. B., *Biochemistry, 3*, 1385 (1964).
23. Marshall, G. R. and Merrifield, R. B., *Biochemistry, 4*, 2394 (1965).
24. Manning, M., *J. Am. Chem. Soc., 90*, 1348 (1968).
25. Merrifield, R. B., Stewart, J. M. and Jernberg, N., *Anal. Chem., 38*, 1905 (1966).
26. Fischer, E., *Ber., 39*, 530 (1906). Translation taken from Greenstein, J. P. and Winitz, M., "Chemistry of the Amino Acids", Vol. 2, John Wiley, 1961, p. 1816b.
27. Hirs, C. H. W., Moore, S. and Stein, W. H., *J. Biol. Chem., 235*, 633 (1960).
28. Kartha, G., Bello, J. and Harker, D., *Nature, 213*, 862 (1967).
29. Gutte, B. and Merrifield, R. B., *J. Biol. Chem., 246*, 1922 (1971).
30. White, F. H. Jr., *J. Biol. Chem., 236*, 1353 (1961).
31. Anfinsen, C. B. and Haber, E., *J. Biol. Chem., 236*, 1361 (1961).
32. Richards, F. M., *Compt. Rend. Trav. Lab. Carlsberg Ser. Chim., 29*, 329 (1955).
33. Finn, F. M. and Hofmann, K., *J. Am. Chem. Soc., 87*, 645 (1965).
34. Rocchi, R., Marchiori, F., Moroder, L., Borin, G. and Scoffone, E., *J. Am. Chem. Soc., 91*, 3927 (1969).
35. Wyckoff, H. W., Hardman, K. D., Allewell, N. M., Inagami, T., Johnson, L. N. and Richards, F. M., *J. Biol. Chem., 242*, 3984 (1967).
36. Lin, M. C., *J. Biol. Chem., 245*, 6726 (1970).
37. Lin, M. C., Gutte, B., Moore, S. and Merrifield, R. B., *J. Biol. Chem., 245*, 5169 (1970).
38. Gutte, B., Lin, M. C., Caldi, D. G. and Merrifield, R. B., *J. Biol. Chem., 247*, 4763 (1972).
39. Hayashi, R., Moore, S. and Merrifield, R. B., *J. Biol. Chem. 248*, 3889 (1973).
40. Lin, M. C., Gutte, B., Caldi, D. G., Moore, S. and Merrifield, R. B., *J. Biol. Chem., 247*, 4768 (1972).
41. Merrifield, R. B. and Hodges, R. S., in "Proceedings Internat. Symp. Macromolecules", E. B. Mano, ed., Elsevier, Amsterdam, 1975, pp. 417–431.
42. Wyckoff, H. W., Tsernoglou, D., Hanson, A. W., Knox, J. R., Lee, B. and Richards, F. M., *J. Biol. Chem., 245*, 305 (1970).
43. Hodges, R. S. and Merrifield, R. B., *J. Biol. Chem., 250*, 1231 (1975).

44. Mitchell, A. R., Erickson, B. W., Ryabtsev, M. N., Hodges, R. S. and Merrifield, R. B., *J. Am. Chem. Soc.*, *98*, 7357 (1976).

45. Mitchell, A. R., Kent, S. B. H., Erickson, B. W. and Merrifield, R. B., *Tetrahedron Lett.*, 3795 (1976).

46. Kent, S. B. H., Mitchell, A. R., Engelhard, M. and Merrifield, R. B., *Proc. Natl. Acad. Sci. USA*, *76*, 2180 (1979).

47. Barany, G. and Merrifield, R. B., *J. Am. Chem. Soc.*, *99*, 7363 (1977).

48. G. Barany (Personal communication).

49. Tam, J. P., Heath, W. F. and Merrifield, R. B., *J. Am. Chem. Soc. 105*, 6442 (1983).

50. Yates, K. and McClelland, R. A., *J. Am. Chem. Soc.*, *89*, 2686 (1967).

51. Barany, G. and Merrifield, R. B., in "The Peptides" Vol. 2, E. Gross and J. Meienhofer, eds., Academic-Press, N.Y., 1979, pp. 1−284.

52. Sarin, V. K., Kent, S. B. H., Tam, J. P. and Merrifield, R. B., *Anal. Biochem.*, *117*, 147 (1981).

53. Laursen, R. A., *J. Am. Chem. Soc.*, *88*, 5344 (1966).

54. Niall, H. D., Tregear, G. W. and Jacobs, J., in "Chemistry and Biology of Peptides", J. Meienhofer, ed., Ann Arbor Press, MI., 1972, pp. 695−699.

55. Manning, J. M. and Moore, S., *J. Biol. Chem.*, *243*, 5591 (1968).

56. Cosand, W. L. and Merrifield, R. B., *Proc. Natl. Acad. Sci. USA*, *74*, 2771 (1977).

57. Wong, T. W. and Merrifield, R. B., *Biochemistry*, *19*, 3233 (1980).

58. Mojsov, S. and Merrifield, R. B., *Biochemistry*, *20*, 2950 (1981).

59. Merrifield, R. B., Vizioli, L. D. and Boman, H. G., *Biochemistry*, 21, 5020 (1982).

60. Andreu, D., Merrifield, R. B., Steiner, H. and Boman, H. G., *Proc. Natl. Acad. Sci. USA*, *80*, 6475 (1983).

61. Manning, M., Balaspiri, L., Acosta, M. and Sawyer, W. H., *J. Med. Chem.*, *16*, 975 (1973).

62. Manning, M., Klis, W. A., Olma, A., Seto, J. and Sawyer, W. H., *J. Med. Chem.*, *25*, 414 (1982).

63. Hughes, J., U.S. Patent 3926−938. December 16, 1975.

64. Lerner, R. A., *Nature*, *299*, 592 (1982).

65. Heath, W. F., unpublished results.

66. Cohen, S., *J. Biol. Chem. 237*, 1555 (1962).

67. Mantei, N., Schwarzstein, M., Streuli, M., Panem, S., Nagata, S. and Weissmann, C., *Gene, 10*, 1 (1980).

68. Levy, W. P., Rubinstein, M., Shively, J., Del Valle, V., Lai, C. Y., Moschere, J., Brink, L., Gerber, L., Stein, S. and Pestka, S., *Proc. Natl. Acad. Sci. USA*, *78*, 6186 (1981).

69. Krim, M., Mecs, I., Merrifield, E. L., Fox, F., Sarin, V. and Merrifield, R. B., TNO-ISIR Meeting, Heidelberg, Federal Republic of Germany, October 21−25, 1984.

Chemistry 1985

HERBERT A. HAUPTMAN and JEROME KARLE

for their outstanding achievements in the development of direct methods for the determination of crystal structures

THE NOBEL PRIZE FOR CHEMISTRY

Speech by Professor INGVAR LINDQVIST of the Royal Academy of Sciences.

Translation from the Swedish text

Your Majesties, Your Royal Highnesses, Ladies and Gentlemen,

The youth of to-day find it quite natural that there are such things as atoms and molecules. They have often seen molecular models in school and experienced the molecule as something obviously existing. There have been people over thousands of years who have come to the conclusion − instinctively, emotionally, or logically−that molecules should exist. These people have also had ideas about the shape and properties of the molecules.

These efforts reached their climax at the end of the 19th century in three extraordinary theories: the idea by van't Hoff about the significance of the tetrahedral carbon atom, the revelation by Kekulé of the structure of benzene and the description by Werner of many metal complexes as having octahedral, tetrahedral or planar square structures. The geniality of these ideas has been strongly confirmed in our century to an extent which proves how epoch-making these achievements were.

It is not, however, until the 20th century that the scientists have created methods which admit a complete determination of the structures of molecules. In this context structure means the geometrical arrangement of atoms as well as the bonding distances between atoms. The most important of these methods is X-ray crystallography.

In such investigations the X-rays are arranged to strike a crystal. The radiation is then scattered in certain directions and the light intensity is measured for each such scattered X-ray. Such an experiment was first made by Max von Laue who obtained the Nobel Prize for physics in 1913 for his discovery. The Braggs, father and son, made the first structure determinations of simple chemical compounds and were awarded the Nobel Prize for physics in 1915.

It was not, however, possible to determine crystal structures without some assumptions or guesses, because the phase differences between the different scattered X-rays were not known. The crystallographers had to use a trial and error method.

Several methodological improvements have since taken place, but it has for a long time been considered a great scientific achievement to determine the molecular structures of organic molecules as large as penicillin or vitamin B_{12}. As late as 1964 Dorothy Hodgkins was awarded the Nobel Prize in chemistry for such structure determinations.

It therefore was met with great interest and much opposition and discussion when Herbert Hauptman and Jerome Karle during the years 1950−56 published a series of papers in which they claimed to have found a general method, a "direct" method for solving the phase problem, thus opening the possibility to determine the structure directly from the experimental results without any further assumptions. Hauptman and Karle built their method on two established facts. One was that the electron density in a molecule can never be negative—there are electrons or there are not. The other fact was that the number of experimental results is large enough to permit application of statistical methods. Recent developments have shown that they were right and the production of modern computers has strongly contributed to the rapidity and efficiency of their methods. These methods are now so efficient that structure determinations for which the Nobel Prize was awarded in 1964 can to-day be made by a clever beginner.

At the same time it has been more and more important for the chemists to know the exact structures of molecules which take part in important chemical and biochemical reactions. One could without exaggeration say that it is only in the last ten years that chemistry has developed into a truly molecular era. Molecules with desired structures and properties can be produced and the molecular mechanism is known for increasingly more reactions.

It is this importance to chemistry which has motivated a Nobel Prize in chemistry to the mathematician Herbert Hauptman and the physicist Jerome Karle. Another way to express it is, that the imagination and ingenuity of the laureates have made it unnecessary to exercise these qualities in normal structure determinations. On the other hand they have increased the possibilities for the chemists to use their imagination and their ingenuity.

Herbert Hauptman and Jerome Karle,

Your basic development of the direct methods för X-ray crystallographic structure determination has given the chemists an efficient tool for faster and more detailed studies of the structures of molecules and therefore also for the study of chemical reactions. On behalf of the Academy I wish to convey to you our warmest congratulations and I now ask you to receive your prizes from the hands of His Majesty the King.

Herbert A. Hauptman

HERBERT A. HAUPTMAN

I was born in New York City on February 14, 1917, the oldest child of Israel Hauptman and Leah Rosenfeld. I have two brothers, Manuel and Robert.

I married Edith Citrynell on November 10, 1940. We have two daughters, Barbara (1947) and Carol (1950).

My interest in most areas of science and mathematics began at an early age, as soon as I had learned to read, and continues to this day. I obtained the B.S. degree in mathematics from the City College of New York (1937) and the M.A. degree in mathematics from Columbia University (1939).

After the war I made the decision to obtain an advanced degree and pursue a career in basic scientific research. In furtherance of these goals I commenced a collaboration with Jerome Karle at the Naval Research Laboratory in Washington, D.C. (1947) and at the same time enrolled in the Ph.D. program at the University of Maryland. The collaboration with Dr. Karle proved to be fruitful because his background in physical chemistry and mine in mathematics complemented each other nicely. Not only did this combination enable us to tackle head-on the phase problem of X-ray crystallography, but this work suggested also the topic of my doctoral dissertation, "An N-Dimensional Euclidean Algorithm". By 1954 I had received my Ph.D. degree and Dr. Karle and I had laid the foundations of the direct methods in X-ray crystallography. Our 1953 monograph, "Solution of the Phase Problem I. The Centrosymmetric Crystal", contains the main ideas, the most important of which was the introduction of probabilistic methods, in particular the joint probability distributions of several structure factors, as the essential tool for phase determination. In this monograph we introduced also the concepts of the structure invariants and seminvariants, special linear combinations of the phases, and used them to devise recipes for origin specification in all the centrosymmetric space groups. The extension to the non-centrosymmetric space groups was made some years later. The notion of the structure invariants and seminvariants proved to be of particular importance because they also serve to link the observed diffraction intensities with the needed phases of the structure factors.

In 1970 I joined the crystallographic group of the Medical Foundation of Buffalo of which I was Research Director in 1972, replacing Dr. Dorita Norton. My work on the phase problem continues to this day. During the early years of this period I formulated the neighborhood principle and extension concept, the latter independently proposed by Giacovazzo under the term "representation theory". These ideas laid the groundwork for the probabilistic theories of the higher order structure invariants and seminvariants which were further developed during the late seventies by myself and others. During the eighties I initiated work on the problem of combining the traditional techniques of direct methods with isomorphous replacement and anomalous dispersion in the attempt to facilitate the solution of macromolecular crystal structures. This work continues to the present time. More recently I have formulated the phase problem of X-ray crystallography as a minimal principle in the attempt to strengthen the existing direct methods techniques. Together with colleagues Charles Weeks, George DeTitta and others, we have made the initial applications with encouraging results.

Honors:
1. Belden Prize (Gold Medal), Mathematics, 1935.
2. RESA Award in Pure Science, 1959.
3. Co-recipient (with Jerome Karle) of the 1984 Patterson Award. Presented at the American Crystallography Association in Lexington, Kentucky, on May 21, 1984.
4. Co-recipient (with Jerome Karle) of the Nobel Prize in Chemistry, 1985.
5. Honorary degree, Doctor of Science, University of Maryland, 1985.
6. Citizen of the Year Award, Buffalo Evening News, April 1986.
7. Inducted into Nobel Hall of Science, Museum of Science and Industry, Chicago, Illinois, April 1986.
8. Recipient of the Norton Medal, SUNY, May 1986.
9. Schoellkopf Award, American Chemical Society, May 1986.
10. Honorary Doctor of Science Degree, CCNY, May 1986.
11. Gold Plate Award, American Academy of Achievement, Salute to Excellence Weekend, Washington, D.C., June 1986.
12. Townsend Harris Medal for 1986, City College of New York, October 1986.
13. Recipient of Medal from Jewish Academy of Arts and Sciences, November 1986.
14. Recipient of the National Library of Medicine Medal, November 1986.
15. Western New York Man of the Year Award, Buffalo Chamber of Commerce, 1986.
16. Honorary Member Phi Beta Kappa, May 1987.
17. Induction as a Fellow of the Jewish Academy of Arts and Sciences.
18. 1987 Honoree, Western New Yorker of the Year, January 1987.
19. Recipient of the Cooke Award, State Univ. of New York at Buffalo, October 1987.
20. Elected to the U.S. National Academy of Sciences, 1988.
21. Honorary Doctorate in Chemistry, Univ. of Parma, Italy 1989.
22. Honorary Doctor of Science Degree, D'Youville College, Buffalo, NY (1989).
23. Elected Member of Townsend Harris Hall of Fame (1989).
24. Honorary Doctor of Science Degree, Honoris Causa, Bar-Ilan Univ., Israel (1990).

25. Honorary Doctor of Science Degree, Honoris Causa, Columbia University, New York, NY, (1990).

26. Dirac Medal for the Advancement of Theoretical Physics, University of New South Wales, Australia, January 1991.

DIRECT METHODS AND ANOMALOUS DISPERSION

Nobel lecture, 9 December, 1985

by

HERBERT A. HAUPTMAN

Medical Foundation of Buffalo, Inc. 73 High Street, BUFFALO, N. Y. 14 203

1. INTRODUCTION

The electron density function, $\varrho(\mathbf{r})$, in a crystal determines its diffraction pattern, i.e. both the magnitudes and phases of its x-ray diffraction maxima, and conversely. If, however, as is always the case, only magnitudes are available from the diffraction experiment, then the density function $\varrho(\mathbf{r})$ cannot be recovered. If one invokes prior structural knowledge, usually that the crystal is composed of discrete atoms of known atomic numbers, then the observed magnitudes are, in general, sufficient to determine the positions of the atoms, i.e. the crystal structure.

It should be noted here that the recognition that observed diffraction data are in general sufficient to determine crystal structures uniquely was an important milestone in the development of the direct methods of crystal structure determination. The erroneous contrary view, that crystal structures could not, even in principle, be deduced from diffraction intensities, had long been held by the crystallographic community prior to c.1950 and constituted a psychological barrier which first had to be removed before real progress could be made.

2. THE TRADITIONAL DIRECT METHODS

2.1. *The phase problem.*
Denote by $\phi_{\mathbf{H}}$ the phase of the structure factor $F_{\mathbf{H}}$:

$$F_{\mathbf{H}} = |F_{\mathbf{H}}| \exp (i\phi_{\mathbf{H}}), \tag{1}$$

where $\mathbf{H}$ is a reciprocal lattice vector (having three integer components) which labels the corresponding diffraction maximum. Then the relationship between the structure factors $F_{\mathbf{H}}$ and the electron density function $\varrho(\mathbf{r})$ is given by

$$F_{\mathbf{H}} = \int_V \rho(\mathbf{r}) \exp(2\pi i \mathbf{H}\cdot\mathbf{r})dV \tag{2}$$

and

$$\rho(\mathbf{r}) = \frac{1}{V} \sum_{\mathbf{H}} F_{\mathbf{H}}\exp(-2\pi i\mathbf{H}\cdot\mathbf{r}) = \frac{1}{V} \sum_{\mathbf{H}}|F_{\mathbf{H}}|\exp i(\phi_{\mathbf{H}}-2\pi\mathbf{H}\cdot\mathbf{r}) \tag{3}$$

in which V represents the unit cell or its volume. Thus the structure factors F_H determine $\varrho(\mathbf{r})$. The x-ray diffraction experiment yields only the magnitudes $|F_H|$ of a finite number of structure factors, but the values of the phases ϕ_H, which are also needed if one is to determine $\varrho(\mathbf{r})$ from (3), cannot be determined experimentally. If arbitrary values for the phases ϕ_H are specified in Eq. (3), then density functions $\varrho(\mathbf{r})$ are defined which, when substituted into (2) yield structure factors F_H the magnitudes of which agree with the observed magnitudes $|F_H|$. It follows that the diffraction experiment does not determine $\varrho(\mathbf{r})$. It was this argument which led crystallographers, prior to 1950, to the erroneous conclusion that diffraction intensities could not, even in principle, determine crystal structures uniquely. What had been overlooked was the fact that the phases ϕ_H could not be arbitrarily specified if (3) is to yield density functions characteristic of real crystals.

Crystals are composed of discrete atoms. One exploits this prior structural knowledge by replacing the real crystal, with continuous electron density $\varrho(\mathbf{r})$, by an ideal one, the unit cell of which consists of N discrete, non-vibrating, point atoms located at the maxima of $\varrho(\mathbf{r})$. Then the structure factor F_H is replaced by the normalized structure factor E_H and (1) to (3) are replaced by

$$E_H = |E_H| \exp(i\phi_H), \tag{4}$$

$$E_H = \frac{1}{\sigma_2^{1/2}} \sum_{j=1}^{N} f_j \exp(2\pi i \mathbf{H} \cdot \mathbf{r}_j), \tag{5}$$

$$\langle E_H \exp(-2\pi i \mathbf{H} \cdot \mathbf{r}) \rangle_H = \frac{1}{\sigma_2^{1/2}} \left\langle \sum_{j=1}^{N} f_j \exp\left[2\pi i \mathbf{H} \cdot (\mathbf{r}_j - \mathbf{r})\right] \right\rangle_H$$

$$= \frac{f_j}{\sigma_2^{1/2}} \text{ if } \mathbf{r} = \mathbf{r}_j$$

$$= 0 \text{ if } \mathbf{r} \neq \mathbf{r}_j \tag{6}$$

where f_j is the zero-angle atomic scattering factor, $\mathbf{r}_j$ is the position vector of the atom labelled j, and

$$\sigma_n = \sum_{j=1}^{N} f_j^n, \quad n = 1, 2, 3, \dots \tag{7}$$

In the x-ray diffraction case the f_j are equal to the atomic numbers Z_j and are presumed to be known. From (6) it follows that the normalized structure factors E_H determine the atomic position vectors $\mathbf{r}_j$, j = 1,2,..., N, i.e. the crystal structure.

In practice a finite number of magnitudes $|E_H|$ of normalized structure factors E_H are obtainable (at least approximately) from the observed magnitudes $|F_H|$ while the phases ϕ_H, as defined by (4) and (5), cannot be determined experimentally. Since one now requires only the 3N components of the N position vectors $\mathbf{r}_j$, rather than the much more complicated electron density function $\varrho(\mathbf{r})$, it turns out that, in general, the known magnitudes are more than sufficient. This is most readily seen by equating the magnitudes of both sides of (5) in order to obtain a system of equations in which the only unknowns are the 3N components of the position vectors $\mathbf{r}_j$. Since the number of such equations, equal to the number of reciprocal lattice vectors $\mathbf{H}$ for which magnitudes $|E_\mathbf{H}|$ are available, usually greatly exceeds the number, 3N, of unknowns, this system is redundant. Thus observed diffraction intensities usually over-determine the crystal structure, i.e. the positions of the atoms in the unit cell. In short, by merely replacing the integral of Eq. (2) by the summation of Eq. (5), i.e. taking Eq. (5) as the starting point of our investigation rather than Eq. (2), one has transformed the problem from an unsolvable one to one which is solvable, at least in principle.

In summary then, the intensities (or magnitudes $|E_\mathbf{H}|$) of a sufficient number of x-ray diffraction maxima determine a crystal structure. The available intensities usually exceed the number of parameters needed to describe the structure. From these intensities a set of numbers $|E_\mathbf{H}|$ can be derived, one corresponding to each intensity. However, the elucidation of the crystal structure requires also a knowledge of the complex numbers $E_\mathbf{H} = |E_\mathbf{H}| \exp(i\phi_\mathbf{H})$, the normalized structure factors, of which only the magnitudes $|E_\mathbf{H}|$ can be determined from experiment. Thus a "phase" $\phi_\mathbf{H}$, unobtainable from the diffraction experiment, must be assigned to each $|E_\mathbf{H}|$, and the problem of determining the phases when only the magnitudes $|E_\mathbf{H}|$ are known is called the "phase problem". Owing to the known atomicity of crystal structures and the redundancy of observed magnitudes $|E_\mathbf{H}|$, the phase problem is solvable in principle.

2.2. *The structure invariants*

Equation (6) implies that the normalized structure factors $E_\mathbf{H}$ determine the crystal structure. However (5) does not imply that, conversely, the crystal structure determines the values of the normalized structure factors $E_\mathbf{H}$ since the position vectors $\mathbf{r}_j$ depend not only on the structure but on the choice of origin as well. It turns out nevertheless that the magnitudes $|E_\mathbf{H}|$ of the normalized structure factors are in fact uniquely determined by the crystal structure and are independent of the choice of origin but that the values of the phases $\phi_\mathbf{H}$ depend also on the choice of origin. Although the values of the individual phases depend on the structure and the choice of origin, there exist certain linear combinations of the phases, the so-called structure invariants, whose values are determined by the structure alone and are independent of the choice of origin.

It follows readily from Eq. (5) that the linear combination of three phases

$$\psi_3 = \phi_\mathbf{H} + \phi_\mathbf{K} + \phi_\mathbf{L} \tag{8}$$

is a structure invariant (triplet) provided that

$$\mathbf{H} + \mathbf{K} + \mathbf{L} = 0; \tag{9}$$

the linear combination of four phases

$$\psi_4 = \phi_H + \phi_K + \phi_L + \phi_M \tag{10}$$

is a structure invariant (quartet) provided that

$$H + K + L + M = 0; \tag{11}$$

etc.

2.3. *The structure seminvariants*

If a crystal possesses elements of symmetry then the origin may not be chosen arbitrarily if the simplifications permitted by the space group symmetries are to be realized. For example, if a crystal has a centre of symmetry it is natural to place the origin at such a centre while if a two-fold screw axis, but no other symmetry element is present, the origin would normally be situated on this symmetry axis. In such cases the permissible origins are greatly restricted and it is therefore plausible to assume that many linear combinations of the phases will remain unchanged in value when the origin is shifted only in the restricted ways allowed by the space group symmetries. One is thus led to the notion of the structure seminvariant, those linear combinations of the phases whose values are independent of the choice of permissible origin.

If the only symmetry element is a centre of symmetry, for example (space group P$\bar{1}$), then it turns out (again from Eq. (5)) that a single phase ϕ_H is a structure seminvariant provided that the three components of the reciprocal lattice vector H are even integers; the linear combination of two phases $\phi_H + \phi_K$ is a structure seminvariant provided that the three components of $H + K$ are even integers; etc.

If the only symmetry element is a two-fold rotation axis (or twofold screw axis) then one finds from Eq. (5) that the single phase ϕ_{hkl} is a structure seminvariant provided that h and l are even integers and k = 0; the linear combination of two phases

$$\phi_{h_1 k_1 l_1} + \phi_{h_2 k_2 l_2}$$

is a structure seminvariant provided that $h_1 + h_2$ and $l_1 + l_2$ are even and $k_1 + k_2 = 0$; etc.

The structure invariants and seminvariants have been tabulated for all the space groups (Hauptman and Karle 1953, 1956, 1959; Karle and Hauptman 1961; Lessinger and Wondratschek 1975). In general the collection of structure invariants is a subset of the collection of structure seminvariants. If no element of symmetry is present, that is the space group is P1, then the two classes coincide.

2.3.1. Origin and enantiomorph specification

The theory of the structure seminvariants leads in a natural way to space group dependent recipes for origin and enantiomorph (i.e. the handedness, right or left) specification.

In general the theory identifies an appropriate set of phases whose values are to be specified in order to fix the origin uniquely. For example, in space group Pl (no elements of symmetry) the values of any three phases

$$\phi_{h_1k_1l_1}, \phi_{h_2k_2l_2}, \phi_{h_3k_3l_3}, \tag{12}$$

for which the determinant $\triangle$ satisfies

$$\Delta = \begin{vmatrix} h_1k_1l_1 \\ h_2k_2l_2 \\ h_3k_3l_3 \end{vmatrix} = \pm 1, \tag{13}$$

may be specified arbitrarily, thus fixing the origin uniquely. Once this is done then the value of any other phase is uniquely determined by the structure alone. For enantiomorph specification it is sufficient to specify arbitrarily the sign of any enantiomorph sensitive structure invariant, i.e. one whose value is different from 0 or π. (See Hauptman 1972, pages 28−52, for further details.)

In the space group P$\bar{1}$ one again specifies arbitrarily the value (0 or π) of three phases (12), but now the condition is that the determinant Δ [defined by (13)] be odd. Similar recipes for all the space groups are now known and are to be found in the literature cited.

2.4. *The fundamental principle of direct methods*

It is known that the values of a sufficiently extensive set of cosine seminvariants (the cosines of the structure seminvariants) lead unambiguously to the values of the individual phases (Hauptman 1972). Magnitudes $|E|$ are capable of yielding estimates of the cosine seminvariants only or, equivalently, the magnitudes of the structure seminvariants; the signs of the structure seminvariants are ambiguous because the two enantiomorphous structures permitted by the observed magnitudes $|E|$ correspond to two values of each structure seminvariant differing only in sign. However, once the enantiomorph has been selected by specifying arbitrarily the sign of a particular enantiomorph sensitive structure seminvariant (i.e. one different from 0 or π), then the magnitudes $|E|$ determine both signs and magnitudes of the structure seminvariants consistent with the chosen enantiomorph. Thus, for fixed enantiomorph, the observed magnitudes $|E|$ determine unique values for the structure seminvariants; the latter, in turn, lead to unique values of the individual phases. In short, the structure seminvariants serve to link the observed magnitudes $|E|$ with the desired phases ϕ (the fundamental principle of direct methods). It is this property of the structure seminvariants which accounts for their importance and which justifies the stress placed on them here.

By the term "direct methods" is meant that class of methods which exploits relationships among the structure factors in order to go directly from the observed magnitudes $|E|$ to the needed phases ϕ.

2.5. *The neighborhood principle*

It has long been known that, for fixed enantiomorph, the value of any structure seminvariant ψ is, in general, uniquely determined by the magnitudes $|E|$ of the normalized structure factors. In recent years it has become clear that, for fixed enantiomorph, there corresponds to ψ one or more small sets of magnitudes |E|, the neighborhoods of ψ, on which, in favorable cases, the value of ψ most sensitively depends; that is to say that, in favorable cases, ψ is primarily determined by the values of |E| in any of its neighborhoods and is relatively insensitive to the values of the great bulk of remaining magnitudes. The conditional probability distribution of ψ, assuming as known the magnitudes |E| in any of its neighborhoods, yields an estimate for ψ which is particularly good in the favorable case that the variance of the distribution happens to be small [the neighborhood principle (Hauptman, 1975a,b)].

The study of appropriate probability distributions (compare § 2.7) leads directly to the definition of the neighborhoods of the structure invariants. Definitions are given here only for the triplet ψ₃ and the quartet ψ₄, but recipes for defining the neighborhoods of all the structure invariants are now known (Hauptman 1977a,b, Fortier & Hauptman, 1977).

2.5.1. The first neighborhood of the triplet ψ₃

Let **H**, **K**, **L** be three reciprocal lattice vectors which satisfy Eq. (9). Then ψ₃, Eq. (8), is a structure invariant and its first neighborhood is defined to consist of the three magnitudes:

$$|E_\mathbf{H}|, |E_\mathbf{K}|, |E_\mathbf{L}|. \tag{14}$$

2.5.2. Neighborhoods of the quartet ψ₄
2.5.2.1. The first neighborhood

Let **H**, **K**, **L**, **M** be four reciprocal lattice vectors which satisfy Eq. (11). Then ψ₄, Eq. (10), is a structure invariant and its first neighborhood is defined to consist of the four magnitudes:

$$|E_\mathbf{H}|, |E_\mathbf{K}|, |E_\mathbf{L}|, |E_\mathbf{M}|. \tag{15}$$

The four magnitudes (15) are said to be the main terms of the quartet ψ₄.

2.5.2.2. The second neighborhood

The second neighborhood of the quartet ψ₄ is defined to consist of the four magnitudes (15) plus the three additional magnitudes:

$$|E_{\mathbf{H}+\mathbf{K}}|, |E_{\mathbf{K}+\mathbf{L}}|, |E_{\mathbf{L}+\mathbf{H}}|, \tag{16}$$

i.e. seven magnitudes |E| in all. The three magnitudes (16) are said to be the cross-terms of the quartet ψ₄.

2.6. *The extension concept*

By embedding the structure seminvariant T and its symmetry related variants in suitable structure invariants Q one obtains the extensions Q of the seminvariant T. Owing to the space group dependent relations among the phases, T is related in a known way to its extensions. In this way the theory of the structure seminvariants is reduced to that of the structure invariants. In particular, the neighborhoods of T are defined in terms of the neighborhoods of its extensions. The procedure will be illustrated in some detail only for the two-phase structure seminvariant in the space group $P\bar{1}$ which serves as the prototype for the structure seminvariants in general, in all space groups, noncentrosymmetric as well as centrosymmetric.

2.6.1. The two-phase structure seminvariant in $P\bar{1}$

It has already been seen (§ 2.3) that the linear combination of two phases

$$T = \phi_H + \phi_K \tag{17}$$

is a structure seminvariant in $P\bar{1}$ if and only if the three components of the reciprocal lattice vector $\mathbf{H} + \mathbf{K}$ are all even. Then the components of each of the four reciprocal lattice vectors $\frac{1}{2}(\pm \mathbf{H} \pm \mathbf{K})$ are all integers. Note also that in this space group the structure factors are real and all phases are O or π.

2.6.2. The extensions of T

One embeds the two-phase structure seminvariant T (17) and its symmetry related variant

$$T_1 = \phi_{-H} + \phi_K \tag{18}$$

in the respective quartets

$$Q = T + \phi_{-\frac{1}{2}(\mathbf{H} + \mathbf{K})} + \phi_{-\frac{1}{2}(\mathbf{H} + \mathbf{K})}, \tag{19}$$

$$Q_1 = T_1 + \phi_{-\frac{1}{2}(-\mathbf{H} + \mathbf{K})} + \phi_{-\frac{1}{2}(-\mathbf{H} + \mathbf{K})} \tag{20}$$

In view of (17) and (18) and the space group-dependent relationships among the phases it is readily verified that Q and Q_1 are in fact (special) four-phase structure invariants (quartets) and

$$T = T_1 = Q = Q_1. \tag{21}$$

The quartets Q and Q_1 are said to be the extensions of the seminvariant T. In this way the theory of the two-phase structure seminvariant T is reduced to that of the quartets. In particular, the neighborhoods of T are defined in terms of the neighborhoods of the quartet.

2.6.3. The first neighborhoods of the extensions

Since two of the phases of the quartet Q (19) are identical, only three of the four main terms are distinct. The first neighborhood of Q is accordingly defined to consist of the three magnitudes

$$\left[\text{since } \left|E_{-\frac{1}{2}(\mathbf{H}+\mathbf{K})}\right| = \left|E_{\frac{1}{2}(\mathbf{H}+\mathbf{K})}\right| \right]:$$

$$|E_{\mathbf{H}}|, \ |E_{\mathbf{K}}|, \ |E_{\frac{1}{2}(\mathbf{H}+\mathbf{K})}|. \tag{22}$$

In a similar way the first neighborhood of the extension Q_1, (20), is defined to consist of the three magnitudes

$$|E_{\mathbf{H}}|, \ |E_{\mathbf{K}}|, \ |E_{\frac{1}{2}(\mathbf{H}-\mathbf{K})}|. \tag{23}$$

2.6.4. The first neighborhood of T

The first neighborhood of the two-phase structure seminvariant T is defined to consist of the set-theoretic union of the first neighborhoods of its extensions, i.e., in view of (22) and (23), of the four magnitudes

$$|E_{\mathbf{H}}|, \ |E_{\mathbf{K}}|, \ |E_{\frac{1}{2}(\mathbf{H}+\mathbf{K})}|, \ |E_{\frac{1}{2}(\mathbf{H}-\mathbf{K})}|. \tag{24}$$

2.7. *The solution strategy*

One starts with the system of equations (5). By equating real and imaginary parts of (5) one obtains two equations for each reciprocal lattice vector $\mathbf{H}$. The magnitudes $|E_{\mathbf{H}}|$ and the atomic scattering factors f_j are presumed to be known. The unknowns are the atomic position vectors r_j and the phases $\phi_{\mathbf{H}}$. Owing to the redundancy of the system (5), one naturally invokes probabilistic techniques in order to eliminate the unknown position vectors r_j, and in this way to obtain relationships among the unknown phases $\phi_{\mathbf{H}}$ having probabilistic validity.

Choose a finite number of reciprocal lattice vectors $\mathbf{H}, \mathbf{K}, \ldots$ in such a way that the linear combination of phases

$$\psi = \phi_{\mathbf{H}} + \phi_{\mathbf{K}} + \ldots \tag{25}$$

is a structure invariant or seminvariant whose value we wish to estimate. Choose satellite reciprocal lattice vectors H', K', ... in such a way that the collection of magnitudes

$$|E_{\mathbf{H}}|, \ |E_{\mathbf{K}}|, \ldots; \ |E_{\mathbf{H}}'|, \ |E_{\mathbf{K}}'|, \ldots \tag{26}$$

constitutes a neighborhood of ψ. The atomic position vectors r_j are assumed to be the primitive random variables which are uniformly and independently distributed. Then the magnitudes $|E_{\mathbf{H}}|, |E_{\mathbf{K}}|, \ldots; |E_{\mathbf{H}}', |E_{\mathbf{K}}'|, \ldots;$ and phases $\phi_{\mathbf{H}}, \phi_{\mathbf{K}}, \ldots; \phi_{\mathbf{H}}', \phi_{\mathbf{K}}', \ldots$ of the complex, normalized structure factors $E_{\mathbf{H}}, E_{\mathbf{K}}, \ldots; E_{\mathbf{H}}', E_{\mathbf{K}}', \ldots$, as functions [(Eq. (5)] of the position vectors r_j, are themselves random variables, and their joint probability distribution P may be obtained. From the distribution P one derives the conditional joint probability distribution

$$P\ (\Phi_H,\ \Phi_K,...\ |\ |E_H|,\ |E_K|,...;\ |E_{H'}|,\ |E_{K'}|,...),\qquad (27)$$

of the phases ϕ_H, ϕ_K,..., given the magnitudes $|E_H|$, $|E_K|$,...; $|E_{H'}|$, $|E_{K'}|$,..., by fixing the known magnitudes, integrating with respect to the unknown phases ϕ_H', ϕ_K',... from O to 2π, and multiplying by a suitable normalizing parameter. The distribution (27) in turn leads directly to the conditional probability distribution

$$P(\Psi|\ |E_H|,\ |E_K|,...;\ |E_{H'}|,\ |E_{K'}|,...)\qquad (28)$$

of the structure invariant or seminvariant ψ assuming as known the magnitudes (26) constituting a neighborhood of ψ. Finally, the distribution (28) yields an estimate for ψ which is particularly good in the favorable case that the variance of (28) happens to be small.

2.8. *Estimating the triplet in P1*

Let the three reciprocal lattice vectors **H**, **K**, and **L** satisfy (9). Refer to § 2.5.1 for the first neighborhood of the triplet ψ_3 (Eq. (8)] and to § 2.7 for the probabilistic background.

Suppose that R_1, R_2, and R_3 are three specified non-negative numbers. Denote by

$$P_{1/3} = P(\Psi|R_1, R_2, R_3)$$

the conditional probability distribution of the triplet ψ_3, given the three magnitudes in its first neighborhood:

$$|E_H| = R_1, |E_K| = R_2, |E_L| = R_3.\qquad (29)$$

Then, carrying out the program described in § 2.7, one finds (Cochran, 1955)

$$P_{1/3} = P(\Psi|R_1, R_2, R_3) \approx \frac{1}{2\pi I_0(A)}\ \exp\ (A \cos \Psi)\qquad (30)$$

where

$$A = \frac{2\sigma_3}{\sigma_2^{3/2}}\ R_1\ R_2\ R_3,\qquad (31)$$

I_O is the modified Bessel function, and σ_n is defined by (7). Since $A>O, P_{1/3}$ has a unique maximum at $\Psi = 0$, and it is clear that the larger the value of A the smaller is the variance of the distribution. See Figure 1, where $A = 2.316$, Figure 2, where $A = 0.731$. Hence in the favorable case that A is large, say, for example, $A>3$, the distribution leads to a reliable estimate of the structure invariant ψ_3, zero in this case:

$$\psi_3 \approx O \text{ if A is large.}\qquad (32)$$

Furthermore, the larger the value of A, the more likely is the probabilistic statement (32). It is remarkable how useful this relationship has proven to be in the applications; and yet (32) is severely limited because it is capable of yielding only the zero estimate for ψ_3, and only those estimates are reliable for which A is large, the favorable cases.

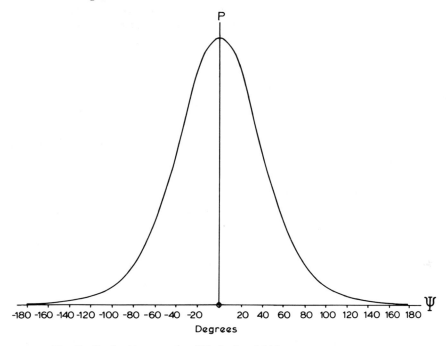

Figure 1. The distribution $P_{1/3}$, equation (30), for A = 2.316.

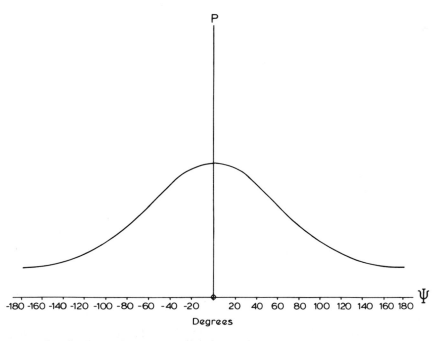

Figure 2. The distribution $P_{1/3}$, equation (30), for A = 0.731

It should be mentioned in passing that a distribution closely related to (30) leads directly to the so-called tangent formula (Karle & Hauptman, 1956) which is universaily used by direct methods practitioners:

$$\tan \phi_h = \frac{\langle |E_K E_{h-K}| \sin (\phi_K + \phi_{h-K}) \rangle_K}{\langle |E_K E_{h-K}| \cos (\phi_K + \phi_{h-K}) \rangle_K} ,\qquad (33)$$

in which **h** is a fixed reciprocal lattice vector, the averages are taken over the same set of vectors **K** in reciprocal space, usually restricted to those vectors **K** for which $|E_K|$ and $|E_{h-K}|$ are both large, and the sign of $\sin \phi_h$ ($\cos \phi_h$) is the same as the sign of the numerator (denominator) on the right hand side. The tangent formula is usually used to refine and extend a basis set of phases, presumed to be known.

2.9. *Estimating the quartet in P1*

Two conditional probability distributions are described, one assuming as known the four magnitudes $|E|$ in the first neighborhood of the quartet, the second assuming as known the seven magnitudes $|E|$ in its second neighborhood.

2.9.1. The first neighborhood

Suppose that **H**, **K**, **L**, and **M** are four reciprocal lattice vectors which satisfy (11). Refer to § 2.5.2.1 for the first neighborhood of the quartet ψ_4 (10) and to § 2.7 for the probabilistic background. Suppose that R_1, R_2, R_3, and R_4 are four specified non-negative numbers. Denote by

$$P_{1/4} = P(\Psi | R_1, R_2, R_3, R_4)$$

the conditional probability distribution of the quartet ψ_4, given the four magnitudes in its first neighborhood:

$$|E_H| = R_1, |E_K| = R_2, |E_L| = R_3, |E_M| = R_4. \qquad (34)$$

Then

$$P_{1/4} = P(\Psi | R_1, R_2, R_3, R_4) \approx \frac{1}{2\pi I_0(B)} \exp (B \cos \Psi) \qquad (35)$$

where

$$B = \frac{2\sigma_4}{\sigma_2^2} R_1 R_2 R_3 R_4, \qquad (36)$$

and σ_n is defined by (7). Thus $P_{1/4}$ is identical with $P_{1/3}$, but B replaces A. Hence similar remarks apply to $P_{1/4}$. In particular, (35) always has a unique maximum at $\Psi = O$ so that the most probable value of ψ_4, given the four magnitudes (34) in its first neighborhood, is zero, and the larger the value of B the more likely it is that $\psi_4 = O$. Since B values, of order $1/N$, tend to be less than A values, of order $1/\sqrt{N}$, at least for large values of N, the estimate (zero) of ψ_4 is in general less reliable than the estimate (zero) of ψ_3. Hence the goal of obtaining a reliable non-zero estimate for a structure invariant is not realized by (35). The decisive step in this direction is made next.

2.9.2. The second neighborhood
Employ the same notation as in § 2.9.1 but refer now to § 2.5.2.2 for the second neighborhood of the quartet ψ_4. Suppose that R_1, R_2, R_3, R_4, R_{12}, R_{23}, and R_{31}, are seven non-negative numbers. Denote by

$$P_{1/7} = P(\Psi|R_1, R_2, R_3, R_4; R_{12}, R_{23}, R_{31})$$

the conditional probability distribution of the quartet ψ_4, given the seven magnitudes in its second neighborhood:

$$|E_H| = R_1, |E_K| = R_2, |E_L| = R_3, |E_M| = R_4; \tag{37}$$

$$|E_{H+K}| = R_{12}, |E_{K+L}| = R_{23}, |E_{L+H}| = R_{31}. \tag{38}$$

Then (Hauptman, 1975 a, b; 1976)

$$P_{1/7} \approx \frac{1}{L} \exp\left(-2B'\cos\Psi\right) I_0\left[\frac{2\sigma_3}{\sigma_2^{3/2}} R_{12}X_{12}\right] I_0\left[\frac{2\sigma_3}{\sigma_2^{3/2}} R_{23}X_{23}\right] \times$$

$$I_0\left[\frac{2\sigma_3}{\sigma_2^{3/2}} R_{31}X_{31}\right], \tag{39}$$

where

$$B' = \frac{1}{\sigma_2^3} (3\sigma_3^2 - \sigma_2\sigma_4) R_1R_2R_3R_4, \tag{40}$$

$$X_{12} = [R_1^2R_2^2 + R_3^2R_4^2 + 2R_1R_2R_3R_4\cos\Psi]^{1/2}, \tag{41}$$

$$X_{23} = [R_2^2R_3^2 + R_1^2R_4^2 + 2R_1R_2R_3R_4\cos\Psi]^{1/2}, \tag{42}$$

$$X_{31} = [R_3^2R_1^2 + R_2^2R_4^2 + 2R_1R_2R_3R_4\cos\Psi]^{1/2}, \tag{43}$$

σ_n is defined by (7), and L is a normalizing parameter, independent of Ψ, which is not needed for the present purpose.

Figures 3–5 show the distribution (39) (solid line ——) for typical values of the seven parameters (37) and (38). For comparison the distribution (35) (broken line – – –) is also shown. Since the magnitudes |E| have been obtained from a real structure with N = 29, comparison with the true value of the quartet is also possible. As already emphasized, the distribution (35) always has a unique maximum at $\Psi = O$. The distribution (39), on the other hand, may have a maximium at $\Psi = O$, or π, or any value between these extremes, as shown by Figures 3–5. Roughly speaking, the maximum of (39) occurs at 0 or π according as the three parameters R_{12}, R_{23}, R_{31} are all large or all small, respectively. These figures also clearly show the improvement which may result when, in addition to the four magnitudes (37), the three magnitudes (38) are also assumed to be known. Finally, in the special case that

$$R_{12} \approx R_{23} \approx R_{31} \approx O \tag{44}$$

the distribution (39) reduces to

$$P_{1/7} \approx \frac{1}{L} \exp\left(-2B'\cos \Psi\right), \tag{45}$$

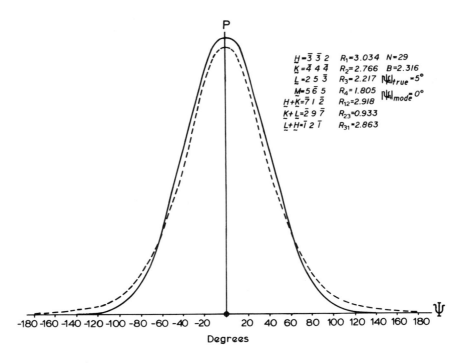

Figure 3. The distribution (39) (——) and (35) (– – –) for the values of the seven parameters (37) and (38) shown. The mode of (39) is O, of (35) always O.

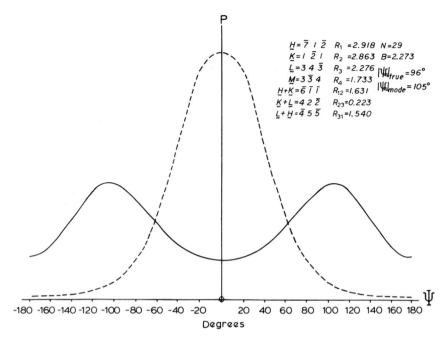

$\underline{H} = \overline{7}\ I\ \overline{2}$ $R_1 = 2.918$ $N = 29$
$\underline{K} = I\ \overline{2}\ I$ $R_2 = 2.863$ $B = 2.273$
$\underline{L} = 3\ 4\ \overline{3}$ $R_3 = 2.276$
$\underline{M} = 3\ \overline{3}\ 4$ $R_4 = 1.733$ $|\Psi|_{true} = 96°$
$\underline{H} + \underline{K} = \overline{6}\ \overline{I}\ \overline{I}$ $R_{12} = 1.631$ $|\Psi|_{mode} = 105°$
$\underline{K} + \underline{L} = 4\ 2\ \overline{2}$ $R_{23} = 0.223$
$\underline{L} + \underline{H} = \overline{4}\ 5\ \overline{5}$ $R_{31} = 1.540$

Figure 4. The distribution (39) (——) and (35) (– – –) for the values of the seven parameters (37) and (38) shown. The mode of (39) is 105°, of (35) always O.

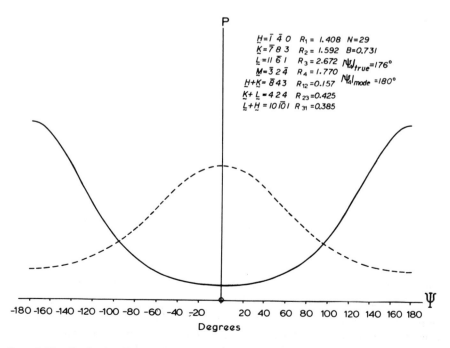

$\underline{H} = \overline{I}\ \overline{4}\ 0$ $R_1 = 1.408$ $N = 29$
$\underline{K} = \overline{7}\ 8\ 3$ $R_2 = 1.592$ $B = 0.731$
$\underline{L} = II\ \overline{6}\ I$ $R_3 = 2.672$ $|\Psi|_{true} = 176°$
$\underline{M} = \overline{3}\ 2\ \overline{4}$ $R_4 = 1.770$ $|\Psi|_{mode} = 180°$
$\underline{H} + \underline{K} = \overline{6}\ 4\ 3$ $R_{12} = 0.157$
$\underline{K} + \underline{L} = 4\ 2\ 4$ $R_{23} = 0.425$
$\underline{L} + \underline{H} = 10\ \overline{I0}\ I$ $R_{31} = 0.385$

Figure 5. The distribution (39) (——) and (35) (– – –) for he values of the seven parameters (37) and (38) shown. The mode of (39) is 180°, of (35) always O.

which has a unique maximum at $\Psi = \pi$ (Fig. 5).

2.10. *Estimating the two-phase structure seminvariant in $P\bar{I}$*

Suppose that $\mathbf{H}$ and $\mathbf{K}$ are two reciprocal lattice vectors such that the three components of $\mathbf{H} + \mathbf{K}$ are even integers. Then the linear combination T of two phases (17) is a structure seminvariant. Refer to § 2.6.4 for the four magnitudes (24) in the first neighborhood of T and to § 2.7 for the probabilistic background. Suppose that R_1, R_2, r_{12}, and $r_{1\bar{2}}$ are four non-negative numbers. In this space group every phase is O or π so that $T = 0$ or π and the conditional probability distribution of T, assuming as known the four magnitudes in its first neighborhood, is discrete. Denote by $P_+(P_-)$ the conditional probability that $T = O$ (π), given the four magnitudes in its first neighborhood:

$$|E_{\mathbf{H}}| = R_1, |E_{\mathbf{K}}| = R_2, \left|E_{\frac{1}{2}(\mathbf{H} + \mathbf{K})}\right| = r_{12}, \left|E_{\frac{1}{2}(\mathbf{H} - \mathbf{K})}\right| = r_{1\bar{2}}. \tag{46}$$

In the special case that all N atoms in the unit cell are identical, the solution strategy described in § 2.7 leads to (Green & Hauptman, 1976)

$$P_{\pm} \approx \frac{1}{M} \exp\left\{ \mp \frac{R_1 R_2 (r_{12}^2 + r_{1\bar{2}}^2)}{2N} \right\} \cosh\left\{ \frac{r_{12} r_{1\bar{2}} (R_1 \pm R_2)}{N^{1/2}} \right\} \tag{47}$$

where upper (lower) signs go together and

$$M = \exp\left\{ -\frac{R_1 R_2 (r_{12}^2 + r_{1\bar{2}}^2)}{2N} \right\} \cosh\left\{ \frac{r_{12} r_{1\bar{2}} (R_1 + R_2)}{N^{1/2}} \right\} +$$

$$\exp\left\{ +\frac{R_1 R_2 (r_{12}^2 + r_{1\bar{2}}^2)}{2N} \right\} \cosh\left\{ \frac{r_{12} r_{1\bar{2}} (R_1 - R_2)}{N^{1/2}} \right\}. \tag{48}$$

It is easily verified that, under the assumption that R_1 and R_2 are both large, $P_+ >> 1/2$ if r_{12} and $r_{1\bar{2}}$ are both large, but $P_+ << 1/2$ if one of r_{12}, $r_{1\bar{2}}$ is large and the other is small. Hence $T \approx O$ or π respectively (the favorable cases of the neighborhood principle for T).

3. COMBINING DIRECT METHODS WITH ANOMALOUS DISPERSION

3.1. *Introduction*

The overview of the traditional direct methods described in §§ 1 and 2 is readily generalized to the case that the atomic scattering factors are arbitrary complex-valued functions of $(\sin \theta)/\lambda$, thus including the special case that one or more anomalous scatterers are present. Once again the neighborhood concept plays an essential role. Final results from the probabilistic theory of the two- and three-phase structure invariants are briefly summarized. In particular, the conditional probability distributions of the two- and three-phase structure invariants, given the magnitudes |E| in their first neighborhoods, are described. The distributions yield estimates for these invariants which are particularly good in those cases that the variances of the distributions happen to be small (the neighborhood principle). It is particularly noteworthy that these estimates are unique in the whole range from $-\pi$ to $+\pi$. An example shows that the method is capable of yielding unique estimates for tens of thousands of structure invariants with unprecedented accuracy, even in the macromolecular case. It thus appears that this fusion of the traditional techniques of direct methods with anomalous dispersion will facilitate the solution of those crystal structures which contain one or more anomalous scatterers.

Most crystal structures containing as many as $80-100$ independent nonhydrogen atoms are more or less routinely solvable nowadays by direct methods. On the other hand, it has been known for a long time (Peerdeman & Bijvoet, 1956; Ramachandran & Raman, 1956; Okaya & Pepinsky, 1956) that the presence of one or more anomalous scatterers facilitates the solution of the phase problem; and some recent work (Kroon, Spek & Krabbendam, 1977; Heinerman, Krabbendam, Kroon & Spek, 1978), employing Bijvoet inequalities and the double Patterson function, leads in a similar way to estimates of the sines of the three-phase structure invariants. Again, some early work of Rossmann (1961), employing the difference synthesis $(|F_{\mathbf{H}}| - |F_{\overline{\mathbf{H}}}|)^2$ in order to locate the anomalous scatterers and recently applied by Hendrickson and Teeter (1981) in their solution of the crambin structure, shows that the presence of anomalous scatterers facilitates the determination of crystal structures. This work strongly suggests that the ability to integrate the techniques of direct methods with anomalous dispersion would lead to improved methods for phase determination. The fusion of these techniques is described here. That the anticipated improvement is in fact realized is shown in Tables 1 and 2 and Fig. 6. Not only do the new formulas lead to improved estimates of the structure invariants but, more important still, because the distributions derived here are unimodal in the whole interval from $-\pi$ to $+\pi$, the twofold ambiguity inherent in all the earlier work is removed. It is believed that this resolution of the twofold ambiguity results from the ability now to make use of the individual magnitudes in the first neighborhood of the structure invariant and the avoidance of explicit dependence on the Bijvoet differences; the explicit use of the Bijvoet differences, as had been done in all previous work, leads apparently to a loss of information resulting in a twofold ambiguity in estimates of the structure invariants. It may be of some interest to observe that in the earlier work with anomalous dispersion only the sine of the invariant may be estimated; in the absence of anomalous scatterers only the cosine of the invariant may be estimated; as a result of the work described here both the sine and the cosine, that is to say the invariant itself, may be estimated. Since, in the presence of anomalous scatterers, the observed intensities are known to determine a unique enantiomorph, and therefore unique values for all the structure seminvariants, formulas of the kind described here should not be unexpected; nevertheless not even their existence appears to have been anticipated.

3.2. *The normalized structure factors*

In the presence of anomalous scatterers the normalized structure factor

$$E_H = |E_H|\exp(i\phi_H) \tag{49}$$

is defined by

$$E_H = \frac{1}{\alpha_H^{1/2}} \sum_{j=1}^{N} f_{jH}\exp(2\pi i H \cdot r_j) \tag{50}$$

$$= \frac{1}{\alpha_H^{1/2}} \sum_{j=1}^{N} |f_{jH}|\exp[i(\delta_{jH} + 2\pi H \cdot r_j)] \tag{51}$$

where

$$f_{jH} = |f_{jH}|\exp(i\delta_{jH}) \tag{52}$$

is the (in general complex) atomic scattering factor (a function of $|H|$ as well as of j) of the atom labeled j, r_j is its position vector, N is the number of atoms in the unit cell, and

$$\alpha_H = \sum_{j=1}^{N} |f_{jH}|^2. \tag{53}$$

For a normal scatter, $\delta_{jH} = O$; for an atom which scatters anomalously, $\delta_{jH} \neq O$. Owing to the presence of the anomalous scatterers, the atomic scattering factors f_{jH}, as functions of $\sin\theta/\lambda$, do not have the same shape for different atoms, even approximately. Hence the dependence of the f_{jH} on $|H|$ cannot be ignored, in contrast to the usual practice when anomalous scatterers are not present. For this reason the subscript H is not suppressed in the symbols f_{jH} and α_H [Eq. (53)].

The reciprocal-lattice vector H is assumed to be fixed, and the primitive random variables are taken to be the atomic position vectors r_j which are assumed to be uniformly and independently distributed. Then E_H, as a function, (Eq. (51)], of the primitive random variables r_j, is itself a random variable and, as it turns out,

$$\langle |E_H|^2 \rangle_{r_j} = 1. \tag{54}$$

3.3. *The two-phase structure invariant*

The two-phase structure invariant, which has no analogue when no anomalous scatterers are present, is defined by

$$\psi = \phi_H + \phi_{\bar{H}}. \tag{55}$$

3.3.1. The first neighborhood

The first neighborhood of the two-phase structure invariant ψ [Eq. (55)] is defined to consist of the two magnitudes

$$|E_H|, |E_{\bar{H}}|, \tag{55}$$

which, because of the breakdown of Friedel's Law, are in general distinct.

3.3.2. Estimating the two-phase structure invariant

Define C_H and S_H by means of

$$C_H = \frac{1}{\alpha_H} \sum_{j=1}^{N} |f_{jH}|^2 \cos 2\,\delta_{jH} \tag{57}$$

$$S_H = \frac{1}{\alpha_H} \sum_{j=1}^{N} |f_{jH}|^2 \sin 2\,\delta_{jH} \tag{58}$$

where f_{jH}, δ_{jH}, and α_H are defined in (52) and (53). Define X and ξ by means of

$$X \cos \xi = C_H, \quad X \sin \xi = -S_H, \tag{59}$$

$$X = \left[C_H^2 + S_H^2 \right]^{1/2}, \quad \tan \xi = -S_H/C_H \tag{60}$$

Suppose that R and $\bar{R}$ are fixed non-negative numbers. In view of (48) to (50) the two-phase structure invariant $\phi_H + \phi_{\bar{H}}$, as a function of the primitive random variables r_j, is itself a random variable. Denote by $P(\Psi|R,\bar{R})$ the conditional probability distribution of the two-phase structure invariant $\phi_H + \phi_{\bar{H}}$, given the two magnitudes in its first neighborhood:

$$|E_H| = R, \quad |E_{\bar{H}}| = \bar{R}. \tag{61}$$

Then (Hauptman, 1982; Giacovazzo, 1983)

$$P(\Psi|R, \bar{R}) \approx \left[2\pi I_0 \left[\frac{2R\bar{R}X}{1-X^2} \right] \right]^{-1} \exp \left\{ \frac{2R\bar{R}X}{1-X^2} \cos (\Psi + \xi) \right\}, \tag{62}$$

where X and ξ, defined by (57)−(60) are seen to be functions of the (complex) atomic scattering factors f_{jH}, which are presumed to be known. It should be noted that the distribution (62) has the same form as (30) but is centered at -ξ instead of *O*. Since (62) has a unique maximum at $\Psi = -\xi$, it follows that

$$\phi_H + \phi_{\bar{H}} \approx -\xi \tag{63}$$

provided that the variance of the distribution is small i.e. provided that

$$A = \frac{2R\bar{R}X}{1-X^2} \text{ is large.} \tag{64}$$

It should be noted that, while A depends on R, $\bar{R}$ and $|\mathbf{H}|$, for a fixed chemical composition ξ depends only on $|\mathbf{H}|$ (or $\sin\theta/\lambda$) and is independent of R and $\bar{R}$.

3.4. *The three-phase structure invariant*
It will be assumed throughout that $\mathbf{H}$, $\mathbf{K}$, and $\mathbf{L}$ are fixed reciprocal-lattice vectors satisfying

$$\mathbf{H} + \mathbf{K} + \mathbf{L} = 0. \tag{65}$$

Owing to the breakdown of Friedel's law there are, in sharp contrast to the case that no anomalous scatterers are present, eight distinct three-phase structure invariants:

$$\psi_0 = \phi_{\mathbf{H}} + \phi_{\mathbf{K}} + \phi_{\mathbf{L}}, \tag{66}$$

$$\psi_1 = -\phi_{\bar{\mathbf{H}}} + \phi_{\mathbf{K}} + \phi_{\mathbf{L}}, \tag{67}$$

$$\psi_2 = \phi_{\mathbf{H}} - \phi_{\bar{\mathbf{K}}} + \phi_{\mathbf{L}}, \tag{68}$$

$$\psi_3 = \phi_{\mathbf{H}} + \phi_{\mathbf{K}} - \phi_{\bar{\mathbf{L}}}, \tag{69}$$

$$\psi_{\bar{0}} = \phi_{\bar{\mathbf{H}}} + \phi_{\bar{\mathbf{K}}} + \phi_{\bar{\mathbf{L}}}, \tag{70}$$

$$\psi_{\bar{1}} = -\phi_{\mathbf{H}} + \phi_{\bar{\mathbf{K}}} + \phi_{\bar{\mathbf{L}}}, \tag{71}$$

$$\psi_{\bar{2}} = \phi_{\bar{\mathbf{H}}} - \phi_{\mathbf{K}} + \phi_{\bar{\mathbf{L}}}, \tag{72}$$

$$\psi_{\bar{3}} = \phi_{\bar{\mathbf{H}}} + \phi_{\bar{\mathbf{K}}} - \phi_{\mathbf{L}}. \tag{73}$$

3.4.1. The first neighborhood
The first neighborhood of each of the three-phase structure invariants (66)–(73) is defined to consist of the six magnitudes:

$$|E_{\mathbf{H}}|, |E_{\mathbf{K}}|, |E_{\mathbf{L}}|, |E_{\bar{\mathbf{H}}}|, |E_{\bar{\mathbf{K}}}|, |E_{\bar{\mathbf{L}}}| \tag{74}$$

which, again owing to the breakdown of Friedel's law, are not in general equal in pairs.

3.4.2. The probabilistic background

Fix the reciprocal-lattice vectors $\mathbf{H}$, $\mathbf{K}$, and $\mathbf{L}$, subject to (65). Suppose that the six non-negative numbers R_1, R_2, R_3, $R_{\bar{1}}$, $R_{\bar{2}}$ and $R_{\bar{3}}$ are also specified. Define the N-fold Cartesian product W to consist of all ordered N-tuples $(\mathbf{r}_1, \mathbf{r}_2,..., \mathbf{r}_N)$, where $\mathbf{r}_1$, $\mathbf{r}_2,...$, $\mathbf{r}_N$ are atomic position vectors. Suppose that the primitive random variable is the N-tuple $(\mathbf{r}_1, \mathbf{r}_2,..., \mathbf{r}_N)$ which is assumed to be uniformly distributed over the subset of W defined by

$$|E_{\mathbf{H}}| = R_1, \ |E_{\mathbf{K}}| = R_2, \ |E_{\mathbf{L}}| = R_3, \tag{75}$$

$$|E_{\overline{\mathbf{H}}}| = R_{\bar{1}}, \ |E_{\overline{\mathbf{K}}}| = R_{\bar{2}}, \ |E_{\overline{\mathbf{L}}}| = R_{\bar{3}}, \tag{76}$$

where the normalized structure factors E are defined by (50). Then the eight structure invariants

$$\psi_j, \ \psi_{\bar{j}}, \ j = 0, 1, 2, 3, \tag{77}$$

$(66)-(73)$, as functions of the primitive random variables $(\mathbf{r}_1, \mathbf{r}_2, ..., \mathbf{r}_N)$, are themselves random variables.

Our major goal is to determine the conditional probability distribution of each of the three-phase structure invariants $(66)-(73)$, given the six magnitudes (75) and (76) in its first neighborhood, which, in the favorable case that the variance of the distribution happens to be small, yields a reliable estimate of the invariant (the neighborhood principle).

3.4.3. Estimating the three-phase structure invariant

Denote by

$$P_j(\Psi | R_1, R_2, R_3, R_{\bar{1}}, R_{\bar{2}}, R_{\bar{3}}) = P_j(\Psi),$$

$$j = 0, 1, 2, 3, \bar{0}, \bar{1}, \bar{2}, \bar{3}, \tag{78}$$

the conditional probability distribution of each ψ_j, assuming as known the six magnitudes (74) in its first neighborhood. Then the final formula, the major result of this article, is simply (Hauptman, 1982; Giacovazzo, 1983)

$$P_j(\Psi) \approx \frac{1}{K_j} \exp\left\{ A_j \cos\left(\Psi - \omega_j\right)\right\},$$

$$j = 0, 1, 2, 3, \bar{0}, \bar{1}, \bar{2}, \bar{3} \tag{79}$$

where the parameters K_j, A_j, and ω_j are expressible in terms of the complex scattering factors $f_{j\mathbf{H}}$, $f_{j\mathbf{K}}$, $f_{j\mathbf{L}}$, presumed to be known, and the observed magnitudes $|E_H|$, $|E_K|$, $|E_L|$, $|E_{\overline{H}}|$, $|E_{\overline{K}}|$, $|E_{\overline{L}}|$ in the first neighborhood of the invariant. Since the K_j's and A_j's are positive, the maximum of (79) occurs at $\Psi = \omega_j$. Hence when the variance of the distribution (79) is small, i.e. when A_j is large, one obtains the reliable estimate

$$\psi_j = \omega_j, j = 0, 1, 2, 3, \overline{0}, \overline{1}, \overline{2}, \overline{3}, \tag{80}$$

for the structure invariant ψ_j. It should be emphasized that the estimate (80) is unique in the whole range from $-\pi$ to $+\pi$. No prior knowledge of the positions of the anomalous scatterers is needed, nor is it required that the anomalous scatterers be identical.

3.4.4. The applications

Using the presumed known coordinates of the $PtCl_4^{2-}$ derivative of the protein Cytochrome c_{550} from *Paracoccus denitrificans* (Timkovich & Dickerson, 1976), molecular weight $M_r \simeq 14,500$, space group $P2_12_12_1$, some 8300 normalized structure factors E were calculated (to a resolution of 2.5Å). In addition to the anomalous scatterers Pt and Cl, this structure contains one Fe and six S atoms which also scatter anomalously at the wavelength used (CuKα). Using the 4000 phases ϕ_{hkl} corresponding to the 4000 largest $|E_{hkl}|$'s with hkl $\neq$ O, the three-phase structure invariants ψ_j, $j = 0, 1, 2, 3, \overline{0}, \overline{1}, \overline{2}, \overline{3}$, $[(66)-(73)]$, were generated and the parameters ω_j and A_j needed to define the distributions (79), were calculated. All calculations were done on the VAX 11/780 computer; double precision (approximately 15 significant digits) was used in order to eliminate round-off errors. The values of the A_j's were arranged in descending order and the first 2000, sampled at intervals of 100, were used in the construction of Table 1; the top 60,000 were used for Table 2.

Table 1 lists 21 values of A_j, sampled as shown from the top 2000, the corresponding estimates ω_j (in degrees) of the invariants ψ_j, the true values of the ψ_j, and the magnitude of the error, $|\omega_j - \psi_j|$. Also listed are the six magnitudes $|E|$ in the first neighborhood of the corresponding invariant.

Table 2 gives the average magnitude of the error,

$$<|\omega_j - \psi_j|>, \tag{81}$$

in the nine cumulative groups shown, for the 60,000 most reliable estimates ω_j of the invariants ψ_j.

Tables 1 and 2 show firstly that, owing to the unexpectedly large number of large values of A_j, our formulas yield reliable (and unique) estimates of tens of thousands of the three-phase structure invariants. Secondly, the invariants which are most reliably estimated lie anywhere in the range from -180° to +180°, and appear to be uniformly distributed in this range (Columns 9 and 10 of Table 1). Finally , in sharp contrast to the case that no anomalous scatterers are present, the most reliable estimates are not necessarily of invariants corresponding to the most intense reflections but of those corresponding instead to reflections of only moderate intensity (Columns 2−7 of Table 1).

Fig. 6 shows a scatter diagram of ω_j *versus* ψ_j for the $PtCl_4^{2-}$ derivative of Cytochrome c_{550}, using 201 invariants sampled at intervals of length ten from the top 2000, as well as the line $\omega_j = \psi_j$. Since the line falls evenly among the points, it appears that the ω_j are unbiased estimates of the invariants ψ_j.

3.4.5. Concluding remarks

In this article the goal of integrating the techniques of direct methods with anomalous dispersion is realized. Specifically, the conditional probability distribution of the three-phase structure invariant, assuming as known the six magnitudes in its first neighborhood, is obtained. In the favorable case that the variance of the distribution happens to be small, the distribution yields a reliable estimate of the invariant (the neighborhood principle). It is particularly noteworthy that, in strong contrast to all previous work, the estimate is unique in the whole interval $(-\pi, \pi)$ and that any estimate in this range is possible (even, for example, in thr vicinity of $\pm\pi/2$ or π). The first applications of this work using error-free diffraction data have been made, and these show that in a typical case some tens of thousands of three-phase structure invariants may be estimated with unprecedented accuracy, even for a macromolecular crystal structure. Some preliminary calculations on a number of structures, not detailed here, show that the accuracy of the estimates depends in some complicated way on the complexity of the crystal structure, the number of anomalous scatterers, the strength of the anomalous signal, and the range of $\sin\theta/\lambda$. With smaller structures the accuracy may be greatly increased, average errors of only three or four degrees for thousands of invariants being not uncommon.

It should be stated in conclusion that the availability of reliable estimates for large numbers of the three-phase structure invariants implies that the traditional machinery of direct methods, in particular the tangent formula [Eq. (33)], suitably modified to accommodate the non-zero estimates of the invariants, may be carried over without essential change to estimate the values of the individual phases and thus to facilitate structure determination *via* anomalous dispersion. In view of the calculations summarized in Tables 1 and 2 and Fig. 6, it seems likely that, in time, even macromolecules will prove to be solvable in this way. It is clear, too, that, owing to the ability to estimate both the sine and cosine invariants, that is to say both the signs and magnitudes of the invariants, the unique enantiomorph determined by the observed intensities is automatically obtained. In fact the first application of these techniques using experimental diffraction data has already facilitated the solution of the unknown macromolecular structure Cd, Zn Metallothionein [Furey, et al., 1986].

Table 1. Twenty-one estimates ω_j (in degrees) of the structure invariants ψ_j sampled from the top 2,000 for the Pt Cl$_4{}^{2-}$ derivative of Cytochrome c$_{550}$.

Serial No.	$|E_H|$	$|E_{\bar{H}}|$	$|E_K|$	$|E_{\bar{K}}|$	$|E_L|$	$|E_{\bar{L}}|$	A_j	Estimated value ω_j of ψ_j	True value of ψ_j	Mag. of the Error $|\omega_j-\psi_j|$
1	2.17	2.04	0.89	1.03	0.85	0.67	6.92	− 58°	− 88°	30°
100	1.91	2.06	1.61	1.49	0.85	0.67	5.62	148	130	18
200	1.91	2.06	1.96	2.06	1.41	1.57	4.83	− 79	−121	42
300	2.36	2.48	1.56	1.69	0.82	0.68	4.52	52	2	50
400	2.17	2.04	1.34	1.48	1.28	1.15	4.31	79	96	17
500	1.85	1.94	0.85	0.67	0.78	0.92	4.21	56	42	14
600	2.17	2.04	0.92	1.04	0.86	0.70	4.10	146	148	2
700	1.39	1.28	0.85	0.67	0.87	0.75	4.02	− 72	− 68	4
800	1.41	1.57	1.61	1.49	0.71	0.85	3.93	70	50	20
900	1.88	1.98	1.28	1.15	0.85	0.67	3.87	104	96	8
1,000	1.29	1.43	0.79	0.71	0.85	0.67	3.80	− 88	−138	50
1,100	1.34	1.48	1.34	1.22	1.25	1.16	3.76	− 72	−126	54
1,200	1.56	1.69	1.41	1.57	0.98	0.90	3.72	73	78	5
1,300	1.98	2.07	2.08	1.94	1.08	1.21	3.68	−161	−124	37
1,400	1.56	1.67	1.41	1.57	1.24	1.33	3.63	− 72	− 3	69
1,500	2.38	2.50	1.91	2.06	0.74	0.64	3.59	84	77	7
1,600	1.91	2.06	1.34	1.22	0.72	0.83	3.55	− 64	− 94	30
1,700	1.91	2.06	2.02	2.12	2.15	2.24	3.51	− 64	− 72	8
1,800	2.38	2.50	1.61	1.49	0.78	0.90	3.46	78	82	4
1,900	2.38	2.50	1.63	1.70	1.81	1.93	3.43	63	123	60
2,000	0.85	0.67	0.97	0.83	1.02	1.09	3.42	− 96	−126	30

Table 2. Average magnitude of the error (in degrees) in the top 60,000 estimated values of the three-phase structure invariants, cumulated in the nine groups shown, for the Pt Cl_4^{2-} derivative of Cytochrome c_{550}.

Group No.	No. in Group	Average Value of A	Average Mag. of Error
1	100	6.01	27.9°
2	500	4.90	29.3
3	1,000	4.44	28.8
4	2,000	4.02	28.0
5	5,000	3.49	31.4
6	10,000	3.09	33.8
7	20,000	2.71	36.1
8	40,000	2.35	38.6
9	60,000	2.15	39.8

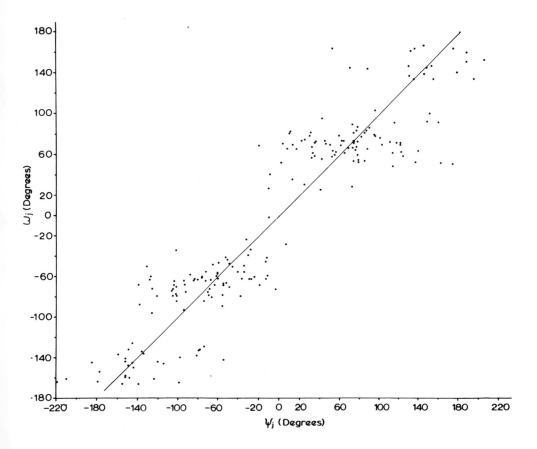

Fig. 6. A scatter diagram of ω_j *versus* ψ_j, using 201 invariants sampled at intervals of length ten from the top 2000, for the $PtCl_4^{2-}$ derivative of Cytochrome c_{550}, as well as the line $\omega_j = \psi_j$.

REFERENCES

Cochran, W. (1955), *Acta Cryst. 8*, 473−478.

Green, E. and Hauptman, H. (1976), *Acta Cryst. A32*, 940−944.

Fortier, S. and Hauptman, H. (1977), *A33*, 694−696.

Furey, W. F., Robbins, A. H., Clancy, L. L., Winge, D. R., Wang, B. C., and Stout, C. D. (1986). Science 231, 704−710.

Giacovazzo, C. (1983), *Acta Cryst. A39*, 585−592.

Hauptman, H. (1972), Crystal Structure Determination: The Role of the Cosine Seminvariants (New York: Plenum Press).

− (1975a), *Acta Cryst. A31*, 671−679.

− (1975b), *Acta Cryst. A31*, 680−687.

− (1976), *Acta Cryst. A32*, 877−882.

− (1977a), *Acta Cryst. A33*, 553−555.

− (1977b), *Acta Cryst. A33*, 568−571.

− (1982), *Acta Cryst. A38*, 632−641.

− and Karle, J. (1953), Solution of the Phase Problem I. The Centrosymmetric Crystal. ACA Monograph No. 3. Polycrystal Book Service.

− and Karle, J. (1956), *Acta Cryst. 9*, 45−55.

− and Karle, J. (1959), *Acta Cryst. 12*, 93−97.

Heinerman, J. J. L., Krabbendam, H., Kroon, J., and Spek, A. L. (1978). *Acta Cryst. A34*, 447−450.

Hendrickson, W. A. and Teeter, M. M. (1981), *Nature* (London) *290*, 107−113.

Karle, J. and Hauptman, H. (1955), *Acta Cryst. 9*, 635−651.

− − (1961), *Acta Cryst. 14*, 217−223.

Kroon, J., Spek, A. L., and Krabbendam, H. (1977), *Acta Cryst. A33*, 382−385.

Lessinger, L. and Wondratschek, H. (1975), *Acta Cryst. A31*, 521.

Okaya, Y. and Pepinsky, R. (1956), *Phys. Rev. 103*, 1645−1647.

Peerdeman, A. F. and Bijvoet, J. M. (1956), *Proc. K. Ned. Akad. Wet. B59*, 312−313.

Ramachandran, G. N. and Raman, S. (1956), *Curr. Sci. 25*, 348−351.

Rossmann, M. G. (1961), *Acta Cryst. 14*, 383−388.

Timkovich, R. and Dickerson, R. E. (1976), *J. Biol. Chem. 251*, 4033−4046.

Jerome Karle

JEROME KARLE

I was born in New York City in 1918 into a family that had a number of artistic people among its members. My father's brother and a sister's husband were probably the best known. The latter, Ivan Olinsky, taught for many years at the Art Students' League in New York City. I have been told that my paternal grandfather professionally made artistic decorations in peoples' homes. The propensity for artistic endeavors extended to my generation and beyond.

My mother was an excellent pianist and organist and it was one of her hopes that I would become a professional pianist. As a youth I was entered into "Music Week" competitions in New York City. I had some modest success, but found at an early age that I had no taste for public performance. On the other hand, I was strongly attracted to science as a lifelong career at an early age.

I had the privilege of attending schools in the New York City public school system. Their standards of education, character building and discipline were very high and I, most certainly, benefited from them. They separated out the more advanced students and permitted them to progress at their own pace. In my case, this occasionally led to some curious circumstances. In my senior year in high school (Abraham Lincoln), the girls would join the boys to practice dancing. I was 14 years old at the time and the girls were the usual 17–18 years old. The physical discrepancy between this 14 year old boy and 17–18 year old girls was considerable. Their first reaction was incredulity but after a while they got used to my presence and even danced with me. I took the chemistry and physics courses that were available, both taught by the same man. He recognized my interests and was very encouraging to me.

I enjoyed a number of sports that I participated in at every opportunity, swimming in the ocean nearby, a game called single-wall handball, played with a little hard black ball and well-known mainly in some metropolitan areas, touch football whose rules eliminate the bruises from tackling and ice-skating that was facilitated by the flooding of a huge parking lot by the local fire department.

I entered the City College of New York in 1933 and, at first, found it to be a bit of a struggle. Their academic standards were very high and they had a concentration of the best students in New York City. In addition, I spent three hours a day traveling on the subway system to and from home. This marked the end of piano practicing. City College had no tuition fee. The only financial requirement was one dollar per year for a library card. At the College, there were broad course requirements for all students that ranged through mathematics, the physical sciences, the social sciences and literature. There were even two years of compulsory public speaking courses. I studied, in addition to the requirements, some additional mathematics, some physics, and much chemistry and biology. The year after graduation from City College was spent at Harvard University in the study of biology, for which I received a master's degree, M.A., in 1938.

After a brief hiatus, I went to work with the New York State Health Department in Albany. While there, I had the opportunity to spend some time again at the piano. At the time I was in Albany, the fluoridation of drinking water was getting underway. I developed a procedure for determining the amount of fluorine in water supplies that became a standard method. This was my first modest contribution to science.

It was my intention to save enough money while at the Health Department to return to graduate school. This I did, and I entered the Chemistry Department of the University of Michigan in 1940 where I met my wife, Isabella Lugoski, whom I married in 1942, at an adjoining laboratory desk the first day that I went to physical chemistry class. We were both attracted to physical chemistry and took our degrees with Professor Lawrence O. Brockway whose speciality was the investigation of gas-phase molecular structure by means of electron diffraction. Although my Ph.D. degree was awarded in 1944, I had completed all my work for it during the summer of 1943 and went off to work on the Manhattan Project at the University of Chicago. Isabella joined me on this project a few months later.

In 1944, we returned to the University of Michigan, I went to work on a project of the Naval Research Laboratory and Isabella as an instructor in the Chemistry Department. While at the University of Michigan, I performed some experiments on the structure of monolayers of long-chain hydrocarbon films involved in the boundary lubrication of metallic surfaces. I also derived a theory that explained the electron diffraction patterns obtained from the oriented monolayers.

In 1946 we both went to work permanently in Washington for the Naval Research Laboratory. Our interest continued in developing the quantitative aspects of gas electron diffraction analysis. The solution of a key problem that arose in such analyses had evident implications for crystal structure analysis and, in fact, other areas of structure determination. At about the time that these matters were developing, Herbert Hauptman joined us at the Naval Research Laboratory and we decided to pursue the implications for crystal structures. This eventually led to the development of the direct methods for crystal structure analysis with the major part of the mathematical foundations and procedural insights established in the early 1950's.

While all this was going on and with hardly missing a step from her research activities, Isabella mothered three children, Louise in 1946, Jean in 1950, and Madeleine in 1955. Louise is a theoretical chemist, Jean an organic chemist and Madeleine is a museum specialist trained in geology.

The initial applications of the procedure for structure determination for centrosymmetric crystals involving probability measures and formulas derived from the joint probability distribution were performed in the middle 1950's in collaboration with colleagues at the U.S. Geological Survey. Then, in the second half of the 1950's, through the efforts of Isabella Karle, an experimental X-ray diffraction facility was established in our own laboratory.

During the 1960's, there was an intensive program in my laboratory to develop a procedure for crystal structure determination of broad applicability that would encompass noncentrosymmetric as well as centrosymmetric crystals. Largely through the efforts of Isabella Karle, such a procedure was developed and called the symbolic addition procedure. This procedure had its origins in the theoretical work and the experience in practical application of the 1950's, but it also required some new procedural insights and some additional theoretical work to make it efficient and broadly applicable and avoid the pitfalls that easily arise when optimal pathways through a procedure must be chosen on the basis of probability measures. The first application of the symbolic addition procedure was published in 1963 and the first essentially equal atom noncentrosymmetric crystal structure to be solved by direct phase determination was published in 1964. This was followed by a number of exciting applications and toward the end of the 1960's many laboratories started to become interested in the potential of the direct method for structure determination.

During the 1960's, I collaborated with Isabella in some of her investigations and derived with her a variance formula that was the basis for applying probability measures to procedures for analyzing noncentrosymmetric crystals. In addition, I also carried out a number of theoretical investigations. Perhaps, the most useful one concerned a procedure for developing a fragment of a structure into a complete one by use of the so-called tangent formula for phase determination.

During the 1950's and 1960's, I maintained an interest in gas electron diffraction and made some experimental and theoretical studies of internal rotation and coherent diffraction associated with excitation processes. The latter was especially interesting, but required extensive experimental development that the resources available to me did not permit.

In the 1970's, I continued theoretical work in crystal structure analysis that included the derivation of a "tangent formula" for phase determination that was based on the more restrictive higher and higher order determinants from the determinantal inequalities. I showed how joint probability distributions relevant to crystallographic quantities could be put into an exponential form and thereby decrease considerably problems with asymptotic convergence. I also derived heuristic joint probability distributions based on the determinants involved in the determinantal inequalities and obtained from them formulas for evaluating triplet phase invariants and, later on, formulas for the expected values of phase invariants and embedded semi-invariants of any order, triplet, quartet, quintet, etc. The utility of phase invariants of high order in phase determination has so far been rather limited, except perhaps collectively in the high order determinants where they have been useful for refining the values of approximately determined phase values.

I participated with Wayne Hendrickson of my laboratory in some refinements of macromolecular structure with the use of the tangent formula and also had some early participation with John Konnert and Wayne Hendrickson in the constrained refinement technique for macromolecules. In collaboration with John Konnert and Peter D'Antonio, procedures were developed for determining atomic arrangements in amorphous materials based on criteria similar to those applied to molecular vapors. Collaborations on structural problems also included Judith Flippen-Anderson, Clifford George, Richard Gilardi and Alfred Lowrey.

At the end of the 1970's Wayne Hendrickson made some valuable advances in the application of anomalous dispersion to the determination of macromolecular structure that rekindled an interest that I formerly had in this subject. I developed an exact, linear algebraic theory that includes any number and type of anomalous scatterer and any number of wavelengths. It can also incorporate information from isomorphous replacement measurements. Exact data give exact values for the unknown quantities that include phase differences. I have also been investigating the evaluation of triplet phase invariants to see what their potential usefulness may be. This activity continues to the present and is greatly facilitated by Stephen Brenner who has performed my programming and computing for me since the early 1960's.

In addition to participating in the development of new analytical methods and their applications, I have taught from time to time, mathematics and physics in the University College of the University of Maryland, I have taken an active role in the affairs of crystallography over the years as, for example, President of the International Union of Crystallography (1981—1984) and have enjoyed having a laboratory that investigates a broad variety of subjects ranging over gaseous molecules, amorphous solids, fibers, crystals and crystalline macromolecules.

During my entire married life I have had the strong support of my wife, both technical and spiritual. I also deeply appreciate the supportive atmosphere provided by the Naval Research Laboratory. This was especially helpful during the early 1950's when a large number of fellow-scientists did not believe a word we said.

Added in 1992:

Since this biography was written in 1985, advances have been made in macromolecular structure analysis by applications of the linear algebraic theory for the multiwavelength anomalous dispersion technique that I published in 1980. A number of such applications have been made by Wayne Hendrickson and colleagues who, along with the applications, developed suitable techniques for the use of synchrotron radiation and relatively weak anomalous scatterers. In recent years, I have been concerned with additional developments in the anomalous dispersion technique and have become interested in some aspects of the solution of nonlinear simultaneous equations, the determination of electron densities in crystals and some new approaches to phase determination in crystal structure problems.

Receipt of the Nobel Prize has given me the opportunity to have contact to an unprecedented degree with young people who look forward to careers in science and other intellectual and artistic pursuits. I have also had many contacts with organizations whose purpose is to improve the quality of life on this planet in a variety of contexts. These contacts have not changed my earlier views, but, in many instances, have perhaps given some of them a sharper focus. I would like to share a few.

Societies must provide a framework of encouragement in which its children can develop their skills fully and an educational system open to all in which this can be achieved. In many societies with which I am familiar, this would require a major change of priorities. Encouragement within the family structure is also very important.

This world has enormous social, economic and political problems, not the least of which concern the environment and natural resources. The degradation of the environment must be brought under control if there is to be a worthwhile and sustainable quality of life for most people. This too will require a reordering of priorities. It is very likely that continued population growth will defeat any attempts to halt environmental degradation and the unconscionable destruction of resources. Everyone has a responsibility in this regard.

Respect for the dignity of all human beings, if widespread, would go very far toward relieving numerous social stresses that much too often lead to societal deterioration and violence.

Our world has a long way to go.

Peace.

RECOVERING PHASE INFORMATION FROM INTENSITY DATA

Nobel lecture, 9 December, 1985

by

JEROME KARLE

Laboratory for the Structure of Matter, Naval Research Laboratory, Washington, D. C. 20375—5000, U.S.A.

The concept of a crystal is that of a solid body in which the atomic or molecular units are so arranged as to form an array having three dimensional periodicity. Because of the periodicity, it is possible to describe the arrangements of the atomic composition by means of Fourier series. The type of Fourier series that is used in crystal structure analysis represents the electron density distribution in a crystal. This is indeed equivalent to representing the structure of a crystal since the atomic locations are represented by the regions of highest electron density in the electron distributions.

The experimental technique used for examining the structure of crystals is called diffraction. In a diffraction experiment, rays are made to impinge on a crystalline substance of interest and, given the proper geometric conditions, the rays are scattered as if they were bouncing off large numbers of different planes imagined to be cutting through the crystal. The collected intensities of scattering (often 5000—10000 in number) are called a scattering pattern or diffraction pattern and comprise the experimental data from which the structure of the crystal of interest is to be elucidated. The most commonly used rays are Roentgen rays or X-rays, as they are usually called. Other rays, composed of neutrons or electrons, also have their purposes.

It is possible to obtain an insight into the character and sense of a diffraction experiment by imagining some experimental circumstances on the macroscopic scale. Let us suppose that we would like to probe the shape of some large object, hidden from view, by using balls that are hurled in a precise way at the object of interest and interact with the surface with essentially perfect restitution. Let us also suppose that it is possible to minimize and correct for the gravitational effects on the impinging balls. We assume that we can observe the results of the bouncing of the balls from the surface of the object. If a large area were scanned perpendicular to the direction in which the balls were hurled and each time the bouncing pattern were essentially parallel but in the opposite direction to that of the impinging balls, we would conclude that the object had a face with a high degree of flatness that was perpendicular to the direction of the impinging balls. Evidently, by varying the orientation of an arbitrarily shaped object and observing the patterns of the bouncing balls or, equivalently, the scattering pattern, knowledge of the shape of the object could be developed in quite some detail. On the submicroscopic scale, the situation is somewhat different. Crystals are quite porous to X-rays and the interactions between the rays and the atomic composition of the crystal are different than that of a ball bouncing off a surface such that, as noted, certain geometric conditions need to be fulfilled before it appears that reflection of the rays has taken place. The nature of the interactions is understood, however, and the relevance of the comparison with the macroscopic thought experiment involving bouncing balls maintains. In the case of X-rays and a crystal, the X-rays replace the bouncing balls and the way the X-rays interact with electron density distributions within the crystal gives rise to a scattering pattern unique to each crystalline substance. The problem that the analyst faces is to be able to take the diffraction pattern and from it determine the atomic architecture of the crystal which cannot be observed directly.

There is a special problem in taking the intensity information from a diffraction pattern and calculating from it the electron density distribution of a crystal by use of the Fourier series. The coefficients in the Fourier series are, in general, complex numbers. Only the magnitudes of the complex numbers appear to be available from the measured intensities of scattering. The required phases of the complex numbers seem to be lost in an ordinary X-ray diffraction experiment. It was therefore generally thought that it was not possible to go directly from a diffraction pattern to a determination of a crystal structure. The impasse was overcome in a series of steps that involved recognition that the required phase information was contained in the experimental intensity information, the derivation of a foundation mathematics that displayed relationships between phases and magnitudes and even among phases alone and, finally, the development of practical procedures for structure determination, strategies that brought together in a more or less optimal fashion the mathematical relationships with suitably adjusted and refined experimental data.

The results of structure determinations have been playing a valuable role in a number of areas of scientific endeavor. Crystallization, for example, is a very common phenomenon and many types of substances form crystals ranging from metals and minerals to huge macromolecules such as viruses. Knowledge of structure allows one to relate structure to function, i.e., understand physical, chemical or biological properties and activities, provides the chemist with useful information for syntheses, modifications and reaction mechanisms and can also be used to identify very small quantities of scarce material. It often provides the theoretical chemist with a starting point for his calculations. Structural research provides a conceptual basis for many associated scientific disciplines and it is the opportunity to interact with a variety of such disciplines that has made structural research particularly appealing to me.

As this article proceeds, it will elaborate on a number of the items discussed in this introductory part, describe some interesting applications and discuss briefly some research paths and opportunities for the future.

ELECTRON DENSITY DISTRIBUTION

The electron density distribution, $\varrho(\mathbf{r})$, is expressed in terms of the three-dimensional Fourier series

$$\varrho\,(\mathbf{r}) = V^{-1} \sum_{\mathbf{h}}^{\infty} F_{\mathbf{h}} \exp\,(-2\pi i \mathbf{h}\cdot\mathbf{r}) \tag{1}$$
$$\scriptstyle -\infty$$

where V is the volume of the unit cell of the crystal, the basic structural unit from which, through three-dimensional periodicity, the crystal is formed. The coefficients

$$F_{\mathbf{h}} = |F_{\mathbf{h}}|\exp(i\phi_{\mathbf{h}}) \tag{2}$$

are the crystal structure factors associated with the planes labeled with the vectors $\mathbf{h}$. The $\mathbf{h}$ have integer components, h, k, and l, the Miller indices, whose values are inversely proportional to the intercepts on the x, y and z axes, respectively, of planes cutting through the crystal. The angle $\phi_{\mathbf{h}}$ is the phase associated with $F_{\mathbf{h}}$ and $\mathbf{r}$ labels the position of any point in the unit cell. $F_{\mathbf{h}}\mathbf{E}$ is the amplitude of the scattered wave associated with the plane labeled by $\mathbf{h}$, where $\mathbf{E}$ is the electric field vector of the incident beam. The measured intensities of X-ray scattering are proportional to $|F_{\mathbf{h}}|^2$. If the values of the $\phi_{\mathbf{h}}$ were also obtained directly from experiment, structures could be immediately calculated from (1). The seeming absence of this information gave rise to the so-called "phase problem".

The Fourier inversion of (1) followed by the replacement of the integral by the sum of contributions from the N discrete atoms in the unit cell gives, for the Fourier coefficient,

$$|F_{\mathbf{h}}|\exp(i\phi_{\mathbf{h}}) = \sum_{j=1}^{N} f_{j\mathbf{h}}\exp(2\pi i \mathbf{h}\cdot\mathbf{r}_j) \tag{3}$$

where $f_{j\mathbf{h}}$ represents the amplitude of scattering of the jth atom in the unit cell and $\mathbf{r}_j$ is its position vector.

OVERDETERMINACY

A system of simultaneous equations is formed by the definition of the crystal structure factors given by (3) since the values of the scattered intensities are measured for a large number of $\mathbf{h}$. The unknown quantities in (3) are the phases $\phi_{\mathbf{h}}$ and the atomic positions $\mathbf{r}_j$. The known quantities are the $|F_{\mathbf{h}}|$ obtained from the measured intensities and the $f_{j\mathbf{h}}$ which differ little from the theoretically calculated atomic scattering factors for free atoms. Since each equation in (3) involves complex quantities, there are really two equations, one for the real and one for the imaginary part. In order to determine the overdeterminacy, a comparison is made of the number of unknown quantities with the number of independent data available. With the use of CuKα radiation, the overdeterminacy can be as great as a factor of about 50 for crystals that have a center of symmetry and about 25 for those that do not. In practice, somewhat fewer than the maximum available data are measured, but the overdeterminacy is still quite high.

SOME ATTEMPTS, SOME SUCCESSES

There were some early attempts to obtain structural information or phase information from the structure factor equations. Ott [1] made use of the structure factor equations (3) to derive relationships among the structure factors and atomic positions and he showed that in some simple cases atomic coordinates could be obtained directly from the relationships. Banerjee [2] devised a trial and error self-consistency routine based on Ott's results for finding the phases of structure factors that are centric and therefore with phases that have values that are limited to zero or π. The number of trials increased rapidly with complexity limiting applications to rather simple structures. Avrami [3] worked with equations that relate intensities to interatomic vectors (4). Solutions to these equations were given in terms of the roots of a polynominal equation whose degree increases rapidly as the complexity of the crystal increases. In all these approaches, the increase in computational demands with complexity, sensitivity to experimental errors, and inherent ambiguities in the results prevented their application to any but the simplest structures. Even though present day computational facility is enormously greater than when this work was done originally, the limitations cannot be suitably overcome even now.

A significant advance in the attempt to obtain structural information from the measured intensities was made in 1934 by A. L. Patterson. He developed a Fourier series which has as its coefficients the magnitude of the square of the structure factors rather than the structure factors themselves. The phases may be eliminated from (3) by multiplying by the corresponding complex conjugates to obtain

$$|F_\mathbf{h}|^2 = \sum_{j=1}^{N} \sum_{k=1}^{N} f_{j\mathbf{h}} f_{k\mathbf{h}} \exp\left[2\pi i \mathbf{h} \cdot (\mathbf{r}_j - \mathbf{r}_k)\right] \tag{4}$$

The Fourier transform of (4) is known as the Patterson function [4,5]

$$P(\mathbf{r}) = \sum_{\mathbf{h}}^{\infty} |F_\mathbf{h}|^2 \exp(-2\pi i \mathbf{h} \cdot \mathbf{r}) \tag{5}$$
$$_{-\infty}$$

The maxima of a Patterson function represent the interatomic vectors in a structure. Evidently the values of the coefficients are directly obtainable from the measured intensities of scattering. This function has been very useful in locating the heavier atoms in a structure, if they are not too numerous, since the interatomic vectors associated with them would predominate in a map computed from (5) and the atomic positions for them could then be readily deduced. The coordinates for the heavy atoms may be used with (3) to compute an initial set of approximate phases. Depending upon the scattering power of the heavy atoms, such a computation may be suitable for structures containing up to a few hundred atoms. There are numerous procedures for developing a complete structure from the initial phase information obtained from the heavy atoms. The use of the Patterson function with structures containing heavy atoms has found widespread application and remains one of the major methods of crystal structure determination.

The difficulty with using the Patterson function with experimental data in a general way in the absence of heavy atoms arises from the lack of resolution that occurs for the $N(N-1)$ interatomic vectors as well as inaccuracies. The Patterson function becomes somewhat accessible when it is used in combination with known atomic groupings [6−8].

The fact that the Patterson function could be used to solve simple structures made a positive contribution to the background atmosphere in which progress in phase determination was made. Once a structure was solved, it was possible to use (3) to calculate values for the phases. In effect then, phase values were determined from the measured intensities. In the case of the Patterson function, this happened through the intermediary step of first determining the structure. It was, however, conceivable that the process could be reversed, namely, to obtain phase information directly from the intensities and from that compute the structure by use of (1). This is indeed what happened. So far, except for special cases involving heavy atoms, it has turned out to be easier to obtain phase values directly from the measured intensities and then compute the structure than to obtain the structure directly from the intensities without the use of phases.

Relationships between phases and magnitudes that anticipated the later developments were the inequalities of Harker and Kasper [9]. They derived a number of inequalities by application of the Schwarz and Cauchy inequalities to the structure factor equations (3) in the presence of crystallographic symmetry. The Harker-Kasper inequalities have provided valuable insights. For example, the simple inequality formulas can provide useful phase information as shown by Kasper, Lucht and Harker [10] in their solution of the structure of decaborane. In addition, work with the inequalities indicated that they may have probabilistic characteristics. Gillis [11], for example, noted that the implication of an inequality was probably correct even when the magnitudes of the structure factors were too small to permit a definitive conclusion to be drawn. Gillis speculated that the smallness of the structure factor magnitudes may have been due to thermal effects and employed an appropriate function to increase the values of the structure factor magnitudes so that the inequalities could be applied. The probabilistic interpretation, however, remained a possible alternative, namely, that although an inequality does not quite determine the value of a phase definitively, it still does so with a high probability that the value is correct. This could be important because it would imply that the inequalities have probabilistic implications that could extend their range of usefulness.

CH₂ CF₂

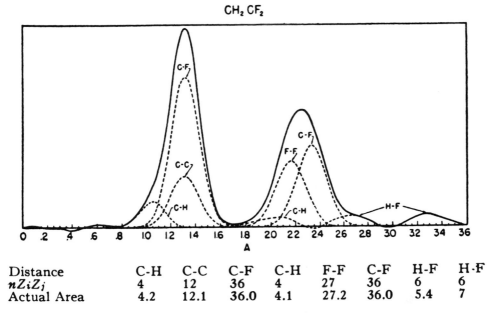

Distance	C-H	C-C	C-F	C-H	F-F	C-F	H-F	H·F
nZ_iZ_j	4	12	36	4	27	36	6	6
Actual Area	4.2	12.1	36.0	4.1	27.2	36.0	5.4	7

Fig. 1. The essentially non-negative radial distribution function (solid curve) for CH₂CF₂, comput-ed from the experimental molecular intensity data extracted by use of a properly formed back-ground intensity curve. The solid curve is a probability density that gives the probability of finding interatomic distances in some distance interval along the horizontal axis. The dashed lines represent the decomposition of the main peaks into their component individual interatomic distances. The individual peaks have a definite width related to the internal vibrations in the molecule.

NON-NEGATIVITY AND GENERAL INEQUALITIES

The initial motivation to investigate the mathematics of crystal structure determination arose from experiences in the development of an analytical procedure for obtaining accurate radial distribution functions for determining the structures of gaseous molecules by electron diffraction. A problem arose, namely, to find an accurate background intensity so that the molecular interfer-ence intensity could be accurately extracted from the total intensity of scatter-ing. The Fourier transform of the molecular intensity can be interpreted as representing the probability of finding interatomic distances in a molecule. Therefore, this transform must be non-negative and the non-negativity im-posed a very useful constraint on the shape of the background intensity [12,13]. Figure 1 shows a radial distribution function for CH₂CF₂ [14] derived from application of the non-negativity constraint and the component distances in the molecule. The attendant accuracy of the result permitted not only equilibrium interatomic distances to be determined but also estimates of the root-mean-square amplitudes of vibration associated with the interatomic distances.

At about the time this work in gas electron diffraction was proceeding, Herbert Hauptman joined our group at the Naval Research Laboratory and, in view of the success of the non-negativity criterion, we decided to explore the possibility that this criterion might be useful in other areas of structural research. This led us to investigate the determination of electron density distributions around free atoms [15] which found a very fine application in the determination of the electron distribution about argon by Bartell and Brockway [16].

We were also quite interested in seeing what the consequences of the application of non-negativity would be for crystal structure analysis since the electron density distribution defined in (1) is constrained not to be negative. This brought in the work of Toeplitz [17] early in this century on non-negative Fourier series and subsequent development by others. We discussed the theory in three-dimensions and wrote it in a form that would have particular relevance to crystallographic data.

The fundamental result was that the necessary and sufficient condition for the electron density distribution in a crystal to be non-negative is that an infinite system of determinants involving the crystal structure factors be non-negative. A typical determinant is [18]

$$
\begin{vmatrix}
F_{000} & F_{-k_1} & F_{-k_2} & \cdots & F_{-h} \\
F_{k_1} & F_{000} & F_{k_1-k_2} & \cdots & F_{k_1-h} \\
F_{k_2} & F_{k_2-k_1} & F_{000} & \cdots & F_{k_2-h} \\
\cdots & \cdots & \cdots & \cdots & \cdots \\
F_{h} & F_{h-k_1} & F_{h-k_2} & \cdots & F_{000}
\end{vmatrix} \geq 0 \tag{6}
$$

The subscripts in the first column start with 0,0,0 but are otherwise arbitrary. The subscripts in the first row are the same but of opposite sign. The subscript of the element in the ith row and jth column is the sum of the subscripts of the elements of the ith row and first column and the first row and jth column. The third order inequality

$$
\begin{vmatrix}
F_{000} & F_{-k} & F_{-h} \\
F_{k} & F_{000} & F_{k-h} \\
F_{h} & F_{h-k} & F_{000}
\end{vmatrix} \geq 0 \tag{7}
$$

contains a relationship among the structure factors that has played a most important role in direct crystal structure analysis. This may be seen by rewriting (7) in the form [18],

$$
\left| F_{h} - \frac{F_{k}F_{h-k}}{F_{000}} \right| \leq \frac{\begin{vmatrix} F_{000} & F_{-k} \\ F_{k} & F_{000} \end{vmatrix}^{1/2} \begin{vmatrix} F_{000} & F_{-h+k} \\ F_{h-k} & F_{000} \end{vmatrix}^{1/2}}{F_{000}} \tag{8}
$$

For structure factors of unusually large magnitude, the right side of (8) becomes quite small and then

$$
F_{h} \sim F_{k}F_{h-k}/F_{000} \tag{9}
$$

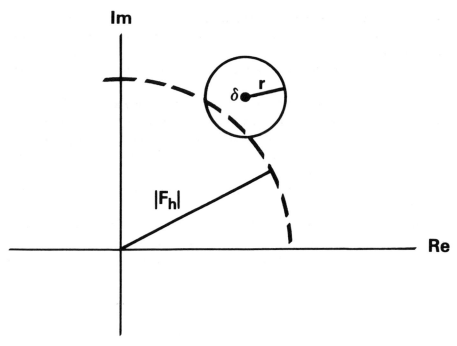

Fig. 2. The determinantal inequalities can be written in the general form $|F_h - \delta| \leq r$. This means that F_h is bounded by a circle of radius r in the complex plane centered at δ. If $|F_h|$ is known, then F_h is confined to a line within the circle.

One evident conclusion from (9) is

$$\phi_h \sim \phi_k + \phi_{h-k} \tag{10}$$

This states that for large structure factor magnitudes, the value of ϕ_h may be defined in terms of the values of two other phases. This may also be seen from a construction based on (8), Fig. 2, in which $\delta = F_k F_{h-k}/F_{000}$ and r is equal to the right side of (8). The form for (8) is then

$$|F_h - \delta| \leq r \tag{11}$$

It can be readily shown [18] that all determinants (6) can be written in the form (11). As the order of the determinants increases, there is a tendency for r to decrease in size, making the determination of ϕ_h rather definitive.

Formula (10) has found wide application beyond the range of usefulness of (8). This is because of the probabilistic characteristics of the inequalities [19] which imply that the most likely value of ϕ_h is that of $\phi_k + \phi_{h-k}$ and the probability decreases the farther the value of ϕ_h deviates from that of $\phi_k + \phi_{h-k}$. Therefore, even when the radius, r, of the bounding circle is large, the most likely value of ϕ_h is $\phi_k + \phi_{h-k}$.

The structure factors in (6) can be replaced by quasi-normalized structure factors, $\mathcal{E}$, that represent point atoms (to an approximation when atoms of unequal atomic number are present) rather than atoms with electron distributions,

$$\mathcal{E}_{\mathbf{h}} = F_{\mathbf{h}}/(\sum_{j=1}^{N} f_{j\mathbf{h}}^2)^{1/2} \tag{12}$$

Structure factors representing point atoms are the type of quantity normally used in phase-determining procedures. Instead of (9), we have

$$\mathcal{E}_{\mathbf{h}} \sim \mathcal{E}_{\mathbf{k}}\mathcal{E}_{\mathbf{h}-\mathbf{k}}/E_{000} \tag{13}$$

For centrosymmetric crystals, we have

$$s\mathcal{E}_{\mathbf{h}} \sim s\mathcal{E}_{\mathbf{k}}\mathcal{E}_{\mathbf{h}-\mathbf{k}} \tag{14}$$

where s means "sign of" a plus sign implying that the phase is equal to zero and a minus sign that it is equal to π. A one-term tangent formula also follows from (13),

$$\tan\phi_{\mathbf{h}} \simeq \frac{|\mathcal{E}_{\mathbf{k}}\mathcal{E}_{\mathbf{h}-\mathbf{k}}|\sin(\phi_{\mathbf{k}} + \phi_{\mathbf{h}-\mathbf{k}})}{|\mathcal{E}_{\mathbf{k}}\mathcal{E}_{\mathbf{h}-\mathbf{k}}|\cos(\phi_{\mathbf{k}} + \phi_{\mathbf{h}-\mathbf{k}})} \tag{15}$$

The tangent formula composed of the sum of terms over $\mathbf{k}$ both in the numerator and the denominator is another formula that has played a major role in the practical applications of the theory for structure determination.

After the set of determinantal inequalities (6) were obtained on the basis of the non-negativity of the electron density distribution in a crystal, it was of interest to investigate their relationship to the inequalities derived by Harker and Kasper [10] from use of the Schwarz and Cauchy inequalities. It was shown [18] that, when the appropriate symmetry was introduced into the third order determinantal inequality by means of certain relationships among the structure factors, the Harker-Kasper inequalities could be derived. Examination of the derivation of the Harker-Kasper inequalities shows, as would be expected, that the non-negativity of the electron density distribution is a requirement for their validity.

The variety of phase determining formulas contained within the determinantal inequalities (6) have their counterpart in probability theory, i.e., similar formulas can be derived with the use of probability theory. Their virtue is that measures of reliability can be attached to them and the judicious use of such measures was an important feature in bridging the gap between mathematical theory and practical application.

It had been pointed out that the determinants have inherent probabilistic characteristics [19] that can, in fact, be directly read out from the form (11). This was not, however, how the first probability formulas were derived. The pursuit of such formulas was motivated by the expectation that the usefulness of the formulas from the inequalities could be extended because of the great overdeterminancy of the structure problem, the expectation that the points within the bounding circle of radius r and center δ, Fig. 2, would not be uniformly probable and the experience of Gillis [11] with structure factor magnitudes that were too small to elicit definitive conclusions from the inequalities but which, when made larger artificially, led to correct conclusions.

In order to characterize the probabilistic aspects of this subject, we initially decided to develop a facility in the use of the random walk [20], but subsequently changed to the joint probability distribution [21, 22] which culminated in a monograph [23] that contained, for the first time, a set of probabilistic formulas and measures for attacking the phase problem, in this case limited to crystals that have a center of symmetry. It was in the monograph [23] also that the theory of invariants and semi-invariants was introduced for the purpose of solving the problem concerning how many and what types of phases to specify to fix the origin in a crystal and what values are permitted. The practical aspects of solving crystals that lack a center of symmetry were developed later on and it was not until 1964 that the structure of the first crystal lacking a center of symmetry was solved [24] by the "direct method" for obtaining phase information by direct use of the measured intensities of scattering. It is interesting that fairly recent developments in the mathematics of the random walk have made this technique quite accurate, stimulating revived interest in its application to the probabilistic aspects of phase determining formulas [25].

FORMULAS FOR PHASE DETERMINATION
The main formulas for phase determination are now listed. They will suffice to characterize the nature of the phase determining procedures. There are additional formulas that play a variety of helpful roles and may be found in the referenced literature of this article.

Centrosymmetric crystals
The Σ_2 formula is [23],

$$sE_h \approx s \sum_{k_r} E_k E_{h-k} \qquad (16)$$

where s means the "sign of" and k_r represents restricted values of k for which the corresponding $|E_k|$ and $|E_{h-k}|$ values are large. A plus sign refers to a phase of zero and a minus sign to a phase of π, the only two values possible for a centrosymmetric crystal when an origin in the crystal is properly chosen. The quantities, E, are normalized structure factors which arise as appropriate quantities to use with probability theory and are the same as the quasi-normalized structure factors, ε, except for a reweighting [26] of certain subsets of the E. The treatment of the intensity data to obtain normalized structure factors [27] arises from the work of Wilson [28,29], the earliest application of probability methods to crystal structure analysis. Formula (16) is the probability equivalent of the set of inequalities (8) as k varies over the set k_r. The appropriate probability function, $P_+(h)$, which represents the probability that the sign of E_h be positive, was given in the monograph [23]. It is conveniently applied in the form derived by use of the central limit theorem by Woolfson [30] and Cochran and Woolfson [31]

$$P_+(\mathbf{h}) \approx \tfrac{1}{2} + \tfrac{1}{2} \tanh \sigma_3 \, \sigma_2^{-3/2} |E_h| \sum_{\mathbf{k}} E_k E_{h-k} \tag{17}$$

where

$$\sigma_n = \sum_{j=1}^{N} Z_j^n \tag{18}$$

and Z_j is the atomic number of the jth atom in the unit cell containing N atoms.

Noncentrosymmetric crystals

The sum of angles and tangent formulas are, respectively,

$$\phi_h \approx <\phi_k + \phi_{h-k}>_{kr} \tag{19}$$

$$\tan\phi_h \simeq \frac{\sum\limits_{k} |E_k E_{h-k}| \sin(\phi_k + \phi_{h-k})}{\sum\limits_{k} |E_k E_{h-k}| \cos(\phi_k + \phi_{h-k})} \tag{20}$$

Formulas (19) and (20) are comparable to (10) and (15), respectively, and result from combining a number of individual terms as $\mathbf{k}$ varies over some chosen set. An appropriate measure of the reliability of (19) and (20) is a variance, V, [32] given by

$$V = \frac{\pi^2}{3} + \left[I_o(\alpha)\right]^{-1} \sum_{n=1}^{\infty} (I_{2n}(\alpha)/n^2) \tag{21}$$

$$-4 \left[I_o(\alpha)\right]^{-1} \sum_{n=0}^{\infty} \left[I_{2n+1}(\alpha)/(2n+1)^2\right]$$

where I_n is a Bessel function of imaginary argument [33]

$$\alpha = \left\{ \left[\sum_{k_r} \kappa\,(\mathbf{h}, \mathbf{k}) \cos\,(\phi_k + \phi_{h-k})\right]^2 \right. \\ \left. + \left[\sum_{k_r} \kappa\,(\mathbf{h}, \mathbf{k}) \sin\,(\phi_k + \phi_{h-k})\right]^2 \right\}^{1/2} \tag{22}$$

and $\qquad \kappa(\mathbf{h}, \mathbf{k}) = 2\sigma_3 \sigma_2^{-3/2} |E_h E_k E_{h-k}| \tag{23}$

Expression (21) gives the variance of ϕ_h as determined from a given set of $\phi_k + \phi_{h-k}$. This variance formula has its origin in a probabilility distribution (in somewhat different notation) of Cochran [34] for ϕ_h, given a particular $\phi_k + \phi_{h-k}$ and the accompanying $|E|$ values. The tangent formula (20) can be derived in many ways. It has arisen, for example, in theoretical investigations of noncentrosymmetric space groups by use of the joint probability distribution [35] and can be shown to occur [32] in a generalization of the Cochran formula [34] for a particular ϕ_h to take into consideration a set consisting of several or more $\phi_k + \phi_{h-k}$ rather than just one [32,34]. The average in (19) is to be taken in the context of maximum clustering i.e. a minimum deviation of the contributions of individual addition pairs, $\phi_k + \phi_{h-k}$ from the average value. All ϕ are kept in a range $-\pi < \phi \leq \pi$ and maximum clustering requires the addition of 0, 2π, or -2π to each addition pair. A practical method for effecting appropriate clustering has been described [32].

PRACTICAL PHASE DETERMINATION

In this part, the various aspects of practical phase determination will be outlined in terms of the first procedure that had broad practical applications to both centrosymmetric and noncentrosymmetric crystals, the symbolic addition procedure [24,32,36,37]. It arose mainly from the efforts of my wife, Dr. Isabella Karle, to bridge the gap between the mathematics of phase determination and the world of experimental data and practical application. At about 1956, we acquired apparatus for carrying out X-ray diffraction experiments with crystals and Isabella Karle taught herself with the aid of a book written by Martin Buerger how to collect and interpret diffraction photographs. Working with crystal diffraction data was quite different then than now; computer-aided collection of diffraction data was essentially nonexistent and, by present standards, computers were very primitive. The thousands of diffraction data collected in the the experiments were measured by eye with the use of a calibrated comparison strip. At the present time, data are collected by automatic diffractometers that measure and record the data after some modest, preliminary human intervention and the largest computers are many orders of magnitude superior in speed and capacity than thirty years ago. Technical advances in the recent past, as exemplified by synchrotron sources of X-radiation and area detectors, promise even greater advances in the near future in terms of the speed and facility with which data can be collected. They provide experimental opportunities that otherwise could not be contemplated.

Once the intensity data are collected, they are transformed into normalized structure factor magnitudes defined by,

$$|E_h| = |F_h|/(\varepsilon \sum_{j=1}^{N} f_{jh}^2)^{1/2} \tag{24}$$

where ε reweights certain subsets of the data [26]. A procedure for doing this is described in International Tables for X-ray Crystallography [38].

It is apparent on examining (16), (19) and (20) that it is necessary to know the values of some phases before additional ones can be evaluated. There are several sources of such information, from certain phase specifications associated with establishing an origin in the crystal [38], the assignment of symbols to some phases for later evaluation, and the use on occasion of auxiliary formulas, such as Σ_1 and Σ_3, that define individual phases in terms of structure factor magnitudes alone [38]. The number and types of phases to be used for specifying the origin in a crystal has been determined by use of the theory of invariants and semi-invariants that was developed for this purpose. Depending upon the type of space group involved, the number can vary from none at all to as many as three. Suitable tables [38] are available for carrying out this task.

The phase determining procedure is a stepwise one with few contributors to (16) or (19) at the start. Use at the start of phases associated with the largest possible values of the normalized structure factor magnitudes, $|E|$, will assure that the probability measures, (17) and (21), will be as large as possible. The large overdeterminacy of the problem helps to ensure that initial probabilities will be large enough to proceed in a stepwise fashion to build up a sufficiently reliable set of phase values to effect a solution to the structure problem. Because the nature of phase determination is inherently probabilistic and contingent in a stepwise and interdependent fashion, the problem of establishing optimal procedures based on experimental data was not at all straightforward. There are a very large number of paths through a phase determination. Among many of them are pitfalls in which there arise, for example, temptations to take a path in which the interconnections between phase evaluations flow easily at the expense somewhat of the probability measures. Such paths are more likely to lead to missteps and cumulative errors that could damage or defeat a phase determination than ones that are based only on the highest values of the probability measures.

There is also an ambiguousness inherent in procedures for phase determination which is controlled by the use of symbols. The symbols can assume more than one value. For centrosymmetric crystals, they can have only two phase values, zero or π. For noncentrosymmetric crystals, experience has shown that whereas phase values for the general reflections can have any value in the range from $-\pi$ to π, it is usually sufficient to use only four possible values for the symbols spaced $\pi/2$ apart. One of the virtues of using symbols is that, as the phase determination develops, relationships develop among the symbols reducing the number to be assigned values. Here again, one must proceed with caution so that reliable relationships are distinguished from those that are not. With centrosymmetric crystals, the entire phase determination is carried through before the remaining symbols are given alternative numerical values and tested to see which set yields a Fourier series that makes good chemical sense and reproduces the measured intensities well. For noncentrosymmetric crystals the symbolic addition procedure initially makes use only of (19). After about 100 phases have been evaluated, the remaining symbols are given alternative numerical values after which (20) is applied to further extend the phase set. Again, the correct phase set is the one whose Fourier series makes good chemical sense and yields a structure that is in fine agreement with the measured intensities. When a satisfactory result is not obtained, it is appropriate to try an alternative path through the phase determination.

An additional specification, whose character also derives from the theory of invariants and semi-invariants, is required for most noncentrosymmetric space groups. In making the specification, a choice is made of enantiomorph or axis direction or both. A good way in which this specification is achieved in the symbolic addition procedure is to find that a symbolic representation of a phase value most likely has a magnitude that differs significantly from 0 and π. The specification is accomplished by assigning a plus or minus sign arbitrarily to the magnitude of the phase. The enantiomorph ambiguity for noncentrosymmetric crystals arises from the fact that such a crystal gives the same diffraction pattern as its mirror image in the absence of detectable anomalous dispersion effects.

In the course of phase determinations, particularly in the case of noncentrosymmetric crystals, it may turn out that only a partial structure will appear in a Fourier map based on the determined phase values. This occurs because the phase determination went somewhat awry but not altogether so. The partial structure would be recognized as possibly being a correct fragment because it would make chemical sense, i.e., the fragment would have connectedness and acceptable interatomic distances and angles. A method has been developed for deriving a complete structure from a partial structure [39] that is based on the use of the tangent formula (20) and has been used in many applications. It may happen that a partial structure is not placed correctly with respect to a proper origin in the unit cell of a crystal. Under such circumstances, the use of a translation function [40−43] to place the fragment properly may be helpful.

The symbolic addition procedure [24,32,36,37] was developed as an outgrowth of the experiences in applying the procedure for phase determination for centrosymmetric crystals described in the 1953 monograph [23]. In the monograph, the procedures proposed for sign determination involved the initial use of auxiliary phase formulas such as Σ_1 and Σ_3 [23]. The application of preliminary phase information obtained from these formulas along with probability measures associated with individual indications from the phase-determining formulas facilitated the use of the Σ_2 formula (16). The combination of auxiliary formulas and probability theory greatly reduced to a practicable level the ambiguousness that would be obtained in the employment of the Σ_2 relation by itself along with some rather insensitive and often misleading criterion such as internal consistency.

In the course of applying the procedures of the monograph [23], two important features were found that ultimately played an important role in the symbolic addition procedure. The first feature was that if probability measures were carefully employed at each step of a phase determination, it was possible to carry out the procedure with a small set of starting phases. It was also apparent that the use of symbols could greatly increase the efficiency of the procedure by carrying along in their alternative values a residual ambiguity that could not be easily overcome. A sufficient number of reliable relationships among the symbols usually developed in the course of a phase determination to reduce the alternative possible sets of phases to consider to just a few. A further reduction could be obtained, if desired, by applying auxiliary phase determining formulas at the end of a phase determination to help evaluate the remaining symbols. If an evaluation were incorrect, as occasionally happened with the auxiliary formulas, no great harm would be done. The alternative value could be tried without the necessity of repeating the phase determination. The symbolic addition procedure for centrosymmetric crystals is thus the procedure of the monograph [23] facilitated by the use of symbols and the application of auxiliary formulas at the end, if needed, instead of at the beginning of a phase determination.

The symbolic addition procedure for centrosymmetric crystals has several features in common with the procedure of Zachariasen [44]. The main distinction from the latter procedure was the application of probability measures to guide each step of the phase determination, especially in the beginning, the consequent use of a minimal number of symbols and the resulting minimization of the ambiguousness of the determination and the optimization of the reliability of the result. There were efforts by others in the 1950's, e.g. Rumanova [45] and Cochran and Douglas [46] that met with some success. Rumanova developed a systematic method for using symmetry relations in centrosymmetric space groups with (16). She used it in connection with Zachariasen's procedure [44]. Cochran and Douglas used a variant of (16) based on a formula derived by Sayre [47] in a procedure that generated a very large number of sets of signs from which the correct one had to be selected under highly ambiguous circumstances.

The procedural features of the symbolic addition procedure for centrosymmetric crystals were extended, in the main by the efforts of Isabella Karle, to noncentrosymmetric crystals [24] with the use of (19), (20), and (21). Several problems arose in developing the technique for phase determination for noncentrosymmetric crystals concerning, for example, the assignment and handling of symbols, the use of the probability measure (21), the number of possible values to assign to the symbols that represent phase values that range continuously from $-\pi$ to π, the combined use of (19) and (20) for phase determination, the proper use of the tangent formula for the processes of phase refinement and phase extension, the development of techniques for specifying an enantiomorph or axis directions or both, and special considerations such as the avoidance of certain troublesome triplet phase invariants involving one and two-dimensional centric reflections. These various aspects of the symbolic addition procedure are to be found in references 32 and 38 and in the papers concerning the applications described further on.

A considerable virtue of the symbolic addition procedure is that, because of its efficiency, a main part of the procedure for phase determination can be carried through by hand. For many years, the procedure for centrosymmetric crystals was carried out in our laboratory completely by hand. In the case of noncentrosymmetric crystals, the first stage, which involved the use of formula (19), was performed by hand until about 100 phases were evaluated and useful relationships developed among the symbols. Only after selected numerical values were assigned to the few remaining symbols was the tangent formula (20) applied with the aid of a computer. The benefits of this aspect of the efficiency of symbolic addition have been the opportunity for those with modest computing facilities to carry out structure determinations, the possibility of close interaction with the phase determination as it progresses, and the educational value for those newly learning about phase determination to be able to witness and carry through the procedure by hand.

As the application of direct phase determination began to increase during the 1960's and structure determination became more and more a part of research programs, there began to be developed at the end of the 1960's "program packages", software for determining structures from X-ray diffraction data. Among the ones that are widely used, alternative numerical values have been used instead of symbolic phases in the case of noncentrosymmetric crystals, although there are some programs that retain the use of symbols for such types of crystals. For those programs that use alternative numerical values for phases, large numbers of alternative phase sets are generated by use of the tangent formula (20) and the selection of the most likely solutions is dependent upon the use of an elaborate set of probability formulas, auxiliary formulas and acceptance criteria. Other computational techniques have also evolved. For example, random sets of phases have been used as starting sets to be refined by application of the tangent formula (20). By considering large numbers of alternative starting sets it is often practicable to obtain a correct answer, although a large amount of computing is involved. There are also special programs for specific purposes such as the development of a structural fragment into a complete structure.

The computer programs are quite successful with centrosymmetric crystals and also do fairly well with noncentrosymmetric crystals having up to 100 independent (nonhydrogen) atoms to be placed in the unit cell. On occasion an answer will not be forthcoming from the use of a program package. In that case, crystallographers may pursue the problem with special techniques and the application of insights and acumen that have been too special to be found in current programs.

Some names that have been associated with the preparation and dissemination of computer programs for various aspects of automated, direct structure determination are Ahmed, Andrianov, Beurskens, Germain, Gilmore, Hall, Main, Schenk, Sheldrick, Stewart, Viterbo and Woolfson. Among them, some of their programs have enjoyed a broader range of popularity than others. Insight into the contents and philosophy of the programs can be obtained from several publications of the Commission on Crystallographic Computing of the International Union of Crystallography [48,49,50,51,52].

APPLICATIONS

This section will be devoted to illustrating the broad range of applications that have been made accessible by the development of direct structure determination. The examples will be mainly taken from my laboratory but will be seen to be representative of activities that now produce thousands of structural investigations each year.

The earliest applications after the publication of the monograph (23) were collaborations with colleagues at the U.S. Geological Survey on colemanite [53] and meyerhofferite [54]. This was followed by the initial investigations based on the experimental work of Isabella Karle, for example, on p,p'-

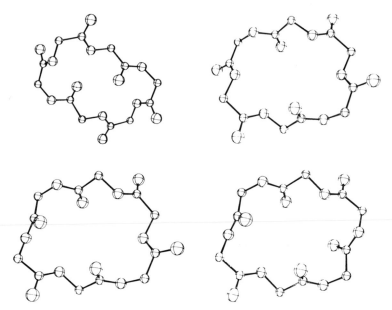

Fig. 3. The four cocrystallizing conformers of cyclohexaglycyl. The conformer at the upper left occurs four times in the unit cell, the one at the lower left occurs twice and the other two occur once.

dimethoxybenzophenone [55] and N-benzyl-1,4-dihydronicotinamide [56] that, in time, led to the symbolic addition procedure.

Early applications of the symbolic addition procedure
The first application of the symbolic addition procedure concerned cyclohexaglycyl, a synthetic polypeptide in which six glycine residues form an eighteen-membered ring. The polypeptide crystallizes as a hemihydrate in the triclinic space group P$\bar{1}$ with eight molecules in the unit cell [36], Fig. 3. An interesting characteristic of this structure is that the eighteen-membered ring occurs in the unit cell in four different conformations. Another feature is that all hydrogen atoms capable of forming hydrogen bonds are so involved.

Noncentrosymmetric crystals are quite common among substances of biochemical interest. The determination of the structure of the amino acid L-arginine dihydrate [24], which crystallizes in space group P$2_1 2_1 2_1$, is the first noncentrosymmetric structure determined by the direct method. As seen in the diagram of the contents of the unit cell, Fig. 4, the hydrogen-bonding indicated by the dotted lines is extensive. As a consequence of this hydrogen-bonding the molecules of arginine form an infinite chain. In addition, water molecules also form hydrogen bonded infinite chains which are perpendicular to the plane of the figure.

The stereochemistry of reserpine, a drug that has been used to control hypertension and nervous disorders, had been determined by chemical means. It was of interest, however, to establish the spatial arrangement of the atoms in the molecule. Reserpine was found to crystallize in the noncentrosymmetric

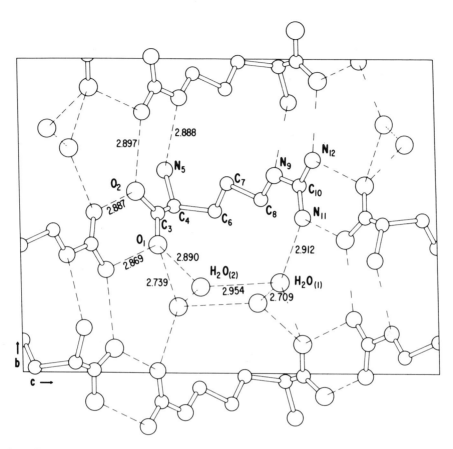

Fig. 4. The structure of L-arginine dihydrate viewed along the *a* axis. Hydrogen bonding is shown by the dashed lines.

space group, P2$_1$ [57]. In this investigation, the symbolic addition procedure yielded a partial structure which was developed into a complete structure, Fig. 5, by use of the recycling procedure [39] involving the tangent formula. Fig. 5 shows sections from an electron density map projected down the b-axis. The trimethoxybenzoxy group at the bottom is nearly perpendicular to the rest of the molecule. The indole group at the top of the diagram is planar and the dihedral angle between least-squares planes for the indole group and the benzoxy group is 82°.

Identification and stereoconfiguration

The problems concerning the determination of molecular formula and stereo-configuration can become especially acute when the amount of sample available is very small, when the chemical linkages are new and unusual or when the number of asymmetric centers is large. Under such circumstances, the use of crystal structure analysis can be not only quite helpful but also essential. An example of this is provided by batrachotoxin, a powerful neurotoxin that can be

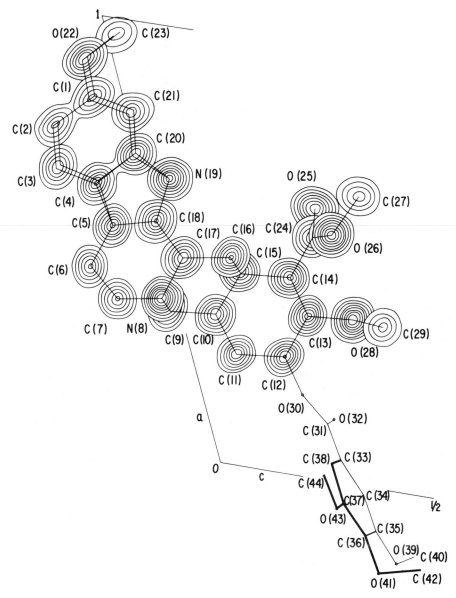

Fig. 5. Sections from a three-dimensional electron density map for the alkaloid reserpine projected down the *b* axis. The electron density contours are equally spaced at le.Å^{-3} starting with the le.Å^{-3} contour.

extracted from the skin of frogs, *Phyllobates aurotaenia*, from tropical America. It is used by native Indians to tip blow darts for hunting. Purified congeners were obtained from ethanol extracts of the skins of about 8000 frogs. The amounts were insufficient to permit the determination of the structural formulae by standard methods. A few, very small crystals of the 0-p-bromobenzoate derivative of batrachotoxinin A, one of the congeners, were grown and one crystal was

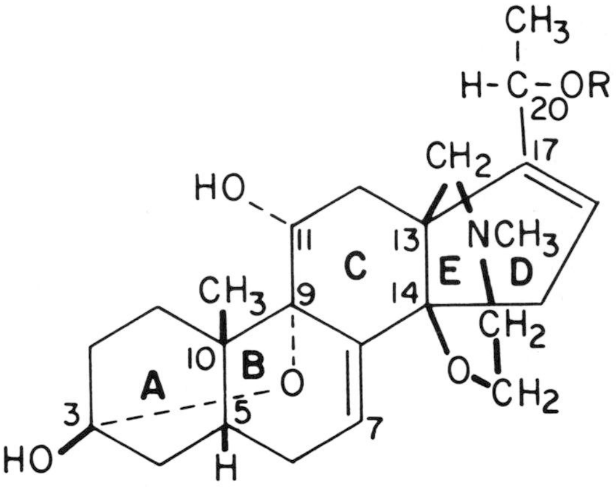

Fig. 6. The molecular formula and absolute stereoconfiguration for batrachotoxinin A (R=H) (heavy solid lines are above plane of paper and dashed ones behind).

selected for structure analysis. It was found to crystallize in space group $P2_12_12_1$. The specialized position of the Br atom (at 1/5, 0,0) and the restricted amount of data available limited the amount of information derivable from knowledge of the location of the Br atom. The structure was determined by use of the tangent formula to develop partial structural information into a complete structure. It showed that batrachotoxinin A was a steroidal alkaloid with several novel features [58], Fig. 6. Other information obtained was the stereoconfiguration at the nine asymmetric centers, the conformations of the six rings including the seven-membered ring containing the alkaloid function, values of bond lengths, bond angles and torsional angles, the location of intermolecular hydrogen bonds, and the packing of the models in the unit cell. The absolute configuration was established by measurement of the anomalous scattering of the Br atom [59]. From the information obtained from the structural analysis, the structure and absolute configuration of batrachotoxin and many other congeners were readily deduced [60].

Another small frog, *Dendrobates aurotaenia*, occurring in Columbia and Ecuador, secretes from its skin defensive substances among which are two major toxic alkaloids, histrionicotoxin and dihydroisohistrionicotoxin. These alkaloids are quite unique having a spiropiperidine system with acetylenic and allenic moieties, Fig. 7. The molecular structures, stereoconfigurations and absolute configurations were established by crystal structure analyses of a hydrochloride and hydrobromide of the histrionicotoxin and a hydrochloride of the dihydroiso compound [61,62]. Space groups C2 and $P2_12_12_1$ were involved in the analyses. The histrionicotoxins appear to offer the first examples of the occurrence of acetylene and allene moieties in the animal kingdom. Other congeners that occur in smaller quantities were shown, subsequently, by means of mass and NMR spectra to differ only in the saturation of the two side chains [63 and references therein]. The spiro ring system, with the internal NH...O hydrogen bond, remained unchanged.

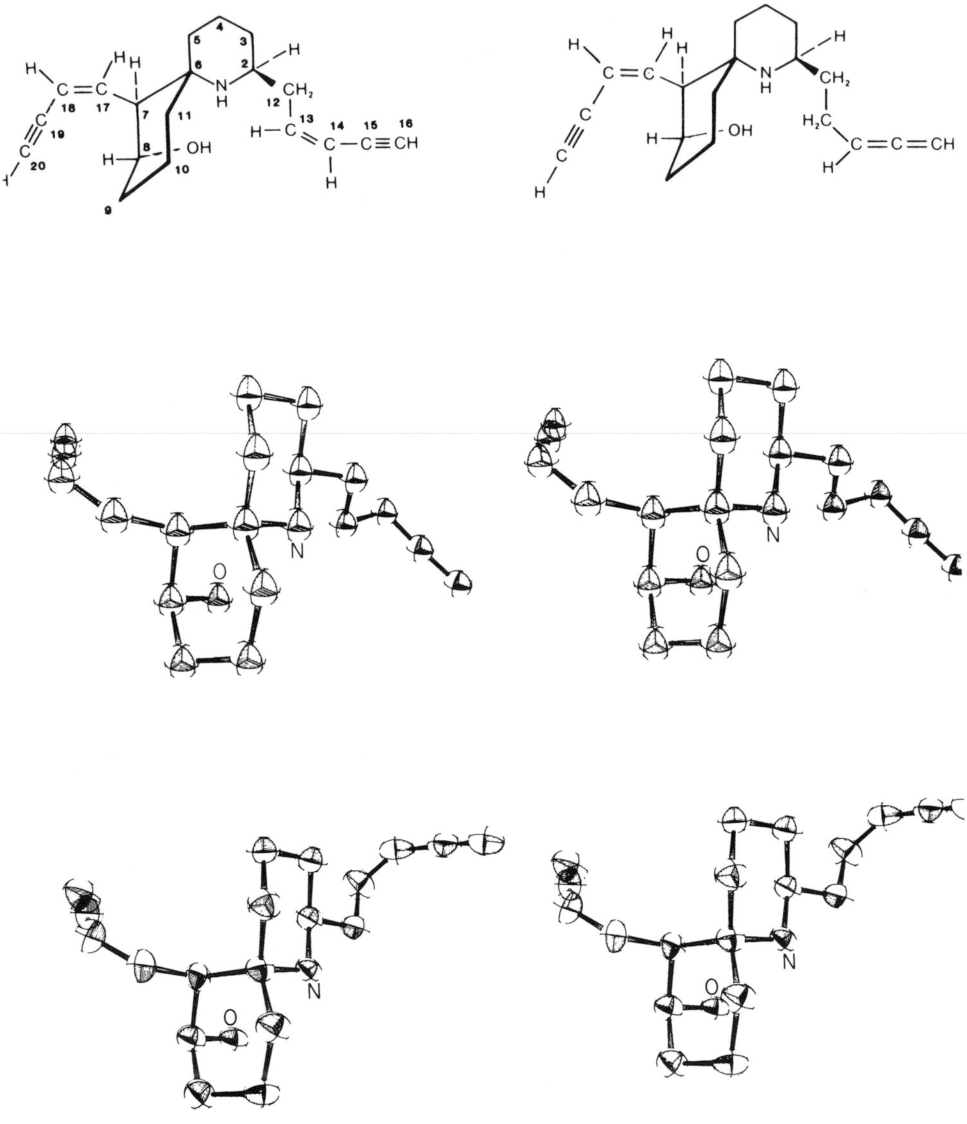

Fig. 7. Histrionicotoxin is shown on the left top and in the corresponding stereodiagram on the bottom. Dihydroisohistrionicotoxin is shown on the right top and in the stereodiagram in the middle.

Another example of the application of X-ray crystal structure analysis to the determination of structural formula and stereoconfiguration of an unknown substance is illustrated by the investigation of an alkaloid derived from an *Ormosia* plant, jamine, which crystallizes in the triclinic space group, P$\overline{1}$ [64]. The configuration determined for the molecule is shown in Fig. 8. The connectivity was found to comprise six six-membered rings, five of which were in the chair configuration and one in the boat configuration. There were six asymmetric carbon atoms and the stereoconfiguration about each was, evidently, readily

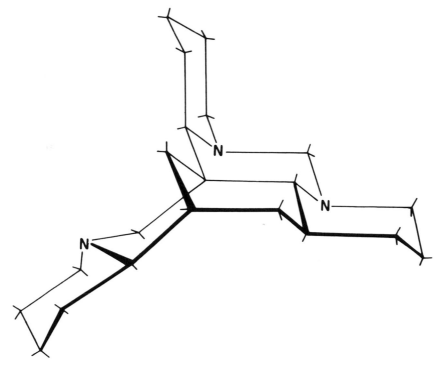

Fig. 8. The stereoconfiguration of jamine.

apparent. The identification of the nitrogen atoms was also readily made in the course of the structure determination.

Brassinolide [65–67] is a very potent plant growth promoter. It is active in very small amounts, 1–10 ng/plant. Brassinolide generates cell enlargement and stimulates cell division in many food plants and plants grown in arid regions that yield oil and other energy related materials. The yield from extraction from plant pollen is very low and therefore large-scale testing awaited chemical synthesis. Evidently, chemical synthesis would be facilitated by structural information. A few crystals of brassinolide became available for crystal structure determination. The crystallization took place in space group $P2_1$ and the analysis [66] established that the steroid nucleus of the molecule had a seven atom B-ring lactone, an unprecedented feature for a natural steroid, Fig. 9. The presence of the lactone appears to be responsible for promoting the plant growth. Subsequent synthesis has shown that compounds analogous to brassinolide are easier to prepare in large quantities and have adequate plant-growth regulating properties. Field evaluations of these regulators are currently underway.

There are very large numbers of examples that could be discussed under the heading of this section. It is noteworthy, however, that the implications of only a small number of definitive structure determinations can be extended far beyond the immediate results of the particular determinations. An outstanding

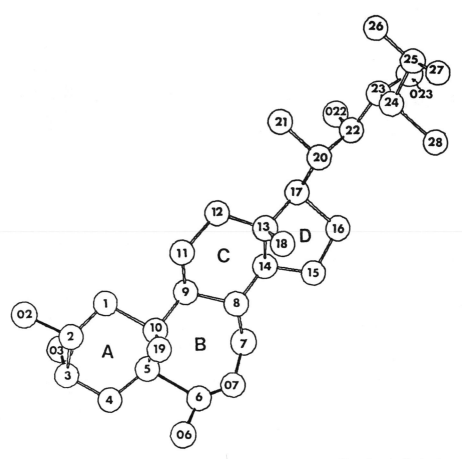

Fig. 9. The molecular formula and stereoconfiguration of brassinolide. Note that the B ring is a seven-membered lactone ring.

example is given by some fundmental structural investigations on frog toxins which have provided the information on which was based the subsequent establishment of molecular formulas and stereoconfigurations of over 200 related frog toxins [68,69].

REARRANGEMENTS

In the case of rearrangement reactions, crystal structure determination can again play a particularly useful role because many rearrangement reactions give products that are the result of vast and unanticipated changes in the starting materials. The following examples provide illustrations of such changes.

A photorearrangement reaction in which a major rearrangement takes place is illustrated by the reaction shown in Fig. 10. A crystal structure investigation of a single optically active crystal, selected from a racemic conglomerate established the structural formula and configuration of the photoproduct [70,71]. The substance crystallized in space group $P2_12_12_1$. The structure

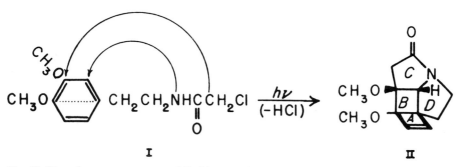

Fig. 10. The photorearrangement of N-chloroacetyldimethoxyphenethylamine to a fused ring system consisting of two four- and two five-membered rings.

analysis showed that the photoproduct consisted of four ring systems, two five-membered and two four-membered rings.

Ultraviolet irradiation of N-chloracetyltyramine, where there is a hydroxyl group on the phenyl ring in contrast to the two methoxy groups in the previous example, yields entirely different photorearrangement products. HCl was eliminated and two unusual photodimers, shown in Fig. 11, were produced. Their molecular formulas and stereoconfigurations have been identified by use of crystal structure analysis [72]. It is interesting to note that dimer (II) is the more stable since it is produced from dimer (I) by use of additional ultraviolet radiation. Dimer (I) crystallizes in space group $P2_1/c$ and dimer (II) crystallizes in space group Pbca. Dimer (I) is seen to have a central cage bounded by four six-membered rings and two four-membered rings. Each four-membered ring is puckered, with the torsion angles around the ring bonds having values of about 20°. The six-membered rings assume distorted boat conformations. Dimer (II) is seen to have a more complex, partially open, cage bounded by one three-, two five-, two six-, and one seven-membered ring. The six-membered rings are again in a distorted boat conformation. Once the structural characteristics of the photoproducts are known, it is possible to consider possible reaction mechanisms that describe the intermediate changes that occur in the rearrangements of the initial materials resulting in the final products. Postulated mechanisms for the formation of dimers (I) and (II) have been presented [73].

Irradiation with ultraviolet light of a solution containing 3-methyl-5,6-diaza-2,4-cyclohexadien-1-one and 2-propenol (I) produces a cyclobutane addition compound (II) which, upon further irradiation, opens and switches bonds to form a tricyclic molecule (III), Fig. 12. Crystal structure analysis [74] of the material which crystallizes in space group P$\bar{1}$ established that the final photoproduct was composed of a fused ring system, one three-membered ring and two five-membered rings.

CONFORMATION

The conformations of molecules can be importantly related to their chemical and physiological behavior. Crystal structure investigations can be helpful in

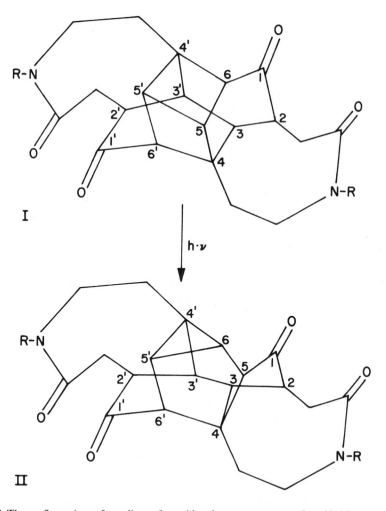

Fig. 11. The configurations of two dimers formed by photorearrangement from N-chloroacetyltyramine. Dimer (II) is seen to be formed from dimer (I) by continued irradiation with ultraviolet light.

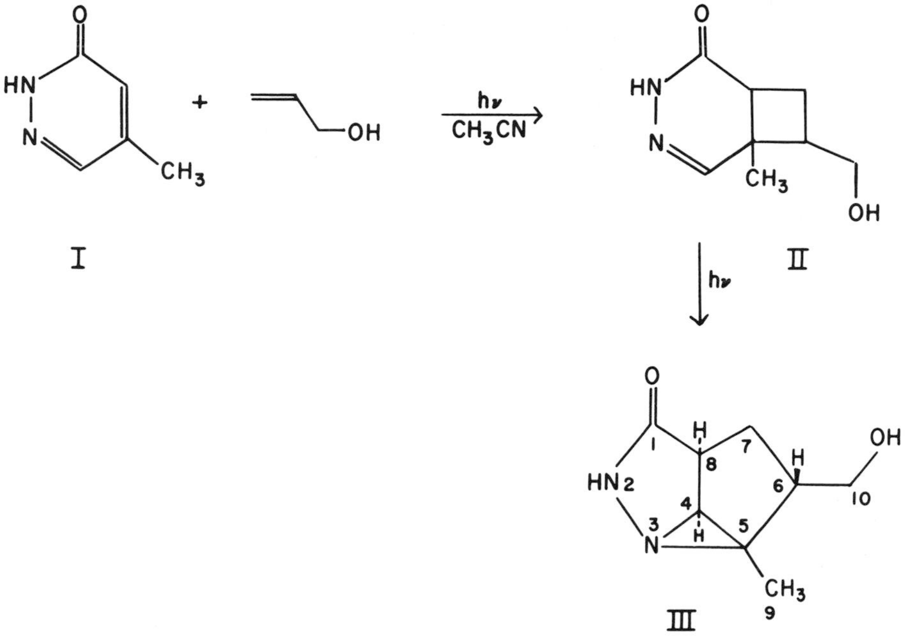

Fig. 12. Irradiation with ultraviolet light of a solution of 3-methyl-5,6-diaza-2,4-cyclohexadien-1-one (I) and 2-propenol leads to a cyclobutane addition product (II). On further irradiation, (II) rearranges to a tricyclic product (III).

providing conformational information. It may be argued, and rightly so, that biologically active materials may assume conformations in the crystalline state that they would not assume in solution. There are, however, numerous instances of conformational studies in which the results of crystal structure analyses are either highly suggestive or rather definitive. The following examples are such instances.

One way in which the crystalline state can imitate the cirumstances found in solution is to include in the crystallization relatively large amounts of solvent. Such a crystal is formed by [Leu⁵]enkephalin (Tyr-Gly-Gly-Phe-Leu) grown from N,N-dimethylformamide (DMFA)/water. Endogenous enkephalin is a linear pentapeptide that occurs in the brain as [Leu⁵]enkephalin and [Met⁵]-enkephalin in varying proportions depending upon the species [75]. Both are quite flexible and the structural study of [leu⁵]enkephalin was undertaken in the hope that insight into probable conformations would be so derived. The crystallization took place in space group P2₁ and it turned out that there were four molecules of the peptide in the asymmetric unit of the unit cell, each having a different conformation [76], Fig. 13. The large amount of solvent surrounding the molecules suggests that the conformations may be relevant to the circumstances in solution. The four conformers with extended backbones form an infinite antiparallel β-sheet. β-sheets related by the twofold screw axis are separated by a 12 Å spacing. Side groups protrude above and below the β-sheets and are entirely immersed in a thick layer of solvent that fills the volume

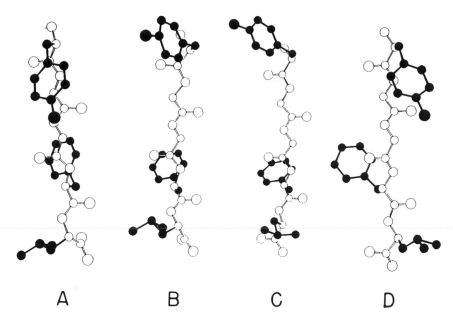

A   B   C   D

Fig. 13. Four conformations of [Leu^5]enkephalin that occurred in N,Ndimethylformamide/water. Note the similarities in the backbone conformations and the differences in the shaded side chains.

between the β-sheets. The crystal was stable only in contact with the mother liquor. Structurally, it consisted of rather rigid sheets of peptide molecules separated by spaces filled with mobile solvent. Many of the solvent molecules appeared from the X-ray analysis to be completely disordered. An asymmetric unit of the unit cell contained four enkephalin molecules, eight water molecules, eight DMFA molecules and an unknown number of disordered solvent molecules.

The crystal of enkephalin contains more than 210 independent C, N and O atoms and is of a size that lies between what is normally considered small-structure crystallography (up to 100 nonhydrogen atoms or so) and protein crystallography (about 500 nonhydrogen atoms or more).

The visual chromophores, 11-*cis* and all-*trans* retinal are present in both rod and cone cells of vertebrate retinas and in the corresponding organs of insects and crustacea [77]. The photochemical isomerization of the 11-*cis* isomer to the *trans* form is an important step in the visual process with the 11-*cis* isomer acting as a photochemical sensor. In the dark, 11-*cis* retinal is covalently linked to proteins in the retina known as opsins. They differ in different species and may differ in different cells of the retina. The details of the geometry of the retinals had not been determined and, in view of their importance to biological function, a crystal structure investigation was undertaken on the 11-*cis* and all-*trans* forms. Special precautions had to be observed with the 11-*cis* retinal. It is unstable to light and oxygen and its stability is favored by lower temperatures. In solution, the *cis* isomer isomerizes to the *trans* form very readily at 20°C. Fortunately, crystals of the cis isomer are more stable than its solution. The

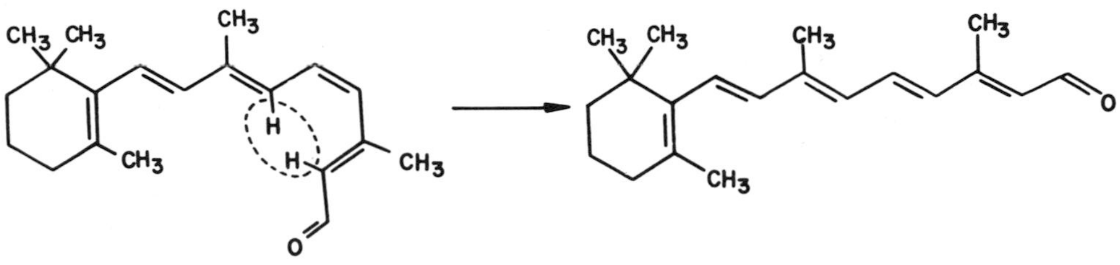

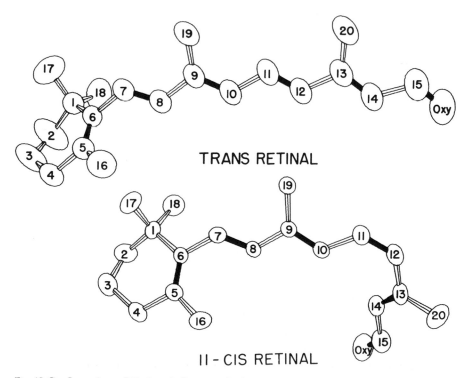

Fig. 14. Two isomers of retinal, the 11-*cis* and all-*trans* forms.

diffraction data for the *cis* form were collected in a light-tight container that was continuously purged with dry nitrogen and held at 16–17°C. Crystal structure analyses [78,79] of the 11-*cis* and all-*trans* forms revealed the detailed geometry of both, Fig. 14. An additional, independent structure analysis of all-*trans* retinal was also reported [80]. Of particular interest was the conformation of the *cis* form as there had been much speculation and theoretical work concerning its conformation. The 11-*cis* retinal crystallized in space group $P2_1/c$ and the *trans* isomer in space group $P2_1/n$. The all-*trans* retinal chain is planar with the six-membered ring inclined to the plane of the chain. The inclination is given by the torsional angle of the 5-6-7-8 segment which is found to be 59°. The torsion angle of a planar *trans* conformation is 180° and that of a planar *cis* conformation is 0°. The conformation for the all-*trans* isomer is typical of

TRANS RETINAL

11 - CIS RETINAL

Fig. 15. Configurations of 11-*cis* and all-*trans* retinal.

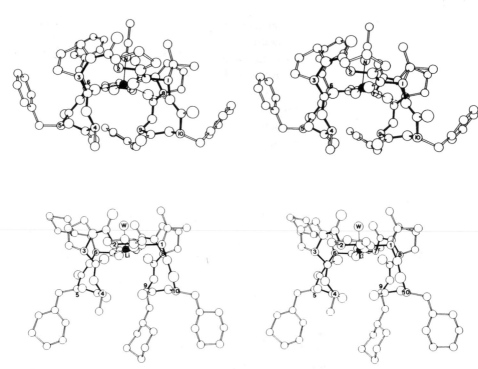

Fig. 16. Phenyl rings folded up around the backbone of a biologically active antamanide complex with Li$^+$ or Na$^+$ providing a lipophilic surface (top pair). In the inactive analogue, Li$^+$·perhydro-antamanide, the cyclohexyl groups are folded down, thus exposing the peptide backbone to the environment (bottom pair).

carotenoids. In forming the 11-*cis* retinal from the all-*trans* form, there is not only a rotation of 180° about the C(11)=C(12) double bond, but also a rather unexpected rotation of 141° about the C(12)-C(13) single bond. This additional rotation relieves the difficulty of too close a contact in the absence of the additional rotation between the hydrogen atom on C(10) and the hydrogen atoms of the methyl group attached to C(13) (see the diagram in Fig. 15). At the time that the manuscript concerning the crystal structure determination of 11-*cis* retinal was being prepared, a theoretical calculation was published by Honig and Karplus [81] in which a ground-state configuration for 11-cis retinal was proposed that closely resembled the result of the crystal structure determination.

The cyclic decapaptide antamanide acts as an antidote to the toxin phalloidin found in the deadly poisonous mushroom *Amanita phalloides*. Antamanide can also be isolated from the same mushroom but occurs in much smaller quantities. The synthetic analog of antamanide in which all four phenylalanyl residues are hydrogenated to cyclohexylalanyl residues (Cha), cyclic(ValPro-ProAlaChaChaProProChaCha), has no antitoxic potency despite its ability to form ion complexes in the same manner as antamanide. A conformational analysis of the hydrogenated antamanide was carried out by means of a crystal structure analysis of Li$^+$·perhydroantamanide which crystallizes in space group P2$_1$2$_1$2$_1$ [82], Fig. 16. The backbone encapsulates a Li$^+$ ion in quite the same fashion as in Li^+·antamanide. In the Li^+·antamanide, however, the four phenyl groups are folded against the backbone, thus providing a hydrophobic surface for the complex, whereas in Li^+·perhydroantamanide the four cyclohexyl moieties are extended away from the folded backbone, with the consequent exposure of large portions of the polar backbone to the surrounding environment. As a result, elements of the backbone that would be otherwise shielded from the environment were found to make hydrogen bonds and ligands that would not occur in the Li^+·antamanide complex. It would appear that the large change of hydrophobicity around the backbone in the perhydroantamanide complex is related to the loss of biological activity.

These few applications represent only a minuscule portion of the broad range of topics and individual studies represented by the thousands of structural investigations performed each year. They do illustrate, however, how structure determination can play a useful and often indispensable role in the progress of many research disciplines.

ANALYSIS OF MACROMOLECULAR STRUCTURE: RECENT INTERESTS

The analysis of structures of macromolecules, molecules that consist of at least about 500 nonhydrogen atoms and more, is facilitated by use of specialized heavy-atom techniques that have the titles anomalous dispersion and isomorphous replacement. Heavy atoms can often be soaked into a macromolecular crystal without seriously disturbing the structure. As a consequence, the techniques of anomalous dispersion and isomorphous replacement have enjoyed considerable success in the determination of the structure of native (unsubstituted) macromolecules.

Anomalous dispersion effects arise from the fact that the atomic scattering factors, f, as would be used in (3), are in general complex numbers defined by

$$f = f^n + f' + if'' \tag{25}$$

where f' and f'' are the real and imaginary parts, respectively, of the correction to f^n and are considered the contributions of anomalous dispersion. The scattering factor f^n is obtained from computations in which it is assumed that the frequency of the radiation is much larger than the absorption frequencies for the constituent atoms. The consequence of having atomic scattering factors with a significant imaginary part, as occurs with the heavier atoms at commonly used wavelengths, is that the intensity measured for an acentric reflection **h** is, in general, different than that for **-h**. The fact that $I_h \neq I_{\bar{h}}$ has been the basis of a technique that has broad applications [83] in the field of structure research.

Crystals that have the same unit cell geometry but differ in chemical composition are called isomorphous. Since it is often possible to substitute heavy atoms into macromolecules with relatively minor perturbations on the structure, the criterion for isomorphicity can be fairly readily achieved to good approximation for such molecules. The method of phase determination based on one substituted isomorphous structure is called the single isomorphous replacement method (SIR) and if several isomorphous structures are available it is called the multiple isomorphous replacement method (MIR). If two isomorphous structures are centrosymmetric, the signs of the structure factors for both crystals can be determined. In practice, some signs would remain indeterminate because of experimental errors and the presence of structure factors of small magnitude. When two noncentrosymmetric crystals form an isomorphous pair, analysis shows that a twofold ambiguity occurs in the evaluation of the phases. It can be readily shown, however, that in the case of noncentrosymmetric crystals many individual phases can be unambiguously evaluated when the structure of the substituted atoms is known [84]. Such phases are either close to the value of the corresponding phases for the substituted atoms or π away from those values. There are a number of ways to resolve the ambiguity from isomorphous replacement. They involve, for example, the use of a second isomorphous substitution or, instead, the additional use of anomalous dispersion data.

Multiple isomorphous substitution with the occasional additional use of anomalous dispersion data has been the technique that has led to the enormous progress in macromolecular structure research of biological importance for the past 25 years. The early theoretical work that supported the advances in this field is that of Patterson [4,5] and Bijvoet [85]. With the types of heavy atoms that are usually substituted into macromolecular structures, there can be significant effects from anomalous dispersion, so that data collection for both anomalous dispersion and isomorphous replacement can be, in effect, combined.

My interest in the techniques of anomalous dispersion and isomorphous replacement was rekindled by a development in my laboratory that demonstrated the considerable potential of the anomalous dispersion technique in structure determination. In an application of the anomalous dispersion technique to the structure determination of the protein, crambin, Hendrickson and Teeter [86] solved the structure by use of the anomalous scattering from six sulfur atoms occurring in three disulfide bridges. The experiment was performed as a single wavelength experiment with CuKα radiation which is far from the absorption edge for S at 5.02 Å. The success with the use of the relatively weak anomalous scattering from the S atoms coupled with the great span in wavelength between that of the absorption edge and the wavelength of CuKα radiation stimulated me to initiate a program to investigate various theoretical aspects of the single and multiple wavelength anomalous dispersion techniques and also, in time, isomorphous replacement.

One investigation concerned an exact algebraic analysis of single and multiple wavelength anomalous dispersion [87] by use of the structure factor equations (3) in which the atomic scattering factors are complex according to (25). The appropriate structure factor equations are treated as simultaneous equations and the key to obtaining the final form in which the equations were expressed was the separation of the contributions from the normal atomic scattering factors f^n from those of the real and imaginary corrections to f^n, f' and f'', respectively.

A simple result of the algebraic analysis that concerns the case when a structure is composed of atoms that scatter normally and one type of atom that scatters anomalously is as follows,

$$|F_{\lambda h}|^2 = |F_{1,\,h}^n|^2 + \left\{ 1 + (f_{\lambda 2,\,h}^a/f_{2,\,h}^n) \left[(f_{\lambda 2,\,h}^a/f_{2,\,h}^n) + 2\cos\delta_{\lambda 2,\,h} \right] \right\} |F_{2,\,h}^n|^2$$

$$+ 2 \left[1 + (f_{\lambda 2,\,h}^a/f_{2,\,h}^n) \cos\delta_{\lambda 2,\,h} \right] |F_{1,\,h}^n||F_{2,\,h}^n|\cos (\phi_{1,\,h}^n - \phi_{2,\,h}^n) \tag{26}$$

$$+ 2 (f_{\lambda 2,\,h}^a/f_{2,\,h}^n) \sin\delta_{\lambda 2,\,h} |F_{1,\,h}^n||F_{2,\,h}^n| \sin (\phi_{1,\,h}^n - \phi_{2,\,h}^n)$$

where $|F_{\lambda h}|^2$ is the measured magnitude squared of the structure factor at the wavelength λ, $|F_{1,\,h}^n|^2$ is the magnitude squared of the structure factor for the nonanomalously scattering atoms and $|F_{2,\,h}^n|^2$ is the magnitude squared of the structure factor for the anomalously scattering atoms, but scattering as if they were doing so normally. The measured $|F_{\lambda h}|^2$ are corrected for vibrational effects and the latter are also absent from $|F_{1,\,h}^n|^2$ and $|F_{2,\,h}^n|^2$. The quantities $f_{\lambda j,\,h}^a$ and $\delta_{\lambda j,\,h}$ are defined for a particular λ,

$$f_{\lambda j,\,h}^a = \left[(f_{\lambda j,\,h}')^2 + (f_{\lambda j,\,h}'')^2 \right]^{1/2} \tag{27}$$

and

$$\delta_{\lambda j,\,h} = \tan^{-1}(f_{\lambda j,\,h}''/f_{\lambda j,\,h}') \tag{28}$$

The f' and f'' are tabulated in International Tables [88]. The phase angles $\phi_{1,\,h}^n$ and $\phi_{2,\,h}^n$ are the angles associated with $|F_{1,\,h}^n|$ and $|F_{2,\,h}^n|$, respectively.

Four equations can be formed from (26) by performing anomalous dispersion experiments at two different wavelengths and making measurements of the intensities of relections associated with $\mathbf{h}$ and $-\mathbf{h}$. The equations are linear if the four quantities, $|F_{1,\,h}^n|^2$, $|F_{2,\,h}^n|^2$, $|F_{1,\,h}^n| |F_{2,\,h}^n| \cos (\phi_{1,\,h}^n - \phi_{2,\,h}^n)$ and $|F_{1,\,h}^n| |F_{2,\,h}^n| \sin (\phi_{1,\,h}^n - \phi_{2,\,h}^n)$ are chosen as the unknowns. It is worthwhile to add the quadratic relationship, $\sin^2\phi + \cos^2\phi = 1$, and treat the defining equations in a least-squares fashion.

A general analysis [87] has also been carried out which is appropriate for any number and type of anomalous scatterers. This general formulation has a number of favorable characteristics that are illustrated by the simple set of simultaneous equations formed from (26) by using multiple wavelengths and measuring the intensities at $\mathbf{h}$ and $-\mathbf{h}$. The unknown quantities are intensities of scattering and phase differences, all associated with the nonanomalous scattering. The intensities are those that would be obtained from individual types of atoms as if each type were present in isolation from all the rest. Knowledge of the intensities for the structure of an individual type of atom can facilitate the determination of the structure associated with this type of atom. Once this is known, the entire structure can be readily determined. The anomalous scattering enters the simultaneous equations as separate factors in terms of known, tabulated quantities. With appropriate definitions of unknown quantities, the simultaneous equations are linear. It is useful to introduce also quadratic relationships that exist among the unknown phase differences. The systems of simultaneous equations involve no approximations and remain exact for any number or type of anomalous scatterers.

There are some additional features of the equations (26). One concerns the fact that the usual analysis of single wavelength anomalous scattering leads to the conclusion that there would be a twofold ambiguity in the evaluation of certain phase differences. It has been recently shown [89] that information that is inherent in the measured intensities $|F_{\lambda h}|^2$ and $|F_{\lambda \bar{h}}|^2$ that gives approximate information concerning $|F^n_{1,\,h}|^2$ and $|F^n_{2,\,h}|^2$ can be used with (26) for $\mathbf{h}$ and $-\mathbf{h}$ to obtain unique or essentially unique values for the phase differences that occur in (26) with potentially useful accuracy. This was specifically shown for a one-wavelength experiment for the case of one type or one predominant type of anomalously scattering atoms in the presence of essentially nonanomalously scattering atoms. It should be more generally true. Hauptman [90] showed earlier the uniqueness of a one-wavelength anomalous dispersion experiment in terms of unique values for triplet phase invariants.

A second feature of equations (26) and their generalization to many kinds of anomalous scatterers concerns the fact that anomalous scatterers in the form of heavy atoms are often substituted into a native substance that scatters essentially nonanomalously. In that case, intensity data for the native material corresponds to $|F^n_{1,\,h}|^2$ in (26), thus reducing the number of unknown quantities.

Another type of theoretical study designed to elicit phase information from anomalous dispersion and isomorphous replacement concerns the development of formulas for evaluating the so-called triplet phase invariants. Triplet phase invariants are sums of three phases whose subscripts add to zero, e.g. $\phi_{\mathbf{h}} + \phi_{\mathbf{k}} + \phi_{\overline{\mathbf{h}}} + \phi_{\overline{\mathbf{k}}}$. In general, the sum of subscripts is carried out by attaching the sign associated with the corresponding phase to each subscript, e.g., $\phi_{\mathbf{h}} + \phi_{\mathbf{k}} - \phi_{\mathbf{h+k}}$ is also an invariant. Note that there are triplet phase invariants in (10) and (19) forming the "sum of angles" formula.

There have been recently two main approaches to the evaluation of triplet phase invariants from isomorphous replacement and anomalous dispersion data. One involves the use of the mathematics of the joint probability distribution by Hauptman for isomorphous replacement [91] and for anomalous dispersion [90], a similar theory for anomalous dispersion by Giacovazzo [92] and a theory for both techniques by Pontenagel, Krabbendam, Peerdeman and Kroon [93]. A recent investigation by Fortier, Moore and Fraser [94] of the evaluation of triplet phase invariants for isomorphous replacement when the heavy-atom structure is known indicates the potential of this approach for good accuracy.

A second approach to the evaluation of triplet phase invariants from isomorphous replacement and anomalous dispersion data has interested me for the past few years. It involved initially the development of formulas by use of approximate but simple algebraic manipulations that were based on certain mathematical and physical characteristics of the two techniques [95 96]. This was followed by algebraic analyses [84,97,98] that indicated a potential for good accuracy for many of a large number of formulas proposed.

The development of optimal strategies for the use of triplet phase invariants awaits practical experience. If the heavy-atom (anomalously scattering) substructure is known, there are a number of techniques for obtaining values for individual phases [84,87,89]. In that case, it is quite possible that the triplet phase invariants would be used mainly for phase extension and refinement. In those circumstances when the heavy-atom (anomalously scattering) substructure is not known and resists analysis, the triplet phase invariants may be able to play a primary role in reaching a solution to the structure since the triplet phase invariants can be evaluated without knowledge of the substructure.

The enhanced opportunities to carry out multiple wavelength anomalous dispersion experiments with high intensity and tunable synchrotron X-ray sources should facilitate the development of applications of the new theoretical formulations. Coupled with the major advances in computing, the new theories have the potential to improve the facility with which macromolecular structure research is performed.

REFERENCES

1. Ott, H. (1928) Z. für Krist. *66* ,136—153.
2. Banerjee, K. (1933) Proc. Roy. Soc. A *141*, 188—193.
3. Avrami, M. (1938) Phys. Rev. *54* 300—303.
4. Patterson, A. L. (1934) Phys. Rev. *46*, 372—376.
5. Patterson, A. L. (1935) Z. für Krist. *90*, 517—542.
6. Hoppe, W. (1957) Z. Elektrochemie *61*, 1076—1083.
7. Nordman, C. E. and Nakatsu, K. (1963) J. Am. Chem. Soc. *85*, 353—354.
8. Huber, R. and Hoppe, W. (1965) Chem. Ber. *98*, 2403—2424.
9. Harker, D. and Kasper, J. S. (1948) Acta Cryst. *1*, 70—75.
10. Kasper, J. S., Lucht, C. M. and Harker, D. (1950) Acta Cryst. *3*, 436—455.
11. Gillis, J. (1948) Acta Cryst. *1*, 174—179.
12. Karle, I. L. and Karle, J. (1949) J. Chem. Phys. *17*, 1052—1058.
13. Karle, J. and Karle, I. L. (1950) J. Chem. Phys. *18*, 957—962.
14. Karle, I. L. and Karle, J. (1950) J. Chem. Phys. *18*, 963—971.
15. Hauptman, H. and Karle, J. (1950) Phys. Rev. *77*, 491—499.
16. Bartell, L. S. and Brockway, L. O. (1953) Phys. Rev. *90*, 833—838.
17. Toeplitz, O. (1911) Rend. Circ. Mat. Palermo, 191—192.
18. Karle, J. and Hauptman, H. (1950) Acta Cryst. *3*, 181—187.
19. Karle, J. (1971) Acta Cryst. B*27*, 2063—2065.
20. Hauptman, H. and Karle, J. (1952) Acta Cryst. *5*, 48—59.
21. Karle, J. and Hauptman, H. (1953) Acta Cryst. *6*, 131—135.
22. Hauptman, H. and Karle, J. (1953) Acta Cryst. *6*, 136—141.
23. Hauptman, H. and Karle, J. (1953) *American Crystallographic Association Monograph No. 3* (Polycrystal Book Service, Western Springs).
24. Karle, I. L. and Karle, J. (1964) Acta Cryst. *17*, 835—841.
25. Shmueli, U., Weiss, G. H., Kiefer, J. E. and Wilson, A. J. C. (1984) Acta Cryst. A*40*, 651—660.
26. Stewart, J. M. and Karle, J. (1976) Acta Cryst. A*32*, 1005—1007.
27. Karle, J., Hauptman, H. and Christ, C. L. (1958) Acta Cryst. *11*, 757—761.
28. Wilson, A. J. C. (1942) Nature *150*, 151—152.
29. Wilson, A. J. C. (1949) Acta Cryst. *2*, 318—321.
30. Woolfson, M. M. (1954) Acta Cryst. *7*, 61—64.
31. Cochran, W. and Woolfson, M. M. (1955) Acta Cryst. *8*, 1—12.
32. Karle, J. and Karle, I. L. (1966) Acta Cryst. *21*, 849—859.
33. Watson, G. N. (1945) *Theory of Bessel Functions*. Cambridge Univ. Press.
34. Cochran, W. (1955) Acta Cryst. *8*, 473—478.
35. Karle, J. and Hauptman, H. (1956) Acta Cryst. *9*, 635—651.
36. Karle, I. L. and Karle, J. (1963) Acta Cryst. *16*, 969—975.
37. Karle, I. L., Dragonette, K. S. and Brenner, S. A. (1965) Acta Cryst. *19*, 713—716.
38. *International Tables for X-ray Crystallography* (1974) Vol. IV, Section 6, editors, J. A. Ibers and W. Hamilton, The Kynoch Press, Birmingham.
39. Karle, J. (1968) Acta Cryst. B*24*, 182—186.
40. Tollin, P. (1966) Acta Cryst. *21*, 613—614.
41. Crowther, R. A. and Blow, D. M. (1967) Acta Cryst. *23*, 544—548.
42. Karle, J. (1972) Acta Cryst. B*28*, 820—824.
43. Langs, D. A. (1985) Acta Cryst A*41*, 578—582.
44. Zachariasen, W. H. (1952) Acta Cryst. *5*, 68—73.
45. Rumanova, I. M. (1954) Doklady Acad. Nauk. U.S.S.R. *98*, 399—402.
46. Cochran, W. and Douglas, A. S. (1955) Proc. Roy. Soc. A *227*, 486—500.
47. Sayre, D. (1952) Acta Cryst. *5*, 60—65.
48. *Crystallographic Computing* (1970) editors, F. R. Ahmed, S. R. Hall and C. P. Huber, Munksgaard, Copenhagen.
49. *Crystallographic Computing Techniques* (1976) editors, F. R. Ahmed, K. Huml and B. Sedlacek, Munksgaard, Copenhagen.

50. *Computing in Crystallography* (1980) editors, R. Diamond, S. Ramaseshan and K. Venkatesan, Indian Academy of Sciences, Bangalore.
51. *Computational Crystallography* (1982) editor, D. Sayre, Clarendon Press, Oxford.
52. *Methods and Applications in Crystallographic Computing* (1984) editors, S. R. Hall and T. Ashida, Clarendon Press, Oxford.
53. Christ, C. L., Clark, J. R. and Evans, H. T. Jr. (1954) Acta Cryst. *7*, 453.
54. Christ, C. L. and Clark, J. R. (1956) Acta Cryst *9*, 830.
55. Karle, I. L., Hauptman, H., Karle, J. and Wing, A. B. (1958) Acta Cryst *11*, 257−263.
56. Karle, I. L. (1961) Acta Cryst. *14*, 497−502.
57. Karle, I. L. and Karle, J. (1968) Acta Cryst. B*24*, 81−91.
58. Karle, I. L. and Karle, J. (1969) Acta Cryst. B*25*, 428−434.
59. Gilardi, R. D. (1970) Acta Cryst. B*26*, 440−441.
60. Tokuyama, T., Daly, J. and Witkop, B. (1969) J. Am. Chem. Soc. *91*, 3931−3938.
61. Daly, J. W., Karle, I., Myers, C. W., Tokuyama, T., Waters, J. A. and Witkop, B. (1971) Proc. Nat. Acad. Sci. USA *68*, 1870−1875.
62. Karle, I. L. (1973) J. Am. Chem. Soc. *95*, 4036−4040.
63. Tokuyama, T., Uenoyama, K., Brown, G., Daly, J. W. and Witkop, B. (1974) Helv. Chim. Acta *57*, 2597−2604.
64. Karle, I. L. and Karle, J. (1964) Acta Cryst. *17*, 1356−1360.
65. Mitchell, J. W., Mandava, N., Worley, J. F., Plimmer, J. R. and Smith, N. V. (1970) Nature *225*, 1065−1066.
66. Mandava, N., Kozempel, M., Worley, J. F., Matthews, D., Warthen, J. D. Jr., Jacobson, N., Steffens, G. L., Kenny, H. and Grove, M. D. (1978) Ind. Eng. Chem., Prod. Res. Dev. *17*, 351−354.
67. Grove M. D., Spencer, G. F., Rohwedder, W. K., Mandava, N., Worley, J. F., Warthen, J. D. Jr., Steffens, G. L., Flippen-Anderson, J. L. and Cook, J. C. Jr. (1979) Nature *281*, 216−217.
68. Witkop, B. and Gössinger, E. (1983) *The Alkaloids*, Vol. XXI, editor, A. Brossi, Academic Press, New York, 139−254.
69. Daly, J. W. (1982) Progress in the Chemistry of Organic Natural Products, Vol. 41, editors, W. Herz, H. Grisebach and G. W. Kirby, Springer-Verlag, New York, 206−340.
70. Yonemitsu, O., Okuno, Y., Kanaoka, Y., Karle, and I. L., Witkop, B. (1968) J. Am. Chem. Soc. *90*, 6522−6523.
71. Karle, I. L., Gibson, J. W. and Karle, J. (1969) Acta Cryst. B*25*, 2034−2039.
72. Jones, D. S. and Karle, I. L. (1974) Acta Cryst. B*30*, 617−623.
73. Iwakuma, T., Nakai, H., Yonemitsu, O., Jones, D. S., Karle, I. L., Witkop, B. (1972) J. Am. Chem. Soc. *94*, 5136−5139.
74. Karle, I. L. (1982) Acta Cryst. B*38*, 1022−1024.
75. Hughes, J., Smith, T. W., Kosterlitz, H. W., Fothergill, L. A., Morgan, B. A. and Morris, H. R. (1975) Nature *258*, 577−579.
76. Karle, I. L., Karle, J., Mastropaolo, D., Camerman, A. and Camerman, N. (1983) Acta Cryst. B*39*, 625−637.
77. Wald, G. (1968) Nature *219*, 800−807.
78. Gilardi, R., Sperling, W., and Karle, I. L., Karle, J. (1971) Nature *232*, 187−188.
79. Gilardi, R. D., Karle, I. L. and Karle, J. (1972) Acta Cryst. B*28*, 2605−2612.
80. Hamanaka, T., Mitsui, T., Ashida, T. and Kakudo, M. (1972) Acta Cryst. B*28*, 214−222.
81. Honig, B. and Karplus, M. (1971) Nature *229*, 558−560.
82. Karle, I. L. (1985) Proc. Natl. Acad. Sci. USA *82*, 7155−7159.
83. Ramaseshan, S. and Abrahams, S. C. (1975) editors, *Anomalous Scattering*, Munksgaard, Copenhagen.
84. Karle, J. (1986) Acta Cryst, A*42* (in press).
85. Bijvoet, J. M. (1954) Nature *173*, 888−891.
86. Hendrickson, W. A. and Teeter, M. (1981) Nature *290*, 107−113.
87. Karle, J. (1980) Int. J. Quantum Chem. Symp. *7*, 357−367.
88. Cromer, D. T. (1974) in *International Tables for X-Ray Crystallography*, Vol. IV, editors, J. A. Ibers and W. Hamilton, The Kynoch Press, Birmingham, pp. 148−151.

89. Karle, J. (1985) Acta Cryst. A*41*, 387—394.
90. Hauptman, H. (1982) Acta Cryst. A*38*, 632—641.
91. Hauptman, H. (1982) Acta Cryst. A*38*, 289—294.
92. Giacovazzo, C. (1983) Acta Cryst. A*39*, 585—592.
93. Pontenagel, W. M. G. F., Krabbendam, H., Peerdeman, A. F. and Kroon J. (1983) Eighth Eur. Crystallogr. Meet., 8—12 Aug., Abstract 4.01P, p. 257.
94. Fortier, S., Moore, N. J. and Fraser, M. E. (1985) Acta Cryst. A*41*, 571—577.
95. Karle, J. (1983) Acta Cryst. A*39*, 800—805.
96. Karle, J. (1984) Acta Cryst. *A40*, 374—379'
97. Karle, J. (1984) Acta Cryst. A*40*, 526—531.
98. Karle, J. (1985) Acta Cryst. A*41*, 182—189.

Chemistry 1986

DUDLEY R. HERSCHBACH, YUAN T. LEE and JOHN C. POLANYI

for their contributions concerning the dynamics of chemical elementary processes

THE NOBEL PRIZE FOR CHEMISTRY

Speech by Professor STURE FORSÉN of the Royal Academy of Sciences.
Translation from the Swedish text

Your Majesties, Your Royal Highnesses, Ladies and Gentlemen,

A burning flame — a little everyday miracle that has astonished and fascinated most of us. A chemical reaction that produces heat and light and that
during historical times has modified the conditions of life for mankind and
made developing civilizations possible even on our northerly latitudes. But at
the same time also a chemical transformation in which the products formed
slowly have modified our atmosphere and most likely will also affect the earth's
climate.

From the point of view of natural science a burning flame is an intriguingly
complex phenomenon. Oxygen molecules in air react with carbon and hydrogen in organic molecules. A manifold of primary products are formed, often
unstable and reactive. Atoms are torn away from their parent molecules. The
products of one reaction become the reactants of another. Dozens or even
hundreds of molecular reactions occur in parallel. It is a scientific challenge to
unravel the details of these transformations.

Let us assume that we choose for our study only one of the many chemical
reactions that take place in a flame. Superficially this reaction may appear very
simple and easy to understand. At a closer look, however, we find that Nature
is elusive and complexity still prevails. Many difficulties await him who would
like to study chemical reactions in their most intimate molecular details. The
reaction event proper is a molecular drama that takes place under an exceedingly short time span — of the order of a millionth of a millionth of a second —
times scientists refer to as a "picosecond". How is it at all possible to obtain
detailed information of what goes on under such short time? Most of our
knowledge has been gained through a deliberate simplification of the reacting
system and through a strict control of the conditions of the reaction. Furthermore our knowledge is to a large extent indirect and based upon a detailed
analysis of the initial conditions as well as of the results of the reaction event. A
reader of detective novels would perhaps like to make a parallell to the concept
of "circumstantial evidence" as a means to prove the guilt of a suspect.

The problem facing the scientist has been compared with that of a spectator
of a drastically shortened version of a classical drama — "Hamlet" say — where
he or she is only shown the opening scenes of the first act and the last scene of
the finale. The main characters are introduced, then the curtain falls for change
of scenery and as it rises again we see on the scene floor a considerable number
of "dead" bodies and a few survivors. Not an easy task for the inexperienced to
unravel what actually took place in between.

This year's Nobel prize winners of chemistry have through their brilliant work in a decisive way enlarged our knowledge of the detailed events in chemical reactions. Reactions between molecules have been studied at low pressures by letting beams of molecules and/or atoms meet at one point in space. The energy of the reacting molecules or atoms has been controlled and the properties of the products formed — their chemical composition, their angular distribution from the place of collision, their speed and their rotational and vibrational energy — have been studied. Through experiments of this kind the prizewinners of this year have been able to paint a very detailed picture of the molecular drama occurring between the opening scene and the finale.

Many of the results obtained have been unexpected and surprising and constitute a rich source of fundamental data for theoreticians to ponder. Their scientific work is truly pure basic chemical research of utmost quality but is nevertheless also of immediate importance for a number of other areas — from basic combustion research to chemistry in the stratosphere and troposphere — areas that are of great concern to mankind.

Professor Herschbach, Professor Lee and Professor Polanyi. Your brilliant research into the finer molecular details of chemical reactions has greatly advanced our knowledge in this central area of chemistry. Your approaches to the problem have differed in detail but your goals have been the same. You have, in a way that is truly admirable, combined extraordinary experimental skill with deep theoretical insight. In recognition of your services to chemistry and to natural science as a whole the Royal Academy of Sciences has decided to confer upon you this years Nobel Prize for Chemistry.

Professor Herschbach, Professor Lee and Professor Polanyi,

To me has been granted the privilege to convey to you the warmest congratulations of the Academy and I now invite you to receive your prize from the hands of His Majesty the King.

Dudley R. Herschbach

DUDLEY R. HERSCHBACH

I was born in San Jose, California on June 18, 1932, the first of six children of Robert and Dorothy Herschbach. My father was then a building contractor and later a rabbit breeder. His family had lived in this part of California for three generations; although our surname comes from a pair of villages in the Rhine Valley, most of his immediate ancestors were of English or Irish origin. My mother's family had moved to San Jose from Illinois when she was a young girl; most of her known ancestors were of German, Dutch, or French origin.

In my boyhood we lived in what was then a rural area of fruit orchards, only a few miles outside San Jose. For years I milked a cow, fed the pigs and chickens, and during summers picked prunes, apricots, and walnuts. From an early age I loved to read but was also very involved in outdoor activities, scouting, and sports. My interest in science was excited at age nine by an article on astronomy in *National Geographic*; the author was Donald Menzel of the Harvard Observatory. For the next few years, I regularly made star maps and snuck out at night to make observations from a locust tree in our back yard.

When I attended Campbell High School, I took all the science and mathematics courses offered. Chemistry I found at first puzzling and then most intriguing, thanks to John Meischke, a superb teacher. At the time, I was at least as interested in football and other sports; perhaps that presaged my later pursuit of molecular collisions. Like most of my classmates, I did not expect to attend college; none of my known relatives had graduated from a university. However, my teachers and coaches presumed I would go. Indeed, I received offers of football scholarships from some universities to which I had not even applied for admission.

I entered Stanford University in 1950 and found a new world with vastly broader intellectual horizons than I'd imagined. Although I gladly played freshman football, I had turned down an athletic scholarship in favor of an academic one. This permitted me to give up varsity football after spring practice, in reaction to a dictum by the head coach that we not take any lab courses during the season. By then the lab and library already were for me much the more exciting playground. My chief mentor at Stanford was Harold Johnston, who imbued me with his passion for chemical kinetics. Many other subjects and professors were also compelling and I took up to ten courses a term. Mathematics was especially appealing; I so admired the teaching of Harold Bacon, George Polya, Gabor Szego, and Bob Weinstock that I simply took all the courses they gave. I received the B.S. in mathematics in 1954 and the M.S. in chemistry in 1955. My Master's thesis, done under the direction of

Harold Johnston, was titled: "Theoretical Pre-exponential Factors for Bimolecular Reactions." It employed the transition-state theory of Henry Eyring
and Michael Polanyi and treated the proportionality factor in the most venerable formula of chemical kinetics, the Arrhenius equation.

My graduate study continued at Harvard, where again I found an exhilerating academic environment. I received the A.M. in Physics in 1956 and the
Ph.D. in Chemical Physics in 1958. My Doctoral Thesis, done under E. Bright
Wilson, Jr., was titled: "Internal Rotation and Microwave Spectroscopy". This
presented theoretical calculations and experiments dealing with hindered internal rotation of methyl groups. The height of the hindering barrier could be
accurately determined because the observed spectra were very sensitive to
tunneling between equivalent potential mimima. Much that shaped my later
research I learned from Bright Wilson and other faculty, especially George
Kistiakowsky and Bill Klemperer, or from fellow students, especially Jerry
Swalen, Victor Laurie and Larry Krisher. My thesis work also benefited from
visits of several months to take spectra at the National Research Council in
Ottawa and to compute Mathieu functions at Los Alamos National Laboratory. During 1957–1959, while a Junior Fellow in the Society of Fellows at
Harvard, I developed plans for molecular beam studies of elementary chemical
reactions.

This work was launched at the University of California at Berkeley, where I
was appointed an Assistant Professor of Chemistry in 1959 and became an
Associate Professor in 1961. The chief experiments dealt with reactions of alkali
atoms with alkyl iodides, systems studied forty years before by Michael Polanyi. Rather simple apparatus sufficed to attain single-collision conditions and
revealed that the product molecules emerged with a preferred range of recoil
angle and translational energy. The possibility of resolving such features of
reaction dynamics encouraged other workers pursuing kindred experiments
and fostered an outburst of new theory. My early work thus interacted particularly with that of Richard Bernstein, Sheldon Datz, Ned Greene, John Polanyi,
John Ross, and Peter Toennies.

This new field developed rapidly after I returned to Harvard in 1963 as
Professor of Chemistry. We studied a wide range of alkali reactions and found
several prototype modes of reaction dynamics which could be correlated with
the electronic structure of the target molecule. Processes involving abrupt,
impulsive bond exchange or formation of a *persistent* complex comprise the two
major categories. In 1967 Yuan Lee joined our group as a postdoctoral fellow
and led the construction of a "supermachine". This employed greatly augmented differential pumping, sophisticated mass spectroscopy using ion counting techniques adapted from nuclear physics, and supersonic beam sources
advocated by enterprising chemical engineers, especially John Fenn and Jim
Anderson. The new machine greatly extended the scope of crossed-beam
experiments, taking us "beyond the alkali age". In particular, we were then
able to study the same reactions elucidated by John Polanyi with his complementary method of infrared chemiluminescence. This much enhanced the
interpretation of reaction dynamics in terms of electronic structure.

The most representative descriptions of the work recognized by the Nobel Prize probably appeared in:

Adv. Chem. Phys. **10**, 319–393 (1966).

Disc. Faraday Soc. **44**, 108–122 (1967).

J. Chem. Phys. **56**, 769–788 (1972).

Faraday Disc. Chem. Soc. **55,** 233–251 (1973).

Pure and Applied Chem. **47**, 61–73 (1976).

Mol. Phys. **35**, 541–573 (1978).

J. Phys. Chem. **87**, 2781–2786 (1983).

In current research we are developing a method for simultaneous measurement of three or four vector properties of reactive collisions, such as reactant or product relative velocities or rotational angular momenta. Theory has shown that data on correlations among these vectors can undo much of the averaging over initial molecular orientations and impact parameters and thereby reveal more incisive information about reaction dynamics. Other studies deal with processes akin to liquid-phase reactions by solvating reactant molecules during a supersonic expansion. We are also examining bulk liquid interactions by means of vibrational frequency shifts induced by high pressure; this offers a way to determine solute-solvent intermolecular forces. In addition to theoretical studies related to these experiments, we are pursuing a new approach to electronic structure calculations which exploits exact solutions obtainable in the limit of one- and infintie-dimension. For two-electron systems this has given high accuracy for the electron correlation energy with far less effort than conventional methods.

Other biographical items pertaining to Harvard include my appointment in 1976 as Frank B. Baird, Jr. Professor of Science; service as Chairman of the Chemical Physics program (1964–1977) and the Chemistry Department (1977–1980), as a member of the Faculty Council (1980–1983), and as Co-Master with my wife of Currier House (1981–1986). At Currier we were in effect reincarnated as undergraduates to preside over an extremely lively community of 400 students and tutors. Typical of many memorable episodes was the night we were summoned to a student's room to meet a seal in the bathtub. My teaching includes graduate courses in quantum mechanics, chemical kinetics, molecular spectroscopy, and collision theory. In recent years I have given undergraduate courses in physical chemistry and especially general chemistry for freshmen, my most challenging assignment.

Away from Harvard, I have been a Visiting Professor at Göttingen University in 1963, a Guggenheim Fellow at Freiburg University in 1968, a Visiting Fellow of the Joint Institute of Laboratory Astrophysics in 1969, and a Sherman Fairchild Scholar at the California Institute of Technology in 1976. I also serve as a consultant to Aerodyne Corporation, the Fluorocarbon Research Panel, and Los Alamos National Laboratory. I was appointed an Exxon Faculty Fellow in 1981 and visit regularly the Corporate Research Laboratory in New Jersey to participate in projects there. I have also served since 1980 as an Associate Editor of the Journal of Physical Chemistry.

Other honors include election to the American Academy of Arts and Sciences in 1964 and to the National Academy of Sciences in 1967; the Pure Chemistry Prize of the American Chemical Society in 1965, the Linus Pauling Medal in 1978, the Michael Polanyi Medal in 1981, and the Irving Langmuir Prize in Chemical Physics in 1983. The University of Toronto bestowed in 1977 the D. Sc., *honoris causa*.

Chemistry also brought my wife Georgene Botyos to Harvard as an organic graduate student. We were married in 1964 and our daughters Lisa and Brenda arrived as harbingers before she received her Ph.D. in 1968. Georgene is now Assistant Dean of Harvard College, a multifaceted position that often requires delicate personal chemistry. Lisa is now a junior in humanities at Stanford, this year enjoying the overseas option at Oxford. Brenda is a junior in chemistry at Harvard, already pursuing research. Our home is in Lincoln, Massachusetts.

MOLECULAR DYNAMICS OF ELEMENTARY CHEMICAL REACTIONS

Nobel Lecture, 8 December 1986

by

DUDLEY R. HERSCHBACH

Department of Chemistry, Harvard University, Cambridge, Massachusetts 02138, U.S.A.

The vast field of Chemical Kinetics embraces three distinct levels of under-standing or abstraction [1]. The broadest level is qualitative description. Here the emphasis is on *substances:* what products are obtained from what reagents under what conditions. The information gathered is encyclopedic, but quite feeble in predictive power. This is because the overall observed transformation in most reactions comprises a more or less complex sequence or network of elementary steps.

At the next level, the aim is to identify the distinct *elementary steps* and to determine quantitatively the rate factors of the Arrhenius equation, k = Aexp $(-E_a/RT)$, a remarkably durable formula now nearly a century old! Some steps involve intermediate species that do not appear among the final products; whenever possible, these intermediates are also characterized by structural, spectral, and thermodynamic properties. Over about the past 70 years, im-mense effort has been devoted to compiling such quantitative data. It is of great practical value because predictions can be made for many reaction systems, on the usually reliable assumption that the rate parameters for each indivisible elementary step can be carried over from one system to another.

As well as mapping the cooperation and competition among elementary processes, kinetic studies at this level also gave rise to many key chemical ideas. These include such concepts as chain reactions carried by free radical interme-diates, collisional energy transfer, and the time-lag for energy flow within an excited molecule. Most fundamental was the idea introduced by Henry Eyring and Michael Polanyi of a transition-state on a potential energy surface. This has guided qualitative reasoning about molecular mechanisms, and when augmented by resourceful semitheoretical approximations for potential param-eters has enabled limited but useful estimates of rate factors. However, these valuable insights were won despite frustrating handicaps. The basic experi-mental variables of concentration and temperature are not incisive enough to allow further progress. Postulated elementary steps or intermediates often prove incomplete or illusory. The observable rate factors, which represent averages over myriad random collisions, are too remote from the molecular interactions.

Emergence of Chemical Dynamics

Over the past 30 years a new level has been attained by study of the intimate *molecular dynamics* of individual reactive collisions. One of the chief experimental approaches is molecular beam scattering. This involves forming the reagent molecules into two collimated beams, each so dilute that collisions within them are negligible. The two beams intersect in a vacuum and the direction and velocity of the product molecules emitted from the collision zone are measured. A host of reactions can now be studied readily in this way, by virtue of an extremely sensitive mass spectrometric detector and the use of supersonic nozzles which generate beams with greatly enhanced intensity and with collision energies much higher than provided by ordinary thermal sources. The coupling of spectroscopic techniques with molecular beams has provided further advances in selectivity and sensitivity, particularly by use of intense laser light sources.

The reaction properties now accessible include the disposal of energy among translation, rotation, and vibration of the product molecules; angles of product emission; angular momentum and its orientation in space; and variation of reaction yield and other attributes with impact energy, closeness of collision, rotational orientation or vibrational excitation of the target molecule. From the earliest stages these experiments have stimulated and responded to a vigorous outburst of theoretical developments. Especially helpful are computer simulations in which dynamical properties are predicted for various postulated forces by Monte Carlo sampling of huge numbers of calculated collisions trajectories. Both the laboratory experiments and computer simulations have prompted a variety of insightful mechanical models as well as very useful diagnostic procedures based on information theory. *Ab initio* electronic structure calculations have also begun to contribute significantly to the exploration of reaction dynamics, although (except for $H + H_2$) satifying overall accuracy has yet to be achieved for potential energy surfaces.

As urged by my students, on this occasion I want both to view our still youthful field of research from a wider perspective and to recount some favorite, instructive episodes from its infancy. I will also briefly discuss several prototype reactions that have served to develop heuristic models and to reveal how electronic structure governs the reaction dynamics. More technical and systematic surveys are abundant [2]. Particularly recommended are recent books by Bernstein and by Levine and Bernstein [3]. The latter, about to appear in its second edition, contains references to 500 review articles in the field of molecular reaction dynamics. Nobody has tried recently to count the research papers in this prolific field; these probably now exceed 5000. This does not include kindred developments in the study of molecular dynamics in solution or the solid phase.

The emergence of chemical dynamics captivated many enterprising scientists and imbued them with a sense of historical imperative. Figure 1 indicates this in the wider context of physical chemistry. The subject began with thermochemistry, still its foundation. The thermochemical era can be considered to reach a pinnacle in 1923, with the publication of the classic text by Lewis and

Randall. This was shortly before the discovery of quantum mechanics ushered in the new structural era (and spawned chemical physics). In turn, the grand pursuit of molecular structure and spectra reached a twin pinnacle in 1951 and 1953 with the monumental discoveries of the alpha-helix by Pauling and the DNA double-helix by Watson and Crick. This was shortly before the early moleular beam and infrared chemiluminescence experiments appeared as har-

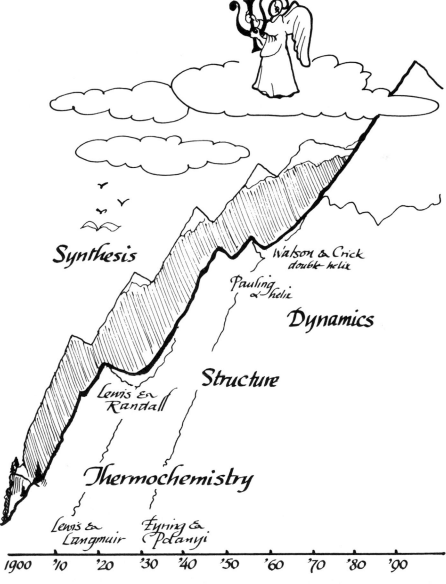

Fig. 1. Historical perspective. Ascending mountains represent three eras, since *thermochemistry* was prerequisite for the *structure* era and both underlie the *dynamics* era. All three connect to a vast range representing *synthetic* chemistry, which draws closest to dynamics as both disappear heavenwards into clouds symbolizing the ultimate triumph of *ab initio* quantum chemistry.

bingers of the dynamics era. It is striking that the prequantum models of Lewis and Langmuir, still valued today, came well before the onset of the structural era. Likewise, the potential surface and transition-state concepts of Eyring and Polanyi anticipated the dynamics era.

Not evident in Fig. 1, and in contrast to the scientific continuity, are cultural chasms between the eras. Linus Pauling once told me of the gulf his students encountered in the late 1920's when, as candidates for an academic position, they presented seminars describing molecular structures they had determined by electron diffraction. All the faculty in the audience had done their Ph. D.'s in thermochemistry, and so imbibed a tradition which emphasized that it did not need to postulate the existence of molecules! Worse, interpretation of the diffraction rings then relied entirely on the so-called "visual method". A densitometer tracing showed only monotonically decreasing intensity curves; the rings could only be seen because the human eye can detect slight changes in intensity. The molecular structures obtained thus depended on an "optical illusion," incomprehensible and reprehensible to many professors of thermodynamics. Forty years later, my own students encountered a similar gulf when presenting seminars describing our early crude molecular beam reactive scattering studies. Almost all the incumbent faculty then had done their Ph. D.'s in structure or spectroscopy. They were dubious about work that depended on drawing velocity vector diagrams to interpret bumps on scattering curves.

Molecular Beams Before Chemical Dynamics

The first molecular beam experiments were carried out 75 years ago, immediately after the invention of the high speed vacuum pump had made it possible to form directed "rays" of neutral molecules at sufficiently low pressures to prevent their disruption by collisions with background gas. Figure 2 indicates the historical progression. Systematic research began with Otto Stern, who developed many aspects of beam techniques. He had retired and was living in Berkeley when I arrived there in 1959. Occasionally Stern attended seminars, and I had the opportunity to hear from him several memorable stories about his work. Two of these are pertinent here.

In his first work, done at Frankfurt in 1919, Stern undertook to test the Maxwell-Boltzmann velocity distribution by analyzing the speeds of a silver atom beam with a rotating device. Gleefully, he described how the money to build the apparatus was supplied by Max Born; the great theoretical physicist was renowed as a public speaker and gave a special lecture series to raise the funds. Stern found that his experimental results were in approximate agreement with the Maxwell-Boltzmann distribution, but deviated from it in a systematic way. He said [4]:

> "After my paper was sent off, I received a letter pointing out that I had overlooked an extra factor of v (the velocity) which enters because the detected atoms must pass through a slit. When this factor was inserted, the agreement became quantitative. That letter came from Albert Einstein!"

The "extra factor" represents the Jacobian for the transformation from number

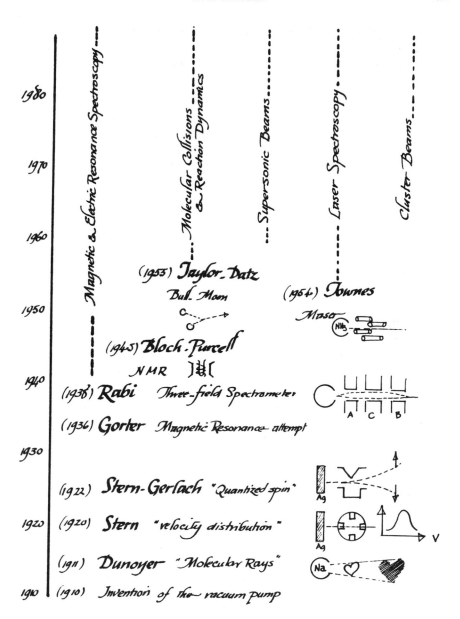

Fig. 2. Evolution of molecular beam and kindred methods.

density (atoms cm^{-3}) to flux density (atoms cm^{-2} sec^{-1}). That same factor and Jacobians for numerous other transformations often appear in chemical dynamics, so I've had reason to tell Stern's story to many bedeviled students. As noted below, some striking effects in molecular collisions are due entirely to a Jacobian factor.

The other favorite story concerns the celebrated Stern-Gerlach experiment, in which a beam of silver atoms was sent through an inhomogeneous magnetic field and discovered to split into two, thereby revealing the space-quantization of the electron spin. This is how Stern told the story [4]:

"The question whether a gas might be magnetically birefringent (in the words we used in those days) was raised at a seminar. The next morning I awoke early, too early to go to the lab. As it was too cold to get out of bed, I lay there thinking about the seminar question and had the idea for the experiment. When I got to the lab, I recruited Gerlach as a collaborator. He was a skilful experimentalist, while I was not. In fact, each part of the apparatus that I constructed had to be remade by Gerlach."

"We were never able to get the apparatus to work before midnight. When finally all seemed to function properly, we had a strange experience. After venting to release the vacuum, Gerlach removed the detector flange. But he could see no trace of the silver atom beam and handed the flange to me. With Gerlach looking over my shoulder as I peered closely at the plate, we were surprised to see gradually emerge two distinct traces of the beam. Several times we repeated the experiment, with the same mysterious result. Finally we realized what it was. I was then the equivalent of an assistant professor. My salary was too low to afford good cigars, so I smoked bad cigars. These had a lot of sulfur in them, so my breath on the plate turned the silver into silver sulfide, which is jet black so easily visible. It was like developing a photographic film."

Although I've rarely tried a cigar, our first reactive scattering work also benefited from lucky contamination of the detector, as described later.

Soon Stern could afford good cigars, as he was appointed professor of physical chemistry at Hamburg. His Institute there conducted a series of remarkable molecular beam studies, including invention of the surface ionization that was to be so vital in the "alkali age" of chemical dynamics. This detector, derived from Langmuir's studies of surface ionization, has since been used in thousands of experiments. It consists simply of a heated filament of a metal such as tungsten and will ionize with nearly 100 % efficiency materials having ionization potentials lower than about 4 or 5 volts, such as alkali atoms. Stern had to abandon his splendid work at Hamburg in 1933 when Hitler took office.

Other epochal molecular beam contributions to physics came from Isidor Rabi. He had also been educated as a chemist and worked with Stern as a postdoctoral fellow before joining the physics faculty at Columbia. In 1937 he invented his magnetic resonance method. This followed a seminar by Gorter from Leiden describing his failed attempt to detect nuclear magnetic resonance in a crystal. It was then not obvious that the tiny nucleus, with dimensions of the order 10^{-13} cm, would interact appreciably with the external atomic electron distribution, with dimensions of the order 10^{-8} cm. If not, the different nuclear spin orientations (favorable: "up" or unfavorable: "down") would remain equally probable even when subjected to an external magnetic field. There would then be no net absorption of radiofrequency radiation in the magnetic resonance experiment. Spin flips induced by the radiation are equally likely up or down, so net absorption cannot occur unless interactions permit the spins to relax toward thermal equilibrium where more will be up than down.

Rabi immediately realized how to exploit beams to escape this constraint. He introduced two opposing Stern-Gerlach fields. A beam traversing the first field (denoted A) is split into its nuclear spin components, but on passing through the second field (denoted B) these are recombined. Between these fields, which act like diverging and converging lenses, Rabi introduced a third field (the C-field). This was homogeneous and so had no lens action, but it served the role of Gorter's field in defining "up" and "down". Since there are no collisions in the beam, there is no means to relax the spin component populations. But now none is needed. Radiofrequency radiation in the C-field region will at resonance flip equal numbers of spins up or down; but now any spin changing its orientation after the A-field will not be refocussed in the B-field. In this ingenious way, Rabi could detect which frequencies produced resonances. He thus created a versatile new spectroscopy with extremely high resolving power. It has provided a wealth of information about nuclear structure. An analogous electric resonance spectroscopy operates with rotational states of polar molecules. This has likewise been very fruitful for molecular structure. It has also been used to fully analyze the quantum states of products in some reactive scattering experiments.

Shortly before inventing his resonance spectroscopy, Rabi undertook some simple elastic scattering studies by observing the attenuation of an alkali atom beam sent through a gas-filled chamber and detected on a surface ionization filament. If Gorter had not come by, perhaps Rabi would have continued on to develop reactive scattering. Another quirk of fate was only revealed in 1945, when nuclear magnetic resonance was successfully done in liquids and solids by Purcell at Harvard and Block at Stanford. Only then did it become clear that Gorter's original experiment would have worked if he had used a material with magnetic impurities which would have induced the necessary relaxation. Yet another epilogue, also of great portent, was the invention of the maser in 1955 by Townes at Columbia. In this device, the precursor of lasers, he sent a beam of ammonia molecules through an electric quadrupole focussing field which selected an energetically unfavorable state. When illuminated with microwave radiation, the molecule reverted to the energetically favorable state, emitting the excess energy as a photon. Thus was born molecular amplifiers and oscillators and the now vast industry of quantum electronics. The idea came to Townes on a sunny park bench, 35 years after Stern woke up early in his warm bed.

My first Encounters with Kinetics and Beams

Within days of my arrival at college, I first learned something of the challenging field of chemical kinetics from my freshman advisor, Harold S. Johnston. He was then a young assistant professor at Stanford University, applying new methods of his own devising to measure rates of fast gas-phase reactions in regimes previously inaccessible. In the summer after my sophomore year, and again after my junior year, Hal hired me as a lab assistant for his research group. One project involved weeks of glassblowing, silver soldering, and plastering with great quantities of asbestos paste (now outlawed) to install a large volume reaction vessel that could be maintained at a high and uniform temperature. I was greatly impressed with the care and attention to detail shown by Hal and his graduate students, and his emphasis on testing theory. He stressed that this could only be done with a reaction known to be an elementary step—and that we could never be entirely certain that all steps and intermediates had been correctly identified.

In my senior year, I took Hal's excellent course in chemical kinetics, and the following year a study of transition-state theory begun there became my master's thesis. In this, we calculated the Arrhenius A-factors for twelve well-characterized bimolecular reactions involving only 4 to 6 atoms. Necessary assumptions about the properties of the transition state were made uniform for the whole set and consistent with similar stable molecules. Internal rotations of the reaction complexes received special attention, thanks to the help of Kenneth Pitzer at Berkeley, who kindly showed me how to design a sensible approximation. Previous tests had been piecemeal and often capricious about internal rotations; Hal expected our test might prove embarassing for the theory. In fact, the agreement with experiment was good. This persuaded Hal that the theory deserved further development; over the next decade he carried this out and produced an exemplary book. Although Hal's contagious enthusiasm had long since convinced me that I wanted to pursue kinetics, the thesis project made me decide to study first the dynamics of stable molecules. Thus I applied to the chemical physics program at Harvard in order to do Ph.D. research with E. Bright Wilson, Jr.

A few weeks before turning in the master's thesis, I first heard of molecular beams. In his course on statistical thermodynamics, Walter Meyerhof briefly decribed Stern's test of the Maxwell-Boltzmann velocity distribution. It was love at first sight. I remember a flush of excitement at the thought that this was *the way* to study elementary reactions, the unequivocal way to know a reaction is elementary and to study directly its properties. This was the spring of 1955. I did not imagine that 5 years later my own first beam apparatus would be nearly complete at Berkeley and I would have met Otto Stern himself.

On arrival at Harvard in the fall, I was delighted to discover that Norman Ramsey was offering a seminar course based on the proof sheets of his book on molecular beams. E. B. Wilson was giving a course using his just published book on molecular vibrations. Roy Glauber provided a rigorous treatment of electromagnetic theory, working through all the mathematical derivations in class with no notes at all. The dessert course of this remarkable academic menu was entitled "Chemical Physics", and given by the legendary Peter Debye, a visitor that year. His course had 40 or more auditors, but only 3 of us took it for credit and thus had individual oral exams with Debye. I still relish the memory of his lucid lectures and the twinkle in his eye as he so often began, "Let me tell you a story ...". Early in Ramsey's course, he too discussed Stern's velocity analysis study and actually announced, in his booming voice: "This would be a wonderful way to do chemistry!" In the spring term I took among others Kenneth Bainbridge's course on nuclear physics to learn something about nuclear reactions; this was indeed to prove very useful later.

Even more exhilerating that year and thereafter was my experience in Bright

Wilson's research group. With microwave spectroscopy, he and his students were developing an elegant new way to study internal rotation by exploiting the tunnel effect. Typical systems had a methyl group attached to an asymmetric molecular frame. The torsion or internal rotation of the methyl group about its axis is hindered by a sinusoidal potential barrier with three equivalent minima. If the barrier is very high, then essentially there are only small torsional oscillations in three separate potential wells, and to good approximation the microwave spectrum is that of a rigid rotor molecule. If the barrier is low enough, however, tunneling between the well occurs and this causes the microwave rotational transitions to split into doublets. These splittings are very sensitive to the barrier height, so an approximate dynamical theory is adequate to extract an accurate value for the barrier. Since I already had seen how prevalent internal rotation is, not only in stable molecules but in transition-states for reaction, I was very excited about this new method. It was such a fresh and powerful approach that almost every day some member of the group had an unanticipated result to report. Because the microwave spectra display a forest of transitions, there was ample need to work out variations on the basic theme and satisfying opportunity for definitive tests.

Although this research was entirely different in character from molecular collisions, I learned invaluable lessons from it and from working with Bright and his extremely able group. A microwave spectroscopist simply must calculate as much as possible before plunging into the forest; for a molecular beamist, the need is less obvious but just as vital. Bright's high standards and absolute integrity had great impact on his students. Thus he put strong emphasis on analyzing and clearly stating what is well-established and what is really based on unproven assumptions. He also gave strong emphasis to blocking out the big questions and on coupling theoretical investigations with experiments.

Microwaves soon led to my first work with beams, in collaboration with an ebullient young instructor, Bill Klemperer. He had developed a technique for high temperature infrared spectroscopy and was using it to study alkali halides (later to become my favorite molecules). Bill invited me to help build a high temperature microwave spectrometer, again a project requiring much asbestos paste. One day Bill came in with a paper just published by Marple and Trischka reporting an electric resonance study of lithium chloride. They had used the venerable tungsten surface ionization filiment as a detector. Depending on resonance field settings, the apparatus could resolve four vibrational states. However, the vibrational frequency derived from the relative intensities was much lower than Bill had measured in the infrared. We examined various possibilities and concluded that most likely was an increase in the efficiency of the surface reaction (dissociation of the salt molecule to ions) with vibrational excitation of the incident molecule [5]. This was near Christmas, 1956. Almost 15 years later, the vibrational excitation mechanism was confirmed and recently it has become a leading issue in surface chemistry.

We were eager to study the surface ionization process experimentally. Since an electric resonance apparatus was not available (at that time only 3 existed, none closer than 300 miles of Cambridge), we could not study alkali halide molecules. Hence we decided to study ionization of alkali atoms as a function of the surface temperature. Ramsey kindly lent us one of his machines over Christmas vacation and we quickly obtained a hundred pages or so of data from which we derived residence times and heats of adsorption for atoms on tungsten and platinum surfaces. Then we discovered that very similar data existed in the Russian literature, so did not publish our results. However, this was a key episode for both Bill and me. He too fell in love with molecular beams, and immediately undertook to build an electric resonance apparatus. Bill and his students have since turned this into a cornocopia for molecular spectroscopy, unprecedented in resolution and chemical scope. Especially revealing have been their studies in recent years of molecules held together by weak van der Waals forces. This work is building an understanding of these ubiquitous forces which may help elucidate the specificity and selectivity of interactions in biomolecules.

In studying the literature on surface ionization, I was elated to discover [6] a paper by Taylor and Datz from Oak Ridge describing an actual crossed beam study of the reaction K + HBr → KBr + H. Although the traditional tungsten surface ionization detector is about equally sensitive to K and KBr, Taylor and Datz found that a platinum alloy is much more effective for K than for KBr. From the difference in the signals read on the two detector filaments, they were able to distinguish the small amount of reactively scattered KBr from the large background of elastically scattered K atoms. Small as it was, the little difference bump was a joyful sight. In a series of classic studies in the 1920's, Michael Polanyi had shown that many alkali reactions proceeded at rates corresponding to "reaction at every collision". With this differential surface ionization detector, there was now prospect that crossed-beam studies of many of these reactions could be made with relatively simple apparatus.

Soon I found another intriguing crossed-beam study, published two years before by Bull and Moon at Birmingham. They bombarded a stream of Cs vapor with a pulsed accelerated beam of CCl_4 produced by swatting with a paddle attached to a high-speed rotor. The intense CCl_4 beam could be monitored with a simple ionization gauge and signal pulses due to scattered Cs or CsCl were detected by surface ionization on a tungsten filament. Although there was no direct means to distinguish between Cs and CsCl, the observed signal pulses appeared to come primarily from reactively scattered CsCl, on the basis of time-of-flight analysis and blank runs with the CCl_4 replaced by Hg vapor. Unjustly, these experiments were long discounted or ignored. This was due to the misconception that elastic scattering would always predominate. The high-speed rotor technique also intimidated other prospective experimenters; thus this fine early work did not have the impact deserved.

During the spring of 1958, while completing my Ph.D thesis, I visited several universities as a faculty candidate. My seminar about internal rotation was always well received, but my plans for beam experiments sometimes produced outright dismay. Yet the response was often enthusiastic from faculty young in either age or spirit. I was delighted to accept an assistant professorship at

Berkeley, especially since Hal Johnston was now there and also Bruce Mahan, who had the year before completed his Ph.D. at Harvard in kinetics with George Kistiakowsky. My appointment was to take effect a year later; meanwhile I continued at Harvard as a Junior Fellow of the Society of Fellows. The intervening year was extremely useful. It gave me time to work out detailed calculations for design of the beam apparatus and a kinematic theory for interpreting crossed-beam experiments. Membership in the Society of Fellows was also a great pleasure because it brought together intense people from all academic disciplines. Monday night dinners included the two dozen Junior Fellows, ten Senior Fellows and typically ten guests, in a "neo-pickwickian atmosphere" with ample wine and cigars too good to have saved the Stern-Gerlach experiment [7].

Just before departing to Berkeley, I went to say goodbye to Kistiakowsky. Like so many others, I revered him as a scientist and a sage as well as an awesome personality. He confirmed a legendary story about a molecular beam experiment he and Slichter had tried in 1951 to test early theoretical calculations on the properties of supersonic expansions [8]. Because of inadequate pumping speed, the experiment was inconclusive and a miserable struggle. Afterwards, Kisty destroyed the apparatus with an ax! He said to me, with his extra heavy Russian accent (conveying whimsey—or so I hoped): "So you have been bitten by the molecular beam bug. Too bad! The trouble is that there are no collisions in one beam, and no collisions in the other beam, so when you cross them there are still no collisions!"

Early Alkali Age of Reactive Scattering

At Berkeley, I was soon joined by two graduate students, George Kwei and Jim Norris. Space was in short supply, and we were content with a corner (only 4m × 6m) of a lab in Lewis Hall. Our first task was to remove a large calorimeter that had long occuppied that corner; this we viewed as a rite of passage, with the historical perspective of Fig. 1 already in mind. The beam apparatus we installed (dubbed *Big Bertha*) was almost rudimentary. As shown in Fig. 3, the beams were formed by thermal effusion from ovens mounted on turnable which could be rotated to sweep the angular distribution of scattered atoms and molecules past the surface ionization detector. Typically, the distance from the scattering center to the alkali oven was 10 cm, to the other oven 1.5 cm, and to the detector 10 cm. Use of a double-chamber oven for the alkali allowed the temperature of the beam emerging from the upper chamber to be varied by about 300 degrees independently of the vapor pressure established in the lower chamber (about 0.1 torr), which governed the beam intensity. The oven for the reactant gas was connected to an external barostat by a supply tube which passed through the support column in the rotatable lid. Cold shields and collimating slits hid both ovens from the scattering center, and a cold shield also surrounded the detector. The entire scattering chamber was enclosed in a copper box attached to a large liquid nitrogen trap. Since the reactants are condensable, the cold shields and trap provided very high pumping speed. This kept the background vacuum in the scattering chamber low

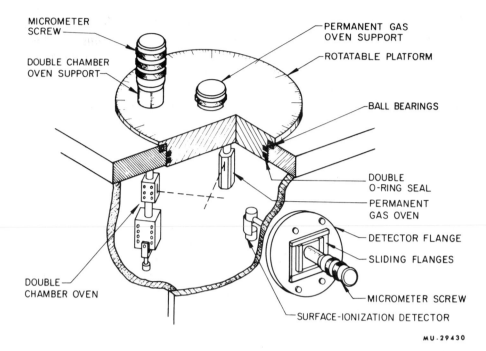

MU·29430

Fig. 3. Crossed-beam apparatus *Big Bertha* for study of alkali atom reactions. Cold shields, collimating slits, shutters to interupt the beams, and other details omitted.

enough (of the order 10^{-7} torr) to ensure that collisions with background gas were negligible for both the reactant beams and the product molecules in flight to the detector.

In typical experiments, the concentration of alkali atoms within the volume defined by the intersection of the beams was about $10^{10}/cm^3$, equivalent to a pressure of 10^{-6} torr, and that of the reactant gas molecules was about 100-fold greater. About 10^{14} alkali atoms/sec entered the reaction volume, of which roughly $0.1-1\%$ reacted to form products while about 10 % underwent elastic scattering. The steady-state concentration of products in the reaction volume was roughly 10^7 to 10^8 molecules/cm^3, the pressure about 10^{-9} to 10^{-10} torr. At the peak of the angular distribution of reactive scattering about 10^{10} to 10^{11} product molecules/cm^2/sec arrive at the detector. Since conversion to ions on a hot tungsten or platinum surface is nearly 100 % efficient for alkali species, such yields gave readily measureable signals of the order of 10^{-13} Amps. For many of the reactions studied more than a month would be required to deposit a monolayer of product molecules, so it was essential to have a detector far more sensitive than Stern's cigar.

Experiments Revealing Product Recoil
Our first experiments were begun in the fall of 1960. On the basis of simple theoretical considerations, we decided to try

$$K + CH_3I \rightarrow KI + CH_3$$

From the angular distribution of the reactively scattered KI we hoped to learn whether there was any preferred direction relating the reactant and product relative velocity vectors and also to get some idea how the reaction energy is partitioned between internal excitation of the product molecules and their relative translational motion. Our rationale was based on the CH_3/KI mass ratio. A simple kinematic calculation (displayed later in Fig. 5) showed that this ratio was large enough to permit the detected product KI to recoil sufficiently far away from the centroid to reveal any directional preference, but small enough to inhibit the KI from spraying out so widely that the reactive scattering became too weak to observe. Happily, even the first experiments worked nicely. Our kinematic rationale was vindicated, but it was not until three years later that we learned how lucky was this choice of reaction system; it turned out that CH_3I played the role of Stern's cigar.

Figure 4 shows typical results. In (a), the parent K beam was attentuated 7 % by the perpendicularly crossed CH_3I beam; that represents chiefly elastic scattering. Readings on the platinum detector (solid points) were normalized to those on tungsten (open points) at the parent beam peak. In this case, the use of differential detection was not crucial, as the KI distribution is displaced

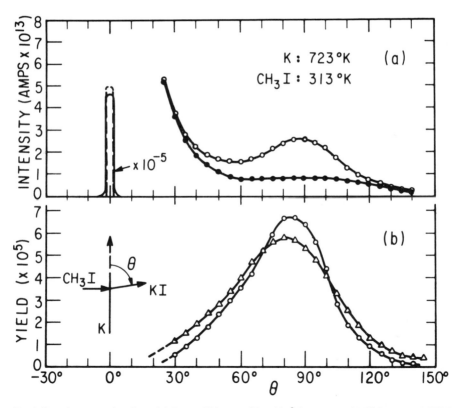

Fig. 4. Reactive scattering data: (a) Parent K beam of 5×10^{-8} A attenuated 7 % by crossed CH_3I beam. Readings on Pt detector (solid circles) normalized to W (open circles) at parent beam peak. (b) KI distributions; circles derived from (a), triangles from a replicate experiment several months later. Area under curves gives collision yield.

far enough from the K beam to appear as a pronounced bump in the signal
from the tungsten detector. In (b), the KI distribution is normalized so that the
area under the curve gives the collision yield (integrated intensity of KI divided
by total K scattered out of the parent beam). This is about 5×10^{-4} and
indicated the reaction cross section (or effective target area) is about 7 A^2.
Measurements at several temperatures of the incident beams indicated that the
activation energy for the reaction is negligibly small, less than 0.3 kcal/mol.
The new dynamical information we were after is contained in the location and
shape of the product angular distribution, however. To extract that information we used a kinematic analysis derived from conservation laws that hold
regardless of the forces governing the collision.

Kinematic Analysis Via Newton Diagrams

Velocity vector diagrams such as Fig. 5 are a convenient aid in the kinematic
analysis. The vectors pertain to the asymptotic initial and final states of the
collision. By virtue of momentum conservation, the center-of-mass velocity
vector **C** remains constant throughout; only motion relative to **C** involves
chemical interactions. The recoil velocity **u** which carries the product **KI** away
from **C** can have any direction but energy conservation limits its magnitude.
Thus, the possible spectrum of recoil vectors **u** is represented by a set of

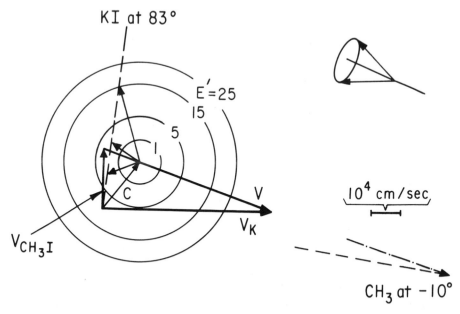

Fig. 5. Newton diagram corresponding to most probable velocitites of reactants in Fig 4. Spheres
indicate range of KI recoil vectors allowed by conservation of energy and linear momentum; each is
labelled by value of E'(kcal/mol), the final relative translational kinetic energy. Distribution of
recoil vectors must have cylindrical symmetry about intial relative velocity vector, by virtue of
"dart board" randomness of collisions. Three sample recoil vectors are shown that correspond to
the KI observed at the peak of the angular distribution. Also shown is estimate of most probable
recoil vector for CH_3 product.

spheres, one for each accessible value of the final relative translational energy of the products. This energy ranges up to a maximum determined by the initial collision energy and the difference in dissociation energy of the new bond formed and the old bond broken.

From the velocity vector diagram we see that the broad peak observed near 83° in the laboratory corresponds to reactive scattering in which an observer riding with the center-of-mass would see the product KI recoil into the backward hemisphere (and the CH_3 forwards) with repect to the incoming K beam. Markedly anisotropic scattering of this kind evidently corresponds to a "hard" collision dominated by repulsion, as the K atom, the CH_3 group, and the center-of-mass of KI must all reverse direction. We called this the *rebound* mechanism. It was later found for many other reactions dominated by repulsion.

Comparison with the velocity vector diagram also permits an estimate of the energy in product relative translation. This affects the displacement of the laboratory angular distribution from that of **C** as computed from the initial conditions. In assessing the displacement, it is necessary to take proper account of a Jacobian factor which enhances the intensity of the laboratory scattering at low recoil velocities. We failed to include this Jacobian in our early work (thereby unwittingly emulating Otto Stern) and hence underestimated the product translation. The correct analysis showed that for the $K + CH_3I$ reaction about half of the available energy appears in recoil of the products. This result, later confirmed by direct measurement of the **KI** velocity distributions, again exhibited the dominant role of repulsive forces in a *rebound* reaction.

For kinematic analysis, Newton's laws suffice. In the asymptotic translational states the beam molecules are too far apart to interact, hence need not be precisely localized in space and can be assigned definite momenta despite Heisenberg's principle. Accordingly, we named our velocity vector constructions "Newton diagrams". These are still much used today, as kinematic analysis is an essential part of the design and interpretation of any collision experiment.

Frustrations and Preparations

We were very happy to find this clear evidence for a preferred range of product recoil and energy and were eager to study other alkali atom reactions to look for variations in the scattering patterns that might be correlated with electronic structure. However, we encountered frustrating difficulties with the surface ionization detector. Many reactant gases (other than alkyl halides) poisoned the Pt filament, inducing spurious and irreproducible responses that precluded measurement of reactive scattering. This difficulty stymied us until the fall of 1963, when we finally eliminated it by means of a procedure due to Touw and Trischka. They demonstrated that the Pt filament could be stabilized in either of two distinct modes. The Pt filament becomes essentially *nondetecting* for alkali compounds (but still detects alkali atoms) if it is preheated while doused with methane or another hydrocarbon. The Pt filament becomes *detecting* for alkali compounds and acts just like a W filament if it is instead preheated while doused with oxygen or merely operated in sufficiently clean vacuum. The nondetecting mode evidently requires a stable carbon layer on the filament. Thus, the differential detection so useful for alkali reactive scattering results from a dirty surface. Before the preheating procedure was developed, the carbon deposit was a matter of luck, supplied either by pump oil vapor if the vacuum were poor or by hydrocarbon-rich reactant species such as methyl iodide; either acted like Stern's cigar.

During the lean period of more than two years before rescue by a dirty detector, our only successful reactive scattering experiments dealt with variants of the $K + CH_3I$ reaction, involving other alkali atoms and alkyl groups with up to seven carbon atoms. However, in this period our research group expanded rapidly and we constructed much new apparatus, experimental and theoretical, that was destined to bring a harvest of results. Jim Kinsey, Ken Cashion, Mark Child and Malcolm Fluendy joined as a postdoctoral fellows, and Phil Brooks, Kent Wilson, John Birely, Ron Herm, Jim Cross and Dick Zare as graduate students. Figure 6 provides a succinct summary, indicating the chief projects pursued by this distinguished roster during our Berkeley era.

In particular, we note among the apparatus construction projects three analyzing devices: rotating slotted-disk velocity selector, inhomogeneous magnetic deflecting field, and inhomogeneous electric deflecting field. These devices are quite similar to those used by Otto Stern; we liked to say that reactive scattering followed the evolutionary principle, "Ontogeny recapitulates Phylogeny". As seen already, velocity analysis was essential to obtain quantitative

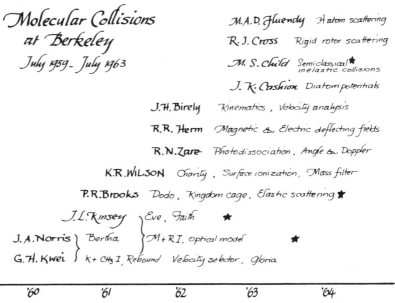

Fig. 6. Roster of graduate students and *postdoctoral fellows* with chief research projects during Berkeley era. Our first apparatus was a helium leak detector called Annie; hence our successive beam machines were named Bertha, Charity, Dodo, Eve, Faith, Gloria, ... Stars indicate completion of Ph. D. theses or postdoctoral terms.

data for the recoil angle and translational energy distributions. Magnetic analysis provided, among other things, a definitive test of our detection technique, immune to poisoning of the filaments. By deflecting away most of the paramagnetic alkali atoms, we could measure directly the distribution of diamagnetic alkali halide molecules produced by reactive scattering. Electric deflection of these polar molecules enabled us to determine the rotational angular momentum. This had special appeal for me because of my background in rotational spectroscopy. Furthermore, since mass, velocity, and position all enter into angular momentum, it was clearly a key property for theoretical models of reaction dynamics.

Among our theoretical studies of this period was a treatment of the velocity vector distribution of fragments from molecular photodissociation. This was undertaken because already we had vague notions of a kinship between chemical reactions and photodissociation. In electronic excitation by photons, the angular dependence is determined simply by the dipole selection rule whereas the exit velocity of the photofragments is determined by the repulsive potential of the excited state. This "half-collision" model, worked out nicely by Dick Zare, provided rather general "form factors" for numerous properties [9]. Subsequently, this has served as the basis for extracting a lode of information about dissociative excited states by means of laser spectroscopy, and the suspected link with reaction dynamics did indeed emerge.

Our Berkeley days also included serveral memorable visits with others captivated by reaction dynamics. One of these occurred in the early summer of 1961, when I visited Don Bunker at Los Alamos. He had undertaken the first realistic Monte Carlo calculations of classical trajectories for chemical reactions, and his initial aim was to interpret our results for the K + CH₃I system in terms of potential surface features. The scene in his office comes back vividly [10]:

> ... He was decorating the walls with strips of recorder chart paper several yards long, on which were plotted a series of undulating, intertwined curves. These were the first results of his Monte Carlo calculations ... The room was bright with sunshine and the varicolored curves seemed to shimmer and dance about, as if choreographed to Don's crisp and witty description. It was an exhilerating moment. Henceforth the mechanics of molecular collisions for any postulated force field could be computed, as he liked to say, "in instructive and entertaining detail."

Such classical trajectory calculations, as developed especially by Martin Karplus and John Polanyi, have indeed proven invaluable for interpreting data and testing concepts of reaction dynamics.

Other inspiring encounters came in April, 1962, at a Faraday Society Discussion held at Cambridge University, where I met many leading workers in chemical kinetics. Equally inspiring was a solo visit to Berkeley a few months later by Michael Polanyi. He witnessed an unsuccessful scattering experiment on the K + Br₂ reaction while he described to us his prescient speculations of 30 years before that it might proceed by a stripping mechanism with high vibrational excitation of the KBr product. He also mentioned that he liked to think that the attacking alkali atom used its valence electron to "harpoon" the halogen. Later I adopted this term to enliven numerous discussions of electron transfer reactions.

From Rebound to Stripping with Rainbows Between

Within a few months my group was discussing an unanticipated transfer to Harvard. We had expected to relocate in a new lab about 100 m away in the basement of the nearly completed Latimer Hall, but in the late summer of 1963 instead moved 5000 km to the basement of Mallinckrodt Laboratory. I went even further that summer, as I'd agreed to present a lecture course on molecular collisions at Göttingen University. There I strolled daily along the path atop the encircling medieval wall, the path taken by many of the pioneers of quantum mechanics during ambulatory debates. But only by communing with the charming little Goose Girl in the Rathaus fountain was I reconciled to the quantum-like dualities of the Harvard/Berkeley choice.

Our new lab was in nearly full swing by Halloween and soon we learned how to avoid poisoning our detector filaments. Thereafter results were rapidly obtained for a wide range of alkali reactions. As illustrated in Fig. 7, many showed striking contrasts to the methyl iodide reaction. We found that for

$$K + Br_2 \rightarrow KBr + Br$$

and similar systems the reaction cross section is remarkably large, $\sim 200 \ A^2$, and most of the alkali halide product recoils into the forward hemisphere with respect to the incident alkali atom. These features indicate dominantly attractive interaction in which the bond exchange can occur for quite large impact parameters. This suggests a *stripping* mechanism of the type familiar in nuclear physics. In an independent study of the Br_2 reaction, Minturn and Datz obtained similar results and many other examples of *stripping* reactions were soon found.

Before long, we had examples for which the reaction cross section is of intermediate size and the product peaks sideways, giving a conical angular distribution about the direction of the initial relative velocity vector. Such a case is the reaction

$$K + CCl_4 \rightarrow KCl + CCl_3.$$

A strong correlation emerged: as the magnitude of the total reaction cross section increases, the preferred recoil direction of the alkali halide product shifts forwards. This is exemplified as $K \rightarrow Rb \rightarrow Cs$ for the CCl_4 reaction, whereas the angular distributions do not change much with the identity of the alkali atom for the CH_3I case or the Br_2 case. Velocity analysis experiments likewise show a nice contrast. The product translational energy is large for the CH_3I case and small for the Br_2 case, but the angular distributions do not change much with the exit energy. For the CCl_4 reaction, the preferred direction of the product moves forward rapidly as the translational recoil energy increases. The form of this strong angle-energy coupling resembles the *rainbows* familiar in elastic scattering of molecules and in sunny but moist skies.

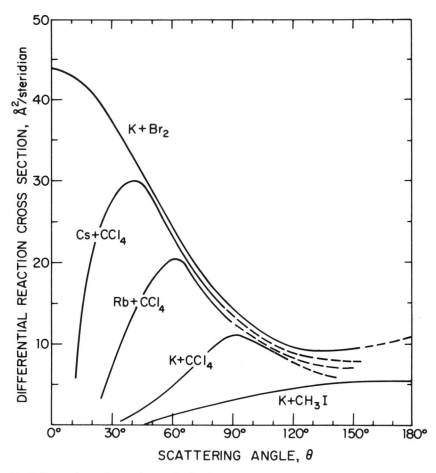

Fig. 7. Comparison of scattering angle distributions for reactions of alkali atoms with Br_2, CCl_4, and CH_3I, illustrating stripping, transitional, and rebound behavior. Ordinate scale indicates approximate normalization to total reaction cross sections. Distributions show angular varation that would be seen by a hypothetical observer traveling with the center-of-mass. Scattering angle indicates deviation of product relative velocity from reactant relative velocity, for $0 - 0°$, the alkali halide emerges in same direction as incident alkali atom.

    The marked anisotropy of the product angular distributions for this family of reactions, ranging between rebound and stripping, indicates that all proceed by a *direct* or *impulsive* mechanism. The duration of the reactive collisions must be so short that most of the transient collision complexes decompose before rotating through 180°. Since the rotational velocities are quite high, roughly half or more of the complexes must decompose within about 5×10^{-13} sec, a time not much longer than a vibrational period.

    The dynamical properties and chemical variations found for these alkali reactions were well suited to the early Monte Carlo trajectory studies and also prompted several insightful heuristic models. Ironically, trajectory studies of product angular distributions suffered at first from an instructive malady akin to the poisoning problem that had held up our experiments. The calculations for atom exchange reactions, $A + BC \rightarrow AB + C$, used plausible forms of potential energy surfaces with empirical parameters. These surfaces were constructed by adding up pairwise interactions modulated by switching functions to weaken each bond as the collision partner draws nigh. Blais and Bunker found that reactive trajectories for such surfaces gave chiefly backward scattering, in satisfying agreement with our first experiments. But this rebound behavior was unduly prevalent; for about two years after the discovery of stripping and rainbow-like reactive scattering, trajectory calculations remained unable to obtain anything but backward scattering. Godfrey and Karplus cured this affliction using potential surfaces derived from approximate quantum mechanical secular equations. This ensured a smooth surface, free from spurious bumps and wrinkles that appeared in surfaces constructed by adding piecewise interactions. It was "contamination by such warts" that induced too much backward scattering. By varying the repulsive or attractive character of the wartless surfaces, the trajectory calculations nicely elucidated the various trends associated with transition from rebound to stripping. Likewise, trajectory studies by John Polanyi and his students mapped out the systematics of energy disposal among product translation, vibration, and rotation.

We now could also relate these general dynamical properties of alkali reactions to the electronic structure of the target molecules. All these reactions involve converting a covalent bond into an ionic bond, and in effect are gas-phase acid-base or ion-recombination processes, via

$$A + BC \rightarrow A^+ + (BC)^- \rightarrow A^+B^- + C.$$

According to Michael Polanyi's *harpooning* model, the attacking alkali atom tosses out its valence electron, hooks the halogen-containing molecule, and hauls it in with the Coulomb force. The basic features of the initial electron transfer had been examined as early as 1940 in an exemplary theoretical study by Magee. As pictured schematically in Fig. 8, the transfer occurs in the vicinity of "curve crossing" between zeroth-order "purely covalent" and "ion-pair" potential surfaces. When the crossing radius r_c is large enough, Coulombic attraction is dominant there and this radius may be estimated from

$$e^2/r_c = IP(A) - EA(BC),$$

the difference in the ionization potential of the alkali atom and the electron affinity of the target molecule. This relates the size of the reaction cross section to the energy required to create the ion-pair.

However, the dynamical properties are also greatly influenced by the exit interaction as the A^+ ion approaches the intermediate molecule-ion BC^-, which is severely distorted by the strong electric field and dissociates. Accordingly, many dynamical features will differ with the shape and particularly the location of the asymptote of the potential curve for the negative molecule ion. Often we could estimate the relevant potential curves for the BC^- ion from data obtained in electron impact experiments or by means of arguments developed by Mulliken for the analysis of charge-transfer spectra. Figure 9 classifies some of the possibilities.

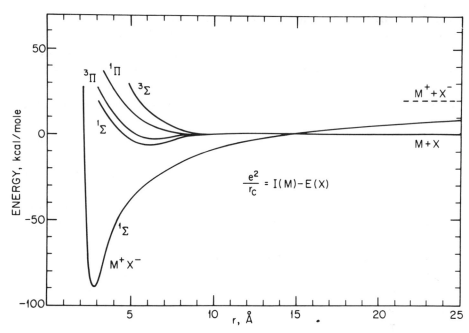

Fig. 8. Potential energy curves for an alkali halide molecule (drawn for KBr) showing the "zeroth-order" crossing of the ionic and covalent states.

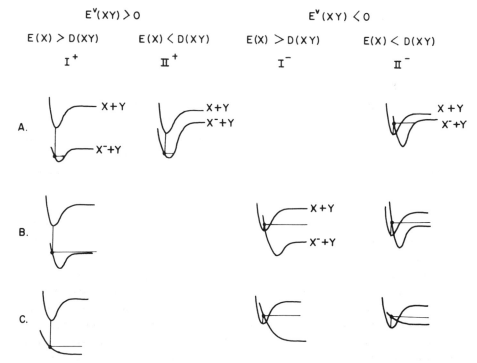

Fig. 9. Classification of electron attachment processes producing negative molecule-ions (XY + e → XY⁻). Here D denotes the bond dissociation energy, E the adiabatic and E^v the vertical electron affinity.

For instance, the case designated I^-C pertains to the CH_3I reaction and many other examples of rebound reactions. The negative ion is then formed in a strongly repulsive state, because the harpooning electron enters a strongly antibonding molecular orbital with a node between the originally bonded atoms. This is usually the same orbital that governs photodissociation of the parent molecule, so we could derive the location and shape of the exit potential curve from its absorption spectrum.

In contrast, the case designated I^+A pertains to the Br_2 reaction and many of the prototype stripping systems. These involve exceptionally large electron affinities, which accounts for the very big reaction cross section and the long-range attraction that gives forward scattering. Here the intermediate negative molecule ion would be stable in the absence of the positive ion, although highly excited vibrationally. But when the ions are within the crossing radius of about $7 A^2$, the Coulomb field exceeds 3×10^9 V/cm and thus the negative ion will be pulled apart in a time comparable to a vibrational period, or $\sim 10^{-13}$ sec. This accounts for the direct character of the reaction. However, the transient attractive coupling between B^- and C in the exit channel has a major role in the energy disposal.

The harpooning model prompted treatments of many other properties, such as the mechanics of curve crossing and dissociative electron attachment, orientation dependence, and orbiting outside a centrifugal barrier at the crossing radius. But in terms of chemical perspective, the most striking aspect was the connection to charge-transfer spectra. In retrospect, the detection-imposed limitation of early beam studies was fortunate, as this alkali family was ideally suited for relating reaction dynamics to the electronic structure of reactant molecules.

Elucidating Polanyi Flame Reactions

As well as demonstrating very large reaction rates, the classic studies of alkali-halogen reactions by Michael Polanyi also revealed intense emission from electronically excited alkali atoms. This had a major role in early discussions about interaction of electronic and nuclear motions, as the Polanyi flames preceded by a few years the Born-Oppenheimer and Franck-Condon approximations, both found in 1925. Some central questions concerning the chemiluminescence remained unresolved 40 years later, however, when we undertook beam experiments to examine each of the postulated elementary steps in the flame mechanism. Again, the results exemplified prototypcial aspects of electronic structure.

One of the key steps producing chemiluminescence involves *vibrational-to-electronic* energy transfer. This we studied by means of a "triple-beam" experiment. Vibrationally excited KBr† was formed at the intersection of crossed beams of K and Br_2 and sent into a second scattering chamber containing a Na crossed beam, where fluorescence from K* was observed. The cross section found for this *reactive transfer* process,

$$Na + KBr\dagger \rightarrow NaBr + K^*$$

is very large, $10-100$ A^2, whereas that for *nonreactive transfer* via

$$K + NaBr\dagger \rightarrow K^* + NaBr$$

is at least tenfold smaller. (The dagger denotes high vibrational energy, the asterisk denotes the lowest ^2P excited electronic state.) This result is interesting because the initial energy distributions of the reactants are nearly the same in the Na + KBr$^+$ and the K + NaBr$^+$ experiments and the same set of potential surfaces is accessible in both cases. On energetic or statistical grounds, both processes might have been expected to form K* + NaBr with the same probability. The interpretation becomes clear when the ionic character of the bonding is considered. If in the intermediate (AB)$^+$X$^-$ complex the (AB)$^+$ ion decomposes rapidly and irreversibly in the field of the X$^-$ ion, configurations in which both A and B are symmetrical with respect to both charge-sharing and interaction with X$^-$ may seldom be traversed. The chemical exchange process $(A + B^+X^- \rightarrow A^+X^- + B^*)$ is then more favorable for electronic excitation because it involves charge-transfer $(B^+ + e \leftrightarrow B^*)$ whereas the nonreactive process $(A + B^+X^- \rightarrow A^* + B^+X^-)$ does not involve charge-transfer. Later, we found *translational-to-electronic* energy transfer for these systems exhibits the same strong preference for reactive over nonreactive transfer, although the cross sections are a thousandfold smaller. Potential surface calculations indicate this preference appears to involve a "stereospecific" orientation dependence. The potential surfaces for the ground and excited states remain widely separated except for nearly collinear configurations with the halogen between the two alkali atoms.

In the other major chemiluminescent step, the same potential surface complex is reached via harpooning, $X + AB \rightarrow (AB)^+X^- \rightarrow A^+X^- + B^*$. This we studied by generating a beam of halogen atoms from thermal dissociation in a graphite oven and a beam of alkali dimers from a supersonic expansion. Contrary to previous indirect evidence, our beam experiments showed that this step occurs with a large cross section, ~ 10 A^2. A "degeneracy-induced" excitation mechanism proposed long before by Magee was thereby vindicated. The reactant halogen atom has three-fold orbital degeneracy, corresponding to location of its valence-shell "hole" in the p_x, p_y, or p_z orbital, whereas the reactant and product molecules are in nondegenerate electronic states. Thus the reactants give rise to three distinct potential energy surfaces but only one of these can lead to the nondegenerate ground state of the product alkali atom and the other two surfaces must lead to electronically excited states.

Many other aspects of the harpooning mechanism were revealed in subsequent studies of electronic excitation processes [11]. Particularly striking was evidence for an "internal reflection" mechanism. This is illustrated in Fig. 10 for K + SO$_2$ collisions. Intense emission of K$^*(^2P)$ fluorescence is seen at collision energies above the endoergicity (1.6 cV) for this excitation. The emission increases strongly up to the threshold (3 eV) for formation of K$^+$ + SO$_2^-$ ion-pairs and then declines steeply. At energies above the A* + X asymptote but below the A$^+$ + X$^-$ asymptote, trajectories which make the $A + X \rightarrow A^+X^-$ crossng and intend to exit via the ion-pair channel find it

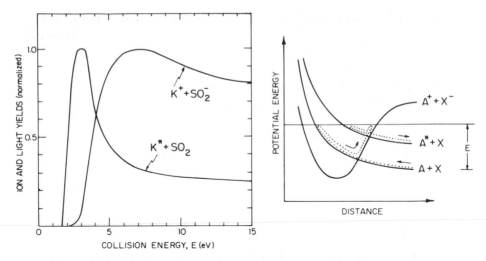

Fig. 10. Cross sections for electronic excitation and ion-pair formation on impact of K atoms with SO$_2$ molecules, as functions of initial relative kinetic energy. Schematic potential curves at right show reaction path illustrating "internal reflection" mechanism.

closed. This flux must be reflected back and redistributed at the $A^+X^- \rightarrow A^* + X$ and $A^+X^- \rightarrow A + X$ crossings until ultimately it escapes via either the excitation channel or the original entrance channel.

Persistent Collision Complexes with Glorys Galore

Until late 1966, the alkali reactions studied in crossed beams or in Polanyi flames all converted a *covalent* bond into an *ionic* bond, and all appeared to involve abrupt, *impulsive* reaction dynamics. Likewise, by then trajectory calculations for A + BC reactions had explored a variety of "mountain-pass" and "ski-run" potential energy surfaces, and found that the lifetime of the transient A-B-C complex is usually too short for rotational or even vibrational motions; the products typically emerge within 10^{-13} sec or less. However, we expected that suitable systems would form a lingering or *persistent* complex with lifetime long compared to rotational as well as vibrational periods. In this regime, the angle and energy distributions of the scattered products would give information about the *unimolecular* decomposition of the complex.

Among alkali reactions likely candidates were the exchange reactions with alkali halides,

$$A + X^-B^+ \rightarrow A^+X^- + B,$$

at ordinary thermal energies, too low to permit electronic excitation. For these *ionic→ionic* reactions, since an ion-pair is present throughout, electron transfer can enhance attraction rather than induce exit repulsion. Furthermore, the near equality of the A^+X^- and B^+X^- bond strengths and the exceptional stability of the diatomic alkali molecule-ions (just then confirmed in the Ph.D. work of Yuan Lee with Bruce Mahan) favor formation of an $(AB)^+X^-$ complex of appreciable lifetime. Electronic structure calculations subsequently demonstrated that the potential energy surfaces for dialkali halides indeed have a pronounced basin, correponding at its minimum to a triangular complex stable by about 10 kcal/mol or more with respect to the separated products.

With the aim of studying these alkali atom + akali halide reactions, before departing Berkeley we had begun constructing an apparatus with a mass spectrometer behind the surface ionization filament, to provide differential detection for both of the product species and both of the reactant species. Walter Miller and Sanford Safron completed the apparatus at Harvard and obtained lovely evidence for persistent collision complexes in more than a dozen of these exchange reactions. In each case, the angular distribution of products has symmetry about 90° and peaks very strongly near 0° and 180°. Furthermore, at wide angles the nonreactive scattering shows a "sticky collision bump" arising from break-up of the complex to reform the reactants rather than proceed to products.

We found the angular distributions could be nicely interpreted in terms of a statistical model adapted from the compound nucleus treatment of nuclear fission. The symmetry about 90° indicates that the complex persists for at least a few rotational periods and hence many vibrational periods. This permits a rough estimate of 5×10^{-12} sec as a lower limit for the mean lifetime of the complex. The strong peaking at 0° and 180° indicates that the typical complex forms and dissociates with centrifugal angular momentum much larger than the rotational momenta of the reactant or product salt molecules. As illustrated in Fig. 11, the products emerge with equal probability at all azimuthal angles about the total angular momentum vector **J** of the complex, like water from a lawn sprinkler. The complete angular distribution is obtained by averaging uniformly over all orientations of the sprinkler about the relative velocity vector **V** of the reactants. The result corresponds to the intersections of circles of latitude and longitude on a globe. The product intensity is low in the equatorial regions but becomes very high in the polar regions, near 0° and 180°. This prominent peaking forward or backward along **V** is called a "glory", after an analogous effect in light scattering from raindrops.

The glory peaks are rounded off when much of the angular momentum **J** appears in rotational tumbling of the reactant or product molecules rather than centrifugal motion. As seen in Fig. 11, this tilts the spinning sprinkler away from the polar regions. The shape of the product angular distribution thus reveals the relative contribution of the centrifugal and tumbling motions. This is illustrated in Fig. 12. In turn, the statistical model links the angular momentum disposal to the moments of inertia of the complex in the transition-states for formation and decomposition. The arrangement of atoms and the rotational motions in the transition-state thereby can leave its imprint in the product angular distribution, even though the complex may dissociate after only a few rotations.

The product energy distributions are more directly related to properties long familiar in the theory of unimolecular reaction rates. The main features are governed primarily by the statistical densities of rotational and vibrational states at the transition-state, as in the Rice-Ramsberger-Kassel-Marcus treat-

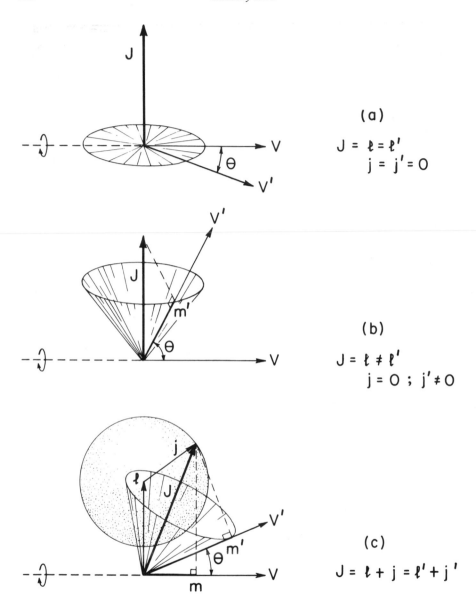

(a)

$$J = \ell = \ell'$$
$$j = j' = 0$$

(b)

$$J = \ell \neq \ell'$$
$$j = 0 \; ; \; j' \neq 0$$

(c)

$$J = \ell + j = \ell' + j'$$

Fig. 11. Relationships among initial and final angular momenta and relative velocity vectors for a long-lived collision complex. In case (a) there is no rotational momentum for either the reactant or product molecules; in (b) it is present only for the products; in (c) for both reactants and products. For fixed magnitudes of the total angular momentum **J** and the projections **m** and **m'** on the relative velocity vectors the product angular distribution is generated by uniform precessions of **V'** about **J** and **J** about **V**.

ment of unimolecular rates. Qualitative aspects are illustrated in Fig. 13. In partitioning the available energy among relative translation, vibration, and rotation of the products, the statistically favored situation puts only a small part into translation, since the vibrational and rotational modes are more numerous. Thus, the probability distribution decline rather rapidly with incre-

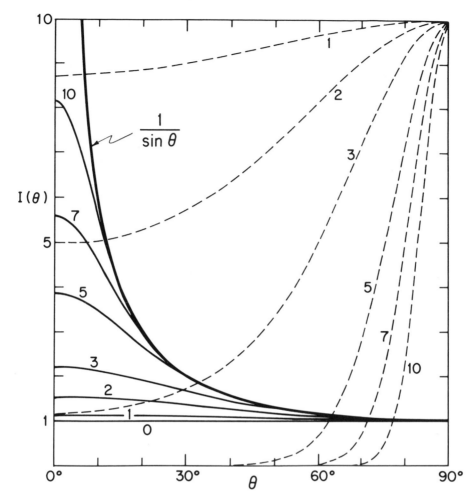

Fig. 12. Angular distributions for the statistical complex model, averaged over distributions of angular momenta appropriate to a prolate complex (solid curves) or an oblate complex (dashed curves), for case (b) of Fig. 11. All the curves are symmetric about 90°. The index gives the ratio of the maximum value of reactant orbital angular momentum 1_{max} to the average value of the projection m' for the dissociating complex; thus the index indicates the relative contribution of centrifugal and tumbling motions.

ase in product translational energy. This decline becomes more rapid as the number of atoms in the complex increases and hence the number of vibrational modes increases. Another effect enters when the complex is rotating, however. The energy in centrifugal motion is not available for statistical distribution among the other modes. On decomposition of the complex this centrifugal energy is converted into relative motion of the emerging product molecules. This changes the shape of the product translational energy distribution. The low energy region is determined by the centrifugal contribution, the high energy region by the statistical contribution.

As outlined here, the statistical model pertains to a "loose" complex defined by the exit and entrance centrifugal barriers associated with long-range attrac-

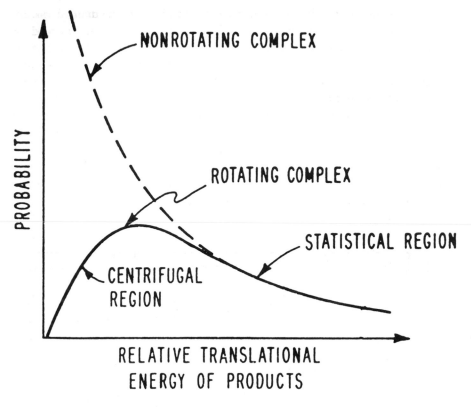

Fig. 13. Schematic distribution of product relative kinetic energy as predicted from statistical theory, without (dashed) and with (solid) the centrifugal contribution that enters for a rotating complex.

tion. Within these barriers, the energy disposal is assumed to be statistical; outside, the collision partners are assumed to rotate freely and travel like point masses subject to two-body forces. This simple model has worked well for many reactions, although various refinements are needed for quantitative analysis. Other basic factors enter for a "tight" complex. The transition-state then occurs at a potential barrier rather than a centrifugal barrier, and bond deformations are required to surmount the barrier. This implies that energy can be exchanged between different degrees of freedom as the system moves from the transition-state to the separated products. With allowance for such effects, statistical treatments akin to the RRKM theory usually prove adequate in the persistent complex realm.

A major, instructive discrepancy appeared for the dialkali halide systems, however. Despite its good agreement with the angle and energy distributions, the statistical model overestimated the ratio of reactive to nonreactive decay of th collision complex, often by a factor of 3 to 5 or more. This striking discrepancy is attributed to preferred reaction geometry. The potential energy surfaces predict that the preferred direction of approach is collinear, with the incoming alkali atom attacking the "wrong end" of the salt molecule. This corresponds to linear $(AB)^+X^-$ and reflects the stability of the dialkali ion. By virtue of the strong long-range attraction, most complexes are formed in collisions with large impact parameters and thus the centrifugal momemtum often restrains the roughly collinear $(AB)^+X^-$ configurations from bending into the triangular configurations required for reaction. Later, we found many similar instances of preferred reaction geometry.

In search of another prototypical deviation from the statistical case, we looked for a collision complex with lifetime comparable to its mean rotational period. This is called an *osculating* complex, a term applied by Wigner to an analogous realm in nuclear reactions. As indicated in Fig. 14, our model calculations for this case assume that during rotation about its total angular momentum **J** the complex is subject to decomposition with a random lifetime

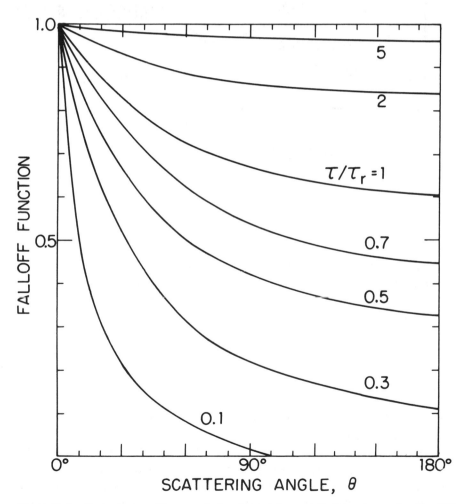

Fig. 14. Ratio of intensity appearing at angle θ to the peak intensity for forward scattering ($\theta = 0°$) as calculated for the *osculating model*. The corresponding angular distribution is obtained by multiplying this "falloff function" by the appropriate curve from Fig. 12. The falloff curves are labelled by the ratio of the mean lifetime of the complex to a typical rotational period, τ/τ_r.

distribution, $\exp(-t/\tau)$. This gives an angle-dependent form factor governed by the ratio τ/τ_r of the mean lifetime to a typical rotational period. The angular distribution is approximated by multiplying that for a persistant complex by this factor. For $\tau \sim 5\tau_r$ we thus expect the forward and backward glory peaks to be about equally strong, whereas for $\tau \sim \tau_r$ the backward peak should be attenuated by about 50%. Reactions of alkali atoms with thallium halides were examined as likely candidates. We expected the lifetime of the complex might be shortened by the higher ionization potential of the Tl atom, which inhibits the sharing of charge, and by the higher reaction exoergicity. Indeed, George Fisk and Doug McDonald found these systems show the glory peaks characteristic of a complex, but the backword peak is substantially weaker, by 30% to 50%. Again, similar behavior was later found for many other reactions. Attractive chemistry often yields not a persistent coupling but just a flirtatious whirl.

Another aspect of unimolecular kinetics explored with alkali reactions is the dependence on the number of degrees of freedom. Even some *covalent→ionic* reactions with sizable exoergicity proved to involve persistent or osculating complexes when six or eight atoms are involved. The most fully characterized example is the $Cs + SF_6$ reaction; in addition to angle and velocity distributions, we measured the CsF rotational excitation by electric deflection and the vibrational populations by electric resonance spectroscopy. Although the available energy is rather large, $40-50$ kcal/mol, the reaction appears throughly statistical. It has symmetrical forward and backward glory peaks and the same effective temperatures for translation, rotation, and vibration, corresponding to equipartition of energy. In this study, a miniature crossed-beam apparatus was joined with Klemperer's electric resonance spectrometer. This enabled quantum states of the reactively scattered CsF to be completely identified with respect to vibrational level, rotational angular momentum, and space orientation of the angular momentum. It also completed another link with the saga of Stern and Rabi.

Beyond the Alkali Age

From the beginning, we spoke of the "early alkali age", a phrase both wistful and whimsical. It was intended to suggest that broader ages would inevitably follow, without predicting how soon. Skeptics discounted alkali reactions as an eccentric, unrepresentative family, but we gladly persisted because they revealed such instructive dynamical variety. Meanwhile, an unduly pessimistic view of prospects for beam studies of nonalkali reactions arose elsewhere. This came from attempts to use mass spectrometric detection which failed to reduce sufficiently the interfering background in the electron bombardment region. In March of 1967 we began designing av new apparatus, to be named *Hope*. Only nine months later it would take us "beyond the alkali age".

Universal Detector and Supersonic Nozzles

Hope was undertaken when Yuan Lee joined our group as a postdoctoral fellow. With Bruce Mahan at Berkeley, he had already built a major apparatus for beam studies of ion-molecule reactions. In this, as in Hope, he made elegant use of ion counting techniques adapted from nuclear physics. Yuan was teamed with two beginning graduate students, Pierre LeBreton and Doug McDonald, and several machinists in our Departmental Shop, headed by George Pisiello. It was a group of extraordinary skill and zeal, inspired by Yuan's genius and fervent sense of mission. Blissfully, as each of the myriad design questions was settled, we all relished his verdict: "Should be all right".

The main features of Hope are shown in Fig. 15. The entire detector unit is mounted on a rotatable lid and the beam sources are fixed, to facilitate changing the source modules and strong pumping of the source chambers. The key component of the detector, the electron bombardment ionizer, is located within three nested chambers, each differentially pumped by ion pumps and cryogenic traps. This fulfills two vital design criteria for the ionization region. (1) The background gas (at the mass of interest) which diffuses in from the scattering chamber must be reduced to a partial pressure comparable to or lower than the product species which enters in free flight. This partial pressure is typically $\sim 10^{-14}$ torr, only tenfold above the vacuum in interstellar space. (2) Background from deposit of product molecules on surfaces near the ionizer

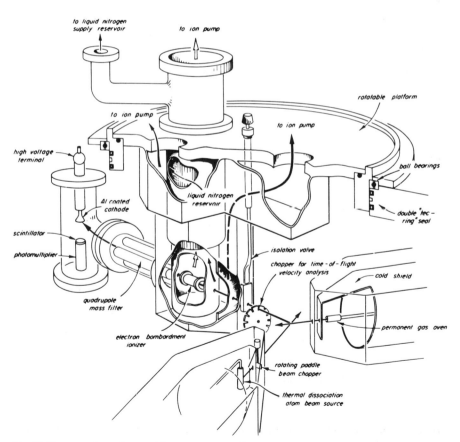

Fig. 15. Cutaway view of crossed beam apparatus *Hope* with mass spectrometric detector, showing beam geometry and arrangement of differential pumping for detector and source chambers.

must be avoided. This is accomplished by the nested design, which permits product molecules that pass through the ionizer without being ionized (the fate of 99.9%!) to fly on to another differentially pumped region before hitting a surface. In practice, Hope and its successors made accessible reactions with much smaller yields as well as greatly extending the chemical scope. Respectable reactive scattering data are now often obtained with product fluxes of only $\sim 10^3$ molecules/sec.

Supersonic nozzles also greatly expanded the variety and scope of beam experiments. Beyond providing very high intensity (typically $\sim 10^{18}$ molecules/steradian/sec), beams generated from such nozzles can be aptly viewed as a "new state of matter", with special properties much different from the old trinity of gas, liquid, and solid. The collisions occurring the high pressure region of a supersonic nozzle organize the beam molecules to a remarkable extent. The exiting crowd of molecules may have mean separations of only 50 diameters, yet have nearly the same velocity and direction and hence suffer almost no collisions. Likewise, the temperatures associated with relative translation and rotation of molecules within the beam are typically very low (of the order of 1 ^0K or less) whereas the vibrational temperature can be kept high or made low by choice of conditions. Another option is translational acceleration by "seeding". The reactant gas is mixed with a large excess (typically 100-fold) of light diluent gas such as helium or hydrogen. Collisions during the supersonic expansion then bring the seeded molecules to the same exit velocity as the diluent gas, and also concentrate the heavier species along the beam axis. Intense beams are thereby obtained with kinetic energy readily variable over a wide range extending well above typical activation energies for chemical reactions.

Like many others since, our adoption of supersonic nozzles was spurred by John Fenn and Jim Anderson, ardent evangelists among the chemical engineers then exploring fluid flow in nozzles. Again, there was a cultural gap to be bridged; at a 1965 ecumenical meeting [12]:

"The engineers spoke only to one another about Reynolds numbers;

the chemists likewise talked just to each other about harpooning electrons." The Kistiakowsky ax story also prompted us to try nozzles and indeed we had already used them in a modest way in alkali studies. Later we extended the scope of the seeding technique by a nozzle design suitable for solid or liquid substances, and others have developed many variants that have found wide application in both collision and spectroscopic experiments. The splinters from Kisty's ax sprouted into a bountiful garden.

Recognizing Covalent and Ionic Cousins
For our first experiments with Hope, we chose to study

$$Cl + Br_2 \rightarrow BrCl + Br,$$

and other exchange reactions of halogen atoms and molecules. These were inviting because the reactant beams were easy to produce and all species could be very effectively pumped cryogenically. Furthermore, trihalogen complexes had long been postulated as intermediates in the mechanism of halogen atom recombination and other processes. Such complexes had not been detected in the gas phase, and the only rate constants available for the exchange reactions had been derived from the always equivocal analysis of multistep processes in photochemical systems.

Our first attempt to study $Cl + Br_2$ gave data of excellent quality and revealed instructive dynamical features. Although the total reaction cross section is $\sim 10 \text{ Å}^2$, smaller than the hard-sphere cross section, the BrCl angular distribution peaks strongly in the forward hemisphere. This indicates that the dominant interaction is short-range and yet attractive. In previous beam experiments and trajectory calculations, the reactive scattering had always peaked backwards when the cross section was smaller than the hard-sphere value. However, despite its much smaller reaction cross section, the $Cl + Br_2$ reaction gives product angle and translational energy distributions quite similar in form to the $K + Br_2$ reaction. We suggested this was evidence for an osculating complex mechanism.

Only a few months after our results for $Cl + Br_2$ were announced, groups at Freiburg and Los Alamos had confirmed our study, using mass spectrometric apparatus of appreciably different design. This illustrates the value of choosing an amenable system to test new apparatus or to extend an experimental domain. Yet the reaction cross sections found in the beam studies indicated that the rate constants derived from the photchemical experiments were about 100-fold too low. Later the photochemical studies were repeated, with results in agreement with the beam work.

In studies of other trihalogen systems, we obtained further evidence for an osculating complex with preferred reaction geometry. The wide-angle reactive scattering indicates complexes containing iodine are more stable than those with only bromine or chlorine, in accord with simple electronic structure arguments. Likewise, for complexes containing different atoms, the most stable configuration has the least electronegative atom in the central position. One of the striking results attributed to this preferred geometry appeared in the Br + ClI case. The yield of BrCl is about 5-fold smaller than otherwise expected and its angular distribution peaks backwards, indicating repulsive interaction. Subsequent studies, including variation of collision energy, brought out other dynamical features. Often these could also be interpreted by analogy to alkali reactions, despite the marked contrast in covalent and ionic bonding.

Next we pursued the reaction that John Polanyi had employed as the prototype in developing his infrared chemiluminescence method,

$$H + Cl_2 \rightarrow HCl + Cl.$$

This system proved to have a gratifying kinship both to photodissociation of Cl_2 and to the $K + CH_3I$ reaction. Angle-velocity contour maps displaying this comparison are shown in Fig. 16. For the Polanyi reaction, the product angular distribution is broad but quite anisotropic, with the HCl recoiling backwards and Cl forwards with respect to the incident H atom. The product velocity is very high, about 1600 m/sec at the peak of the distribution. This corresponds to

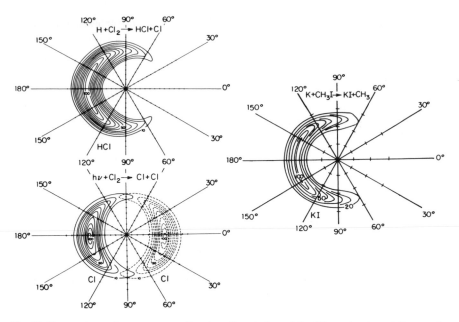

Fig. 16. Comparison of contour maps for photodissociation of the Cl_2 molecule and for reactive scattering of $H + Cl_2$ and $K + CH_3I$. Map for the latter case from work of A.M. Rulis and R.B. Bernstein [3]. For each map, origin is at center-of-mass and horizontal axis is along reactant relative velocity vector, with direction of incident atom or photon designated $\theta = 0°$. Tic marks along radial lines indicate velocity internals of 200 m/sec.

release of about half of the available energy (reaction exoergicity of ~45 kcal/mol plus initial collision energy of ~10 kcal/mol) in the translational recoil of HCl and Cl. The rest appears in vibrational and rotational excitation of HCl, which is observed in the infrared luminescence. The form of the angular distribution indicates collinear H-Cl-Cl as the preferred reaction geometry and the high recoil energy shows that strong repulsive forces are abruptly released.

The contour map for photodissociation was constructed from the continuous absorption spectrum of Cl_2, which shows directly the distribution of relative translational energy of the fragment Cl atoms and hence the repulsive energy release. The spectrum can be closely approximated by simply "reflecting" the Gaussian vibrational distribution of the ground electronic state from the steep (~7 eV/A) repulsive wall of the dissociative excited electronic state. The angular distribution is governed by the dipole selection rule for absorption, which makes the transition probability vary as the square of the cosine of the angle between the Cl-Cl axis and the photon beam direction. We see that $H + Cl_2$ and $h\nu + Cl_2$ give remarkably similar angle-velocity maps.

More remarkable still is the close resemblance of $H + Cl_2$ and $K + CH_3I$; except for a change of scale, the contour maps are almost congruent. In terms of electronic structure, as already noted, the analogy to photodissociation is obvious in the $K + CH_3I$ case because the harpooning electron enters the same strongly antibonding molecular orbital excited in photodissociation. In the $H + Cl_2$ case the analogy is not obvious, since the very high ionization potential of

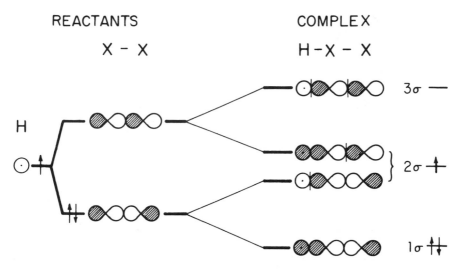

Fig. 17. Schematic construction of the three σ molecular orbitals of collinear H-X-X. The *frontier orbital, 2σ,* may be regarded as a superposition of two components, one H-X antibonding and X-X bonding, the other vice-versa.

the H atom prohibits electron transfer. However, we found a molecular orbital rationale using the "frontier orbital" concept of Fukui. This is indicated in Fig. 17. Collinear approach of the H atom generates three σ orbitals for the H-Cl-Cl complex. The middle one of these, 2σ, is expected to be the highest occupied orbital in the reaction complex. This frontier orbital has one node, resulting from the superposition of two components, one H-Cl antibonding and Cl-Cl bonding, the other vice versa. Simple calculations based on empirical data indicate the latter component is dominant. Thus, the frontier node lies roughly midway between the chlorine atoms, just as in the lowest-lying unoccupied orbital of Cl_2, the orbital excited during photodissociation. The similarity in the contour maps stems from the congruence in frontier nodes which govern the repulsive energy release in both reaction and photodissociation.

It is uncanny that the H + Cl_2 and K + CH_3I reactions, seminal in developing the infrared chemiluminescence and molecular beam methods, proved to be so closely related; not only are both rebound reactions, they are even "kissing cousins." The nineteenth century notation still used to write down chemical reactions gives no hint of such kinships, whereas electronic structure interpretations often bring out the underlying simplicity and broad scope of reaction dynamics.

Other hydrogen atom + halogen reactions provide instructive contrasts. As $Cl_2 \rightarrow Br_2 \rightarrow I_2$, the repulsive energy release becomes a smaller fraction of that in photodissociation and the hydrogen halide angular distribution shifts from backwards to sideways with respect to the H atom direction. These aspects are illustrated in Fig. 18. The molecular orbital treatment relates both trends to the decrease in halogen electronegativity. This enhances the p-character of hybrid orbitals involving the central atom and thereby favors a bent configuration.

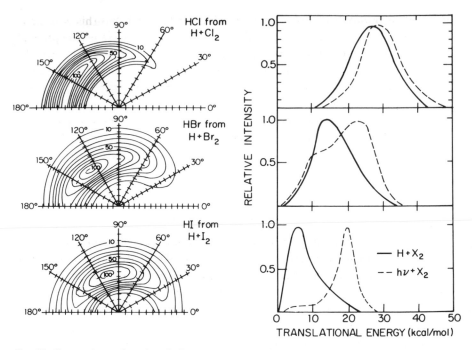

Fig. 18. Comparison of angle-velocity contour maps for reactions of H atoms with halogen molecules. Since maps must be symmetric with respect to the horiozontal axis, only upper halves are shown. Tic marks along radial lines indicate velocity intervals of 200 m/sec for Cl_2 case, 100 m/ sec for Br_2 and I_2. Panels at right compare distributions of product relative translational energy (solid curves) with continuous absorption spectra of halogen molecules (dashed curves). Abscissa scales for spectra are shifted to place origin at the dissociation asymptote, and thus show directly the repulsive energy release in photodissociation.

The frontier node also shifts from about midway between the halogen atoms in H-Cl-Cl to close to the central atom in H-I-I. Many analogous stable molecules are known which have one more or one less valence electron than these H-X-X systems. The molecules withone more electron are linear or nearly so; those having one less electron are strongly bent, with bond angles of 90° to 110°. The shift of the frontier node might be expected to make H-I-I resemble a case with one less electron. Thus it is plausible that decreasing the halogen electronegativity fosters bent reaction geometry and reduces the repulsive energy release.

The H + ICl reaction is especially interesting. On an energetic or statistical basis, reaction at the "Cl-end" would be more favorable than at the "I-end", since the H-Cl bond strength (102 kcal/mol) is much larger than H-I (70 kcal/ mol). The molecular orbitals suggest the H atom should prefer to attack the I- end, however. As a consequence of the electronegativity difference, in ICl both the highest occupied orbital (π^*) and the lowest unoccupied orbital (σ^*) are predominantly I atom orbitals. The beam studies find the HI yield is at least comparable to HCl and probably substantially higher (although the effect of some experimental factors that discriminate against HCl remains unresolved). The angular distribution of HI peaks sideways, the HCl backwards. The infrared chemiluminescence detects only HCl and not HI, but this is compatible with the beam results because the dipole derivative of HI is exceptionally small and hence the infrared emission is very weak. The HCl energy distribution has a very unusual bimodal form. Comparison with trajectory calculations suggests that some of the HCl results from direct attack at the Cl-end, but most comes from indirect reaction after initial attack at the I-end. These two reaction modes produce HCl in low and high rotation-vibration states, respectively. Analogous steric preferences for many other reactions can likewise be attributed to *orbital asymmetry* arising from differences in electronegativity.

Further aspects of this theme were examined by studying reactions of oxygen atoms with halogen molecules. We found the reaction

$$O + Br_2 \rightarrow BrO + Br$$

goes via a persistent complex with large yield and no activation energy. The product contour map shows prominent glory peaks. For other reactions with such maps, the intermediate complex usually corresponds to a stable molecular species which correlates with the ground-state reactants and products. Here, however, different spin states are involved. The reactants approach on a triplet surface, whereas the products can depart on either a singlet or triplet surface. The known stable OBr_2 molecule has a symmetric, strongly bent geometry and a singlet ground state. Qualitative electronic structure arguments suggest that the reaction goes predominantly via a less stable, triplet O-Br-Br potential surface without transition to the more stable, singlet Br-O-Br surface. Likewise, the O + ICl reaction would be expected to give primarily IO + Cl rather than ClO + I, although the latter path is much more exoergic. We indeed found a large yield of IO but no detectable ClO, nice evidence that the reaction involves end-on attack rather than insertion. These results led us to the unorthodox prediction that the $O + F_2$ reaction should prefer the F-O-F geometry rather than O-F-F, since oxygen is less electronegative than fluorine. This implies a relatively high activation energy, associated with the switch to an insertion mechanism, despite the large reaction exoergicity and the notorious chemical personalities of the reactants. Subsequent experiments [13] indeed confirmed that $O + F_2$ is inhibited by a large activation energy.

Migratory Atoms and Bonds
With the mass spectrometric detector, we could now also proceed to "the organic age" and thereby pursue many further aspects of reaction dynamics. Particularly inviting were unimolecular reactions involving isomerizations or rearrangements. Fig. 19 shows the contour maps obtained for a favorite pair of examples, the displacement reactions of chlorine atoms with vinyl bromide and with allyl bromide. In these systems,

$$Cl + RBr \rightarrow ClRBr \rightarrow RCl + Br,$$

where R denotes $CH_2=CH$ or $CH_2=CHCH_2$, respectively, the intermediate chlorobromoalkyl radical is a known, stable species. It is vibrationally excited by ~30 kcal/mol, the sum of the initial collision energy and thermal excitation

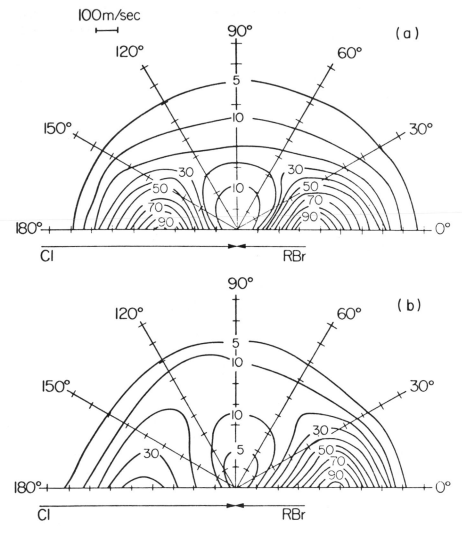

Fig. 19. Comparison of contour maps for chlorolefin products from reactions of chlorine atoms with (a) vinyl bromide and (b) allyl bromide. Tic marks along radial lines indicate velocity intervals of 100 m/sec.

of the reactant olefin, the loss in bond strength in converting the double bond to a single bond (~57 kcal/mol), and the gain in forming the new C-Cl bond (80 kcal/mol). Under our single-collision conditions, the excited radical cannot be deactivated by a subsequent collision and thus must undergo unimolecular decomposition. The net energy available to the products on reforming the double bond and releasing the Br atom is ~19 kcal/mol. These energetic aspects are the same for the vinylic and the allylic reactions.

The customary theory of unimolecular processes predicts that the allylic reaction should proceed more slowly and hence show more nearly statistical behavior, since it involves more atoms and therefore the excitation energy shuffles among more vibrational modes than in the vinylic reaction. On the contrary, the contour map for the vinylic reaction indicates a persistent complex, that for the allylic reaction an osculating complex. Some special feature must intervene to make the vinylic reaction more nearly statistical than the allylic reaction. Indeed, this was anticipated in choosing these systems for study. According to a large body of work on organic reaction mechanisms, the initial stage in both reactions should involve addition of the Cl atom to the carbon atom "most distant" from that with the Br atom. As pictured in Fig. 20, this implies that a "free valence" appears on the carbon linked to Br in the vinylic case, but not in the allylic case. Thus, the vinylic reaction seems likely to proceed via a 1,2-chlorine *atom migration*, whereas the allylic reaction can go via a 1,3-*bond migration*. Since the heavy atom migration would be much slower than the bond migration, this might provide a rate-limiting process which makes the vinylic reaction more statistical than the allylic one.

These mechanisms require the product chlorolefin to have Cl on the carbon to which Br was originally bonded in the vinylic case, but to remain on the "most distant" carbon in the allylic case. We verified this by analysis of the fragment ion mass spectra of chlorolefins formed in corresponding reactions with a methyl group added to "label" one or another carbon atom. The methyl substitutions also produced revealing changes in the reactive scattering. For the vinylic cases, the angular distributions show variations which reflect changes in the rotational motions caused by the methyl group. For the allylic cases, the product translational energy distributions become statistical, as if the methyl group diverts the intramolecular energy flow and thereby makes it more random.

Further tests of the migration mechanisms were later obtained by Rowland using radioactive tracers [14]. In this work, the same vinylic and allylic reactions examined in the beam experiments were studied under "bulb" conditions. At low pressures, the products obtained were those expected for the atom- and bond-migration mechanisms, respectively. At high pressures, how-

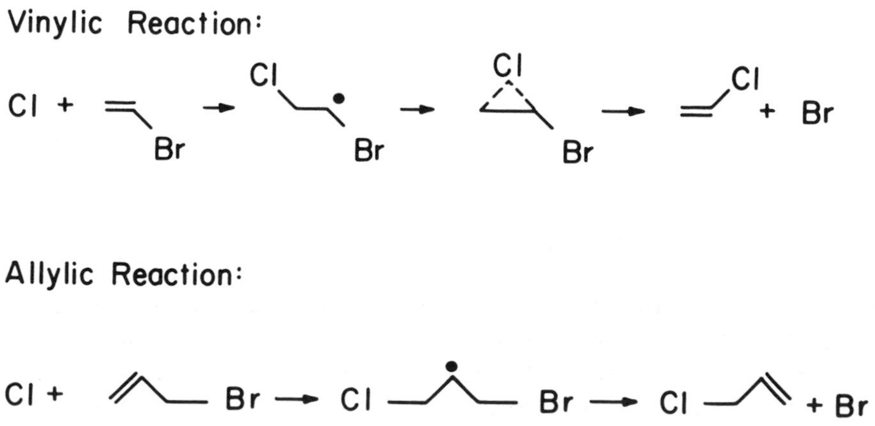

Vinylic Reaction:

Allylic Reaction:

Fig. 20. Stereoselective mechanisms for 1,2-chlorine *atom migration* in vinylic reaction and 1,3-*bond migration* in allylic reaction.

ever, different products were obtained. These corresponded to the species expected from collisional stabilization of the intermediate chlorobromoalkyl radicals, followed by standard reactions of those radicals.

A host of other organic reactions have now been studied by beam methods, especially by Yuan Lee and his students. Often the intermediates and initial elementary steps have proven to be quite different than those postulated in conventional mechanistic studies. The wide range of bonding and stereochemical situations in organic systems ensures vast scope for dynamical detective work in this domain.

Facile Molecular Reactions

The reaction systems discussed so far all involve attack by an open-shell atom or free radical. One of the chief criteria invoked in postulating elementary steps in reaction mechanisms is that only such processes have very low activation energies, less than 5 or 10 kcal/mol. The advent of the Woodward-Hoffmann rules focussed attention on reactions of molecules with molecules, which involve the concerted making and breaking of two or three pairs of bonds. These processes typically have activation energies above 20 kcal/mol when fully "allowed" by the rules and above about 40 kcal/mol when "forbidden." In this context, we sought instructive examples of facile bimolecular and termolecular reactions of diatomic molecules.

The simplest case is the exchange reaction of two alkali halides such as

$$Cs^+Cl^- + K^+I^- \rightarrow Cs^+I^- + K^+Cl^-.$$

This might be termed a "no-electron" reaction, since the salt molecules are essentially closed-shell ion-pairs. Accordingly, no restraints are imposed by the molecular orbital correlations involved in the Woodward-Hoffmann rules. Yet a contemporary theoretical paper had concluded that even this alkali halide reaction would be subject to a high activation energy of ∼50 kcal/mol. This seemed unlikely in view of the strong long-range dipole-dipole attraction. Since alkali halides form rhombic quadrupolar dimers with dissociation energy ∼30−50 kcal/mol, the potential surface for the exchange reaction has a deep basin. Our beam experiments indeed found the reaction proceeds via a persistent complex. There is no activation energy and the reaction cross section is extremely large, corresponding to formation of the complex in collisions with impact parameters up to ∼8−9 A. We therefore spoke of a "vacuum cleaner" potential. The product angle and velocity distributions conform nicely to the usual statistical model, but the ratio of nonreactive to reactive decay of the complex was 2 or 3 times larger than statistical. As in the analogous atom + salt case, this can be attributed to geometrical isomerism. Ionic model calculations predict less stabel, linear chain isomers exist in addition to the rhombic dimer. These linear chain isomers may often dissociate nonreactively rather than rearrange to the cyclic form required for the exchange process, especially when the centrifugal angular momentum in the collision keeps the chain "ends" apart. It is like two pairs of ice skaters playing "crack the whip."

Extensive trajectory calculations by Brumer and Karplus confirmed this striking role of geometrical isomerism.

Since the four-center ionic + ionic reaction proved facile whereas the typical covalent + covalent case was forbidden according to Hoffmann's obital correlations, we examined several reactions involving ionic + covalent reactions. These also proved facile at thermal collision energies; for example,

$$Cs^+Br^- + ICl \rightarrow Cs^+Cl^- + IBr.$$

This reaction very likely involves formation of an alkali trihalide salt, Cs^+ $(BrICl)^-$, and charge migration within the trihalide anion. Although apparently unknown in the gas phase, trihalides have been much studied in solution and in the solid state. In agreement with molecular orbital theory, the trihalide anions are linear or nearly linear, the middle atom is always the least electronegative (I, in this case) and it acquires a small positive charge whereas the end atoms share the negative charge. Clear evidence for this structure appears in the reactive scattering. There is no observable yield of $Cs^+ \cdot I^- + BrCl$, even at collision energies more than 20 kcal/mol above the energetic threshold for this channel. Furthermore, as seen in Fig. 21, the IBr angle-velocity contour map has a very unusual skewed shape. The lefthand product peak has distinctly higher intensity and velocity than the righthand peak. This shows that collisions from which IBr and CsCl rebound backwards with respect to the incident ICl and CsBr, respectively, are more probable and involve more repulsive energy release than collisions from which IBr and CsCl emerge in the same direction as the incident ICl and CsBr, respectively. These properties are

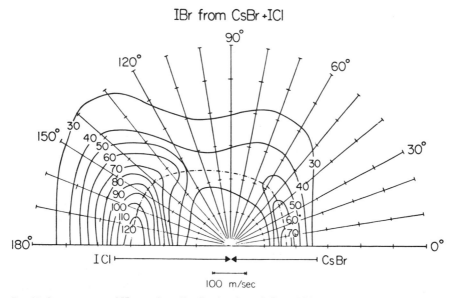

Fig. 21. Contour map of IBr product distribution from CsBr + ICl reaction at collision energy of 3.9 kcal/mol.

consistent with the expectation that in Cs^+Br^- + ICl reactive configurations, Br^- tends to be collinear with ICl while Cs^+ is likely to collide with the central I atom. The positive charge acquired by I as the trihalide forms then repels Cs^+, which picks up the emerging Cl^- and departs quickly in the direction opposite to the incident salt. Similar ionic + covalent reactions such as CsF + HCl and NaO + HCl [15] likewise go with near zero activation energy.

In the domain of four-center covalent + covalent reactions, the landmark is the classic work of Sullivan on the favorite "textbook" case,

$$H_2 + I_2 \rightarrow 2HI.$$

In 1967 he showed this does not occur as a four-center reaction but involves dissociation or near-dissociation of I_2 followed by I + H_2 + I. Soon after, Jaffe and Anderson studied HI + DI using the seeded supersonic beam technique and found no detectable HD yield at collision energies far above the empirical activation energy. This is certainly an allowed reaction, at least as the reverse of I + HD + I, but apparently vibrational rather than translational activation is required. Likewise, in our laboratory David Dixon and David King found no evidence for the four-center exchange reaction

$$Br_2 + Cl_2 \rightarrow 2BrCl.$$

Their experiment employed two supersonic nozzles and scanned the collision energy up to ~25 kcal/mol. Formation of BrCl is readily observed on mixing the reactants in a bulb, and rate studies had indicated a relatively low activation energy, only ~15 kcal/mol. However, this was suspect because of "possible catalysis by moisture or surfaces," a traditional lament in kinetic studies. Hoffmann's molecular orbital correlation diagram predicts a much larger activation barrier, comparable to the promotion energy of two electrons from a bonding to an antibonding orbital and thus above the dissociation energy of the weaker reactant bond 45 kcal/mol.

Since a six-center, termolecular reaction is allowed by the orbital correlations, we decided to look for

$$Br_2 + Cl_2... Cl_2 \rightarrow 2BrCl + Cl_2.$$

Scattering experiments using three crossed beams are utterly impractial, but appreciable fluxes of reactant dimer molecules, linked by a weak van der Waals bond, can readily be generated by a supersonic expansion. We were encouraged by tantalizing evidence obtained by Noyes for a third order reaction of halogen molecules in solution. Yet analogy to other allowed six-center cases (such as Diels-Alder reactions) suggested that the activation barrier might well be ~20 kcal/mol, and it might also require chiefly vibrational rather than translational excitation. Thus, we were startled when King and Dixon found large yields of BrCl that appeared to come from the termolecular process even at thermal collision energies of only ~3 kcal/mol. However, velocity analysis data for this system and the analogous HI reaction provided several kinematic consistency tests which even indicated that three sequential bond scissions can be resolved in the termolecular exchange process. Soon another example of a facile molecular reaction fostered by a van der Waals dimer was provided by Durana and McDonald [16]. They found that F_2 + $(HI)_2$ yields intense HF infrared chemiluminescence whereas F_2 + HI gives none. Whether or not these processes actually involve six-center transition-states, the weak van der Waals link is remarkably effective in promoting a reaction that otherwise does not go.

Subsequently, we undertook potential surface calculations to examine termolecular six-center bond exchange poroceses. For Cl_6 a traditional semiempirical treatment [17] indicates facile reaction but the approximations involved are dubious. For hexagonal H_6 *ab initio* calculations of high quality were carried out by Dixon and Stevens [18] and show that a termolecular path is indeed accessible without breaking an H-H bond whereas there appears to be no such path for a bimolecular, four-center exchange. The results for H_4 and H_6 thus conform to the rule, amply demonstrated in organic chemistry, that cycloaddition reactions involving 4m electrons are forbidden as concerted processes while those involving 4m + 2 electrons are allowed. But it turns out that H_6 obeys this rule only through configuration interaction, not by virtue of the usually invoked nodal properties.

The special role of H_6 was elucidated further by comparison with several previous calculations for larger H_n polygons, available for n = 4m + 2 extending from n = 14 to 62. These calculations, pertinent to model treatments of metallic hydrogen, found the bond length and cohesive energy (binding per atom) to be nearly constant for the whole series. Since our values for H_6 proved to be practically the same, we can reliably estimate the stability of H_{10} and higher polygons by taking the cohesive energy as constant. This shows that H_6 is the only H_{4m+2} polygon that is stable with respect to dissociation of an H_2 bond and hence the only polygon that can serve as transition-state for a concerted bond exchange. The customary orbital symmetry criterion thus fails for all the higher 4m + 2 polygons; it must be supplemented by an energetic criterion in order to predict whether a reaction is concerted.

The unique stability of H_6 relative to dissociation of an H_2 bond also suggests that solid molecular hydrogen might undergo a high pressure transition to form a new phase involving termolecular complexes before transistion to the long-sought atomic or metallic phase (predicted in 1948 by Wigner). This possibility was examined by Rich LeSar by means of lattice-energy calculations employing approximate pair potentials [19]. Although uncertainties in the potentials prevent a definite conclusion, the results obtained for a range of parameters indicate that a phase comprising a "partly dissociated" form of the six-center transition state may be stable at pressures above a few hundred kilobars. As shown in Fig. 22, repulsive forces from the neighboring units in the crystal prevent dissociation of this complex to diatoms. A marked drop in the H_2 stretching frequency observed at high pressure is qualitatively consistent with formation of molecular clusters but by no means a clear test. In any case, these calculations serve to illustrate how pursuit of reaction dynamics can lead to far distant domains!

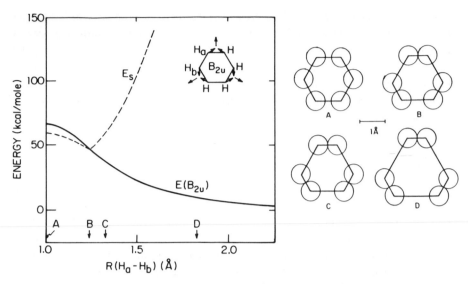

Fig. 22. Potential energy profile for dissociation of a hexagonal H_6 complex in free space (solid curve) and in a hexagonal closest-packed crystal (dashed curve, computed for pressure of 400 kbar) in which repulsive interactions with neighbors prevent the dissociation. At right are shown configurations of the termolecular complexes corresponding to positions marked along abcissa. The circles for each H atom have a radius of half of the free H_2 bond length.

Van der Waals Dimers and Clusters

Supersonic expansions make accessible a vast range of species and processes involving feeble bonds. In effect, the very low internal temperaturs in supersonic beams circumvent the Second Law, since the entropy term in the free energy, $\Delta H - T\Delta S$, can be made negligible. Even a weakly favorable enthalpy term then suffices to produce large yields of dimers or bigger clusters. This permits studies involving "solvation" of reactant species and other interactions akin to condensed phase or surface chemistry. In the past few years such work has created a major new field, with particular emphasis on metal clusters [20]. Here we describe some of our experiments to illustrate a few favorite themes.

In solution kinetics, much attention has been devoted to photodissociation and recombination processes. These are usually interpreted in terms of "the cage effect," in which the solvent inhibits diffusive separations of the photofragments. A different perspective has now emerged from beam experiments [21]. Complexes of I_2 with various solvent species such as Ar, N_2, benzene, ... were formed by supersonic expansion and excited with a laser in a spectral region above the dissociation threshold for the bare I_2 molecule. This produced intense fluorescence, showing that instead of dissociating much of the iodine relaxes into bound vibrational levels of the eletronically excited B state. Energy balance is maintained by breaking the van der Waals bond(s) of the original solvent complex and releasing some repulsion into relative translation of the product fragments. In solution photodissociation must likewise by mollified by energy transfer to the solvent and the surviving excited iodine molecules may account for features previously attributed to recombination of caged atoms.

Another typical role of solvation is exemplified in reactions of ammonia clusters with hydrogen halides. For unclustered ammonia proton transfer is endoergic to form the inoic salt, $NH_4^+X^-$, increasingly so for $X = I \rightarrow Br \rightarrow Cl$. In crossed-beam scattering with clustered ammonia, new mass peaks appear that correspond to $(NH_3)_n\, HX$ adducts with n as large as 15 or more. Framentation of these peaks appears to decrease markedly on completion of the first solvation shell. For sufficiently large clusters the complex formation probably involves proton transfer and is driven by solvation of the resulting $NH_4^+X^-$ ion - pair by the "extra" ammonia molecules in the reactant cluster.

Although solvated ions govern a host of solution phenomena, the individual molecular species and processes often cannot be resolved or characterized in condensed phases. Thus studies of molecular cluster ions in the gas phase have become another major field; for instance, the elusive hydrated electron has now been produced in beams of unfragmented water clusters [22]. In related work, we studied collisional electron transfer from alkali atoms to molecular clusters. Since the process is endoergic, fast atoms are generated from a seeded supersonic expansion. Because the electron transfer can only occur within the crossing radius for the ionic-covalent potential curves, the time-scale becomes very short, typically ~0.3 psec. This limits use of the method for electron affinity studies but also provides a means to probe the size of cluster subunits involved in impulsive interactions [23].

For small clusters, particularly dimers, it is feasible to study the exchange of van der Waals bonds by the same methods employed for chemical bonds. Fig. 23 shows an angle-velocity contour map obtained by Worsnop and Buelow [24] for the $Xe + Ar_2$ reaction. Since thermal collision energies are typically several times larger than the dimer well depth, conservation of energy and momentum constrain the scattered diatom product to a narrow circular band centered on the center-of-mass velocity; only within that band can XeAr be formed with low enough internal excitation to hang together. However, several other fea-

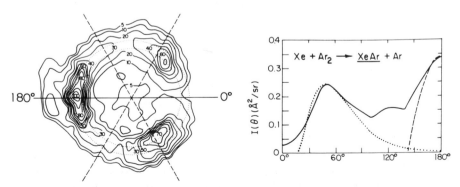

Fig. 23. Contour map for XeAr from $Xe + Ar_2$ reaction at collision energy of 1.3 kcal/mol. The dotted circular band shows the kinematically allowed scattering region for nominal parent beam velocities. At right angular distribution derived from experimental map (solid curve) is compared with predictions from the hard-sphere model, including *knock-out* (dashed curve) and *sequential impulse* (dotted curve) contributions.

tures provide dynamical information. These include the pronounced "hole" in the reactive scattering in the forward direction, which is accompanied by strong peaking near 50°, and an even stronger backward scattering.

This pattern is remarkably similar to that found for some ion-molecule reactions, such as the $O^+ + HD \rightarrow OH^+ + D$ reaction, for which the collision energies and bond strengths are more than 100-fold higher. As illustrated in Fig. 23, we find several properties are in nearly quantitative agreement with an impulsive model based on pairwise hard sphere interactions. Two distinct collision modes give the chief contributions. In the mode that accounts for the forward pitched scattering, $A + BC$ interact via *sequential* hard sphere elastic collisions (A off B, then B off C); the exchange reaction occurs only when the final relative velocity of A and B corresponds to an energy less than the AB bond strength. For this mode, the angular distribution may be found from a geometric construction devised by Mahan for high energy ion-molecule reactions [25]. The other mode gives strong backward scattering by a process familiar in billiards; A *knocks-out* B and thereby comes nearly to rest with respect to the C atom, so the resulting AC molecule moves briskly backward with respect to the center-of-mass. This study of an exchange of feeble bonds thus helped develop a more comprehensive asymptotic model applicable to any atom transfer reaction when the ratio of collision energy to bond strength becomes large.

Pursuit of Vector Correlations

In addition to the product angular distribution, many other directional or vector properties of reactions are now accessible in beam experiments, especially by means of laser-induced fluorescence techniques developed by Zare which exploit the orientation dependence of the dipole selection rule for electronic excitation [26]. These vector properties offer much information not provided by energetic or other scalar properties, and their study is now emerging as a vigorous field referred to as dynamical stereochemistry or *stereodynamics*. For instance, as indicated in Fig. 24, of special interest is the *triple-vector correlation* among the initial and final relative velocities and the product rotational angular momentum. In principle, this vector correlation offers a means to undo the "dart-board" averaging over the random azimuthal orientations of initial impact parameters. The distrilbutions of both $\mathbf{k}'$ and $\mathbf{j}'$ must have azimuthal symmetry about $\mathbf{k}$, but when a subset is selected of $\mathbf{k}'$ vectors with particular $\mathbf{j}'$ (or vice-versa), this subset in general will not have azimuthal symmetry about $\mathbf{k}$. Model calulations and an early electric deflection study by Hsu and McClelland of the $Cs + CH_3I$ reaction revealed marked asymmetry; as the freshly formed CsI recoils from the collision, its internuclear axis tends to rotate in or near the plane of the initial and final relative velocity vectors. Far more detailed information of this kind is in prospect from studies using laser-induced fluorescence, especially since the great sensitivity of the method may allow the alignment of individual rotaton-vibration states to be measured as a function of the scattering angle.

We may even be optimistic about measuring some properties of the four

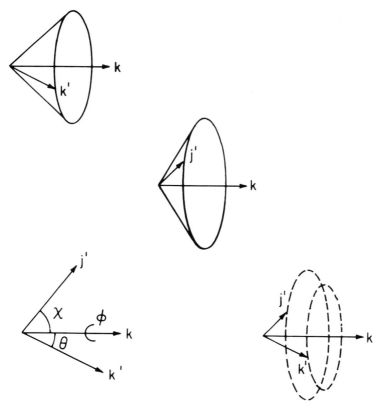

Fig. 24. Three-vector correlation among initial and final relative veloticy vectors, here denoted by **k** and **k′**, and product rotational angular momentum vector **j′**. Upper pair of diagrams indicate the azimuthal symmetry about **k** of the **k′** and **j′** product vectors inherent when these are observed separately, as in the two-vector correlations (**k,k′**) and (**k,j′**). Lower pair of diagrams indicates how the three-vector correction (**k,k′,j′**) can give information about the dihedral angle φ, in effect undoing the azimuthal averaging about the initial velocity.

vector correlation involving both the velocity vectors and the reactant as well as product rotation. This correlation of course contains still more information than found in the six two-vector and four three-vector correlations involving pairs or triads of the four vectors. As exhibited in a model calculation [27], such information includes the azimuthal asymmetry with respect to both **k** and **k′**, the preferred rotational orientation of both the reactant and product molecules with respect to the **k, k′** plane, and even the relative sense of rotation (parallel or contrary). Several averages implicit in lower order vector correlations are thus undone. The loss of information otherwise imposed by the random aspects of the initial conditions resembles the celebrated "phase problem" encountered in X-ray scattering. Thus the complications required to unravel phases for a molecular structure determination are heuristically analogous to undoing impact parameter averaging by observing one or two "extra angles" in collision experiments. In such ways, a further level of molecular resolution awaits reaction dynamics.

ACKNOWLEDGEMENTS

This paper is a tour through a family album. The frequent mentions of work by "we" pertain to my colleagues listed in Table 1, a roster of 51 graduate students and 35 postdoctoral fellows (in *italics*) listed in order of their "graduation" from our research group. Their work totals about 250 years of research, 150 of that in reaction dynamics. The investment of human resources is far greater, however. Beyond the direct contributions by the machine shop and other staff, all of us received vital help along the way from our teachers, our relatives, and many friends and colleagues. Sustained financial support was likewise vital, including many scholarship and fellowship awards as well as research grants from several agencies, especially the National Science Foundation. To all who made our research possible, I am deeply grateful.

In revisiting the album, I have lingered fondly on the early episodes because, as in dynamics generally, the initial conditions are often as important as the force field. I hope that students and young researchers starting their own work might be encouraged to see how simple and naive were our first steps. I hope also that some research administrators and funding agencies might be encouraged to take a longer-range view of quixotic projects. In our work the essential impetus was the evangelical fervor of young scientists captivated by new vistas. But to pursue such distant vistas we have freedom and support. In many quarters this seems less fortcoming today. In expressing thanks for our good fortune, I must urge renewed efforts to foster enterprising work on fundamental problems.

The family album contains far more than shown here, including many intriguing chapters from other laboratories. Again I recommend an admirable pair of texts [3] which survey much work deserving of honor at this forum. Over 30 years of reaction dynamics, the fellowship of striving and joy of discovery have created an intense sense of community. Perhaps in emulation of our molecular friends, recruits to this field seem unusually excitable, zestful, and generous. Working in such a community in pursuit of new insights is a splendid prize, enhanced in the sharing. Thereby we offer thanks for the privilege of studying our ever mysterious atoms and molecules.

Table 1. Roster of Graduate Students and Postdoctoral Fellows

Vintage	Alumni
'62	*J.L. Kinsey, M.S. Child*
'63	J.A. Norris, P.R. Brooks
'64	R.N. Zare, K.R. Wilson, *M.A.D. Fluendy, R.J. McNeal*
'65	M.C. Moulton, R.J. Cross, R.R. Herm, *R.M. Martin, J. Chryusochoos*
'66	J.H. Birely, E.A. Gislason, *M. Cosandey, J.R. Jordan, A. Niehaus*
'67	G.H. Kwei, R. Grice, E.A. Entemann, *H.D. Cohen, P.M. Strudler, P.B. Empedocles*
'68	W.C. Stwalley, R.W. Anderson, *Y.T. Lee, Ch. Ottinger, G.A. Fisk, J.F. Wilson, V. Aquilanti*
'69	S.A. Safron, W.B. Miller, P.E. Siska, H.L. Kramer, C. Maltz, *T. Kitagawa*
'70	P.R. LeBreton, R.J. Gordon, S.J. Riley, R.H. Harris, S.M. Freund, *K. Lacmann*
'71	J.D. McDonald, *R.M. Düren*

'72   N.D. Weinstein, S.A. Adelman, W.S. Struve, *H.J. Loesch*
'73   R.P. Mariella, *D.L. McFadden, D.D. Parrish, J.R. Krenos*
'74   D.L. King, D.S.Y. Hsu, J.T. Cheung, R.A. Larsen, D.M. Lindsay, P.E. McNamee
'75   S.K. Neoh, D.A. Dixon
'76   D.A. Case, *R.A. Sanders*
'77   K.H. Bowen
'78   L.D. Trowbridge, *G.W. Liesegang, J.J. Valentini*
'79   G.M. McClelland, *W. Lee*
'80   E.L. Quitevis, *R. Naaman*
'81   K.L. Saenger, R.A. LeSar, *N. Agmon, O. Cheshnovsky*
'82   D.R. Worsnop, *S. Raynor, A. Yokozeki*
'83   S.J. Buelow, J.D. Barnwell
'84   J.G. Loeser, *C.J. Sandroff*
'86   M.R. Zakin, D.J. Doren, D.Z. Goodson, *S.G. Grubb*

REFERENCES

[1]   H.S. Johnston, *Gas Phase Reaction Rate Theory*, Ronald Press, New York, 1966.

[2]   For most of the work described here, references to the original literature may be obtained readily from a few reviews. Early chemical applications are discussed by S. Datz and E.H. Taylor, in *Recent Research in Molecular Beams*, I. Estermann, Ed., Academic Press, New York, 1959. We have surveyed the development of our studies and allied work in: Disc. Faraday Soc. **33,** 149 (1962); App. Optics Suppl. *(Chemical Lasers)*, 128 (1965); Adv. Chem. Phys. **10**, 319 (1966); Faraday Disc. Chem. Soc. **55,** 233 (1973); Pure and Applied Chem. **47,** 61 (1976). Usually we will omit specific citation except for papers not referenced in these reviews.

[3]   R.B. Bernstein, *Chemical Dynamics via Molecular Beam and Laser Techniques*, Clarendon Press, Oxford, 1982; R.D. Levine and R.B. Bernstein, *Molecular Reaction Dynamics*, 2nd Ed., Oxford University Press, New York, 1987.

[4]   Of course, these are not Stern's exact words, although my memory of him is vivid. I have presented his stories in first person as an attempt to capture his way of telling them.

[5]   W. Klemperer and D.R. Herschbach, Proc. Natl. Acad. Sci. (U.S.) **43,** 429 (1957). The response to vibrational excitation was shown by K.T. Gillen and R.B. Bernstein, Chem. Phys. Letts. **5,** 275 (1970).

[6]   This had been published about a year before but somehow I had missed seeing that issue of the *Journal of Chemical Physics*. In the same volume I also found an exceptional paper on transition-state theory by another author new to me: J.C. Polanyi!

[7]   The Pickwickian phrase is a typically Polanyian comment by John, delivered when he and his father Michael visited the Society of Fellows in 1966.

[8]   G.B. Kistiakowsky and W.P. Slichter, Rev. Sci.Inst. **22,** 333 (1951).

[9]   R.N. Zare and D.R. Herschbach, Proc. I.E.E.E. **51,** 173 (1963); R.N. Zare, Mol. Photochem. **4,** 1 (1972); C.H. Greene and R.N. Zare, Ann. Rev. Phys. Chem. **33,** 119 (1982).

[10]   D.R. Herschbach, J. Phys. Chem. **83**, 4A (1979).

[11]   K. Lacmann and D.R. Herschbach, Chem, Phys. Letts. **6,** 106 (1970).

[12]   C.E. Kolb and D.R. Herschbach, J. Phys. Chem. **88,** 4447 (1984).

[13]   R.H. Krech, G.J. Diebold, and D.L. McFadden, J. Am. Chem. Soc. **99,** 4605 (1977).

[14]   R.S. Iyer and F.S. Rowland, Chem. Phys. Letts **103,** 213 (1983). See also Faraday Disc. Chem. Soc. **67,** 250 (1979). See also J. Phys. Chem. **89,** 2042, 3730 (1985) for evidence suggesting less stereospecific addition as well as Br atom migration.

[15] J.A. Silver and C.E. Kolb, J. Phys. Chem. **90,** 3267 (1986).

[16] J.F. Durana and J.D. McDonald, J. Am. Chem. Soc. **98,** 1289 (1976).

[17] D.A. Dixon and D.R. Herschbach, Faraday Disc. Chem. Soc. **62,** 162 (1977); D.L. Thompson and H.H. Suzukawa, J. Am. Chem. Soc. **99,** 3614 (1977).

[18] D.A. Dixon, R.M. Stevens, and D.R. Herschbach, Faraday Disc. Chem. Soc. **62,** 110 (1977).

[19] R. LeSar and D.R. Herschbach, J. Phys. Chem. **85,** 3787 (1981).

[20] See, for example, F. Trager and G. zu Putliz, Eds., *Metal Clusters,* Z. f. Phys. **D 3** (1986).

[21] K.L. Saenger, G.M. McClelland, and D.R. Herschbach. J. Phys. Chem. **85,** 3333 (1981); J.J. Valentini and J.B. Cross, J. Chem. Phys. **77,** 572 (1982).

22] H.Haberlund, H.-G. Schindler, and D.R. Worsnop, J. Phys. Chem. **88,** 390 (1984).

[23] K.H. Bowen G.W. Liesegang, B.S. Sanders and D.R. Herschbach, J. Phys. Chem. **87,** 557 (1983).

[24] D.R. Worsnop, S.J. Buelow, and D.R. Herschbach, J. Phys. Chem. **90,** 5121 (1986).

[25] B.H. Mahan, W.E.W. Ruska, and J.S. Winn, J. Chem. Phys. **65,** 3888 (1976).

[26] D.A. Case, G.M. McClelland, and D.R. Herschbach, Mol. Phys. **35,** 541 (1978).

[27] J.D. Barnwell, J.G. Loeser, and D.R. Herschbach. J. Phys. Chem. **87,** 2781 (1983).

Yuan T. Lee

YUAN T. LEE

Yuan Tseh Lee was born on November 19, 1936 in Hsinchu, Taiwan. His father is an accomplished artist and his mother a school teacher.

He started his early education while Taiwan was under Japanese occupation—a result of a war between China and Japan in 1894. His elementary education was disrupted soon after it started during World War II while the city populace was relocated to the mountains to avoid the daily bombing by the Allies. It was not until after the war when Taiwan was returned to China that he was able to attend school normally as a third year student in grade school.

His elementary and secondary education in Hsinchu was rather colorful and full of fun. In elementary school, he was the second baseman on the school's baseball team as well as a member of the ping-pong team which won the little league championship in Taiwan. In high school he played on the tennis team besides playing trombone in the marching band.

Besides his interest in sports during this time, he was also an avid and serious reader of a wide variety of books covering science, literature, and social science. The biography of Madame Curie made a strong impact on him at a young age. It was Madame Curie's beautiful life as a wonderful human being, her dedication toward science, her selflessness, idealism that made him decide to be a scientist.

In 1955, with his excellent academic performance in high school, Lee was admitted to the National Taiwan University without having to take the entrance examination, a practice the Universities took to admit the best students. By the end of his freshman year he had decided chemistry was to be his chosen field. Although the facilities in the Taiwan University were less than ideal, the free and exciting atmosphere, the dedication of some professors, and the camaraderie among fellow students in a way made up for it. He worked under Professor Hua-sheng Cheng on his B.S. thesis which was on the separation of Sr and Ba using the paper electrophoresis method.

After graduation in 1959, he went on to the National Tsinghua University to do his graduate work. He received his Master's degree on the studies of the natural radioisotopes contained in Hukutolite, a mineral of hot spring sediment under Professor H. Hamaguchi's guidance. After receiving his M.S. he stayed on at Tsinghua University as a research assistant of Professor C. H. Wong and carried out the x-ray structure determination of tricyclopentadienyl samarium.

He entered the University of California at Berkeley as a graduate student in 1962. He worked under the late Professor Bruce Mahan for his thesis research on chemiionization processes of electronically excited alkali atoms. During his graduate student years, he developed an interest in ion-molecule reactions and the dynamics of molecular scattering, especially the crossed molecular beam studies of reaction dynamics.

Upon receiving his Ph.D. degree in 1965, he stayed on in Mahan's group and started to work on ion molecule reactive scattering experiments with Ron Gentry using ion beam techniques measuring energy and angular distributions. In a period of about a year he learned the art of designing and constructing a very powerful scattering apparatus and carried out successful experiments on $N_2^+ + H_2 \rightarrow N_2H^+ + H$ and obtained a complete product distribution contour map, a remarkable accomplishment at that time.

In February 1967, he joined Professor Dudley Herschbach at Harvard University as a post-doctoral fellow. He spent half his time working with Robert Gordon on the reactions of hydrogen atoms and diatomic alkali molecules and the other half of his time on the construction of a universal crossed molecular beams apparatus with Doug McDonald and Pierre LeBreton. Time was certainly ripe to move the crossed molecular beams method beyond the alkali age. With tremendous effort and valuable assistance from the machine shop foreman, George Pisiello, the machine was completed in ten months and the first successful non alkali neutral beam experiment on $Cl + Br_2 - \rightarrow BrCl + Br$ was carried out in late 1967.

He accepted the position as an assistant professor in the Department of Chemistry and the James Franck Institute of the University of Chicago in October 1968. There he started an illustrious academic career. His further development as a creative scientist and his construction of a new generation state-of-the-art crossed molecular beams apparatus enabled him to carry out numerous exciting and pioneering experiments with his students. He was promoted to associate professor in October 1971 and professor in January 1973.

In 1974, he returned to Berkeley as professor of chemistry and principal investigator at the Lawrence Berkeley Laboratory of the University of California. He became an American citizen the same year.

In the ensuing years, his scientific efforts blossomed and the scope expanded. His world leading laboratory now contains seven very sophisticated molecular beams apparati which were specially designed to pursue problems associated with reaction dynamics, photochemical processes, and molecular spectroscopy. His laboratory has always attracted bright scientists from all over the world and they always seem to enjoy working together. He takes great pride in the fact that more than fifteen of his former associates are serving as professors in major universities, and many others are making great contributions at the national laboratories and in the private sector.

Lee and his wife, Bernice Wu, whom he first met in elementary school have two sons, Ted (born in 1963), Sidney (born in 1966) and a daughter, Charlotte (born in 1969).

Among some of the awards and recognitions he has received over the years include:

Alfred P. Sloan Fellow, 1969–1971
Camille and Henry Dreyfus Foundation Teacher Scholar Grant, Recipient 1971–1974.

Fellow, American Academy of Arts and Science, 1975.

Fellow, American Physical Society, 1976.

John Simon Guggenheim Fellow, 1976−1977.

Member, National Academy of Sciences, 1979.

Member, Academia Sinica, Taiwan, China, 1980.

Honorary Professor, Institute of Chemistry, Chinese Academy of Science, Beijing, China, 1980.

Honorary Professor, Fudan University, Shanghai, China, 1980.

Miller Professorship, University of California, Berkeley, California, 1981−1982.

Ernest O. Lawrence Award, U.S. Department of Energy, 1981.

Sherman Fairchild Distinguished Scholar, California Institute of Technology, 1983.

Harrison Howe Award, Rochester Section, American Chemical Society, 1983.

Peter Debye Award of Physical Chemistry, American Chemical Society, 1986.

National Medal of Science, 1986.

Honorary Professor, Chinese University of Science and Technology, Hofei, Anhuei, China, 1986.

Honorary Doctor of Science Degree, University of Waterloo, 1986.

MOLECULAR BEAM STUDIES OF ELEMENTARY CHEMICAL PROCESSES

Nobel lecture, 8 December, 1986

by

YUAN TSEH LEE

Lawrence Berkeley Laboratory and Department of Chemistry, University of California, Berkeley, CA 94720, USA

Chemistry is the study of material transformations. Yet a knowledge of the rate, or time dependence, of chemical change is of critical importance for the successful synthesis of new materials and for the utilization of the energy generated by a reaction. During the past century it has become clear that all macroscopic chemical processes consist of many elementary chemical reactions that are themselves simply a series of encounters between atomic or molecular species. In order to understand the time dependence of chemical reactions, chemical kineticists have traditionally focused on sorting out all of the elementary chemical reactions involved in a macroscopic chemical process and determining their respective rates.

Our basic understanding of the relation between reactive molecular encounters and rates of reactions (formulated in terms of activation energies, E_a, and pre-exponential factors, A, as elucidated by Arrhenius in his rate constant expression, $k = A\exp(-E_a/kT)$), was deepened some fifty years ago following the discovery of quantum mechanics. Since a chemical reaction is fundamentally a mechanical event, involving the rearrangement of atoms and molecules during a collision, detailed information on the dynamics of simple chemical reactions could be obtained by first carrying our extensive quantum mechanical calculations of the interaction potential as a function of interatomic distances and then computing classical trajectories based on this potential energy surface [1]. Although these initial theoretical studies were only qualitative, they heralded a new era in the field of chemical kinetics; the chemist could now, in principle, predict the dynamical course of a chemical reaction.

During the past three decades, with the development of many sophisticated experimental techniques, it has become possible to study the dynamics of elementary chemical reactions in the laboratory. For example, detailed information on the nascent quantum state distributions of simple products for some chemical reactions can be derived from the chemiluminescence spectra of reaction products obtained under single collision conditions [2], the analysis of the threshold operating conditions of a chemical laser [3] or the spectra obtained using various linear or non-linear laser spectroscopic techniques [4, 5]. However, when one desires to (1) control the energies of the reagents, (2) understand the dependence of chemical reactivity on molecular orientation, (3) explore the nature of reaction initermediates and their subsequent decay dynamics, and (4) identify complex reaction mechanisms involving polyatomic radical products, the crossed molecular beams technique is most suitable [6, 7].

Information derived from the measurements of angular and velocity distributions of reaction products played a crucial role in the advancement of our understanding of the dynamics of elementary chemical reactions. This and the more general investigations of chemical reactions under single collision conditions in crossed molecular beams will be the subject of this lecture.

Crossed Molecular Beams Experiments: Measurements of Angular and Velocity Distributions of Products.

If the motion of individual atoms were observable during reactive collisions between molecules, it would be possible to understand exactly how a chemical reaction takes place by just following the motion of these atoms. Unfortunately, despite recent advances in microscope technology that allow us to observe the static arrangement of atoms in a solid, we are still far from being able to follow the motion of atoms in the gas phase in real time. The idea of crossed molecular beams experiments is in a sense to "visualize" the details of a chemical reaction by tracing the trajectories of the reaction products. This is done by first defining the velocities, approach angle, and other initial conditions of the reactants, and then measuring the velocity and angular distribution of the products. For example, in the investigation of the $F + D_2 \rightarrow DF + D$ reaction [8], if we let F atoms and D_2 molecules collide at a relative energy of 1.82 kcal/mol and then measrue the angular and velocity distributions of DF products, we will obtain the results shown in Fig. 1. This contour map shows the probability of DF products appearing at specific angles and velocities and reveals a great deal about the dynamics of the reaction. 0° corresponds to the initial direction of the F atom beam and the distance between any point and the center is the center-of-mass velocity. The strong backward peaking of DF products with respect to the initial direction of F atoms indicates that not all the collisions between F atoms and D_2 molecules produce DF product. Only those collisions in which the F atom and the two D atoms are nearly linear will react and produce DF. Apparently, if an F atom collides with a D_2 molecule from a direction perpendicular to the molecular axis of the D_2, the F atom will only bounce off elastically. The appearance of DF in several velocity bands is due to the fact that DF molecules are produced in several vibrational states with different recoil velocities as indicated in the figure. Since the total energy released in every reactive encounter between F and D_2 is the same, the maximum energy available for translational motion will depend on the vibrational quantum state of DF. Because the rotational energy spread of DF products is less than the spacings of the vibrational energy levels, the recoil velocities of various vibrational states of DF products are well separated and can be identified easily.

If a crossed molecular beams study of the $F + D_2 \rightarrow DF + D$ reaction is carried out using the an experimental arrangement shown in Fig. 2, the rate of

F+D$_2$→DF+D, 1.82 kcal/mole

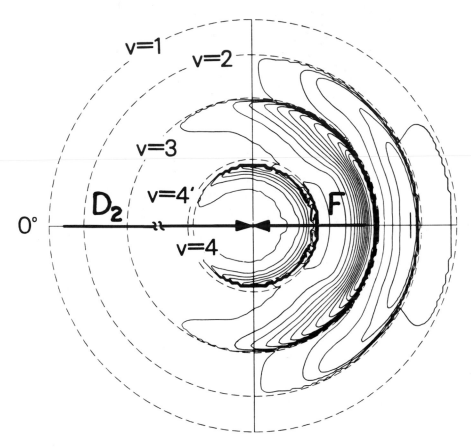

Fig. 1. Center-of-mass velocity flux contour map for the F + D$_2$ → DF + D reaction. F atoms and D$_2$ molecules move towards each other at a collision energy of 1.82 kcal/mol, with the F atoms moving from right to left.

production of DF products, dN$_{DF}$/dt, in the scattering volume defined by the crossing of two beams can be estimated from the follwing equation:

$$\frac{dN_{DF}}{dt} = n_F n_{D2} \sigma g \Delta V$$

where n$_f$ and n$_{D2}$ are the number densities of F atoms and D$_2$ molecules in the scattering region, σ, g, and ΔV are the reaction cross section, the relative velocity between F and D$_2$ and the scattering volume, respectively. In a experiment using a velocity selected effusive F atom source and a supersonic beam of D$_2$, the values of n$_F$, n$_{D2}$, and ΔV are typically 10^{10} molecules/cc, 10^{12} molecules/cc, and 10^{-2} cc. If the relative velocity between F and D$_2$ is 10^5 cm/sec and the reactive cross section is 10^{-15} cm^2, the dN$_{HF}$/dt will have a value of

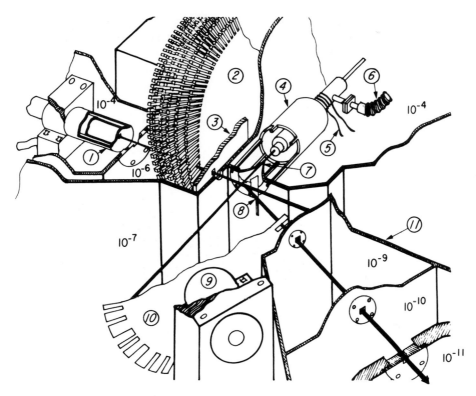

Fig.2. Experimental arrangement for F + D$_2$ → DF + D and F + H$_2$ → HF + H reactive scattering. Pressures (in torr) for each region are indicated. Components shown by numbers are: (1) effusive F atom beam source made of nickel, resistively heated; (2) velocity selector; (3) liquid nitrogen cooled cold trap; (4) D$_2$ or H$_2$ beam source, supersonic expansion, (5) heater, (6) liquid nitrogen feed line, (7) skimmer, (8) tuning fork chopper, (9) synchoronous motor, (10) cross correlation chopper for time-of-flight velocity analysis, (11) ultrahigh vacuum, triply differentially pumped, mass spectrometric detector chamber.

10^{10} molecules/sec. These DF products with various recoil velocities will scatter into a range of laboratory angles. If the DF is scattered fairly evenly within 1 steradian of solid angle in the laboratory and if the movable detector which scans the angular distribution has an acceptance solid angle of 1/3000 steradian (approximately an angular width of 1° in both directions from the detector axis), the detector will receive ~3 × 10^6 DF molecules per second.

This would certainly constitute a large product signal, assuming we were able to count all of these molecules. Indeed, in a reactive scattering experiment using a beam of alkali atoms, surface ionization could be used to detect the alkali containing product with nearly 100 percent efficiency and with high specificity. Therefore, even in the presence of a billion times more background molecules very good signal to noise ratios can be obtained in a short time.

To detect the DF products, however, first it is necessary to ionize DF to DF$^+$ by electron bombardment. The product ion can then be mass filtered and counted. The typical ionization efficiency for a molecule during the short transit time through the ionizer is about 10^{-4}. 3 × 10^6 DF/sec reaching the detector will yield only 300 DF^+ ions/sec. However, this is a large enough number to allow reliable measurements of angular and velocity distributions in a relatively short time if the background count rate is not much greater. Indeed, the success of a crossed molecular beams study of such a chemical reaction depends entirely on whether the background in the mass spectrometric detector can be reduced sufficiently [9].

There are two sources of background molecules in the detector that one has to deal with in a crossed molecular beams experiment: the inherent background in the detector chamber and the background caused by the effusion of molecules from the collision chamber into the detector when the beams are on. The former is mainly due to outgassing from the materials used for the construction of the chamber and to limitations imposed by the performance of the ultrahigh vacuum pumping equipment. Reduction of the latter requires many stages of differential pumping using buffer chamber. 3×10^6 molecules/sec entering the detector with a speed of 10^5 cm/sec through a 0.3 cm^2 aperture will establish a steady state density of only 100 molecules/cc, which is equivalent to a DF pressure of about 3×10^{-15} torr. This is four orders of magnitude lower than the pressure attainable using conventional ultrahigh vacuum techniques. Since none of the chemical compounds found in a vacuum system give ions with a mass-to-charge ratio (m/e) of 21 (DF^+) in the ionization process, inherent background will not be a problem for the investigation of the $F + D_2 \rightarrow DF + D$ reaction even if the ultimate pressure of the detector chamber is around 10^{-11} torr.

Suppose the partial pressure of DF background molecules in the collision chamber after the introduction of beams of F atoms and D_2 molecules reaches $\sim 10^{-9}$ torr. Then three stages of differential pumping will be required to obtain a partial pressure of $\sim 10^{-15}$ in the ionizer chamber if the partial pressure of DF is reduced by a factor of 100 in each separately pumped buffer chamber. As long as the inherent background in the detector does not contain the species to be detected, extensive differential pumping appears to be the only thing needed to make reactive scattering experiments feasible. This conclusion is unforunately not quite correct. In order to detect the scattered products, the defining apertures, which are located on the walls of the buffer chambers of the detector, must be perfectly aligned. This limits the reduction of background that would be possible through many stages of differential pumping since some of the DF molecules in the main chamber moving along the axis of the detector will pass straight through all the defining apertures and into the detector chamber. It is important to understand that no matter how many stages of differential pumping are arranged between the collision chamber and the detector chamber, the number of these "straight through" molecules can not be reduced.

If all the apertures on the walls of the buffer chambers and the detector chamber have the same area, A, the steady state density of "straight through" molecules in the detector chamber, n', at a distance d from the entrance aperture of the first buffer chamber can be calculated from the following relation

$$n' = \frac{nA}{4\pi d^2}$$

where n is the number density of background molecules in the collision chamber. If d is 20 cm and A is 0.3 cm^2

$$n' = \frac{0.3}{4\pi \cdot (20)^2} n \approx 6 \times 10^{-5} n.$$

If the partial pressure of DF background molecules in the collision chamber is 10^{-9} torr, the "straight through" molecules will create a steady state density of 6×10^{-14} torr, which is 60 times larger than what one hopes to accomplish with 3 stages of differential pumping. Of course, reduction of the partial pressure of DF molecules in the collision chamber will also reduce the "straight through" background, but increasing the pumping speed in the collision chamber to reduce the partial pressure of DF by several orders of magnitude is simply not a practical solution.

Fortunately there is a way to reduce this background without substantially increasing the pumping speed in the collision chamber. Recognizing that at a pressure of 10^{-7} torr the mean free path between molecular collisions in the collision chamber is more than 100 meters, which is two orders of magnitude larger than the size of a typical scattering apparatus, one realizes that almost all of the "straight through" background will come from those molecules that bounce off the surface which is in the line-of-sight of the detector, and not from gas phase collisions that occur in the viewing window of the detector. Installing a small liquid helium cooled surface opposite the detector and behind the collision region, such that the detector line-of-sight always faces a cold surface, will help to eliminate this background since the surface will trap essentially all condensable molecules that impinge on it.

Since the mid-1960's, many "universal" crossed molecular beams apparati have been constructed in various laboratories. Since ultrahigh vacuum equipment available twenty years ago could attain an ultimate pressure of only $\sim 10^{-10}$ torr, and only two stages of differential pumping were needed to reduce the pressure from 10^{-7} torr in the collision chamber to 10^{-10} torr in the detector chamber, almost all mass spectrometric detectors with electron impact ionization were constructed with no more than two stages of differential pumping, the principal exceptions being those built in our laboratory [10]. The failure to recognize that for many chemical species the partial pressure in the detector chamber could be reduced well below the ultimate total pressure through additional differential pumping was part of the reason why many of these apparatuses did not perform optimally.

Direct Experimental Probing of Potential Energy Surfaces

For gaseous rare gas systems, if the interaction potentials between the atoms are accurately known, all bulk properties and transport phenomena can be predicted theoretically. Similarly, for a simple atom-molecule reaction, the potential energy surface, which describes the interaction potential as a function of the coordinates of the atoms, will be the basis for the understanding of the detailed dynamics of a chemical reaction.

One of the systems which has attracted extensive attention in both experimental and theoretical efforts during the last fifteen years is the reaction of $F+H_2 \rightarrow HF+H$. In the early 1970's, using quasi-classical trajectory calculations, Muckerman derived a semiempirical potential energy surface, known as the Muckerman V surface, which gave results in agreement with all experimental data available at that time [11]. These results included rate constants, vibrational-rotational state distributions obtained from chemical laser and chemiluminescence experiments, as well as product angular distributions obtained from $F+D_2 \rightarrow DF+D$ experiments as shown in Fig. 1. The potential energy surface obtained from the ab initio quantum mechanical calculations [12] was still rather limited at that time, but it did show many important features which were in good qualitative agreement with the Muckerman V surface.

If the Muckerman V surface were sufficiently accurate, it would be possible to carry out scattering calculations using this surface under conditions which could not be easily arranged in the laboratory. This would significantly expand the scope of our understanding of the dynamics of this system. However, the accuracy of the Muckerman V surface depends not only on the reliability of the experimental input used in the derivation of the surface, but also on the applicability of classical mechanics in treating the $F+H_2 \rightarrow HF+H$ reaction. This is certainly a major concern for a H atom transfer reaction.

One dimensional quantum calculations on the $F+H_2$ reaction, although not necessarily realistic, had in fact shown the inadequacy of classical mechanics in handling this reaction [13, 14]. Quantum effects were, indeed, very important, and in all these calculations strong "resonances" were found in the dependence of reaction probability on collision energy [15]. These resonances were later shown to be due to the formation of "quasi-bound" states in the F-H-H reaction intermediate [16, 17]. The $F+H_2$ surface has a barrier in the entrance channel, but there is no attractive well in the intimate region near the transition state. The quasi-bound states in the $F+H_2$ reaction are entirely dynamical in nature. Loosely speaking, the first dynamic resonance is due to the formation of a bound state which is a superposition of $F+H_2$ (v = 0) and HF (v=3)+H in the intimate region of chemical interaction.

The discovery of dynamical resonances in the collinear quantal calculations of the $F+H_2$ system provided new possibilities for probing the critical region of the potential energy surface more directly. In contrast to most other microscopic experiments, in which the influence of the potential energy surface on the final distribution of products is assessed, the experimental observation of resonances is almost equivalent to carrying out vibrational spectroscopy directly on the reaction intermediate. Thus it should offer a more stringent test of the details of the calculated potential energy surface [16].

In a three dimensional quantum scattering calculation of $F + H_2$ on the Muckerman V surface, Wyatt et al. [18] have shown that as the collision energy is increased beyond the one dimensional resonance energy, the reso-

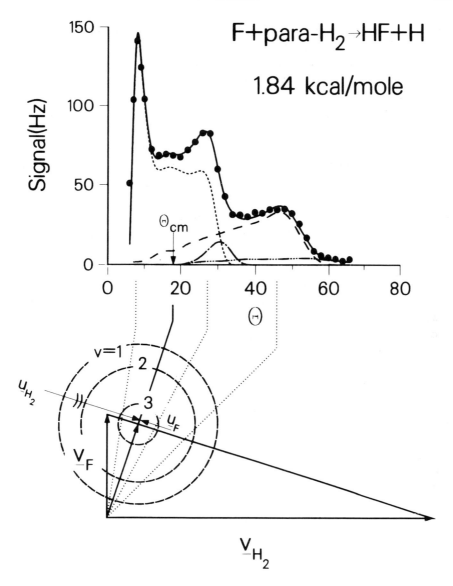

Fig 3. Laboratory angular distribution for F + para-H_2, 1.84 kcal/mol, velocity diagram shown below. Both the data and calculated laboratory distributions are shown (● data, ——— total calculated, — ·· — v=1, --- v=2, ------ v=3 — · — v=3, (from $H_2(J=2)$)).

nance does not just disappear but occurs at increasingly larger impact parameters. Consequently, resonances cannot be observed in an experiment in which the reaction cross section is measured as a function of collision energy. On the other hand, if the experiment is carried out at a fixed collision energy, and if the reaction probability is measured as a function of impact parameter, the resonance should be observable. Unfortunately, one has no hope of controlling or

F+p-H$_2$→HF+H, 1.84 kcal/mole

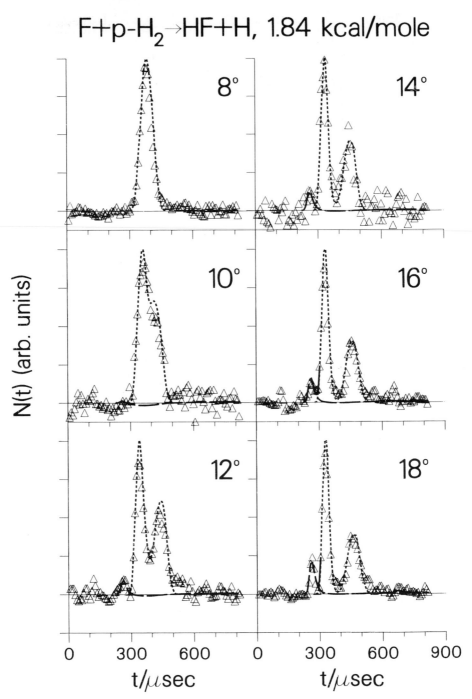

Fig. 4(a). Time-of-flight spectra for F + para-H$_2$, 1.84 kcal/mol. ($\triangle$ data, ———total calculated, —···— v=1, – – – v=2, ······ v=3, —·— v=3′.

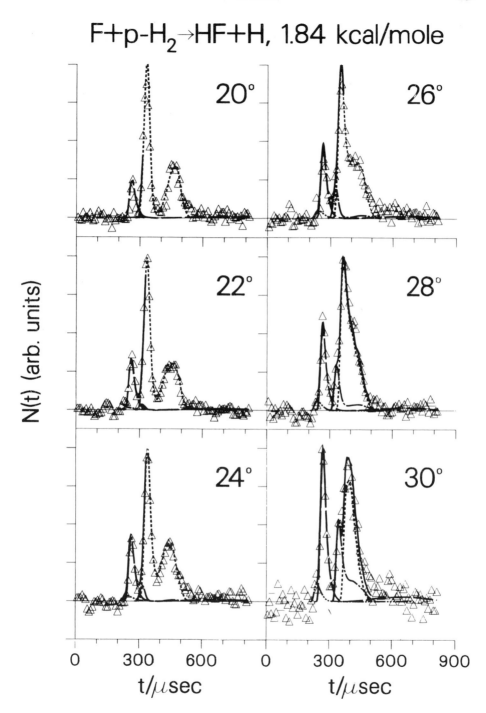

$F+p-H_2 \rightarrow HF+H$, 1.84 kcal/mole

N(t) (arb. units)

t/μsec   t/μsec

XBL 841-38

Fig. 4(b)

F+p-H$_2$→HF+H, 1.84 kcal/mole

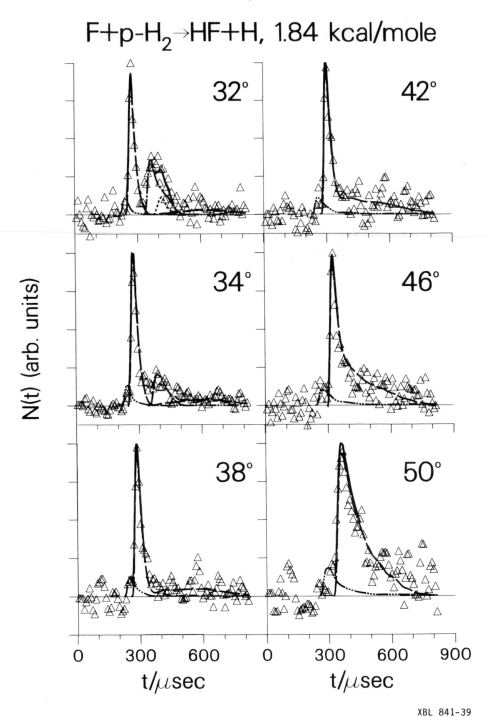

XBL 841-39

Fig. 4(c)

F+p-H₂→HF+H, 1.84 kcal/mole

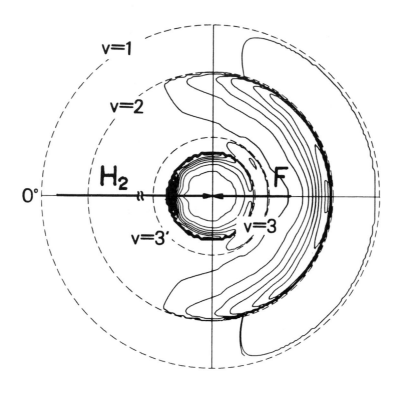

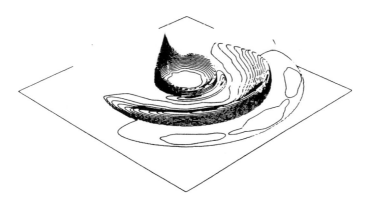

Fig. 5. Center-of-mass velocity flux contour map for F + para-H₂, 1.84 kcal/mol, shown in three-dimensional perspective.

measuring the impact parameter in a scattering experiment. But, for $F + H_2 \rightarrow$ HF + H, in which the collinear configuration dominates the reactive scattering at lower collision energies, the scattering angle of HF should depend on the impact parameter. In particular, when a quasi-bound state is formed, if the average lifetime of the F-H-H intermediate is an appreciable fraction of its rotational period, HF produced from the decay of the F-H-H quasi-bound state is expected to scatter in a more forward direction compared to the strongly backward peaked HF produced by direct reaction. One of the unique and most important aspects of the measurement of product angular distributions is that one can use the rotational period of the reaction intermediate, typically $10^{-12}-10^{-13}$ sec to judge the lifetime of the reaction intermediate [6]. If the average lifetime of the intermediate is much longer than the rotational period, the angular distribution of products will show forward-backward symmetry. On the other hand, if the lifetime is much shorter, the asymmetric angular distribution reveals the preferred molecular orientation for the chemical reaction to occur.

Experimental measurements of the laboratory angular distribution and time-of-flight velocity distributions of HF products at a collision energy of 1.84 kcal/ mole, using the experimental arrangement shown in Fig. 2, are shown in Figs. 3 and 4. The velocity and angular distributions in the center-of-mass coordinate system derived from these experimental results are shown in Fig. 5 [19]. The enhanced forward peaking in the angular distribution of HF in v=3 is a strong indication that quasi-bond states are indeed formed in the $F + H_2 \rightarrow HF + H$ reaction at this energy, and that they seem to decay exclusively into HF in the v=3 state.

For the reaction of $F + HD \rightarrow HF + D$, quantal collinear calculations give a very striking result. There is a sharp spike in the HF(v=2) reaction probability near threshold and virtually no other product is formed at higher collision energies up to 0.2 eV [15]. The collinear calculations therefore indicate that the formation of HF in this reaction is dominated by resonant scattering while the DF product is formed by direct scattering. As shown in Fig. 6, the product angular distribution of HF measured at a collision energy of 1.98 kcal/mol indeed shows that most of the signal is in the forward direction as expected, in strong contrast to the DF signal whoch is formed through direct scattering and is therefore mainly scattered in the backward direction. Again, the forward peaked HF products are found to be in the v=3 state, rather than v=2, as was observed in the quantal calculations on $F + H_2 \rightarrow HF + F$. This disagreement is certainly due to the shortcoming in the M5 surface. These vibrationally state specific angular distributions obtained at various collision energies for $F + H_2$, HD and D_2 reactions provide a very stringent test for the ever improving potential energy surfaces obtained from ab initio quantum mechanical calculations.

There is no doubt that through meaningful comparisons with experimental results, more sophisticated and reliable ab initio quantum mechanical calculation techniques will emerge. In the near future, ab initio calculations of potential energy surfaces and exact scattering calculations on these systems will

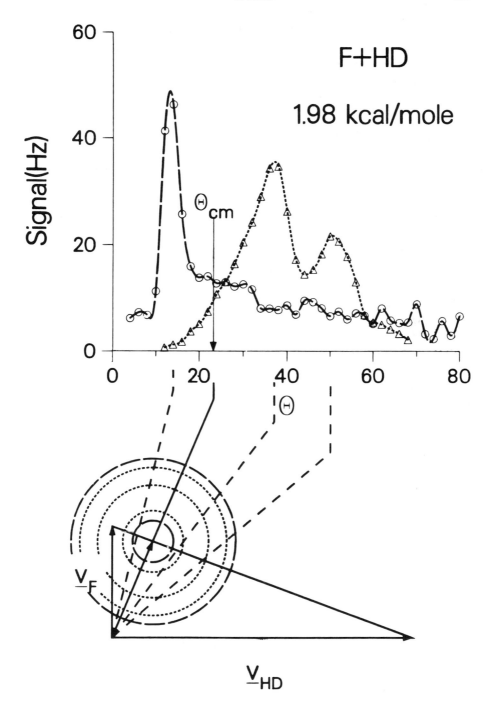

Fig. 6. Laboratory angular distribution for F + HD, at a collision energy of 1.98 kcal/mol (– – – ○ – – – ○ – – – HF product, ...△....△ ..... DF product). The Newton circles corresponding to HF and DF product are drawn with the same texture as the lines in the angular distributions. The HF (v=3) and v=2 circles are shown, as are the v=4, 3, and 2 circles for DF.

likely provide more detailed and accurate information in simple reactive systems such as $F + H_2$ than one could possibly learn in the laboratory. The fruitful interplay of theory and experiment will then extend to more complicated systems, making chemistry a more exact science.

Exploration of new chemistry under single collision conditions
There are many mysterious phenomena in nature which have thus far defied explanation. The mystery is often due to the fact that a certain phenomenon cannot be understood based on our established knowledge or common sense.

The ease with which F_2 and I_2 react to produce electronically excited IF molecules which relax through photon emission was a mystery a dozen years ago [20]. A molecule-molecule reaction is supposed to have a high energy barrier and the four-center reaction producing two IF molecules, with either both in the ground state or one of them in an excited state, is a symmetry forbidden process. The text book mechanism has either I_2 or F_2 molecules first dissociating into atoms followed by the radical chain reactions $F + I_2 \rightarrow IF + I$ and $I + F_2 \rightarrow IF + F$. However, neither of these reactions is exothermic enough to produce electronically excited IF.

The clue that something new might be happening in this reaction was actually discovered in a crossed molecular beams study of the $F + CH_3I \rightarrow IF + CH_3$ reaction [21]. When we found that this reaction proceeded through the formation of a long lived complex, we began to increase the collision energy to see whether it was possible to shorten the lifetime enough to make it comparable to the rotational period of the CH_3IF complex. If we could estimate the lifetime of the collision complex using the rotational period as a clock, it would be possible to evaluate the stability of this reaction intermediate using statistical theories for the unimolecular decomposition rate constants. At higher collision energies, the angular distribution of products monitored at $m/e = 146$ (IF^+) showed a peculiar feature which could not possibly be due to IF produced from the $F + CH_3I$ reaction. This was later shown to be from stable CH_3IF produced in the collision volume of the two beams which yielded additional IF^+ signal after dissociative ionization.

The stable CH_3IF was in fact formed by the reaction of undissociated F_2 in our F atom beam with CH_3I:

$$CH_3I + F_2 \rightarrow CH_3IF + F.$$

The threshold for this reaction was found to be 11 kcal/mol as shown in Fig. 7. The product angular distribution measured at a collision energy of 25.1 kcal/mol is shown in Fig. 8. Since the dissociation energy of F_2 is 37 kcal/mol, the dissociation energy of $CH_3IF \rightarrow CH_3I + F$ could be as high as 26 kcal/mol (Fig. 9). This was certainly a surprising result and was entirely unsuspected.

In the reaction of I_2 and F_2, it was not surprising that the stability of the I_2F radical is the driving force for the reaction

$$I_2 + F_2 \rightarrow I_2F + F$$

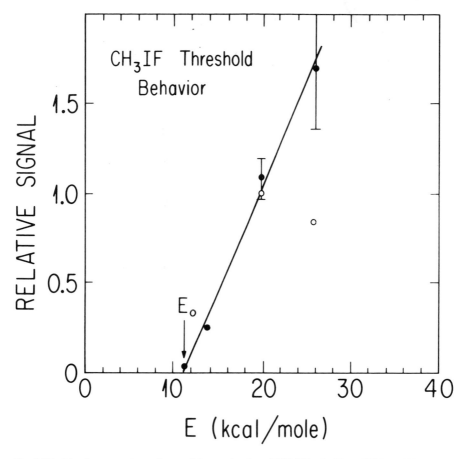

Fig. 7. The kinetic energy dependence of the production of CH_3IF in the $F_2 + CH_3I \rightarrow CH_3IF + F$ reaction.

to proceed. But, what amazed us most was that this reaction had a threshold of only 4 kcal/mol, and that at 7 kcal/mol the reaction

$$I_2 + F_2 \rightarrow I + IF + F$$

was observed [22]. The production of I and IF in this reaction is most likely through the secondary decomposition of vibrationally excited I_2F radicals. Later, a careful investigation showed that the threshold energy for producing electronically excited iodofluoride, IF* [23],

$$I_2 + F_2 \rightarrow IF^* + IF$$

is identical to that for $I_2F + F$ formation. However, the formation of electronically excited IF* is only a minor channel compared to $I_2F + F$ formation. Apparently, it is a secondary encounter between the departing F atom and the terminal I atom in I_2F which produces IF*. A relatively rare sequential process during a binary collision between F_2 and I_2 is responsible for the production of electronically excited IF, not the symmetry forbidden four-center reaction which breaks and forms two bonds simultaneously.

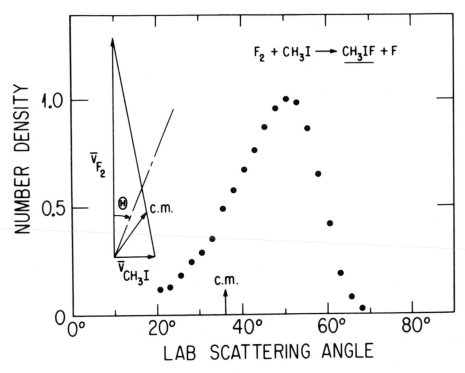

Fig. 8. CH₃IF angular distribution, obtained in the reaction of $F_2 + CH_3IF \rightarrow CH_3IF + F$ at a collision energy of 25.1 kcal/mol.

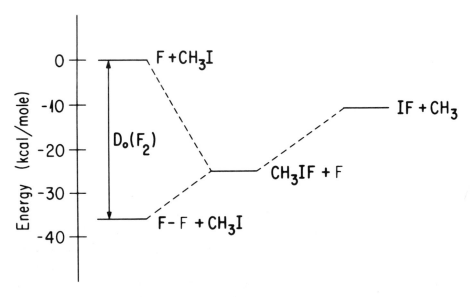

Fig. 9. Energy diagram showing the relative energy CH₃IF intermediate in the reaction of F + CH₃I → CH₃ + IF and as a product of the endothermic $F_2 + CH_3I \rightarrow CH_3IF + F$ reaction.

Fig. 10. An acrobat bounced off the plank converts his kinetic energy into potential energy on his way to forming a delicate three-man formation. Many delicate radicals which can not be synthesized through exothermic channels were synthesized by this method.

The fact that one can control kinetic energy precisely and carry out a synthetic study of delicate new radicals through endothermic reactions is certainly among the most dramatic features of crossed molecular beams experiments. Successful studies of the stabilities of a series of I-F containing radicals such as HIF, ClIF and I_2F were carried out by transferring the correct amount of kinetic energy into potential energy, just like an acrobatic performance in a circus in which an acrobat is bounced off of a plank and lands gently on the shoulder of a second acrobat who is standing on top of a third to form a fragile three acrobat formation (Fig. 10).

The development of the seeded supersonic beam source has been largely responsible for making crossed molecular beams experiments at higher collision energies possible [24]. If a gaseous mixture is expanded into a vacuum chamber through a small nozzle with a sufficiently high stagnation pressure, all molecules, regardless of their molecular weights, attain the same average terminal speed. Consequently, the kinetic energies of molecules in the beam will be proportional to their molecular weights, and for heavier atoms or

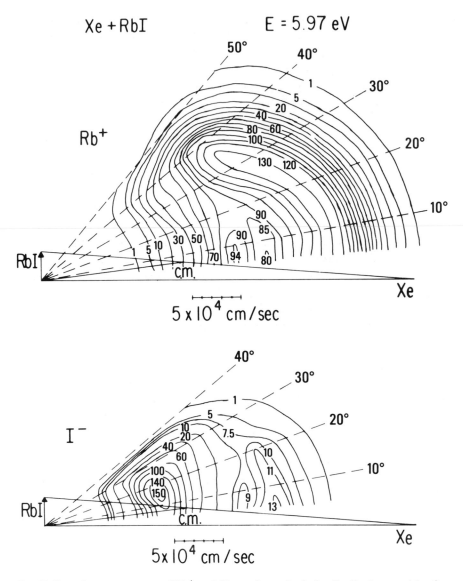

Fig. 11. Cartesian contour map of Rb$^+$ and I$^-$ angular and velocity distributions resulting from dissociative Xe + RbI collisions at a most probable relative collision energy of 5.97 eV.

molecules a very high kinetic energy can be obtained by seeding a small fraction of heavy particles in a very light carrier gas.

Using this aerodynamic acceleration for heavier particles many interesting experiments have been carried out in our laboratory. In the collision induced dissociation of alkali halides by rare gas atoms [25], it was found from classical trajectory simulations of velocity and angular distributions of the products that for most dissociative collisions at energies near the dissociation threshold, the most efficient means of transferring translational energy into internal energy is through initial bond compression in near collinear collisions. The experimental angular and velocity distributions of Rb^+ and I^- from the reaction $Xe + RbI \rightarrow Xe + Rb^+ + I^-$ at a collision energy of 5.97 eV is shown in Fig. 11. The amount of energy transferred as measured from the final translational energy distributions of dissociated atoms agrees with the estimated initial momentum transfer to one of the atoms in the diatomic molecule using the impulse approximation.

In a recent series of investigations of substantially endothermic reactions of Br atoms eith ortho-, meta- and para-chlorotoluenes, a beam of energetic Br atoms was used to study of reactivity and dynamics of Cl atom substitution by Br atoms [26]. The intermediates of these reactions are expected to have potential wells which are much shallower than the endothermicity of reaction. From the measurements of the translational energy dependence of the reaction cross sections and the product translational energy distributions, the extent of energy randomization among various vibrational degrees of freedom was found to be rather limited. Despite the fact that ortho- and para-chlorotoluenes react easily, no substitution was observed for meta-chlorotoluene indicating that the eletron density distribution on the benzene ring strongly influences the reactivity, even though dynamic factors are expected to be more important in endothermic substitution reactions.

Elucidation of Reaction Mechanisms from Product Angular and Velocity Distributions.
In elementary chemical reactions involving complicated polyatomic molecules, the unravelling of the reaction mechanism is often the most important issue. Without the positive identification of primary products, it is not possible to discuss reaction dynamics in a meaningful way. In bulk experiments, the identification of primary products has often been complicated by fast secondary reactions of primary products. Recently, advances in sensitive detection methods and time resolved laser techniques have allowed single collision experiments to become possible even in the bulk, and complications caused by secondary collisions can be avoided. However, the positive identification of internally excited polyatomic radicals produced under single collision conditions is still a very difficult problem. Spectroscopic techniques which are so powerful in providing state resolved detection of atoms or diatomic molecules are often not very useful, either because of the lack of spectroscopic information or simply because huge numbers of states are involved. The more general mass spectrometric technique, which depends heavily on "fingerprints" of fragment ions for positive identification, also suffers from the fact that fragmentation patterns for vibrationally excited polyatomic radical products in electron bombardment ionization are not known. This problem is especially serious because many radicals do not yield parent ions. Even if stable molecules are formed as products, the change in fragmentation patterns with increasing internal energy can be so drastic that erroneous conclusions are often reached. For example, at room temperature both ethanol (C_2H_5OH) and acetaldehyde (CH_3CHO) will yield $C_2H_5OH^+$ and CH_3CHO^+ as major ions by electron bombardment ionization. However, since both these ions contain a very weak bond and most

of the vibrational energy is retained in the ionization process, when highly
vibrationally excited C_2H_5OH and CH_3CHO are ionized, even if parent ions
are initially produced, they will further dissociate into $C_2H_5O^+$ and CH_3CO^+
by ejecting an H atom [27].

The problem of product identification caused by the fragmentation of primary products during the ionization process can be overcome if product angular
and velocity distributions are measured carefully in high resolution crossed
molecular beams experiments. For example, the reaction between $=O(^3P)$ and
C_2H_4 under single collision conditions using a mass spectrometer to detect the
products generates signal at $m/e=43,42,29,27$, and 15. The fact that $m/e=15$
(CH^+_3) and 29 (HCO^+) are the most intense signals suggests that CH_3 +
HCO is the major reaction channel. This conclusion is in agreement with
previous studies of the reaction of $O(^3P)$ with C_2H_4 carried out by Cvetanovic
[28], Pruss et al. [29], and Blumenberg et al. [20]. From the analysis of final
products in a bulk experiment using photoionization detection of products with
hydrogen Lyman-α (10.2 eV) radiation and electron bombardment ionization
mass spectrometry, it was concluded that formation of CH_3 and HCO, resulting from 1,2 migration of a hydrogen atom in the reaction intermediate and
subsequent C-C rupture, as shown below, provides 90 percent of the products.

$$O\,(^3P) + C_2H_4 \rightarrow \left[\; H\overset{\dot{O}}{\underset{H}{\diagup}}C - \dot{C}\overset{H}{\underset{H}{\diagdown}}\;\right]^{+\!+} \rightarrow \left[\; \overset{O}{\underset{H}{\diagup}}C - C\overset{H}{\underset{H}{\diagdown}H}\;\right]^{+\!+}$$

$$\rightarrow HCO + CH_3$$

The remaining 10 percent is ketene formed by a three center elimination of an
H_2 molecule from the reaction intermediate.

$$O\,(^3P) + C_2H_4 \rightarrow \left[\; H\overset{\dot{O}}{\underset{H}{\diagup}}C - \dot{C}\overset{H}{\underset{H}{\diagdown}}\;\right] \rightarrow \left[\; H\text{----}\overset{\dot{O}}{\underset{H}{\diagup}}C - \dot{C}\overset{H}{\underset{H}{\diagdown}}\;\right]$$

$$\rightarrow H_2 + CH_2CO$$

The measurements of product angular distributions in a crossed molecular
beams experiment [31], as shown in Fig. 12, gave strong evidence that the
above conclusion was not quite correct. The fact that the intense $m/e=42$
signal and the weak $m/e=43$ signal (not shown) have the same angular
distributions indicates that the substitution reaction forming vinyloxy radical,
CH_2CHO + H, occurs. The $m/e=42$ signal $(C_2H_2O^+)$ results from dissocia-

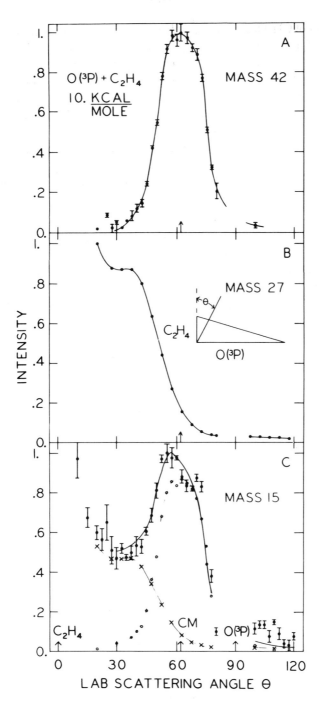

Fig. 12. Angular distributions from the reaction $0 + C_2H_4$ at 10.7 kcal/mol collision energy. (A) CH_2CHO product, (B) elastic scattering of C_2H_4 monitored at m/e=27 ($C_2H_3^+$), (C) m/e=15 (CH_3^+), contributions from C_2H_4 and CH_2CHO are indicated by x and o respectively.

tive ionization of CH_2CHO product rather than from the formation of CH_2CO and H_2. The formation of CH_2CO through the three center elimination of a hydrogen molecule is expected to release a larger amount of recoil energy and the fact that CH_2CO is recoiling from H_2 rather than from an H atom will cause the laboratory angular distribution of CH_2CO to be much broader than that of CH_2CHO. The $m/e = 15$ (CH_3^+) angular distribution clearly shows that in addition to reaction products, elastically scattered C_2H_4 molecules also produce CH_3^+ ions during ionization. The angular distribution of scattered C_2H_4 can be unambiguously measured at $m/e = 27$ ($C_2H_3^+$). After substracting the contribution from elastically scattered C_2H_4 from the angular distribution at $m/e = 15$, it is quite clear that the residual angular distribution of reactively scattered CH_3^+ has an identical angular distribution to that measured at m/e 43 and 42. Apparently, most of the CH_3^+ from reactive scattering are also daughter ions of vinyloxy radicals, CH_2CHO. If the product channel $CH_3 + HCO$ were dominant the angular distribution of CH_3^+ would be much broader. Without the measurement of product angular or velocity distributions which reveal the parent-daughter relations one would not have suspected that the simple substitution reaction forming vinyloxy radical

$$O\,(^3P) + C_2H_4 \rightarrow \left[H \overset{\displaystyle \dot{O}}{\underset{H}{\diagdown}} C \underline{\quad\quad} \dot{C} \overset{H}{\underset{H}{\diagup}} \right] \rightarrow CH_2CHO + H$$

is in fact the major channel.

This was certainly a shocking discovery to chemical kineticists, since the reaction mechanism, $O(^3P) + C_2H_4 \rightarrow CH_3 + HCO$, was thought to be well established. The important role played by the $CH_2CHO + H$ channel was never suspected. Several recent bulk studies using various time resolved spectroscopic techniques [32−36] have verified the major role played by the hydrogen subsitution channel indicated by the crossed molecular beams experiments. These were not strictly single collision experiments, but all showed that the reaction channel $CH_2CHO + H$ accounted for at least half of the products.

For the reaction of oxygen atoms with benzene the story is quite similar [37]. In earlier mass spectrometric studies of the reaction products under single collision conditions, it was concluded that in addition to the formation of a stable addition product, CO elimination from the reaction intermediate to form C_5H_6 was another major pathway. The CO elimination mechanism was mostly based on the experimental observation that $m/e = 66$ and 65 ($C_5H_6^+$ and $C_5H_5^+$) were the most intense signals. However, the angular distribution of products monitored at $m/e = 66$ and 65 in the crossed molecular beams experiments clearly show that they are different from each other but very similar to those monitored at $m/e = 94$ and 93 [$C_6H_5OH^+$ and $C_6H_5O^+$], respectively. Apparently, $C_5H_5^+$ ions observed are not from neutral C_5H_5 product, but are actually daughter ions of the phenoxy radical C_6H_5O. The fact that the very weak signals at $m/e = 94$ and 93 have different angular distributions, as also

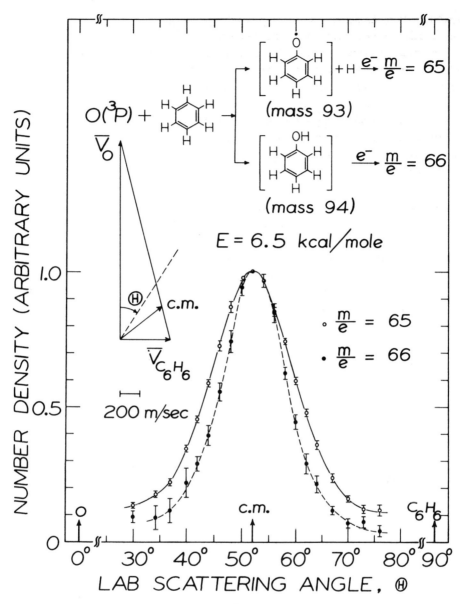

Fig. 13. Angular distribtuions from the reaction $O(^3P) + C_6H_6$, at a collision energy of 6.5 kcal/mol. The primary reaction products formed were C_6H_5O and C_6H_5OH, which subsequently fragmented during electron bombardment ionization.

reflected in the angular distributions of m/e = 66 and 65 shown in Fig. 13, is convincing evidence that the m/e = 93 signal ($C_6H_5O^+$) is not entirely from the dissociative ionization of the addition product, C_6H_5OH. It is the substitution reaction, in which an oxygen atom replaces a hydrogen atom in the benzene molecule, which causes the angular distribution of the phenoxy product, C_6H_5O, to be broader than that of the adduct, C_6H_5OH. The benzene ring does not seem to open up after the initial attack of an oxygen atom. The subsequent decomposition of phenoxy radicals appears to be the important ring opening step. Crossed beams studies of substitution reactions of oxygen atoms with a series of halogenated benzenes [38], indeed showed that very highly vibrationally excited phenoxy radicals, produced by substituting bromine and iodine atoms in bromo- and iodobenzene with oxygen atoms, undergo decomposition to eliminate CO.

The fact that each product in a crossed molecular beams experiment has a unique angular and velocity distribution and the requirements that total mass number in a chemical reaction be conserved and that a pair of products from a given channel have the ratio of their center-of-mass recoil velocities inversely proportional to their mass ratio in order to conserve linear momentum are three of the main reasons why measurements of product angular and velocity distributions are so useful in the positive identification of reaction products, even in those cases where none of the products yield parent ions in mass spectrometric detection [9]. In fact, there is no other general method more useful in elucidating complex gas phase reaction mechanisms and providing information on the energetics and dynamics.

Molecular beam studies of photochemical processes
In the investigation of reaction dynamics, lasers have become increasingly important. Not only are they used extensively for the preparation of reagents and quantum state specific detection of products, but they have also become indispensable for the investigation of the dynamics and mechanisms of photochemical processes.

One of the more exciting application of lasers in crossed molecular beams experiments is the control of the alignment and orientation of electronically excited orbitals before a reactive encounter. For example, in the reaction of Na with O_2 [39, 40], if lineary polarized dye lasers are used to sequentially excite Na atoms from the 3S to 3P to 4D states, the electronically excited 4d orbital can be aligned along the polarization direction of the electric field vector of the lasers. Consequently, the effect of the alignment of the excited orbital on chemical reactivity can be studied in detail by simply rotating the polarization of the lasers with respect to the relative velocity vector.

For many atom-molecule reactions that proceed directly without forming long-lived complexes, for example, $K + CH_3I$, $F + H_2$ and D_2, and $Na(4D) + O_2$, the dependence of chemical reactivity on the molecular orientation can be obtained from measurements of product angular distributions. For symmetric top moleculars, control of molecular orientation in the laboratory frame is possible, and careful investigations of the orientation dependence of chemical reactivity have been carried out for many systems. The combination of laser induced alignment of excited atomic orbitals and measurements of product angular distributions provide the first oppotunity for the detailed experimental probing of the correlation between the alignment of the excited atomic orbital and the orientation of the molecule in a reactive encounter between an atom and a molecule.

The experimental arrangement for the reactive scattering of electronically

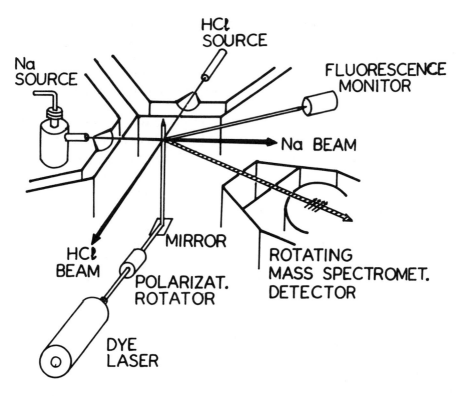

Fig. 14. Cut-out view the experimental apparatus for the reactive scattering of electronically excited sodium atoms with various molecules.

excited Na with simple molecules is schematically shown in Fig. 14. Because the radiative lifetimes of electronically excited Na are short, excitation and alignment have to be carried out in the intersection region of the two beams.

The reaction of ground state Na atoms with O_2

$$Na(3S) + O_2 \rightarrow NaO + O$$

is substantially endothermic. Even if Na is electronically excited to the 3P state, it is still slightly endothermic, and excess translational energy in the reactants was not found to promote NaO formation in our recent experiments. Further excitation of Na from 3P to either the 4D or 5S state, which require comparable excitation energies, makes NaO formation highly exothermic, but only Na(4D) reacts with O_2 and then only at collision energies greater than 18 kcal/mol. The NaO produced is sharply backward peaked with respect to the Na atom beam. As in the low energy reactive scattering of $F + D_2 \rightarrow DF + D$, the Na(4D) and two O atoms must be lined up collinearly in order for chemical reaction to take place. Such a strict entrance channel configuration with a high threshold energy for $Na(4D) + O_2 \rightarrow NaO + O$ and the lack of chemical reactivity for Na(5S) are quite astonishing for a system in which electron transfer from Na to O_2 is expected to take place at a relatively large separation.

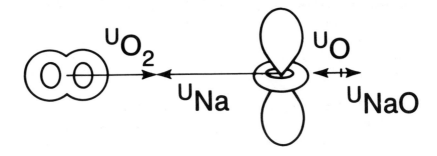

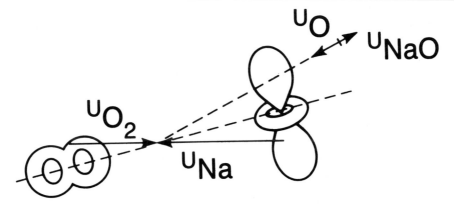

Fig. 15. From the measurements of polarization dependence at various angles, the required geometry for reaction was shown to have the Na-O-O intermediate collinear, so that with increasing impact parameter, the Na-O-O axis must be tilted with respect to the relative velocity vector. The Na($4d_{z^2}$) orbital remains perpendicular to the Na-O-O axis as shown.

The 4D state of Na prepared by sequential excitation using linearly polarized lasers has an electron density distribution similar to that of the d_{z^2} orbital in a H atom, if we take the laser polarization axis to be the Z-axis. Rotation of this excited 4D orbital with respect to the relative velocity vector was found to cause a strong variation in reactivity. The reactively scattered signal reaches a maximum when the d_{z^2} orbital approaches O_2 perpendicularly to the relative velocity vector as shown in Fig. 15. The polarization dependence of products appearing at different scattering angles reflects the strong preference for the d_{z^2} orbital to be aligned perpendicularly to the O_2 molecular axis as shown in the bottom frame of Fig. 15.

These experimental observations are contrary to what one would expect from simple theoretical considerations. Because O_2 has a finite electron affinity, the Na(4D)$-O_2$ potential energy surface is expected to cross the Na$^+$$-O_2^-$ surface at at relatively large internuclear distance and the long range electron transfer from Na(4D) to O_2, to form a Na$^+O_2^-$ intermediate, should play an important role. If this chemically activated $Na^+O_2^-$ complex is indeed responsible for the formation of an ionic NaO product, the reaction should proceed with a large cross section at low collision energies. Also, because the most stable structure of $Na^+O_2^-$ is an isosceles triangle, the angular distribution of NaO product should show either forward peaking or forward-backward symmetry.

Apparently, this long range electron transfer, in spite of its importance, is not the mechanism by which NaO product is formed, and may lead only to the quenching of electronically excited Na(4D) through an inelastic scattering process. It appears that only those collinear collisions that have the d_z2 orbital of Na(4D) aligned perpendicularly to the molecular axis can effectively avoid the long range electron transfer. Then, with this configuration and sufficient collision energy, Na and O_2 could follow a covalent surface to reach a very short Na-O distance where the electron from Na(4D) is transferred to an electronically excited orbital of O_2 after which the complex can separate as Na^+O^- and O.

Na(4D) + NO_2 → NaO + NO is a substantially more exothermic reaction, but it has many features which are similar to those found in the
Na (4D) + O_2 → NaO + O reaction [41]. First of all, the high translational energy requirement, >18 kcal/mol, for NaO product formation again indicates that the entrance channel is very restricted and is likely to be along an O−N bond. If a Na(4D) atom must approach an NO_2 molecule along an O−N bond at a high translational energy in order for a chemical reaction to occur, the orbital angular momentum between Na(4D) and NO_2 will overwhelm the molecular angular momentum of NO_2, coplanar scattering will dominate, and NaO product will be scattered in the plane of the NO_2, which also contains the relative velocity vector. In other words, when Na(4D) approaches NO_2 along the NO bond, all the forces between the interacting atoms will lie in the plane of the NO_2, and the scattered NaO will be confined to that plane. Thus the detector, which rotates in a plane containing both the Na and NO_2 beams, can only detect those NaO products which are produced from NO_2 molecules lying in this plane at the instant when the reactions take place. In contrast to the collinear approach of Na and O_2, there is no cylindrical symmetry about the O−N axis when Na approaches NO_2 along that axis. Because of this, the reaction will depend not only on the d_z2 orbital alignment in the plane defined by the beams and the detector, but also on the alignment of the d_z2 about the relative velocity vector. This is exactly what we have observed in the laboratory. The reactivity of Na(4D) + NO_2 → NaO + NO as a function of the d_z2 orbital alignment with respect to the NO_2 molecule is shown in Fig. 16. The most favorable approach has the d_z2 orbital approaching the O−N axis perpendicularly and lying in the plane of the NO_2. When the alignment of the d_z2 orbital is rotated in the plane of the NO_2 the reactivity is reduced as the d_z2 orbital comes closer to being collinear with the O−N axis. When the d_z2 orbital is rotated out of the NO_2 plane from this collinear configuration, the reactivity decreases further and reaches a minimum when the orbital perpendicular to the NO_2 plane.

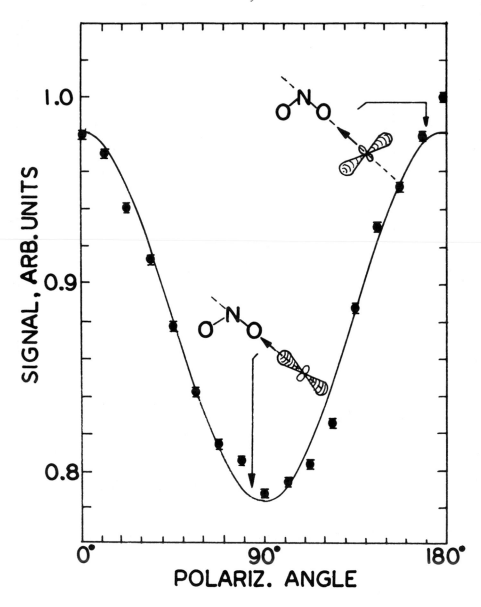

Fig. 16. Polarization dependence of NaO signal from the Na(4D) + NO$_2$ → NaO + NO reaction. As the (4d$_z^2$) orbital of Na was rotated in the plane which contains both beams and the detector, the signal was found to reach a maximum when the (4d$_z^2$) orbital was perpendicular to the relative velocity vector and reached a minimum when it was parallel.

The reaction of ground state Na atoms with HCl is endothermic by 5.6 kcal/mol. Figure 17 shows the product NaCl angular distributions for Na(3S, 3P, 4D) at an average collision energy of 5 kcal/mol. These angular distributions were measured at m/e = 23 because most of the NaCl fragments yield Na$^+$ in the electron bombardment ionizer. The rising signal at low angles is due to elastically scattered Na atoms. The reactive cross section increases with in-

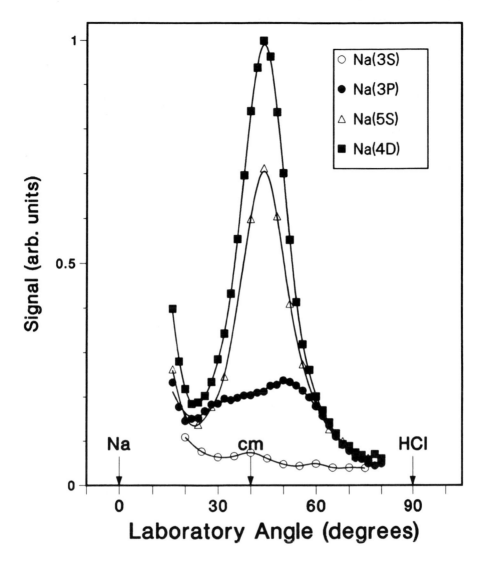

Fig. 17. NaC1 angular distributions for Na (3S,3P,4D,5S) + HC1 at a collision energy of 5.6 kcal/ mol.

creasing electronic energy. At the collision energy shown, the Na(3S) ground state atoms react because the high velocity components of each beam just barely overcome the endothermicity of the reaction. For the reaction of the Na(3P) atoms, NaCl product is observed over the full laboratory angular range possible allowing for the conservation of the momentum and total energy of the system. This implies a broad range of product translational energies, a conclusion which is supported by product velocity measurements. The same is not true of the reaction of the Na(4D) and Na(5S) states, in which the NaCl is scattered over a narrower angular range than that produced from the Na(3P) state, indicating less translational energy despite an additional 2 eV of excess

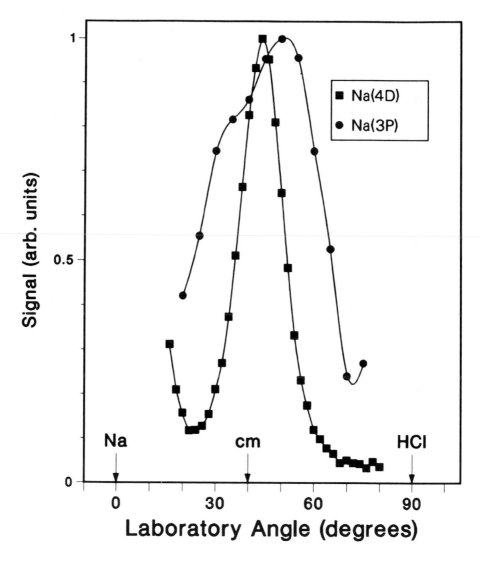

Fig. 18. NaCl angular distributions for Na(3P,4D) + HCl at a collision energy of 5.6 kcal/mol from Fig. 17. The peak intensity of the distribution arising from the reaction of Na(3P) is normalized to that arising from Na(4D) allowing comparison of the angular widths of the two distributions.

energy. This is illustrated in Fig. 18 in which the Na(3P) and Na(4D) angular distributions from Fig. 17 have both been normalized.

These interesting results can be explained by invoking electron transfer followed by repulsion the H atom and NaCl products. HCl is known to be dissociated by slow electrons, and has a *negative* vertical electron affinity of approximately 1 eV. For the reaction of the Na(3P) atoms this electron transfer becomes energetically possible at 3.5 Å. This is, incidentally, the peak of the Na(3P) orbital density. What the departing H atom feels is the repulsion from the Cl atom of the fully developed closed shell NaCl molecule, and a significant impulse is given to it, In the case of the Na(4D) atoms, the crossing of the neutral and ionic potential curves (the initial point of electron transfer) moves out to 7.7 Å. Thus, an electron transfers over, HCl^- dissociates, the H atom departs and the Na^+ and Cl^- are drawn together. The highly vibrationally excited NaCl cannot get rid of any of its energy as the H atom is already gone. The H atom has only felt the repulsion of the loosely bound or highly vibrationally excited NaCl. This interpretation is borne out by the polarization measurements in which the favored alignment of the Na(4D) orbital for reactive signal at any laboratory detector angle is along the relative velocity vector of the system. This corresponds to pointing the 4d orbital towards the HCl, because at long range the relative velocity vector in the laboratory is from the Na to the HCl.

Such a detailed study of the dependence of reactivity on the orbital alignment and the molecular orientation is possible only by combining the crossed molecular beams method with laser excitation.

Concluding remarks

The experimental investigation of elementary chemical reactions is presently in a very exciting period. The advancement in modern microscopic experimental methods, especially crossed molecular beams and laser technology, had made it possible to explore the dynamics and mechanisms of important elementary chemical reactions in great detail. Through the continued accumulation of detailed and reliable knowledge about elementary reactions, we will be in a better position to understand, predict and control many time-dependent macroscopic chemical processes which are important in nature or to human society.

In addition, because of recent improvements in the accuracy of theoretical predictions based on large scale ab initio quantum mechanical calculations, meaningful comparisons between theoretical and experimental findings have become possible. In the remaining years of the twentieth century, there is no doubt that the experimental investigation of the dynamics and mechanisms of elementary chemical reactions will play a very important role in bridging the gap between the basic laws of mechanics and the real world of chemistry.

The experimental investigations described in this article would not have been possible without the dedicated efforts of my brilliant and enthusiastic coworkers during the past twenty years. I enjoyed working with them immensely and sharing the excitement of carrying out research together.

I entered the field of reaction dynamics in 1965 as a post-doctoral fellow in the late Bruce Mahan's group at Berkeley, and learned a lot about the art of designing and assembling a complex experimental apparatus from many scientists and supporting staff at the Lawrence Berkeley Laboratory while studying ion-molecule scattering. In February of 1967, I joined Dudley Herschbach's group at Harvard as a post-doctoral fellow. There, I was exposed to the excitement of the crossed molecular beams research and participated in the construction of an universal crossed molecular beams apparatus. Dudley's contagious enthusiasm and spectacular insight motivated not only me, but a whole generation of chemical dynamicists.

Fig. 19. Part of the new molecular beam laboratory at the University of California, Berkeley.

Molecular collision dynamics has been a wonderful area of research for all practitioners. This is especially true for those who were following the footsteps of pioneers and leaders of the field twenty years ago. In my early years, I was also inspired by the pioneering research work of Sheldon Datz and Ellison Taylor, Richard Bernstein, John Ross, and Ned Green, as well as the "supersonic" John Fenn. They have been most generous and caring scientists and all of us admire them. Their work is the main reason why the field of molecular beam scattering has attracted many of the best minds in the world and made it a most exciting and rewarding field.

My associations with the University of Chicago (1968–74) and with the University of California, Berkeley (1974–) have been very rewarding. I could not ask for a better environment to pursue an academic career. The stimulating colleagues and excellent facilities as shown in Fig. 19 are what made these institutions so wonderful.

Throughout all these years, my scientific research activities have been supported continuously by the Office of Basic Energy Sciences of the Department of Energy and the Office of Naval Research. The stable and continuing support and the confidence they have shown in my research have been most important and are gratefully appreciated.

REFERENCES

1. H. Eyring and M. Polanyi, Z. Phys. Chem. *B12* (1931) 279.
2. J.C. Polanyi and D.C. Tardy, J. Chem. Phys. *51* (1969) 5717.
3. J.H. Parker and G.C. Pimentel, J. Chem. Phys. *51* (1969) 91.
4. H.W. Cruse, P.J. Dagdigian, and R.N. Zare, Farady Discussion of the Chemical Society, *55* (1973) p. 277.
5. D.P. Gerrity and J.J. Valentini, J. Chem. Phys. *82* (1985) 1323.
6. D.R. Herschbach, in *Les Prix Nobel* 1986, Nobel Foundation, 1987.
7. R.B. Bernstein, *Chemical Dynamics via Molecular Beam and Laser Techniques*, Oxford University Press, 1982.
8. D.M. Neumark, A.M. Wodtke, G.N. Robinson, C.C. Hayden, K. Shobatake, R.K. Sparks, T.P. Schaefer, and Y.T. Lee, J. Chem. Phys. *82* (1985) 3067.
9. Y.T. Lee, in *Atomic and Molecular Beam Method*, edited by G. Scoles and U. Buck, Oxford University Press, 1986.
10. Y.T. Lee, J.D. McDonald, P.R. LeBreton and D.R. Herschbach, Rev. Sci. Instr. *40* (1969) 1402.
11. J.T. Muckerman, in *Theoretical Chemistry—Advances and Perspectives*, edited by H. Eyring and D. Henderson (Academic Press, New York, 1981), Vol. 6A, 1—77.
12. C.F. Bender, P.K. Pearson, S.V. O'Neill, and H.F. Schaefer, J. Chem. Phys. *56* (1972) 4626; Science, *176* (1972) 1412.
13. D.G. Truhlar and A. Kuppermann, J. Chem. Phys. *52* (1970) 384; *56* (1972) 2232.
14. S.F. Wu and R.D. Levine, Mol. Phys. *22* (1971) 991.
15. G.C. Schatz, J.M. Bowman, and A. Kuppermann, J. Chem. Phys. *63* (1975) 674.
16. A. Kuppermann, in *Potential Energy Surfaces and Dynamics Calculations*, edited by D.G. Truhlar (Penum Publishing Corporation, New York, 1981).
17. J.M. Launay and M. LeDourneuf, J. Chem. Phys. *B15* (1982) L455.
18. R.E. Wyatt, J.F. McNutt, and M.J. Redmon, Ber. Bunsenges. Chem. Phys. *86* (1982) 437.
19. D.M. Neumark, A.M. Wodtke, G.N. Robinson, C.C. Hayden, and Y. T. Lee, J. Chem. Phys. *82* (1985) 3045.
20. J.W. Birks, S.D. Gabelnick, and H.S. Johnston, J. Mol. Spectr. *57* (1975) 23.
21. J.M. Farrar, and Y.T. Lee, J. Chem. Phys. *63* (1975) 3639.
22. M.J. Coggiola, J.J. Valentini, and Y.T. Lee, Int. J. Chem. Kin. *8* (1976) 605.
23. C.C. Kahler and Y.T. Lee, J. Chem. Phys. *73* (1980) 5122.
24. N. Abuaf, J.B. Anderson, R.P. Andres, J.B. Fenn, and D.R. Miller *Rarefied Gas Dynamics*, Fifth Symposium 1317—1336 (1967)
25. F.P. Tully, N.H. Cheung, H. Haberland, and Y.T. Lee, J. Chem. Phys. *73* (1980) 4460.
26. G.N. Robinson, R.E. Continetti, and Y.T. Lee, to be published Faraday Disc. Chem. Soc. *84* (1987).
27. F. Huisken, Krajnovich, Z. Zhang, Y.R. Shen, and Y.T. Lee, J. Chem. Phys. *78* 3806 (1983).
28. R.J. Cvetanovic, Can. J. Chem. *33* (1955) 1684.
29. F.J. Pruss, J.R. Slagle, and D. Gutman, J. Phys. Chem. *78* (1974) 663.
30. B. Blumenberg, K. Hoyerman and R. Sievert, Proc. 16th Int. Symp. on Combution, The Combustion Institute, 1977, p. 841.
31. R.J. Buss, R.J. Baseman, G. He, and Y.T. Lee, J. Photochem. *17* (1981) 389.
32. G. Inoue and H. Akimoto, J. Chem. Phys. *74* (1981) 425.
33. H.E. Hunziker, H. Kneppe, and H.R. Wendt, J. Photochem. *17* (1981) 377.
34. F. Temps and H.G. Wagner, Max Planck Inst. Stromungsforsch. Ber. 18 (1982).
35. U.C. Sridharan and F. Kaufman, Chem. Phys. Lett. *102* (1983) 45.
36. Y. Endo, S. Tsuchiya, C. Yamada, and E. Hirota, J. Chem. Phys. *85*, (1986) 4446.
37. S.J. Sibener, R.J. Buss, P. Casavecchia, T. Hirooka, and Y.T. Lee, J. Chem. Phys. *72* (1980) 4341.

38. R.J. Brudzynski, A.M. Schmoltner, P. Chu, and Y.T. Lee, to be published in J. Chem. Phys. 1987.
39. H. Schmidt, P.S. Weiss, J.M. Mestdagh, M.H. Covinsky, and Y. T. Lee Chem. Phys. Lett. *118* (1985) 539.
40. P.S. Weiss PH.D. Thesis, University of Calfornia, 1986.
41. B.A. Balko, H. Schmidt, C.P. Schulz, M.H. Covinsky, J.M. Mestdagh, and Y.T. Lee, to be published Faraday Disc. Chem. Soc. *84* (1987).

John C. Polanyi

JOHN C. POLANYI

John Charles Polanyi was born in 1929 in Berlin, Germany, of Hungarian parents, Michael and Magda Elizabeth Polanyi. The family moved to England in 1933 where he received his education.

His University training was at Manchester University, where he obtained his B.Sc. in 1949, and his Ph.D. in 1952.

From 1952–1954, he was a Postdoctoral Fellow at the National Research Council Laboratories in Ottawa, Canada, and from 1954–1956 Research Associate at Princeton University.

In 1956, John Polanyi was appointed as a Lecturer at the University of Toronto where he was successively Assistant Professor (1957–1960), Associate Professor (1960–1962) and Professor (1962–present). He was given the (honorific) title University Professor in January 1974.

In 1958, he married Anne (Sue) Ferrar Davidson. They have two children, Margaret Alexandra (born 1961), and Michael Ferrar (born 1963).

He serves on the Board of the Ontario Laser and Lightwave Research Centre, Canada (1988–present), is a Member of the Board of the Steacie Institute for Molecular Sciences, Canada (1991–present), and Member of the Science Advisory Board, Max Planck Institute for Quantum Optics, Germany (1982–present), and is Honorary Consultant to the Institute for Molecular Science, Okazaki, Japan (1989–1992). He was a Founding Member and is currently President of the Canadian Committee of Scientists and Scholars, and also was a Founding Member of The Royal Society of Canada Committee on Scholarly Freedom, a Member of the American Academy of Arts and Science Committee on International Security Studies, and a Member of the Board of the Canadian Centre for Arms Control and Disarmament to which he is currently an Advisor.

He was awarded the Marlow Medal of the Faraday Society 1962, Centenary Medal of the British Chemical Society 1965, the Steacie Prize for Natural Sciences (shared with N. Bartlett) 1965, the Noranda Award of the Chemical Institute of Canada 1967, the Henry Marshall Tory Medal of the Royal Society of Canada 1977, the Wolf Prize in Chemistry (shared with G. Pimentel) 1982, the Izaak Walton Killam Memorial Prize 1988, the Royal Medal of the Royal Society of London 1989, and the John C. Polanyi Lecture Award of the Canadian Society for Chemistry 1992.

He is a Fellow of the Royal Society of Canada (1966), and the Royal Society of London (1971), a Member of the American Academy of Arts and Sciences, (1976), the U.S. National Academy of Sciences (1978), the

Pontifical Academy of Rome (1986), a Fellow of the Royal Society of Edinburgh (1988), an Honorary Fellow of the Royal Society of Chemistry of the United Kingdom (1991), and of the Chemical Institute of Canada (1991).

He has been the recipient of honorary degrees from the Universities of Waterloo 1970; Memorial 1976; McMaster 1977; Trent 1977; Carleton 1981; Harvard 1982; Dalhousie 1983; Rensselaer 1984; Brock 1984; St. Francis Xavier 1984; Lethbridge 1987; Victoria 1987; Ottawa 1987; Sherbrooke 1987; Laval 1987; York 1988; Manchester, England 1988; Montreal 1989, Acadia 1989; Weizmann Institute, Israel 1989; Bari, Italy 1990; British Columbia 1990; Concordia 1990, McGill 1990 and Queen's 1992.

He was made an Officer of the Order of Canada in 1974, and a Companion of the Order of Canada in 1979.

In addition to his scientific papers he has published approximately one hundred articles on science policy, on the control of armaments and the impact of science on society. He has produced a film 'Concepts in Reaction Dynamics' (1970), and has co-edited a book, 'The Dangers of Nuclear War' (1979).

SOME CONCEPTS IN REACTION DYNAMICS

Nobel lecture, December 8, 1986.

by

JOHN C. POLANYI

Department of Chemistry, University of Toronto, Toronto M5S 1Al. Canada

The objective in this work has been one which I have shared with the two other 1986 Nobel lectures, D. R. Herschbach and Y. T. Lee, as well as with a wide group of colleagues and co-workers who have been responsible for bringing this field to its current state. That state is summarized in the title; we now have some concepts relevant to the motions of atoms and molecules in simple reactions, and some examples of the application of these concepts. We are, however, richer in vocabulary than in literature. The great epics of reaction dynamics remain to be written. I shall confine myself to some simple stories.

1. EXPERIMENTAL AND THEORETICAL APPROACHES.

In this section I shall say something about the experimental and theoretical tools we have used, and give some indication of their origins.

Our principal experimental method has been the study of infrared chemiluminescence. The most valuable theoretical tool has been, and remains, the computer-integration of the classical equations of motion. Though both of these methods belong to the modern period of reaction dynamics, both have clear antecedents in earlier times.

Infrared chemiluminescence derives from the presence of vibrationally excited molecules in the products of reaction. Indirect evidence for the existence of such species was obtained by M. Polanyi and his co-workers in 1928 in the course of studies of the reactions of alkali metal atoms and halogens [1]. Following a suggestion by Bates and Nicolet [2], and by Herzberg [3], McKinley, Garvin, and Boudart [4] looked for and found visible emission from vibrationally excited hydroxyl radicals formed in the reaction of atomic hydrogen with ozone. This finding was followed soon after by the identification, by absorption spectroscopy, of vibrational excitation in the products of the reactions of atomic oxygen with NO_2 and ClO_2 [5].

Reference [5] appeared in the year that, in collaboration with J. K. Cashion, we undertook to look for infrared chemiluminescence from vibrationally excited hydrogen halides (and other hydrogen containing compounds) formed in simple exchange and addition reactions. The first reaction studied was that of atomic hydrogen with molecular halogens [6, 7]. Infrared emission was observed in the region 1.5—4.5 μm arising from the low-pressure ($\sim 10^{-1}$ torr) room temperature reaction,

$$H + X_2 \rightarrow HX\dagger + X \tag{1}$$

where X was Cl or Br.

At that date there was lively dicussion of the possibility of a visible analogue of the maser, that would operate on population-inverted electronic states (a working laser was still in the future) [8]. Given our interests it was natural to speculate about the properties of a vibrational laser. In our communication ("Proposal for an Infrared Laser Dependent on Vibrational Excitation") [9] we pointed to a number of virtues of such a device, of which we shall note two here.

(1) Provided that the vibrational temperature, T_V, sufficiently exceeds the rotational temperature, T_R, (both temperatures being positive, in contrast to the negative temperatures associated with population inversion in the discussions of that date) a large number of P-branch transitions will exhibit population inversion (fig. 1). We gave the name *'partial population inversion'* to this phenomenon, and noted that lasing based on partial population inversion could ensue shortly after a thermal pulse (arising from a pulsed arc, or shock wave), since 'partial cooling' would ensure $T_V \gg T_R$. (2) A chemical reaction could be used to obtain either 'partial' or 'complete' population inversion. We were struck by the fact that the upper atmosphere could constitute a natural laser in the infrared favoured by its long path length and large number of partially population inverted transitions [9, 10].

Vibrational lasing was first achieved by Patel and co-workers [11]. Chemical reaction as the working material in a vibrational laser followed shortly afterward, through the work of Pimentel and his associates [12]. In both cases the major contribution to lasing came from partial population inversion, as indicated by the predominance of P-branch emission.

In our first communication regarding infrared chemiluminescence, we expressed the view that the method "promises to provide for the first time information concerning the distribution of vibrational and possibly rotational energy among the products of a three-centre reaction" [6]. This outcome was not so readily achieved.

By 1962 it was evident that reaction (1), despite the fact that it had proved to be a ready source of infrared chemiluminescence, converted its heat of reaction into vibration with the modest efficiency of $\leqslant 50\%$ [13]. This finding met with scepticism, since it was pointed out, correctly, that it implied for the reverse endothermic process that one could have so much vibration in the reagents that the reaction probability would be markedly diminished [14]. The somewhat lame reply at that date was that "though odd, it could still be the case" [15].

Today we would surmise that it is the case since for the endothermic process $Cl + HCl \rightarrow Cl_2 + H$ a heavy particle (Cl) must approach during the brief time that the H-Cl bond remains extended; too much vibrational excitation can have the consequence that the $Cl + HCl$ interaction is averaged over many vibrational periods, resulting in an 'adiabaticity' with respect to the flow of energy into the bond to be broken. At that date the importance of the relative timing of molecular motions (though recognised for other types of inelastic encounters) had not yet been documented for chemical reactions.

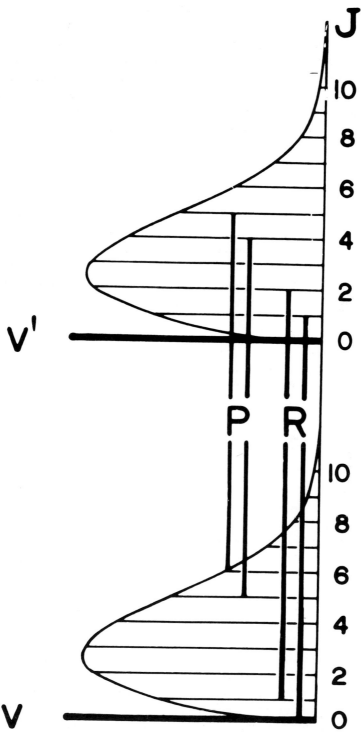

Figure 1. 'Partial population-inversion'; $T_V \gg T_R (>O)$ ensures that the population in the upper state, N_u, exceeds that in the optically linked lower state, N_l for the indicated transitions (among others). The condition for lasing, $(N_u/g_u) > (N_l/g_l)$ where g is degeneracy, is only met for the P-branch. From J.C. Polanyi, Appl. Opt. Suppl. *2*, 109 (1965); (also J. Chem. Phys. *34*, 347 (1961)).

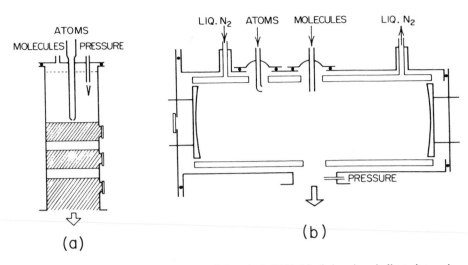

Figure 2. (a) Reaction vessel for 'Measured Relaxation' (MR) (shaded regions indicate internal reflective coating; three sapphire windows at successive sampling points are indicated at the right); (b) Reaction vessel for 'Arrested Relaxation' (AR) shown schematically (mirrors at either end of the vessel collect its radiation and bring it to the sapphire window at the left). From K.G. Anlauf et al., Disc. Faraday Soc. *44*, 183 (1967).

It was not until 1967, with the development of the methods of "measured relaxation" (MR) [16] and "arrested relaxation" (AR) [16] that the goals established nine years earlier were truly achieved. The impediment to obtaining fully quantitative data by infrared chemiluminescence, or any other means, in regard to what we have termed the "detailed rate constants", $k(V', R')$ (V' is the product vibrational excitation, and R' the product rotational excitation) has been the presence of an undetermined amount of vibrational and rotational relaxation of the reaction product prior to observation.

In the MR approach [16, 17−19], this was addressed by measuring the vibrational relaxation at points along the direction of flow, and hence correcting for this effect; see fig. 2 (a). In the AR method [16, 18−21], relaxation rather than being measured was arrested to the fullest extent possible, by the rapid removal of excited products. This was achieved by complete deactivation at a surface, usually cooled in the range 20−77K; fig. 1 (b). Arrest of relaxation yielded detailed rate constants at the level $k(V',R',T')$, where T' is the translational energy distribution in the newly-formed product, obtained by subtracting the initial vibrational plus rotational distributions from the total available energy.

Concurrently with the development of these infrared chemiluminescence techniques, the crossed molecular beam method was established as a quantitative tool, particularly by Herschbach and co-workers. This method obviated problems of relaxation by conducting reaction under single-collision conditions. Since the prime measurables were product angular and translational energy distributions, by 1967 a degree of overlap existed between the approaches. It was not, however, until the incorporation of universal detectors by Lee and Herschbach, following what we have come to know as the 'alkali age' of beam chemistry, that the same systems could be studied by both infrared chemiluminescence and molecular beam scattering.

By the time that the first quantitative data began to appear, powerful theoretical tools were available which could be used to link the potential field operating between atoms A, B, and C in a reaction

$$A + BC \rightarrow AB + C \tag{2}$$

to the motions of the particles. In the original work on the London equation [22] by Eyring and Polanyi [23] an attempt was made, in collaboration with Wigner, to solve the classical equations of motion for the reactive system A + BC. This work was taken up by Hirschfelder and Wigner [24], but it was not until the first computer calculations of reaction dynamics by Wall, Hiller, and Mazur [25] that the full power of the approach became evident.

In parallel with Blais and Bunker [26] our laboratory [27] employed Wall's, Hiller's and Mazur's approach in an attempt to map out the major determinants of A + BC reaction dynamics. Both groups were inspired by the proposal made by M. G. Evans and M. Polanyi [28], over two decades earlier, that vibration in a newly-formed bond might originate from the release of reaction energy as the reagents approached. Detailed computation indicated that this suggestion embodied an important kernel of truth, though the full story, as might have been expected, revealed a substantially richer range of scenarios.

2. ENERGY DISTRIBUTION AMONG REACTION PRODUCTS

2.1 *Experiment*

In the MR method (see previous section), several observation windows for recording infrared (ir) chemiluminescence were located along the line of streaming flow. Provided that the observation windows were situated at distances corresponding to times during which relaxation was moderate, a simple graphical extrapolation back to zero time yielded adequate values for the relative $k(v')$ (v' is the vibrational quantum number corresponding to the product vibrational energy V') [16, 17]. Detailed numerical analyses were also made, which included the effects of reaction, diffusion, flow, radiation, and collisional deactivation [17]. A model of this sort permitted, in effect, a more intelligent extrapolation to time $t=0$.

Subsequently the flow-tube approach was applied to the determination of vibrational energy distributions in a substantial number of infrared chemiluminescent reactions, particularly by Setser and his co-workers [29] and by Kaufman [30]. Infrared chemiluminescence experiments in flow tubes have also been extended to the elucidation of product energy-distributions from ion-molecule reactions [31].

In the A. R. method two uncollimated beams of reagent met in the centre of a vessel, exhibiting a background pressure, with reagents flowing, of $\sim 10^{-4} - 10^{-5}$ torr. Reaction occurred at the intersection of the beams (as demonstrated in ref. [18]). The products of reaction were transferred, following a few secondary

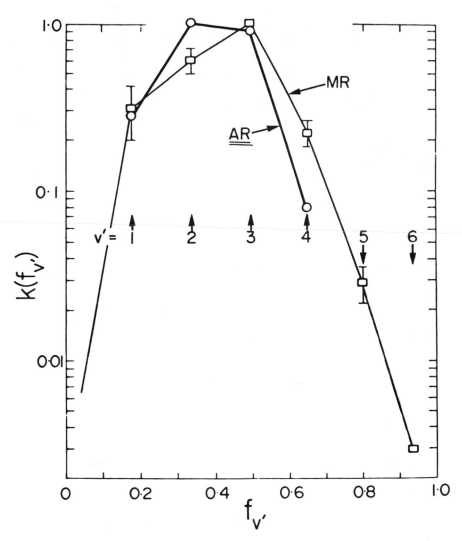

Figure 3. Semi-logarithmic plot of the probability that a fraction f_v, of the energy in the products of the reaction $H+Cl_2 \rightarrow HCl+Cl$ enters vibration in the newly-formed bond. The data were obtained independently by the MR and AR methods, as indicated. The vibrational quantum states $v' = 1-6$ correspond to the values of f_v, indicated by arrows across the centre of the figure. From K.G. Anlauf et al., Disc. Faraday Soc. *44*, 183 (1967).

collisions, to a deactivating surface which surrounded the reaction zone. Depending on the reagents, this surface acted as a cryo-pump, as a getter pump (trapping non-condensibles in excess condensible), or simply as an adsorber capable of bringing about complete vibrational deactivation (adsorbing product for a time sufficient to deactivate the material to its ground vibrational state, $v=0$).

The infrared chemiluminescence stemming from the small fraction of molecules that emitted early in their lifetimes, was collected by gold front-surface

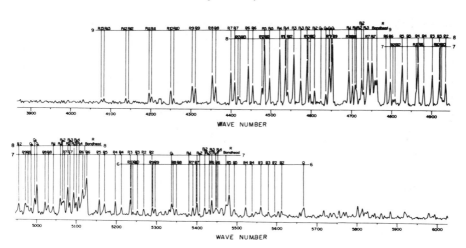

WAVE NUMBER

WAVE NUMBER

Figure 4. First-overtone $(\Delta v' = -1)$ spectrum of OH formed in the reaction $H + O_3 \rightarrow OH + O_2$, $-\Delta H^\circ_o = 77$ kcal/mole. Under arrested relaxation conditions a continuous flow of reagents yields a steady-state distribution of $OH(v')$ that increases with v' up to the highest accessible vibrational level, $v' = 9$ $(E(v' = 9) = 75$ kcal/mole, i.e. 3.25 eV). This spectrum was recorded by using a Fourier transform spectrometer and AR geometry; emission is from $v' = 6-9$ as indicated. From P.E. Charters et al., J. Appl. Optics, *10*, 1747 (1971); (also J.C. Polanyi and J.J. Sloan, Int. J. Chem. Kin. Symposium *1*, 51 (1975).

mirrors, freshly coated for each experiment; this procedure increased the effective solid angle being viewed by almost 50x [32]. By this means $\sim 10^{-9}$ torr of vibrationally excited product could be detected in one v, J (vibrational) state, using the liquid-nitrogen cooled PbS semiconductor detectors that became available in the late 1950's. The emission spectrum (fig. 3) was recorded in early work on a (NaCl, then LiF) prism spectrometer, later on a grating spectrometer and (from the early 1970's, for e.g. [33]) on a variety of Fourier transform interferometric spectrometers.

The AR method was adopted by several laboratories, notably N. Jonathan's [34], D.W. Setser's [35], and J.D. McDonald's [36]. It is characterised by molecular flow of the reactants and hence low densities in the reaction zone, permitting observation of reaction products with highly non-Boltzmann rotational distribution, as well as non-Boltzmann vibrational distributions.

The effectiveness of the combination of molecular flow plus rapid removal of reaction products in reducing relaxation to an insignificant amount in the majority of systems studied was indicated by the fact that the steady-state distribution observed under these conditions was in satisfactory agreement with the initial vibrational distribution obtained by the MR method, (fig. 4) and the observed rotational distribution was largely unrelaxed, (fig. 5) indicating that collisional deactivation of vibrators—generally a less efficient process— was insignificant. In the case of a few very fast reactions (notably $F + HBr \rightarrow HF + Br$ [29]) $k(v')$ from MR were at variance with those from AR. At the low flows used in the AR machine, it appears that reaction and deactivation were

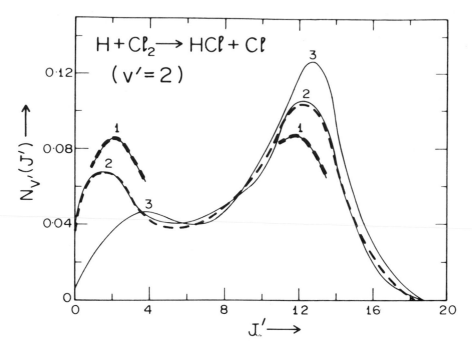

Figure 5. Solid lines record the decreasing contribution of the peak at low J' to the observed rotational distribution (for H+Cl$_2$→HCl(v' = 2, J')+Cl) as the pressure in the beam-crossing region is decreased (in the sequence 3→2→1) in an A.R. vessel. The broken lines superimposed on the solid lines give relaxed distributions computed from 3. Based on fig. 1 of J.C. Polanyi and K:B. Woodall, J. Chem. Phys. *56*, 1563 (1972).

occurring within the inlets, invalidating the results [37]; in such a case the AR rotational distributions were also found to have been relaxed [38].

The observed rotational distributions were typically bimodal, comprising a non-thermal distribution at high rotational quantum number (high J') and a thermal distribution peaking at low J'. The contribution of the peak at low J' relative to that at high J' decreased with diminishing pressure in the reaction zone (fig. 5). Accordingly the low J' peak was ascribed to collisional relaxation, the relaxation mechanism being analyzed in detail [39].

Some detailed rate constants, k(v'), for the reaction

$$F + H_2 \rightarrow HF + H \qquad\qquad (3)$$

obtained by different procedures and by different workers are given in Table 1. Included among the methods used is one due to Pimentel and co-workers which made use of stimulated, rather than spontaneous, infrared emission [40]. This approach was an outgrowth of the first chemical laser, due to the same laboratory [12]. The laser approach [40] gave quantitative data concerning the population in the vibrational ground-state v'=0, which previously had been

Table I. Detailed rate constants, k(v') for $F+H_2 \rightarrow (HF(v')+H$; comparison data from various studies (with some variation in reagent collision energy).

	Reagt.Colln Energy	v'=1	v'=2	v'=3
Parker & Pimentel[1](1969)	539°K	0.18	1.00	–
Polanyi & Tardy[2](1969)	300°K	0.29	1.00	0.47
Jonathan et al.[3](1971)	300°K	0.28±0.02	1.00	058±0.12
Polanyi & Woodall[4](1972)	300°K	0.31	1.00	0.47
Coombe & Pimentel[5](1973)	236°K	–	1.00	0.505±0.01
	298°K	–	1.00	0.478±0.005
	364°K	–	1.00	0.463±0.006
	172°K	0.222±0.01	1.00	–
	432°K	0.345±0.03	1.00	–
Chang & Setser[6](1973)	300°K	0.29	1.00	0.56
Berry[7](1973)	297°K	0.294±0.01	1.00	0.63±0.04
Perry & Polanyi[8](1976)	279°K	0.28	1.00	0.55
	718°K	0.38	1.00	0.55
	1315°K	0.44	1.00	0.55
Neumark et al.[9](1984)	1.84 kcal/mole	0.21	1.00	0.67

[1] J.H. Parker and G.C. Pimentel, J. Chem. Phys. *51*, 91 (1969).
[2] J.C. Polanyi and D.C. Tardy, ibid. *51*, 5717 (1969).
[3] N. Jonathan, C.M. Melliar-Smith, S. Okuda, D.H. Slater, and D. Timlin, Mol. Phys. *22*, 561 (1971).
[4] J.C. Polanyi and K.B. Woodall, J. Chem. Phys. *57*, 1574 (1972).
[5] R.D. Coombe and G.C. Pimentel, ibid. *59*, 251 (1973).
[6] H.W. Chang and D.W. Setser, ibid. *58*, 2298 (1973).
[7] M.J. Berry, ibid. *59*, 6229 (1979).
[8] D.S. Perry and J.C. Polanyi, Chem. Phys. *12*, 419 (1976).
[9] D.M. Neumark, A.M. Wodtke, G.N. Robinson, C.C. Hayden, and Y.T. Lee, J. Phys. *82*, 3045 (1985).

unobservable. Also included are the findings from crossed-molecular beam studies performed in Y.T. Lee's laboratory. In these demanding experiments the vibrational distribution was obtained from structure in the product translation energy-distribution, which was the measured quantity (the findings are for somewhat different reagent energy distributions than for the other experiments tabulated). The reaction $F + H_2 \rightarrow HF + H$ is of considerable fundamental interest since it involves only 11 electrons, and should therefore be amenable to *ab initio* computation (see below). The agreement between determinations of k(v') is satisfactory.

When these experiments and those described below were undertaken, quantitative data regarding k(v') did not exist. The prior observation of highly vibrationally excited reaction products noted in section 1, since it was a qualitative finding, left open the question of the form of k(v') which could have peaked anywhere from v'=0 to v'=v'$_{max}$ (the highest accessible vibrational state). In fact the distributions observed by ir chemiluminescence were found

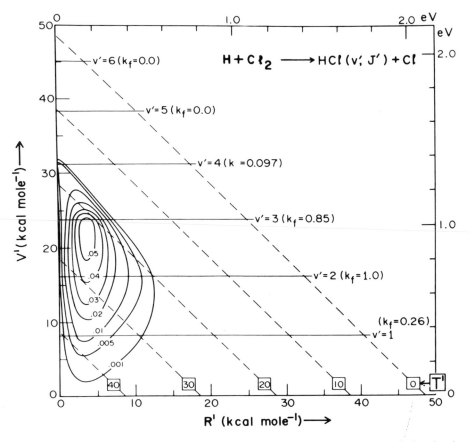

Figure 6. 'Triangle plot' for $H+Cl_2 \rightarrow HCl+Cl$ reaction showing the distribution of vibrational, rotational, and translational energy as contours of equal detailed rate constant $k(V',R',T')$. The ordinate is product vibrational energy, and the abscissa is the product rotational energy. Since the total available energy is approximately constant ($E'_{tot} = 48.4$ kcal/mole), translational energy, T', given by the broken diagonal lines, increases to 48.4 kcal/mole at $V' = 0$, $R' = 0$. The rate constant $k_f \equiv k(v') = \Sigma_j k(v'J')$; the subscript f denotes the 'forward', exothermic, direction. Contours have been normalized to $k(v') = 1.00$ where v'is the most-populated vibrational level. From K.G.Anlauf et al., J. Chem. Phys. *57*, 1561 (1972).

to peak at intermediate v', so that the mean fraction of the available energy entering vibration in the new bond ranged from $< f'_v > = 0.39$ for $H + Cl_2$, $< f'_v > = 0.55$ for $H + Br_2$, $< f'_v > = 0.53$ for $H + F_2$, $< f'_v > = 0.66$ for $F + H_2$ up to $< f'_v > = 0.71$ for $Cl + H1$ and $< f'_v > \approx 0.9$ for $H + O_3 \rightarrow OH + O_2$.

An early finding was that the product vibrational energy distribution $k(v')$ altered dramatically with isotopic substitution, but was roughly invariant if represented in the reduced form $k(f'_v)$ [41]. This is what one would expect if classical mechanics applied rather than conservation laws involving quantum states. The finding gave us courage in applying classical mechanics to these reactions (despite the fact that they were characterised by widely spaced

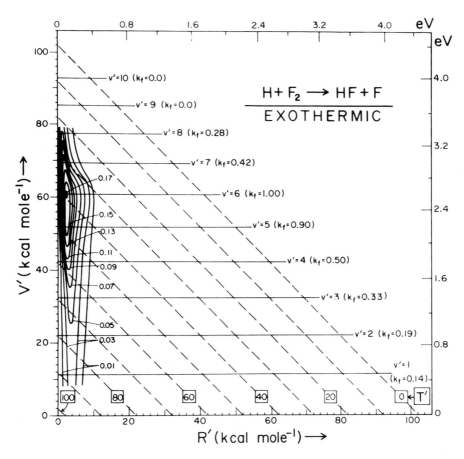

Figure 7. Triangle plot for $H+F_2 \to HF+F$ (see caption to fig. 6). $E'_{tot} = 102$ kcal/mole for this reaction. From J.C. Polanyi and J.J. Sloan, J. Chem. Phys. *57*, 4988 (1972).

vibrational and rotational states), and also encouraged us to record our results in a graphical form that rode rough-shod over the quantised nature of $k(v'J')$.

Four examples of the graphs that we have used (as a supplement to the tabulation of the actual $k(v'J')$ values) as an aid to the visualisation of the experimentally determined product energy distributions, are shown in figs. 6–9. By interpolation between permitted combinations of V', R', and T', we obtain contours of equal $k(V',R',T')$. In these "triangle plots" contours are shown in V',R',T' space. They delineate a "hill" of detailed rate constant in that space. Portions of the hill that lie between vibrational energy states are merely included to guide the eye as to the breadth and shape of the product energy distribution over the axis' corresponding to V',R' and T'. Since classical mechanics is widely used in order to interpret energy distributions, these classical "fingerprints" of differing chemical reactions help us to picture contrasting types of behaviour.

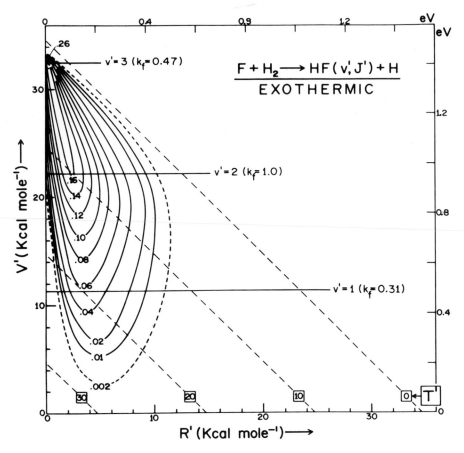

Figure 8. Triangle plot for F+H$_2$→HF+H (see caption to fig. 6). E'$_{tot}$ = 34.7 kcal/mole for this reaction. From J.C. Polanyi and K.B. Woodall, J. Chem. Phys. *57*, 1574 (1972).

It is readily apparent that the first two reactions, H + Cl$_2$ and H + F$_2$ (figs. 6 and 7), though they liberate markedly different energies, both exhibit inefficient vibrational and rotational excitation and consequently efficient translational excitation. The third reaction, F + H$_2$ (fig. 8), gives rise to highly efficient vibrational excitation together with inefficent rotational excitation. The fourth reaction pictured, Cl + HI (fig. 9), combines efficient vibrational excitation with exceptionally efficient rotational excitation ($< f_R' > = 0.13$ for Cl + HI in contrast to $< f_R' > = 0.03$ for H+F$_2$). In concurrent work, alluded to in the following section, we linked these changes in dynamics to the release of repulsive energy in systems having differing mass combination and preferred geometries of the intermediate.

The ridge of the "detailed rate constant hill" is almost vertical in the case of the first reactions, moving out to higher R' at lower V' in the fourth of the reactions pictured (Cl + HI), (for a tabulation of mean $< f_v' > < f_R' >$, $< f_T' >$, and the slope of the ridge $\triangle R'/\triangle V'$ for these reactions see table IV of

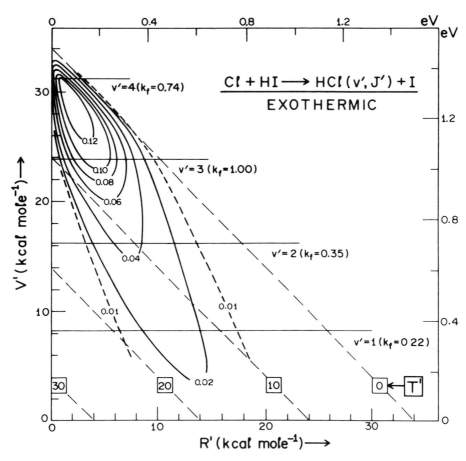

Figure 9. Triangle plot for Cl+HI→HCl+I. See caption to fig. 7. E'$_{tot}$ = 34 kcal/mole. From D.H. Maylotte et al., J. Chem. Phys. *57*, 1547 (1972).

ref. [21]) leaving aside the dynamical implications of this distinctive behaviour (see 2.2 below). It is noteworthy that successively lower vibrational states exhibit marked increases in *translational* energy in the first three cases, but a somewhat overlapped "translational energy spectrum" in the fourth case. This finding was seen to have interesting implications for the crossed molecular beam chemistry in which T' (as already noted) is the prime measurable. Figure 10 gives sample distributions over f$_T$' (blending rotational states into a continuum, since they cannot be resolved in molecular beam time-of-flight (tof) studies). The reaction H + F$_2$, not pictured here, gives rise to eight well-resolved translational peaks (fig. 5 of [21]). In the case of the F + H$_2$ reaction and its isotopic analogues in addition to resolving these peaks by tof, Y.T. Lee's laboratory have in recent times fully characterised the angular distribution of each [42].

The triangle plots, exemplified in figs. 6–9, constitute one way of "compacting" the substantial data embodied in k(v'J',T'). The information-theory,

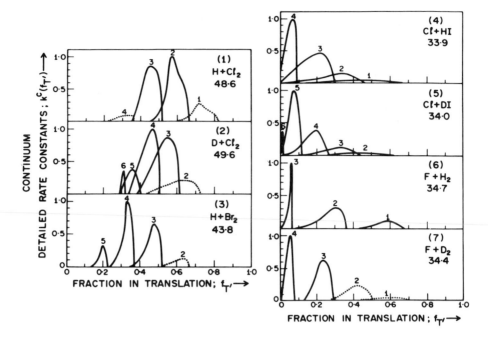

Figure 10. Translational energy-distribution in the products of some exchange reactions
A+BC→AB+C obtained by subtracting the vibrational excitation measured by its chemiluminescence from the total product energies (indicated in the figure). The vibrational states that contribute to the translational peaks are recorded above each peak. From K.G. Anlauf et al. J. Chem.
Phys. *53*, 4091 (1970).

pioneered by Bernstein and Levine [43], has provided a widely-used alternative method of compaction in terms of parameters which describe the deviation
of the observed vibrational and rotational distributions from a statistical outcome. These deviations can be described as vibrational and rotational "surpris-
als". In those cases for which the surprisal is found to be a linear function of f'_v
or f'_R, not only can the data be reduced to only two constants, but extrapolations can be made. We have made use of this in our work in order to estimate
the (small)-fraction of product formed in the non-emitting state, $v'=0$.

Figure 11 contrasts the triangle plots for HF (fig. 11(a)) and DF (fig. 11 (b))
formed in the thermal reaction $F+HD \rightarrow HF(v'J')$ or $DF(v'J')$ (+D or +H)
[44]. These data are included since they shed light on a source of product
rotational excitation. The forces operating are very similar but are clearly
substantially more effective in rotationally exciting HF (mean fraction of the
available energy entering rotation is $<f'_R>= 0.125$) than $DF(<f'_R>=0.066)$.
We postpone discussion to section 2.2 below.

It is interesting to note that for the family of reactions $F+HD \rightarrow HF$, $F+HD$
$\rightarrow DF$, $F+H_2$, $F+D_2$ studied by ir chemiluminescence, the first reaction in the
list exhibited vibrational excitation that was anomalously low as compared
with classical calculation. This (together with the reagent-energy dependence

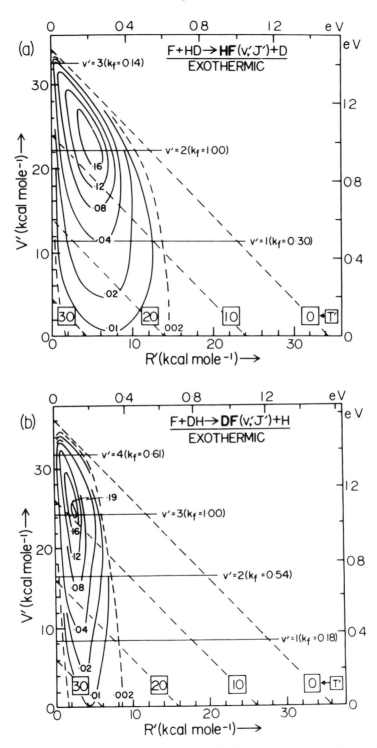

Figure 11. Triangle plots for both branches of the F+HD exchange reaction. The total available energy in the products was taken to be $E'_{tot} = 34.3$ kcal/mole for the HF product, and $E'_{tot} = 36.1$ kcal/mole for DF, in calculating product translation T'. From D.S. Perry and J.C. Polanyi, Chem. Phys. *12.* 419 (1976).

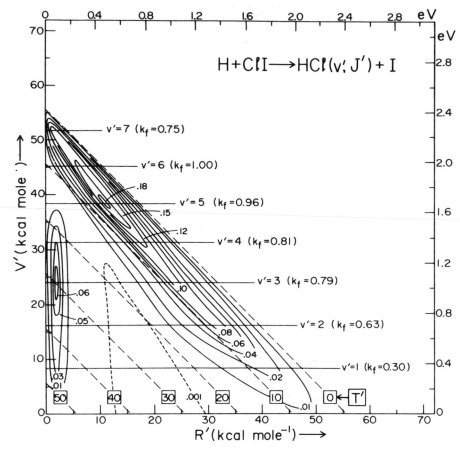

Figure 12. Triangle plot for the reaction H+ClI at 300K showing bimodality (18 % of the HCl is in the low -J' mode, 82 % in the high -J'). From M.A. Nazar, J.C. Polanyi, and W.J. Skrlac, Chem. Phys. Letts. *29*, 473 (1974). See also J.C. Polanyi and W.J. Skrlac, Chem. Phys. *23*, 167 (1977).

of k(v'=3)) was taken to be evidence of a quantum effect at the threshold energy for formation of HF(v'=3). Molecular beam experiments in Y.T. Lee's laboratory have shown the importance of quantum effects on the dynamics of the first member of this family of reactions [42].

Figure 12 shows the dramatic influence the atom C can have on the dynamics of a reaction A+BC → AB+C. Superficially the reaction illustrated resembles H+ClY→ HCl+Y, for which the detailed rate constant k(V',R',T') is given in fig. 6; the difference is that the second halogen atom, Y, has been changed from Y=Cl to Y=I, with the effect on k(V',R',T'), recorded in fig. 12. Bimodal product energy distributions were also observed in a number of other studies [45−48] and were regarded as indicative of the existence of two types of **reaction dynamics leading to the identical product** (HCl in the case illustrated). **This phenomenon was termed 'microscopic' branching,** to distinguish it from the 'macroscopic' branching leading to chemically different products (see

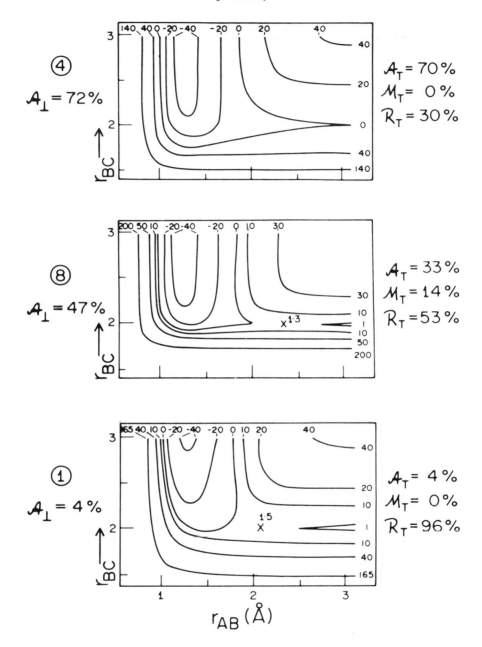

Figure 13. Three potential-energy surfaces (pes) illustrative of the changes in potential energy as reagents A+BC (lower right) pass through collinear transition states A−B−C to form products AB+C (upper left). Contour energies are in kcal/mole; all three reactions have the same exothermicity. The pes, designated 1, 8, and 4, are arranged vertically in order of increasing "attractiveness". The pes are indexed at the left according to the percentage attractive energy-release taken from a perpendicular path across the surface ($A\perp$ = 4, 47, and 72 % for surfaces 1, 8, and 4). The same pes are described at the right according to the percentage attractive, mixed and repulsive energy release, A_T, M_T, and R_T, along a single collinear trajectory; this designation depends upon mass combination, which was L+HH (L = 1 amu, H = 80 amu). The barrier crest, designated x, shifts to "earlier" locations along r_{AB} as the surfaces become more attractive. Excerpted from P.J. Kuntz et al., *J. Chem. Phys. 44*, 1168 (1966).

for example fig. 11). As we shall indicate in section 2.2, these two types of branching are linked.

It is interesting to consider the strengths and weaknesses of ir spectroscopy as a tool in the study of reaction dynamics. The major limitations have stemmed from the low transition probabilities in the ir, the insensitivity of ir detectors, and the lack of a sufficiently precise timescale against which to measure the vibrational and especially rotational relaxation. Substantial improvements in sensitivity have been achieved using the important new methods of laser-induced fluorescence (LIF) [48] and resonantly enhanced multiphoton ionization (REMPI) [49]. However, since these make use of electronic transitions, they cannot so readily scan the full manifold of product vibrotational energy states to yield data as complete as that exemplified in figs. 7–9, which remains the most complete available.

Tunable ir lasers, combined with adequate time-resolution, could give a new lease of life to the vibrotational spectroscopy of reaction products. A promising development employing ir chemiluminescence is the introduction of time-resolved fourier transform spectroscopy following the pulsed photolytic initiation of exchange reaction [50].

2.2 *Theory*

The origin of these theoretical studies is outlined in section 1, above. In the early 1960's two groups [26, 27] became interested in using the computer-based classical trajectory method [25] (often called the Monte Carlo approach) to explore, for the first time, a suggestion made over two decades earlier [28], that energy released as reagents approach is responsible for vibration in the newly-formed bond.

Figure 13 shows three potential-energy surfaces (pes), all of a modified London, Eyring, Polanyi, Sato (LEPS) variety [27, 52, 53], selected from a wider group examined in ref. [53]. The energy released as the reagents approached (termed the attractive energy release, and defined as $A_\perp$ or A_T [53, 54]; see the caption of fig. 13) increases as one moves upwards in the figure. There is, in related families of reaction, a concurrent diminution in the height of the energy-barrier, and a shift of the barrier to earlier locations along the coordinate of approach with increased A. It has proved possible to give an empirical expression to these correlations [55, 56].

In fig. 14 it can be seen that for a group of 8 pes moderate percentages of energy release along the coordinate of approach (% $A\perp$) are converted quantitatively into product vibration for a particular reagent mass combination (**L+HH**; see caption). For high % $A\perp$ the reaction moves into a new regime in which product vibration decreases with increased attractive energy release. Examination of the trajectories indicates that a transition is occurring from predominantly *direct* reaction (products once they start to separate continue to do so [57]) to *indirect* reaction in which subsequent "clutching" and "clouting" secondary encounters drain the incipient vibration out of the new bond into rotation and translation [58, 59]. What we are seeing is the onset of statistical

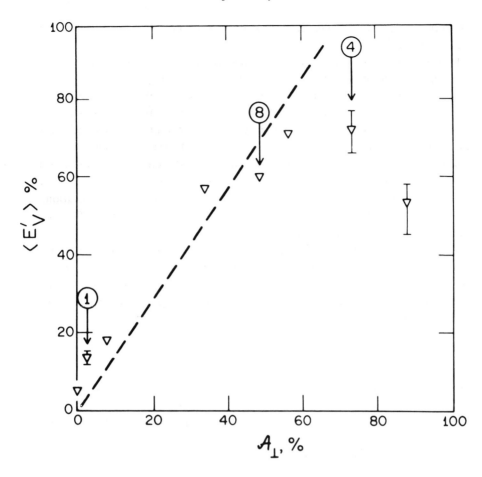

Figure 14. Mean vibrational excitation in the products of 3D L+HH reaction on 8 potential energy hypersurfaces (pes) plotted against the perpendicular attractive energy-release, $A\perp$. The collinear cuts through pes 1, 8, and 4 are to be found in the previous figure. The fall-off in $<E'_v>$ at high $A\perp$ is due to the prevalence of secondary encounters (see text). From P.J. Kuntz et al., J. Chem. Phys. *44*, 1168 (1966).

behaviour, in the limit of which product vibration will receive no more than its modest equipartition allocation.

At the other end of the scale in fig. 14, where repulsive energy-release predominates, the low vibrational excitation shown in the figure is by no means the rule; the observed vibrational excitation is trongly dependant on the chosen mass-combination. The **L+HH** mass combination shown is in fact anomalous in fulfilling the initial expectation that repulsive energy-rclcase R (energy released along the coordinate of separation of the pes) will be inefficiently channelled into vibration. For **L+HH** the attacking atom A approaches BC rapidly up to the normal bonding separation before the repulsion is released (dynamics in which r_{AB} decreases to its equilibrium separation, r^0_{AB}, and only then does r_{BC} increase, correspond on the pes to a rectilinear path). Since the repulsion operates on an already existent AB bond, vibrational excitation is indeed inefficient (left end of fig. 14). This behaviour is sufficiently atypical to be described as the "light-atom anomaly" [53].

For the more general case of a heavier attacking atom relative to the molecule under attack, the new bond is still extendend ($r_{AB} > r^0_{AB}$) at the time that the bulk of the repulsion is released; we can symbolise this as $A- -B \cdot C$. In this case repulsion between the separating atoms B and C results in recoil of B, rather than of AB as a whole; i.e. it gives rise to efficient internal excitation of the new bond, AB.

Since the solution of the collinear equations of motion can be pictured in terms of the motion of a sliding mass across a suitably scaled and skewed representation of the pes, it is instructive to consider the characteristic path across the pes in the latter more general case. The fact that r_{AB} decreases at the same time as r_{BC} increases (we term the energy-released in this phase, "mixed"; M) means that the sliding mass, rather than following a rectilinear path, is cutting the corner of the pes. Instead, therefore, of approaching the exit valley from its head, it approaches the valley from the side; consequently it oscillates from side to side of the exit valley indicating that the new bond is vibrationally-excited.

The proportion of the energy released during corner-cutting is clearly relevant to the efficiency with which the repulsive energy is channelled into vibration. We have determined the extent of mixed energy-release, M_T, from a specimen collinear trajectory employing the appropriate mass-combination and the pes in question [10, 53, 60]. Figure 13 shows that % M_T is insignificant for **L+HH**. For other mass-combinations it is so great that the points on a plot of the type shown in fig. 14 begin at substantial $< E'_V>$, and the region of linearity prior to the fall-off in $< E'_V >$ due to the onset of indirect dynamics becomes a very short region (see the case of the heavy attacking atom in fig. 7 of [53]).

Turning to the experimental findings of 2.1, it was evident that the moderate mean fractional conversion of reaction energy into product vibration, $< f'_V >$, could be explained either by a highly attractive interaction leading to secondary encounters, or by a predominantly repulsive energy-release. The first of these alternatives appeared implausible since it implied a substantially broader distribution over product vibrational and also rotational states than was observed [16]. Instead the evidence favoured a strongly repulsive pes, with the light-atom anomaly explaining the markedly reduced $< f'_V >$ for $H+X_2$ as compared with $X+H_2$ or $X+HY$, and the lower barrier leading to a slightly increased $A_\perp$ in $H+Br_2$ as compared with $H+Cl_2$ (cf. the correlation noted above) accounting for the greater $< f'_V >$ observed for $H+Br_2$ than for $H+Cl_2$ [16].

The success of a strongly repulsive pes in accounting for the general form of the triangle plot for $Cl+HI \rightarrow HCl + I$ is illustrated, by way of an example, in fig. 15 [61], which should be compared with fig. 9 above. Similar success has been obtained for the reaction $F+H_2 \rightarrow HF+H$ (fig. 8) above using a strongly

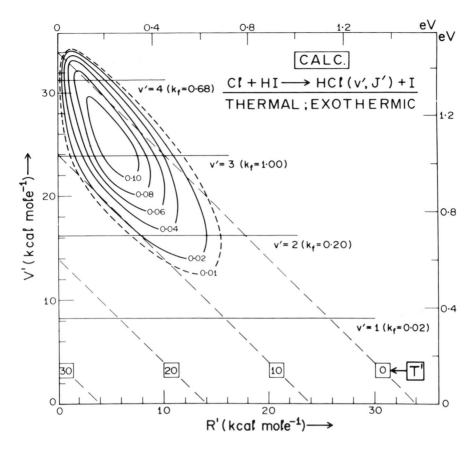

Figure 15. Computed triangle plot for the reaction Cl+HI at thermal collision energies (300K). The 3D classical trajectory method was applied to a highly repulsive pes (proposed by Anlauf et al., J. Chem. Phys. *49*, 5189 (1968)). From C.A. Parr et al., J. Chem. Phys. *58*, 5 (1973).

repulsive pes [62–64]. In this case there is now dependable evidence from *ab initio* variational treatments of FHH [65, 66] that the energy-release is indeed substantially repulsive. The success of a repulsive pes in describing **L+HH** dynamics will be demonstrated (for $H+F_2 \rightarrow HF+F$; cf. fig. 7 above) in section 3.2 below.

The discussion of the previous paragraphs as it relates to repulsive energy-release is summarized pictorially in fig. 16. Though visualisation of reaction dynamics is generally, and often adequately, based on the collinear pes to which the sliding-mass analysis applies, tests of the validity of pes' are made by 3D trajectories.

Product rotational excitation is eliminated from the picture in the collinear world. Though this is a minor constituent of the product energy, it is revealing of the dynamics. In the visualisation of fig. 16, we have included the effect of

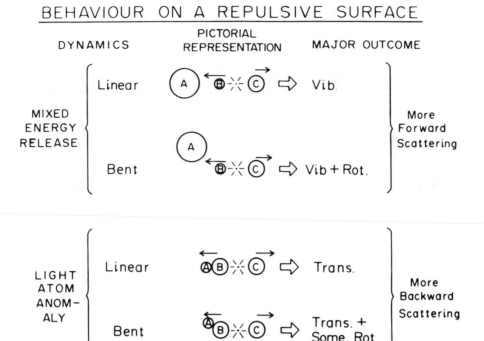

Figure 16. Pictorial representation of contrasting types of reaction dynamics on a repulsive energy surface. Type (a), "mixed energy release", is commonly observed. Type (b), termed "the light-atom anomaly", is observed if the attacking atom is very light. From J. C. Polanyi, Accounts Chem. Res *5*, 161 (1972).

repulsive energy release in bent configurations as one source of product rotation. The experimental data in the triangle plots of fig. 11 give persuasive evidence of the significance of this effect in the important reaction $F+H_2$. Product repulsion would be expected [47, 67] on the basis of momentum conservation to give rise to decreasing $<f'_R>$ in the series $FH \cdot D > FH \cdot H > FD \cdot D > FD \cdot H$ (the dot indicated the locus of the repulsion); this is found theoretically, in 3D trajectory studies, and also experimentally. The triangle plots of fig. 11 correspond to the extremes of this range of isotopic mass-combinations; $<f'_R>$ for the $FH \cdot D$ pathway substantially exceeds that for $FD \cdot H$.

Qualitative pictures of the type given in fig. 16 suggest simple models of the reactive event. Several such models are indicated in fig. 17, all of which stress the role of the forces operating along the coordinate of separation of the pes. The implications of each of these models are discussed in ref. [59]. Despite their crudity simple models have the important virtue that their "moving parts" are open to inspection. Model (c), called the Simple Harmonic model, has been

PRODUCT − FORCE MODELS

$$A + BC \longrightarrow \underline{\underline{A\text{-}\text{-}B\!\times\!C}} \longrightarrow AB + C$$

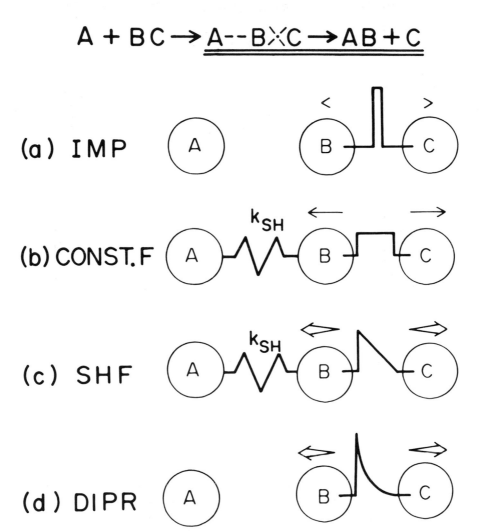

(a) IMP

(b) CONST. F

(c) SHF

(d) DIPR

(e) FOTO

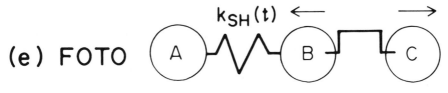

Figure 17. Five types of simple "product-force models", that interpret product energy-distributions in terms of the mode of relaxation of ABC in its retreat from the activated state→AB+C. A repulsive force is assumed to be located between the separating atoms B·C. Various assumptions are made regarding the bond A−B which is forming. Model (a) is the Impulsive model, (b) is the Constant Force model, (c) is the Simple Harmonic model, (d) is the DIPR (Direct Interaction with Product Repulsion) model, and (e) is the FOTO (Forced Oscillation in a Tightening Oscillator) model. From J.C. Polanyi, Faraday Disc. Chem. Soc., *55*, 389 (1973).

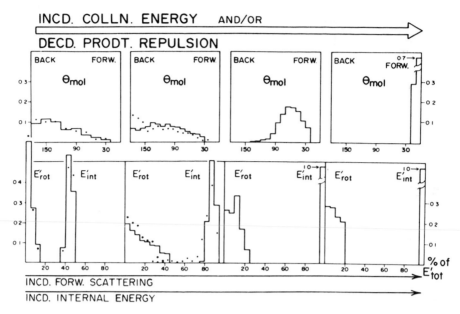

Figure 18. Effect of repulsive energy release on product angular distribution (top row) and energy distribution (bottom row; E'_{yot} is product rotational excitation and E'_{int} is vibration plus rotation). The solid lines show the results of 3D (three-dimensional) DIPR model calculations (masses $K+Br_2$, energy-release $E'_{\text{tot}} = 53$ kcal/mole). The total product repulsion used in the DIPR model calculations was decreased in four stages, from left to right in the figure, simulating decreased % *R*, implying increased % *A*. According to the DIPR model this is mathematically equivalent to increasing the reagent collision energy, hence the dual label of the arrow at the top. The *consequences* of these changes are summarized, qualitatively, by the arrows at the foot of the figure. The dots record the distributions obtained from 3D trajectories on potential-energy hypersurfaces with comparable partitioning of the energy-release. Adapted from P.J. Kuntz et al., J. Chem. Phys. *50*, 4623 (1969).

used to improve our understanding of the mechanism of mixed energy-release, in which a repulsive force operating between particles B and C drives an oscillator joining A to B that is under tension [60]. The model shows that for a wide range of conditions mixed energy release can play a significant role in channelling product repulsion into vibration.

Model (d), the DIPR (Direct Interaction with Product Repulsion) model [68], sidesteps consideration of forces in the new bond by obtaining the product vibration, V', from $E'_{\text{tot}} - (R'+T')$. From the direction of recoil of atom C relative to the direction of approach of A, angular distributions can readily be generated in 3D. The model provides a simple means to expose the link between the angular distribution of reaction products and their internal excitation, as well as the connection between these two quantities and the collision energy.

Findings for the DIPR model are summarised in fig. 18 (solid lines). Reading from left to right the interactions become less repulsive (more attractive). This

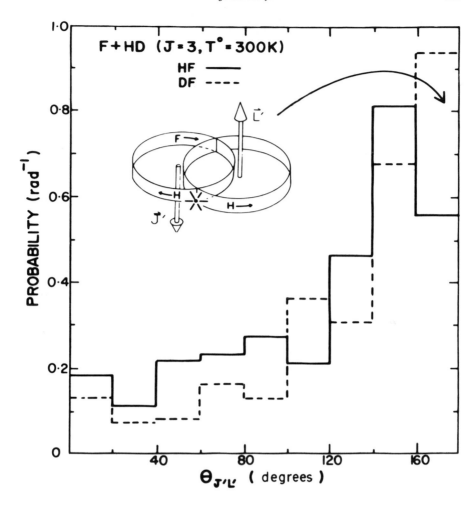

Figure 19. Computed distribution, $\theta_{J'L'}$, of the angle between the product rotational angular momentum, J', and orbital angular momentum, L', following repulsive energy-release between the products of the reaction $F + HD \rightarrow HF + D$ (solid line) and $\rightarrow DF + H$ (broken line). At the extreme left $\theta_{J'L'} = 0$ corresponds to J' parallel to L', while at the extreme right $\theta_{J'L'} = 180°$ corresponds to the favoured outcome of J' anti-parallel to L'.

is achieved by systematically decreasing the repulsive impulse (the time integral of the force, $F(t)$, between B and C) which alone governs the dynamics. Since a measure of $\int F(t) \, dt$ can be obtained by a variety of means for any actual reactive pes [59, 68], the model can be applied to cases for which the results of 3D trajectory computation are known. The agreement—see the points for the more repulsive cases in fig. 18—is excellent, so long as the assumption of direct interaction remains valid.

The lessons that can be learned from fig. 18 are that strongly repulsive energy-release will tend to give backward scattering of the molecular product

MICROSCOPIC BRANCHING

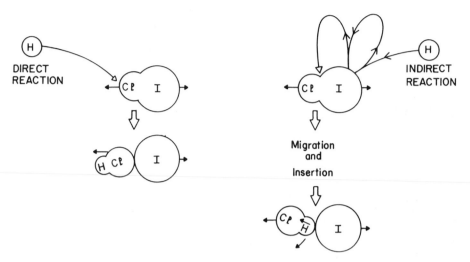

Figure 20. Pictorial representation of the alternative dynamics—direct and indirect—in the microscopic branching underlying the bimodality of product energy distribution exemplified in fig. 12. (Based on a classical trajectory study; J.C. Polanyi et al., Faraday Disc. Chem. Soc. *67*, 66 (1979)).

(θ_{mol} is substantial at 180°) coupled with moderate internal excitation, E'_{int}. As the repulsion is decreased θ_{mol} shifts forward, and E'_{int} increases. The same effect can be achieved by increasing the reactant collision energy, since this too has the effect of decreasing the repulsive impulse $\int F(t)dt$.

Herschbach and co-workers have injected physical content into the DIPR model in their DIPR-DIP extension, in which the repulsive impulse is assumed to be "Distributed as In Photodissociation" [69]. The model then accounts nicely for product distributions observed in a number of reactions. More recently Zare's laboratory has shown how the DIPR approach can be used to understand the plane of rotation of new-born reaction products [70].

The FOTO (Forced Oscillation in a Tightening Oscillator) model [71], item (e) of fig. 17, gives a fuller rendition of the forces on a collinear pes; the B·C repulsion operates on an AB oscillator whose force constant is in the process of increasing, and whose equilibrium separation, r^0_{AB}, is decreasing. The model, which has been applied to ten reactions for which experimental or theoretical data exist, is sufficiently complete to embody analogues of the "attractive", "mixed", and "repulsive" phases of energy-release.

To the extent that the rotation in a newly-formed reaction product AB originates in repulsion AB·C, the rotational motion should be coplanar with the repulsive force, and B in AB should recoil away from C. This implies that the angle between the product rotational angular momentum vector, J', and the product orbital angular momentum vector, L', should be $\theta_{J'L'} \approx 180°$.

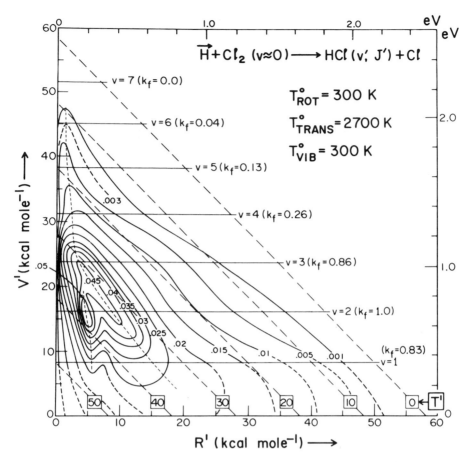

Figure 21. Effect of enhanced reagent collision energy on the product energy distribution, i. e. experimental k(V',R',T'), for H + Cl$_2$(v = 0) → HCI(v'J') + Cl. Mean collision energy, <T> ≈ 10.6 kcal/mole. Compare k(V',R',T') for room temperature reaction, in fig. 6. From A.M. Ding et al., Faraday Disc. Chem. Soc., *55*, 252 (1973).

Analyses of product distributions from 3D trajectory studies [64,72] bear this out. Experimental studies of correlations between such vector attributes will add materially to the mosaic from which our picture of reaction dynamics is composed.

The richer the detail in the rate constants k(V', R', T'), the clearer the message. The striking bimodality in the product energy-distribution from H + ClI ↦ HCI(v', J') + I recorded in fig. 12 can be ascribed to "microscopic branching"; two patterns of molecular dynamics result in the formation of the same reaction product (HCl in this example). The dynamics have been explored in 3D. They are shown schematically in fig 20. In the case of H + ClI the migratory *microscopic* pathway (at the right of fig. 20) is thought to dominate. This is for the same reason that *macroscopic* branching favours formation of HI rather than HCl, namely the existence of a smaller energy barrier for approach from the I end of ICl (see [73] for the greater stability of ClIH than HClI). In collisions of H with the I end of ICl that fails to yield HI, the H atom can migrate to the Cl end of the molecule to yield highly internally-excited HCl. The HCl product with low internal excitation comes from H that have reacted directly at the Cl end of ClI. Since the barrier to this mode of reaction is higher, the yield (at normal reagent energies) is lower. It is evident that the relative yields of HCl and HI in *macroscopic* branching are linked to the yields of low E'_{int} and high E'_{int} HCl by way of *microscopic* branching.

Microscopic branching is thought to constitute more than merely an interesting curiosity, since it can contribute to the dynamics in any case where macroscopic branching is possible—a large class of reactions. As would be expected, the importance of microscopic branching, in common with macroscopic branching, depends sensitively on the reagent energy [49, 50, 74].

3. CHANNELLING REAGENT ENERGY INTO PRODUCTS

3.1 *Experiment*

In the experiments that will now be briefly described reagent translational, vibrational, and rotational energies were altered by simple means in order to explore the effect on the product energy-distribution. To increase the collision energy we formed the atomic reagent (Cl for Cl + HI [75], F for F + HCl [50] or F + H$_2$ [50,67], and H for H + Cl$_2$ [50]) by pyrolysis, rather than using room temperature atoms as in the work described in the previous section. The effect is shown for H + Cl$_2$ in the triangle plot of fig. 21. Comparison with the earlier 300K triangle plot (fig. 6) shows marked changes. The *additional* reagent translation, $\triangle T$, in excess of the barrier to reaction, has been channelled principally into additional product translation.

A second repository for this additional reagent energy was product rotation. (It was interesting to observe in both Cl + HI [75] and H + Cl$_2$ [50] that this enhanced product rotation was associated with the appearance of a second maximum in the product rotational distribution, $k(J')_{v,}$, as if enhanced collision energy had led to *induced microscopic branching*). The general finding regarding the disposition of additional reagent translation could be summarised as,

$$\triangle T \rightarrow \triangle T' + \triangle R' \tag{4}$$

This describes the average behaviour; the breadth of the product energy distribution increased markedly over V', R', and T'.

By heating the molecular species under attack and recording the change in product energy distribution (see fig. 22), we found evidence of a similiar adiabaticity, on the average, in regard to the conversion of reagent vibration into product excitation,

$$\triangle V \rightarrow \triangle V' \tag{5}$$

where the symbol $\triangle$ denotes energy in excess of that required for barrier crossing. As before, the breadths of the product energy distributions increased, that over V' appearing to be bimodal.

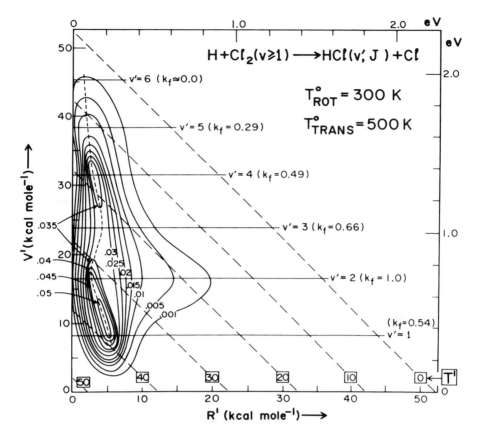

Figure 22. Effect of enhanced reagent internal energy on the product energy distribution, i. e., experimental k(V′,R′,T′), for H + Cl$_2$(v≥1) → HCl(v′,J′) + Cl. This result is approximate since it was obtained by subtracting the detailed rate constants for H + Cl$_2$(v ≈ 0) → HCl + Cl from H + Cl$_2$(v≥0) → HCl + Cl. Mean vibrational energy, <V> ≈ 3.3 kcal/mole. Compare with fig. 6. From A.M.G. Ding et al, Faraday Disc. Chem. Soc. *55*, 252 (1973).

A more stringent test of this tendency toward vibrational adiabaticity was obtained using the method of ir chemiluminescence depletion (CD; section 4.1 below) and ir emission. Vibrationally excited reagent, OH(v), was formed in a selected "pre-reaction". Pulses of a further reagent—Cl in the present case— were introduced, and a record was made of the depletion of chemiluminescence (at the pulsing frequency) in OH(v), as well as of the appearance of chemiluminescence in reaction product HCl(v′). The reaction responsible for both emissions was

$$Cl + OH(v) \rightarrow HCl(v') + 0 \qquad (6)$$

Using two different pre-reactions (see caption fig. 23) the reagent vibrational energy could be given mean values of <V> = 72.1 kcal/mole or 20.5 kcal/ mole. For these high vibrational energies over 90 % of the reagent vibration

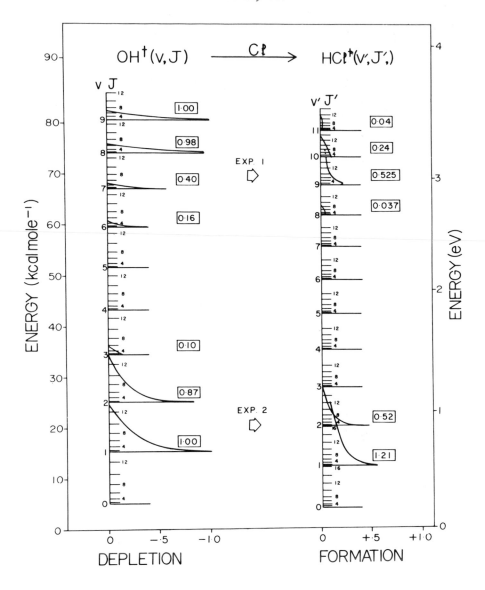

Figure 23. Experimental evidence for the conversion of high reagent vibrational excitation into the corresponding degree of freedom of the product in the reaction $Cl + OH(v,J) \rightarrow HCl(v'J') + 0$. The curves at the left show the *decrease* in population of each OH vibrotational level resulting from the introduction of Cl atoms; the curves at the right were taken from a concurrent record of the amount of HCl formed in each vibrational level. In experiment 1 (top of figure), the OH was formed in $v = 6-9$ by the pre-reaction $H + O_3$. In experiment 2 (lower part of figure), the OH was formed in $v = 1-3$ by the pre-reaction $H + NO_2$. The numbers in the boxes at the left give the relative fractional depletions of each OH v-level (summed over J), and those at the right give the corresponding information for the product HCl. From B.A. Blackwell et al., Chem. Phys. *24*, 25 (1977).

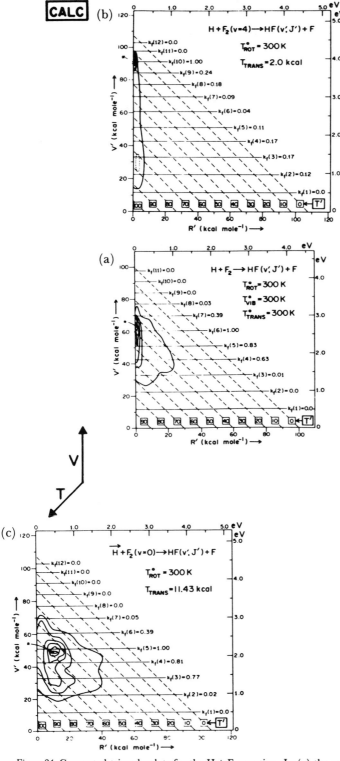

Figure 24. Computed triangle plots for the H + F₂ reaction. In (a) the reagents were Monte Carlo selected from a 300K Boltzmann distribution. In (b) the F₂ reagent vibrational excitation, V, was increased to v = 4 (reagent collision energy 2.0 kcal/mole), and in (c) the reagent relative translational energy, T, was increased to 11.43 kcal/mole (reagent vibrational energy corresponded to v = 0). From J.C. Polanyi et al., *Chem. Phys*, *9*, 403 (1975).

was found to be channelled into product vibration. Once again $\triangle V \rightarrow \triangle V'$ [76].

Evidence regarding the fate of reagent rotation was more fragmentary, coming from studies of the reaction $F + H_2(J)$ with J varied systematically from $0 \rightarrow 1 \rightarrow 2$ [77]. A small but significant decrease was observed in the fraction of the total energy entering product vibration when $J = 0 \rightarrow 1$, followed by an increase in this fraction as $J = 1 \rightarrow 2$. This finding was in accord with an earlier less detailed study by Coombe and Pimentel [78]. As is often the case at present with the effects of reagent rotation (see 4.2 below), detailed understanding is lacking.

3.2 *Theory*

The nature and the origins of the approximate adiabaticity relations (4) and (5) have been discussed in a number of places [49,50,54,61,67,79−82]. The success of the 3D classical trajectory method in describing the types of change in detailed rate constant observed experimentally (see figs. 21 and 22) can be judged from fig. 24 [81]. This figure shows a computed triangle plot for the 300K $H + F_2 \rightarrow HF(v'J') + F$ reaction at the centre (see fig. 7 for the experimental counterpart), and the predictions for enhanced reagent translation (ref. [82] fig. 3 has the experimental counterpart) and vibration, below and above respectively. It is evident from inspection of fig. 24 that $\triangle T \rightarrow \triangle T' + \triangle R'$, and $\triangle V \rightarrow \triangle V'$. In addition the increased breadths of product energy distributions over T' and R' in the former case, and over V' in the latter are apparent, as also is the bimodality over V' resulting from enhanced V. (See [50] for comparable trajectory studies using parameters chosen for $H + Cl_2$).

The origin of this adiabaticity is evident from an inspection of trajectories with and without the additional reagent energy. The effect of enhanced translation is to shift the characteristic pathway across the collinear pes (used as a diagnostic of 3D behaviour) toward more-compressed configurations−the sliding mass impelled by its enhanced momentum along r_{AB} caroms into the corner of the pes in the region of ① in fig. 25. The compressed intermediate $A \cdot B \cdot C$ if collinear then flies apart to give translation in $AB + C$, and if bent to give enhanced rotation in AB. For enhanced reagent vibration the most common paths to product are shifted to the region ② in fig. 25; the effect is to favour a more stretched intermediate $A-B-C$ that pulls together along r_{AB} to give $AB(v \gg 0) + C$. The opposite phase of vibration to that pictured in fig. 25, drives the intermediate into region ① giving rise to the low V' peak in the product distribution. Since energy released along r_{BC} has been characterised as repulsive and that along r_{AB} as attractive, we have termed these changes in dynamics *induced repulsive energy-release* in the first case, and *induced attractive energy-release* in the second. For a strongly exothermic reaction such as $H + F_2$ (fig. 24), these shifts affect not only the channelling of the additional reagent energy, but also the channelling of the exothermicity into product excitation, so that $(\triangle T' + \triangle R')/\triangle T$ and $\triangle V'/\triangle V$ exceed unity [81].

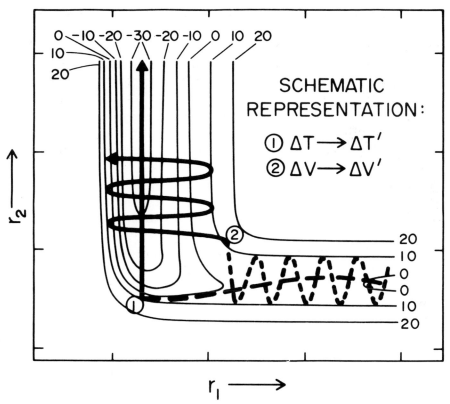

Figure 25. Schematic representation (on a typical collinear exothermic pes) of the mechanism (1) by which "additional" reagent translation ($\Delta T \gg E_c$, where E_c is the barrier height) is channelled into "additional" product translation ($\Delta T'$), and also (2) the mechanism by which "additional" reagent vibration ($\Delta V \gg E_c$) becomes "additional" product vibration ($\Delta V'$). (These are not actual trajectories since they show neither the exothermic energy-release nor the effect on that energy-release of changing reagent energy; the intention, which is artificial, is to show the effect of ΔT and ΔV in isolation). From A.M.G. Ding et al., Faraday Disc. Chem. Soc. *55, 252* (1973).

4. SURMOUNTING THE ENERGY BARRIER

4.1 *Experiment*

The most detailed information regarding the relative efficiency of various types of reagent motions in surmounting an energy barrier, comes from the application of microscopic reversibility to detailed rate constants for forward reaction, $k_f(V',R',T')$[83]. Detailed rate constants for the reverse reaction obtained in this way, $k_r(V',R',T')$, are recorded in the triangle plot of fig. 26, where V',R',T', are now the vibrational, rotational and translational energies of the *reagents* for the endothermic reaction $HF(v'J') + H \rightarrow F + H_2$[84]. The total energy available for distribution is fixed by the nature of the experiment, being equal to the energy made available by the forward exothermic reaction; 34.7 kcal/mole in the case illustrated. This approach has been tested numerically by applying it to a case where $k_r(V',R',T')$ could be obtained directly from **3D**

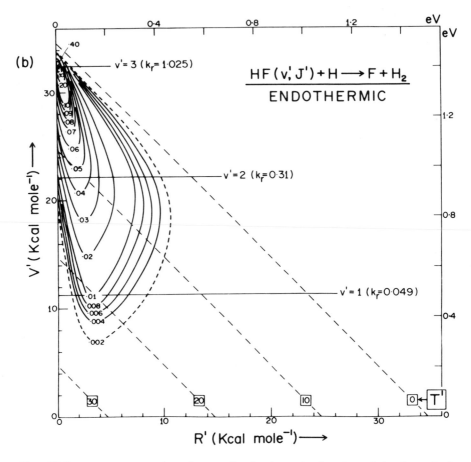

Figure 26. Contours represent values of *reagent* vibrational, rotational, and translational energies for the endothermic reaction $HF(v',J') + H \rightarrow F + H_2$ which correspond to equal $k_r(\equiv k_{endo})$. The data were obtained by application of microscopic reversibility to the $k_f(\equiv k_{exo})$ given in fig. 8. From J.C. Polanyi and D.C Tardy, J. Chem. Phys. *51*, 5717 (1969).

trajectories or by application of microscopic reversibility; the procedures agreed with one another to approx. 10% [85].

Inspection of fig. 26 indicates that a redistribution of 30 kcal/mole from T′ into V′ has the dramatic effect of increasing by $\sim 10^3 x$ the rate of reaction in the endothermic direction [58,84]. Endothermic triangle plots of this type are to be found in, for example, references [67,83,84]. In a few cases, in other laboratories, direct measurement has been made of the relative efficiency of vibration and translation in surmounting an energy barrier ([86−88]); in one of these cases [88] the barrier corresponded to "substantially" [67] endothermic reaction: the reaction was $HF(v') + K \rightarrow H + KF$ (17 kcal/mole endothermic) and the effect of transferring 11 kcal/mole from reagent translation to vibration was to give $> 10^2 x$ increase in reaction rate.

The infrared "chemiluminescence depletion" (CD) method, already alluded

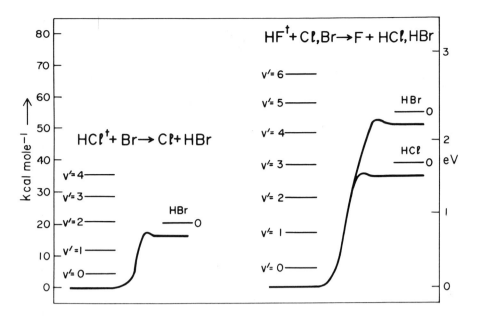

Figure 27. Endothermicities of three reactions HX + Y → X + HY; showing kcal/mole at left, eV at right, and the vibrational spacing of the relevant HX within the figure. Energy profiles are schematic. From D.J. Douglas et al., Chem. Phys. *13*, 15 (1976).

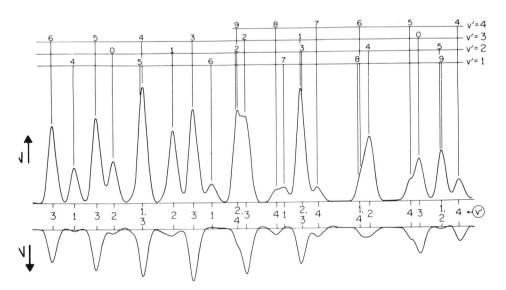

Figure 28. A portion (one fifth) of the spectrum recorded in an ir chemiluminescence depletion (CD) study of the endothermic reaction HCl(v' = 1−4) + Br → Cl + HBr. At the top (labelled N) is the emission from the HCl(v' = 1−4) formed in a pre-reaction. Below (△N) is the concurrent record of the depletion, due to pulses of Br (trace enlarged 2.5x). The fraction △N(v')/N(v')>0 for v'⩾2. (Excursions of △N above the baseline can be identified with product emission (HBr(v = 1,2) rather than reagent depletion.) From D.J. Douglas et al., J. Chem. Phys., *59*, 6679 (1973).

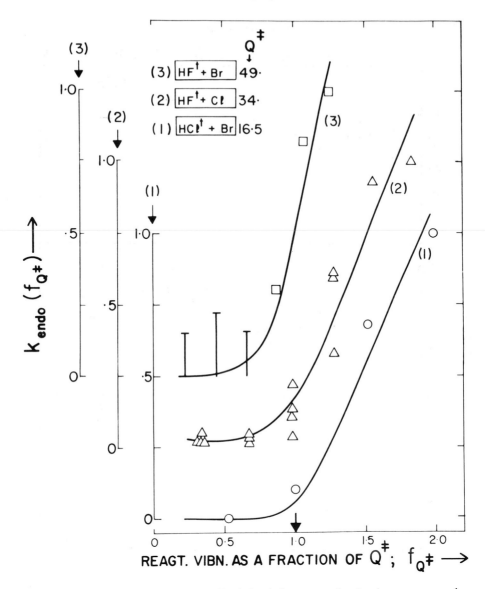

Figure 29. Endothermic rate constant $k_{endo}(f_{Q^{\ddagger}})$; $f_{Q^{\ddagger}}$ is the reagent vibrational energy expressed as a fraction of the endothermicity $Q^{\ddagger}$ for reactions (1)−(3). From D.J. Douglas et al., Chem. Phys. *13*, 15 (1976).

to in section 3.1, was used to obtain a measure of the efficiency of a high degree of vibrational excitation in promoting endothermic reaction. In contrast to the microscopic reversibility procedure outlined above, by CD the vibrational energy could be varied independently of the collision energy and could be made much larger than the endothermic barrier height. The energies of one family of endothermic reactions investigated in this fashion is shown in fig. 27; HX† + Y → X + HY, where X = Cl or F, Y = Br or Cl and the dagger indi-

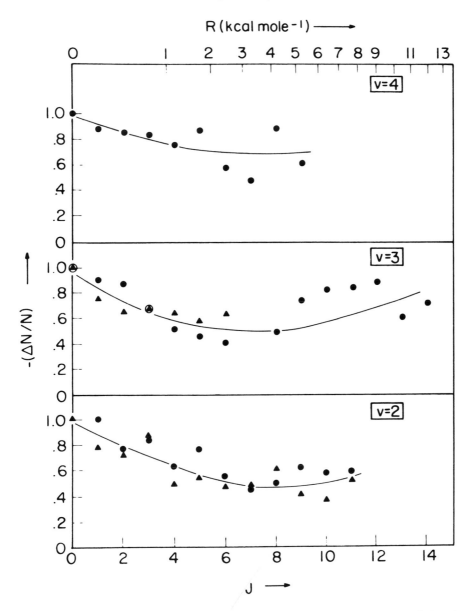

Figure 30. Fractional depletions of individual J states of v = 2−4 in a CD study of the endothermic reaction HF† + Na. From B.A. Blackwell et al., Chem. Phys. *30*, 299 (1978).

cates vibrational excitation [89]. An analogous study has been made of the endothermic reactions HX† + Na → H + NaX, X = Cl,F[90]. For CD experiments, the vibrationally excited reagent HX† was formed in a variety of pre-reactions or in a variety of partially-relaxed vibrational distributions from single prereaction. The atomic reagent, Y or Na, was pulsed and a concurrent record was made of the ir chemiluminescence from HX† (cw) and of the

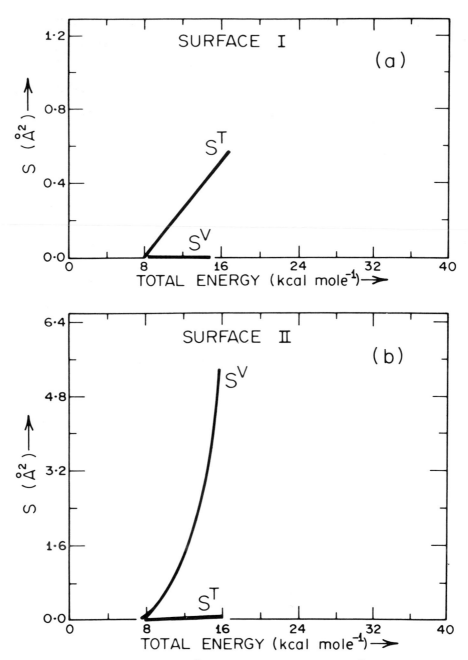

Figure 32. (a) Reactive cross-sections S^T (reagent energy in translation) S^V (reagent energy in vibration, + 1.5 kcal/mole of translation) on the "early barrier" surface depicted at the left of fig. 31. (b) Reactive cross-sections S^T and S^V on the "late barrier" surface shown at the right of fig. 31. Three equal masses. From J.C. Polanyi and W.H. Wong, J. Chem. Phys., *51*, 1439 (1969).

to apply to surface II, fig. 32; reagent vibrational energy (with a small amount of translation to bring the reagents together) was highly effective in giving reaction, whereas translation was without effect.

As we have remarked, these correlations are of more than empirical value; by focussing attention on certain simple attributes of the pes, we can shed light upon dominant features of the dynamics. In terms of the dynamics of the reaction A + BC the correlations noted in the preceding paragraph reduce to the proposition that a barrier predominantly along the coordinate of approach, r_{AB}, is best traversed by motion in that same coordinate, namely reagent translation, whereas a barrier predominantly along the coordinate of separation, r_{BC}, is best traversed by motion in that coordinate, i.e., vibration in the bond under attack. The trajectories in fig. 31 (obtained by solution of the equatons of motion, but capable of being pictured as the motion of a sliding mass across the pes) illustrate this; the reactive cases are (a) and (c) (early and late barriers), the nonreactive are (b) and (d) (early and late barriers once more). The effect illustrated is sufficiently striking that it may also have relevance to the dynamics of reactions involving many atoms (cf. [99]), provided that these occur by a direct pathway.

The interest of these correlations is greatly increased if we can link the form of the pes to the nature of the chemical reaction. Though unrecognised by reaction dynamicists, Hammond's postulate [100] that for endothermic reactions the transition state resembles reaction products might have pointed the way to the desired connection.

Instead a systematic study of a considerable number of exchange reactions A + BC → AB + C in terms of both the London, Eyring, Polanyi, Sato (LEPS), and the Bond-Energy Bond-Order (BEBO) methods led to the conclusion that for substantially exothermic reactions the barrier-crest is located in the entrance valley ("early barrier") of the pes, and for substantially endothermic reactions it is in the exit valley ("late barrier") [55]. From this, and the dynamical study of early and late barriers, it followed (as previously surmised [101]) that the favoured degree of freedom for barrier-crossing in substantially exothermic reactions would be translation, whereas reagent vibration would be most effective in giving rise to substantially endothermic reaction. (In the BEBO approximation "substantially" has the meaning ≥ 10 kcal/mole [67]).

Behaviour of the type illustrated in fig. 32(a), computed for the well-studied exothermic reaction F + H_2 → HF + H is shown in ref. [64]. Behaviour of the type illustrate in fig. 32(b) is in accord with the evidence of section 4.1 above. There is, nonetheless, a clear need for comparisons of what we have termed (fig. 32) S^T with S^V for a range of reactions, and within related series of reactions.

If barrier location is to be made to any degree a quantitative index of the relative efficacy of translation and vibration in bring about barrier-crossing, the location of the barrier must be considered as it appears on a surface skewed and scaled for the appropriate masses m_A, m_B, and m_C, so that a sliding mass will correctly solve the equations of (collinear) motion [98]. The mass combination **H + HL** accentuates surface I behaviour while diminishing the surface II effect, whereas **L + HH** has the converse effect. The qualitative behaviour noted above remains unaltered.

A further significant refinement in understanding the controlling features of the pes takes note of the fact that in the case of a late barrier-crest some part of the energy-barrier is located along the coordinate of approach. This further element was included in the picture by distinguishing between type "IIS" and "IIG" barriers, where S and G differentiate late-barriers approached by a Sudden or Gradual potential rise [97]. The gradual rise implies the existence of a significant fraction of the type II barrier in the coordinate of approach, with a corresponding requirement for a minimum reagent translational energy along with the (dominant) vibrational contribution.

A further dividend that accrues from the inclusion of this refinement is that it has the effect of tying in the considerations raised in this section regarding the favoured type of reagent energy for endothermic barrier crossing, with those described in section 2.2 in regard to the channelling of exothermic energy into product degrees of freedom. The portion of the (late) barrier to endothermic reaction that extends into the coordinate of approach for 'reverse' reaction, is the portion of the exothermic energy release that occurs in the coordinate of separation for 'forward' reaction; we are viewing the same hill from below and above. Similarly the reagent translational energy required for endothermic reaction is related to the translational energy derived from repulsive energy-release in the corresponding exothermic reaction. Since the pes is a single whole, it is to be hoped that the various indices of reactive behaviour bear an evident relationship to one another.

The "indices of reactive behaviour" referred to in the previous paragraph (see 2.2, 3.2 and 4.2) make frequent use of the collinear "diagnostic" pes. Since this allows for neither reagent nor product angular momenta, it gives no insight into the sources of product rotation (section 2.2), nor into the channelling of reagent rotation into motion across the barrier.

The experimentally observed dependence of reaction probability on J can plausibly be explained as follows [91]. In the great majority of reactions there will be a preferred orientation for the required close-approach of the attacking atom to the molecule under attack. With increased J, the time that the system spends in this orientation will decrease, as will the detailed rate constant, $k(J)$. With further increase in J, the preferred orientation will be recovered at short intervals during the approach of A to BC, resulting in a reversal of the downward trend in $k(J)$. At high J, the attacking atom responds to an average interaction around BC; $k(J)$ would be expected to level off. There is, however, a further contributing factor at J sufficiently high that the rotational energy is significant in terms of the energy required for barrier-crossing; this is the effect of rotation in carrying the system across the barrier to reaction. If the barrier is "late", this would be expected to occur largely through vibration-rotation interaction; if "early" through the component of force along r_{AB} as BC rotates into A (see [72b] for the geometry of this collision).

These scenarios are speculative. What can be said with assurance is that the ratio of time spent along the approach coordinate to the rotational period of the molecule under attack will play an important role in the outcome of a potential-ly-reactive encounter, in addition to the energy vested in rotation.

5. FUTURE DIRECTIONS

This talk has described the development of one type of "state specified" chemistry, from which some simple lessons regarding reaction dynamics can be drawn. What we see obscurely today, we shall in the future see much more clearly through the more rigorous application of state-selection, an increase in the range of attributes being selected, and also an extension in the range of reactions amenable to study. All of this will add up to a maturation of the field of state-to-state chemistry.

In closing I mention two further approaches which could assist materially in the quest for understanding of the choreography of chemical reaction. In the first, attempts are being made to observe the molecular partners while they are, so to speak, on the stage, rather than immediately prior to and following the reactive dance. This we term "transition state spectroscopy" (TSS); it is a young but burgeoning field. In the second novel approach, the intention, stated a little grandiosely, is to have a hand in writing the script according to which the dynamics occurs; the reagents are aligned by the forces at a crystal surface and held in a fixed arrangement immediately prior to the initiation of the reaction by light, thus restricting the subsequent pattern of motion. This field of "surface aligned photochemistry" (SAP) has only very recently made its appearance. SAP may initiate reaction part way across the pes and hence is related to TSS, since the adsorbed reactants can be held together in configurations that locate them at the outset part way up the reactive barrier. In the SAP experiments performed till now the forces holding the reactants to the surface have been physisorptive.

These two approaches, TSS and SAP, will be briefly reviewed.

The impetus for transition state spectroscopy, TSS, came from a series of theoretical studies of the effect of laser fields on reactive encounters [102]. A variety of intriguing alterations in dynamics were predicted, exemplifying in the reactive domain the effects being explored at that date for non-reactive encounters under the heading of "laser assisted collisions". However, the laser powers thought to be required lay in the gigawatt to terawatt range (powers associated with the electrical breakdown of gases) so that the outlook for success appeared bleak.

It was pointed out that a simple alternative would be to look for emission from electronically-excited transition-state configurations (ABC‡*) [103]. (In reaction rate theory "transition states" are sets of intermediate configurations variously defined; in the spectroscopic context we use the term to denote the full range of configurations intermediate between reactants $A + BC$ and products $AB + C$). An example is shown in fig. 33. In collisional line-broadening of $Na^* + NaX$ the wings of the atomic line emission would be due to the approach of Na^* to NaX (moving from right to left in the figure) followed by emission of frequency ν_{em}. In TSS comparable wings should be observed due to the fact that every Na^* is formed by separation from NaX (moving from left to right in the figure); separation can be preceded by emission at frequency ν_{em}. A simple calculation showed that the emission intensity at a ν_{em}, of ~ 1 cm^{-1} bandwidth, would total $\sim 10^{-6}x$ the chemiluminescent intensity of Na^*, and that this

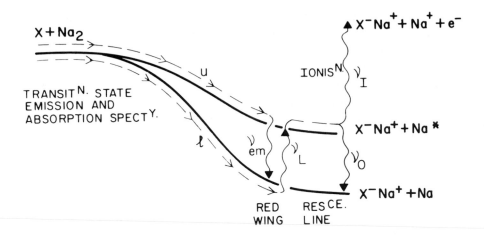

Figure 33. Illustration of transition state emission and absorption spectroscopy, in analogy to collisional line-broadening. The reaction is $X + Na_2 \rightarrow NaX + Na(*)$ (X = halogen). Reaction on the upper surface, u, leads to emission as the products separate (v_{em} constitutes a red-shifted "wing" on the D-line at v_o). Reaction on the lower surface, l, may be detected by laser absorption as products separate ($v_L < v_o$ in the illustration); absorption at v_L may be evidenced by subsequent emission of Na* at v_o, or by laser ionization of Na* using a second laser tuned to v_I. From J.C. Polanyi, Faraday Disc. Chem. Soc., *67*, 129 (1979).

should be measurable. Observation of wings of this intensity, having the appropriate kinetics, followed shortly thereafter [104]. Preliminary reports of similar measurements in absorption [105–107] have led to more definitive successes [108] at increasing levels of spectroscopic detail [109].

The relative intensity of the TSS at various v_{em} or v_L (corrected for changing transition moment, where warranted) will yield information concerning the relative times that the absorber, ABC‡, spends at successive configurations *en route* from reagents to products [110–112]. This is precisely the information that reaction dynamicists have sought to infer from a knowledge of reagent and product motions. The more direct approach of TSS suffers from the drawback that it is a spectroscopy of two pes rather than one, namely the reactive pes and the optically-linked pes (preferably a bound state) to which emission or absorption occurs. This complication need no more be fatal to TSS than it was to the study of stable molecules by electronic emission or absorption spectroscopy in decades past. It does however mean that as soon as our experimental virtuosity permits, we should turn our attention to the TSS of the simplest reactive systems, such as that involving only hydrogen atoms $(H + H_2 \rightarrow H_3‡ \rightarrow H_2 + H)$ [111, 112].

In regard to SAP (Surface Aligned Photochemistry), we can be brief. In the experiments performed to date [113], a submonolayer of H_2S was adsorbed in ultrahigh vacuum (UHV) on a single crystal of LiF(001), and was irradiated with ultraviolet (uv) at 193 and 222 nm coming from an excimer laser (5mJ/ pulse). At low coverage (~0.01 monolayers) fastmoving H photofragments were observed to leave the surface. From their energy and angular distribution these were clearly due to the uv photolysis of H_2S still held at the surface at the instant of photofragmentation. With increasing coverage ($\geqslant 0.1$ monolayers) the photoproducts included fast-moving H_2 formed in the reaction $H + H_2S \rightarrow H_2 + HS$. The yield of H_2 increased with coverage at a rate and to an extent that could only be explained by the reaction of photorecoiling H with *adsorbed* H_2S; accordingly the reaction is ascribed to SAP. The H_2 product is fast and directional.

This is a small first step along the way to inducing reactions under conditions where the range of reagent angles-of-approach and impact parameters is restricted by surface alignment and (by choice of crystal, crystal face and coverage) is controllable. A trajectory study [113] illustrates the simplified link between product attributes and reactive geometries that applies under these more disciplined conditions.

Even in the world of molecules the civilising influence of modest restraints is a cause for rejoicing.

ACKNOWLEDGMENTS

The research described here in so far as it relates to the work of my laboratory was performed over a thirty year period at the University of Toronto. It is a pleasure to express my indebtedness to my colleagues at this University—most especially to the late Prof. D.J. LeRoy who fostered this work from its inception—and to my students and postdoctoral associates whose talents, generosity and friendship have made this undertaking possible under fulfilling.

REFERENCES

1. M. Beutler and M. Polanyi, Z. Phys. Chem B*1*, 3 (1928); St. v. Bogdandy and M. Polanyi ibid. *1*, 21 (1928); M. Polanyi and G. Schay, ibid. *1*, 30 (1928).
2. D.R. Bates and M. Nicolet, J. Geophys. Res. *55*, 301 (1950).
3. G. Herzberg, J. Roy. Astron. Soc. Canada *45*, 100 (1951).
4. J.D. McKinley, D. Garvin and M. Boudart, J. Chem. Phys. *23*, 784 (1955); T.M. Cawthorn and J.D. McKinley, ibid. *25*, 583 (1956).
5. F.J. Lipscomb, R.G.W. Norrish and B.A. Thrush, Proc. Roy. Soc. A *233*, 455 (1956).
6. J.K. Cashion and J.C. Polanyi, J. Chem. Phys. *29*, 455 (1958).
7. J.K. Cashion and J.C. Polanyi, J. Chem. Phys., *30*, 1097 (1959); same authors Proc. Roy. Soc. (London) *A258*, 529, 564, 570 (1960).
8. A.L. Schawlow and C.H. Townes, Phys. Rev. *112* 1940 (1958); A.M. Prokhorov, Zh. Eksptl. i. Teor. Fiz. *34* 1658 (1958) and Sov. Phys.–JETP *7*, 1140 (1958).
9. J.C. Polanyi, Proc. Roy. Soc. (Canada) *54(C)*, 25 (1960); J. Chem. Phys. *34*, 347 (1961).
10. J.C. Polanyi, Applied Optics Supplement *2*, 109 (1965).
11. C.K.N. Patel, W.L Faust, and R.A. McFarlane, Bull. Am. Phys. Soc. *119*, 500 (1964); C.K.N. Patel and R.J. Kerl, Appl. Phys. Letts. *5*, 81 (1964); C.K. N. Patel, Phys. Rev. Letts. *12*. 588 (1964).
12. J.V.V. Kasper and G.C. Pimentel, Phys. Rev. Lett., *14*, 352 (1965); G.C. Pimentel, Sci. Am., *214*, 32 (1966).
13. P.E. Charters and J.C. Polanyi, Disc. Faraday. Soc *33*. 107 (1962).
14. H.O. Pritchard, Trans. Faraday Soc., *33*, 278 (1962).
15. J.C. Polanyi, Trans. Faraday Soc., *33*, 279 (1962).
16. K.G. Anlauf, P.J. Kuntz, D.H. Maylotte, P.D. Pacey and J.C. Polanyi, Disc. Faraday Soc. *44*, 183 (1967).
17. P.D. Pacey and J.C. Polanyi, J. Applied Optics, *10*, 1725 (1971).
18. D.H. Maylotte, J.C. Polanyi and K.B. Woodall, J. Chem. Phys. *57*, 1547 (1972).
19. K.G. Anlauf, D.S. Horne, R.G. Macdonald, J.C. Polanyi and K.B. Woodall, J. Chem. Phys. *57* 1561 (1972).
20. J.D. Polanyi and K.B. Woodall, J. Chem. Phys. *57*, 1574 (1972).
21. J.C. Polanyi and J.J. Sloan, J. Chem. Phys. *57*, 4988 (1972).
22. F. London, *Problem der Modernen Physik* (Sommerfeld Festschrift) p. 104 (1928); Z. Elektrochem. *35*, 552 (1924).
23. H. Eyring and M. Polanyi, Z. Physik. Chem. *B12*, 279 (1931).
24. J.O. Hirschfelder and E. Wigner, J. Chem. Phys. *7*, 616 (1929).
25. F.T.Wall, L.A. Hiller, and J. Mazur, J. Chem. Phys. *29*, 255 (1958).; ibid. *35*, 1284 (1961).
26. N.C. Blais and D.L. Bunker, J. Chem. Phys., *37*, 2713 (1962); ibid. *39*, 315 (1962).
27. J.C. Polanyi in *Transfert d'Energie dans les Gaz*, Ed. R. Stoops, p. 177–182, 526–528, Interscience Publishers, N.Y. (1962); J.C. Polanyi and S.D. Rosner, J. Chem. Phys. *38*, 1028 (1963).
28. M.G. Evans and M. Polanyi, Trans. Faraday Soc., *35*, 178 (1939).
29. For a review see B.S. Agrawala and D.W. Setser in *'Gas Phase Chemiluminescence and Chemi-ionization'*, Ed. A. Fontijn, p. 157, Elsevier (1985).
30. B.M. Berquist, J.W. Bozzelli, L.S. Dzelzkalns, L.G. Piper and F. Kaufman, J. Chem. Phys. 76, 2976 (1982); B.M. Berquist, L.S. Dzelzkalns, F. Kaufman, ibid. 76, 2984 (1982); L.S. Dzelzkalns, and F. Kaufman, ibid. 77, 3508 (1982); L.S. Dzelzkalns and F. Kaufman, ibid. 79, 3836 (1983).
31. J.C. Weisshaar, T.S. Zwier and S.R. Leone, J. Chem. Phys. *75*, 4873 (1981); for a review see C.E. Hamilton and S.R. Leone in *'Gas Phase Chemiluminescence and Chemi-ionization'*, Ed. A. Fontijn, p. 139, Elsevier (1985).

32. H.L. Welsh, C. Cumming, and E.J. Stansbury. J. Opt. Soc. Am. *41*, 712 (1951); H.L. Welsh, E.J. Stansbury, J. Romanko, and T. Feldman, J. Opt. Soc. Am. *45*, 338 (1955).

33. P.E. Charters, R.G. Macdonald, and J.C. Polanyi, Appl. Optics, *10*, 1747 (1971).

34. N. Jonathan, C.M. Melliar-Smith, and D.H. Slater, Mol. Phys. *20*, 93 (1971); N. Jonathan, C.M. Melliar-Smith, S. Okuda, D.M. Slater, and D. Timlin, Mol. Phys. *22*, 561 (1971).

35. H.W. Chang and D.W. Setser, J. Chem. Phys. *58*, 2298 (1973); D.J. Bogan and D.W. Setser, ibid. *64*, 586 (1976).

36. J.C. Moehlmann and J.D. McDonald, J. Chem. Phys. *62*, 3061 (1975); J.W. Hudgens and J.D. McDonald, ibid. *67*, 3401 (1977).

37. P.M. Aker, D.J. Donaldson, and J.J. Sloan, J. Phys. Chem., *40*, 3110 (1986).

38. D. Brandt, L.W. Dickson, L.N.Y. Kwan, and J.C. Polanyi, Chem. Phys. *39*, 189 (1979); L.W. Dickson, Ph.D Thesis, Univ. of Toronto, 1982.

39. J.C. Polanyi and K.B. Woodall, J. Chem. Phys. *56*, 1563 (1872).

40. J.H. Parker and G.C. Pimentel, J. Chem. Phys. *51*, 91 (1969); R.D. Coombe and G.C. Pimentel, ibid. *59*, 251, 1535 (1973); M.J. Berry, ibid. *59*, 6229 (1973).

41. K.G. Anlauf, J.C. Polanyi, W.H. Wong, and K.B. Woodall, J. Chem. Phys., *49*, 5189 (1968).

42. D.M. Neumark, A.M. Wodtke, G.N. Robinson, C.C. Hayden, and Y.T. Lee, J. Chem. Phys. *82*, 3045 (1985); D.M. Neumark, A.M. Wodtke, G.N. Robinson, C.C. Hayden, K. Shobotake, R.K. Sparks, T.P. Schaefer, and Y.T. Lee, J. Chem. Phys. *82*, 3967 (1985).

43. R.B. Bernstein and R.D. Levine, J. Chem. Phys. *57*, 434 (1972); R.B. Bernstein and R.D. Levine 'Advances in Atomic and Molecular Physics', Vol. 11 (Academic Press, New York, 1975); R.B. Bernstein and R.D. Levine in 'Modern Theoretical Chemistry', Vol. 3, Dynamics of Molecular Collisions, Part B, Ed. W.H. Miller (Plenum Press, New York, 1976) Chapter 7.

44. R.N. Zare and P.J. Dagdigian, Science, *185*, 739 (1974).

45. "Multiphoton Spectroscopy of Molecules", by S.H. Lin, Y. Fujimura, H.J. Neusser, and E.W. Schlag (Academic Press, 1984).

46. P.M. Aker and J.J. Sloan, J. Chem. Phys. *85*, 1412 (1986); P.M. Aker, J.J. Sloan, and J.S. Wright, Chem. Phys., In press (1986).

47. D.S. Perry and J.C. Polanyi, Chem. Phys. *12*, 419 (1976).

48. K.G. Anlauf, P.E. Charters, D.S. Horne, R.G. Macdonald, D.H. Maylotte, J.C. Polanyi, W.J. Skrlac, D.C. Tardy, and K.B. Woodall, J. Chem. Phys. *53*, 4091 (1970); H. Heydtmann and J.C. Polanyi, J. Appl. Optics, *10*, *1738 (1971);* M.A. Nazar, J.C. Polanyi, and W.J. Skrlac, Chem. Phys. Letts. *29*, 473 (1974); J.C. Polanyi and W.J. Skrlac, Chem. Phys. *23*, 167 (1977); D. Brandt and J.C. Polanyi, Chem. Phys. *35*, 23 (1978); ibid. *45*, 65 (1980).

49. L.T. Cowley, D.S. Horne, and J.C. Polanyi, Chem. Phys. Letts., *12*, 144 (1971).

50. A.M.G. Ding. L.J. Kirsch, D.S. Perry, J.C. Polanyi, and J.L. Schreiber, Faraday Disc. Chem. Soc., *55*, 252 (1973).

51. J.C. Polanyi, J.L. Schreiber, and W.J. Skrlac, Faraday Disc. Chem. Soc. *67*, 66 (1979).

52. J.C. Polanyi, J. Quant. Spectrosc. Radiat. Transfer *3*, 471 (1963).

53. P.J. Kuntz, E.M. Nemeth, J.C. Polanyi, S.D. Rosner, and C.E. Young, J. Chem. Phys. *44*, 1168 (1966).

54. D.S. Perry, J.C. Polanyi, and C. Woodrow Wilson Jr., Chem. Phys. *3*. 317 (1974).

55. M.H. Mok and J.C. Polanyi, J. Chem. Phys. *51*, 1451 (1969).

56. T.H. Dunning Jr, J. Phys. Chem. *88*, 2469 (1984).

57. J.C. Polanyi, Disc. Faraday Soc. *44*, 293 (1967).

58. For a review see: J.C. Polanyi, Accounts, Chem. Res. *5*, 161 (1972).

59. For a review see: J.C. Polanyi and J.L. Schreiber, Physical Chemistry — An Advanced Treatise, Vol. VIA. *Kinetics of Gas Reactions*, Eds., H. Eyring, W. Jost and D. Henderson (Academic Press, New York, 1974) Chap. 6, p. 383.

60. P.J. Kuntz, E.M. Nemeth, and J.C. Polanyi, J. Chem. Phys. *50*, 4607 (1969).

61. C.A. Parr, J.C. Polanyi, and W.H. Wong, J. Chem. Phys. *58* 5 (1973).

62. J.T. Muckerman, J. Chem. Phys. *54*, 1155 (1971); ibid. *56*, 2997 (1972); ibid. *57*, 3388 (1972).

63. J.C. Polanyi and J.L. Schreiber, Chem. Phys. Letts. *29*, 319 (1974).

64. J.C. Polanyi and J.L. Schreiber, Faraday Disc. Chem. Soc. *62*, 267 (1977).

65. C.F. Bender, S.V. O'Neill, P.K. Pearson, and H.F. Schaefer III, Science *176*, 1412 (1972).

66. H.F. Schaefer III, J. Phys. Chem. *89*, 5336 (1985).

67. D.S. Perry and J.C. Polanyi, Chem. Phys. *12*, 419 (1976).

68. P.J. Kuntz, M.H. Mok, and J.C. Polanyi, J. Chem. Phys. *50*, 4623 (1969).

69. D.R. Herschbach, Faraday Disc. Chem. Soc., *55, 233 (1973)*.

70. M.G. Prisant, C.T. Rettner, and R.N. Zare, J. Chem. Phys. *81, 2699 (1984)*.

71. M.D. Pattengill and J.C. Polanyi, Chem. Phys. *3*, 1 (1974).

72. (a) N.H. Hijazi and J.C. Polanyi, J. Chem. Phys. *63*, 2249 (1975);
    (b) N.H. Hijazi and J.C. Polanyi, Chem. Phys. *11*, 1 (1975).

73. J.J. Valentini, M.J. Coggiola, and Y.T. Lee, J. Am. Chem. Soc. *98*, 853 (1976); Faraday Disc., Chem. Soc., *62*, 232 (1977).

74. J.W. Hudgens and J.D. McDonald, J. Chem. Phys. *67*, 3401 (1977).

75. L.T. Cowley, D.S. Horne, and J.C. Polanyi, Chem. Phys. Letts. *12*, 144 (1971).

76. B.A. Blackwell, J.C. Polanyi, and J.J. Sloan, Chem. Phys. *24*, 25 (1977).

77. D.J. Douglas and J.C. Polanyi, Chem. Phys. *16*, 1 (1976).

78. R.D. Coombe and G.C. Pimentel, J. Chem. Phys. *59*, 1535 (1973).

79. D.L. Thompson, J. Chem. Phys. *56*, 3570 (1972).

80. J.C. Polanyi, Faraday Disc. Chem. Soc. *55*, 389 (1973).

81. J.C. Polanyi, J.L. Schreiber, and J.J. Sloan, Chem. Phys. *9*, 403 (1975).

82. J.C. Polanyi, J.J. Sloan, and J. Wanner, Chem. Phys. *13*, 1 (1976).

83. K.G. Anlauf, D.H. Maylotte, J.C. Polanyi, and R.B. Bernstein, J. Chem. Phys. *51*, 5716 (1969).

84. J.C. Polanyi, and D.C. Tardy, J. Chem. Phys. *51*, 5717 (1969).

85. D.S. Perry, J.C. Polanyi and C. Woodrow Wilson Jr., Chem. Phys. Letts. *24*, 484 (1974).

86. T.J. Odiorne, P.R. Brooks, and J.V.V. Kasper, J. Chem. Phys. *55*, 1980 (1971; J.G. Pruett, F.R. Grabiner, and P.R. Brooks, ibid. *63*, 3335 (1974); J.G. Pruett, F.R. Grabiner, and P.R. Brooks, ibid. *63*, 1173 (1975).

87. A Gupta, D.S. Perry, and R.N. Zare, J. Chem. Phys. *72*, 6237, 6250 (1980).

88. F. Heismann and H.J. Loesch, Chem. Phys. *64*, 43 (1982).

89. D.J. Douglas, J.C. Polanyi, and J.J. Sloan, J. Chem. Phys. *59*, 6679 (1973).

90. F.E. Bartozek, B.A. Blackwell, J.C. Polanyi, and J.J. Sloan, J. Chem. Phys *74* 3400 (1981).

91. B.A.Blackwell, J.C. Polanyi, and J.J. Sloan, Chem. Phys. *30*, 299 (1978).

92. H.H. Dispert, M.W. Geis, and P.R. Brooks, J. Chem. Phys. *70*, 5317 (1979).

93. M.Hoffmeister, L. Potthast, and H.J. Loesch, Book of Abstracts, XII International Conf. on the Physics of Elektronic and Atomic Collisions, Gatlinburg (1981).

94. J.C. Polanyi and W.H. Wong, J. Chem. Phys. *51*, 1439 (1969).

95. J.W. Duff and D.G. Truhlar, J. Chem. Phys. *62*, 2477 (1975).

96. G.L. Hofacker and R.D. Levine, Chem. Phys. Letts. *9*, 617 (1971).

97. J.C. Polanyi and N. Sathyamurthy, Chem. Phys. *33*, 287 (1978).

98. B.A. Hodgson and J.C. Polanyi, J. Chem. Phys. *55*, 4745 (1971).

99. M.H. Mok and J.C. Polanyi, J. Chem. Phys. *53*, 4588 (1970).

100. G.S. Hammond, J. Amer. Chem. Soc. *77*, 334 (1955).

101. J. C. Polanyi, J. Phys. Chem. *31,* 1338 (1959).

102. (a) T.F. George, J. Phys. Chem. *86*, 10 (1982) and references therein; (b) A.M.F. Lau, Phys. Rev. *A13*, 139 (1976); *14*, 279 (1976); Phys. Rev. Lett. *43*, 1009 (1978);

Phys. Rev A*22*, 614 (1980); (c) V.S. Dubov, L.I. Gudzenko, L.V. Gurvich and S.I. Yakovlenko, Chem. Phys. Lett. *45*, 330 (1977); S.I. Yakovlenko, Sov. J. Quantum Electron, *8*, 151 (1977); (d) A.E. Orel and W.H. Miller, Chem. Phys. Lett. *57*, 362 (1979); J. Chem. Phys. *70*, 4393 (1979); (e) J.C. Light and A. Altenberger-Siczek, ibid. *70*, 4108 (1979).

103. J.C. Polanyi, Faraday Disc. Chem. Soc. *67*, 129 (1979).
104. P. Arrowsmith, F.E. Bartoszek, S.H.P. Bly, T. Carrington, Jr., P.E. Charters, and J.C. Polanyi, J. Chem. Phys. *73*, 5895 (1980); H.-J.. Foth, J.C. Polanyi, and H.H. Telle, J. Phys. Chem. *86*, 5027 (1982); P. Arrowsmith, S.H.P. Bly, P.E. Charters, and J.C. Polanyi, J. Chem. Phys. *79*, 283 (1983).
105. P. Hering, P.R. Brooks, R.F. Curl Jr., R.S. Judson, and R.S. Lowe, Phys. Rev. Lett. *44*, 657 (1980); P.R. Brooks, R.F. Curl, and T.C. Maguire, Ber. Bunsenges. Phys. Chem. *86*, 401 (1982).
106. H.P. Grieneisen, H. Xue-Jing, and K.L. Kompa, Chem. Phys. Letts. *82*, 421 (1981).
107. J.K. Ku, G. Inoue, and D.W. Setser, J. Phys. Chem. *87*, 2989 (1983).
108. T.C. Maguire, P.R. Brooks R.F. Curl, J.H. Spence, and S. Ulrick, J. Chem. Phys. *85*, 844 (1986); T.C. Maguire, P.R. Brooks, R.F. Curl, J.H. Spence, and S. Ulrick, Phys. Rev. A*34*, 4418 (1986).
109. P.D. Kleiber, A.M. Lyyra, K.M. Sando, S.P. Heneghan, and W.C. Stwalley, Phys. Rev. Letts. *54*, 2003 (1985); P.D. Kleiber, A.M. Lyyra, K.M. Sando, V. Zafiropolos, and W.C. Stwalley, J. Chem. Phys. *85*, 5493 (1986).
110. J.C. Polanyi and R.J. Wolf, J. Chem. Phys. *75*, 5951 (1981).
111. H.R. Mayne, R.A. Poirier, and J.C. Polanyi, J. Chem. Phys. *80*, 4025 (1984); H.R. Mayne, J.C. Polanyi, N. Sathyamurthy, and S. Raynor, ibid. *88*, 4064 (1984).
112. V. Engel, Z. Bacic, R. Schinke, and M. Shapiro, J. Chem. Phys. *82*, 4844 (1985); V. Engel and R. Schinke, Chem. Phys. Letts. *122*, 103 (1985).
113. E.B.D. Bourdon, J.P. Cowin, I. Harrison, J.C. Polanyi, J. Segner, C.D. Stanners, and P.A. Young, J. Phys. Chem. *88*, 6100 (1984); E.B.D. Bourdon, P. Das, I. Harrison, J.C. Polanyi, J. Segner, C.D. Stanners, R.J. Williams, and P.A. Young, Faraday Disc. Chem. Soc. *82*, 343 (1986).

Chemistry 1987

**DONALD J. CRAM, JEAN-MARIE LEHN and
CHARLES J. PEDERSEN**

*for their development and use of molecules with structure-specific interactions of
high selectivity*

THE NOBEL PRIZE FOR CHEMISTRY

Speech by Professor SALO GRONOWITZ of the Royal Academy of Sciences.
Translation form the Swedish text

Your Majesties, your Royal Highnesses, Ladies and Gentlemen,

A prerequisite for all life processes is that molecules recognize each other and bind to each other in order to be able to react. The molecules are said to form complexes. The proteins in our food bind to other proteins, called enzymes, which catalyze− − − that is accelerate − − − their breakdown. Other compounds recognize ions, such as sodium and potassium, and transport them in and out of the living cell. Our biological defence against intruders is based on the formation of antibodies, which recognize the enemy, the antigen, and disarm him by forming harmless complexes. Our whole life, our consciousness, our instincts are governed by signal substances that are recognized by various receptors.

This biological recognition is very specific and selective. As the great chemist Emil Fischer, who received the Nobel Prize in chemistry in 1902, formulated it: The two molecules have to fit like a key in a lock. However, the lock is always a very complicated molecule of high molecular weight, a protein or a nucleic acid.

Organic chemists have been wondering for a long time, how much of the very large biomolecule is really necessary for achieving the desired result. How much of the molecule is really needed to form the hole or cleft in which the key has to fit for the reaction to take place. Organic chemists have dreamt about preparing molecules that mimic biomolecules in the laboratory, molecules that for instance have the same catalytic effects as enzymes, which chemical evolution has taken millions of years to develop.

A breakthrough towards this goal occurred when Charles Pedersen in the 1960's prepared cyclic compounds that contained carbon and oxygen atoms in rings of 18 up to 40 atoms. As their form was similar to a royal crown, he called these new compounds crown ethers. He showed that these compounds had many peculiar and unexpected properties. For example, they formed strong complexes with the alkali metal ions, lithium, sodium, potassium, rubidium and cesium, which previously had been difficult to bind. Of these spherical ions, lithium is the smallest and cesium the largest. Depending upon the size of the synthesized crown, potassium, for example, could fit inside, while the cesium ion was too large to be trapped.

The explosive development of the art of organic synthesis during the last decades, which is reflected in the Nobel Prizes in Chemistry to Robert Woodward in 1965, to Herbert Brown and Georg Wittig in 1979 and to Bruce Merrifield in 1984, has made it possible for Donald Cram and Jean-Marie Lehn to very skilfully design more complex molecules containing holes and clefts which bind inorganic as well as organic positive ions even more selectively. Other synthetic molecules can bind various negative ions as well as neutral molecules. Even molecules which can differentiate between mirror image forms of a molecule have been prepared. It has also been possible to synthesize molecules which enzymes in their mode of action and strongly accelerate various types of chemical reactions. Other organic compounds have been prepared which transport ions through biological membranes. By means of their penetrating investigations the laureates have also elucidated the factors that govern complex formation, and the changes that occur in both partners' chemical and physical properties, and how this can be used in practice for different purposes.

Through their work the laureates laid the foundation to what today is one of the most expansive chemical research areas, for which Cram has coined the term host-guest chemistry, while Lehn calls it supramolecular chemistry. Their research has been of enormous importance for the development of coordination chemistry, organic synthesis, analytical chemistry, bioinorganic and bioorganic chemistry, it is no longer science fiction to prepare supermolecules which are better and more versatile catalysts than the highly specialized enzymes. The dream may soon become reality. Through their work Cram, Lehn and Pedersen have shown the way.

Professor Cram, Professor Lehn, Mister Pedersen, in these few minutes I have tried to explain your fundamental work in the field of molecular recognition and the great importance and consequences your results have in many branches of chemistry, such as analytical chemistry, organic synthesis, coordination chemistry and bioinorganic and bioorganic chemistry. You have indeed shown, as the father of organic synthesis, Marcelin Berthelot, Professor at Collège de France, formulated it more than a century ago "La Chimie crée ses objets".

In recognition of your important contributions to chemistry the Royal Academy of Sciences has decided to confer upon you this year's Nobel Prize for Chemistry.

It is an honour and a pleasure for me to extend to you the congratulations of the Royal Academy of Sciences, and to ask you to receive your prizes from the hands of his Majesty the King.

Donald J. Cram

DONALD J. CRAM

The beginning is distant, and was a time when we as a people were without much of the fruits of science that now refine our lives. But it was also a good time, when family and town were the domain of our existence. My father of Scottish, and mother of German extraction, migrated with their three children from Ontario, Canada, to rural Chester, Vermont, USA, where I was born in the spring of 1919, as the Cram's fourth and only male child. Two years later, the family moved to Brattleboro, Vermont.

My mother, Joanna, was high-spirited throughout her 94 years, starting with a girlhood rebellion against the strict Mennonite faith in which she was raised. My father, William, was a romantic, a cavalry officer, later working alternately as a successful lawyer and unsuccessful farmer. He died of pneumonia at 53, leaving my mother with a set of Victorian upper-class English values, and the task of providing for and raising my sisters and me, then aged four.

According to my oldest sister, Elizabeth, I was as a child precocious, curious, and constantly in, or causing, trouble. This character trait started at birth; I weighed over ten pounds, and had an unusually large head! Determined to walk at seven months, I pulled a pan of fresh eggs down from a table onto that head. At three years, I broke my first window. My father took me directly to our neighbor, Mr. Mason, to apologize. Reputedly, I said to him, "Sorry, you nasty Mason".

By the time I was four and a half, I was reading children's books. My mother, steeped in English literature, cultivated incentive by reading to me only the *beginnings* of tales that involved heroes, heroines, hypocrites, and villains. When we reached the exciting part, she left me with the story to finish by myself. My childhood was adventuresome and idyllic in the things that mattered.

Elementary schooling for me was series of multiclass single-room buildings, where the young and very young witnessed each other being taught. On report card days, I faced searching questions and criticism during which "character grades" were stressed over academic accomplishments. I usually was marked "A" in attitude and accomplishment, "B" in effort, and "C" in obedience.

What I call my real *education* occurred principally outside of the classroom, in a private world of books and brooks. All of Dickens, Kipling, Scott, Shaw, and much of Shakespeare were read. I learned how to run up spring-fed brooks, jumping from one glacier-polished stone to another. I carried firewood, emptied ashes, and shoveled snow for music lessons. I picked apples to be paid in apples, strawberries to be paid in strawberries; and I moved the lawns of a large estate belonging to the dentist who filled my cavities and pulled my teeth.

The rate of exchange of my time for his was fifty to one. During the Christmas rush, I sold ties, shirts, shoes, gloves, and jackets in return for the same. By the time I was fourteen, my mother's barter arrangements were replaced by a flat fee of 15 cents an hour working for neighbors, raking leaves, hoeing corn, digging potatoes, pitching hay, and delivering newspapers. The last job taught me about debts, dogs, and bicycle bags.

At sixteen, I left my home in Brattleboro, Vermont, a handsome old town on the Connecticut River. By then I had learned how to adapt to eighteen different employers, and had played high school varsity tennis, football, and ice hockey. I also had my full growth of 195 pounds, and was 6 feet tall.

My family dispersed at that time, and I drove two elderly ladies to Florida in a Model-A Ford, without benefit of a license, and in return for transportation. Stopping at Lake Worth, Florida, I worked in an ice-cream shop and weeded lawns in return for room and board. There, I continued my secondary schooling while suffering an acute case of homesickness for New England. Nine months later, I hitchhiked north again to Massachusetts, and passed the summer as a house painter and roofer. My twelfth grade was spent at Winwood, a little private school on Long Island, New York, working as a factotum for my tuition and board. While there, I did three things significant to my future. I took a course in chemistry, taught myself solid geometry from a book, and won a $6,000 four-year National Rollins College Honor Scholarship.

While at Rollins, in the resort town of Winter Park, Florida, I worked at chemistry and played at philosophy (four courses!). I read Dostoevski, Spengler, and Tolstoy, and sang in the choir and in a barbershop quartet. In four halcyon years, I obtained an airplane pilot's license, acted in plays, produced-announced a minor radio program, and, while my fellow students complained, dined on the best food I had yet encountered.

During the summers of 1938 to 1941, I worked for the National Biscuit Company in New York City, at first as a salesman covering an area from 144th Street to 78th Street on the tough East Side. The largest city in the country was also its best teacher. This was my first exposure to ghetto slums, youth gang warfare, drugs, prostitution, and petty thievery. The northern end of my territory was dominated by Jewish delicatessens, out of which at first (but not later) I was urged to leave. I got to know everybody, and everybody's business. In the Irish-run grocery stores, we would conduct "wakes" for the last fig bar in the bin. Harlem was a place where each street was a playground, and each store a small fort. The Puerto Rican district was full of street vendors of all sorts. I started that summer weighing 195 pounds, and ended it at 155. My $ 15 per week provided me with cash, and a crash course in ethnic groups and big city street-life. The other summers involved analyzing cheeses for moisture and fat content in the National Biscuit Laboratories.

From these routine jobs, I extracted pleasure by making them into games. They taught me self-discipline, and illustrated how I did not want to spend my life. But this period provided me with a keen interest in the differences between people, and an overwhelming dislike of repetitive activities. When the word "research" entered my vocabulary, it had a magic ring, suggesting the search for new phenomena. Chemical research became my god, and the conducting of it, my act of prayer, from 1938 to the present. When told by my first college chemistry professor, Dr. Guy Waddington, that he thought I would make a good industrial investigator—but probably not a good academic one—I determined upon an academic research career in chemistry.

Out of 17 applications for teaching assistantships to go to graduate school, three offers came. I accepted the University of Nebraska's, where an MS was granted in 1942. My thesis research there was done under the supervision of Dr. Norman O. Cromwell. By then, World War II was upon us, so I went to work for Merck & Co., ultimately on their penicillin project, where my search for excellence was symbolized by Dr. Max Tishler. Immediately after the war ended in 1945, Max arranged for me to attend Harvard University, working for Professor L. F. Fieser. The work for my Ph.D. degree on a National Research Council Fellowship was in hand in eighteen months.

At Harvard, scientific excellence was personified for me by Professors Paul D. Bartlett and Robert B. Woodward. After three months at M.I.T., working for Professor John D. Roberts, I set out for the University of California at Los Angeles on August 1, 1947, and have taught and researched there ever since, after 1985 as the S. Winstein Professor of Chemistry.

In retrospect, I judge that my father's death early in my life forced me to construct a model for my character composed of pieces taken from many different individuals, some being people I studied, and others lifted from books. The late Professor Saul Winstein, my colleague, friend, and competitor, contributed much to this model. It was almost complete by the time I was 35 years of age. Thus, through the early death of my father, I had an opportunity — indeed the necessity — to animate that father image that was slowly maturing in my mind's eye. And, at last, I realized who in fact this figure was. It was I.

The times and environment have been very good to me during my forty-six years of chemical research. I entered the profession at a period when physical, organic, and biochemistry were being integrated, when new spectroscopic windows on chemical structures were being opened, and when UCLA, a fine new university campus, was growing from a provincial to a world-class institution. My over 200 co-workers have shared with me the miseries of many failures and the pleasures of some triumphs. Their careers are my finest monument. My countrymen have supported our research without mandating its character. Jean Turner Cram, as my first wife, sacrificed for my career from 1940 to 1968. Dr. Jane Maxwell Cram, my second wife, acted as foil, unsparing but inspiring critic and research strategist in ways beyond mention.

My fellow scientists have generously honored my research program with three American Chemical Society awards: for Creative Work in Synthetic Organic Chemistry; the Arthur C. Cope Award for Distinguished Achievement in Organic Chemistry; and the Roger Adams Award in Organic Chemistry. Local sections of the same society awarded me the Willard Gibbs and Tolman Medals. I was elected to membership in the National Academy of Science (1961), to become the 1974 California Scientist of the Year, and the 1976 Chemistry Lecturer and Medalist of the Royal Institute of Chemistry (UK). In

1977, I was given an Honorary Doctor's degree from Sweden's Uppsala University, and in 1983 a similar one from the University of Southern California.

I have contributed directly to the teaching of organic chemistry—about 12,000 undergraduate students—and, indirectly, by writing three textbooks: Organic Chemistry (with G. S. Hammond; translated into twelve languages), Elements of Organic Chemistry (with D. H. Richards and G. S. Hammond; three translations), and Essence of Organic Chemistry (with J. M. Cram; one translation), plus the monograph, "Fundamentals of Carbanion Chemistry" (one translation). I enjoy skiing and surfboarding, playing tennis, and playing the guitar as an accompaniment to my singing folk songs. The award of a Nobel Prize at the age of 68 years was ideally timed to enhance rather than divert my research career.

In the four years that have elapsed since I shared a Nobel Prize in Chemistry (1987), the effect of receiving this honor on my life has been profound. Most importantly, the Prize has extended my career by enough years to allow me to obtain the most exciting results of my 50 years of carrying out research. The Prize has also broadened the range of my experiences, most of which have been both interesting and educational. Finally, the research field of molecular recognition in organic chemistry gained much impetus by being recognized by the Nobel Prize. I am grateful that our research results were chosen as a vehicle for honoring those who know the joys of carrying out organic chemical research.

THE DESIGN OF MOLECULAR HOSTS, GUESTS, AND THEIR COMPLEXES

Nobel Lecture, 8 December 1987

by

DONALD J. CRAM

Department of Chemistry and Biochemistry, University of California at Los Angeles, Los Angeles, California 90024, U.S.A.

Origins

Few scientists acquainted with the chemistry of biological systems at the molecular level can avoid being inspired. Evolution has produced chemical compounds exquisitely organized to accomplish the most complicated and delicate of tasks. Many organic chemists viewing crystal structures of enzyme systems or nucleic aids and knowing the marvels of specificity of the immune systems must dream of designing and synthesizing simpler organic compounds that imitate working features of these naturally occurring compounds. We had that ambition in the late 1950's. At that time, we were investigating *pi*-complexes of the larger [m.n.]paracyclophanes with $(NC)_2C=C(CN)_2$, and envisioned structures in which the *pi*-acid was sandwiched by two benzene rings. Although no intercalated structures were observed [1,2], we recognized that investigations of *highly* structured complexes would be central to simulation of enzymes by relatively simple organic compounds.

In 1967, Pedersen's first papers appeared [3,4] which reported that alkali metal ions bind crown ethers to form highly structured complexes. We immediately recognized this work as an entree into a general field. The 1969 papers on the design, synthesis, and binding properties of the cryptands by J.-M. Lehn, J.-P. Sauvage, and B. Dietrich [5,6] further demonstrated the attractions and opportunities of complexation chemistry. Although we tried to interest graduate students in synthesizing *chiral crown ethers* from 1968 on, the efforts were unsuccessful. In 1970 we insisted that several postdoctoral co-workers enter the field. During 1973, we published five Communications on the subject [7,11]. In 1974 with Jane M. Cram, we published a general article entitled "Host-Guest Chemistry", which defined our approach to this research [12].

Aeschylus, the Athenian Poet-Dramatist, wrote 2 500 years ago, "Pleasantist of all ties is the tie of host and guest" [13]. Our research of the past 17 years has dealt with the pleasant tie between host and guest and the organic molecular level. The terms *host, guest, complex,* and their binding forces were defined in 1977 as follows [14]. "Complexes are composed of two or more molecules or ions held together in unique structural relationships by electrostatic forces other than those of full covalent bonds ... molecular complexes are usually held together by hydrogen bonding, by ion pairing, by *pi*-acid to *pi*-base interactions, by metal to ligand binding, by van der Waals attractive forces, by solvent reorganizing, and by partially made and broken covalent bonds (transition states) ... high structural organization is usually produced only through multiple binding sites ... a highly structured molecular complex is composed of at least one host and one guest component ... a host-guest relationship involves a complementary stereoelectronic arrangement of binding sites in host and guest ... the host component is defined as an organic molecule or ion whose *binding sites converge* in the complex ... the guest component is defined as any molecule or ion whose *binding sites diverge* in the complex ..." In these definitions, hosts are synthetic counterparts of the receptor sites of biological chemistry, and guests, the counterparts of substrates, inhibitors, or cofactors. These terms and concepts have gained broad international acceptance [15]. A new field requires new terms which, if properly defined, facilitate the reasoning by analogy on which research thrives.

From the beginning, we used Corey-Pauling-Koltun (CPK) molecular models [16], which served as a compass on an otherwise uncharted sea full of synthesizable target complexes. We have spent hundreds of hours building CPK models of potential complexes, and grading them for desirability as research targets. Hosts were then prepared by my co-workers to see if they possessed the anticipated guest-binding properties. Crystal structures of the hosts and their complexes were then determined to compare what was anticipated by model examination with what was experimentally observed. By the end of 1986, Drs. K. N. Trueblood, C. B. Knobler, E. F. Maverick, and I. Goldberg, working at UCLA, had determined the crystal structures of over 50 complexes, and those of another 25 hosts. These crystal structures turned our faith into confidence. Chart I traces the steps involved in linking the structures of *biotic complexes* of evolutionary chemistry with our *abiotic complexes* designed with the aid of CPK molecular models [17].

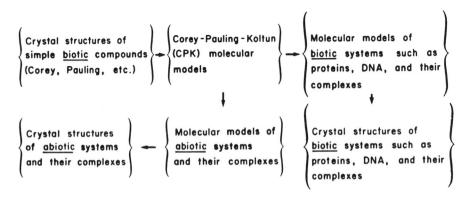

Chart I. Crystal structures of biotic compounds are correlated with those of abiotic compounds through CPK models.

In molecular modeling, we made extensive use of the self-evident principle of complementarity: "to complex, hosts must have binding sites which cooperatively contact and attract binding sites of guests without generating strong nonbonded repulsions" [18]. Complexes were visualized as having three types of common shapes: 1) perching complexes, resembling a bird perching on a limb, an egg protruding from an egg cup, or a scoop of ice cream sitting on a cone; 2) nesting complexes, similar to an egg resting in a nest, a baby lying in its cradle, or a sword sheathed in its scabbard; 3) capsular complexes, not unlike a nut in its shell, a bean in its pod, or a larva in its cocoon. Chart II provides a comparison of CPK models of the three types of complexes (**1, 2, and 3**) and their actual crystal structures [19,20].

Molecular model structures   Crystal structures

Perching complex (**1**)

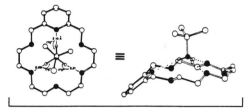

Nesting complex (**2**)

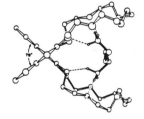

Capsular complex (**3**)

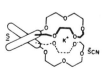

Chart II. Three types of complexes.

Principle of Preorganization

Crystal structures of Pedersen's 18-crown-6 [21] and Lehn's [2.2.2] cryptand [22,23] show that in their uncomplexed states, they contain neither cavities nor convergently-arranged binding sites. Comparisons of the crystal structure of host **4** with that of its K$^+$ complex **5**, and of host **6** with that of its K$^+$ complex **7** indicate that the complexing act must be accompanied by host reorganization and desolvation.

4   **5**   **6**   **7**

With the help of CPK molecular models, we designed ligand system 8, whose oxygens have no choice but to be octahedrally arranged around an enforced spherical cavity complementary to Li$^+$ and Na$^+$ ions. We have given the family name, *spherand*, to completely preorganized ligand systems, and the name, *spheraplex*, to their complexes, which like **7**, are capsular [24]. The syntheses and crystal structures of **8, 9** and **10**, have been reported [25]. As expected, the crystal structure of **11** contains a hole lined with 24 electrons, which are shielded from solvation by six methyl groups. The snowflake-like structures of **11** and of spheraplexes **12** and **13** are nearly identical. Thus **8** is the first ligand system to be designed and synthesized which was completely organized for complexation during synthesis, rather than during complexation.

9   **8**   **10**

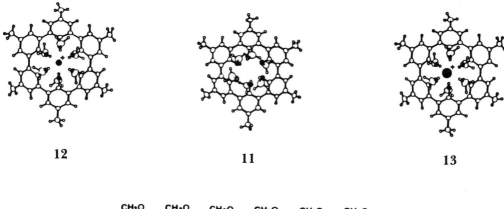

12　　　　**11**　　　　**13**

14

A method was developed of determining the binding free energies of lipophilic hosts toward guest picrate salts of Li^+, Na^+, K^+, Rb^+, Cs^+, NH_4^+, $CH_3NH_3^+$, and t-BuNH$_3^+$. The guest salts were distributed between $CDCl_3$ and D_2O at 25 °C in the presence and absence of host. From the results, K_a (mol^{-1}) and $-\Delta G°$ values (kcal mol^{-1}) were calculated (equations (1)). This method was rapid and convenient for obtaining

$$H + GPic \underset{k_{-1}}{\overset{k_1}{\rightleftarrows}} H \cdot G \cdot Pic \quad K_a = k_1/k_{-1} \quad -\Delta G° = -RTlnK_a \tag{1}$$

$-\Delta G°$ values at 25 °C ranging from about 6 to 16 kcal mol^{-1} in $CDCl_3$ saturated with D_2O [26]. Higher values (up to 22 kcal mol^{-1}) were obtained by equilibration experiments between complexes of known and those of unknown $-\Delta G°$ values [18, 27, 28]. Others were determined from measured k_{-1} and k_1 values, all in the same medium at 25 °C [18]. Spherand **8** binds LiPic with >23 kcal mol^{-1}, NaPic with 19.3 kcal mol^{-1}, and totally rejects the other standard ions, as well as a wide variety of other di- and trivalent ions [18]. The open-chain counterpart of **8**, podand **14**, binds LiPic and NaPic with $-\Delta G° <6$ kcal mol^{-1} [29]. Podand is the family name given to acyclic hosts [15].

Podand **14** differs constitutionally from spherand **8** only in the sense that **14** contains two hydrogen atoms in place of one Ar-Ar bond in **8**. The two hosts differ radically in their conformational structures and states of solvation. The spherand possesses a single conformation ideally arranged for binding Li^+ and Na^+. Its oxygens are deeply buried within a hydrocarbon shell. The orbitals of their unshared electron pairs are in a microenvironment whose dielectric properties are between those of a vacuum and of a hydrocarbon. No solvent can approach these six oxygens, which remain unsolvated. The free energy costs of organizing the spherand into a single conformation and of desolvating its six oxygens were paid for during its synthesis. Thus spherand **8** is preorganized for binding [30]. The podand in principle can exist in over 1 000 conformations, only two of which can bind metal ions octahedrally. The free energy for organizing the podand into a binding conformation and desolvating its six oxygens must come out of its complexation free energy. Thus the podand is not preorganized for binding, but is randomized to maximize the entropy of mixing of its conformers, and to maximize the attractions between solvent and its molecular parts.

The difference in $-\Delta G°$ values for spherand **8** and podand **14** binding Li^+ is >17 kcal mol^{-1}, corresponding to a difference in K_a of a factor of $>10^{12}$. The difference in $-\Delta G°$ values for **8** and **14** binding Na^+ is >13 kcal mol^{-1}, corresponding to a difference in K_a of a factor of $>10^{10}$. These differences are dramatically larger than any we have encountered that are associated with other effects on binding power toward alkali metal ion guests. We conclude that *preorganization is a central determinant of binding power.* We formalized this conclusion in terms of what we call the principle of preorganization, which states that "the more highly hosts and guests are organized for binding and low solvation prior to their complexation, the more stable will be their complexes." Both enthalpic and entropic components are involved in preorganization, since solvation contains both components [29]. Furthermore, binding conformations are sometimes enthalpically rich. For example, the benzene rings in spherand **8** and spheraplexes **9** and **10** are somewhat folded from their normal planar structures to accommodate the spacial requirements of the six methoxyl groups [30]. The anisyl group is an intrinsically poor ligand [31, 32]. That **8** is such a strong binder provides an extreme example of the power of preorganization.

Families of hosts generally fall into the order of their listing in Chart III when arranged according to their $-\Delta G°$ values with which they bind their most complementary guests: spherands > cryptaspherands > cryptands > hemispherands > corands > podands. Corand is the family name given to modified crown ethers [33]. Spheraplex **8**·Li^+ provides a $-\Delta G°$ value of >23 kcal mol^{-1}. Cryptaspheraplexes **15**·Na^+, **16**·Na^+, and **17**·Cs^+ [34] give values of 20.6, 21.0, and 21.7 kcal mol^{-1}, respectively [27]. Cryptaplexes **18**·Li^+, **19**·Na^+, and **6**·K^+ give respective values of 16.6, 17.7, and 18.0 kcal mol^{-1} [27]. Hemispheraplexes **20**·Na^+, **21**·Na^+, and **22**·K^+ are bound by 12.2, 13.5, and 11.6 kcal mol^{-1} [35, 36]. Coraplex **23**·K^+ has a $-\Delta G°$ value of 11.4 [26, 31] and podaplexes **14**·M^+ values of <6 kcal mol^{-1} [29]. Although the numbers of binding sites and their characters certainly influence these values, the degree of preorganization appears to be dominant in providing this order.

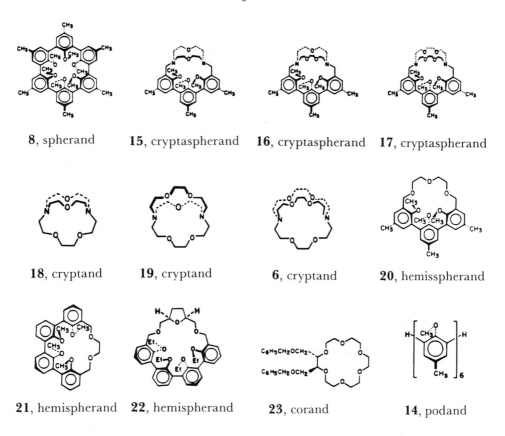

8, spherand   **15**, cryptaspherand   **16**, cryptaspherand   **17**, cryptaspherand

18, cryptand   **19**, cryptand   **6**, cryptand   **20**, hemisspherand

21, hemispherand   **22**, hemispherand   **23**, corand   **14**, podand

Chart III. Host structures arranged in the order of decreasing $-\Delta G°$ values for binding their most complementary guest picrate salts at 25 °C in $CDCl_3$ saturated with D_2O.

Structural Recognition

Just as preorganization is the central determinant of binding power, complementarity is the central determinant of structural recognition. The binding energy at a single contact site is at most a few kilocalories per mole, much lower than that of a covalent bond. Contacts at several sites between hosts and guests are required for structuring of complexes. Such contacts depend on complementary placements of binding sites in the complexing partners.

The most extensive correlations of structural recognition with host-guest structure involve the K_a values with which the spherands, cryptaspherands, cryptands, and hemispherands associate with the various alkali metal picrate salts at 25 °C in $CDCl_3$ saturated with D_2O. Chart IV lists the $K_a^A/K_a^{A'}$ ratios for various hosts binding two alkali metal ions A and A' that are adjacent to one another in the periodic table [33]. Notice that factors as high as $>10^{10}$ are observed for the spherands binding Na^+ better than K^+. Cryptaspherand **15** provides a factor of 13,000. The highest factors for hosts binding K^+ better than Na^+ are observed for cryptaspherand **17** (11,000) and hemispherand **22** (2000). The highest factors for a host binding Li^+ over Na^+ are found for cryptand **18** (4,800). These particular selectivities are important because of the

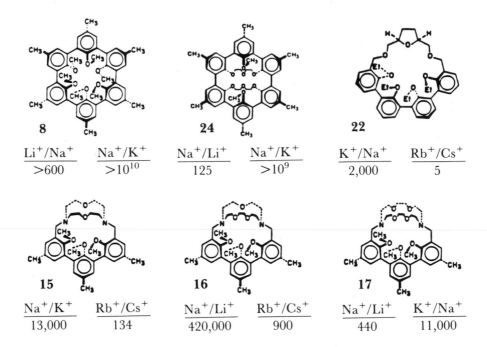

8		**24**		**22**	
Li^+/Na^+	Na^+/K^+	Na^+/Li^+	Na^+/K^+	K^+/Na^+	Rb^+/Cs^+
>600	>10^{10}	125	>10^9	2,000	5

15		**16**		**17**	
Na^+/K^+	Rb^+/Cs^+	Na^+/Li^+	Rb^+/Cs^+	Na^+/Li^+	K^+/Na^+
13,000	134	420,000	900	440	11,000

Chart IV. Structural recognition measured by $K_a^A/K_a^{A'}$ values for alkali metal picrates at 25 °C in $CDCl^3$ saturated with water.

physiological importance of these ions. These hosts, or modifications of them, are being developed for commercial use in the medical diagnostics industry.

Chart V provides stereoviews of crystal structures of capsular complexes $15 \cdot Na^+$, $17 \cdot Na^+$, and $17 \cdot K^+$. Notice that in $15 \cdot Na^+$ and $17 \cdot K^+$ the metal ions contact all of the heteroatoms, whereas in $17 \cdot Na^+$, the Na^+ ion does not. Here is a visual example of complementarity vs. noncomplementarity. The $K_a^A/K_a^{A'}$ ratio for $17 \cdot K^+/17 \cdot Na^+ = 11,000$ [34].

Arrangement of the classes of hosts in decreasing order of their ability to select between the alkali metal ion guests provides spherands > cryptaspherands ~ cryptands > hemispherands > corands > podands. This order is similar but less rigidly followed than that for host preorganization. In some cases, rather small changes in structure provide a substantial spread in $-\Delta G°$ values for binding under our standard conditions [33].

Chiral recognition in complexation is a fundamental aspect of structural recognition in complexation in the biotic world. We synthesized host **25** in an enantiomerically pure form to study its ability to distinguish between enantiomers in complexation of amino acids and ester salts in solution. We were careful to design a system containing at least one C_2 axis of symmetry, a tactic that made the hosts *nonsided* with respect to perching guests. A $CDCl_3$ solution of (R, R)-**25** in $CDCl_3$ at 0 °C was used to extract D_2O solutions of racemic amino acid or ester salts. As predicted in advance by CPK molecular models, the (D)-enantiomers were extracted preferentially into the organic layer. Chiral

15 · Na⁺

17 · Na⁺

17 · K⁺ · H₂O

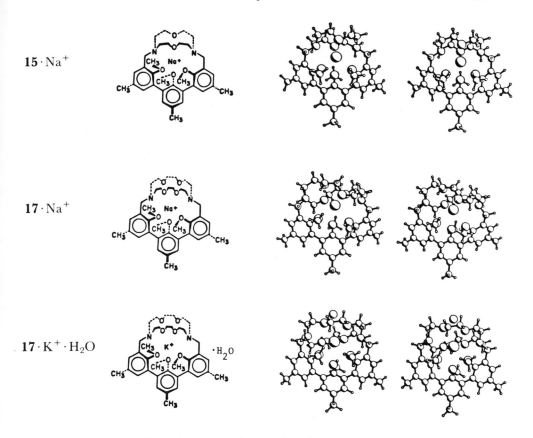

Chart V. Stereoviews of crystal structures of cryptaspheraplexes.

recognition factors ranged from a high of 31 for $C_6H_5CH(CO_2CH_3)NH_3PF_6$ to a low of 2.3 with $CH_3CH(CO_2H)NH_3ClO_4$. These factors represent free energy differences between diastereomeric complexes of 1.9 kcal mol⁻¹ and 0.42 kcal mol⁻¹, respectively. Other amino acid and ester salt guests ranged between these values. We interpreted these results in terms of the complementarity between host and guest of the (R, R)-(D)-configurations as visualized in the complex **26,** and the lack of complementarity in those of the $(R,R)-(L)$-configurations, which were designed not to form [38, 39].

An amino acid and ester resolving machine was designed, built, and tested, which is pictured in Figure 1. It made use of chiral recognition in transport of amino acid or ester salts through lipophilic liquid membranes. From the central reservoir of the W-tube containing an aqueous solution of racemic salt, the (L)-enantiomer was picked up by (S,S)-**25** in the left hand chloroform reservoir and delivered to the left hand aqueous layer, while the (D)-enantiomer was transported by (R,R)-**25** in the right hand chloroform reservoir and delivered to the right hand aqueous layer. The thermodynamic driving force for the machine's operation involved exchange of an energy-lowering entropy of dilution of each enantiomer for an energy-lowering entropy of mixing. To maintain the concentration gradients down which the enantiomers traveled in

25

26

STABLER COMPLEX

each arm of the W-tube, fresh racemic guest was continuously added to the central reservoir and (*L*)- and (*D*)-C$_6$H$_5$CH(CO$_2$CH$_3$)NH$_3$PF$_6$ of 86–90 % enantiomeric excess were continuously removed from the left and right hand aqueous reservoirs, respectively [40].

   In another experiment, we covalently attached the working part of (*R,R*)-**25** at a remote position of the molecule to a macroreticular resin (polystyrene-divinylbenzene) to give immobilized host of ~18,000 mass units per average

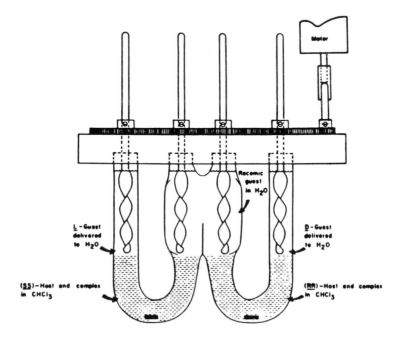

Figure 1. Enantiomer resolving machine.

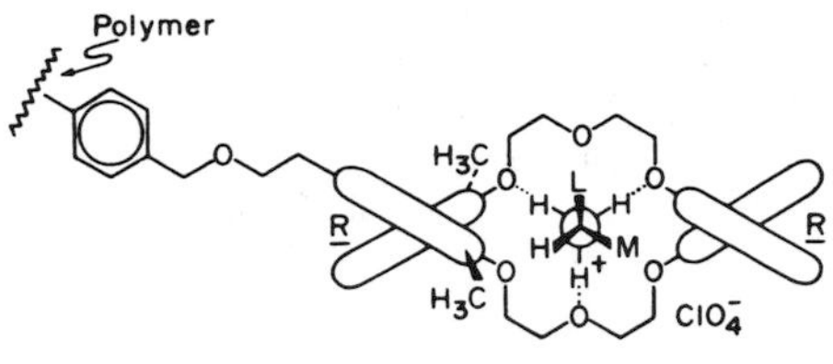

27

active site. This material (the host part of **27**) was used to give complete
enantiomeric resolution of several amino acid salts. The behavior in the chromatographic resolution paralleled that observed in the extraction and transport
experiments, and was useful both analytically and preparatively. Separation
factors ranged from 26 to 1.4, the complexes of the $(R,R)-(D)$- or $(S,S)-(L)$-
configurations always being the more stable. The structure envisioned for the
more stable complex is formulated in **27** [41].

Partial Transacylase Mimics

The design and synthesis of enzyme-mimicking host compounds remains one of
the most challenging and stimulating problems of organic chemistry. We chose
to examine transacylase mimics first because the mechanism of action of these
enzymes had been so thoroughly studied.

The active site of chymotrypsin combines a binding site, a nucleophilic
hydroxyl, an imidazole, and a carboxyl group in an array preorganized largely
by hydrogen bonds as indicated in **28**. With the help of molecular models, we
designed **29** as an "ultimate target" host possessing roughly the same organization of groups as that of **28**.

Compound **29** is much too complicated to synthesize without getting encouragement from simpler model compounds. An incremental approach to **29** was
employed. We first prepared **30,** and found that it binds t-BuNH$_3$Pic in CDCI$_3$
saturated with D$_2$O with $-\Delta G^\circ = 13.2$ kcal mol^{-1}. The complex,
30·(CH$_3$)$_3$CNH$_3^+$, had the expected crystal structure [42]. Accordingly, **31**
was prepared, and found to bind CH$_3$NH$_3$Pic and NaPic under our standard

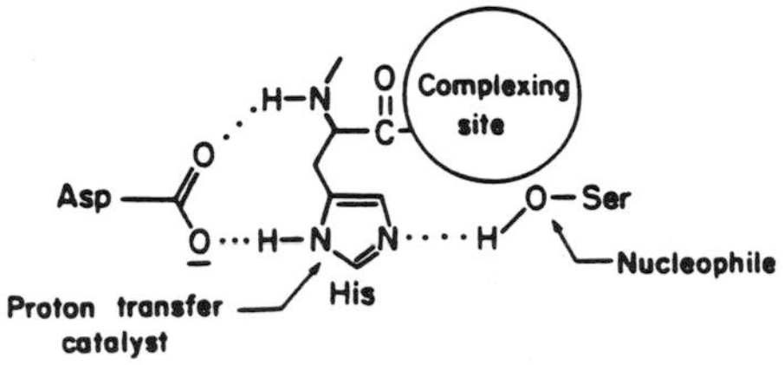

28

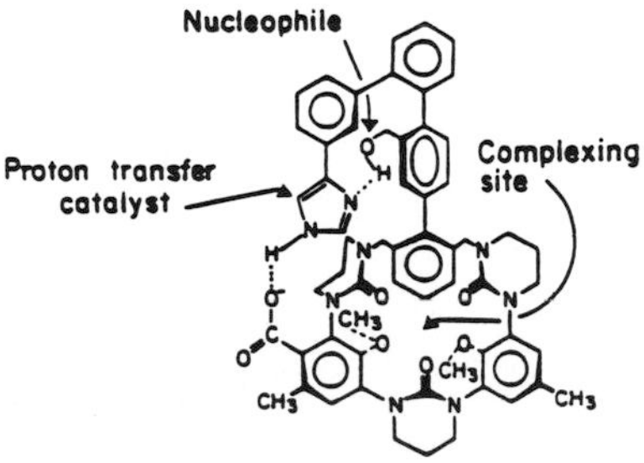

29

30   **30** · (CH₃)₃CNH₃⁺

30 · $(CH_3)_3CNH_3^+$

conditions with $-\Delta G° = 12.7$ and 13.6 kcal mol⁻¹, respectively [43]. Host **31** was acylated by **32** to give **33** and *p*-nitrophenol. The kinetics of formation of **33** were measured in CHCl₃, and found to be first order in added Et₃N/ Et₃NHClO₄ buffer ratio. Thus the alkoxide ion is the nucleophile. The rate constant for acylation of **31** by **32** was calculated to be ∼10¹¹ higher valued than the rate constant for the noncomplexed model compound, 3-phenylbenzyl alcohol [44]. This high factor demonstrates that collecting and orienting reactants through highly structured complexation can result in an enormous rate acceleration. When NaClO₄ was added to the medium, the acylation rate of **31** was depressed by several powers of ten. Thus the acylation of **31,** like that of the serine esterases, is subject to competitive inhibition.

31   **32**   **33**

34   32   half-life ≤2 min   35   half-life 4.1 hour   36

37   38   39   40

A thirty-step synthesis of **34** was then devised, and about 0.5 g of the compound prepared [45]. This compound combines the binding site, the nucleophilic hydroxyl, and the imidazole proton-transfer agent in the same molecule, lacking only the carboxyl group of final target compound **29.** Compound **34** complexed CH_3NH_3Pic and NaPic with respective $-\Delta G°$ values of 11.4 and 13.6 kcal mol^{-1} in $CDCl_3$ saturated with D_2O at 25° C. In pyridine-chloroform, amino ester salt **32** instantaneously acylated the imidazole group of **34** to give **35,** which more slowly gave **36.** In $CHCl_3$ in the absence of any added base, the observed rate constant for acylation of **34** by **32** was higher by a factor of 10^5 than that for acylation of an equal molar mixture of noncomplexing model compounds **39** or **40** under the same conditions. The same ratio was obtained when **37** was substituted for **34.** Thus the imidazole groups of **34** and **37** are the sites of acylation. Introduction of $NaClO_4$ into the medium as a competitive inhibitor of complexation destroyed much of the rate acceleration. When **32** added to **38** was substituted for **34,** the resulting complex acylated imidazole **40** with a 10 rate-constant factor increase. Thus complexed **32** is a better acylating agent than **32** alone.

The disadvantages of comparing rate constants for reactions with different molecularities are avoided by referring to uncomplexed **34** or **37,** noncomplexing imidazole **40,** and uncomplexed acylating agent **32** as standard starting states, and the rate-limiting transition states for transacylation as standard final states. This treatment introduces K_a into the second order rate constant expression when complexation precedes acylation. The resulting second order rate constants for **32** acylating **34** or **37** are higher by factors of 10^{10} or 10^{11} than the second order rate constant for **32** acylating **40.** This work clearly demonstrates that complexation of the transition states for transacylation can greatly stabilize those transition states to produce large rate factor increases over comparable noncomplexed transition states [46]. Others have shown that the imidazole of chymotrypsin is acylated first by esters of nonspecific substrates [47].

These investigations demonstrate that totally synthetic systems can be designed and prepared which mimic the following properties of enzymes: the ability to use complexation to vastly enhance reaction rates and the sensitivity to competitive inhibition. In a different, chiral system, we demonstrated that a synthetic host was capable of distinguishing between enantiomeric reactants [48, 49]. We anticipate that as the field matures, many of the other remarkable properties of enzyme systems will be observed in designed, synthetic systems. Our results illustrate some of the strategies and methods that might be applied in this expanding field of research.

Cavitands — Synthetic Molecular Vessels
Although enforced cavities of molecular dimensions are frequently encountered in enzyme systems, RNA, or DNA, they are almost unknown among the seven million synthetic organic compounds. In biological chemistry such cavities play the important role of providing concave surfaces to which are attached convergent functional groups which bind substrates and catalyze their reactions. If synthetic biomimetic systems are to be designed and investigated, simple means must be found of synthesizing compounds containing enforced concave surfaces of dimensions large enough to embrace simple molecules or ions. We applied the name *cavitand* to this class of compound [50].

Cavitands designed and studied include compounds **42—45,** many of which were prepared from **41.** The structure and conformational mobility of **41** had been established by A. G. S. Högberg [51]. The substance is prepared in good yield by treatment of resorcinol with acetaldehyde and acid. We rigidified **41** and its derivatives by closing four additional rings to produce **42—45** [50, 52].

As anticipated by molecular model examinations, **42—45** crystallize only as solvates because these rigid molecules taken alone are incapable of filling their voids either intermolecularly or intramolecularly. They are shaped like bowls of differing depth supported on four methyl "feet." Compound **42** forms crystallates with SO_2, CH_3CN, and CH_2Cl_2, molecules to which it is complementary (molecular model examination). Cavitand **43,** whose cavity is deeper, crystallizes with a mole of $CHCl_3$. Crystal structures of **42**·CH_2Cl_2 and **43**·$CHCl_3$ show they are caviplexes, as predicted [53]. Cavitand **44** is vase-shaped. It crystallizes with one mole of $(CH_3)_2NCHO$, which is just small enough to fit into the interior of **44** in models. Although the amide cannot be removed at high temperature and low pressure, it is easily displaced with $CHCl_3$, one and one-half moles of which appear to take the place of the $(CH_3)_2NCHO$ in the crystallate [50].

Treatment of octol **41** with R_2SiCl_2 gave a series of cavitands, of which **45** is typical. In molecular models, **45** has a well-shaped cavity, defined by the bottoms of four aryls and by four inward-turned methyl groups. In molecular

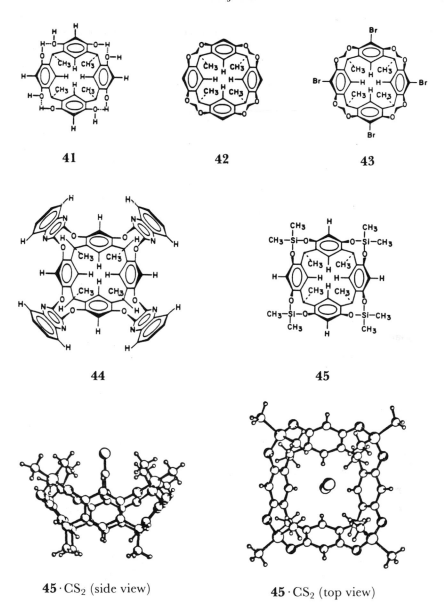

41

42

43

44

45

45·CS₂ (side view)

45·CS₂ (top view)

models, this well is complementary to small, cylindrical molecules such as S=C=S, CH₃C≡CH, and O=O, but not to larger compounds such as CDCl₃ or C₆D₆. Cavitand **45** and its analogues when dissolved in CDCl₃ or C₆D₆ complex guests such as those mentioned above, whose external surfaces are complementary to the internal surface of the host cavity. Association constants were determined for **45** and its analogues binding S=C=S. Values of $-\triangle G°$ as high as 2 kcal mol⁻¹ have been observed. A crystal structure of **45·CS₂** shows that CS₂ occupies the well in the expected manner. Compound **45** in CDCl₃ was also shown to bind dioxygen reversibly [52]. Dissolution of **45** in solvents such as CDCl₃ or C₆D₆ is the equivalent of dissolving "holes" in a medium into which appropriately shaped solutes fall. The discrimination shown by the holes for the guests exemplifies the principle of complementarity as applied to cavitand complexation.

The next steps in research on these cavitands is to append to them water-solubilizing and catalytic groups. The former will provide them with hydrophobic driving forces to complex nonpolar guests, and the latter to catalyze reactions of such guests.

Carcerands—Synthetic Molecular Cells

Absent among the millions of organic compounds hitherto reported are closed-surface hosts with enforced interiors large enough to imprison behind covalent bars, guests the size of ordinary solvent molecules. After much thought and molecular model examination, we chose **48** as the target for synthesis of the first molecular cell. The term *carcerand* was applied to this class of compound. The synthesis involved treating Cs_2CO_3 with a solution in $(CH_3)_2NCHO-$ $(CH_2)_4O$ of equal molar amounts of cavitands **46** and **47** under an atmosphere of argon. The first question to be answered was: what guest compounds would be trapped inside during the shell closure? This question is akin to asking whether two soup bowls closed rim-to-rim under the surface of a kettle of stew would net any stew. The answer was that **48** "contained" essentially every kind of component of the medium present during ring closure [54].

The product (**48** and guests) was very insoluble in all media, and was purified by extracting it with the most powerful solvents of each type. The remaining material was subjected to elemental analysis for C, H, S, O, N, Cl, and Cs. Nitrogen analysis and an IR spectrum of the substance revealed that $(CH_3)_2NCHO$ had been entrapped. The presence of equivalent amounts of Cs and Cl demonstrated that one or the other ion or both had to be encapsulated in the host.

A fast atom bombardment mass spectrum of **48·G** showed the presence of the following host-guest combinations, the species trapped in the interior of **48** being enclosed by parentheses:
48·no guest; **48**·(Cs^+) ·Cl^-; **48**· ((CH_3)$_2NCHO$)): **48**·($Cs^+ + H_2O$)·Cl^-; **48**·((CH_2)$_4O + H_2O$); **48**·((CH_3)$_2NCHO + Cs^+$)·Cl^-; **48**·($Cs^+ + Ar$)·Cl^-; **48**·($Cs^+ + H_2O + Cs^+$)·$Cl_2^=$; **48**·($Cs^+ + Cl^-$), and **48**·($Cs^+ + Cs^+ + Cl^-$)·Cl^-.

46   47   48

No peaks were found at molecular masses above that of the last carcaplex listed. None were observed that could not be interpreted in terms of appropriate host-guest combinations. When highly dried **48** was boiled with D_2O, the **48**·($Cs^+ + H_2O$) peak was substantially replaced by a **48**·($Cs^+ + D_2O$) peak. Models suggest that **48** has two small portals lined with methyl groups through which molecules as small as H_2O can pass.

Molecular models of **48** show that its interior surface is complementary to the outer surface of anti-$ClCF_2CF_2Cl$. Shell closure of **46** and **47** in the presence of this Freon resulted in entrapment of a small amount of this gas in the interior of **48**.

The FAB-MS coupled with the elemental analyses indicated that about 5 % of the mixture was noncomplexed **48,** about 60 % encapsulated Cs^+, about 45 % encapsulated $(CH_3)_2NCHO$, 15 % encapsulated $(CH_2)_4O$, but only 1 – 2 % encapsulated Cl^-. Thus Cs^+ was mainly inside and Cl^- mainly outside the carcaplex. Models show that if the final covalent bond leading to **48**·G involves an intramolecular SN_2 linear transition state as in **48,** any Cs^+ ion-paired to the S^- is trapped inside the cavity and the Cl^- must be external to the cavity [54].

$$Cs^+S^-\overset{\frown}{}CH_2\overset{\frown}{}Cl$$

49

We anticipate that unusual physical and chemical properties will provide unusual uses for carcaplexes, particularly when their design renders them soluble and separable.

We warmly thank the following co-workers for carrying out the research described here: L. A. Singer, R. H. Bauer, M. G. Siegel, J. M. Timko, K. Madan, S. S. Moore, T. L. Tarnowski, G. M. Lein, J. L. Toner, J. M. Mayer, S. P. Ho, M. P. de Grandpre, S. P. Artz, G. D. Y. Sogah, S. C. Peacock, L. A. Domeier, H. E. Katz, I. B. Dicker, J. R. Moran, and K. D. Stewart as graduate students; E. P. Kyba, L. R. Sousa, K. Koga, R. C. Helgeson, G. W. Gokel, D. M. Walba, J. M. Cram, T. Kaneda, S. B. Brown, K. E. Koenig, P. Stückler, G. D. Y. Sogah, G. R. Weisman, Y. Chao, F. C. A. Gaeta, M. Newcomb, P. Y. S. Lam, S. Karbach, A. G. S. Högberg, Y. H. Kim, and M. Lauer as postdoctoral fellows. The crystal structure work of colleagues K. N. Trueblood, C. B. Knobler, E. F. Maverick and I. Goldberg was indispensable.

We gratefully acknowledge the financial support of the following granting agencies: the Division of Basic Energy Sciences of the Department of Energy for the work on the metal ion binding; the National Science Foundation for the work on structural recognition; the National Institutes of Health for research on catalysis. We warmly thank all former and present co-workers, over 200 in number, and the many others whose results and discussions have stimulated and instructed us over the years. My long-time colleague, Roger C. Helgeson, has provided us not only with excellent ideas and results, but also with continuity. The artwork displayed here and in my slides and publications for the last twelve years was done by Mrs. June Hendrix, to whom we are much indebted.

REFERENCES AND NOTES

1. Cram, D. J.; Bauer, R. H. *J. Am. Chem. Soc.* **1959**, *81*, 5971–5977.
2. Singer, L. A.; Cram, D. J. *J. Am. Chem. Soc.* **1963**, *85*, 1080–1084
3. Pedersen, C. J. *J. Am. Chem. Soc.* **1967**, *89*, 2495–2496.
4. Pedersen, C. J. *J. Am. Chem. Soc.* **1967**, *89*, 7017–7036.
5. Dietrich, B.; Lehn, J.-M.; Sauvage, J.-P. *Tetrahedron Lett.* **1969**, 2885–2888.
6. Dietrich, B.; Lehn, J.-M.; Sauvage, J.-P. *Tetrahedron Lett.* **1969**, 2889–2892.
7. Kyba, E. P.; Siegel, M. G.; Sousa, L. R.; Sogah, G. D. Y.; Cram, D. J. *J. Am. Chem. Soc.* **1973**, *95*, 2691–2692.
8. Kyba, E. P.; Koga, K.; Sousa, L. R.; Siegel, M. G.; Cram, D. J. *J. Am. Chem. Soc.* **1973**, *95*, 2692–2693.
9. Helgeson, R. C.; Koga, K.; Timko, J. M.; Cram, D. J. *J. Am. Chem. Soc.* **1973**, *95*, 3021–3023.
10. Helgeson, R. C.; Timko, J. M.; Cram, D. J. *J. Am. Chem. Soc.* **1973**, *95*, 3023–3025.
11. Gokel, G. W.; Cram, D. J. *J. C. S. Chem. Commun.* **1973**, *521*, 481–482.
12. Cram, D. J.; Cram, J. M. *Science,* **1974**, *183*, 803–809.
13. Aeschylus [525–456 B. C.], *The Choëphoroe,* Translated by Sir Gilbert Murry, taken from J. Bartlett, *Familiar Quotations,* 11th. Edition, C. Morley and L. D. Everett editors, Garden City Publishing Co., Garden City, New York, 1944, p. 963.
14. Kyba, E. P.; Helgeson, R. C.; Madan, K.; Gokel, G. W.; Tarnowski, T. L.; Moore, S. S.; Cram, D. J. *J. Am. Chem. Soc.* **1977**, *99*, 2564–2571.
15. Topics in Current Chemistry, "Host-Guest Complex", Volumes I–III, ed. E. L. Boschke, Springer Verlag, Berlin, 1982–1984.
16. Koltun, W. L. *Biopolymers* **1965**, *3*, 665–679.
17. The crystal structures and references to them up to 1980 are gathered in Cram, D. J.; Trueblood, K. N. *Topics in Current Chemistry,* **1981**, *98*, 43–106.
18. Cram, D. J.; Lein, G. M. *J. Am. Chem. Soc.* **1985**, *107*, 3657–3668.
19. Timko, J. M.; Moore, S. S.; Walba, D. M.; Hiberty, P. C.; Cram, D. J. *J. Am. Chem. Soc.* **1977**, *99*, 4207–4219.
20. Helgeson, R. C.; Tarnowski, T. L.; Cram, D. J. *J. Org. Chem.* **1979**, *44*, 2538–2550.
21. Dunitz, J. D.; Dobler, M.; Seiler, P.; Phizackerly, R. P. *Acta Crystallogr. Sect. B,* **1974**, *30*, 2733 and following papers to 2750.
22. Weiss, R.; Metz, B.; Moras, D. *Proc. Int. Conf. Coord. Chem.* 13th. **1970**, *2*, 85–86.
23. Metz, B.; Moras, D.; Weiss, R. *Acta Crystallogr. Sect. B,* **1973**, *29*, 1377–1381.
24. Cram, D. J.; Kaneda, T.; Helgeson, R. C.; Lein, G. M. *J. Am. Chem. Soc.* **1979**, *101*, 6752–6754.
25. Trueblood, K. N.; Knobler, C. B.; Maverick, E.; Helgeson, R. C.; Brown, S. B.; Cram, D. J. *J. Am. Chem. Soc.* **1981**, *103*, 5594–5596.
26. Helgeson, R. C.; Weisman, G. R.; Toner, J. L.; Tarnowski, T. L.; Chao, Y.; Mayer, J. M.; Cram, D. J. *J. Am. Chem. Soc.* **1979**, *101*, 4928–4941.
27. Cram, D. J.; Ho, S. P. *J. Am. Chem. Soc.* **1985**, *107*, 2998–3005.
28. Cram, D. J. *Science* **1983**, *219*, 1177–1183.
29. Cram, D. J.; deGrandpre, M. P.; Knobler, C. B.; Trueblood, K. N. *J. Am. Chem. Soc.* **1984**, *106*, 3286–3292.
30. Cram, D. J.; Kaneda, T.; Helgeson, R. C.; Brown, S. B.; Knobler, C. B.; Maverick, E.; Trueblood, K. N. *J. Am. Chem. Soc.* **1985**, *107*, 3645–3657.
31. Mitsky, J.; Jaris, L.; Taft, R. W. *J. Am. Chem. Soc.* **1972**, *94*, 3442–3445.
32. Aitken, H. W.; Gilkerson, W. R. *J. Am. Chem. Soc.* **1973**, *95*, 8551–8559.
33. Cram, D. J. *Angew, Chemie Int. Ed.,* 1986, 25, 1039–1057.

34. Cram, D. J.; Ho, S. P.; Knobler, C. B.; Maverick, E.; Trueblood, K. N. *J. Am. Chem. Soc.* **1985**, *107*, 2989−2998.
35. Koenig, K. E.; Lein, G. M.; Stückler, P.; Kaneda, T.; Cram, D. J. *J. Am. Chem. Soc.* **1979**, *101*, 3553−3566.
36. Artz, S. P.; Cram, D. J. *J. Am. Chem. Soc.* **1984**, *106*, 2160−2171.
37. The *cis*-isomer gave a lower-−$\Delta G°_{av}$ value than the *trans*-isomer by 0.7 kcal mol $^{-1}$.
38. Peacock, S. C.; Domeier, L. A.; Gaeta, F. C. A.; Helgeson, R. C.; Timko, J. M.; Cram, D. J. *J. Am. Chem. Soc.* **1978**, *100*, 8190−8202.
39. Peacock, S. C.; Walba, D. M.; Gaeta, F. C. A.; Helgeson, R. C.; Cram, D. J. *J. Am. Chem. Soc.* **1980**, *102*, 2043−2052.
40. Newcomb, M.; Toner, J. L.; Helgeson, R. C.; Cram, D. J. *J. Am. Chem. Soc.* **1979**, *101*, 4941−4947.
41. Sogah, G. D. Y.; Cram, D. J. *J. Am. Chem. Soc.* **1979**, *101*, 3035 3042
42. Cram, D. J.; Dicker, I. B.; Lauer, M.; Knobler, C. B.; Trueblood, K. N. *J. Am. Chem. Soc.* **1984**, *106*, 7150−7167.
43. Katz, H. E.; Cram, D. J. *J. Am. Chem. Soc.* **1983**, *105*, 135−137.
44. Cram, D. J.; Katz, H. E.; Dicker, I. B. *J. Am. Chem. Soc.* **1984**, *106*, 4987−5000.
45. Cram, D. J.; Lam, P. Y. S. *Tetrahedron Symposium-in-Print*, **1986**, *42*, 1607−1615.
46. Cram, D. J.; Lam, P. Y. S.; Ho, S. P. *J. Am. Chem. Soc.* **1986**, *108*, 839−841.
47. Hubbard, C. D.; Kirsch, J. F. *Biochemistry*, **1972**, *11*, 2483−2493.
48. Chao, Y.; Cram, D. J. *J. Am. Chem. Soc.* **1976**, *98*, 1015−1017.
49. Chao, Y.; Weisman, G. R.; Sogah, G. D. Y.; Cram, D. J. *J. Am. Chem. Soc.* **1979**, *101*, 4948−4958.
50. Moran, J. R.; Karbach, S.; Cram, D. J. *J. Am. Chem. Soc.* **1982**, *104*, 5826−5828.
51. Högberg, A. G. S. *J. Am. Chem. Soc.* **1980**, *102*, 6046−6050.
52. Cram, D. J.; Stewart, K. D.; Goldberg, I.; Trueblood, K. N. *J. Am. Chem. Soc.* **1985**, *107*, 2574−2575.
53. Cram, D. J.; Cram, J. M. "Designed Complexes−Science and Applications", Chapter in Monograph "Selectivity; A goal for Synthetic Efficiency", W. Bartmann and B. M. Trost, Ed., Workshop Conference Hoechst, *14*, *Verlag* Chemie, Weinheim, Germany, **1983**, 42−64.
54. Cram, D. J.; Karbach, S.; Kim, Y. H.; Baczynskyj, L.; Kalleymeyn, G. W. *J. Am. Chem. Soc.* **1985**, *107*, 2575−2576.

Jean-Marie Lehn

JEAN-MARIE LEHN

I was born on September 30 1939 in Rosheim, a small medieval city of Alsace in France. My father, Pierre Lehn, then a baker, was very interested in music, played the piano and the organ and became later, having given up the bakery, the organist of the city. My mother Marie kept the house and the shop. I was the eldest of four sons and helped out in the shop with my first brother. I grew up in Rosheim during the years of the second world war, went to primary school after the war and, at age eleven, I entered high school, the Collège Freppel, located in Obernai, a small city about five kilometers from Rosheim. During these years I began to play the piano and the organ, and with time music has become my major interest outside science. My high school studies from 1950 to 1957 were in classics, with latin, greek, german, and english languages, french literature and, during the last year, philosophy, on which I was especially keen. However, I also became interested in sciences, especially chemistry, so that I obtained the baccalauréat in Philosophy in July 1957 and in Experimental Sciences in September of the same year.

I envisaged to study philosophy at the University of Strasbourg, but being still undecided, I began with first year courses in physical, chemical and natural sciences (SPCN). During this year 1957/58, I was impressed by the coherent and rigorous structure of organic chemistry. I was particularly receptive to the experimental power of organic chemistry, which was able to convert at will, it seemed, complicated substances into one another following well defined rules and routes. I bought myself compounds and glassware and began performing laboratory practice experiments at my parents home. The seed was sown, so that when, the next year, I followed the stimulating lectures of a newly appointed young professor, Guy Ourisson, it became clear to me that I wanted to do research in organic chemistry.

After having obtained the degree of Licencié-ès-Sciences (Bachelor), I entered Ourisson's laboratory in October of 1960, as a junior member of the Centre National de la Recherche Scientifique in order to work towards a Ph.D. degree. This was the first decisive stage of my training. My work was concerned with conformational and physico-chemical properties of triterpenes. Being in charge of our first NMR spectrometer, I was led to penetrate more deeply into the arcanes of this very powerful physical method; this was to be of much importance for later studies. My first scientific paper in 1961 reported an additivity rule for substituent induced shifts of proton NMR signals in steroid derivatives.

Having obtained my degree of Docteur ès Sciences (Ph.D.) in June of 1963, I spent a year in the laboratory of Robert Burns Woodward at Harvard University, where I took part in the immense enterprise of the total synthesis of Vitamin B_{12}. This was the second decisive stage of my life as a researcher. I also followed a course in quantum mechanics and performed my first computations with Roald Hoffmann. I had the chance to witness in 1964 the initial stages of what was to become the Woodward-Hoffmann rules.

After my return to Strasbourg, I began to work in the area of physical organic chemistry, where I could combine the knowledge acquired in organic chemistry, in quantum theory and on physical methods. It was clear that, in order to be able to better analyze physical properties of molecules, a powerful means was to synthesize compounds that would be especially well suited for revealing a given property and its relationships to structure. This orientation characterized the years 1965-1970 of my activities and of my young laboratory, newly established after my appointment in 1966 as maître de conférences (assistant professor) at the Chemistry Department of the Univerity of Strasbourg. Our main research topics were concerned with NMR studies of conformational rate processes, nitrogen inversion, quadrupolar relaxation, molecular motions and liquid structure, as well as ab initio quantum chemical computations of inversion barriers, of electronic structures and later on, of stereoelectronic effects.

While pursuing these projects, my interest for the processes occurring in the nervous system (stemming diffusely from the first year courses in biology as well as from my earlier inclination towards philosophy), led me to wonder how a chemist might contribute to their study. The electrical phenomena in nerve cells depend on sodium and potassium ion distributions across membranes. A possible entry into the field was to try to affect the processes which allow ion transport and gradients to be established. I related this to the then very recent observations that natural antibiotics were able to make membranes permeable to cations. It thus appeared possible to devise chemical substances that would display similar properties. The search for such compounds led to the design of cation cryptates, on which work was started in October 1967 This area of research expanded rapidly, taking up eventually the major part of my group and developing into what I later on termed "supramolecular chemistry". Organic, inorganic and biological aspects of this field were explored and investigations are continuing. In 1976 another line of research was started in the area of artificial photosynthesis and the storage and chemical conversion of solar energy; it was first concerned with the photolysis of water and later with the photoreduction of carbon dioxide.

I was promoted associate professor in early 1970 and full professor in October of the same year. I spent the two spring semesters of 1972 and 1974 as visiting professor at Harvard University giving lectures and directing a research project. This relationship extended on a loose basis to 1980. In 1979, I was elected to the chair of "Chimie des Interactions Moléculaires" at the Collège de France in Paris. I took over the chemistry laboratory of the Collège de France when Alain Horeau retired in 1980 and thereafter divided my time between the two laboratories in Strasbourg and in Paris, a situation continuing up to the present. New lines of research developed, in particular on combining the recognition, transport and catalytic properties displayed by supramolecular species with the features of organized phases, the long range goal being to design and realize "molecular devices", molecular components that would eventually be able to perform signal and information processing at the molecular level. A major research effort is presently also devoted to supramolecular self-organisation, the design and properties of "programmed" supramolecular systems.

The scientific work, performed over twenty years with about 150 collaborators from over twenty countries, has been described in about 400 publications and review papers. Over the years I was visiting professor at other institutions, the E.T.H. in Zürich, the Universities of Cambridge, Barcelona, Frankfurt.

In 1965 I married Sylvie Lederer and we have two sons, David (born 1966) and Mathias (born 1969).

Honors

Bronze (1963), Silver (1972) and Gold (1981) Medals of the CNRS; Adrian Prize (1968) and Raymond Berr Prize (1978) of the French Chemical Society; Gold Medal of the Pontifical Academy of Sciences (1981); Parcelsus Prize (1982, Swiss Chemical Society); Prize of the CEA (1984, French Academy of Sciences); Alexander von Humboldt (1983), Rolf-Sammet (1985, University of Frankfurt), George Kenner (1987, University of Liverpool) Prizes; Nobel Prize of Chemistry (1987); Sigillum Magnum (1988, University of Bologna); Vermeil Medal (1989, Ville de Paris); Gold Medal (1989, Société d'Encouragement au Progrés); Karl-Ziegler Prize (1989, Gesellschaft Deutscher Chemiker).

Member of the Académie des Sciences, Institut de France (1985); foreign member of the National Academy of Sciences of the USA (1980), the American Academy of Arts and Sciences (1980), the Royal Academy of Arts and Sciences of the Netherlands (1983), the Deutsche Akademie der Naturforscher Leopoldina (1985), the Accademia Nazionale dei Lincei (1985), the American Philosophical Society (1987); honorary member of the Royal Society of Chemistry of Belgium and of Great-Britain (1987); member of the Academia Europaea (1988), the Académie d'Alsace (1989); foreign associate of the Akademie der Wissenschaften und der Literatur-Mainz (1989); honorary member of the Yugoslav Academy of Sciences and Arts (1990); member of the Akademie der Wissenschaften of Göttingen (1990); associate member of the Koninklijke Academie voor Wetenschappen, Letteren en Schone Kunsten van België (1990); honorary fellow of the Indian Academy of Sciences (1991); foreign member of the Polish Academy of Sciences (1991), the Academy of Arts and Sciences of Puerto Rico (1991), the Ukrainian Academy of Sciences (1992).

Honorary Doctorates from the Hebrew University of Jerusalem (1984), the Universidad Autonoma of Madrid 1985), the Georg-August University of Göttingen (1987), the Université Libre of Bruxelles (1987), the Iraklion University (1989), the Università degli Studi di Bologna (1989), the Charles

University of Prague (1990), the University of Sheffield (1991) and the University of Twente (1991),

Chevalier dans l'Ordre National du Périte (1976), de la Légion d'Honneur (1983); officier de la Légion d'Honneur (1988); Chevalier dans l'Ordre des Palmes Académiques (1989); member of the Order "Pour le Mérite" for Sciences and Arts (1990, Allemagne).

Review papers in the area of Supramolecular Chemistry

- "Design of organic complexing agents. Strategies towards properties", Structure and Bonding 16, 1, 1973.
- "Cryptates: The chemistry of macropolycyclic inclusion complexes", Accounts of Chem. Res. 11, 49, 1978.
- "Cryptates: Inclusion complexes of macropolycyclic receptor molecules", Pure & Appl. Chem. 50, 871, 1978.
- "Macrocyclic receptor molecules: Aspects of chemical reactivity. Investigations into molecular catalysis and transport processes", Pure & Appl. Chem. 51, 979, 1979.
- "Physicochemical studies of crown and cryptate complexes", in Coordination Chemistry of Macrocyclic Compounds, Plenum Publishing Corporation, Ch. 9, 1979 (with A.I. Popov).
- "Cryptate inclusion Complexes. Effects on solute-solute and solute-solvent interactions and on ionic reactivity", Pure & Appl. Chem. 52, 2303, 1980.
- "Dinuclear cryptates: Dimetallic macropolycyclic inclusion complexes. Concepts − Design − Prospects", Pure & Appl. Chem. 52, 2441, 1980.
- Inaugural Lecture. Chair of "Chimie des Interactions Moléculaires", Collège de France, 7 Mars 1980.
- "Macropolycyclic structures for bio- and abio-inorganic chemistry from dinucleating to polynucleating cryptands", IUPAC Frontiers of Chemistry, Pergamon Press, p.265, 1982.
- "Supramolecular organic chemistry − From molecular receptors to coreceptors", in Biomimetic Chemistry, p.163, Elsevier (Amsterdam), Kodansha Ltd (Tokyo), 1983.
- "Chemistry of transport processes − Design of synthetic carrier molecules", in Physical Chemistry of Transmembrane Ion Motions, G. Spach Ed., p. 181, Elsevier, 1983.
- "Supramolecular chemistry: Receptors, catalysts and carriers", Science 227, 849, 1985.
- "Recent studies of supramolecular catalysis and transport processes", in International Symposium on Bioorganic Chemistry, Annals of the New York Academy of Sciences, Vol. 471, p.41, 1986.
- "Molecular recognition: Design of abiotic receptor molecules", in Design and Synthesis of Organic Molecules Based on Molecular Recognition, Ed. G. van Binst, Springer-Verlag, p.173, 1986.

— "Design of cation and anion receptors, catalysts, and carriers", Vol. 4 of Synthesis of Macrocycles: The Design of Selective Complexing Agents, Eds. R.M. Izatt & J.J. Christensen, John Wiley & Sons, Inc. P. 167, 1987 (with P.G. Potvin).
— "Multidentate Macrocyclic and Macropolycyclic Ligands" in Comprehensive Coordination Chemistry, Ed. G. Wilkinson, Pergamon Press, Vol. 1, Part 2, Ch. 21.3, p.915, 1987 (with K.B. Mertes).
— "Photophysical and photochemical aspects of supramolecular chemistry", in Supramolecular Photochemistry, Ed. V. Balzani, D. Reidel Publishing Company, p. 29, 1987.
— "Perspectives in Supramolecular Chemistry – From Molecular Recognition towards Molecular Information Processing and Self-Organization", Angew. Chem. Int. Ed. 29, 7304, 1991.

N.B. The references cited in the text of the Nobel Lecture give access to the original papers on our work in this area.

SUPRAMOLECULAR CHEMISTRY — SCOPE AND PERSPECTIVES MOLECULES — SUPERMOLECULES — MOLECULAR DEVICES

Nobel lecture, December 8, 1987

by

JEAN-MARIE LEHN

Institut Le Bel, Université Louis Pasteur, 4, rue Blaise Pascal, 67000 Strasbourg and Collège de France, 11 Place Marcelin Berthelot, 75005 Paris.

Abstract

Supramolecular chemistry is the chemistry of the intermolecular bond, covering the structures and functions of the entities formed by association of two or more chemical species. Molecular recognition in the supermolecules formed by receptor-substrate binding rests on the principles of molecular complementarity, as found in spherical and tetrahedral recognition, linear recognition by coreceptors, metalloreceptors, amphiphilic receptors, anion coordination. Supramolecular catalysis by receptors bearing reactive groups effects bond cleavage reactions as well as synthetic, bond formation via cocatalysis. Lipophilic receptor molecules act as selective carriers for various substrates and allow to set up coupled transport processes linked to electron and proton gradients or to light. Whereas endo-receptors bind substrates in molecular cavities by convergent interactions, exo-receptors rely on interactions between the surfaces of the receptor and the substrate; thus new types of receptors such as the metallonucleates may be designed. In combination with polymolecular assemblies, receptors, carriers and catalysts may lead to molecular and supramolecular devices, defined as structurally organized and functionally integrated chemical systems built on supramolecular architectures. Their recognition, transfer and transformation features are analyzed specifically from the point of view of molecular devices that would operate via photons, electrons or ions, thus defining fields of molecular photonics, electronics and ionics. Introduction of photosensitive groups yields photoactive receptors for the design of light conversion and charge separation centres. Redox active polyolefinic chains represent molecular wires for electron transfer through membranes. Tubular mesophases formed by stacking of suitable macrocyclic receptors may lead to ion channels. Molecular selfassembling occurs with acyclic ligands that form complexes of double helical structure. Such developments in molecular and supramolecular design and engineering open perspectives towards the realization of molecular photonic, electronic and ionic devices, that would perform highly selective recognition, reaction and transfer operations for signal and information processing at the molecular level.

1. From Molecular to Supramolecular Chemistry

Molecular chemistry, the chemistry of the covalent bond, is concerned with uncovering and mastering the rules that govern the structures, properties and transformations of molecular species.

Supramolecular chemistry may be defined as "chemistry beyond the molecule", bearing on the organized entities of higher complexity that result from the association of two or more chemical species held together by intermolecular forces. Its development requires the use of all ressources of molecular chemistry combined with the designed manipulation of non-covalent interactions so as to form supramolecular entities, supermolecules possessing features as well defined as those of molecules themselves. One may say that supermolecules are to molecules and the intermolecular bond what molecules are to atoms and the covalent bond.

Basic concepts, terminology and definitions of supramolecular chemistry were introduced earlier [1−3] and will only be summarized here. Section 2.3. below provides a brief account on the origins and initial developments of our work which led to the formulation of supramolecular chemistry. Molecular associations have been recognized and studied for a long time [4] and the term "übermoleküle", i.e. supermolecules, was introduced already in the mid-1930's to describe entities of higher organization resulting from the association of coordinatively saturated species [5]. The partners of a supramolecular species have been named *molecular receptor* and *substrate* [1, 2, 65], the substrate being usually the smaller component whose binding is being sought. This terminology conveys the relation to biological receptors and substrates for which *Paul Ehrlich* stated that molecules do not act if they are not bound ("Corpora non agunt nisi fixata"). The widely employed term of *ligand* seemed less appropriate in view of its many unspecific uses for either partner in a complex. Molecular interactions form the basis of the highly specific recognition, reaction, transport, regulation etc. processes that occur in biology such as substrate binding to a receptor protein, enzymatic reactions, assembling of protein-protein complexes, immunological antigen-antibody association, intermolecular reading, translation and transcription of the genetic code, signal induction by neurotransmitters, cellular recognition, etc. The design of artificial, abiotic, receptor molecules capable of displaying processes of highest efficiency and selectivity requires the correct manipulation of the energetic and stereochemical features of the non-covalent, intermolecular forces (electrostatic interactions, hydrogen bonding, Van der Waals forces etc.) within a defined molecular architecture. In doing so, the chemist may find inspiration in the ingenuity of biological events and encouragement in their demonstration that such high efficiencies, selectivities and rates can indeed be attained. However chemistry is not limited to systems similar to those found in biology, but is free to invent novel species and processes.

Binding of a substrate σ to its receptor ρ yields the supermolecule and

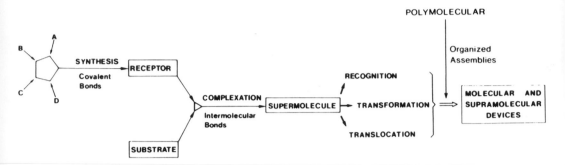

Scheme 1. From molecular to supramolecular chemistry: molecules supermolecules, molecular and supramolecular devices.

involves a molecular recognition process. If, in addition to binding sites, the receptor also bears reactive functions it may effect a chemical transformation on the bound substrate, thus behaving as a supramolecular reagent or catalyst. A lipophilic, membrane soluble receptor may act as a carrier effecting the translocation of the bound substrate. Thus, *molecular recognition, transformation* and *translocation* represent the basic functions of supramolecular species. More complex functions may result from the interplay of several binding subunits in a polytopic coreceptor. In association with organized polymolecular assemblies and phases (layers, membranes, vesicles, liquid crystals, etc.), functional supermolecules may lead to the development of *molecular devices*. The present text describes these various aspects of supramolecular chemistry (diagrammatically shown in Scheme 1) and sketches some lines of future development (for earlier general presentations see [1–3, 6–9]). The results discussed here, taken mainly from our own work, have been completed by references to other studies, in order to draw a broader picture of this rapidly evolving field of research. Emphasis will bear on conceptual framework, classes of compounds and types of processes. Considering the vast literature that has developed, the topics of various meetings and symposia, etc., there is no possibility here to do justice to the numerous results obtained, all the more to provide an exhaustive account of this field of science. Supramolecular chemistry, the designed chemistry of the intermolecular bond, is rapidly expanding at the frontiers of molecular science with physical and biological phenomena.

2. Molecular Recognition

2.1. *Recognition — Information — Complementarity*

Molecular recognition has been defined as a process involving *both binding* and *selection* of substrate(s) by a given receptor molecule, as well as possibly a specific *function* [1]. Mere binding is not recognition, although it is often taken as such. One may say that recognition is binding with a purpose, like receptors are ligands with a purpose. It implies a structurally well defined pattern of intermolecular interactions.

Binding of σ to ϱ forms a supermolecule characterized by its thermodynamic and kinetic stability and selectivity, i.e. by the amount of energy and of information brought into operation. Molecular recognition thus is a question of *information storage* and *read out* at the supramolecular level. Information may be stored in the architecture of the ligand, in its binding sites (nature, number, arrangement) and in the ligand layer surrounding bound σ; it is read out at the rate of formation and dissociation of the supermolecule. Molecular recognition thus corresponds to optimal information content of ϱ for a given σ [1, 3]. This amounts to a generalized *double complementarity principle* extending over energetical (electronic) as well as geometrical features, the celebrated "lock and key", steric fit concept enunciated by *Emil Fischer* in 1894 [10]. Enhanced recognition beyond that provided by a single equilibrium step may be achieved by multistep recognition and coupling to an irreversible process [11].

The ideas of molecular recognition and of receptor chemistry have been penetrating chemistry more and more over the last fifteen years, namely in view of its bioorganic implications, but more generally for its significance in intermolecular chemistry and in chemical selectivity [1−3, 6−9, 12−21].

2.2. *Molecular Receptors — Design Principles*

Receptor chemistry, the chemistry of artificial receptor molecules may be considered a generalized coordination chemistry, not limited to transition metal ions but extending to of all types of substrates: cationic, anionic or neutral species of organic, inorganic or biological nature.

In order to achieve high recognition it is desirable that receptor and substrate be in contact over a large area. This occurs when ϱ is able to wrap around its guest so as to establish numerous non covalent binding interactions and to sense its molecular size, shape and architecture. It is the case for receptor molecules that contain intramolecular cavities into which the substrate may fit, thus yielding an inclusion complex, a *cryptate.* In such concave receptors the cavity is lined with binding sites directed towards the bound species; they are endopolarophilic [1] and *convergent,* and may be termed *endoreceptors* (see also below).

Macropolycyclic structures meet the requirements for designing artificial receptors: − they are large (macro) and may therefore contain cavities and clefts of appropriate size and shape; − they possess numerous branches, bridges and connections (polycyclic) that allow to construct a given architecture endowed with desired dynamic features; − they allow the arrangement of structural groups, binding sites and reactive functions.

The balance between rigidity and flexibility is of particular importance for the dynamic properties of ϱ and of σ. Although high recognition may be achieved with rigidly organized receptors, processes of exchange, regulation, cooperativity and allostery require a built-in flexibility so that may adapt and respond to changes. Flexibility is of great importance in biological receptor-substrate interactions where adaptation is often required for regulation to occur. Such designed dynamics are more difficult to control than mere rigidity and recent developments in molecular design methods allowing to explore both structural and dynamical features may greatly help [22]. Receptor design thus covers both static and dynamic features of macropolycyclic structures.

The stability and selectivity of σ binding result from the set of interaction sites in ρ and may be structurally translated into *accumulation* (or collection) + *organization* (or orientation) i.e. bringing together binding sites and arranging them in a suitable pattern. Model computations on $(NH_3)_n$ clusters of different geometries have shown that collection involves appreciably larger energies than changes in orientation [23]. One may note that these intersite repulsions are built into a polydentate ligand in the course of synthesis [1].

We have studied receptors belonging to various classes of macropolycyclic structures (macrocycles, macrobicycles, cylindrical and spherical macrotricycles, etc.) expanding progressively our initial work on macrobicyclic cationic cryptates into the investigation of the structures and functions of supermolecules presenting molecular recognition, catalysis and transport processes.

2.3. *Initial Studies. Spherical Recognition in Cryptate Complexes.*

The simplest recognition process is that of spherical substrates; these are either positively charged metal cations (alkali, alkaline-earth, lanthanide ions) or the negative halide anions.

During the last 20 years, the complexation chemistry of alkali cations developed rapidly with the discovery of several classes of more or less powerful and selective ligands: natural [24] or synthetic [25, 26] macrocycles (such as valinomycin, 18-crown-6, spherands) as well as macropolycyclic cryptands and crypto-spherands [1, 6, 9, 26−29]. It is the design and study of alkali metal cryptates that started our work which developed into supramolecular chemistry.

It may be suitable at this stage to recount briefly the *origins of our work*, trying to trace the initial motivations and the emergence of the first lines of research. In the course of the year 1966, my interest for the processes occurring in the nervous system, led me to wonder how a chemist might contribute to the study of these highest biological functions. The electrical events in nerve cells rest on changes in the distributions of sodium and potassium ions accross the membrane. This seemed a possible entry into the field, since it had just been shown that the cyclodepsipeptide valinomycin [24c], whose structure and synthesis had been reported [24d], was able to mediate potassium ion transport in mitochondria [24e]. These results [24d,e] made me think that suitably designed synthetic cyclopeptides or analogues could provide means of monitoring cation distribution and transport across membranes. Such properties were also displayed by other neutral antibiotics [24f] of the enniatin and actin [24g] groups, and were found to be due to selective complex formation with alkali metal cations [24h-1], thus making these substances *ionophores* [24m]. However, since cation complexation might also represent a means of increasing the reactivity of the counteranion (anion activation) [6, 35], it became desirable to

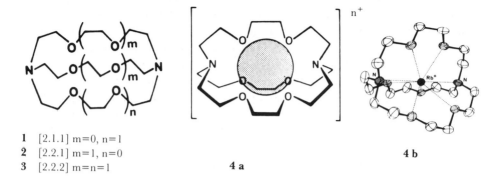

1   [2.1.1] m=0, n=1
2   [2.2.1] m=1, n=0
3   [2.2.2] m=n=1

4 a

4 b

envisage molecules which would be chemically less reactive than cyclic peptides [a]. Thus, when the cation binding properties of macrocyclic polyethers (crown ethers) were reported by Charles Pedersen [25a], these substances were perceived as combining the complexing ability of the macrocyclic antibiotics with the chemical stability of ether functions. Meanwhile, it had also become clear that compounds containing a three-dimensional, spheroidal cavity surrounding entirely the bound ion, should form stronger complexes than the rather flat shaped macrocycles; thus emerged the idea of designing macro*bicyclic* ligands.

Work started in October 1967 yielded the first such ligand [2.2.2] **3** in September of 1968; its very strong binding of potassium ions was noted at once and a cryptate structure was assigned to the complex obtained, allowing also to envisage its potential use for anion activation and for cation transport [29a] [b]. Other ligands such as **1** and **2** or larger ones were synthesized and numerous cryptates were obtained [29b]. Their structure was confirmed by crystal structure determinations of a number of complexes, such as the rubidium cryptate of **3, 4b** [29c] and their stability constants were measured [28].

The problem of *spherical recognition* is that of selecting a given spherical ion among a collection of different spheres of same charge. Thus, the macrobicyclic cryptands **1**—**3** form highly stable and selective cryptates $[M^{n+} \subset (\text{cryptand})]$ such as **4**, with the cation whose size is complementary to the size of the cavity i.e. Li^+, Na^+ and K^+ for **1**, **2** and **3** respectively [28a, 29a]. Others display high selectivity for alkali versus alkaline-earth cations [28b]. Thus, recognition features equal to or higher than those of natural macrocyclic ligands may be achieved. The spherical macrotricyclic cryptand **5** binds strongly and selectively the larger spherical cations, giving a strong Cs^+ complex, as in **6** [30].

[a] Earlier observations had suggested that polyethers interact with alkali cations. See for instance in H.C. Brown, E.J. Mead, P.A. Tierney, J. Am. Chem. Soc. *79* (1957) 5400; J.L. Down, J. Lewis, B. Moore, G. Wilkinson, J. Chem. Soc. *1959*, 3767; suggestions had also been made for the design of organic ligands, see in R.J.P. Williams, The Analyst *78* (1953) 586, Quaterly Rev. *24* (1970) 331.

[b] To name this new class of chemical entities, a term rooted in greek and latin, and which would also be equally suggestive in French, English, German and possibly (!) other languages was sought; "cryptates" appeared particularly suitable for designating a complex in which the cation was contained inside the molecular cavity, the crypt, of the ligand termed "cryptand".

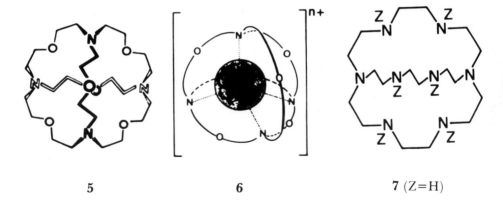

5   **6**   **7** (Z=H)

Anion cryptates are formed by the protonated polyamines **7** [31] and **5** [32] with the spherical halide anions F^- and Cl^- respectively. **5**-4H$^+$ binds Cl^- very strongly and very selectively with respect to Br^- and other types of anions, giving the [$Cl^- \subset$ (**2**-4H$^+$)] cryptate **8**. Quaternary ammonium derivatives of such type of macrotricycles also bind spherical anions [33].

Thus, cryptands **1**−**3** and **5** as well as related compounds display *spherical recognition* of appropriate cations and anions. Their complexation properties result from their macropolycyclic nature and define a *cryptate effect* characterized by high stability and selectivity, slow exchange rates and efficient shielding of the bound substrate from the environment.

As a consequence of these features, cryptate formation strongly influences physical properties and chemical reactivity. Numerous effects have been brought about and studied in detail, such as: stabilization of alkalides and electrides [34], dissociation of ion pairs, anion activation, isotope separation, toxic metal binding, etc. These results will not be described here and reviews may be found in [6, 35−38].

2.4. Tetrahedral Recognition

Selective binding of a tetrahedral substrate requires the construction of a receptor molecule possessing a tetrahedral recognition site, as realized in the macrotricycle **5** that contains four nitrogen and six oxygen binding sites located respectively at the corners of tetrahedron and of an octahedron [30].

Indeed, **5** forms an exceptionnally stable and selective cryptate [NH$_4^+ \subset$ **5**], **9**, with the tetrahedral NH$_4^+$ cation, due to the high degree of structural and energetical complementarity. NH$_4^+$ has the size and shape for fitting into the cavity of **5** and forming a tetrahedral array of $^+$N-H...N hydrogen bonds with the four nitrogen sites [39]. As a result of its very strong binding, the pK$_a$ of the NH$_4^+$ cryptate is about six units higher than that of free NH$_4^+$ indicating how much strong binding may affect the properties of the substrate. It also indicates that similar effects exist in enzyme active sites and in biological receptor-substrate binding.

The unusual protonation features of **5** in aqueous solution (high pK$_a$ for double protonation, very slow exchange) and ^{17}O-NMR studies led to the

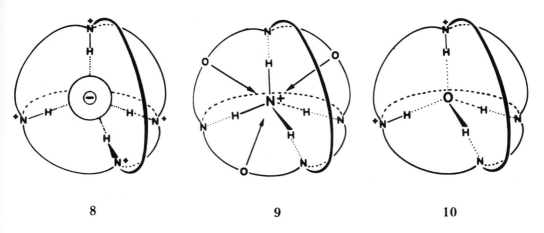

8   **9**   **10**

formulation of a water cryptate [$H_2O \subset (5\text{-}2H^+)$] **10** with the diprotonated macrotricycle [2, 6, 40]. The facilitation of the second protonation of **5** represents a *positive cooperativity*, in which the first proton and the effector molecule water set the stage both structurally and energetically for the fixation of a second proton.

Considering together the three cryptates [$NH_4^+ \subset 5$] **9**, [$H_2O \subset (5\text{-}2H^+)$] **10** and [$Cl^- \subset (5\text{-}4H^+)$] **8**, it is seen that the spherical macrotricycle **5** is a molecular receptor possessing a *tetrahedral recognition site* in which the substrates are bound in a tetrahedral array of hydrogen bonds. It represents a state of the art illustration of the molecular engineering involved in abiotic receptor chemistry. Since it binds a tetrahedral cation NH_4^+, a bent neutral molecule H_2O or a spherical anion Cl^- when respectively unprotonated, diprotonated and tetraprotonated, the macrotricyclic cryptand **5** behaves like a sort of molecular chameleon responding to pH changes in the medium!

The macrobicycle **3** also binds NH_4^+ forming cryptate **11**. The dynamic properties of **11** with respect to **9** reflect the receptor-substrate binding complementarity: whereas NH_4^+ is firmly held inside the cavity in **9**, it undergoes internal rotation in **11** [41].

2.5. *Recognition of Ammonium Ions and Related Substrates*

In view of the important role played by substituted ammonium ions in chemistry and in biology, the development of receptor molecules capable of recognizing such substrates is of special interest. Macrocyclic polyethers bind primary ammonium ions by anchoring the $-NH_3^+$ into their circular cavity via three $^+N\text{-}H \ldots O$ hydrogen bonds as shown in **12a** [12−15,25,42]; however they complex alkali cations such as K^+ more strongly. Selective binding of $R\text{-}NH_3^+$ may be achieved by extending the results obtained for NH_4^+ complexation by **5** and making use of the aza-oxa macrocycles [15,43] developed in the course of the synthesis of cryptands. Indeed, the triaza-macrocycle [18]-N_3O_3 which forms a complementary array of three $^+N\text{-}H \ldots N$ bonds **13**, selects $R\text{-}NH_3^+$ over K^+ and is thus a receptor unit for this functional group [43].

11

12 a X=H
   b X=CO$_2^-$
   c X=CONYY'

13

A great variety of macrocyclic polyethers have been shown to bind R-NH$_3^+$ molecules with structural and chiral selectivity [12,13,42]. Particularly strong binding is shown by the tetracarboxylate **12b** which conserves the desirable basic [18]-O$_6$ ring and adds electrostatic interactions, thus forming the most stable metal ion and ammonium complexes of any polyether macrocycle [44]. Very marked *central discrimination* is observed in favour of primary ammonium ions with respect to more highly substituted ones; it allows preferential binding of biologically active ions such as noradrenaline or norephedrine with respect to their N-methylated derivatives adrenaline and ephedrine [44].

Modulation of the complexation features of **12** by varying the side groups X so as to make use of specific interactions (electrostatic, H-bonding, charge transfer, lipophilic) between X and the R group of the centrally bound R-NH$_3^+$ substrate, brings about *lateral discrimination* effects. This also represents a general way of modeling interactions present in biological receptor-substrate complexes, such as that occurring between nicotinamide and tryptophane [45]. One may thus attach to **12** amino-acid residues, leading to "parallel peptides" [44] as in **12c**, nucleic bases or nucleosides, saccharides, etc.

Binding of metal-amine complexes M(NH$_3$)$_n^{m+}$ to macrocyclic polyethers via N-H...O interactions with the NH$_3$ groups, leads to a variety of supramolecular species of "supercomplex type" by second sphere coordination [46]. As with R-NH$_3^+$ substrate, binding to aza-oxa or polyaza macrocycles (see **13**) may also be expected. Strong complexation by macrocycles bearing negative charges (such as **12b** or the hexacarboxylate in **14** [47]), should allow to induce various processes between centrally bound metal-amine species and lateral groups X in **12** (energy and electron transfer, chemical reaction, etc.).

Receptor sites for secondary and tertiary ammonium groups are also of interest. R$_2$NH$_2^+$ ions bind to the [12]-N-$_2$O$_2$ macrocycle via two hydrogen bonds [48]. The case of quaternary ammonium ions will be considered below.

The guanidinium cation binds to [27]-O$_9$ macrocycles through an array of six H-bonds [49] yielding a particularly stable complex **14** with a hexacarboxylate receptor, that also binds the imidazolium ion [49a].

3. Anion Coordination Chemistry and the Recognition of Anionic Substrates

Although anionic species play a very important role in chemistry and in biology, their complexation chemistry went unrecognised as a specific field of research, while the complexation of metal ions and, more recently, of cationic molecules was extensively studied. The coordination chemistry of anions may be expected to yield a great variety of novel structures and properties of both chemical and biological significance [2, 6, 32]. To this end, anion receptor molecules and binding subunits for anionic functional groups have to be devised. Research has been increasingly active along these lines in recent years and anion coordination chemistry is progressively building up [8, 9, 50].

Positively charged or neutral electron deficient groups may serve as interaction sites for anion binding. Ammonium and guanidinium units which form $^+$N-H...X$^-$ bonds have mainly been used, but neutral polar hydrogen bonds (e.g. with -NHCO- or -COOH functions), electron deficient centres (boron, tin, etc.) or metal ion centres in complexes, also interact with anions.

Polyammonium macrocycles and macropolycycles have been studied most extensively as anion receptor molecules. They bind a variety of anionic species (inorganic anions, carboxylates, phosphates, etc.) with stabilities and selectivities resulting from both electrostatic and structural effects.

Strong and selective complexes of the spherical *halide anions* are formed by macrobicyclic and by spherical macrotricyclic polyammonium receptors such as the protonated forms of **5** [32] (see **8**), of bis-tren **15** [51] and of related compounds [50, 52].

The hexaprotonated form of bis-tren, **15**-6H$^+$ complexes various monoatomic and polyatomic anions [51]. The crystal structures of four such anion

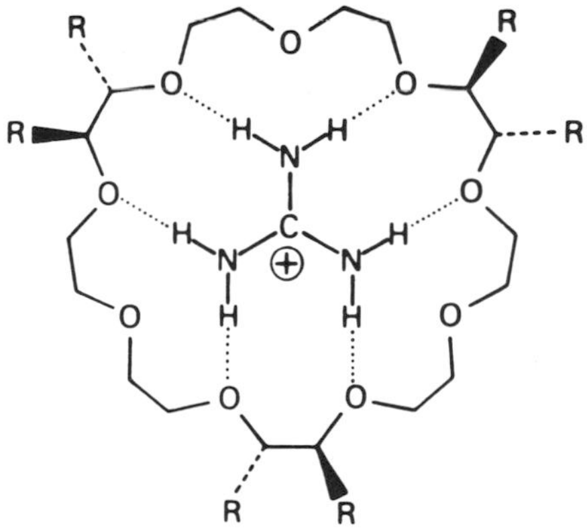

14 (R-CO$_2^-$)

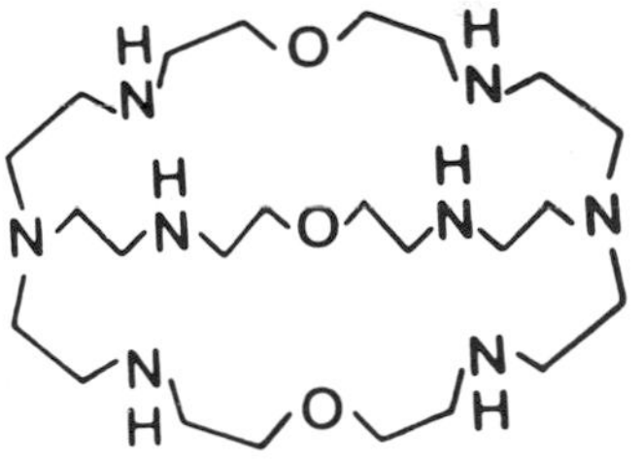

15

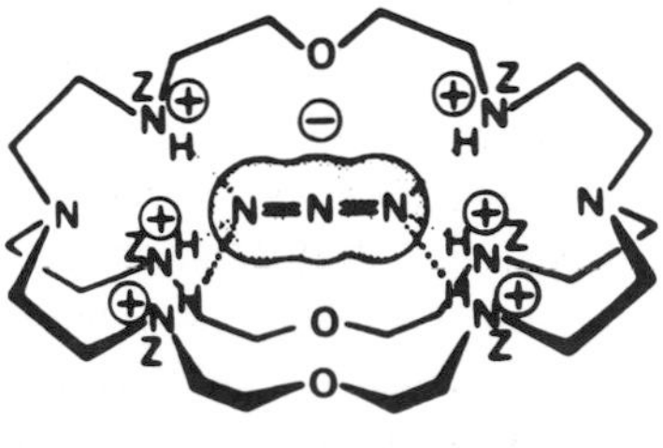

16

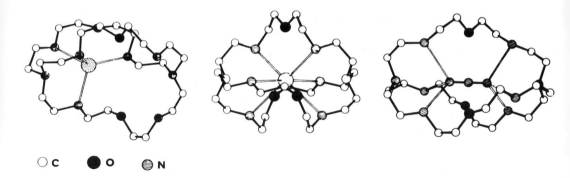

Fig. 1. Crystal structure of the anion cryptates formed by the hexaprotonated receptor molecule **15**-$6H^+$ with fluoride (left), chloride (centre) and azide (right) anions [51b].

cryptates provide a unique series of anion coordination patterns [51b]. The spherical halide ions are not complementary to the ellipsoidal receptor cavity and distort the structure, F^- being bound in a tetrahedral array of H-bonds and Cl^- and Br^- having octahedral coordination. The linear triatomic anion N_3^- has a shape and size complementary to the cavity of **15**-$6H^+$ and is bound inside by a pyramidal array of three H-bonds to each terminal nitrogen, forming the cryptate $[N_3^- \subset (\mathbf{15}\text{-}6H^+)]$, **16** (Figure 1). Thus, **15**-$6H^+$ is a molecular receptor recognizing linear triatomic species such as N_3^-, which is indeed bound much more strongly than other singly charged anions.

Carboxylates and *phosphates* bind to polyammonium macrocycles with stabilities and selectivities determined by structure and charge of the two partners [50, 51, 53−57]. The design of receptor units for these functional groups is of much interest since they serve as anchoring sites for numerous biological substrates. Thus, strong complexes are obtained with macrobicyclic polyammonium pockets in which carboxylate (formate, acetate, oxalate, etc.) and phosphate groups interact with several ammonium sites [51]. The guanidinium group, which serves as binding site in biological receptors, may form two H-bonds with carboxylate and phosphate functions and has been introduced into acyclic [58] and macrocyclic [59] structures. Binding units mimicking that of vancomycin are being sought [60].

Complexation of complex anions of transition metals such as the hexacyanides $M(CN)_6^{n-}$ yields second coordination sphere complexes, *supercomplexes* [53a] and affects markedly their electrochemical [61, 62] and photochemical [63] properties. Of special interest is the strong binding of adenosine mono-, di- and triphosphate (AMP, ADP and ATP) and related compounds that play a very important biological role [55−57].

Cascade type binding and *recognition* [64] of anionic species occurs when a ligand first binds metal ions which then serve as interaction sites for an anion. Such processes occur for instance in lipophilic cation/anion pairs [65] and with Cu(II) complexes of bis-tren **15** and of macrocyclic polyamines [66].

Heteronuclear NMR studies give information about the electronic effects induced by anion complexation as found for chloride cryptates [67].

Complexation of various molecular anions by other types of macrocyclic ligands have been reported [50], in particular with cyclophane type compounds. Two such receptors of defined binding geometries are represented by the protonated forms of the macropolycycles **17** [68] and **18** [69].

Anion coordination chemistry has thus made very significant progress in recent years. The development of other receptor molecules possessing well defined geometrical and binding features will allow to further refine the requirements for anion recognition, so as to yield highly stable and selective anion complexes with characteristic coordination patterns. Theoretical studies may be of much help in the design of anion receptors and in the *a priori* estimation of binding features, as recently illustrated by the calculation of the relative affinity of **5**-4H$^+$ for chloride and bromide ions [70].

4. Coreceptor Molecules and Multiple Recognition

Once binding units for specific groups have been identified, one may consider combining several of them within the same macropolycyclic architecture. Thus are formed polytopic coreceptor molecules containing several discrete binding subunits which may cooperate for the simultaneous complexation of several substrates or of a multiply bound (polyhapto) polyfunctional species. Suitable modification would yield cocatalysts or cocarriers performing a reaction or a transport on the bound substrate(s). Furthermore, because of their ability to perform multiple recognition and of the mutual effects of binding site occupation, such coreceptors provide entries into higher forms of molecular behaviour such as cooperativity, allostery, regulation as well as communication or signal transfer, if a species is released or taken up. Basic ideas and definitions concerning coreceptor molecules have been presented in more detail elsewhere [7].

The simplest class of coreceptors are those containing two binding subunits, ditopic coreceptors, which may belong to different structural types. Combination of chelating, tripodal and macrocyclic fragments yields macrocyclic, axial or lateral, macrobicyclic, or cylindrical macrotricyclic structures (Fig. 2). Depending on the nature of these units the resulting coreceptors may bind metal ions, organic molecules or both.

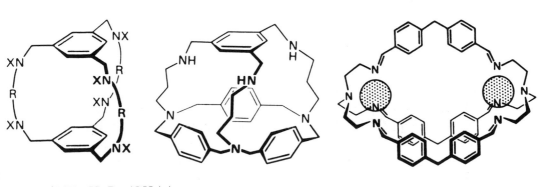

17 (X=H, R=(CH$_2$)$_3$)   **18**   **19**

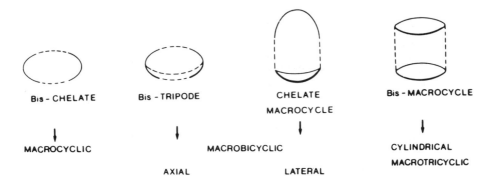

BIS - CHELATE   BIS - TRIPODE   CHELATE   BIS - MACROCYCLE
                                              MACROCYCLE

MACROCYCLIC   MACROBICYCLIC   CYLINDRICAL
                                                             MACROTRICYCLIC
                    AXIAL   LATERAL

Fig. 2. Combination of chelating, tripodal and cyclic subunits into ditopic coreceptors of macrocyclic, axial and lateral macrobicyclic and cylindrical macrotricyclic types (from left to right).

4.1. *Dinuclear and Polynuclear Metal-Ion Cryptates*

Coreceptor molecules containing two or more binding subunits for metal ions form dinuclear or polynuclear cryptates in which the arrangement of the metal ions is determined by the macropolycyclic structure. Such complexes may present a multitude of new properties, such as interactions between cations, electrochemical and photochemical processes, fixation of bridging substrates etc., that are of interest both for bioinorganic modeling and for multicentermultielectron reactions and catalysis.

Dinuclear cryptates of ligands belonging to all structural types shown in Fig. 2 have been obtained. This vast area will only be illustrated here by a few recent examples, (for more details and references see earlier reviews [64, 71].

Axial macrobicyclic ligands give dinuclear cryptates such as the bisCu(I) complex **19** formed by a large hexaamine structure obtained in a one step multiple condensation reaction; its crystal structure is shown in **20** [72].

Lateral macrobicycles are dissymmetric by construction and allow to arrange metal centres of different properties in the same ligand. Thus, complexes of type **21** combine a redox centre and a Lewis acid centre for activation of a bound substrate [73].

20   21 (m, n=0,1)   22

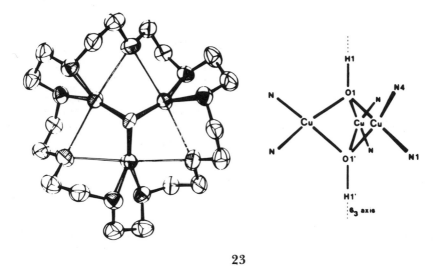

23

"Cluster cryptates" may be formed by assembling of metal ions and bridging species inside the molecular cavity of polytopic receptors. Thus, in the trinuclear Cu(II) complex **22** (crystal structure **23**) a [tris Cu(II), bisμ_3-hydroxo] group is bound in the cavity of a tritopic macrocycle [74]. Modeling of biological iron-sulfur cluster sites may employ inclusion into appropriate macrocyclic cavities [75].

This inorganic aspect of supramolecular species represents in itself a field of research in which many novel structures and reactivities await to be discovered.

4.2. Linear Recognition of Molecular Length by Ditopic Coreceptors

Receptor molecules possessing two binding subunits located at the two poles of the structure will complex preferentially substrates bearing two appropriate functional groups at a distance compatible with the separation of the subunits. This distance complementarity amounts to a recognition of molecular length of the substrate by the receptor. Such *linear molecular recognition* of dicationic and dianionic substrates corresponds to the binding modes illustrated by **24** and **25**.

Incorporation of macrocyclic subunits that bind -NH$_3^+$ groups (see above) into cylindrical macrotricyclic [76] and macrotetracyclic [77] structures, yields

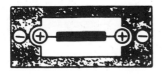

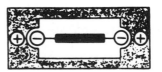

24   **25**

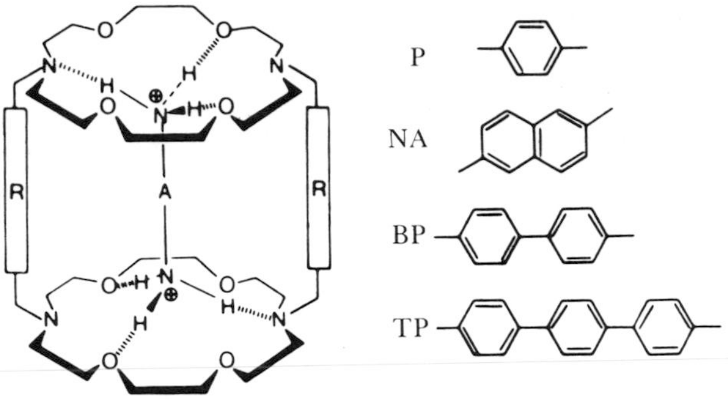

26 (R=P, NA, BP, TP)

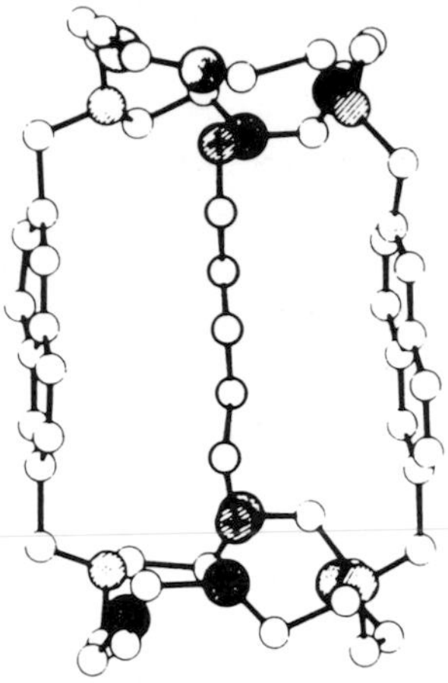

27

ditopic coreceptors that form molecular cryptates such as **26** with terminal *diammonium cations* ^+H_3N-$(CH_2)_n$-NH_3^+. In the resulting supermolecules the substrate is located in the central molecular cavity and anchored by its two -NH_3^+ in the macrocyclic binding sites, as shown by the crystal structure **27** (**26** with R=NA and A=$(CH_2)_5$) [78]. Changing the length of the bridges R in **26** modifies the binding selectivity in favour of the substrate of complementary length. NMR relaxation data have also shown that optimal partners present similar molecular motions in the receptor-substrate pair. Thus, complementarity in the supramolecular species expresses itself in both steric *and* dynamic fit [79].

Dianionic substrates, the dicarboxylates ^-O_2C-$(CH_2)_n$-CO_2^-, are bound with length discrimination by ditopic macrocycles such as **28**. These receptors contain two triammonium groups as binding subunits interacting with the terminal carboxylate functions, via a pattern schematically shown in **29** [80].

Thus, for both the terminal diammonium and dicarboxylate substrates, selective binding by the appropriate receptors describes a linear recognition

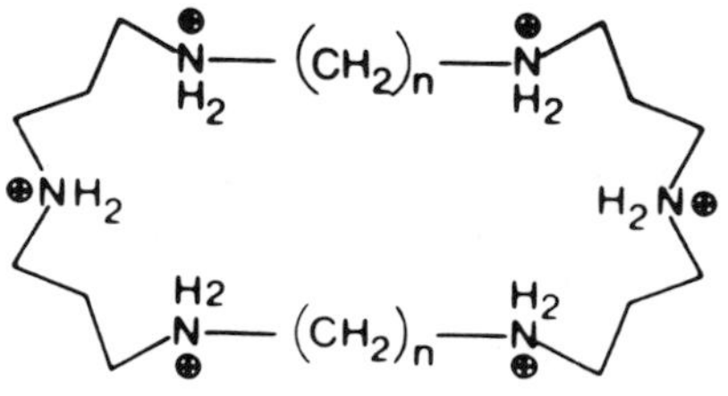

28 (*n*=7, 10)

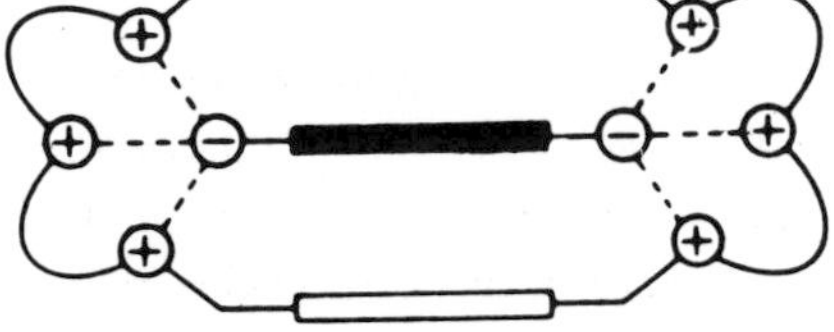

29

process based on length complementarity in a ditopic binding mode. Important biological species such as polyamines, amino-acid and peptide diamines or dicarboxylates, etc. may also be bound selectively.

Numerous variations in the nature of the binding subunits or of the bridges linking them, are conceavable and may be tailored to specific complexation properties (see for instance [15, 81]). The development of *heterotopic receptors* may allow to bind ion pairs [82] or zwitterionic species [83].

Studies on *dynamic coupling* [79, 84] between a receptor and a substrate are of much interest. Dynamic features of supermolecules correspond on the intermolecular level to the internal conformational motions present in molecules themselves and define molecular recognition processes by their dynamics in addition to their structural aspects.

4.3. *Heterotopic Coreceptors. Speleands, Amphiphilic Receptors*

Combination of binding subunits of different nature yields heterotopic receptors that may bind substrates by interacting simultaneously with cationic, anionic or neutral sites, making use of electrostatic and Van der Waals forces as well as of solvophobic effects.

The natural cyclodextrins were the first receptor molecules whose binding properties towards organic molecules, yielded a wealth of results on physical and chemical features of molecular complexation [21, 85].

Numerous types of synthetic macrocyclic receptors that contain various organic groups and polar functions, have been developed in recent years. They complex both charged and uncharged organic substrates. Although the results obtained often describe mere binding rather than actual recognition, they have provided a large body of data that allow to analyze the basic features of molecular complexation and the properties of structural fragments to be used in receptor design. We describe here mainly some of our own results in this area, referring the reader to specific reviews of the subject [20, 86]

Synergetical operation of electrostatic and hydrophobic effects may occur in *amphiphilic receptors* combining charged polar sites with organic residues which shield the polar sites from solvation and increase electrostatic forces. Such macropolycyclic structures containing polar binding subunits maintained by apolar shaping components, termed *speleands*, yield molecular cryptates, *(speleates)*, by substrate binding [87].

Thus, macrocycle **30** incorporating four carboxylate groups and two diphenylmethane units [88], not only forms very stable complexes with primary ammonium ions, but also strongly binds secondary, tertiary and quaternary ammonium substrates. In particular, it complexes *acetylcholine*, giving information about the type of interactions that may play a role in biological acetylcholine receptors, such as the combination of negative charges with hydrophobic walls. Similar effects operate in other anionic receptors complexing quaternary ammonium cations [86a, 89]. Extensive studies have been conducted on the complexation of heterocyclic ammonium ions such as diquat by macrocyclic polyether receptors [90].

The $CH_3-NH_3^+$ cation forms a selective speleate **31** by binding to the [18]-

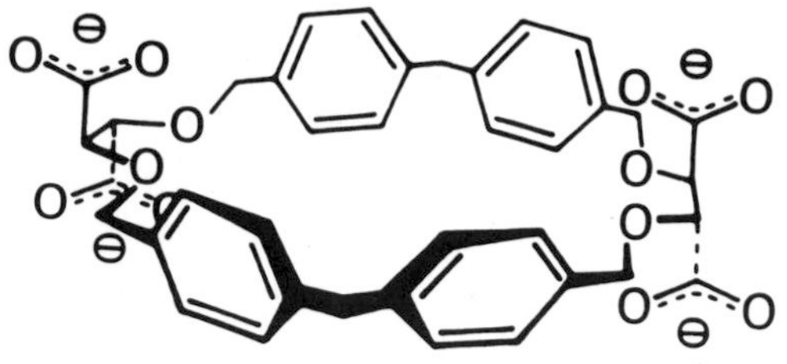

30

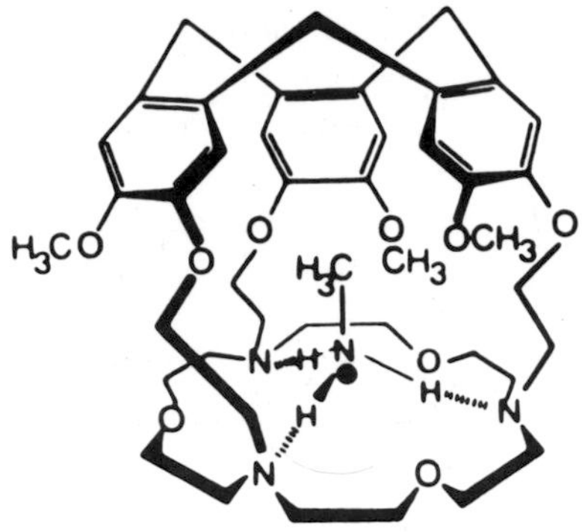

31

N_3O_3 subunit of a macropolycycle maintained by a cyclotriveratrylene shaping component. The tight intramolecular cavity efficiently excludes larger substrates [87, 91].

Amphiphilic type of binding also occurs for molecular anionic substrates [20, 86, 92]. Charged heterocyclic rings systems such as those derived from the pyridinium group represent an efficient way to introduce simultaneously electrostatic interactions, hydrophobic effects, structure and rigidity in a molecular receptor; in addition they may be electroactive and photoactive [93]. Even single planar units such as diaza-pyrenium dications bind flat organic anions remarkably well in aqueous solution, using electrostatic interactions as well as hydrophobic stacking [93]. A macrocycle containing four pyridinium sites was found to strongly complex organic anions [94].

Receptors of *cyclointercaland* type, that incorporate intercalating units into a macrocyclic system, are of interest for both the binding of small molecules and their own (selective) interaction with nucleic acids. A *cyclo-bis-intercaland* has been found to form an intercalative molecular cryptate **32** in which a nitrobenzene molecule is inserted between the two planar subunits of the receptor [95]. Such receptors are well suited for the recognition of substrates presenting flat shapes and become of special interest if intercalating dyes are incorporated [96].

Fitting the macrocyclic polyamine **33** with a side chain bearing a 9a-minoacridine group yields a coreceptor that may display both anion binding via the polyammonium subunit and stacking interaction by the intercalating dye. It interacts with both the triphosphate and the adenine groups of ATP and provides in addition a catalytic site for its hydrolysis (see below) [97].

Flat aromatic heterocyclic units bearing lateral acid and amid functional groups, function as receptors that perform size and shape recognition of complementary substrates within their molecular cleft [17b]. Receptor units containing heterocyclic groups such as 2,6-diaminopyridine [98a] or a nucleic base combined with an intercalator [98b] may lead to recognition of nucleotides via base pairing [98c].

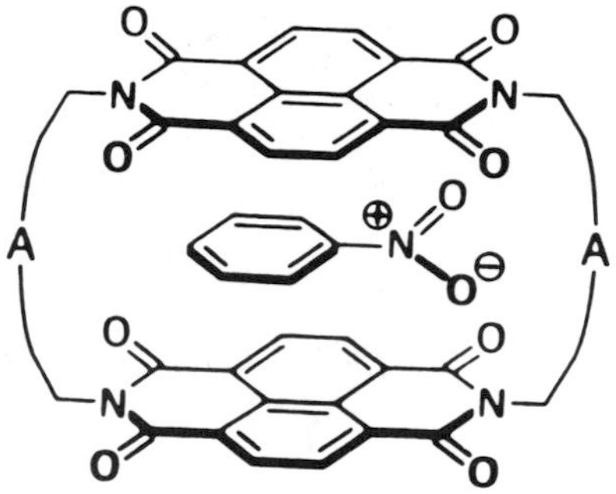

32 (A=(CH$_2$)$_8$)

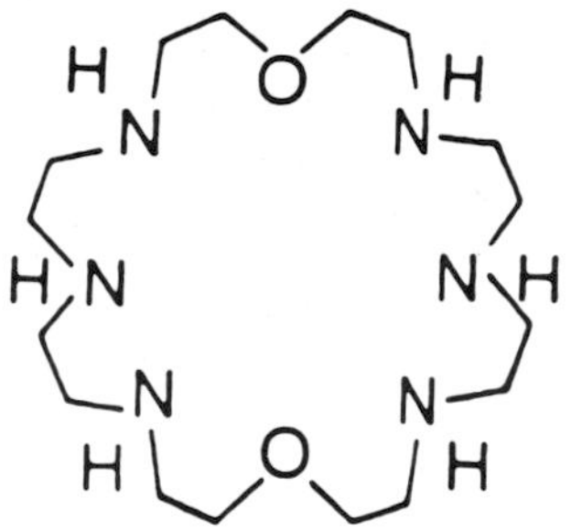

33

The spherically shaped *cryptophanes* allow to study recognition between neutral receptors and substrates, and in particular the effect of molecular shape and volume complementarity on selectivity [99].

4.4. Multiple Recognition in Metallo-Receptors

Metalloreceptors are heterotopic coreceptors that are able to bind both metal ions and organic molecules by means of substrate-specific units.

Porphyrin and α,α'-bipyridine (bipy) groups have been introduced as metal ion binding units in macropolycyclic coreceptors containing also macrocyclic sites for anchoring -NH$_3^+$ groups [64, 71, 100]. These receptors form mixed-substrate supermolecules by simultaneously binding metal ions and diammonium cations as shown in **34** [101]. Metalloreceptors and the supermolecules which they form, thus open up a vast area for the study of interactions and reactions between co-bound organic and inorganic species. In view of the number of metal ion complexes known and of the various potential molecular substrates to the bound, numerous types of metalloreceptors may be imagined which would be of interest as abiotic chemical species or as bioinorganic model systems.

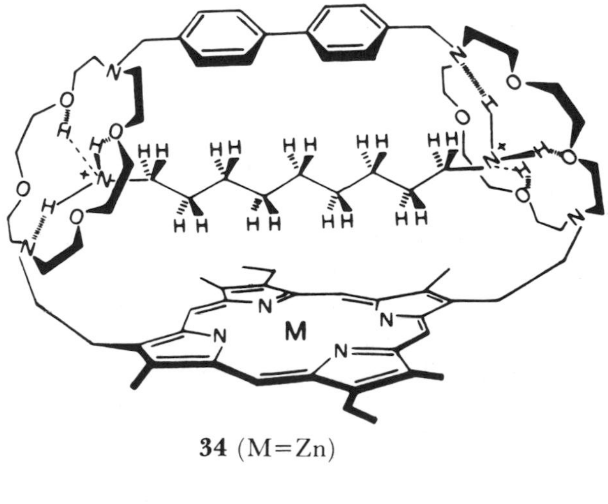

34 (M=Zn)

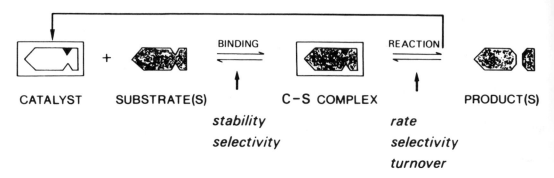

Fig. 3. Schematic representation of the supramolecular catalysis process.

5. Supramolecular Reactivity and Catalysis

Reactive and catalysis represent major features of the functional properties of supramolecular systems. Molecular receptors bearing appropriate reactive groups in addition to binding sites, may complex a substrate (with given stability, selectivity and kinetic features), react with it (with given rate, selectivity, turnover) and release the products, thus regenerating the reagent for a new cycle (Fig. 3).

Supramolecular reactivity and catalysis thus involve two main steps: *binding* which selects the substrate, followed by — *transformation* of the bound species into products within the supermolecule formed. Both steps take part in the molecular *recognition* of the *productive* substrate and require the correct molecular information in the reactive receptor [1]. Compared to molecular catalysis, a binding step is involved that selects the substrate and precedes the reaction itself.

The design of efficient and selective supramolecular reagents and catalysts may give mechanistic insight into the elementary steps of catalysis, provide new types of chemical reagents and effect reactions that reveal factors contributing to enzymatic catalysis. This led to numerous investigations, that made use mainly of reagents based on functionalized α- cyclodextrin, macrocyclic polyethers and cyclophanes [84, 85, 102, 103]

5.1. *Catalysis by Reactive Cation Receptor Molecules*

Ester cleavage processes have been most frequently investigated in enzyme model studies. Macrocyclic polyethers fitted with side chains bearing thiol groups cleave activated esters with marked rate enhancements and chiral discrimination between optically active substrates [104−106]. The tetra-(L)-cysteinyl derivative of macrocycle **12c** binds *p*-nitrophenyl (PNP) esters of amino-acids and peptides, and reacts with the bound species, releasing *p*-nitrophenol as shown in **35** [105]. The reaction displays i) substrate selectivity with ii) marked rate enhancements in favour of dipeptide ester substrates, iii) inhibition by complexable metal cations that displace the bound substrate, iv) high chiral recognition between enantiomeric dipeptide esters, v) slow but definite catalytic turnover.

35 (R=CH$_3$)   **36** (X=CONHnBu)

Binding of pyridinium substrates to a macrocycle of type **12c** bearing 1,4-dihydropyridyl side chains led to enhanced rates of *hydrogen transfer* from dihydropyridine to pyridinium within the supramolecular species **36** formed. The first order intracomplex reaction was inhibited and became bimolecular on displacement of the bound substrate by complexable cations [107].

Activation and orientation by binding was observed for the hydrolysis of 0-acetylhydroxylamine. CH$_3$COONH$_3^+$ forms such a stable complex with the macrocyclic tetracarboxylate **12b** [44], that it remains protonated and bound even at neutral pH, despite the low pK$_a$ of the free species ($\sim$ 2.15). As a consequence, its hydrolysis is accelerated and exclusively gives acetate and hydroxylamine, whereas in presence of K$^+$ ions, which displace the substrate, the latter rearranges to acetylhydroxamic acid CH$_3$CONH-OH ($\sim$ 50%) [108]. Thus, strong binding may be sufficient for markedly accelerating a reaction and affecting its course, a result that also bears on enzyme catalyzed reactions.

5.2. *Catalysis by Reactive Anion Receptor Molecules*

The development of anion coordination chemistry and of anion receptor molecules has made possible to perform molecular catalysis on anionic substrates of chemical and biochemical interest [50], such as adenosine triphosphate (ATP).

ATP hydrolysis was found to be catalyzed by a number of protonated macrocyclic polyamines. In particular [24]-N$_6$O$_2$, **33**, strongly binds ATP and markedly accelerates its hydrolysis to ADP and inorganic phosphate over a wide pH range [109]. The reaction presents first-order kinetics and is catalytic with turnover. It proceeds via initial formation of a complex between ATP and protonated **33**, followed by an intracomplex reaction which may involve a

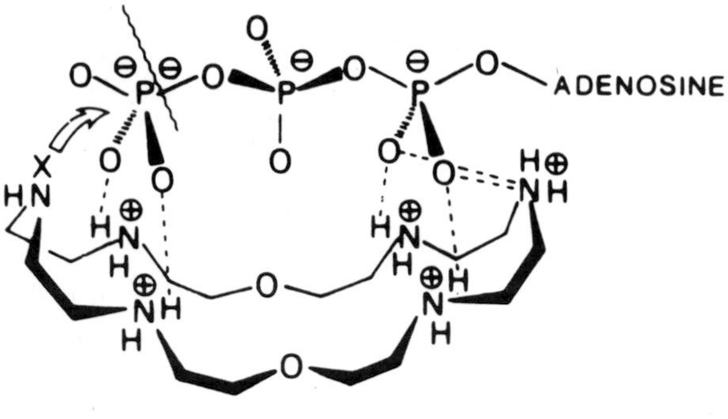

37

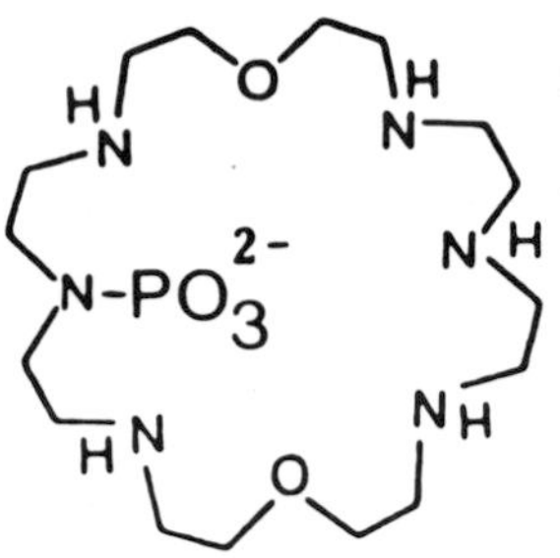

38

combination of acid, electrostatic, and nucleophilic catalysis. Structure **37** represents one possible binding mode of the ATP-**33** complex and indicates how cleavage of the terminal phosphoryl groups might take place. A transient intermediate identified as phosphoramidate **38**, is formed by phosphorylation of the macrocycle by ATP and is subsequently hydrolyzed. Studies with analogues of ATP indicated that the mechanism was dissociative in character within a pre-associative scheme resulting from receptor-substrate binding [110]. In this process catalyst **33** presents prototypical ATPase activity, *i.e.* it behaves as a proto-ATPase.

5.3. Cocatalysis: Catalysis of synthetic reactions

A further step lies in the design of systems capable of inducing *bond formation* rather than bond cleavage, thus effecting *synthetic* reactions as compared to degradative ones. To this end, the presence of several binding and reactive groups is essential. Such is the case for coreceptor molecules in which subunits may cooperate for substrate binding and transformation [7]. They should be able to perform *cocatalysis* by bringing together substrate(s) and cofactor(s) and mediating reactions between them within the supramolecular structure (Fig. 4).

A process of this type has been realized recently [111]. Indeed, when the same macrocycle **33** used in the studies of ATP hydrolysis was employed as catalyst for the hydrolysis of acetylphosphate (AcP=$CH_3COOPO_3^{2-}$), it was found to mediate the *synthesis of pyrophosphate* from AcP. Substrate consumption was accelerated and catalytic with turnover. The results obtained agree with a catalytic cycle involving the following steps: i) substrate AcP binding by the protonated molecular catalyst **33**; ii) phosphorylation of **33** within the supramolecular complex, giving the phosphorylated intermediate PN **38**; iii) binding of the substrate HPO_4^{2-} (P); iv) phosphoryl transfer from PN to P with formation of pyrophosphate PP (Fig. 5); v) release of the product and of the free catalyst for a new cycle.

The fact that **33** is a ditopic coreceptor containing two diethylenetriamine subunits is of special significance for both PN and PP formation. These subunits may cooperate in binding AcP and activating it for phosphoryl transfer via

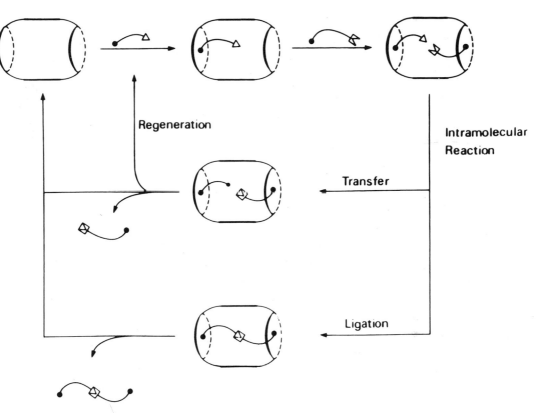

Fig. 4. Schematic illustration of cocatalysis processes: group transfer and ligation reactions occurring within the supramolecular complex formed by the binding of substrates to the two macrocyclic subunits of a macrotricyclic coreceptor molecule.

the ammonium sites, in providing an unprotonated nitrogen site for PN formation, as well as in mediating phosphoryl transfer from PN to P. Thus **33** would combine electrostatic and nucleophilic catalysis in a defined structural arrangement suitable for PP synthesis via two successive phosphoryl transfers, displaying protokinase type activity (Fig. 5). This bond-making process extends supramolecular reactivity to cocatalysis, mediating *synthetic reactions* within the supramolecular entities formed by coreceptor molecules. The formation of PP when ATP is hydrolyzed by **33** in presence of divalent metal ions has also been reported [112].

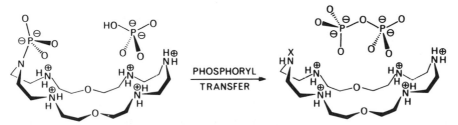

Fig. 5. Cocatalysis: pyrophosphate synthesis by phosphoryl transfer mediated by macrocycle **33** via the phosphorylated intermediate **38**.

Functionalized macrocyclic polyethers were used for peptide bond formation
in two successive intra-complex steps [113] and a thiazolium bearing macrobicyclic cyclophane was shown to effect supramolecular catalysis of the benzoin
condensation of two benzaldehyde molecules [114].

The systems described above possess the properties that define supramolecular reactivity and catalysis: substrate recognition, reaction within the supermolecule, rate acceleration, inhibition by competitively bound species, structural
and chiral selectivity, catalytic turnover. Many other types of processes may be
imagined. Thus, supramolecular catalysis of the hydrolysis of unactivated
esters and of amides presents a challenge [115] that chemistry has met in
natural enzymatic reagents but not yet in abiotic catalysts. Designing modified
enzymes by chemical mutation [116], or by protein engineering [117] and
producing catalytic proteins by antibody induction [118] represent biochemical approaches to artificial catalysts. Of particular interest is the development
of supramolecular catalysts performing synthetic reactions that create new
bonds rather than cleave them. By virtue of their multiple binding features
coreceptors open the way to the design of cocatalysts for ligation, metallocata-
lysis, cofactor reactions, that act on two or more co-bound and spatially
oriented substrates.

Supramolecular catalysts are by nature *abiotic* reagents, chemical catalysts,
that may perform the same *overall* processes as enzymes, without following the
detailed way in which the enzymes actually realize them. This chemistry may
develop reagents that effect highly efficient and selective processes that enzymes do not perform or realize enzymatic ones in conditions in which enzymes
do not operate.

6. Transport Processes and Carrier Design

The organic chemistry of membrane transport processes and of carrier molecules has only recently been developed, although the physico-chemical features
and the biological importance of transport processes have long been recognized. The design and synthesis of receptor molecules binding selectively
organic and inorganic substrates made available a range of compounds which,
if made membrane soluble, could become carrier molecules and induce selective transport by rendering membranes permeable to the bound species. Thus,
transport represents one of the basic functional features of supramolecular
species together with recognition and catalysis [2, 103].

The chemistry of transport systems comprises three main aspects: to design
transport effectors, to devise transport processes, to investigate their applications in chemistry and in biology. Selective membrane permeability may be
induced either by *carrier molecules* or by *transmembrane channels* (Fig. 6).

6.1. *Carrier-mediated Transport*

Carrier-mediated transport consists in the transfer of a substrate accross a membrane, facilitated by a carrier molecule. The four step cyclic process (association, dissociation, forward and back-diffusion) (Fig. 6) is a *physical catalysis*
operating a translocation on the substrate like chemical catalysis effects a

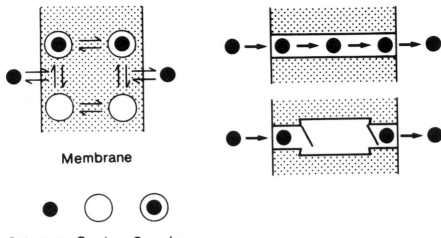

Membrane

Substrate Carrier Complex

Fig. 6. Transport processes: carrier mediated (left), channel (top right), gated channel (bottom right).

transformation into products. The carrier is the transport catalyst and the active species is the carrier-substrate supermolecule. Transport is a three-phase process, whereas homogeneous chemical and phase transfer catalyses are respectively single phase and two-phase.

Carrier design is the major feature of the organic chemistry of membrane transport since the carrier determines the nature of the substrate, the physico-chemical features (rate, selectivity) and the type of process (facilitated diffusion, coupling to gradients and flows of other species, active transport). The carrier must be highly selective, present appropriate exchange rates and lipophilic/hydrophilic balance, bear functional groups suitable for flow coupling. The transport process depends also on the nature of the membrane, the concentrations in the three phases, the other species present. More detailed considerations on these internal and external factors that affect transport processes may be found in earlier reports [1, 103, 120, 121].

Our initial work on the *transport of amino-acids*, dipeptides and acetylcholine through a liquid membrane employed simple lipophilic surfactant type carriers. It was aimed at the physical organic chemistry of transport processes, exploring various situations of transport coupled to flows of protons, cations or anions in concentration and pH gradients [122].

Selective *transport of metal cations*, mainly of alkali cations, has been a major field of investigation, spurred by the numerous cation receptors of natural or synthetic origin that are able to function as cation carriers [24, 103, 120, 121, 123, 124]. It was one of the initial motivations of our work [29a].

Cryptands of type **1—3** and derivatives thereof carry alkali cations [125], even under conditions in which natural or synthetic macrocycles are inefficient. The selectivities observed depend on the structure of the ligand the nature of the cation and the type of co-transported counter anion. Designed structural changes allow to transform a cation receptor into a cation carrier [120, 125].

The results obtained with cryptands indicated that there was an optimal complex stability and phase transfer equilibrium for highest transport rates [125]. Combined with data for various other carriers and cations, they gave a bell-shaped dependence of transport rates on extraction equilibrium (see Fig. 3 in [120]), with low rates for two small or too large (carrier saturation) extraction and highest rates for half-filled carriers [120, 125]. Kinetic analyses allowed to relate the experimental results to the dependence of transport rates and selectivities on carrier properties [124, 126, 127] Detailed studies of the transport of K^+ and Na^+ by lipophilic derivatives of **2** and **3** in vesicles bore on the efficiency, the selectivity and the mechanism of the processes [128].

Modifying the nature of the binding sites allows to selectively transport other cations such as toxic metal ions. Macrocyclic polyethers carry organic primary ammonium cations, in particular physiologically active ones [129]. The nature of the counterion and the concentrations of species strongly influence rates of transport and may affect its selectivity [120, 125].

Anion transport may be effected by lipophilic ammonium ions or by metal complexes [130] acting as anion receptors. Progress in anion coordination chemistry should provide a range of anion carriers that will help to develop this area of transport chemistry. The selective transport of carboxylates and phosphates is of great interest. Some results, references and suggestions have been given earlier [120].

Cation-anion cotransport was effected by a chiral macrotricyclic cryptand that carried simultaneously an alkali cation and a mandelate anion, see **39** [82]. Employing together a cation and an anion carrier should give rise to *synergetic transport* with double selection, by facilitating the flow of both components of a salt (see the electron-cation symport, below).

Selective transport of *amino-acids* occurs with a macrotricycle containing an internal phosphoric acid group [83b] and with a convergent dicarboxylic acid receptor [83c]. *Neutral* molecules are carried between two organic phases through a water layer by water soluble receptors containing a lipophilic cavity [132].

Z = BENZYL

39   **40**

It is clear that numerous facilitated transport processes may still be set up, especially for anions, salts or neutral molecules and that the active research in receptor chemistry will make available a variety of carrier molecules. Of special interest are those transport effectors derived from coreceptors, that allow coupled transport, *cotransport* to be performed.

6.2. *Coupled Transport Processes*

A major goal in transport chemistry is to design carriers and processes that involve the coupled flow of two (or more species) either in the same (symport) or in opposite (antiport) direction. Such parallel or antiparallel vectorial processes allow to set up pumped systems in which a species is carried in the potential created by physico-chemical gradients of electrons (redox gradient), protons (pH gradient) or other species (concentration gradient). To this end, either two or more individual carriers for different species may be used simultaneously or the appropriate subunits may be introduced into a single species, a *cocarrier*.

6.2.1. *Electron Coupled Transport in a Redox Gradient*

Electron-Cation symport has been realized in a double carrier process where the coupled parallel transport of electrons and metal cations was mediated by an electron carrier and a selective cation carrier [133]. The transport of electrons by a nickel complex in a redox gradient was the electron pump for driving the selective transport of K^+ ions by a macrocyclic polyether (Fig. 7). The process has the following features: − active K^+ transport and coupled electron flow; − two cooperating carriers acting synergetically; − a redox pump; − a selection

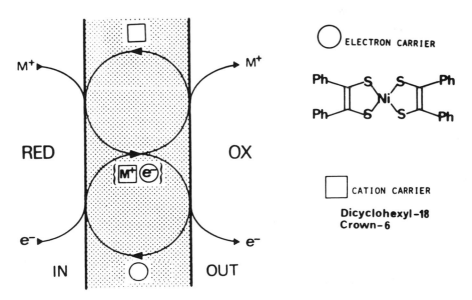

Fig. 7. Electron-cation coupled transport: redox driven electron-cation symport via an electron carrier (nickel complex) and a selective cation carrier (macrocyclic polyether) RED: potassium dithionite, OX potassium ferricyanide [133].

process by the cation carrier; — regulation by the cation/carrier pair. This system represents a prototype for the design of other multicarrier coupled transport processes. Electron transport with quinone carriers involves a ($2e^-$, $2H^+$) symport [134, 135].

Electron-anion antiport has been realized for instance with carriers such as ferrocene or alklylviologens [136]. The latter have been used extensively in light driven systems and in studies on solar energy conversion.

6.2.2. *Proton Coupled Transport in a pH Gradient*

Carriers bearing negatively charged groups may effect cation antiport accross membranes so that if one cation is a proton, a proton pump may be set up in a pH gradient. This has been realized for alkali cations with natural or synthetic carboxylate bearing ionophores [137].

A case of special interest is that of the transport of divalent ions such as calcium, versus monovalent ones. The lipophilic carrier **40** containing a *single* cation receptor site and *two* ionizable carboxylic acid groups, was found to transport selectively Ca^{2+} in the dicarboxylate form and K^+ when monoionized, thus allowing pH control of the process. This striking change in transport features involves *pH regulation* of Ca^{2+}/K^+ selectivity in a competitive (Ca^{2+}, K^+) symport coupled to (Ca^{2+}, $2H^+$) and (K^+, H^+) antiport in a pH gradient, which provides a proton pump (Fig. 8) [138].

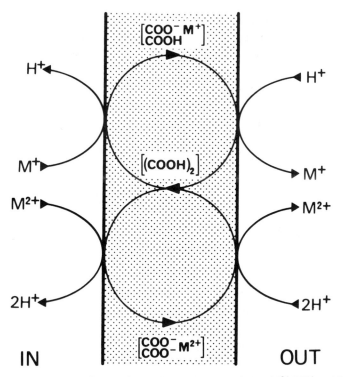

Fig. 8. Competitive divalent/monovalent cation symport coupled to $M^{2+}/2H^+$ and M^+/H^+ antiport in a pH gradient by a macrocyclic carrier such as **40**; the state of the carrier is indicated as diprotonated [$(CO_2H)_2$] or complexed [$(CO_2^-)_2$, M^{2+}] and [$(CO_2H)(CO_2^-)$, M^+].

This system demonstrates how carrier design allows to endow transport processes with regulation of rates and selectivity as well as coupling to energy sources, for transport of a species against its own concentration gradient.

6.2.3. *Light Coupled Transport Processes*

Light driven transport may be brought about by photogeneration of a species that will induce the process or perturb it.

This has been achieved in a *light induced electron transport* process involving the proflavine sensitized photogeneration of reduced methylviologen MV^+ which transfers electrons to a quinone type carrier contained in the membrane [135]. Light driven (electron, cation) symport occurs when combining this system with the (nickel complex, macrocycle) process described above [139]. The increased lipophilicity of reduced viologens facilitates electron and phase transfer [140]. Such processes are also of interest for solar energy storage systems where electron permeable membranes separate the reductive and oxidation components. Photocontrol of ion extraction and transport has been realized with macrocyclic ligands undergoing a structural change under irradiation, between two forms having different ion affinities [141].

The results obtained on coupled transport processes stress the role of cocarrier systems capable of transporting several substrates with coupling to physical and chemical energy sources.

6.3. *Transfert via Transmembrane Channels*

Transmembrane channels represent a special type of multiple-unit effector allowing the passage of ions or of molecules through membranes by a flow or site-to-site hopping mechanism. They play an important role in biological transport. Natural and synthetic peptide channels (gramicidine A, alamethicin) for cations have been studied [142, 143].

Artificial *cation channels* would provide fundamental information on the mechanism of cation flow and channel conduction. A solid state model of cation transfert inside a channel is provided by the crystal structure of the KBr complex of **12c** (Y=Y'=CH$_3$) which contains stacks of macrocycles with cations located alternatively inside and above a macrocyclic unit, like a frozen picture of cation propagation through the "channel" defined by the stack [144].

A polymeric stack of macrocycles has been synthesized [145] and a cyclodextrin based model of a half-channel has been reported [146]. A derivative of monensin, an acyclic polyether ionophore, forms lithium channels in vesicles [147] which may be sealed by diammonium salts [148].

Cylindrical macrotricycles such as **26** (substrate removed) represent the basic unit of a cation channel based on stacks of linked macrocycles; they bind alkali cations [149] and cation jumping processes between the top and bottom rings have been observed [150]. Theoretical studies give insight into the molecular dynamics [151] of ion transport and the energy profiles [152] in cation channels.

Electron channels, as transmembrane wires, represent the channel type counterpart to the mobile electron carriers discussed above and will be considered below. *Anion channels* may also be envisaged.

Thus, several types of studies directed towards the development of artificial ion channels and the understanding of ion motion in channels are engaged. These effectors desserve and will receive increased attention, in view also of their potential role in molecular ionics (see below).

A further step in channel design must involve the introduction of (proton, ion, redox or light activated) *gates and control elements* for regulating opening and closing, rates and selectivity. In this respect, one may note that ionizable groups on mobile carriers (as in **40**, see above) are, for these effectors, the counterpart of gating mechanisms in channels.

In conclusion, transport studies open ways to numerous developments of this chemically and biologically most important function: effector design, analysis of elementary steps and mechanisms, coupling to chemical potentials, energy and signal transduction, models of biological transport processes, construction of vesicular microreactors and artificial cells, etc., with a variety of possible applications, for instance in separation and purification or in batteries and systems for artificial photosynthesis.

By the introduction of polytopic features, which may include selective binding subunits as well as gating components for flow regulation, the design of cocarriers and of artificial molecular channels should add another dimension to the chemistry of the effectors and mechanisms of transport processes.

7. From Endo-receptors to Exo-receptors. Exo-Supramolecular Chemistry

The design of receptor molecules has mainly relied on macrocyclic or macropolycyclic architectures (see above) and/or on rigid spacers or templates (see for instance [17b, 26, 76−79, 81b, 83b] that allow to position binding sites on the walls of molecular cavities or clefts in such a way that they converge towards the bound substrate. The latter is more or less completely surrounded by the receptor, forming an *inclusion* complex, a *cryptate*. This widely used principle of *convergence* defines a convergent or *endo-supramolecular chemistry* with endoreceptors effecting endo-recognition. Biological analogies are found in the active sites of enzymes where a small substrate binds inside a cavity of a large protein molecule.

The opposite point of view would be to make use of an external *surface* with protuberances and depressions, rather than an internal cavity, as substrate receiving site. This would amount to the passage from a convergent to a *divergent* or *exo-supramolecular chemistry* and from endo- to *exo- receptors*. Receptor-substrate binding then occurs by surface-to-surface interaction, that may be termed *affixion* and symbolized by // or by the mathematical symbol of intersection (if there is notable interpenetration of the surfaces) [153], [ϱ// σ] and [ϱ $\cap$ σ]respectively. *Exo-recognition* with strong and selective binding requires a large enough contact area and a sufficient number of interactions as well as geometrical and site (electronic) complementarity between the surfaces of ϱ and σ. Such a mode of binding finds biological analogies in protein-protein

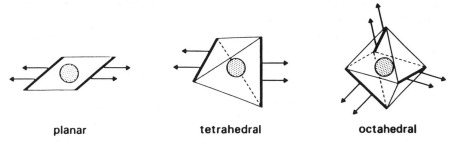

planar　　　　**tetrahedral**　　　　**octahedral**

Fig. 9. Schematic representation of the arrangement of external interaction sites (represented by
arrows) around a metal ion of given coordination geometry in mononuclear metallo-exoreceptors.

interactions, for instance at the antibody-antigen interface where the immunological recognition processes occur [154].

One may note that exo-recognition includes recognition between (rather
large) bodies of similar size as well as recognition at interfaces with monolayers, films, membranes, cell walls, etc.

7.1. *Metallo-exoreceptors. Metallonucleates*

In order to reinforce the binding strength one may think of introducing one or
more interaction *poles*, for instance electrical charges, generating strong electrostatic forces. Such could be the case for a metal complex whose central cation
would be the electrostatic pole and whose external surface could bear functional groups containing the molecular information required for recognizing the
partner. In addition, the metal cation provides a further way of organizing the
structure by virtue of its coordination geometry (tetrahedral, square-planar,
octahedral, etc.) which leads to a given arrangement of the external interaction
sites (Fig. 9).

As a first approach to receptor design along these lines, specific groups,
selected for their information content, have been attached to functionalized
α,α'-bipyridines, giving ligands of type **41** that form metal complexes of
defined geometry and physico-chemical properties. With nucleosides, species
such as the bis-adenosine derivative **42** are obtained [155]. The resulting
positively charged, *metallonucleate* complexes should interact strongly and selectively with the negatively charged oligonucleotides and nucleic acids, the
fixation site depending on the nature and disposition of the external nucleosides. Double helical metallonucleates [155] may be derived from the helicates
described below. Numerous variations may be imagined for the external groups
(intercalators, amino-acids, oligo-peptides, reactive functions) so as to yield
metal complexes displaying selective fixation and reactivity (chemical, photochemical) determined by the nature of the attached sites.

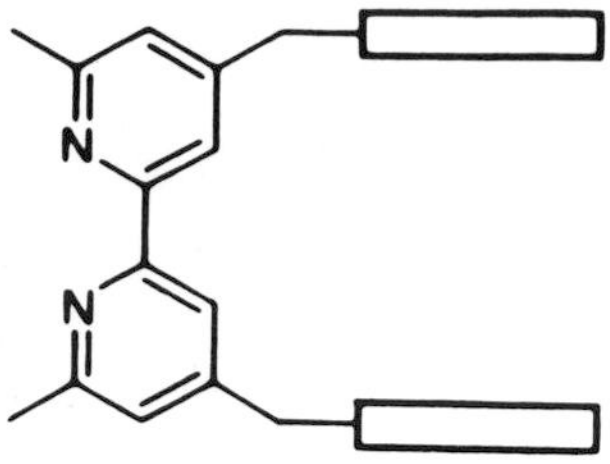

41

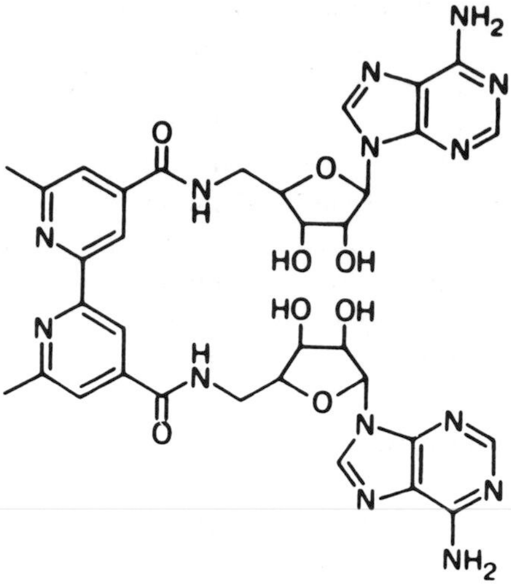

42

7.2. *Molecular Morphogenesis*

The design of exo-receptors requires means of generating defined external molecular shapes. This may be achieved in a number of ways, namely by stepwise synthesis of molecular architectures or by assembling on metal templates that impose a given coordination geometry, as noted above.

Another approach to such controlled molecular morphogenesis, is provided by the generation of globular molecules such as the "starburst dendrimers" [156] and the "arborols" [157], based on branched structures formed via cascade processes. Starting from a given core, it may be possible to produce a molecule of given size and external shape. On the polymolecular level, this may be related to the generation of shapes and structures in two-dimensional lipid monolayers [158] and in aggregates of amphilic molecules [159] (see also below).

8. From Supermolecules to Polymolecular Assemblies

Molecular chemistry is the domain of (more or less) independent *single* molecules. *Supramolecular chemistry* may be divided into two broad, partially overlapping areas: — *supermolecules* are well-defined *oligo*molecular species that result from the intermolecular association of a few components (a receptor and its substrate(s)) following a built-in Aufbau scheme based on the principles of molecular recognition; — *molecular assemblies* are polymolecular systems that result from the spontaneous association of a non-defined number of components into a specific phase having more or less well-defined microscopic organization and macroscopic characteristics depending on its nature (layers, membranes, vesicles, micelles, mesomorphic phases... [160]).

Continuous progress is being made in the design of synthetic molecular assemblies, based on a growing understanding of the relations between the features of the molecular components (structure, sites for intermolecular binding, etc.), the characteristics of the processes which lead to their association and the supramolecular properties of the resulting polymolecular assembly

Molecular organization, self-assembling and cooperation, construction of multilayer films [161−164], generation of defined aggregate morphologies [159], etc. allow to build up supramolecular architectures. The polymerization of the molecular subunits has been a major step in increasing control over the structural properties of the polymolecular system [165, 166]. Incorporating suitable groups or components may yield functional assemblies capable of performing operations such as energy, electron or ion transfer, information storage, signal transduction, etc. [160, 162, 167, 168]. The combination of receptors, carriers and catalysts, handling electrons and ions as discussed above, with polymolecular organized assemblies, opens the way to the design of what may be termed *molecular* and *supramolecular devices*, and to the elaboration of chemical microreactors and artificial cells.

One may also note that polymolecular assemblies define *surfaces* on which and through which processes occur, a feature that again stresses the interest of designing exoreceptors operating at the interfaces, in addition to endoreceptors embedded in the bulk of the membranes.

9. Molecular and Supramolecular Devices

Molecular devices may be defined as structurally organized and functionally integrated chemical systems built into supramolecular architectures. The development of such devices requires the design of molecular components (effectors) performing a given function and suitable for incorporation into an organized array such as that provided by the different types of polymolecular assemblies (see above). The components may be photo-, electro-, iono-, magneto-, thermo-, mecano-, chemo-active depending on whether they handle photons, electrons or ions, respond to magnetic fields or to heat, undergo changes in mechanical properties or perform a chemical reaction. A major requirement would be that these components and the devices that they bring about, perform their function(s) at the *molecular* and *supramolecular* levels as distinct from the bulk material.

Molecular receptors, reagents, catalysts, carriers and channels are potential effectors that may generate, detect, process and transfer signals by making use of the three-dimensional information storage and read-out capacity operating in molecular recognition and of the substrate transformation and translocation processes conveyed by reactivity and transport. Coupling and regulation may occur if the effectors contain several subunits that can interact, influence each other and respond to external stimuli such as light, electricity, heat, pressure, etc.

The nature of the mediator (substrate) on which molecular devices operate defines fields of molecular photonics, molecular electronics and molecular ionics. Their development requires to design effectors that handle these mediators and to examine their potential use as components of molecular devices. We shall now analyze specific features of molecular receptors, carriers and reagents from this point of view.

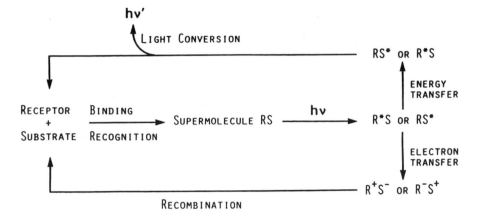

Fig. 10. Representation of the processes involved in supramolecular photochemistry. R*S, RS*, R^+S^- or R^-S^+ may be followed by a chemical reaction.

9.1. *Supramolecular Photochemistry and Molecular Photonics*

The formation of supramolecular entities from photoactive and/or electroactive components may be expected to perturb the ground state and excited state properties of the individual species, giving rise to novel properties that define a *supramolecular photochemistry* and *electrochemistry*.

Thus, a number of processes may take place within supramolecular systems, modulated by the arrangement of the bound units as determined by the organizing receptor: photoinduced energy migration, charge separation by electron or proton transfer, perturbation of optical transitions and polarisabilities, modification of redox potentials in ground or excited states, photoregulation of binding properties, selective photochemical reactions, etc.

Supramolecular photochemistry, like catalysis, may involve three steps: binding of substrate and receptor, mediating a photochemical process followed by restoration of the initial state for a new cycle or by a chemical reaction [169] (Fig. 10). The photophysical and photochemical features of supramolecular entities form a vast area of investigation into processes occurring at a level of intermolecular organization.

9.1.1. *Light Conversion by Energy Transfer*

A light conversion molecular device may be realized by an AbsorptionEnergy Transfer-Emission *A-ET-E* process in which light absorption by a receptor molecule is followed by intramolecular energy transfer to a bound substrate which then emits. This occurs in the europium(III) and terbium(III) cryptates of the macrobicyclic ligand [bipy.bipy.bipy] **43** [170]. UV light absorbed by the bipy groups is transferred to the lanthanide cation bound in the molecular cavity, and released in the form of visible lanthanide emission via an A-ET-E process, as shown in **44** [171]. These Eu(III) and Tb(III) complexes display a bright luminescence in aqueous solution, whereas the free ions do not emit in the same conditions. Numerous applications may be envisaged for such substances, in

43

44

particular as luminescent probes for monoclonal antibodies, nucleic acids, membranes, etc.

Photophysical processes occurring in macrocyclic [172], cryptate [173] and molecular [174, 175] complexes have been investigated.

9.1.2. *Photoinduced Electron Transfer in Photoactive Coreceptor Molecules*

The photogeneration of charge separated states is of interest both for inducing photocatalytic reactions (*e.g.* for artificial photosynthesis) and for the transfer of photo-signals, for instance through a membrane. It may be realized in D-PS-A systems in which excitation of a photosensitizer PS, followed by two electron transfers from a donor D and to and to an acceptor A, yields D^+-PS-A^-. Numerous systems of this type are being studied in many laboratories from the standpoint of modelling photosynthetic centres.

The D,A units may be metal coordination centres. Thus, binding of silver ions to the lateral macrocycles of coreceptors containing porphyrin groups as photosensitizers, introduces electron acceptor sites, as shown in **45**. This

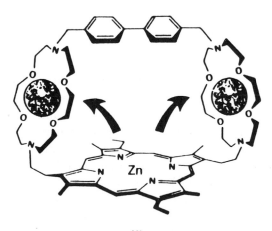

45

results in quenching of the singlet excited state of the Zn-porphyrin centre by an efficient intra-complex electron transfer, leading to charge separation and generating a phorphyrinium cation of long half-life [176].

Systems performing photoinduced electron transfer processes represent components for *light to electron conversion devices*. The light activated electron transport mentioned above [135] also belongs to this general type.

9.1.3. *Photoinduced Reactions in Supramolecular Species*

The binding of a substrate to a receptor molecule may affect the photochemical reactivity of either or both species, orienting the course of a reaction or giving rise to novel transformations.

Complexation of coordination compounds may allow to control their photochemical behaviour via the structure of the supramolecular species formed. Thus, the photoaquation of the $Co(CN)_6^{3-}$ anion is markedly affected by binding to polyammonium macrocycles. The results agree with the formation of supramolecular species, in which binding to the receptor hinders some CN^- groups from escaping when the Co-CN bonds are temporarily broken following light excitation [63]. It thus appears possible to orient the photosubstitution reactions of transition metal complexes by using appropriate receptor molecules. Such effects may be general, applying to complex cations as well as to complex anions, and providing an approach to the control of photochemical reactions via formation of defined supramolecular structures [169, 177, 178].

Structural or conformational changes photoproduced in receptor molecules affect their binding properties thus causing release or uptake of a species, as occurs in macrocyclic ligands containing light sensisitive groups [141, 172]. Such effects may be used to generate protonic or ionic photosignals in *light- to-ion conversion* devices.

9.1.4. *Non-linear Optical Properties of Supramolecular Species*

Substances presenting a large electronic polarisability are likely to yield materials displaying large macroscopic optical non-linearities. This is the case for push-pull compounds possessing electron donor and acceptor units. The potential non-linear optical (NLO) properties of materials based on metal complexes and supramolecular species have been pointed out recently [179a]. The D-PS-A system investigated for their ability to yield charge separated states possess such features; they may be metallic or molecular complexes.

Ion dependent optical changes produced by indicator ligands [180] might lead to cation control of non-linear optical properties.

Molecular electron-donor-acceptor complexes could present NLO effects since they possess polarized ground states and undergo (partial) intermolecular charge separation on excitation. It is possible to more or less finely tune their polarization, polarizability, extent of charge transfer, absorption bands, etc. by many variations in basic structural types as well as in substituents (see for instance [174, 179b]).

The results discussed above provide illustrations and incentives for further studies of photo-effects brought about by the formation of supramolecular species.

In a broader perspective, such investigations may lead to the development of *photoactive molecular devices*, based on photoinduced energy migration, electron transfer, substrate release or chemical transformation in supermolecules. Thus would be brought together molecular design, intermolecular bonding and supramolecular architectures with photophysical, photochemical and optical properties, building up a kind of *molecular photonics*.

9.2. Molecular Electronic Devices

More and more attention is being given to *molecular electronics* and to the possibility of developping electronic devices that would operate at the molecular level [3, 181]. Molecular rectifiers, transistors and photodiodes have been envisaged, the latter requiring features such as those of charge transfer states in organic molecules, in metal complexes and in D-PS-A systems discussed above (see ref. in [182]).

Among the basic components of electronic circuitry at the molecular level, a unit of fundamental importance is a connector or junction allowing electron flow to take place between different parts of the system, *i.e.*, a *molecular wire*.

An approach to such a unit is represented by the *caroviologens*, vinylogous derivatives of methylviologen that combine the features of the carotenoids and of the viologens [182]. They present features required for a molecular wire: a conjugated polyene chain for electron conduction; terminal electroactive and hydrosoluble pyridinium groups for reversible electron exchange; a length sufficient for spanning typical molecular supporting elements such as a monolayer or bilayer membrane.

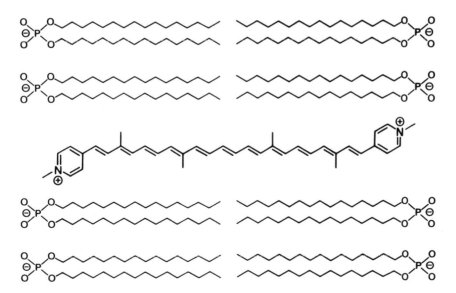

Fig. 11. Transmembrane incorporation of a caroviologen into sodium dihexadecylphosphate vesicles.

Caroviologens have been incorporated into sodium dihexadecyl phosphate vesicles. The data obtained agree with a structural model in which the caroviologens of sufficient length span the bilayer membrane, the pyridinium sites being close to the negatively charged outer and inner surfaces of the vesicles and the polyene chain crossing the lipidic interior of the membrane (Fig. 11) These and other functionalized membranes are being tested in processes in which the caroviologen would function as a continuous, transmembrane electron channel, *i.e.*, a genuine molecular wire. With respect to electron transport by means of redox active mobile carrier molecules (see above), the caroviologens represent electron channels, like there are cation channels and cation carriers. They portray a first approach to molecular wires and several modifications concerning the end-groups and the chain may be envisaged. For instance, coupling with photoactive groups could yield photoresponsive electron channels, as well as charge separation and signal transfer devices [183]. Combination with other polymolecular supports, such as polymeric layers or mesomorphic phases, and with other components, might allow to assemble nanocircuits and more complex molecular electronic systems.

Redox modification of an electroactive receptor or carrier molecule leads to changes in binding and transport properties, thus causing release or uptake of the substrate, in a way analogous to the photo-effects discussed above. To this end redox couples such as disulfide/dithiol [184, 185a] and quinone/hydroquinone [185] have been introduced into receptor molecules and shown to modify their substrate binding features. Complexation of metal hexacyanides $M(CN)_6^{n-}$ by polyammonium macrocycles markedly perturbs their redox properties [61]. These electrochemical features resulting from receptor-substrate association define a *supramolecular electrochemistry*. The mutual effects between redox changes and binding strength in a receptor-substrate pair, may allow to achieve electrocontrol of complexation and, conversely, to modify redox properties by binding (see also below).

9.3. *Molecular Ionic Devices*

The numerous receptor, reagent and carrier molecules capable of handling inorganic and organic ions are potential components of molecular and supramolecular *ionic devices* that would function via highly selective recognition, reaction and transport processes with coupling to external factors and regulation. Such components and the devices that they may build up form the basis of a field of *molecular ionics*, the field of systems operating with ionic species as support for signal and information storage, processing and transfer. In view of the size and mass of ions, ionic devices may be expected to perform more slowly than electronic devices. However ions have a very high information content by virtue of their multiple molecular (charge, size, shape, structure) and supramolecular (binding geometry, strength and selectivity) features. Molecular ionics appear a promising field of research which may already draw from a vast amount of knowledge and data on ion processing by natural and synthetic receptors and carriers.

Selective ion receptors represent basic units for ionic transmitters or detectors, selective ion carriers correspond to ionic transducers. These units may be fitted with triggers and switches sensitive to external physical (light, electricity, electric or magnetic field, heat, pressure) or chemical (other binding species, regulating sites) stimuli for connection and activation.

Binding or transport and triggering may be performed by separate species having each a specific function, as in multiple carrier transport systems (see above) [133]. This allows a variety of combinations between photo- or electroactive components and different receptors or carriers. On the other hand, light and redox sensitive groups have been incorporated into receptors and carriers and shown to affect binding and transport properties (see [141, 172, 184, 185] and references therein). Coreceptors and cocarriers provide means for regulation via cofactors, co-bound species that modulate the interaction with the substrate. Thus, a simple ionizable group such as a carboxylic acid function represents a *proton switch* and leads to gated receptors and carriers responding to pH changes, as seen for instance in the regulation of transport selectivity by **40** [138].

The main problem is the generally unsufficient changes brought about in most systems by the switching process. It is worth stressing that proton triggered yes/no or $+/-$ switches are potentially contained in the ability of polyamine receptors and carriers to bind and transport cations when unprotonated and anions when protonated; also, zwitterions such as amino-acids may change from bound to unbound or conversely when they undergo charge inversion as a function of pH. Molecular *protonic devices* thus represent a particularly interesting special case of ionic devices. A proton conducting channel would be a *proton wire*.

Functional molecular assemblies provide ways of organizing molecular devices and of introducing regulation and cooperative processes which may induce a much steeper response of the system to the stimuli. Vesicles have been fitted with functional units [168], for instance with a Li^+ ion channel that may

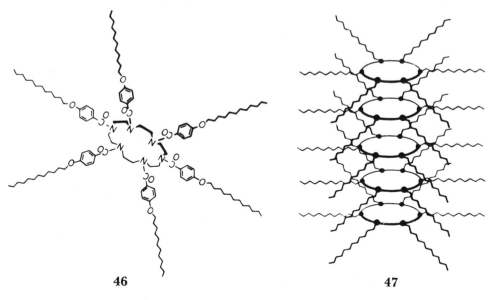

46   **47**

be sealed [147, 148]. Liquid crystals consisting of macrocyclic molecules bearing mesogenic groups, such as **46**, form *tubular mesophases*, composed of stacked rings as in **47**, that lead to the development of phase dependent ion conducting channels [186, 187]. Cooperativity and self-assembly may also be designed directly into the molecular components (see below).

Biological information and signals are carried by ionic and molecular species (Na^+, K^+, Ca^{2+}, acetylcholine, cyclic AMP, etc.). Polytopic metalloproteins such as calcium binding proteins [188] perform complicated tasks of detecting, processing and transferring ionic signals. These biological functions give confidence and inspiration for the development of molecular ionics.

One may note that ion receptors, channels, switches and gates have been considered within a potential European research programme on "Ionic Adaptative Computers". Applications such as selective chemical sensors based on molecular recognition are already well advanced [121].

9.4. *Molecular Self-Assembly*

Self-assembling and multiple binding with positive cooperativity are processes of spontaneous molecular organization that also allow to envisage *amplification molecular devices*. Such phenomena are well documented in biology, but much less so in chemistry. By virtue of their multiple binding subunits, polytopic coreceptors may display self-assembling if substrate binding to one receptor molecule generates binding sites that induce association with another one.

Positive cooperativity in substrate binding is more difficult to set up than negative cooperativity. However, both effects are of interest for regulation processes. Cooperativity is found in the unusual facilitation of the second protonation of **5** by a water effector molecule as in **10** (see above) [40]. Allosteric effects on cation binding to macrocyclic polyethers have been studied by inducing conformational changes via a remote site [17a] and subunit cooperativity occurs in a dimeric porphyrin [189].

The spontaneous formation of the double helix of nucleic acids represents the self-assembling of a supramolecular structure induced by the pattern of intermolecular interactions provided by the complementary nucleic bases. It involves recognition and positive cooperativity in base pairing [190].

Such self-assembling has recently been shown to occur in repetitive chain ligands, acyclic coreceptors containing several identical binding sites arranged linearly. Based on earlier work with a quaterpyridine ligand, oligo-bipyridine chain ligands incorporating two to five bipy groups were designed. By treatment with Cu(I) ions they underwent a spontaneous assembling into *double stranded helicates* containing two ligand molecules and one Cu(I) ion per bipy site of each ligand, the two receptor strands being wrapped around the metal ions which hold them together. Thus, the tris-bipy ligand **48** forms the trinuclear complex **49** whose crystal structure has been determined, confirming that it is indeed an *inorganic double helix* [191]. The results obtained indicate that the process occurs probably with positive cooperativity, binding of a Cu(I) ion with two ligands facilitating complexation of the next one. Furthermore, it appears that in a mixture of ligands containing different numbers of bipy units,

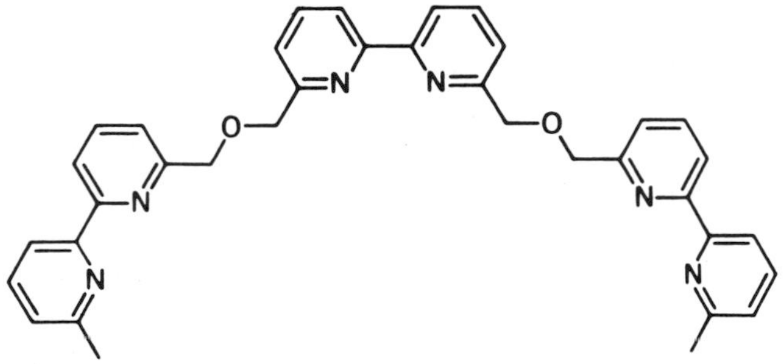

48

49

pairing occurs preferentially between the *same* ligands, thus performing *self-self recognition*. This spontaneous formation of an organized structure of intermolecular type opens ways to the design and study of self-assembling systems presenting cooperativity, regulation and amplification features. Catalytic reactions, gated channels, phases changes represent other processes that might be used for inducing amplification effects.

Various further developments may be envisaged along organic, inorganic and biochemical lines. Thus, if substituents are introduced at the para positions of the six pyridine units in **48**, treatment with Cu(I), should form an inorganic double helix bearing twelve outside directed functional groups [155].

Designed self-assembling rests on the elaboration of molecular components that will spontaneously undergo organization into a desired supramolecular architecture. Such control of self-organization at the molecular level is a field of major interrest in molecular design and engineering, that may be expected to become subject of increasing activity.

9.5. *Chemionics*

Components and molecular devices such as molecular wires, channels, resistors, rectifiers, diodes, photosensitive elements, etc. might be assembled into nanocircuits and combined with organized polymolecular assemblies to yield systems capable ultimately of performing functions of detection, storage, processing, amplification and transfer of signals and information by means of various mediators (photons, electrons, protons, metal cations, anions, molecules) with coupling and regulation [3, 8, 181, 182, 192].

Molecular photonics, electronics and ionics represent three areas of this intriguing and rather futuristic field of chemistry which may be termed "*chemionics*" [3,8] − the design and operation of photonic, electronic and ionic components, devices, circuitry and systems for signal and information treatment at the molecular level. Such perspectives lie, of course, in the long range

[193], but along the way they could yield numerous spin-offs and they do represent ultimate goals towards which work may already be planned and realized.

10. Conclusion

The present text was aimed at presenting the scope, providing illustration and exploring perspectives of supramolecular chemistry. Its conceptual framework has been progressively laid down and the very active research on molecular recognition, catalysis and transport, together with extension to molecular surfaces and polymolecular assemblies, is building up a vast body of knowledge on molecular behaviour at the supramolecular level. It is clear that much basic chemistry remains to be done on the design and realization of numerous other systems and processes that await to be imagined. The results obtained may also be analyzed in the view of developing components for molecular devices that would perform highly selective functions of recognition, transformation, transfer, regulation and communication, and allow signal and information processing. This implies operation via intermolecular interactions and incorporation of the *time* dimension into recognition events. One may note that such functions have analogies with features of expert systems, thus linking processes of artificial intelligence and molecular behaviour.

These developments in molecular and supramolecular science and engineering offer exciting perspectives at the frontiers of chemistry with physics and biology. Of course, even with past achievements and present activities, extrapolations and predictions can only be tentative; yet, on such an occasion for the celebration of chemistry, it appears justified to try looking out into the future for "He (or she) who sits at the bottom of a well to contemplate the sky, will find it small" (*Han Yu*, 768–824).

Acknowledgements. I wish to thank very warmly my collaborators at the Université Louis Pasteur in Strasbourg and at the Collège de France in Paris whose skill, dedication and enthusiasm, allowed the work described here to be realized. Starting with B. Dietrich et J.-P. Sauvage, they are too numerous to be named here, but they all have contributed to the common goal. I also wish to acknowledge the fruitful collaboration with a number of laboratories in various countries, beginning with the crystal structure determinations of cryptates by Raymond Weiss and his group and extending over various areas of structural and coordination chemistry, electrochemistry, photochemistry, biochemistry. Finally, I thank the Université Louis Pasteur, the Collège de France and the Centre National de la Recherche Scientifique for providing the intellectual environment and the financial support for our work.

REFERENCES

1. J.-M. Lehn, Struct. Bonding (Berlin) *16* (1973) 1.
2. J.-M. Lehn, Pure Appl. Chem. *50* (1978) 871.
3. J.-M. Lehn, Leçon Inaugurale, Collège de France, Paris 1980.
4. R. Pfeiffer "Organische Molekülerverbindungen", Stuttgart, 1927.
5. K.L. Wolf, F. Frahm, H. Harms, Z. Phys. Chem. Abt. B *36* (1937) 17; K.L. Wolf,

H. Dunken, K. Merkel, ibid. *46* (1940) 287; K.L. Wolf, R. Wolff, Angew. Chem. 61 (1949) 191.

6. J.-M. Lehn, Acc. Chem. Res. *11* (1978) 49.
7. J.-M. Lehn, in Z.I. Yoshida and N. Ise (Eds.): "Biomimetic Chemistry", Kodan-sha, Tokyo, Elsevier, Amsterdam 1983, p. 163.
8. J.-M. Lehn, Science *227* (1985) 849.
9. P.G. Potvin, J.-M. Lehn in R.M. Izatt, J.J. Christensen (Eds.): "Synthesis of Macrocycles: The Design of Selective Complexing Agents"; Progress in Macrocyclic Chemistry vol. 3, Wiley, New York, 1987, p. 167.
10. E. Fischer, Ber. Deutsch. Chem. Gesell. *27* (1894) 2985.
11. F. Cramer, W. Freist, Acc. Chem. Res. *20* (1987) 79.
12. J.F. Stoddart, Chem. Soc. Rev. *8* (1979) 85; Annual Reports B, Roy. Soc. Chem. (1983) p'353'
13. D.J. Cram, J.M. Cram, Acc. Chem. Res. *11* (1978) 8.
14. R.C. Hayward, Chem. Soc. Rev. *12* (1983) 285.
15. I.O. Sutherland, Chem. Soc. Rev. *15* (1986) 63.
16. G. van Binst (Ed.) "Design and Synthesis of Organic Molecules based on Molecular Recognition", Springer, Berlin 1986.
17. a) J. Rebek, Jr., Acc. Chem. Res. *17* (1984) 258; b) Science *235* (1987) 1478.
18. a) P.B. Dervan, R.S. Youngquist, J.P. Sluka, in W. Bartmann, K.B. Sharpless (Eds.) "Stereochemistry of Organic and Bioorganic Transformations" Verlag Chemie, Heidelberg 1987, p.221; b) W.C. Still, ibid. p.235.
19. R.M. Izatt, J.J. Christensen (Eds.), "Progress in Macrocyclic Chemistry", Wiley, New York, vol. *1* (1979), vol. *2* (1981), vol. *3* (1987).
20. F. Vögtle (Ed.) "Host Guest Chemistry", Topics Curr. Chem. 98 (1981), 101 (1982); F. Vögtle, E. Weber (Eds.), ibid., *121* (1984).
21. J.L. Atwood, J.E.D. Davies, D.D. MacNicol, "Inclusion Compounds", Academic Press, London, vol. *1, 2, 3* (1984).
22. G. Wipff, P.K. Kollman, J.-M. Lehn, J. Mol. Struct. *93* (1983) 153; G. Ranghino, S. Romano, J.-M. Lehn, G. Wipff, J. Am. Chem. Soc. *107* (1985) 7873; G. Wipff, P. Kollman, Nouv. J. Chim. *9* (1985) 457; "Structure and Dynamics of Macromolecules", S. Lifson, M. Levitt (Eds.) Israel J. Chem. *27* (1986) N°2.
23. An idea about the respective role of collection and orientation may be gained from examining the energies calculated for a series of $Li(NH_3)_n^+$ complexes and of the corresponding $(NH_3)_n$ units in the identical geometry. In presence of Li^+ the formation energies of the complexes are obtained at the optimized $Li^+ \ldots NH_3$ distances. When Li^+ is removed and the NH_3 molecules are kept at the same position the energies calculated are for the formation of the coordination shell alone. These energies represent the repulsion between the NH_3 groups; they are a measure of the intersite repulsive energy for bringing together two, three or four amine binding sites into a polydentate ligand of same coordination g-eometry. *Results,* $(NH_3)_n$, geometry (repulsive energy, kcal/mole): $(NH_3)_2$, linear (-2.4), bent (-3.8); $(NH_3)_3$, trigonal (-9.1), pyramidal (-10.8); $(NH_3)_4$, tetrahedral (-20.8). Thus, the total collection energies are appreciably larger than the organization energies represented by the changes from one geometry to another, linear to bent (1.4 kcal/mole) or trigonal to pyramidal (1.7 kcal/mole). *Ab initio* computations performed with a set of gaussian type basis functions, contracted into a double set with polarisation; J.-M. Lehn, R. Ventavoli, unpublished results; see also: R. Ventavoli, 3e Cycle Thesis, Université Louis Pasteur, Strasbourg, 1972.
24. (a) Yu. A. Ovchinnikov, V.T. Ivanov, A.M. Skrob, "Membrane Active Complex-ones", Elsevier, New York (1974); (b) B.C. Pressman, Ann. Rev. Biochem. *45* (1976) 501; (c) H. Brockmann, H. Geeren, Annalen *603* (1957) 217; (d) M.M. Shemyakin, N.A. Aldanova, E.I. Vinogradova, M.Yu. Feigina, Tetrahedron Lett. *1963*, 1921; (e) C. Moore, B.C. Pressman, Biochem. Biophys. Res. Commun. *15* (1964) 562; (f) B.C. Pressman, Proc. Natl. Acad. Sci. USA 53 (1965) 1077; (g) J.

Beck, H. Gerlach, V. Prelog, W. Voser, Helv. Chim. Acta 74 (1962) 620; (h) Z. Stefanac, W. Simon, Chimia 20 (1966) 436 and Microchem. J. 12 (1967) 125; (i) K.T. Kilbourn, J.D. Dunitz, L.A.R. Pioda, W. Simon, J. Mol. Biol. 30 (1967) 559; (j) P. Mueller, D.O. Rudin, Biochem. Biophys. Res. Commun. 26 (1967) 398; (k) T.E. Andreoli, M. Tieffenberg, D.C. Tosteson, J. Gen. Biol. 50 (1967) 2527; (l) M.M. Shemyakin, Yu.A. Ovchinnikov, V.T. Ivanov, V.K. Antonov, A.M. Skrob, I.I. Mikhaleva, A.V. Evstratov, G.G. Malenkov, ibid. 29 (1967) 834; (m) B.C. Pressman, E.J. Harris, W.S. Jagger, J.H. Johnson, Proc. Natl. Acad. Sci. USA 58 (1967) 1949.

25. (a) C.J. Pedersen, J. Am. Chem. Soc. 89 (1967) 7017; (b) C.J. Pedersen, H.K. Frensdorff, Angew. Chem. Int. Ed. Engl. 11 (1972) 16.

26. D.J. Cram, Angew. Chem. Int. Ed. Engl. 25 (1986) 1039.

27. D. Parker, Adv. Inorg. Chem. Radiochem. 27 (1983) 1; B. Dietrich, J. Chem. Ed. 62 (1985) 954. For the sepulchrate type of encapsulated metal ions, see: A.M. Sargeson, Pure Appl. Chem. 56 (1984) 1603.

28. (a) J.-M. Lehn, J.-P. Sauvage, J. Am. Chem. Soc. 97 (1975) 6700; (b) B. Dietrich, J.-M. Lehn, J.-P. Sauvage, J. Chem. Soc., Chem. Commun. 1973, 15.

29. (a) B. Dietrich, J.-M. Lehn, J.-P. Sauvage, Tetrahedron Lett. 1969, 2885 and 2889; (b) B. Dietrich, J.-M. Lehn, J.-P. Sauvage, J. Blanzat, Tetrahedron 29 (1973) 1629; B. Dietrich, J.-M. Lehn, J.-P. Sauvage, ibid. 29 (1973) 1647; (c) B. Metz, D. Moras, R. Weiss, J. Chem. Soc., Chem. Commun. 1970, 217; F. Mathieu, B. Metz, D. Moras, R. Weiss, J. Am. Chem. Soc. 100 (1978) 4412 and references therein.

30. E. Graf, J.-M. Lehn, J. Am. Chem. Soc. 97 (1975) 5022; Helv. Chim. Acta 64 (1981) 1040.

31. B. Dietrich, J.-M. Lehn, unpublished results.

32. E. Graf, J.-M. Lehn, J. Am. Chem. Soc. 98 (1976) 6403.

33. F. Schmidtchen, G. Muller, J. Chem. Soc., Chem. Commun. 1984, 1115.

34. J.L. Dye, Angew. Chem. Int. Ed. Engl. 18 (1979) 587; J.L. Dye, M.G. DeBacker, Ann. Rev. Phys. Chem. 38 (1987) 271.

35. J.-M. Lehn, Pure Appl. Chem. 52 (1980) 2303.

36. A.I. Popov, J.-M. Lehn in G.A. Melson(Ed.): "Coordination Chemistry of Macro-cyclic Compounds" Plenum Press, New York 1979.

37. I.M. Kolthoff, Anal. Chem. 51 (1979) lR.

38. F. Montanari, D. Landini, F. Rolla, Topics Curr. Chem. 101 (1982) 203; E. Blasius, K.-P. Janzen, ibid. 98 (1981) 163.

39. E. Graf, J.-M. Lehn, J. LeMoigne, J. Am. Chem. Soc. 104 (1982) 1672.

40. E. Graf, J.-P. Kintzinger, J.-M. Lehn, unpublished results.

41. B. Dietrich, J.-P. Kintzinger, J.-M. Lehn, B. Metz, A. Zahidi, J. Phys. Chem. 91 (1987) 6600.

42. a) D.J. Cram, K.N. Trueblood, Topics Curr. Chem. 98 (1981) 43; b) F. De Jong, D.N. Reinhoudt in V. Gold, D. Bethell (Eds.) Adv. Phys. Org. Chem. 17 (1980) 219, Academic Press, New York.

43. J.-M. Lehn, P. Vierling, Tetrahedron Lett. 1980, 1323.

44. J.-P. Behr, J.-M. Lehn, P. Vierling, J. Chem. Soc., Chem. Commun. 1976, 621; Helv. Chim. Acta 65 (1982) 1853.

45. J.-P. Behr, J.-M. Lehn, Helv. Chim. Acta 63 (1980) 2112.

46. H.M. Colquhoun, J.F. Stoddart, D.J. Williams, Angew. Chem. Inter. Ed. Engl. 25 (1986) 487.

47. Stability constants of about 800 and 10^5 lmol^{-1} have been obtained for binding of Ru(NH$_3$)$_6$$^{3+}$ to **12b** and to the hexacarboxylate in **14**, respectively (aqueous solution, pH=7.3); J.-M. Lehn, P. Vierling, unpublished results.

48. J.C. Metcalfe, J.F. Stoddart, G. Jones, J. Am. Chem. Soc. 99 (1977) 8317; J. Krane, O. Aune, Acta Chem. Scand. B34 (1980) 397.

49. a) J.-M. Lehn, P. Vierling, R.C. Hayward, J. Chem. Soc., Chem. Commun. 1979, 296; b) see also: K. Madan, D.J. Cram, ibid. 1975, 427; J.W.H.M. Uiterwijk, S. Harkema, J. Geevers, D.N. Reinhoudt, ibid. 1982, 200.

50. a) F. Vögtle, H. Sieger, W.M. Muller, Topics Curr. Chem. *98* (1981) 107; b) K. Saigo, Kagaku to Kogyo *35* (1982) 90; c) J.-L. Pierre, P. Baret, Bull. Soc. Chim. Fr. II *1983*, 367; d) E. Kimura, Topics Curr. Chem. *128* (1985) 113; e) F.P. Schmidtchen, ibid. *132* (1986) 101.

51. a) J.-M. Lehn, E. Sonveaux, A.K. Willard, J. Am. Chem. Soc. 100 (1978) 4914; b) B. Dietrich, J. Guilhem, J.-M. Lehn, C. Pascard, E. Sonveaux, Helv. Chim. Acta *67* (1984) 91; c) for other macrobicyclic receptors see also M.W. Hosseini, Thèse de Doctorat-ès-Sciences, Université Louis Pasteur, Strasbourg 1983.

52. C.H. Park, H. Simmons, J. Am. Chem. Soc. *90* (1968) 2431.

53. a) B. Dietrich, M.W. Hosseini, J;-M. Lehn, R.B. Sessions, J. Am. Chem. Soc. *103* (1981) 1282; b) Helv. Chim. Acta *68* (1985) 289.

54. J. Cullinane, R.I. Gelb, T.N. Margulis, L.J. Zompa, J. Am. Chem. Soc. 104 (1982) 3048, E. Suet, H. Handel, Tetrahedron Lett. *1984*, 645.

55. M.W. Hosseini, J.-M. Lehn, M.P. Mertes, Helv. Chim. Acta *66* (1983) 2454.

56. E. Kimura, M. Kodama, T. Yatsunami, J. Am. Chem. Soc. *104* (1982) 3182; J.F. Marecek, C.J. Burrows, Tetrahedron Lett. *1986*, T943.

57. M.W. Hosseini, J.-M. Lehn, Helv. Chim. Acta *70* (1987) 1312; see also H.R. Wilson, R.J.P. Williams, J. Chem. Soc., Faraday Trans. I *83* (1987) 1885.

58. B. Dietrich, D.L. Fyles, T.M. Fyles, J.-M. Lehn, Helv. Chim. Acta *62* (1979) 2763.

59. B. Dietrich, T.M. Fyles, J.-M. Lehn, L.G. Pease, D.L. Fyles, J. Chem. Soc., Chem. Commun. *1978*, 934.

60. M.J. Mann, N. Pant, A.D. Hamilton, J. Chem. Soc., Chem. Commun. *1986*, 158.

61. F. Peter, M. Gross, M.W. Hosseini, J.-M. Lehn, J. Electroanal. Chem. 144 (1983) 279.

62. E. Garcia-Espana, M. Micheloni, P. Paoletti, A. Bianchi, Inorg. Chim. Acta *102* (1985) L9; A. Bianchi, E. Garcia-Espana, S. Mangani, M. Micheloni, P. Orioli, P. Paoletti, J. Chem. Soc., Chem. Commun. *1987*, 729.

63. M.F. Manfrin, L. Moggi, V. Castelvetro, V. Balzani, M.W. Hosseini, J.-M. Lehn, J. Am. Chem. Soc. *107* (1985) 6888.

64. J.-M. Lehn, Pure Appl. Chem. *52* (1980) 2441.

65. J.-M. Lehn, J. Simon, J. Wagner, Angew. Chem. Int. Ed. Engl. *12* (1973) 578, 579.

66. R.J. Motekaitis, A.E. Martell, B. Dietrich, J.-M. Lehn, Inorg. Chem. 23 (1984) 1588; R.J. Motekaitis, A.E. Martell, I. Murase, ibid. 25 (1986) 938; A. Evers, R.D. Hancock, I. Murase, ibid. *25* (1986) 2160; D.E. Whitmoyer, D.P. Rillema, G. Ferraudi, J. Chem. Soc., Chem. Commun. *1986*, 677.

67. J.-P. Kintzinger, J.-M. Lehn, E. Kauffmann, J.L. Dye, A.I. Popov, J. Am. Chem. Soc. *105* (1983) 7549.

68. D. Heyer, J.-M. Lehn, Tetrahedron Lett. *1986*, 5869.

69. T. Fujita, J.-M. Lehn, Tetrahedron Lett. *1988*, 1709.

70. T.P. Lybrand, J.A. McCammon, G. Wipff, Proc. Natl. Acad. Sci. USA *83* (1986) 833.

71. J.-M. Lehn, in K.J. Laidler (Ed.): "IUPAC Frontiers of Chemistry" Pergamon Press, Oxford 1982, p 265.

72. J. Jazwinski, J.-M. Lehn, D. Lilienbaum, R. Ziessel, J. Guilhem, C. Pascard, J. Chem. Soc., Chem. Commun. *1987*, 1691.

73. A. Carroy, J.-M. Lehn, J. Chem. Soc., Chem. Commun. *1986*, 1232.

74. J. Comarmond, B. Dietrich, J.-M. Lehn, R. Louis, J. Chem. Soc., Chem. Commun. *1985*, 74.

75. Y. Okuno, K. Uoto, O. Yonemitsu, T. Tomohiro, J. Chem. Soc., Chem. Commun. *1987*, 1018; J.R. Holmes, J.-M. Lehn, work in progress.

76. F. Kotzyba-Hibert, J.-M. Lehn, P. Vierling, Tetrahedron Lett. *1980*, 941.

77. F. Kotzyba-Hibert, J.-M. Lehn, K. Saigo, J. Am. Chem. Soc. *103* (1981) 4266.

78. C. Pascard, C. Riche, M. Cesario, F. Kotzyba-Hibert, J.-M. Lehn, J. Chem. Soc., Chem. Commun. *1982*, 557.

79. J.-P. Kintzinger, F. Kotzyba-Hibert, J.-M. Lehn, A. Pagelot, K. Saigo, J. Chem.

Soc., Chem. Commun. *1981*, 833.

80. M.W. Hosseini, J.-M. Lehn, J. Am. Chem. Soc. *104* (1982) 3525; Helv. Chim. Acta *69* (1986) 587.

81. a) F.P. Schmidtchen, J. Am. Chem. Soc. *108* (1986) 8249; b) J. Rebek, Jr. D. Nemeth, P. Ballester, F.-T. Lin, ibid. *109* (1987) 3474.

82. J.-M. Lehn, J. Simon, A. Moradpour, Helv. Chim. Acta 61 (1978) 2407.

83. a) F.P. Schmidtchen, J. Org. Chem. *51* (1986) 5161; b) J. Simon, Thèse de Doctorat d'Etat, Université Louis Pasteur, Strasbourg 1976; see also structure **28** in ref. [14] p. 305; c) J. Rebek, Jr., B. Askew, 0. Nemeth, K. Parris, J. Am. Chem. Soc. *109* (1987) 2432.

84. J.-P. Behr, J.-M. Lehn, J. Am. Chem. Soc. *98* (1976) 1743.

85. F. Cramer, "Einschlussverbindungen", Springer Verlag, Berlin 1954; M.L. Bender, M. Komiyama, "Cyclodextrin Chemistry" Springer Verlag, Berlin 1978.

86. a) J. Franke, F. Vögtle, Topics Curr. Chem. *132* (1986) 137; b) F. Vögtle, W.M. Müller, W.H. Watson, ibid. *125* (1984) 131; c) for calixarenes, see C.D. Gutsche, ibid. *123* (1984), 1; Acc. Chem. Res. *16* (1983) 161; d) for cavitands, see: D.J. Cram, Science *219* (1983) 1177.

87. J. Canceill, A. Collet, J. Gabard, F. Kotzyba-Hibert, J.-M. Lehn, Helv. Chim. Acta *65* (1982) 1894.

88. M. Dhaenens, L. Lacombe, J.-M. Lehn, J.-P. Vigneron, J. Chem. Soc., Chem. Commun. *1984*, 1097.

89. H.-J. Schneider, D. Guttes, U. Schneider, Angew. Chem. Int. Ed. Engl. *25* (1986) 647.

90. B.L. Atwood, F.H. Kohnke, J.F. Stoddart, D.J. Williams, Angew. Chem. Int. Ed. Engl. *24* (1985) 581.

91. For dissymmetric cylindrical macrotricyclic coreceptors that bind ammonium ions see: A.D. Hamilton, P. Kazanjian, Tetrahedron Lett. *1985*, 5735; K. Saigo, R.-J. Lin, M. Kubo, A. Youda, M. Hasegawa, Chem. Lett. *1986*, 519.

92. For macrocyclic cyclophane type receptors see also K. Odashima, T. Soga, K. Koga, Tetrahedron Lett. (1980) 5311; F. Diederich, K. Dick, ibid. *106* (1984) 8024 and references therein.

93. A.J. Blacker, J. Jazwinski, J.-M. Lehn, Helv. Chim. Acta *70* (1987) 1.

94. I. Bidd, B. Dilworth, J.M. Lehn, unpublished work.

95. J. Jazwinski, A.J. Blacker, J.-M. Lehn, M. Cesario, J. Guilhem, C. Pascard, Tetrahedron Lett. *1987*, 6057.

96. J.-M. Lehn, F. Schmidt, J.-P. Vigneron, work in progress.

97. M.W. Hosseini, A.J. Blacker, J.-M. Lehn, J. Chem. Soc., Chem. Commun. *1988*, 596.

98. a) J.-P. Behr, J.-M. Lehn, unpublished work; b) J.-M. Lehn I. Stibor work in progress; c) for recent examples see: J. Rebek Jr. B Askew P Ballester, C. Buhr, S. Jones, D. Nemeth, K. Williams, J. Am. Chem. Soc. *109* (1987) 5033; A.D. Hamilton, D. Van Engen, ibid. *109* (1987) 5035.

99. J. Canceill, M. Cesario, A. Collet, C. Riche, C. Pascard, J. Chem. Soc., Chem. Commun. *1986*, 339; J. Canceill, L. Lacombe, A. Collet, C.R. Acad. Sc. Paris *304*, II (1987) 815.

100. A.D. Hamilton, J.-M. Lehn, J.L. Sessler, J. Chem. Soc., Chem. Commun. *1984*, 311; J. Am. Chem. Soc. *108* (1986) 5158.

101. For other metalloreceptor type species see for instance references in [15]; N.M. Richards, I.O. Sutherland, P. Camilleri, J.A. Pape, Tetrahedron Lett. *1985*, 3739; M.C. Gonzalez, A.C. Weedon, Can. J. Chem. *63* (1985) 602; D.H. Busch, C. Cairns, in [9] p. 1; V. Thanabal, V. Krishnan, J. Am. Chem. Soc. *104* (1982) 3643; G.B. Maiya, V. Krishnan, Inorg. Chem. 24 (1985) 3253.

102. a) R. Breslow, Science *218* (1982) 532 and in ref. [21] vol. *3*, p. 473; b) R.M. Kellogg, Topics Curr. Chem. *101* (1982) 111; c) I. Tabushi, K. Yamamura, ibid. *113* (1983) 145; d) Y. Murakami, ibid. *115* (1983) 107; e) C. Sirlin, Bull. Soc. Chim.

France II (1984) 5; f) R.M. Kellogg, Angew. Chem. Int. Ed. Engl. *23* (1984) 782; g) V.T. D'Souza, M. Bender, Acc. Chem. Res. *20* (1987) 146.

103. J.-M. Lehn, Pure Appl. Chem. *51* (1979) 979; Ann. N.Y. Acad. Sci. *471* (1986) 41.
104. Y. Chao, G.R. Weisman, G.D.Y. Sogah, D.J. Cram, J. Am. Chem. Soc. *101* (1979) 4948.
105. J.-M. Lehn, C. Sirlin, J. Chem. Soc., Chem. Commun. *1978*, 949; Nouv. J. Chim. *11* (1987) 693.
106. S. Sasaki, K. Koga, Heterocycles 12 (1979) 1305.
107. J.-P. Behr, J.-M. Lehn, J. Chem. Soc., Chem. Commun. *1978*, 143.
108. J.-M. Lehn, T. Nishiya, Chem. Lett. *1987*, 215.
109. M.W. Hosseini, J.-M. Lehn, M.P. Mertes, Helv. Chim. Acta *66* (1983) 2454; M.W. Hosseini, J.-M. Lehn, L. Maggiora, K.B. Mertes, M.P. Mertes, J. Am. Chem. Soc. *109* (1987) 537.
110. G.M. Blackburn, G.R.J. Thatcher, M.W. Hosseini, J.-M. Lehn, Tetrahedron Lett. *1987*, 2779.
111. M.W. Hosseini, J.-M. Lehn, J. Chem. Soc., Chem. Commun. *1985*, 1155; J. Am. Chem. Soc. *109* (1987) 7047; J. Chem. Soc., Chem. Commun. *1988*, 397.
112. P.G. Yohannes, M.P. Mertes, K.B. Mertes, J. Am. Chem. Soc. *107* (1985) 8288.
113. S. Sasaki, M. Shionoya, K. Koga, J. Am. Chem. Soc. *107* (1985) 3371.
114. H.-D. Lutter, F. Diederich, Angew. Chem. Int. Ed. Engl. *25* (1986) 1125.
115. F.M. Menger, M. Ladika, J. Am. Chem. Soc. *109* (1987) 3145.
116. E.T. Kaiser, D.S. Lawrence, Science *226* (1984) 505.
117. See for instance: J.A. Gerlt, Chem. Rev. *87* (1987) 1079; A.J. Russell, A.R. Fersht, Nature *328* (1987) 496.
118. A. Tramontano, K.D. Janda, R.A. Lerner, Science *234* (1986) 1566; S.J. Pollack, J.W. Jacobs, P.G. Schultz, ibid. *234* (1986) 1570; R.A. Lerner, A. Tramontano, TIBS *12* (1987) 427.
119. B. Dietrich, J.-M. Lehn, J.-P. Sauvage, Tetrahedron Lett. *1969*, 2889.
120. J.-M. Lehn, in G. Spach (Ed.) "Physical Chemistry of Transmembrane Ion Motions", Elsevier, Amsterdam, 1983, p. 181.
121. W. Simon, W.E. Morf, P.Ch. Meier, Structure Bonding *16* (1973) 113; W.E. Morf, D. Amman, R. Bissig, E. Pretsch, W. Simon, in ref. [19], vol.1, p.l.
122. J.-P. Behr, J.-M. Lehn, J. Am. Chem. Soc. *95* (1973) 6108.
123. B.C. Pressman, Ann. Rev. Biochem. *45* (1976) 501.
124. J.D. Lamb, J.J. Christensen in ref [21] vol. *3*, p. 571.
125. M. Kirch, J.-M. Lehn, Angew. Chem. Int. Ed. Engl. *14* (1975) 555; M. Kirch, Thèse de Doctorat-ès-Sciences, Université Louis Pasteur, Strasbourg, 1980.
126. J.-P. Behr, M. Kirch, J.-M. Lehn, J. Am. Chem. Soc. *107* (1985) 241.
127. T.M. Fyles, Can. J. Chem. *65* (1987) 884.
128. M. Castaing, F. Morel, J.-M. Lehn, J. Membrane Biol. *89* (1986) 251; M. Castaing, J.-M. Lehn, ibid. *97* (1987) 79.
129. E. Bacon, L. Jung, J.-M. Lehn, J. Chem. Res. (S) *1980*, 136.
130. H. Tsukube, Angew. Chem. Int. Ed. Engl. *21* (1982) 304.
131. Anion transport with anion cryptands has been observed recently: B. Dietrich, T. M. Fyles, M. W. Hosseini, J.-M. Lehn, K. C. Kaye, J. Chem. Soc., Chem. Commun. *1988*, 691.
132. F. Diederich, K. Dick, J. Am. Chem. Soc. *106* (1984) 8024; A. Harada, S. Takahashi, J. Chem. Soc., Chem. Commun. *1987*, 527.
133. J.J. Grimaldi, J.-M. Lehn, J. Am. Chem. Soc. *101* (1979) 1333.
134. S.S. Anderson, I.G. Lyle, R. Paterson, Nature *259* (1976) 147.
135. J.J. Grimaldi, S. Boileau, J.-M. Lehn, Nature *265* (1977) 229.
136. J.K. Hurst, D.H.P. Thompson, J. Membrane Sci. *28* (1986) and references therein; I. Tabushi, S.-i. Kugimiya, Tetrahedron Lett. *1984*, 3723.
137. M. Okahara, Y. Nakatsuji, Topics Curr. Chem. *128* (1985) 37.
138. A. Hriciga, J.-M. Lehn, Proc. Natl. Acad. Sci. USA *80* (1983) 6426.

139. R. Frank, H. Rau, Z. Natur'forsch. *37a* (1982) 1253.
140. I. Tabushi, S.-i. Kugimiya, J. Am. Chem. Soc. *107* (1985) 1859.
141. S. Shinkai, O. Manabe, Topics Curr. Chem. *121* (1984) 67.
142. D.W. Urry, Topics Curr. Chem. *128* (1985) 175.
143. R. Nagaraj, P. Balaram, Acc. Chem. Res. *14* (1981) 356; R.O. Fox, Jr., F.M. Richards, Nature *300* (1982) 325.
144. J.-P. Behr, J.-M. Lehn, A.-C. Dock, D. Moras, Nature *295* (1982) 526.
145. U.F. Kragten, M.F.M. Roks, R.J.M. Nolte, J. Chem. Soc., Chem. Commun. 1985, 1275.
146. I. Tabushi, Y. Kuroda, K. Yokota, Tetrahedron Lett. *1982*, 4601.
147. J.-H. Fuhrhop, U. Liman, J. Am. Chem. Soc.*106* (1984) 4643.
148. J.-H. Fuhrhop, U. Liman, H.H. David, Angew. Chem. Int. Ed. Engl. *24* (1985) 339.
149. J.-M. Lehn, J. Simon, Helv. Chim. Acta *60* (1977) 141.
150. J.-M. Lehn, M.E. Stubbs, J. Am. Chem. Soc. 96 (1974) 4011.
151. W. Fischer, J. Brickmann, P. Lauger, Biophys. Chem. *13* (1981) 105.
152. C. Etchebest, S. Ranganathan, A. Pullman, FEBS Letters *173* (1984) 301.
153. For an earlier use of the intersection sign, see: E. Kauffmann, J.L. Dye, J.-M. Lehn, A.I. Popov, J. Am. Chem. Soc. *102* (1980) 2274.
154. A.G. Amit, R.A. Mariuzza, S.E.V. Phillips, R.J. Poljak, Science *233* (1986) 747; H.M. Geysen, J.A. Tainer, S.J. Rodda, T.J. Mason, H. Alexander, E.D. Getzoff, R.A. Lerner, ibid. *235* (1987) 1184.
155. M.M. Harding, J.-M. Lehn, unpublished work.
156. D.A. Tomalia, M. Hall, D.M. Hedstrand, J. Am. Chem. Soc. *109* (1987) 1601 and references therein.
157. G.R. Newkome, Z.-q. Yao, G.R. Baker, V.K. Gupta, P.S. Russo, M.J. Saunders, J. Am. Chem. Soc. *108* (1986) 849; G.R. Newkome, G.R. Baker, M.J. Saunders, P.S. Russo, V.K. Gupta, Z.-q. Yao, J.E. Miller, K. Bouillion, J. Chem. Soc., Chem. Commun. *1986*, 752.
158. H.M. McConnell, L.K. Tamm, R.M. Weiss, Proc. Natl. Acad. Sci. USA 81 (1984) 3249; R.M. Weiss, H.M. McConnell, Nature *310* (1984) 47.
159. T. Kunitake, Y. Okahata, M. Shimomura, S.-i. Yasunami, K. Takarabe, J. Am. Chem. Soc. *103* (1981) 5401; N. Nakashima, S. Asakuma, T. Kunitake, ibid. *107* (1985) 509.
160. J.H. Fendler, "Membrane Mimetic Chemistry" Wiley, New York, 1982.
161. H. Kuhn, D. Moebius, Angew. Chem. Int. Ed. Engl. *10* (1971) 620.
162. D. Moebius, Acc. Chem. Res. *14* (1981) 63; Ber. Bunsenges Phys. Chem. *82* (1978) 848; Z. Physik. Chem. Neue Folge *154* (1987) 121.
163. J.A. Hayward (Ed.) "New technological applications of phospholipid bilayers, thin films and vesicles", Tenerife, January 6-9, 1986, Plenum Press.
164. J. Sagiv, in [163]; L. Netzer, J. Sagiv, J. Am. Chem. Soc. 105 (1983) 674.
165. H.-H. Hub, B. Hupfer, H. Koch, H. Ringsdorf, Angew. Chem. Int. Ed. Engl. *19* (1980) 938; L. Gros, H. Ringsdorf, H. Schupp, ibid. *20* (1981) 305.
166. G. Wegner, Chimia *36* (1982) 63; C.M. Paleos, Chem. Soc. Rev. *14* (1985) 45.
167. H. Kuhn, Pure Appl. Chem. *53* (1981) 2105.
168. a) J.-H. Fuhrhop, J. Mathieu, Angew. Chem. Int. Ed. Engl. *23* (1984) 100; b) J.-H. Fuhrhop, D. Fritsch, Acc. Chem. Res. *19* (1986) 130; c) Y. Okahata, Acc. Chem. Res. 19 (1986) 57.
169. a) V. Balzani (Ed.) "Supramolecular Photochemistry", D. Reidel Publ. Co., Dordrecht, Holland, 1987; b) J.-M. Lehn, p. 29–43 therein.
170. J.-C. Rodriguez-Ubis, B. Alpha, D. Plancherel, J.-M. Lehn, Helv. Chim. Acta *67* (1984) 2264.
171. B. Alpha, J.-M. Lehn, G. Mathis, Angew. Chem. Int. Ed. Engl. *26* (1987) 266; B. Alpha, V. Balzani, J.-M. Lehn, S. Perathoner, N. Sabbatini, Angew. Chem. Int. Ed. Engl. *26* (1987), 1266; N. Sabbatini, S. Perathoner, V. Balzani, B. Alpha, J.-

M. Lehn, in ref. [169a], p. 187.

172. a) H. Bouas-Laurent, A. Castellan, J.-P. Desvergne, Pure Appl. Chem. *52* (1980) 2633; b) H. Bouas-Laurent, A. Castellan, M. Daney, J.-P. Desvergne, G. Guinand, P. Marsau, M.-H. Riffaud, J. Am. Chem. Soc. *108* (1986) 315.

173. J.P. Konopelski, F. Kotzyba-Hibert, J.-M. Lehn, J.-P. Desvergne, F. Fagès, A. Castellan, H. Bouas-Laurent, J. Chem. Soc., Chem. Commun. *1985*, 433.

174. D.F. Eaton, Tetrahedron *43* (1987) 1551 and references therein.

175. A. Guarino, J. Photochem. *35* (1986) 1.

176. M. Gubelmann, J.-M. Lehn, J.L. Sessler, A. Harriman, J. Chem. Soc., Chem. Commun. *1988*, 77.

177. V. Balzani, N. Sabbatini, F. Scandola, Chem. Rev. *86* (1986) 319.

178. Photoactive units may photooxidize complexed substrates [93] and effect DNA photocleavage: A.J. Blacker, J. Jazwinski, J.-M. Lehn, F.-X. Wilhelm, J. Chem. Soc., Chem. Commun. *1986*, 1035.

179. a) J.-M. Lehn, in D.S. Chemla, J. Zyss (Eds.), "Non-linear Optical Properties of Organic Molecules and Crystals", Academic Press, New York, vol. 1987, p. 215; b) J.F. Nicoud, R.J. Twieg, ibid. p. 221.

180. M. Takagi, K. Ueno, Topics Curr. Chem. *121* (1984) 39; H.-G. Lohr, F. Vögtle, Acc. Chem. Res. *18* (1985) 65; R. Klink, D. Bodart, J.-M. Lehn, B. Helfert, R. Bitsch, Merck GmbH, Eur. Pat. Apl. 83100281.1 (14.01.1983).

181. R.C. Haddon, A.A. Lamola, Proc. Natl. Acad. Sci. USA *82* (1985) 1874; R.W. Munn, Chem. Britain *1984*, 518; J. Simon, J.-J. André, A. Skoulios, Nouv. J. Chim. *10* (1986) 295; J. Simon, F. Tournilhac, J.-J. André, ibid. *11* (1987) 383; and references therein.

182. T.S. Arrhenius, M. Blanchard-Desce, M. Dvolaitzky, J.-M. Lehn, J. Malthête, Proc. Natl. Acad. Sci. USA *83* (1986) 5355.

183. a) Biological effectors may also be sought: I. Tabushi, T. Nishiya, M. Shimomura, T. Kunitake, H. Inokuchi, T. Yagi, J. Am. Chem. Soc. *106* (1984) 219; b) for other related work see for instance: J.K. Nagle, J.S. Bernstein, R.C. Young, T.J. Meyer, Inorg. Chem: *20* (1981) 1760; E.T.T. Jones, O.M. Chyan, M.S. Wrighton, J. Am. Chem. Soc. *109* (1987) 5526.

184. R. Schwyzer, A. Tun-Kyi, M. Caviezel, P. Moser, Helv. Chim. Acta *53* (1970) 15; Experientia *26* (1970) 577; see also ref. [1] p. 19.

185. a) S. Shinkai, Pure Appl. Chem. *59* (1987) 425; b) D.A. Gustowski, M. Delgado, V.J. Gatto, L. Echegoyen, G.W. Gokel, J. Am. Chem. Soc. *108* (1986) 7553 and references therein.

186. J.-M. Lehn, J. Malthête, A.-M. Levelut, J. Chem. Soc., Chem. Commun. *1985*, 1794.

187. For phthalocyanine derived columnar mesophases see D. Masurel, C. Sirlin, J. Simon, New. J. Chem. *11* (1987) 455 and references therein.

188. E. Carafoli, J.T. Penniston, Sci. Amer. November (1985) 50; T. Hiraoki, H.J. Vogel, J. Cardiovasc. Pharm. *10* (Suppl. 1) (1987) S14.

189. I. Tabushi, S.-i. Kugimiya, T. Sasaki, J. Am. Chem. Soc. *107* (1985) 5159 and references therein.

190. D. Pörschke, M. Eigen, J. Mol. Biol. *62* (1971) 361 and references therein.

191. J.-M. Lehn, A. Rigault, J. Siegel, J. Harrowfield, B. Chevrier, D. Moras, Proc. Natl. Acad. Sci. USA *84* (2565) 1987; J.-M. Lehn, A. Rigault, Angew. Chem. Int. Ed. Engl. *27* (1988), in press.

192. K.E. Drexler, Proc. Natl. Acad. Sci. USA *78* (1981) 5275; C. Joachim, J.-P. Launay, Nouv. J. Chim. *8* (1984) 723.

193. G.C. Pimentel, Chairman "Opportunities in Chemistry", Natl. Acad. Sci., Washington, DC, pp. 219–220.

Charles J. Pedersen

CHARLES J. PEDERSEN

I was born in Fusan, Korea, on October 3, 1904. My father Brede Pedersen, was a Norwegian marine engineer who left home as a young man and shipped out as an engineer on a steam freighter to the Far East. He eventually arrived in Korea and joined the fleet of the Korean customs service, which was administered by the British. Later, he abandoned seafaring and became a mechanical engineer at the Unsan Mines in what is now the northwestern section of present-day North Korea.

My mother, Takino Yasui, was born in 1874 in Japan. She had accompanied her family to Korea when they decided to enter large-scale trade in soybeans and silkworms. They established headquarters not far from the Unsan Mines, where she met my father. I had a sister, Astrid, five years my senior, and an elder brother who died in childhood prior to my birth.

The Unsan Mines were an American gold and lumber concession, 500 square miles in area. Because the mines were administered by Americans, there was an effort to make life there as American as possible. English was the spoken language, and it was the language I learned as a child. Foreign language schools did not exist in Korea at that time and so at the age of 8 years I was sent to Japan to attend a convent school in Nagasaki. When I was 10 years old my mother took me to Yokohama, and I began my studies at St. Joseph College. St. Joseph's was a preparatory school run by a Roman Catholic religious order of priests and brothers called the Society of Mary, also known as the Marianists. There I received a general secondary education and took my first course in chemistry.

When it came time for university, I chose, with my father's encouragement, to study in America. I selected the University of Dayton because it was in Ohio were we had family and friends and because it too was run by the Society of Mary. After taking a bachelor's degree in chemical engineering at the University of Dayton, I went to the Massachusetts Institute of Technology where I obtained a master's degree in organic chemistry. I did not remain at MIT to take a Ph.D.; I was still being supported by my father, and I was anxious to begin working. In 1927, I obtained employment at the Du Pont Company in Wilmington, Delaware, through the good offices of Professor James F. Norris, a very prominent professor and my research advisor. At Du Pont, I was fortunate enough to be directed to research at Jackson Laboratory by William S. Calcott. I remained at Du Pont for my entire 42-year career as a chemist.

As a new scientist I was initially set to work on a series of typical problems, which I solved successfully. After a while, I began to search for oil-solvable precipitants for copper, and I found the first good metal deactivator for petroleum products. As a result of this work, I developed a great interest in the affects of various ligands on the catalytic properties of copper and the transition elements generally and worked in the field for several years.

I next expanded my interests in the oxidative degradation of the substrates I was working on, namely petroleum products and rubber. By the mid-1940s I was in full career, having established myself in the field of antioxidants and independent in terms of the problems I might choose. In 1947, I was appointed research associate, then the highest title that a Du Pont Company researcher could attain. Also at that time, I married Susan Ault and settled in the town of Salem, New Jersey, where I have lived ever since.

During the late '40s and '50s my scientific interests became more varied. I became interested in the photochemistry of some new phthalocyanine adducts and of quinoneimine dioxides. I developed polymerization initiators and even made some novel polymers. In 1960 I returned to investigations in coordination chemistry, and decided to study the effects of bi- and multidentate phenolic ligands on the catalytic properties of the vanadyl group, VO. In the course of these investigations one of my experiments yielded an unexpected small quantity of unknown white crystals which I eventually identified as dibenzo-18-crown-6, first crown ether. The last nine years of my career were spent in the further study of crown ethers. I retired from Du Pont in 1969. During my retirement, I have pursued interests in fishing, gardening, bird study and poetry.

THE DISCOVERY OF CROWN ETHERS

Nobel lecture, December 8, 1987

by

CHARLES J. PEDERSEN

E. I. du Pont de Nemours and Company, Wilmington, Delaware 19898

Ladies and Gentlemen, Dear Colleagues,

This is a wonderful day in my life, and I am looking forward to sharing my thoughts with you.

Before I begin, I would like to convey the warm greetings of the people of Salem County, New Jersey—where I have lived for many years—to the people of Sweden. Salem County is where a very early Swedish settlement was established in 1643. Next year we will join with the people of our neighboring state of Delaware to celebrate the 350th anniversary of the first landing of Swedes in the New World at The Rocks in Wilmington, Delaware. We look forward to the visit of His Majesty King Carl XVI Gustaf and Her Majesty Queen Silvia and others from Sweden to our celebration next April.

Now I would like to discuss the discovery of the crown ethers. I will divide my lecture into three parts.

First, because every discovery takes place in more than a scientific context, I would like to touch on my life and background. In the weeks since it was announced that I would share this year's prize in chemistry, people have expressed as much interest in my early life as they have in my later work. So I think it appropriate to express myself on the matter. It may also be that details of my past have more than casual bearing on my work.

Second, I would like to describe for you my research program and some of the specific events that led to the discovery of the first crown ether. Since I am the only one who knows at firsthand the excitement and pleasure of the discovery, I will devote a portion of my time to sharing this experience with you.

And third, I would like to discuss the properties and preparation of crown ethers. In doing so, I hope I will convey to you that I was always a "hands-on" chemist; I took satisfaction from what I did in the laboratory. Also, I was very much an industrial chemist and was always interested in the potential application of my work. In fact, when I submitted my first major paper on the discovery of the crown ethers, the editor of the *Journal of the American Chemical Society*, Marshall Gates, remarked that my descriptions were replete with industrial jargon. Fortunately he published the paper anyway.

Personal background

Let me start then with how I began life and went on to discover the crown ethers.

My father, Brede Pedersen, was born in Norway in 1865 and trained as a marine engineer. Due to sibling disharmony, he left home for good as a young man and shipped out as an engineer on a steam freighter to the Far East. He eventually arrived in Korea and joined the fleet of the Korean customs, which was administered by the British. He rose in rank and later joined one of the largest Japanese steamship lines and became a chief engineer. Then a tragedy occurred that changed the course of his life. A childhood disease took the life of my elder brother while my father was away from home on a long journey. He abandoned the sea and became a mechanical engineer at the Unsan Mines in what is now the northwestern section of present-day North Korea.

My mother, Takino Yasui, was born in 1874 in Japan. She had accompanied her family to Korea when they decided to enter a large-scale trade in soybeans and silkworms. They established headquarters not far from the Unsan Mines, where she met my father.

The Unsan Mines were an American gold and lumber concession, 500 square miles in area. It had been granted by the Emperor of Korea to an American merchant named James R. Morse prior to 1870. I was conceived there in mid-winter just before the start of the Russo-Japanese war. Frequent incursions by Cossacks across the Yalu River into the region of the mines were considered to endanger my mother, so she and several American ladies were sent south by carriage to the railhead for safety. I was thus born on October 3, 1904, in the southern Port of Fusan, the largest in Korea. My arrival was doubly welcomed because mother was still grieving the loss of her firstborn. She devoted the next 10 years to overseeing my education and that of my sister, Astrid, five years my senior, in foreign language schools.

I spent my first and last winter at the mines when I was 4 years old. The region was known for severe weather due to the confluence of the Siberian steppes, Mongolian Gobi Desert and the mountains of Korea. Large Siberian tigers still roamed the countryside and were frightened away with bells on the pony harnesses. Wolves killed children during the cold winter nights, and foxes slept on roofs against the chimneys to keep warm.

Because the Unsan Mines were an American enclave—the top management being all Americans—great emphasis was placed on making life as American as possible. The country club was the center of social activities and life was considerably more gentle than at the typical gold mine of the legendary American West. So my contacts with Americans began early, and I spoke English which was the common language at the mines.

I do not know if such an environment had a lifelong influence on me, but I can speculate that perhaps it did. Freedom of the Americans to administer their affairs in taking care of themselves in the wilds where things could not be ordered for overnight delivery no doubt taught a certain independent approach to problem solving. As for chemistry, I recall that the gold was recovered by the cyanide process, and the monthly cleanup day was marked by the pervasive odor of the process. The pouring of the molten gold was always a beautiful sight, and that might have started my interest in chemistry. Also, my sister claimed that I loved to play with a collection of colorful Siberian minerals.

Foreign language schools did not exist in Korea then, and so at the age of 8, I was sent to a convent school in Nagasaki. When I was 10 years old, my mother took me to Yokohama where she remained with me for a year as I began my studies at St. Joseph College. St. Joseph was a preparatory school run by a Roman Catholic religious order of priests and brothers called the Society of Mary. There I received a general secondary education and took my first course in chemistry.

When it came time for me to start my higher education, there was no question of where it would be obtained. I had lived among Americans and had determined, with my father's encouragement, to study in America. I selected the University of Dayton in Ohio for two reasons: First, we had family friends in Ohio, and secondly, the same organization, the Society of Mary, ran both St. Joseph College and the University of Dayton.

My four years in Dayton and a year in graduate school at Massachusetts Institute of Technology were pleasant and taken up with activities that made me into an American. This perhaps also molded my scientific character and represented something of a personal metamorphosis. The sequence—Dayton first and then MIT—was also good, making a false start by a young man much less likely. The University of Dayton was a college of 400 men, most of them living in dormitories under strict monastic regimen. Training of the spirit was considered as important as training of the body and soul. I enjoyed all phases of the training. I became vice president of my graduating class, won letters in tennis and track and a gold medal for excellence that reflected my four years of performance there. Excellence in general was encouraged; I was even awarded a gold medal for conduct.

MIT was another matter. Boston, where I lived, is an old city of great charm and a center of the arts. I did not apply myself to my courses as I should, but my extracurricular activities contributed to the formation of my ultimate character. It was while studying at MIT that I first felt the exhilaration of utter freedom. MIT was considered deficient in the humanities, but with a little effort that deficiency could be remedied delightfully by visiting second-hand book stores. Why second-hand books appealed to me more than library books still remains a mystery—though it possibly was the prospect of finding unexpected treasures. I celebrated my graduation from MIT as a chemist by taking a walking tour of the Presidential Range in New Hampshire.

In spite of the urging of James F. Norris— a very prominent professor and my research advisor—I did not remain at MIT to take a Ph.D. My bills were still being paid by my father, and I was anxious to begin supporting myself. In 1927, I obtained employment at Du Pont through the good offices of Professor Norris, and I was fortunate enough to be directed to research at Jackson Laboratory by William S. Calcott. My career of 42 years had begun.

The research environment at Du Pont during those years was not altogether typical of industrial laboratories of the time. The company had formed the

nucleus of a basic research department that in a few years' time would have
scientists such as Wallace Carothers and the young Paul Flory working on the
polymer studies that led to nylon and other breakthroughs. And in general, Du
Pont was a productive center of research where many interesting and important problems were being solved. For example, one day while visiting Julian
Hill at the Du Pont Experimental Station in Wilmington, Delaware, I observed
him pull the first oriented fiber of a polyester. On another occasion, at Jackson
Laboratory, across the Delaware River in New Jersey where I worked, I
noticed commotion in the laboratory of Roy Plunkett, which was across the hall
from my own. I investigated and witnessed the sawing open of a cylinder from
which was obtained the first sample of Teflon® fluoropolymer. At Jackson
Laboratory, during that time, other important advances were taking place in
tetraethyl lead and new petroleum chemicals, new elastomers, and a new series
of fluorocarbons for refrigeration and aerosols. The atmosphere was vibrant
and exciting, and success was expected. It was in this atmosphere I began my
career.

As a new scientist I was initially set to work on a series of typical problems,
the successful solution of which buoyed my research career (Ref. 1–5). After a
while, I began to search for oil-soluble precipitants for copper, and I found the
first good metal deactivator for petroleum products (Ref. 6–8). As a result of
this work, I developed a great interest in the effects of various ligands on the
catalytic properties of copper and the transition elements generally, and I
worked in that field for several years. I noticed a very unusual synergistic affect
wherein a metal deactivator greatly increased the efficacy of antioxidants (Ref.
9–10).

So more and more, I became interested in the oxidative degradation of the
substrates themselves, particularly petroleum products and rubber. As my
interests moved in that direction, I left off working on metal deactivators and
coordination chemistry. By the mid-1940s, I was in full career, having established myself in the field of oxidative degradation and stabilization (Ref. 11–
13). I was independent in terms of the problems I might choose and had
achieved the highest non-management title then available to a scientist at Du
Pont. During the 1940s and 1950s, my interests became more varied. For
example, I became interested in the photochemistry of new phthalocyanine
adducts and of quinoneimine dioxides. I found some polymerization initiators,
discovered that ferrocene was a good antiknock agent for gasoline, and made
some novel polymers (Ref. 14–23).

Discovery of the crown ethers

But then there arose a challenging opportunity that led me back to ligand
chemistry. In response to my desire to contribute to the elastomer field, my
colleague Herman Schroeder suggested that there was an interesting problem
in the coordination chemistry of vanadium. This sparked my curiosity, and I
began work with the initial goal of understanding factors which govern catalytic activity of vanadium in oxidation and polymerization. This was a relatively
unexplored area, and previous work had been empirical. It was my work in this

Copper Complex of N,N′-(1,2-Propylenebis) (Salicylideneimine)

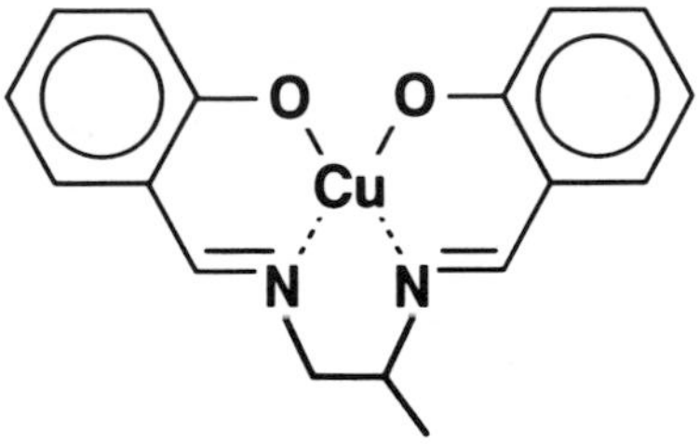
O
O
Cu
N
N

Fig. 1.

area that led to the discovery of crown ethers, which I will now describe.

As I have related, I studied for many years the autoxidation of petroleum products and rubber and its retardation by antioxidants. Autoxidation is greatly catalyzed by trace metals, such as copper and vanadium. Hence, I had developed the compounds referred to earlier, namely the "metal deactivators" which suppress the catalytic activity of the metal salts by converting them into inactive multidentate complexes. The first of these was N,N'-(1,2-propylene-bis) (salicylideneimine) shown in Figure 1—an excellent deactivator for copper which has been used industrially for many years.

In 1960 when I returned to investigations in coordination chemistry, I decided to study the effects of bi- and multidentate phenolic ligands on the

Synthesis of Bis[2-(o-Hydroxyphenoxy)Ethyl]Ether

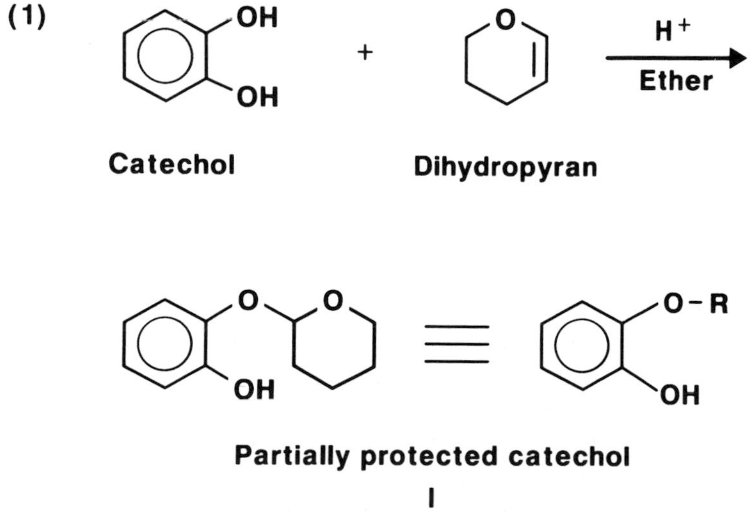
(1)
OH
OH
+
O
H+
Ether
Catechol
Dihydropyran
O
O
OH
O-R
OH
Partially protected catechol
I

Fig. 2.

catalytic properties of the vanadyl group, VO (Ref. 24). The multidendate
ligand I selected is the bis[2-(o-hydroxyphenoxy)ethyl] ether whose synthesis
is depicted in Figure 2. As I proceeded, I knew that the partially protected
catechol was contaminated with about 10 percent unreacted catechol. But I
decided to use this mixture for the second step anyway since purification would
be required at the end. The reactions were carried out as outlined and gave a
product mixture in the form of an unattractive goo. Initial attempts at purification gave a small quantity (about 0.4 percent yield) of white crystals which
drew attention by their silky, fibrous structure and apparent insolubility in
hydroxylic solvents.

The appearance of the small quantity of the unknown should have put me in
a quandary. I probably was not the target compound because that would be
obtained in a higher yield. My objective was to prepare and test a particular
compound for a particular purpose. Had I followed this line, I would have
doomed the crown ethers to oblivion until such a time as another investigator
would retrace my steps and make the better choice at the critical moment.
Crown ethers, however, were in no danger, because of my natural curiosity.
Without hesitation, I began study of the unknown.

It was fortunate that I used an ultraviolet spectrophotometer to follow the
reactions of the phenols. These compounds and their ethers in neutral methanol solutions absorb in the region of 275 millimicrons. On treatment with
alkali, the absorption curve is not significantly altered if all the hydroxyl groups
are covered, but it is shifted to longer wavelengths and higher absorption if one
or more hydroxyl groups are still free, as shown by the dashed curve in Figure
3.

Synthesis of Bis[2-(o-Hydroxyphenoxy)Ethyl]Ether

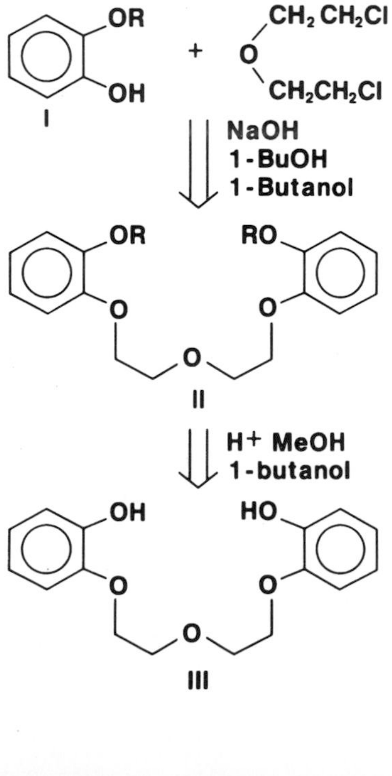

Fig. 2 (continued).

Effect of NaOH on Ultraviolet Spectrum

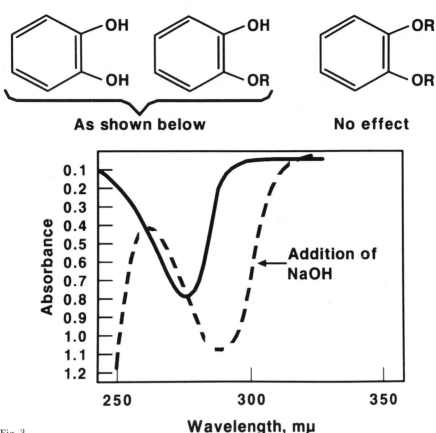

Fig. 3.

The unknown product was very little soluble in methanol, and the neutral solution gave an absorption curve characteristic for a phenolic compound. The solution was made alkaline with sodium hydroxide with the expectations that the curve would either be unaffected or shifted to longer wavelengths. The resulting spectrum, however, showed neither effect, but rather the one shown in Figure 4. At the same time, I noticed that the fibrous crystals were freely soluble in methanol in the presence of sodium hydroxide. This seemed strange since the compound did not contain a free phenolic group, a fact confirmed by its infrared and NMR spectra. I then found that the compound was soluble in methanol containing any soluble sodium salt. Thus, the increased solubility was due not to alkalinity but to sodium ions. But there was no obvious explanation for the behavior of the compound because its elementary analysis corresponded with that for a 2,3-benzo-1,4,7-trioxacyclononane, (Figure 5) a plausible product from the reaction of catechol and bis(2-chloroethyl)ether in the presence of sodium hydroxide. However, the moment of revelation came when I learned that its molecular weight was exactly twice that of the above compound. The true structure was that of an 18-membered ring, dibenzo-18-

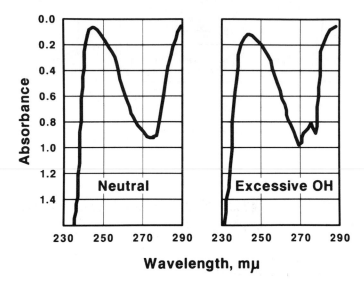

Effect of Sodium Hydroxide on the Ultraviolet Spectrum of Dibenzo-18-Crown-6 in Methanol

Fig. 4.

crown-6, the first and most versatile of the aromatic crown compounds, depicted in Figure 6. The shape is that of a torus or a doughnut.

It seemed clear to me now that the sodium ion had fallen into the hole in the center of the molecule and was held there by the electrostatic attraction between its positive charge and the negative dipolar charge on the six oxygen atoms symmetrically arranged around it in the polyether ring. Tests showed that other alkali metal ions and ammonium ion behaved like the sodium ion so that, at long last, a neutral compound had been synthesized which formed stable complexes with alkali metal ions. Up to that point, no one had ever found a synthetic compound that formed stable complexes with sodium and potassium.

2,3-Benzo-1,4,7-Trioxacyclononane

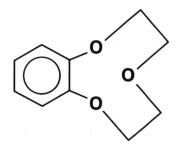

Fig. 5.

Dibenzo-18-Crown-6

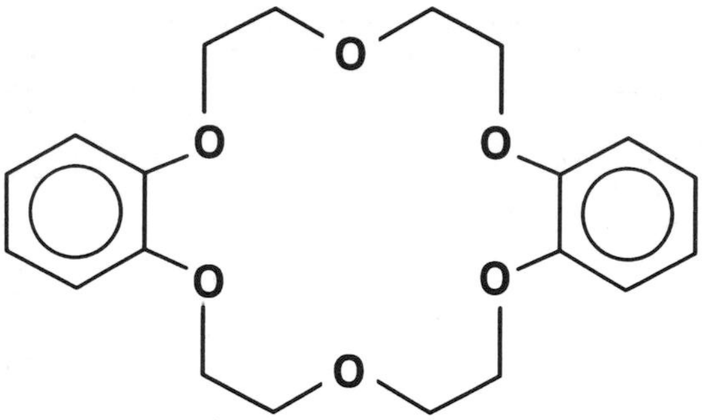

2,3,11,12-Dibenzo-1,4,7,10,13,16-Hexaoxacyclo-octadeca-2,11-Diene

Fig. 6.

My excitement, which had been rising during this investigation, now reached its peak and ideas swarmed in my brain. One of my first actions was motivated by esthetics more than science. I derived great esthetic pleasure from the three-dimensional structure as portrayed in the computer-simulated model in Figure 7. What a simple, elegant and effective means for the trapping of hitherto recalcitrant alkali cations! I applied the epithet "crown" to the first member of this class of macrocyclic polyethers because its molecular model looked like one and, with it, cations could be crowned and uncrowned without physical damage to either as shown for the potassium complex in Figure 8. As my studies progressed, I created the system of crown nomenclature chiefly because the official names of the crown ethers were so complex and hard for me to remember. It is a source of special satisfaction to me that this system of abbreviated names, devised solely for the ready identification of the macrocyclic polyethers, has been retained by the scientific establishment. In Figure 9 I have illustrated how the nomenclature system is made up of the side-ring substituents, the total number of oxygen atoms in the main ring and the size of the ring.

Another aspect of this discovery filled me with wonder. In ordinary organic reactions only rings of 5, 6, or 7 members form easily. Here a ring of 18 atoms had been formed in a single operation by the reaction of two molecules of catechol, which was present as a minor impurity, with two molecules of bis(2-chloroethyl)ether. Further experiments revealed that dibenzo-18-crown-6 can be synthesized from these intermediates in a 45 percent yield without resorting to high dilution techniques. This was most unexpected and some good reason must exist for such an unusual result. I concluded that the ring-closing step, either by a second molecule of catechol or a second molecule of bis(2-chloroethyl)ether, was facilitated by the sodium ion which, by ion-dipole interaction, "wrapped" the molecular pieces around itself to form a three-quarter circle and

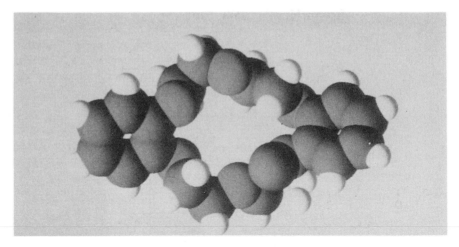

Fig. 7.

disposed them for the final ring closure in much the same fashion as is involved
in the synthesis of the porphyrins and phthalocyanines. Later experiments
appear to support this hypothesis. The yields of dibenzo-18-crown-6 are higher
when it is prepared with sodium or potassium hydroxide than when lithium or
tetramethylammonium hydroxide is used. Lithium and the quaternary ammonium ions are not strongly complexed by the polyether. The best complexing
agents are rings of 15 to 24 atoms including 5 to 8 oxygen atoms. They are
formed in higher yields than smaller or larger rings, or rings of equal sizes with
only four oxygen atoms. Finally, even open-chain polyethers such as 3,4,12,13-
diebenzo-2,5,8,11,14-pentaoxapentadeca-3,12-diene (Figure 10) were found to
form complexes with sodium and potassium ions.

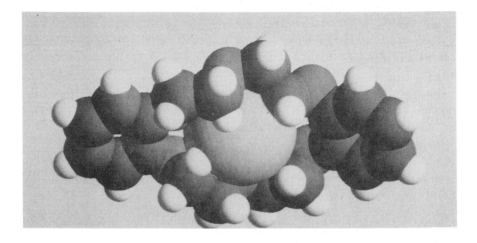

Fig. 8.

Some Macrocyclic Polyethers

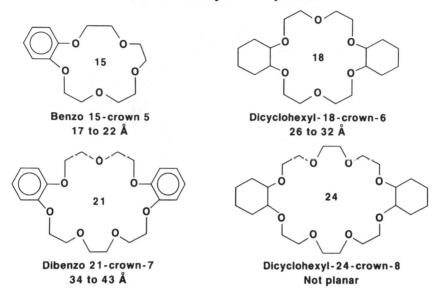

Benzo 15-crown 5
17 to 22 Å

Dicyclohexyl-18-crown-6
26 to 32 Å

Dibenzo 21-crown-7
34 to 43 Å

Dicyclohexyl-24-crown-8
Not planar

The numbers within the diagrams are the numbers of atoms in the polyether rings. The numbers under the names are the estimated diameters of the holes in Å.

Fig. 9.

Thus did I discover dibenzo-18-crown-6, the first crown ether and the first neutral synthetic compound capable of complexing the alkali metal cations (Ref. 25–26, 36).

With the realization that I had something very unusual and with the utmost curiosity and anticipation, I devoted all my energies over the next several years to the study of this fascinating class of ligands and their interaction with inorganic cations. Every successful experiment produced a significantly novel result and led to new thoughts on what to synthesize and also as to the many potential uses of these extraordinary substances.

I was especially interested in the stability of the "complexes" and the reason for their behavior. For example, I found that for maximum stability of its salt complex, each cation has an optimum size of the ring of the polyether. A complex can form even if the fit is not the best by forming a sandwich consisting of two molecules of polyether per cation. The thermal stability of some salt complexes, for example, that with KCNS, is attested to by their having melting points higher than those of the components.

Preparation and properties of macrocyclic polyethers

Spurred by curiosity regarding the factors involved in the stability of the salt complexes (such as the relative sizes of the hole and the cation, and the number and symmetrical arrangement of the oxygen atoms in the polyether ring), I initiated an extensive program of syntheses. Ultimately, about 60 macrocyclic polyethers were prepared containing 12 to 60 atoms to a polyether ring including 4 to 10 oxygen atoms and some with nitrogen and sulfur atoms. Many of these compounds were found to be useless as complexing agents, but they

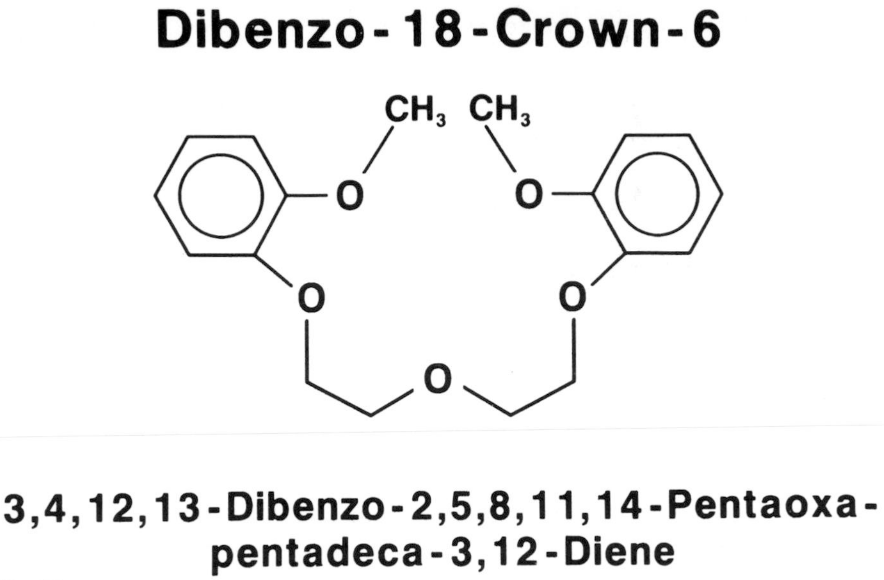

Dibenzo-18-Crown-6

3,4,12,13-Dibenzo-2,5,8,11,14-Pentaoxa-pentadeca-3,12-Diene

Fig. 10.

served to define the effective ones which are compounds containing 5 to 10 oxygen atoms in the ring, each separated from the next by 2 carbon atoms. I also noted that even whole molecules such as the thioureas formed complexes with some crown compounds. I accomplished all this working alone with the help of my able technician, Ted Malinowski.

Some of the general properties of the aromatic macrocyclic polyethers are as follows: They are neutral, colorless compounds with sharp melting points, and are little soluble in water and alcohols, fairly soluble in aromatic solvents, and very soluble in methylene chloride and chloroform. They undergo substitution reactions characteristic for aromatic ethers (halogenation, nitration, etc.), and form formaldehyde resins when treated with paraformaldehyde under acid conditions. They are decomposed by reactions which cause the scission of ethers.

The saturated macrocyclic polyethers are obtained most simply by catalytically hydrogenating the aromatic compounds using ruthenium catalyst. Bridge-bond isomers are obtained from compounds containing two or more aromatic side-ring substituents. For example, dibenzo-18-crown-6 gives a mixture of stereoisomers of dicyclohexyl-18-crown-6. The saturated polyethers are colorless, viscous ills or solids of low melting points. They are thermally stable but, like the aromatic compounds, must be protected from oxygen at high temperatures. They are, as a group, very much more soluble than the aromatic compounds in all solvents, and most of them are even soluble in petroleum ether.

The unique property of the macrocyclic polyethers as complexing agents is their preference for alkali metal ions, which do not form complexes with the numerous ligands used for the transition metal ions. The crown compounds form stable crystalline complexes and solutions of the complexes with some or

Diameters of Holes in Ångström Units

Macrocyclic Polyethers	Diameters
All 14-crown-4	1.2-1.5
All 15-crown-5	1.7-2.2
All 18-crown-6	2.6-3.2
All 21-crown-7	3.4-4.3

Table 1.

all of the cations of alkali and alkaline earth metals plus ammonium ions and others. Some of them, for example, dicyclohexyl-18-crown-6, also form complexes with Co(II), and some other transition metal ions. The saturated compounds are better complexing agents than the corresponding aromatic compounds.

Three criteria have been used for the formation of complexes between macrocyclic polyethers and salts: (a) isolation of the complexes as crystals; (b) characteristic changes in the ultraviolet spectra of the aromatic compounds; and (c) changes in the solubilities of the polyethers and salts in different solvents.

As is evident from Table 1, these compounds have holes of different diameters in the center of the polyether rings. The uncomplexed cations also differ in size, given in Table 2 in Ångströms units: sodium 1.94, potassium 2.66, ammonium 2.86, rubidium 2.94, and cesium 3.34. Depending, therefore, on the relative sizes of the hole and the cation, crystalline complexes with polyether/cation ratios of 1:1, 3:2, and 2:1 have been prepared as illustrated in Table 3. The aromatic macrocyclic polyethers tend to give high melting complexes which are not readily soluble in aprotic solvents, while the saturated compounds give lower melting complexes which are more soluble. Most of the pure complexes are decomposed by water, the rate and extent of decomposition depending on the proportion of water and the temperature.

It was postulated from the beginning that complexes of macrocyclic polyethers containing less than seven oxygen atoms consisted of a cation surrounded by the oxygen atoms arranged symmetrically in a single plane. The essential correctness of this view of the structure has been confirmed by Professor M. R. Truter and her collaborators who have been the first to determine the structures of a number of crystalline salt complexes of crown compounds by X-ray diffraction methods (Ref. 27).

All macrocyclic polyethers containing one or more benzo groups have a characteristic absorption maximum at 275 millimicrons in methanol, and the shapes of the curves are altered by the addition of complexable salts as was shown in *Figure 4*. The spectral evidence is nearly always confirmed by the other two criteria.

Complexable Cations and Their Diameters
in Ångström Units

Group I		Group II		Group III		Group IV	
Li	1.36						
Na	1.94						
K	2.66	Ca	1.98				
Cu(I)	1.92	Zn	1.48				
Rb	2.94	Sr	2.26				
Ag	2.52	Cd	1.94				
Cs	3.34	Ba	2.68	La	2.30		
Au(I)	2.88	Hg(II)	2.20	Tl(I)	2.80	Pb(II)	2.40
Fr	3.52	Ra	2.80				
NH$_4$	2.86						

Table 2.

Macrocyclic polyethers and complexable salts mutually increase their solubilities in solvents wherein the complexes are soluble. Sometimes these effects are spectacular, for instance, the solubility of the potassium thiocyanate complex is about a tenth of a mole per liter, a 100-fold increase. Some of the saturated polyethers, such as dicyclohexyl-18-crown-6, have the useful property of solubilizing alkali metal salts, particularly those of potassium, in aprotic solvents. Crystals of potassium permanganate, potassium tertiary-butoxide, and potassium palladous tetrachloride (PdCl2+2KCl) can be made to dissolve in liquid aromatic hydrocarbons merely by adding dicyclohexyl-18-crown-6. This is dramatic for the crown complex of potassium permanganate which colors toluene purple. Benzylpotassium is rendered soluble in n-heptane by the polyether, but the polyether ring is gradually decomposed by this organometallic compound. The solubilizing power of the saturated macrocyclic polyethers permits ionic reactions to occur in aprotic media. It is expected that this property will find practical use in catalysis, enhancement of chemical reactivity, separation and recovery of salts, electrochemistry, and in analytical chemistry.

The complexing efficiencies of saturated macrocyclic ethers can be ranked numerically by measuring the relative distribution of a colored alkali metal salt (such as picrate) between an immiscible organic solvent and water in the presence of the crown ether as depicted. If the polyether is ineffective, the organic phase will be colorless; if the polyether is very powerful, most of the color will be in the organic phase. The efficiencies of the polyether will lie between these two limits as shown in Table 4 (Ref. 28—35).

Dr. H. K. Frensdorff has determined the stability constants for 1:1 complexes of many macrocyclic polyethers with alkali metal ions by potentiometry with

Crystalline Complexes of Polyethers

Crystalline Complex	Mole Ratio[a]
Benzo-15-crown-5	
NaI	1:1
KCNS	2:1
Dibenzo-15-crown-5	
KCNS	2:1
Dibenzo-18-crown-6	
KCNS	1:1
NH_4CNS	1:1
RbCNS	1:1
RbCNS	2:1
CsCNS	2:1
CsCNS	3:2
Dicyclohexyl-18-crown-6	
KI_3	1:1
CsI_3	3:2
Dibenzo-24-crown-8	
KCNS	1:1
KCNS	1:2
Dibenzo-30-crown-10	
KCNS	1:1
NaCNS	1:2

[a] (Polyether):(salt).

Table 3.

cation-selective electrodes. Selectivity toward the different cations varies with polyether ring size, the optimum ring size being such that the cation just fits into the hole, that is 15—18 for sodium ion, 18 for potassium ion, and 18—21 for cesium ion (Ref. 33).

That concludes my remarks on the discovery, properties and preparation of the crown ethers. It remains only for me to mention certain individuals who contributed to the success of my research and to add a few words concerning my interest and hope for the future of research in this area.

First, I want to remember on this occasion my wife Susan who died in 1983. It would have been wonderful to share with her all that has happened to me of late as we shared everything else during our marriage of 36 years.

Next, I would like to thank the Du Pont Company. They encouraged me to pursue my research on crown ethers, even when it was evident that, at least initially, my work might not have a significant practical impact. At another company, I might not have met with such encouragement and latitude.

Within the company I received support from certain individuals. I appreciate the advice and counsel of my close friend, Dr. Herman Schroeder, who was

Extraction Results[a]

Polyether	Picrate Extracted (%)			
	Li$^+$	Na$^+$	K$^+$	Cs$^+$
Dicyclohexyl-14-crown-4	1.1	0	0	0
Cyclohexyl-15-crown-5	1.6	19.7	8.7	4.0
Dibenzo-18-crown-6	0	1.6	25.2	5.8
Dicyclohexyl-18-crown-6	3.3	25.6	77.8	44.2
Dicyclohexyl-21-crown-7	3.1	22.6	51.3	49.7
Dicyclohexyl-24-crown-8	2.9	8.9	20.1	18.1

[a]Two-phase liquid extraction: methylene chloride and water.

Table 4.

always interested in my research and whose companionship has meant so much to me during the many years we have known each other. I also thank my friend Dr. Rudolph Pariser, who has been tireless in his efforts to assure recognition for my accomplishments.

Finally, I want to thank the analytical groups of the company for making all their resources available to me; my technical colleagues for their scientific consultation; and our academic friends for their interest.

Of course, I must mention my respect and admiration for the two scientists with whom I share this year's prize. If I may use an analogy reflecting my youth at the Unsan gold mines, I see the discovery of the crown ethers as comparable to the finding of a new field with a lot of action in it. Professor Cram and Professor Lehn staked claims to particular veins of rich ore and went on to discover gold mines of their own.

I know that the crown ethers continue to create great interest among biologists for studying the mechanism of transport of ions across cell membranes (Ref. 36). But whether it be in biology or some other field, it is my fervent wish that before too long it matters not by whom the crown ethers were discovered but rather that something of great benefit to mankind will be developed about which it will be said that were it not for the crown compounds it could not be.

REFERENCES

1. F. B. Downing, A. E. Parmalee and C. J. Pedersen, U.S.P. 2,004,160 (6/11/35) to Du Pont.
2. F. B. Downing and C. J. Pedersen U.S.P. 2,008,753 (7/23/35) to Du Pont; also 2,087,103 (7/13/37).
3. R. G. Clarkson and C. J. Pedersen, U.S.P. 2,054,282 (9/15/36) to Du Pont.
4. L. Spiegler and C. J. Pedersen, U.S.P. 2,087,098 (7/13/37) to Du Pont.
5. F. B. Downing and C. J. Pedersen, U.S.P. 2,121,397 (6/21/38) to Du Pont.
6. F. B. Downing and C. J. Pedersen, U.S.P. 2,181,121 (11/28/39) to Du Pont.
7. C. J. Pedersen, *Oil & Gas Journal*, p. 97, July 27, (1939).
8. C. J. Pedersen, *Ind. & Eng. Chem.*, *41*, 824, (1949).
9. C. J. Pedersen, *Delaware Chemical Symposium*, Dec. 1, (1948). Prooxidant Catalytic Activity of Metal Chelates.
10. C. J. Pedersen, *Symposium on Chelate Chemistry*. Centenary Celebration of Brooklyn Polytechnic Institute, New York, N.Y. Published in *Advances in Chelate Chemistry*, p. 113 (1954).
11. C. J. Pedersen, *Delaware Chemical Symposium*, Jan. 21, (1950). Mechanism of Decomposition of Perbenzoic Acid Compared with Benzoyl Peroxide.
12. C. J. Pedersen, Antioxidants, *Encyclopedia Britannica*, (1953).
13. C. J. Pedersen, *J. Org. Chem.*, *22*, 127 (1957); U.S.P. 2,662,895-7 (12/15/53); U.S.P. 2,681,347 (6/15/54); U.S.P. 2,741,531 (5/12/56); U.S.P. 2,831,805 (4/22/58) all to Du Pont.
14. C. J. Pedersen, *Ind. & Eng. Chem.*, *48*, 1881 (1956).
15. C. J. Pedersen, *J. Am. Chem. Soc.*, *79*, 2295 (1957).
16. C. J. Pedersen, *J. Am. Chem. Soc.*, *79*, 5014 (1957); U.S.P. 2,681,918 (6/22/54); U.S.P. 2,741,625 (4/10/56); U.S.P. 2,831,805 (4/22/58) all to Du Pont.
17. C. J. Pedersen, U.S.P. 2,867,516 (1/6/59) to Du Pont.
18. C. J. Pedersen, U.S.P. 3,341,311 (9/12/67) to Du Pont.
19. C. J. Pedersen, U.S.P. 3,038,299−300 (6/12/62) to Du Pont.
20. C. J. Pedersen, *J. Org. Chem.*, *23*, 252 & 255 (1958).
21. J. Diekmann and C. J. Pedersen, *J. Org. Chem.*, *28*, 2879 (1963). See also *Chem. Rev.*, *67*, 611 (1967, p. 617).
22. C. J. Pedersen, U.S.P. 3,232,914 (2/1/66) to Du Pont.
23. C. J. Pedersen, U.S.P. 3,320,214 (5/16/67) to Du Pont.
24. C. J. Pedersen, U.S.P. 3,361,778 (1/2/68) to Du Pont.
25. C. J. Pedersen, *J. Am. Chem. Soc.*, *89*, 2495, 7017 (1967).
26. C. J. Pedersen, *Aldrichimica Acta*, *(4) 1*, 1 (1971).
27. M. R. Truter and C. J. Pedersen, Endeavor, XXX (111), 142 (1971).
28. C. J. Pedersen, *Fed. Proc., Fed. Am. Soc. Exp. Biol.*, *27*, 1305 (1968).
29. C. J. Pedersen, *J. Am. Chem. Soc.*, *92*, 386 (1970).
30. C. J. Pedersen, *J. Am. Chem. Soc.*, *92*, 391 (1970).
31. C. J. Pedersen, *J. Org. Chem.*, *36*, 254 (1971).
32. C. J. Pedersen, *J. Org. Chem.*, *36*, 1690 (1971).
33. C. J. Pedersen and H. K. Frensdorff, *Angew. Chem.*, *84*, 16 (1972); *ibid. (int. Ed.)*, *11*, 16 (1972).
34. C. J. Pedersen, *Org. Syn.*, *52*, 66 (1972).
35. C. J. Pedersen, U.S.P. 3,562,295, (2/9/71); U.S.P. 3,622,577 (11/23/71); U.S.P. 3,686,225 (8/22/72); U.S.P. 3,687,978 (8/29/72); with M. Bromeis U.S.P. 3,847,949 (11/12/74); U.S.P. 3,856,813 (12/24/74); U.S.P. 3,873,569 (3/25/75); U.S.P. 3,987,061 (10/19/76); U.S.P. 3,998,838 (12/21/76); with M. Bromeis U.S.P. 4,031,111 (6/21/77); all to Du Pont.
36. C. J. Pedersen in Current Topics in Macrocyclic Chemistry in Japan, Ed. E. Kimura, Hiroshima Univ. School of Medicine (1987), p. 1. H. E. Schroeder 5.

Chemistry 1988

**JOHANN DEISENHOFER, HARTMUT MICHEL and
ROBERT HUBER**

*for the determination of the three-dimensional structure of a photosynthetic reaction
centre*

THE NOBEL PRIZE IN CHEMISTRY

Speech by Professor Bo G. Malmström of the Royal Swedish Academy of Sciences.
Translation from the Swedish text.

Your Majesties, Your Royal Highnesses, Ladies and Gentlemen,

In the beginning there was light. Light played an important role in the origin of life on earth, and the radiation of the sun is an absolute prerequisite for the forms of life which today inhabit our planet. In the green leaves of plants solar light is converted to chemical energy, which is used as nutrition not only by the green plants themselves but also, for example, by cows who eat green grass, by ourselves who eat the meat of the cow and drink its milk, and so on through the nutritional chain.

The energy required for life processes is to a large extent liberated in the combustion of sugar and fat by the oxygen of the air. This process can, however, continue idefinitely thanks to the fact only that the nutritional substances used up are re-manufactured in the photosynthesis of green plants. In photosynthesis the plants use solar energy to build up complicated nutritional substances from two simple molecules, carbon dioxide and water, with concomitant liberation of oxygen. In the respiration in the cells of living organisms this nutrition is then reconverted to carbon dioxide and water, so that there is a continous cyclic process driven by the sun.

In respiration as well as in photosynthesis, electrons fall from a higher to a lower energy level, somewhat like an electric current. They do not, however, pass through an electric wire but are transferred between a number of complicated proteins, which often contain metals, e. g. iron. The principles of electron transfer between simple metal compounds has been analyzed in detail by Henry Taube, the Nobel Prize winner for chemistry in 1983. An important goal in the chemical research of today is to extend these contributions in order to explain the more complicated biochemical processes.

The proteins mediating the electron transfer are organized in large molecular aggregates which are bound to biological membranes. In the electron transfer energy is liberated, and this is used to make ATP, the universal energy storage molecule of living cells. The ATP formation takes place according to a mechanism formulated by the Nobel Prize winner for chemistry in 1978, Peter Mitchell.

For a long time it has been impossible to prepare membrane-bound proteins in a form allowing the determination of the detailed structure in three dimensions. Before 1984, there were only rather fuzzy structural pictures available for a few membrane proteins. These had been derived with the aid of an electron microscopic method developed by the Englishman Aaron Klug, who was awarded the Nobel Prize in chemistry in 1982 for this achievment. But the situation had actually drastically changed in 1982, when Hartmut Michel thanks to systematic experiments succeeded in preparing highly ordered crystals of a photosynthetic reaction center from a bacterium. With these crystals he could in the period 1982—1985, in collaboration with Johann Deisenhofer and Robert Huber, determine the structure of the reaction center in atomic detail.

The structural determination awarded has led to a giant leap in our understanding of fundamental reactions in photosynthesis, the most important chemical reaction in the biosphere of our earth. But it has also consequences far outside the field of photosynthesis research. Not only photosynthesis and respiration are associated with membrane-bound proteins but also many other central biological functions, e. g. the transport of nutrients into cells, hormone action or nerve impulses. Proteins participating in these processes must span biological membranes, and the structure of the reaction center has delineated the structural principles for such proteins. Michel's methodological contribution has, in addition, the consequence that there is now hope that we can determine detailed structures also for many other membrane proteins. Not least important is the fact that the reaction center structure has given theoretical chemists an indispensable tool in their efforts to understand how biologic electron transfer over very large distances on a molecular scale can occur as rapidly as in one billionth (American English, trillionth) of a second. In a longer perspective it is possible that such research can lead to important energy technology in the form of artificial photosynthesis.

Drs. Deisenhofer, Huber and Michel,

I have tried to describe — in Swedish — how your determination of the structure of a photosynthetic reaction center has lead to a leap in our understanding of the perhaps most important chemical reaction on earth. But it has also major implications far outside the field of photosynthesis by clarifying the structural principles for membrane proteins and by providing theoretical chemists with an important tool for understanding the basis of rapid electron transfer in biological systems. It is for these fundamental contributions that the Royal Swedish Academy of Sciences has decided to award to you this year's Nobel Prize in chemistry.

On behalf of the Academy I wish to convey to you our warmest congratulations, and I now ask you to receive the Prize from the hands of His Majesty the King.

JOHANN DEISENHOFER

I was born on September 30, 1943 in Zusamaltheim, Bavaria, now Federal Republic of Germany, as the first son of Thekla and Johann Deisenhofer. After my father's return from military service my parents ran the family farm. In 1948, our family grew to its final size with the birth of my only sister, Antonie.

My early youth was influenced by the environment provided by a little village that, after World War II, tried to find its way back to some kind of a normal life. Nevertheless, it was a most enjoyable place for a little boy. In 1949, I entered elementary school at Zusamaltheim, and continued to attend until 1956. According to the local custom, the oldest son was designated to take over the family's farm. However, to their great dissappointment, my parents early noticed my lack of interest in farming, and made the difficult decision to send me away to school. My way to higher education started in 1956 at the "Knabenmittelschule Hl. Kreuz", Donauwoerth, and continued 1957 to 1959 at the "Staatliche Realschule Wertingen" and 1959 to 1963 at the "Holbéin Gymnasium", Augsburg. There, in 1963, I underwent the "Abitur" examination that allowed me to go to a university. I was awarded the "Stipendium fuer besonders Begabte" of the "Bayerische Staatsministerium fuer Unterricht und Kultus" which helped to lower the financial burden on my parents for my education.

In the fall of 1965, after 18 months of military service in the German Bundeswehr, where I did not exceed the rank of private, I began to study physics at the Technische Universitaet Muenchen. A major reason for choosing physics was an interest in physical, especially astronomical problems, aroused by popular books on this subject. The book I most clearly remember was a popular review of the state of astronomy by Fred Hoyle, describing the impact of modern physics on astronomy, and the recent achievements and open questions in that field. The Technische Universitaet Muenchen (TUM) was the obvious choice because Rudolf L. Moessbauer had just accepted a professorship at the TUM; moreover Munich is only about 100 km from Zusamaltheim.

During my time at the TUM, I learned that physics was quite different from what I expected; also, my interest slowly shifted to solid state physics. Together with a couple of colleagues I started my Diplomarbeit in this field in the laboratory of Klaus Dransfeld. As it turned out, Klaus Dransfeld was a person almost as shy as myself, so that we could not establish a good personal contact at the time. Nevertheless, the experimental work I did under the supervision of Karl-Friedrich Renk in Dransfeld's lab was very successful, and led to a publication in Physical Review Letters in 1971; this was my first scientific publication. During my time in his lab, Klaus Drans-feld transmitted his interest in biophysical problems to many students. This had direct consequences for my career because it made me look for a suitable institution to get a Ph. D. in this field. From a friend I heard about a new group at the Max-Planck-Institut fuer Eiweiss- und Lederforschung whose head, Robert Huber, was looking for students. After a brief interview with Robert Huber, it was agreed that I could start my work in June 1971, following the final examination for my physics diploma at the TUM. In 1972, the institute moved a few kilometers from Munich to Martinsried, and became the Max-"Planck-Institut fuer Biochemie". The work I did together with Wolfgang Steigemann (also one of Huber's Ph.D. students at that time) on the crystallographic refinement of the structure of bovine pancreatic trypsin inhibitor was a success, and our 1975 paper in Acta Crystallographica has been cited ever since.

At the end of 1974, when I had obtained my Ph.D. degree, Robert Huber offered me a postdoctoral position for two years which I accepted. This position was converted into a permanent position in 1976. I joined Peter M. Colman, then a postdoctoral fellow in Huber's lab, and Walter Palm from the University of Graz, Austria, in their work on the human myeloma protein Kol. After the solution of this interesting structure, I continued, together with Robert Huber, Peter Colman's work on the human Fc-fragment, and its complex with an Fc-binding fragment from protein A from *Staphylococcus aureus*. The refinement of these structures was finished in 1980. In the following two years I joined several projects in Robert Huber's lab: human C3a, citrate synthase, and α1-proteinase inhibitor. During all my time in Martinsried I enjoyed working with computers, and developing and maintaining crystallographic software.

In 1982, Hartmut Michel, who had come to Martinsried together with Dieter Oesterhelt, reported in one of Huber's group seminars about his spectacular success with the crystallization of the photosynthetic reaction center from *Rhodopseudomonas viridis*. After discussions between Hartmut and myself, and after Robert Huber had given his agreement, I joined the reaction center project in order to determine the three-dimensional struc-ture of this molecule. Shortly afterwards Kunio Miki, a post-doctoral fellow from Osaka University, arrived in Martinsried, and helped us until Septem-ber 1983. Later Otto Epp, a colleague and friend since I joined the Max-Planck-Institute, made most valuable contributions to the project.

In a surprisingly short time, at the end of 1983, we came to a point where the success of the project was at the horizon. It still took almost two years until we had worked out the complete structure, and two more years to refine the model at 2.3Å resolution.

The work on the photosynthetic reaction center changed my life in many ways. It was a special privilege to belong to the very small group of people who saw the structural model of this molecule grow on the screen of a computer workstation, and it is hard to describe the excitement I felt during this period of the work. Soon after the news of our success spread through the interested scientific community, we received many invitations to report our results during scientific meetings, in seminars, and even in TV shows. The wide recognition of our work also opened the possibility for me to move to a new place, and to build a research group of my own. The best of several opportunities was an offer from the University of Texas Southwestern Medical Center at Dallas which I joined in March 1988 to become Professor of Biochemistry, and Investigator in the Howard Hughes Medical Institute. Almost immediately after my arrival I fell in love with Kirsten Fischer Lindahl, Professor of Microbiology and Biochemistry and Investigator in the Howard Hughes Medical Institute; we got married in 1989.

For the determination of the three-dimensional structure of the reaction center Hartmut Michel and I received the 1986 Biological Physics Prize of the American Physical Society, and the 1988 Otto Bayer Prize. The 1988 Nobel Prize in Chemistry was followed by several non-scientific honors such as honorary citizenships of my home town Zusamaltheim and of my current residence Dallas, and a high order of the Federal Republic of Germany. I am a member of the Academia Europaea, and a Fellow of the American Association for the Advancement of Science.

HARTMUT MICHEL

I was born in Ludwigsburg, Württemberg, in the southwestern part of the Federal Republic of Germany on July 18, 1948, as the elder son of Karl and Frieda Michel. My ancestors lived in that area for generations, mainly as farmers. There the inherited land is equally divided among sisters and brothers, and not enough land was left for one family's living during my grandparents' generation. During the day my father worked in a factory as a joiner, my mother at home as a dressmaker, in the evenings and on Saturdays care had to be taken of the huge gardens.

As a child I liked to play outside, to stroll through the fields, and I was an active member of the local children's gang, frequently being chased by field guards and building supervisors. Nevertheless, my performance at school was very good, and mainly due to the influence of my mother I was allowed to attend high school. At age eleven I became a member of the circulating library of my home town. From there on I was rarely seen outside, but was reading two to four books per week, the subjects ranging from archaeology over ethnology and geography to zoology. Needless to say that I did not do much homework. At school my favorite subjects were history, biology, chemistry and physics. Especially the teaching in physics was excellent. Most of my understanding of it I got at high school, not at the university.

In parallel, my interest in molecular biology rose. In 1969 — after the obligatory military service — I applied to study biochemistry at the University of Tübingen. At that time Tübingen was the only place in Germany, where one could study biochemistry from the first year, and I was happy to be accepted. Studying biochemistry meant that one had to take part in nearly the same amount of lectures and courses as chemistry students in addition to numerous lectures and courses in biology. The atmosphere between senior teachers and students was impersonal, and the only time I talked to the full professor of biochemistry was during the final examination. However, the possibility existed to work for one year in the various biochemistry labs at the University of Munich and the Max-Planck-Institut für Biochemie instead of attending lab courses in Tübingen. I took that chance in 1972/1973, and at the end I was convinced that academic research was what I wanted to do.

After the examination in Tübingen in 1974 I did the experimental part of my biochemistry diploma in Dieter Oesterhelt's lab at the Friedrich Miescher-Laboratorium of the Max-Planck-Gesellschaft in Tübingen. In cooperation with Walter Stockenius, Dieter Oesterhelt had discovered bacteriorhodopsin in halobacteria and later proposed that it acts as a light-driven proton pump in the framework of Peter Mitchell's chemiosmotic theory. During my diploma work I characterized the ATPase-activity of halobacteria. In 1975, Dieter Oesterhelt moved to Würzburg. I joined him, and as a thesis I correlated the intracellular levels of adenosine di- and triphosphate with the electrochemical proton gradient across the halobacterial cell membrane. Having received the doctorate in June 1977 I tried to fuse delipidated bacteriorhodopsin with bacterial vesicles in order to achieve light-driven amino acid uptake. Upon storage in the freezer the delipidated bacteriorhodopsin yielded solid, glass-like aggregates. On the basis of this observation I was convinced that it should be possible to crystallize membrane proteins like bacteriorhodopsin, which was considered to be impossible at that time. With Oesterhelt's help I started the experiments, and already four weeks later we obtained a new two-dimensional membrane crystal of bacteriorhodopsin. It was not the three-dimensional crystal we wanted, but allowed me to travel to the MRC at Cambridge, England, and to do electron microscopical studies together with Richard Henderson. Back in Würzburg, we observed the first real three-dimensional crystals of bacteriorhodopsin in April 1979. The success led me to cancel my plans to do post-doctoral studies with Susumu Ohno, Duarte, California, on sexual differentiation in mammals. Instead of this, I moved with Dieter Oesterhelt again, this time to the Max-Planck-Institut für Biochemie at Martinsried near Munich, where he became a department head and director. Before moving to Munich, Ilona Leger became my wife. Her understanding and patience helped me a lot.

A promising aspect of the move to Martinsried was the possibility of a cooperation with Robert Huber and colleagues, who at the Max-Planck-Institut had established a very productive department for X-ray crystallographic protein structure analysis. Our bacteriorhodopsin crystals were found to diffract X-rays, but to be too small and too disordered for a structural analysis. We tried to improve size and quality of the crystals. Since all the X-ray crystallographers had beautifully diffracting crystals of soluble proteins, I, understandably, had very limited access to the X-ray equipment at Martinsried. As a consequence, I spent four months at the MRC in Cambridge, England, together with Richard Henderson in 1980, in order to perform X-ray experiments. This period was essential for improving the crystallization method. After my return Dieter Oesterhelt decided to buy an X-ray generator for the ongoing work with bacteriorhodopsin. The generator was installed in Robert Huber's department and guaranteed us continued access to the equipment, and the know how, of the X-ray crystallographers. Later on, I used this generator for the work with the reaction centres.

Frustrated from the lack of the final success with bacteriorhodopsin, I tried to crystallize several other membrane proteins, mainly photosynthetic ones. After developing a new isolation procedure I obtained the first crystals of the photosynthetic reaction centre from the purple bacterium *Rhodopseudomonas viridis* at the end of July 1981. One week later our daughter Andrea was born. During September 1981 the first reaction centre crystal was X-rayed by Wolfram Bode and myself, and turned out to be of excellent quality. Therefore 1981 was the happiest and most successful year of my life.

Dieter Oesterhelt immediately agreed that the reaction centre should be a project of the young people. In February 1982, I started the data collection for the X-ray structure analysis. In April or May I gave a seminar in Robert Huber's department and asked officially for collaboration. After some internal discussions Robert Huber agreed that Johann ("Hans") Deisenhofer, who was the partner of my choice, should take part in the reaction centre project. During the work Hans and I became the best friends. In August 1982, Hans and Kunio Miki, a Japanese post-doctoral research associate in Robert Huber's department, started to evaluate the pile of X-ray films. I continued with the experimental work, occasionally helped by Robert Huber, who showed me how the diffraction pattern of a promising derivative should look like. Not only the X-ray work, but also the entire biochemical characterization and sequence determination had to be done. After the preliminary tracing of the peptide chains by Johann Deisenhofer, the sequence determination, which was performed by Karl A. Weyer, Heidi Gruenberg and myself with Dieter Oesterhelt's support and help, turned out to be the bottle neck for our progress. During that period of heavy work our son Robert Joachim was born in 1984.

As one of the results of the success I received many offers. I accepted the one to become a department head and director at the Max-Planck-Institut für Biophysik in Frankfurt/Main, West Germany, where I am since October 1987.

For the success with the crystallization of membrane proteins and the elucidation of the three-dimensional structure of the photosynthetic reaction centre from the purple bacterium *Rhodopseudomonas viridis* I received various prizes and awards. Among these are the Biophysics Prize of the American Physical Society (together with J. Deisenhofer), the "Chemiedozentenstipendium" of the "Fonds der Chemischen Industrie", the "Otto-Klung-Preis" for chemistry, the Leibniz-Preis of the Deutsche Forschungsgemeinschaft, the "Otto-Bayer-Preis" (together with J. Deisenhofer) and now the Nobel Prize (together with J. Deisenhofer and R. Huber).

THE PHOTOSYNTHETIC REACTION CENTRE FROM THE PURPLE BACTERIUM *RHODOPSEUDOMONAS VIRIDIS*

Nobel Lecture, December 8, 1988

by

JOHANN DEISENHOFER* and HARTMUT MICHEL**

* Howard Hughes Medical Institute and Department of Biochemistry, University of Texas Southwestern Medical Center, 5323 Harry Hines Blvd., Dallas, Texas 75235, U.S.A

** Max-Planck-Institut für Biophysik, Heinrich-Hoffmann-Str. 7, D-6000 Frankfurt/M 71, West Germany

In our lectures we first describe the history and methods of membrane protein crystallization, before we show how the structure of the photosynthetic reaction centre from the purple bacterium *Rhodopseudomonas viridis* was solved. Then the structure of this membrane protein complex is correlated with its function as a light-driven electron pump across the photosynthetic membrane. Finally, we draw conclusions on the structure of the photosystem II reaction centre from plants and discuss the aspects of membrane protein structure. Paragraphs 1 (crystallization), 4 (conclusions on the structure of photosystem II reaction centre and evolutionary aspects) and 5 (aspects of membrane protein structure) were presented and written by H. M., paragraphs 2 (determination of the structure) and 3 (structure and function) by J. D. We arranged the manuscript in this way in order to facilitate continuous reading.

1. THE CRYSTALLIZATION

1.1. *The background*

As in many instances of new scientific developments and technical inventions an accidental observation caused the beginning of the experiments, which ultimately resulted in the elucidation of the three-dimensional structure of a photosynthetic reaction centre. This initiating observation was the formation of solid, most likely glass-like aggregates in August 1978, when bacteriorhodopsin delipidated according to Happe and Overath (1976) was stored in the freezer. These aggregates are shown in *Figure 1a*. From there on I was convinced that it should be possible not only to obtain these solid bodies but also to produce three-dimensional crystals. The availability of well-ordered three-dimensional crystals is the prerequisite for a high resolution X-ray crystallographic analysis, which — despite the progress made by

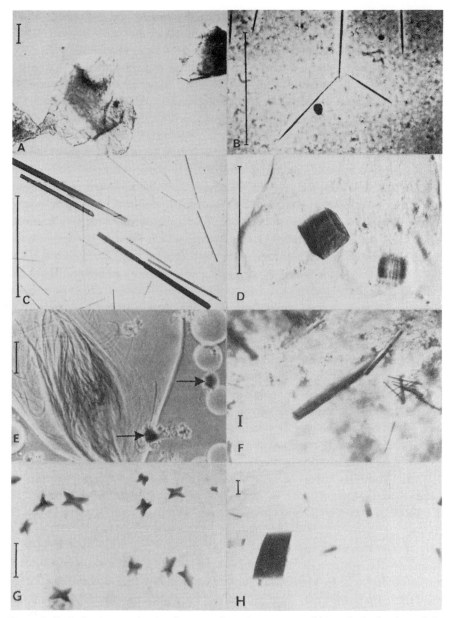

Figure 1: Optical micrographs showing crystals and aggregates of bacteriorhodopsin and the photosynthetic reaction centre from *Rhodopseudomonas viridis:* (A) "glass-like" aggregates of bacteriorhodopsin obtained after freezing of delipidated bacteriorhodopsin; (B) rolled up sheets of the two-dimensionally crystalline orthorhombic form of purple membrane (taken from Michel et al., 1980); (C) needle-like crystals of bacteriorhodopsin obtained with sodium phosphate as precipitant; (D) cube-like crystals of bacteriorhodopsin obtained with ammonium sulphate as precipitant; (E) filamentous aggregates of bacteriorhodopsin and a few cubes (arrows) obtained with ammonium sulphate as precipitant (taken from Michel, 1982a); (F) hexagonal columns of bacteriorhodopsin obtained in the presence of 3% heptane-1,2,3-triol with ammonium sulphate as precipitant; (G) star-like reaction centre crystals obtained within two days (starting conditions: 1 mg protein/ml, 3% heptane-1,2,3-triol, 1.5 M ammonium sulphate) by vapor diffusion against 3 M ammonium sulfate (taken from Michel, 1982b); (H) tetragonal crystals of the reaction centre obtained within three weeks (starting conditions as under G) by vapor diffusion against 2.4 M ammonium sulfate (taken from Michel, 1982b). The bar indicates 0.1 mm in all photographs.

Henderson and Unwin (1975) with electron microscopy and electron diffraction on bacteriorhodopsin — was and still is the only way to obtain detailed structural knowledge of large biological macromolecules.

I was working at the University of Würzburg as a post-doc in D. Oesterhelt's lab, who in collaboration with Walter Stoeckenius had discovered bacteriorhodopsin (Oesterhelt and Stoeckenius, 1971) and was later the first to propose its function (Oesterhelt, 1972). My intention to try to produce well-ordered three-dimensional crystals of bacteriorhodopsin received his immediate support. It turned out that he already tried to crystallize a modified form of bacteriorhodopsin in organic solvents.

Bacteriorhodopsin, the protein component of the so-called purple membrane resembles the visual pigment rhodopsin and acts as a light energy converting system. It is part of a simple "photosynthetic" system in halobacteria. It is an integral membrane protein, which forms two-dimensional crystals in the so called purple membrane. At that time the general belief was that it was impossible to crystallize membrane proteins. With the exception of bacteriorhodopsin there was no information about the three-dimensional structure of membrane proteins, which might have helped to understand their various functions, e.g. as carriers, energy converters, receptors or channels.

The first attempts were to decrease the negative surface charge of purple membrane by addition of long chain amines and to add some Triton X100, a detergent, in order to allow rearrangements of the bacteriorhodopsin molecules, which were partly solubilized by the detergent. This procedure may be a way to obtain the type I crystals described below. Within four weeks the "needles" presented in *Figure 1b* were obtained. Electronmicroscopic studies done in collaboration with Richard Henderson in Cambridge showed that the "needles" were a new two-dimensionally crystalline membrane form of bacteriorhodopsin. In this new form the membranes are rolled up like tobacco leaves in a cigar (Michel et al., 1980).

1.2. *A more systematic approach*

Based on the properties of membrane proteins, a new strategy was developed (Michel, 1983). Membrane proteins are embedded into the electrically insulating lipid bilayers. The difficulties in handling membrane proteins reside in the amphipathic nature of their surface. They possess a hydrophobic surface where in the membrane they are in contact with the alkane chains of the lipids and they have a polar surface where they are in contact with the aqueous phases on both sides of the membrane and the polar headgroups of the lipids (see fig. 2). As a result, membrane proteins are not soluble in aqueous buffers or in organic solvents of low dielectric constant. In order to solubilize membrane proteins one has to add detergents. Detergents are amphiphilic molecules which form micelles above a certain concentration, the so-called critical micellar concentration. The detergent micelles take up the membrane proteins and shield the hydrophobic surface parts of the membrane' protein from contact with water. A schematic

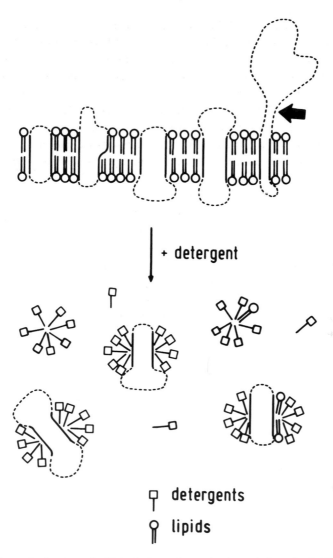

+ detergent

detergents

lipids

Figure 2: Schematic drawing of a biological membrane (top) consisting of a lipid bilayer and membrane proteins embedded into it, and its solubilization by detergents (bottom). The polar part of the membrane protein surface is indicated by broken lines, (modified after Michel, 1983).

drawing of a biological membrane and its solubilization with detergents is shown as *Figure 2*. The membrane protein in the detergent micelle then has to be purified by the various chromatographic procedures.

Once the protein has been isolated and is available in large quantities, one can try to crystallize it. For membrane proteins, which are merely anchored in the membrane, the most promising approach is to remove the membrane anchor by proteases or to use genetically modified material where the part of the gene coding for the membrane anchor has been deleted. At present there are already four examples where the structures of the hydrophilic domains have been reported at high resolution: cytochrome b_5 (Mathews et al., 1972), haemagglutinin (Wilson et al., 1981) and neuraminidase (Varghese et al., 1983) from influenza virus, and the human class I histocompatibility antigen, HLA-A2 (Bjorkman et al., 1987). For really integral membrane proteins, two possibilities exist to arrange them in the form of true three-dimensional crystals:

I) One could think to form stacks of two-dimensional crystals of membrane proteins. In the third dimension the two-dimensional crystals must be ordered with respect to translation, rotation and up and down orientation during or after their formation. In most cases the lipids might still be present in the form of bilayers and compensate the hydrophobicity of the intramembranous protein surface. Hydrophobic and polar interactions would stabilize the crystals in the membrane planes, whereas polar interactions would dominate in the third dimension. In a reasonable crystallization procedure one would have to increase both types of interaction at the same time. This seems to be difficult to achieve.

II) The alternative is to crystallize the membrane proteins within the detergent micelles. The crystal lattice will be formed by the membrane proteins via polar interactions between polar surface parts. *Figure 3* (bottom) shows one example of such a crystal. It is immediately clear that membrane proteins with large extramembranous domains should form this type of crystal much easier than those with small polar domains. The size of the detergent micelle plays a crucial role. A large detergent micelle might prevent the required close contact between the polar surface domains of the membrane proteins. One way to achieve a small detergent micelle is to use small linear detergents like octylglucopyranoside. However, a general experience of membrane biochemists is that membrane proteins in micelles formed by a detergent with a short alkyl chain are not very stable. An increase of the alkyl chain length by one methylene group frequently leads to an increase of the stability by a factor of two to three. One therefore has to find a compromise.

The advantage of the type II crystals is that basically the same procedures to induce supersaturation of the membrane protein solution can be used as for soluble proteins, namely vapor diffusion or dialysis with salts or polymers like polyethylene glycol as precipitating agents. As a serious complication caused by the detergents frequently a viscous detergent phase is formed, which seems to consist of precipitated detergent micelles (see e.g. Zulauf et al., 1985). Membrane proteins are enriched in the detergent phase, and frequently undergo denaturation. In several examples, crystals which were already formed are redissolved.

Bacteriorhodopsin, solubilized in octylglucopyranoside, forms needle-like crystals (*Figure 1c*) when phosphate is used as precipitant (*Figure 1c*) and cubes when ammonium sulphate (*Figure 1d*) is used (Michel and Oesterhelt, 1980). The cubes are not the most stable material, and a conversion into a hairy, thread-like material (*Figure 1e*) is found after several weeks (Michel, 1982a). In this hairy material bacteriorhodopsin probably forms membranes again.

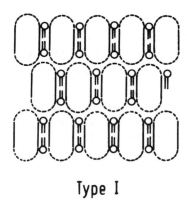

Type I

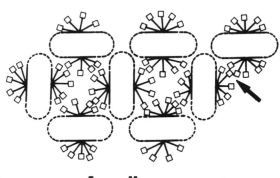

Type II

Figure 3: The two basic types of membrane protein crystals. Type I: stacks of membranes containing two-dimensionally crystalline membrane proteins, which are then ordered in the third dimension. Type II: A membrane protein crystallized with detergents bound to its hydrophobic surface. The polar surface part of the membrane proteins is indicated by broken lines. The symbols for lipids and detergents are the same as in *Figure 2* (Taken from Michel, 1983).

OmpF-porin, an outer membrane protein from *Escherichia coli,* was also crystallized after solubilization in octylglucopyranoside by Garavito and Rosenbusch (1980). We received knowledge of this parallel development when D. Oesterhelt and J. P. Rosenbusch met in China at the end of 1979.

1.3. *The improvement*

My feeling for the lack of the final success with bacteriorhodopsin was always that the detergent micelles still were too large. The use of even smaller detergents was impossible due to the insufficient stability of bacteriorhodopsin in detergents with a shorter alkyl chain or a smaller polar head group. One way out was to add small amphiphilic molecules (Michel, 1982a, 1983) for several reasons: (i) These molecules, might displace detergent molecules which were too large to fit perfectly into the protein's crystal lattice in certain positions. (ii) The small amphiphilic molecules are too small to form micelles themselves, but they are incorporated into the detergent micelles. These mixed micelles are smaller than the pure detergent micelles and possess a different curvature of their surface. As a result the proteins could come closer together. (iii) Their polar head group is smaller than that of the detergent and less of the proteins polar surface would be covered by the polar part of the mixed small amphiphile/detergent micelle.

I had a look through the catalogues of the major chemical companies and ordered nearly everything which was hydrophilic at one end and hydrophobic at the other. In addition, I synthesized about 20 amphiphilic compounds, mainly alkylpolyols and alkyl-n-oxides. These compounds were added during our attempts to crystallize bacteriorhodopsin. Several of the compounds had the effect that hexagonal columns (see *Figure 1f*) were obtained, whereas cubes (*Figure 1d*) had been obtained without the additives. The most effective compound was heptane-1,2,3-triol, but it had a slightly denaturing effect on bacteriorhodopsin. The diffraction quality of the bacteriorhodopsin crystals was improved: Using synchroton radition H. Bartunik, D. Oesterhelt and myself found that they occasionally diffracted to 3 Å resolution, but only in one direction.

1.4. *The turn to classical photosynthesis*

Frustrated from the lack of the final breakthrough with bacteriorhodopsin, which is partly due to the absence of large extramembranous domains in this protein, I looked for more promising membrane proteins to be crystallized. My choices were the photosynthetic reaction centres from the purple bacteria *Rhodospirillum rubrum* and *Rhodopseudomonas viridis* and the light-harvesting chlorophyll a/b protein from spinach. It was influenced by the fact that these proteins (or protein complexes) were said to be part of a two-dimensional crystalline array already in their native environment. As additional benefits, they were available in large quantities, could easily be isolated, were colored and denaturation of the proteins was indicated by color changes.

I learnt about the *Rhodopseudomonas viridis* system when E. Wehrli from the ETH Zürich presented the results of electron microscopical studies during a workshop at Burg Gemen, Germany, in June 1979 (Baumeister and Vogell, 1980). Initially, I received some isolated photosynthetic membranes from him in December 1980. At that time I had moved with D. Oesterhelt to the Max-Planck-Institut für Biochemie at Martinsried near Munich and I was just back from a stay at the MRC in Cambridge where we had done X-ray diffraction experiments on the bacteriorhodopsin crystals. I isolated the reaction centres using hydroxyapatite chromatography according to a published procedure (Clayton and Clayton, 1978) and tried to crystallize it, without success. I developed a new isolation procedure using only molecular sieve chromatography, tried it again and met immediate success (Michel, 1982b). The conditions were nearly identical to those found to be optimal

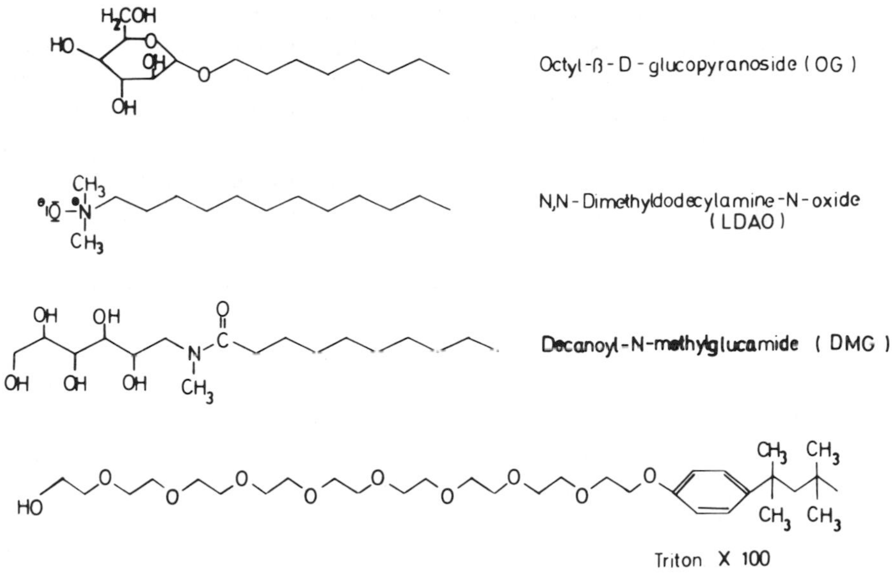

Octyl-β-D-glucopyranoside (OG)

N,N-Dimethyldodecylamine-N-oxide (LDAO)

Decanoyl-N-methylglucamide (DMG)

Triton X 100

Figure 4: Structural formulas of commonly used detergents: octylglucopyranoside, N,N-dimethyldodecylamine-N-oxide, decanoyl-N-methylglucamide are promising for membrane protein crystallization, whereas Triton X100 is not.

for bacteriorhodopsin. The exception was that I could use N,N-dimethyldodecylamine-N-oxide as detergent instead of octylglucopyranoside (see *Figure 4*). In the presence of 3 % heptane-1, 2, 3-triol (high melting point isomer) and 1.5 to 1.8 M ammonium sulphate, star-like crystals are obtained upon vapor diffusion against 2.5 to 3 M ammonium sulphate in two days, more regular tetragonal columns with a length of up to 2 mm upon vapor diffusion against 2.2 to 2.4 M ammonium sulphate in two to three weeks (see *Figure 1g, 1h*). The much smaller polar head group of N,N-dimethyldodecylamine-N-oxide is certainly of importance. Unfortunately, this detergent denatures bacteriorhodopsin. D. Oesterhelt generously considered the reaction centre as my project.

The crystals turned out to be of excellent quality from the beginning. After a scaling up of the isolation procedure, a continuous supply of crystals was guaranteed. I could then start collecting the X-ray data with the initial help of W. Bode and R. Huber. *Figure 5* shows a rotation photograph similar to that used for data collection.

2. DETERMINATION OF THE STRUCTURE

In spring 1982, I (J.D.) joined H.M. in order to determine the three-dimensional structure of the reaction centre. The tetragonal crystals have unit cell dimensions of a = b = 223.5Å, c = 113.6Å, and the symmetry of space group P4$_3$2$_1$2 (Michel, 1982; Deisenhofer et al., 1984). As it turned out, there is one reaction centre with a "molecular weight" of 145 000 Daltons in the asymmetric unit.

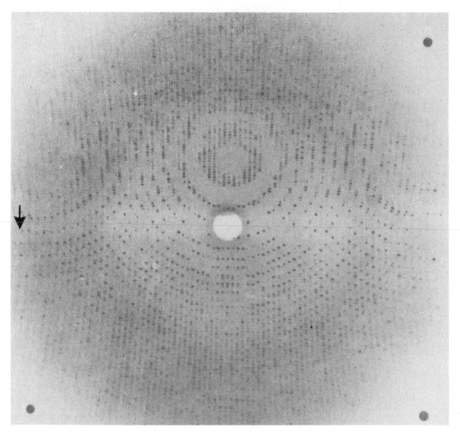

Figure 5: X-ray diffraction pattern of a single reaction centre crystal (1° rotation). Exposure time; 20 h, Cu-Kα-radiation, crystal to film distance: 100 mm. The arrow indicates 3.0 Å resolution (Taken from Michel, 1982b).

2.1. *Collection of X-ray reflection intensity data*

For data collection we used the rotation method with a rotating anode X-ray generator as the source, and photographic film as the detector (Deisenhofer et al., 1984). The large unit cell of the reaction centre crystals, in combination with the resolution limit of the diffraction pattern at 2.9Å limited the rotation interval per film exposure to 0.5°, so that more than two thirds of the reflections on any given film were partially recorded. However, the long lifetime of the crystals in the X-ray beam at about 0°C, and their positional stability allowed to add up partially recorded reflections from successive exposures, so that their treatment did not present a serious problem. Nevertheless, it took about three to four months to collect a complete data set. Data collection for the heavy atom derivatives was speeded up by choosing a rotation interval of 0.6° per exposure. A later re-collection of the native data set at the HASYLAB facilities of DESY in Hamburg was done at 2.3Å resolution in rotation intervals of 0.4° by Irmgard Sinning, Gebhard

Schertler, and H.M. The most tedious and time consuming task in this type of data collection was the processing of films. Kunio Miki and later Otto Epp provided most valuable help during that period of the work. We used the computer programs FILME (Schwager et al., 1975; Jones et al., 1977), and OSC (Rossmann, 1979; Schmid et al., 1981) for film evaluation, and PROTEIN (principal author: W. Steigemann) for scaling and merging data.

2.2 *Solution of the phase problem*
To solve the phase problem for the reaction centre crystal structure we used the method of isomorphous replacement with heavy atom compounds. The experimental part was performed by H.M., the film evaluation and data analysis by myself with support from Kunio Miki and Otto Epp. In order to find the heavy atom derivatives, crystals were soaked for three days in 1mM solutions of the respective heavy atom compounds in a soak buffer similar to the mother liquor. A number of compounds like K_2PtBr_4, $K_2Pt(CN)_4$, $KHg(CN)_2$, K_2HgI_4, and $EuCl_3$ could not be used since they induced the phase separation of the soak buffer into the viscous detergent phase and the aqueous phase. At the beginning, large heavy atom compounds like $(C_6H_5)_3PbNO_3$ or C_6H_5HgCl completely abolished the diffraction, whereas the smaller homologues $(CH_3)_3PbCl$ or C_2H_5HgCl decreased the diffraction to about 6Å resolution. However, after additional purification of the reaction centres prior to the crystallization the diffraction quality of the crystals was unchanged by the small heavy atom compounds. One compound ($KAuCl_4$) caused a shrinkage of the c-axis. Rotation photographs (1°) showing a large part of the 1,k,l lattice plane were taken and inspected visually for changes in the diffraction pattern. For promising candidates, ca. 50% complete data sets were collected and evaluated.

On the average, each heavy atom derivative had nine heavy atom binding sites (Deisenhofer et al., 1984). The major binding sites were found with the automatic search procedure in the PROTEIN program package. Using five different heavy atom derivatives, we could calculate phases to 3.0Å resolution, and an electron density map (Deisenhofer et al., 1984). Phases and map were further improved by solvent flattening (Wang, 1985).

2.3 *Model building*
Map interpretation and model building were done in three stages: At first the prosthetic groups in the reaction centre were identified. We found the four heme groups, the four bacteriochlorophyll-b, the two bacteriopheophytin-b, and one quinone (Deisenhofer et al., 1984). Next, the polypeptide chains were built with polyalanine sequence, except in the amino terminal regions of the subunits, L, M, and cytochrome where partial amino acid sequences were known (Michel et al., 1983), and could be used to distinguish between the subunits. At that stage, some use was made of the local symmetry of the subunits L and M. Finally, as the gene sequences of the reaction centre subunits were determined (Michel et al., 1985; Michel et al., 1986a; Weyer et al., 1987b), the model of the protein subunits was complet-

ed. The sequence information led to an overall verification but also to a
number of minor corrections of the polypeptide backbone model, since in
the previous model building stage the electron density was not always clear
enough to allow determination of the correct number of amino acids.

Our tools for model building were interactive graphics display systems: a
black and white Vector General 3400 system, and later a color Evans &
Sutherland PS 300. On both systems we used Alwyn Jones' program package FRODO (Jones, 1978). The model library of this package was extended
to include bacteriochlorophyll-b, bacteriopheophytin-b, menaquinone-7,
and ubiquinone-1. Frequent use was made of the real-space-refinement
facility in FRODO which allowed to correctly place long stretches of helical
structure into the electron density.

2.4. *Model refinement*

The reaction centre model, with about half of the side chains of the
cytochrome subunit still missing, already had the rather low crystallographic
R-value of 0.359 at 2.9Å resolution ($R = \Sigma(\parallel F_{obs} \mid - \mid F_{calc} \parallel)/\Sigma \mid F_{obs} \mid$; F_{obs}
and F_{calc} are observed and calculated structure factors, respectively). Crys-
tallographic refinement of the model was started at 2.9Å resolution, and
continued at 2.3Å resolution. The program packages used for refinement
were PROTEIN, EREF (Jack & Levitt, 1978; Deisenhofer et al., 1985b),
TNT (Tronrud et al., 1987), and again FRODO.

As a result of the refinement the R-value was brought down to 0.193 for
95762 unique reflections at 2.3Å resolution, the refined model consists of
10288 non-hydrogen atoms. Erros in the initial model, e.g. peptide groups
and side chains with wrong orientations were removed. New features were
added to the model: a partially ordered carotenoid molecule, a ubiquinone
in the partially occupied Q_B binding pocket, a complete detergent molecule
(LDAO, see Fig. 4), a candidate for a partially ordered LDAO or similar
molecule, seven candidates for negative ions, and 201 ordered water molecules. The upper limit of the mean coordinate error was estimated (Luzzati,
1952) to be 0.26Å. A detailed description of refinement and refined model
of the photosynthetic reaction centre from *Rhodopseudomonas viridis* will be
given elsewhere (Deisenhofer, J., Epp, O., Sinning, I., Michel, H., to be
published).

3. STRUCTURE AND FUNCTION

3.1. *Structure overview*

An overall view of the structure of the photosynthetic reaction centre from
Rhodopseudomonas viridis is shown in *Figure 6*. It is a complex of four protein
subunits, and of 14 cofactors. The protein subunits are called H (heavy), M
(medium), L (light), and cytochrome; the names H, M, and L were chosen
according to the apparent molecular weights of the subunits, as determined
by electrophoresis. The core of the complex is formed by the subunits L and
M, and their associated cofactors: four bacteriochlorophyll-b (BChl-b), two

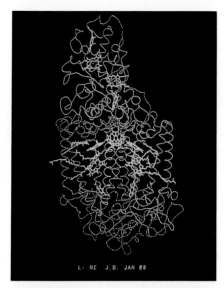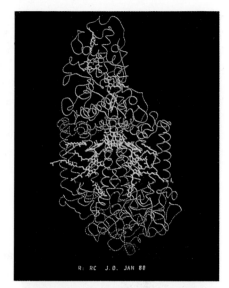

Figure 6: Stereo pair showing an overall view of the reaction centre structure. Protein chains are represented as smoothed backbone drawings; green: cytochrome, blue: M-subunit, brown: L-subunit, purple: H-subunit. Cofactors are drawn in bright atom-colors; yellow: carbons, blue: nitrogens, red: oxygens green: magnesium. Smoothed backbone representations of polypeptide chains were produced following an idea of Richard J. Feldman, with help from Marius G. Clore.

bacteriopheophytin-b (BPh-b), one nonheme iron, two quinones, one carotenoid. Structural properties, e.g. the hydrophobic nature of the protein surface, and functional considerations strongly indicate that the subunits L and M span the bacterial membrane. This aspect of the structure will be discussed in detail below. Each of the subunits L and M contains five membrane spanning polypeptide segments, folded into long helices. The polypeptide segments connecting the transmembrane helices form flat surfaces parallel to the membrane surfaces.

The H subunit contributes another membrane spanning helix with its N-terminus near the periplasmic membrane surface. The C-terminal half of the H subunit forms a globular domain that is bound to the L-M complex near the cytoplasmic membrane surface. On the opposite side of the membrane, the cytochrome subunit with its four covalently bound heme groups is attached to the L-M complex. Both the cytochrome subunit and the globular domain of the H-subunit have surface properties typical for water soluble proteins.

The total length of the reaction centre, from the tip of the cytochrome to the H subunit is about 130Å. The core has an elliptical cross section with axes of 70Å and 30Å.

The photosynthetic reaction centres from purple bacteria are the best characterized among all photosynthetic organisms (for reviews, see (Feher & Okamura, 1978; Okamura et al., 1982). All of them contain the three subunits H, M, and L; some bacteria lack the tightly bound cytochrome subunit. An example of a reaction centre without a bound cytochrome subunit is that from *Rhodobacter sphaeroides* which was crystallized (Allen and

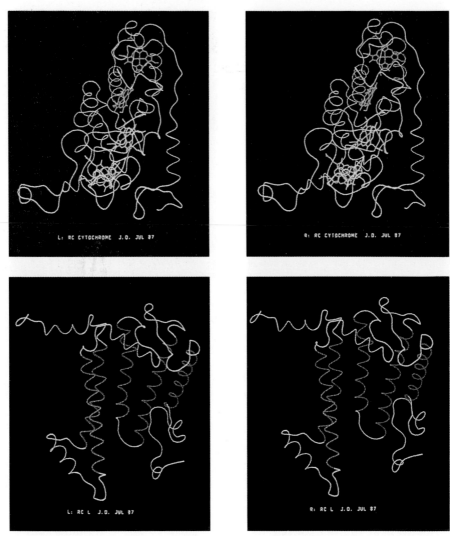

Figure 7: Stereo pairs showing smoothed backbone representations of the protein subunits. Secondary structure is indicated by colors. Yellow: no apparent secondary structure; red: transmembrane helices, purple: other helices; blue: antiparallel β-sheets. a) cytochrome (with the four heme groups); b) L-subunit; c) M-subunit; d) H-subunit. N-termini are marked blue, C-termini are marked red.

Feher, 1984; Chang et al., 1985); its structure has been shown to be very similar to the reaction centre from *Rhodopseudomonas viridis* (Allen et al., 1986; Chang et al., 1986).

3.2. *Subunit structure*

Schematic drawings of the polypeptide chain folding of the four reaction centre subunits are shown in *Figure 7*. As mentioned above, major elements of secondary structure in the subunits L and M are the five membrane spanning helices. A comparison of the polypeptide chain folding in both subunits shows a high degree of similarity. Structurally similar segments

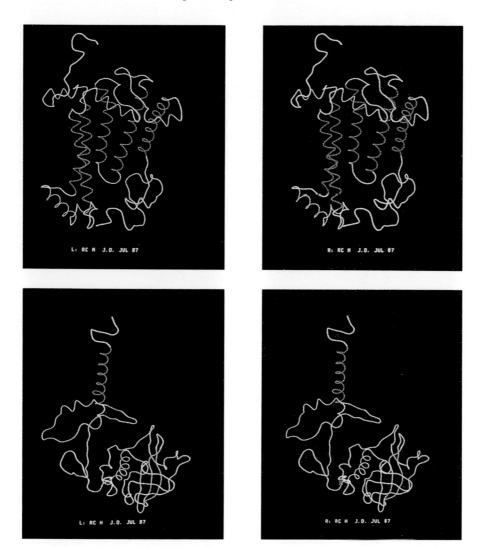

include the transmembrane helices and a large fraction of the connections. In total, 216 α-carbons from the M-subunit can be superimposed onto corresponding α-carbons of the L-subunit with an r.m.s. deviation of only 1.22Å. The superposition of the subunits is done by a rotation of about 180° around an axis running perpendicular to the membrane surface; we call this axis the central local symmetry axis. *Table I* lists the helices in both subunits; *Table II* lists the structurally similar regions in both subunits. Besides the transmembrane helices, called LA, LB, LC, LD, LE, and MA, MB, MC, MD, ME, with lengths between 21 and 28 residues, there are shorter helices in the connecting segments, notably helix de (between transmembrane helices D and E), and helix cd. Subunit M (323 residues) is 50 residues longer than L (273 residues). The insertions in M, with respect to L, are located near the N-terminus, (20 residues), in the connection between the helices MA and MB (7 residues), in the connection between

Table I: Helical segments in subunits L and M

helix	segment (length)	
	subunit L	subunit M
transmembrane		
A	L33-L53 (21)	M52-M76 (25)
B	L84-L111 (28)	M111-M137 (27)
C	L116-L139 (24)	M143-M166 (24)
D	L171-L198 (28)	M198-M223 (26)
E	L226-L249 (24)	M260-M284 (25)
periplasmic		
	– – –	M81-M87 (7)
cd	L152-162 (11)	M179-M190 (12)
ect	L259-L267 (9)	M292-298 (7)
cytoplasmic		
	– – –	M232-M237 (6)
de	L209-L220 (12)	M241-M254 (14)

Table II: Regions with similar polypeptide chain folding in subunits L and M

subunit M		subunit L	length
M49-76	->	L29-56	28
M88-96	->	L61-69	9
M100-224	->	L73-197	125
M243-290	->	L209-256	48
M291-296	->	L258-263	6

MD and ME (7 residues), and at the C-terminus (16 residues). The insertions at the N- and C- termini make the M-subunit dominate the contacts with the peripheral subunits. The insertion between MD and ME, containing another small helix (see *Table I*) is of importance for the different conformations of the quinone binding sites in L and M, and for binding of the non-heme iron (see below).

The H-subunit with 258 residues can be divided into 3 structural regions with different characteristics (see *Figure 7*). The N-terminal segment, beginning with formyl-methionine (Michel et al., 1985), contains the only transmembrane helix of subunit H; it includes 24 residues from H12 to H35. Near the end of the transmembrane helix the sequence shows seven consecutive charged residues (H33 to H39). Residues H47 to H53 are disordered in the crystal, so that no significant electron density can be found for them.

Following the disordered region the H chain forms an extended structure along the surface of the L-M complex, apparently deriving structural stability from that contact. The surface region contains a short helix and two two-stranded antiparallel β-sheets.

The third structural segment of the H-subunit, starting at about H105 forms a globular domain. This domain contains an extended system of antiparallel and parallel β-sheets between residues H134 and H203, and an α-helix (resudues H232 to H248). The β-sheet region, the only larger one in the whole reaction centre, forms a pocket with highly hydrophobic interior walls. This structural property reminds of transport proteins like retinol binding protein (Newcomer et al., 1984), bilin binding protein (Huber et al., 1987), and others; however the strand topology is different. So far, no evidence for a ligand has been found.

With 336 residues (Weyer et al., 1987b) the cytochrome is the largest subunit in the reaction centre complex. Its last four residues, C333 to C 336, are disordered. Also disordered is the lipid molecule bound to the N-terminal cysteine residue (Weyer et al., 1987b). The complicated structure of the cytochrome can be summarized as follows: The structure consists of an N-terminal segment, two pairs of heme binding segments, and a segment connecting the two pairs. Each heme binding segment consists of a helix with an average length of 17 residues, followed by a turn and the Cys-X-Y-Cys-His sequence typical for c-type cytochromes.

The hemes are connected to the cysteine residues via thioether linkages. This arrangement leads to the heme planes being parallel to the helix axes. The sixth ligands to the heme irons are in three of the four cases methionine residues within the helices. The iron of heme 4 has histidine C124, located in a different part of the structure, as a sixth ligand. The two pairs of heme binding segments, containing hemes 1 and 2, and 3 and 4, respectively, are related by a local twofold symmetry. From each pair 65 residues obey this local symmetry with an r.m.s. deviation between corresponding α-carbon atoms of 0.93Å. The local symmetry of the cytochrome is not related to the central local symmetry.

3.3 Arrangement of cofactors

Figure 8 shows the arrangement of the 14 cofactors associated with the

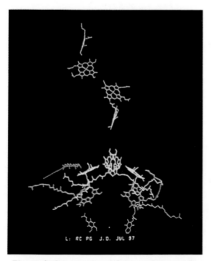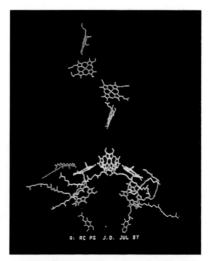

Figure 8: Stereo view of the cofactors. Brown: heme groups; yellow bacteriochlorophyll-bs; light blue: bacteriopheophytin-bs, blue: carotenoid (dihydro-neurosporene; I. Sinning, H. Michel, unpublished results); purple: quinones (right: Q_A, left: Q_B; red dot: non-heme iron.

reaction centre protein subunits. The four heme groups of the cytochrome, numbered according to the order of attachment to the protein, form a linear chain that points to a closely associated pair of BChl-bs. This pair, the so-called "special pair" is the origin of two branches of cofactors, each consisting of another BChl-b (the "accessory" BChl-b), a BPh-b, and a quinone. The non-heme iron sits between the quinones. The tetrapyrrole rings of BChl-bs, BPh-bs, and quinones approximately follow the same local symmetry that is displayed by the L- and M-chains. The branches of cofactors from the special pair to the BPh-bs can be clearly associated with subunits L or M, so that we speak of an L-branch and an M-branch.

This is the basis for our nomenclature: BChl-bs and BPh-bs are called BC_{XY} and BP_X, respectively, where X denotes the branch (L or M), and Y is P for "special pair" or A for "accessory". At the level of the quinones the situation is more complicated because the subunits interpenetrate here, and the quinone at the end of the L-branch is actually bound in a pocket of the M-subunit and vice versa. Therefore, we prefer the nomenclature Q_A and Q_B with Q_A at the end of the L-branch. Q_A is menaquinone-9, and Q_B is ubiquinone-9 (Gast et al., 1985). The local symmetry is violated by the phytyl chains of BChl-bs and BPh-bs, by the different chemical nature and different occupancy of the quinones, and by the presence of a carotenoid molecule near the accessory BChl-b of the M-branch.

3.4 *Functional overview*
The current understanding of the function of the reaction centre was developed by combining structural information with information from

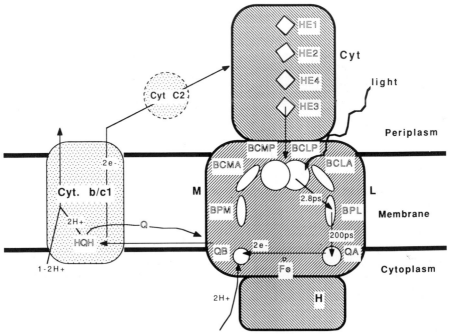

Figure 9: Schematic view of the reaction centre, showing the light-driven cyclic electron flow.

other experimental techniques, notably spectroscopy, as described in recent reviews (Parson & Ke, 1982; Parson, 1987; Kirmaier & Holten, 1987). *Figure 9* shows a schematic view of the reaction centre with its cofactors in the bacterial membrane. The special pair, P, is the starting point for a light driven electron transfer reaction across the membrane. Absorption of a photon, or energy transfer from light harvesting complexes in the membrane puts P into an excited state, P*. From P* an electron is transferred to the BPh-b on the L- branch, BP_L with a time constant of 2.8 ps (Breton et al., 1986; Fleming et al., 1988). The distinction between the two BPh-bs was possible because they absorb at slightly different wavelengths, and, with the knowledge of the crystal structure, linear dichroism absorption experiments could distinguish between the two chromophores (Zinth et al., 1983; Zinth el al., 1985; Knapp et al., 1985).

From BP_L the electron is transferred to Q_A with a time constant of $\sim$ 200ps. At this point the electron has crossed most of the membrane. Both these electron transfer steps function at very low temperatures ($\sim$ 1K) with time constants even shorter than at room temperature (Kirmaier et al., 1985b; Kirmaier et al., 1985a).

From Q_A the electron moves on to Q_B within about 100 μs. The non-heme iron does not seem to play an essential role in this step (Debus et al., 1986).

Q_B can pick up 2 electrons and, subsequently, 2 protons (Dracheva et al., 1988). In the $Q_B H_2$ state it dissociates from the reaction centre, and the Q_B site is re-filled from a pool of quinones dissolved in the membrane. Electrons and protons on $Q_B H_2$ are transferred back through the membrane by the cytochrome b/c_1, complex. The electrons are shuttled via a soluble cytochrome c_2 to the reaction centre's cytochrome from which P+ had been reduced with a time constant of $\sim$ 270μs. This time constant increases with decreasing temperature down to $\sim$ 100K, and remains constant for lower temperatures. The whole process can be described as a light driven cyclic electron flow, the net effect of which is the generation of a proton gradient across the membrane that is used to synthesize adenosine triphosphate, as described by P. Mitchell's chemiosmotic theory. Complete understanding of the reaction centre's function still meets with a number of problems. The nature of electron transfer along the stages described above, its speed and temperature dependence has not yet been explained theoretically. The first step, with the question of the role of the bridging BC_{LA} is a matter of fascinating debate.

One of the major surprises from the structural work was the symmetry of the core structure, raising the question of the factors leading to the use of only the L-branch of cofactors and of the significance of the apparently unused branch. Further open questions relate to electron transfer between Q_A and Q_B, the role of the non-heme iron, and the function of Q_B as two-electron gate and proton acceptor. Finally, the purpose of the cytochrome, as well as details of electron transfer from the soluble cytochrome, and among the 4 hemes, is as yet not completely explained.

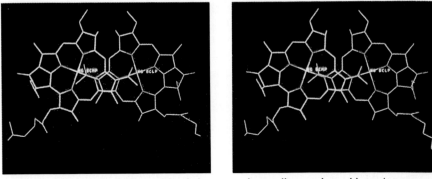

Figure 10: Stereo view of the special pair in atom-colors: yellow: carbons, blue: nitrogens, red: oxygens, green: magnesium.

3.5 *Structural details in relation to function*

Here I describe the arrangement of the cofactors, and their environment in some detail. Observations relating to open functional questions are emphasized.

Figure 10 shows the BChl-b ring systems of the special pair, the primary electron donor of the photosynthetic light reaction. On the basis of spin resonance experiments the existence of a special pair had been postulated a long time ago (Norris et al., 1971). The two molecules overlap with their pyrrole rings I in such a way that, when looking in a direction perpendicular to the ring planes, the atoms of these rings eclipse each other. The orientation of the rings leads to a close proximity between the ring I acetyl groups, and the Mg^{2+} ions; however, the acetyl groups do not act as ligands to the Mg^{2+}. The pyrrole rings I of both BChl-bs are nearly parallel, and ~ 3.2Å apart. Both tetrapyrrole rings, however, are nonplanar; planes through the pyrrole nitrogens of each BChl-b form an angle of $11.3°$.

The special pair BChl-bs are arranged with a nearly perfect twofold symmetry. This is illustrated also in *Figure 11*, which shows a view along the twofold axis (Deisenhofer & Michel, 1988). The BChl-b rings of the special pair are nearly parallel to the symmetry axis. Further objects shown in *Figure 11* that obey the central local twofold symmetry are the histidine residues (L173, M200) acting as ligands to the special pair Mg^{2+} ions, the rings of the accessory BChl-bs, the water molecules H-bonded between histidine nitrogens and ring V carbonyl groups of the accessory BChl-bs, and the transmembrane helices of subunits L and M. The carotenoid molecule in contact with the accessory BChl-b BC_{MA}, the side chains of the accessory BChl-bs, and the trans-membrane helix of the H-subunit are examples of structural elements that brake the twofold symmetry. A more subtle deviation from symmetry is the different degree of non-planarity of the two BChl-b ring systems of the special pair. The BC_{MP} ring is considerably more deformed than that of BC_{LP}. This can cause an unequal charge distribution between the two components of the special pair, which in turn can be part of the reason for unidirectional electron transfer (Michel-Beyerle et al., 1988).

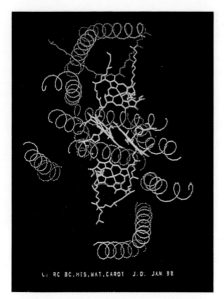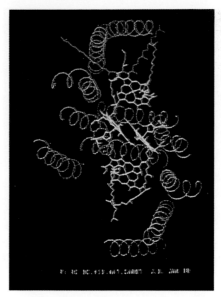

Figure 11: Stereo pair: view along the central local twofold axis showing in atom colors: special pair with histidine ligands, accessory bacteriochlorophyll-bs (BC_{LA} bottom, BC_{MA} top), two waters; the transmembrane helices of subunits L (brown), M (blue), and H (purple) are shown in smoothed backbone representation.

Even though the tetrapyrrole rings of the BChl-bs and BPh-bs of the L- and M-branches can be rotated on top of each other using a single transformation with the reasonably low r.m.s. deviation of 0.38Å between the positions of equivalent atoms, a closer inspection shows considerable differences between the local symmetry operations of special pair, accessory BChl-bs, and BPh-bs. Optimum superposition of the tetrapyrrole rings of the special pair alone is achieved by a rotation of 179.7°, for the accessory BChl-bs by a rotation of $-175.8°$, and for the BPh-bs by a rotation of $-173.2°$. This deviation from twofold symmetry is illustrated in *Figure 12*, where the cofactors of the M-branch were rotated using the transformation that optimally superimposes the special pair tetrapyrrole rings. It is clear that, due to the imperfect symmetry, interatomic distances, and interplanar angles are different in both branches. For example, the closest distance of atoms involved in double bonds in the special pair, and in BP_L is shorter by 0.7Å than the corresponding distance between special pair and BP_M. Another example is the angles between the tetrapyrrole rings of the special pair, and those of the accessory BChl-bs: the angles for BP_L are about 6° smaller than for BP_M. These structural differences lead to differences in overlap of electronic orbitals, and are expected to lead to different electron transfer properties in both branches. This may be another contribution to the unidirectional charge separation in the reaction centre.

Yet another observation that may relate to the different electronic properties of the L- and M-branches is the different degree of structural order. The amount of disordered structure, measured by the number of atoms

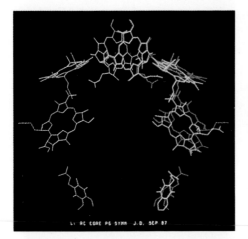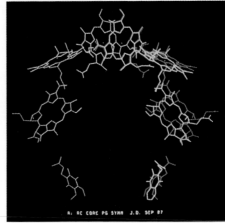

Figure 12: Stereo pair showing cofactors of M-branch (purple), and of L-branch (green); phytyl chains are omitted for clarity. Red: cofactors of M-branch, rotated using a transformation that optimally superimposes the tetrapyrrole ring of BC_{MP} onto that of BC_{LP}.

without significant electron density, is larger in the constituents of the M-branch than in those of the L-branch. Both phytyl side chains of BC_{MA} and BP_M are partially disordered at their ends; the phytyl chains of BC_{LA} and BP_l have a different conformation and are well ordered. The carotenoid near BC_{MA} may contribute to this difference in phytyl chain structure since its presence prevents an identical arrangement of phytyl chains on both sides.

A measure of the rigidity of the structure is the atomic B-values obtained during crystallographic refinement. These values are higher in the M-branch than in the L-branch. An example is the tetrapyrrole ring of BP_M with an average B of $21.1Å^2$, as compared to $10.3Å^2$ for BP_L.

A major source of asymmetry is the protein subunits L and M surrounding the core pigments. Their overall sequence homology is only 25% (Michel et al., 1986a). Although key residues like the histidines that are ligands to the Mg^{2+} ions of the BChl-bs, and to the non-heme iron are strictly conserved, most of the residues in contact with the core pigments are different between the two branches.

I now describe details of the protein environment of the pigments along the pathway of the electron, and mention additional differences between the branches that may be functionally important. *Figure 13* shows a close view of the structures that are directly involved in the first step of the light driven electron transfer reaction: the special pair, the accessory BChl-b BC_{LA}, and the first electron acceptor, BP_L. In addition, a few amino acid residues in close contact to these pigments are shown. BC_{LA} is in van der Waals contact to both the special pair and BP_L. The closest approach between the tetrapyrrole rings of the special pair and BP_L is 10Å (atoms in double bonds). The phytyl chain of BC_{LP} follows a cleft formed by BC_{LA} and BP_L; it is in van der Waals contact to both tetrapyrrole rings. At first glance this arrangement suggests that the electron should follow the path P $\longrightarrow$

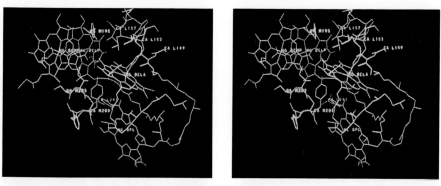

Figure 13: Special pair, BC_{LA}, BP_L, and selected residues in atom-colors.

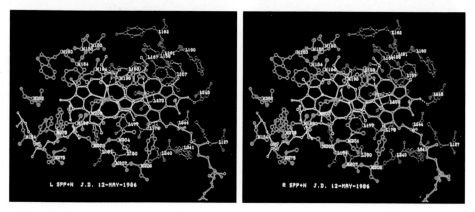

Figure 14: Stereo pair, showing the special pair, and its protein environment (Michel et al., 1986b). Brown: residues from the L-subunit; blue: residues from the M-subunit; green: BC_{LP}; yellow: BC_{MP}; hydrogen bonds are indicated in purple. The hydrogen bond between serine M203 and BC_{MP} is no longer present in the refined model.

$BC_{LA} \longrightarrow BP_L$. However, attempts to observe bleaching of the absorption bands of BC_{LA} due to transient reduction failed. Spectroscopic experiments done with ultrafast laser systems indicated direct reduction of BP_L from P* without intermediate steps (Breton et al., 1986; Fleming et al., 1988; Kirmaier et al., 1985b). This result has initiated an intense debate on the mechanism of electron transfer from P to BP_L, and on the role of BC_{LA} in this process. As indicated in *Figure 13* with the example of tyrosine M208, it seems plausible that the protein plays an important role, not only as a scaffold to keep pigments in place, but also in influencing functional properties.

Numerous protein-pigment interactions are apparent also for the special pair itself (Michel et al., 1986b), as shown in *Figure 14*. These interactions include bonds between Nε atoms of histidines L173 and M200 to the Mg^{2+} ions of BC_{LP} and BC_{MP}, respectively. Both acetyl groups of the special pair are hydrogen bonded: BC_{LP} to histidine L168, and BC_{MP} to tyrosine M195. A further hydrogen bond is found between the ring V keto carbonyl oxygen and threonine L248; there is no equivalent hydrogen bond for BC_{MP}.

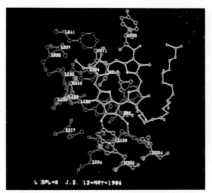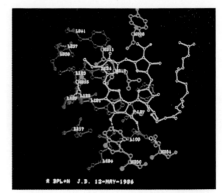

Figure 15: Stereo pair, showing BP$_L$ (yellow), and its protein environment, colored as in *Figure 14.*

The special pair environment is rich in aromatic residues: five phenylalanines, three tyrosines, and three tryptophans are in direct contact with the tetrapyrrole rings of the special pair. Tyrosine L162 is located between the special pair and the closest heme group (HE3) of the cytochrome, and may play a role during reduction of P+ by the cytochrome (Michel et al., 1986b).

Figure 15 shows BP$_L$, the first electron acceptor, with its protein environment (Michel et al., 1986b). The BPh-bs are held in their places by noncovalent interactions only. In the positions where histidine ligands of BChl-bs would be expected, we find leucine M212 for BP$_L$ (see *Figure 15*), and methionine L184 for BP$_M$. BP$_L$ forms two hydrogen bonds with the protein. The one between the ring V ester carbonyl group and tryptophan L100 has an equivalent in a hydrogen bond between BP$_M$ and tryptophan M127. The other hydrogen bond, between the ring V keto carbonyl oxygen and glutamic acid L104 is unique for the L-branch; the residue on the M-side corresponding to glutamic acid L104 is valine M131. Glutamic acid L104 is conserved in all currently known sequences of reaction centre L-subunits from purple bacteria. Its position in the electron transfer pathway strongly suggests that it is protonated; otherwise, the negative charge of the ionized glutamic acid side chain would make electron transfer to BP$_L$ energetically highly unfavorable.

As for the special pair, aromatic residues are found in the neighborhood of the BPh-bs; the neighborhood of BP$_L$ is richer in aromatic residues than that of BP$_M$. An especially noteworthy aromatic residue is tryptophan M250, whose side chain forms a bridge between BP$_L$ and the next electron acceptor, Q$_A$. The M-branch residue equivalent to tryptophan M250 is phenylalanine L216 which, due to the smaller side chain, cannot perform a similar bridging function between BP$_M$ and Q$_B$.

The environment of the quinones, and of the non-heme iron (Michel et al., 1986b) is shown in *Figure 16*. Instead of Q$_B$, the figure shows the herbicide terbutryn in the Q$_B$ binding pocket. The non-heme iron appears in the centre of the drawing, between the binding sites of Q$_A$ and Q$_B$, very near the central local twofold symmetry axis. It is bound by five protein side

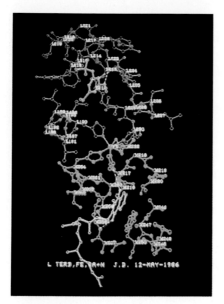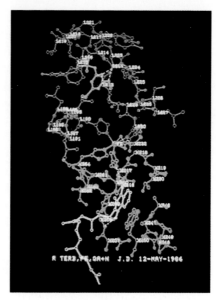

Figure 16: Stereo pair, showing Q_A, the non-heme iron, the herbicide terbutryn in the Q_B binding pocket, and the protein environment of these cofactors, colored as in *Figure 14* (Michel et al., 1986b).

chains, four histidines (L190, L230, M217, M264), and glutamic acid M232, whose carboxylate group acts as a bidentate ligand. The iron sits in a distorted octahedral environment with the axial ligands histidine L230 and histidine M264, and equatorial ligands histidine L190, histidine M217, and glutamic acid M232. Histidine L190 and histidine M217 also contribute significantly to the binding of Q_B and Q_A, respectively. The location of the iron, and its binding to residues from subunits L and M immediately suggests that the part of iron's role is to increase the structural stability of the reaction centre. It is surprising that its role in electron transfer between the quinones seems to be relatively minor (Kirmaier et al., 1986).

The head group of Q_A is bound in a highly hydrophobic pocket; its carbonyl oxygens are hydrogen bonded to the peptide NH of Ala M258, and to the Nδ of the iron ligand histidine M217. As mentioned above, tryptophan M250 forms part of the Q_A's binding pocket; its indole ring is nearly parallel to the head group of Q_A at a distance of 3.1Å. The isoprenoid side chain of Q_A is folded along the surface of the L-M complex; the last three isoprenoid units are disordered in the crystal. The Q_A binding pocket is well shielded from the cytoplasm by the globular domain of the H-subunit. Since the Q_B binding site in the reaction centre crystals is only partially occupied, the Q_B model is less reliable than the other parts of the structural model discussed above. Nevertheless, the crystallographic data suggested a highly plausible arrangement of the Q_B head group in its pocket; the Q_B side chain remained undefined. It appears that Q_B, similar to Q_A, forms hydrogen bonds to the protein with its two carbonyl oxygens: one to Nδ atom of the iron ligand histidine L190, and a bifurcated hydrogen bond to Oγ of serine L223, and to NH of glycine L225. As tryptophan

M250 for Q_A, phenylalanine L216 forms a significant part of the Q_B binding pocket. Major differences between the binding sites of Q_A and Q_B are the more polar nature of the Q_B site, and the presence of pathways through the protein, through which protons may enter the Q_B site. The bottom of the Q_B site is formed to a large part by the side chain of glutamic acid L212. Protons can move from the cytoplasm along a path marked by charged or polar residues to glutamic acid L212 and from there, by an, as yet, unknown mechanism, to the doubly reduced Q_B^{2-}.

Some herbicides are competitive inhibitors of Q_B binding to reaction centres of purple bacteria. Crystallographic binding studies with the herbicide terbutryn (see *Figure 16*), and with o-phenanthroline (Deisenhofer et al., 1985a; Michel et al., 1986b) demonstrated binding of these molecules in the Q_B binding pocket, and provided a structural basis for understanding mutations that render *Rps. viridis* herbicide resistant (Sinning & Michel, 1987b; Sinning & Michel, 1987a). The fact that herbicides, which were developed to inhibit photosystem II reaction centres of green plants, can also inhibit reaction centres of purple bacteria is one of the many indications of a close structural similarity between these kinds of photosynthetic reaction centres (see paragraph 4 and Michel & Deisenhofer, 1988).

4. THE RELATION TO PHOTOSYSTEM II AND EVOLUTIONARY ASPECTS

4.1 *Conclusions on the structure of photosystem II reaction centre*

The most surprising result of the X-ray structure analysis was the discovery of the nearly symmetric arrangement of the reaction centre core formed by the homologous L- and M-subunits together with the pigments. Primary electron donor, as well as the ferrous non-heme iron atom, are found at the interface between both subunits. Both subunits are needed to establish the reaction centre.

During the X-ray structure analysis the following results suggesting a close relation between the reaction centres from purple bacteria and photosystem II were or became available. i) Photosystem II reaction centre and the reaction centre from purple bacteria both possess two pheophytin molecules (Omata et al., 1984; Feher and Okamura, 1978). Upon removal of the quinones or prereduction of them, it is possible to trap one electron on one of them. (Tiede et al., 1976, Shuvalov & Klimov, 1976). ii) Both reaction centres possess a magnetically coupled Q_A- Fe- Q_B complex. iii) The L-subunit of the purple reaction centre and the D1 protein (which is the product of the psbA-gene and also called Q_B protein, 32 kD protein or herbicide binding protein) bind the herbicide azidoatrazine upon photoaffinity labelling (de Vitry & Diner, 1984; Pfister et al. 1981). (iv) Weak but significant sequence homologies between the L- and M-subunits of the purple bacteria (Williams et al., 1983, 1984; Youvan et al., 1984; Michel et al., 1986a), the D1 (Zurawski et al., 1981) and later on also D2 proteins (Alt

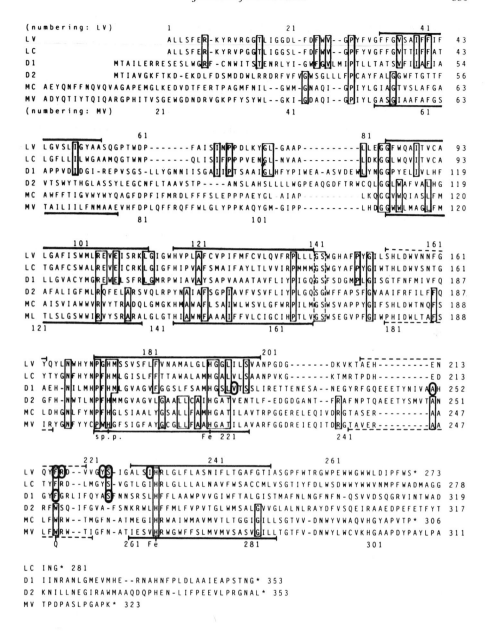

LC ING* 281
D1 IINRANLGMEVMHE--RNAHNFPLDLAAIEAPSTNG* 353
D2 KNILLNEGIRAWMAAQDQPHEN-LIFPEEVLPRGNAL* 353
MV TPDPASLPGAPK* 323

Figure 17: The amino acid sequences of the L and M subunits from the purple bacteria *Rps. viridis* (LV, MV; Michel et al., 1986a) and *Rb. capsulatus* (LC, MC; Youvan et al., 1984). Amino acids common to all six subunits or the L subunits and D1, or the M subunits and D2 are boxed.. The position of the transmembrane α-helices in the *Rps. viridis* reaction centre is indicated by bars above the sequences of the L subunits and below the sequences of the M subunits. The positions of the short α-helices in the connections of transmembrane α-helices C and D, as well as D and E, are indicated by dashed lines. The histidine ligands of the special pair bacteriochlorophylls and of the non-heme iron atom are marked by sp.p. or Fe. Circles show amino acids known to be mutated in herbicide-resistant reaction centres from the purple bacteria or from photosystem II (Taken from Michel & Deisenhofer, 1988).

et al., 1984, Holschuh et al., 1984; Rasmussen et al., 1984, Rochaix et al., 1984) of photosystem II were discovered.

The meaning of the results was obvious: The reaction centre of photosystem II from plants and algae had to be expected to be formed by the D1 and D2 proteins with D1 corresponding to the L-subunit and D2 corresponding to the M-subunit. This proposal was at variance with the accepted view that the so-called CP47, a chlorophyll-binding protein with apparent molecular weight of 47000, is the apoprotein of the photosystem II reaction centre (Nakatani et al., 1984).

Figure 17 compares the amino acid sequences of the L- and M-subunits from two purple bacteria with the D1 and D2 proteins from spinach chloroplast. Significant sequence homology starts with the glycine-glycine pair (L83,84, 110,111) at the beginning of the seconds transmembrane helices. Mainly amino acids of structural importance (such as glycines, prolines, and arginines) are conserved. Part of the amino acids involved in the binding of the pigments and co-factors are also conserved: The histidine-ligands to the magnesium atoms of the special pair chlorophylls (L173, M200) and to the non-heme iron atom. In the L-subunit and the D1 protein a phenylalanine residue (L216, D1-255) and a serine residue (L223, D1-264) are found in the corresponding sequence positions. These residues are involved in the binding of the *s*-triazine herbicides like atrazine and terbutryn which presumably act by competing with the secondary quinone Q_B for its binding site. Mutations of these amino acids cause herbicide resistance in the purple bacteria, in plants, and algae. The phenylalanines L216 and D1-255 correspond to tryptophans M250 and D2-254 which form the major part of the binding site of the primary quinone Q_A (see 3.5).

Several important differences exist between the reaction centres of photosystem II and the purple bacteria: the amino acids involved in the binding of the accessory bacteriochlorophylls in the purple bacteria, and a glutamic acid, which is a bidentate ligand to the ferrous non-heme iron are not conserved. There is no hint for the existence of an analogue to the H-subunit in photosystem II reaction centre. The overall structure of the photosystem II reaction centre core, however, must be very similar to the reaction centre from purple bacteria formed by the L- and M-subunits. Figure 18 shows those helices which are presumably conserved between the reaction centre cores of the purple bacteria and photosystem II and the position of the amino acids conserved between the L- and M-subunits and the D1 and D2 proteins. Identities of amino acids, which are found specifically in the L-subunits and the D1 proteins, or specifically in the M subunits and D2 proteins, and involved in the quinone binding, might be the result of convergent evolution. Their location is also shown.

The rereduction of the photooxidized primary electron donor occurs from the cytochrome subunit in the reaction centre from *Rhodopseudomonas viridis*. In the position equivalent to the cytochrome subunit we have to expect the water soluble proteins forming part of the manganese-containing oxygen evolving complex in the photosystem II reaction centre. Experi-

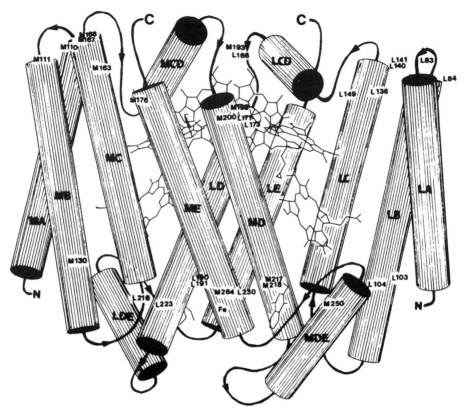

Figure 18: Column model for the core of the reaction centre from *Rps. viridis.* Only helices which are presumably conserved in photosystem II reaction centres are shown. The connections of the helices are only indicated schematically. The transmembrane helices of the L (M) subunit are labelled by LA-LE (MA-ME) and the major helices in the connections by LCD (MCD) and LDE (MDE). The special pair bacteriochlorophylls are at the interface of the L and M subunits between the D and E helices, the bacteriopheophytins near the L helices. The binding site for Q_A is between the LDE and LD helices. The location of the amino acids conserved between all L and M subunits and the D1 and D2 proteins, as well as those forming the quinone binding sites, is indicated by their sequence numbers (Taken from Michel & Deisenhofer 1988).

mental proof for the existence of a similar reaction centre core in photosystem II was the recent isolation of a complex consisting of the proteins D1, D2 and cytochrome b559 from spinach chloroplasts which contained four to five chlorophylls and two pheophytins (Nanba & Satoh, 1987). It has been shown to be active in electron transport to the pheophytins. Recently, evidence has been presented by two groups that a tyrosine residue located on the D1 subunit in the third transmembrane helix is an intermediate electron carrier between the primary electron donor of photosystem II and the oxygen evolving manganese cluster (Debus et al., 1988, Vermaas et al., 1988, Hoganson & Babcock, 1988). At present it is speculated if even the manganese cluster is bound to the D1 and D2 proteins.

As a result of the work on the bacterial photosynthetic reaction centre our entire view on the photosystem II reaction centres from plants and algae has changed.

4.2. *Evolutionary aspects*

The sequence similarities discussed above suggest that the reaction centres from purple bacteria and photosystem II are evolutionary related. A common ancestor possessed an entirely symmetric reaction centre with two parallel electron transporting pigment branches across the membrane. In this view the symmetric reaction centre was formed by two copies of the same protein subunit encoded by one gene, i. e. was a homodimer. After a gene duplication and subsequent mutations the formation of the asymmetric dimer ("heterodimer") and the use of only one pigment branch for electron transfer became possible. It is an open question if in evolution this gene duplication occurred only once, before the lineages leading to the purple bacteria and the photosystem II containing organisms split, or twice, after the splitting into these two lineages. In the latter case, the specific sequence similarities between L and D1, as well as those between M and D2, would be the result of convergent evolution, whereas the identities of the structurally important amino acids would date back to the original symmetric dimer. Sequence comparisons are in favour of the latter possibility (see Williams et al., 1986): The sequence identity between the D1 and D2 proteins is much higher than that between the L- and M-subunits. This observation possibly indicates that the gene duplication giving rise to separate D1 and D2 proteins occurred later during evolution than the gene duplication leading to the L- and M-subunits. On the other hand, due to more and stronger interactions with neighbouring proteins, the D1 and D2 proteins had less freedom to mutate than the L- and the M-subunits. As a result sequence comparisons might be misleading.

The evolutionary relations also indicate that there must be an advantage for reaction centres possessing only one active electron transport chain with two quinones acting in series. There might be rather trivial explanations for the use of only one branch, e. g. an asymmetry in the protein environment can cause an asymmetry in the distribution of electrons in the excited state and subsequently lead to a preferred release of an electron only in one direction. This existing polarity might lead to a faster rate of the first electron transfer step, a minimization of competing reactions, and thus a higher quantum yield for the electron transfer.

It is a clear advantage in the present day's reaction centres that the two quinones act in series, and only the released secondary quinone, Q_B, is a two-electron carrier. Consider the situation of the ancient symmetric reaction centre: Upon the first excitation the electron is transferred to the quinone at the end of one pigment branch. The resulting semiquinone is not stable and its electron is lost in the time range of seconds. Only if it receives a second electron can it be protonated and energy is stored in the form of the quinol. With two identical parallel electron transfer chains the probability for the second electron to be funneled into the same chain to the same quinone as the first electron is only 50%. A possible electrostatic repulsion by the negatively charged semiquinone might even decrease this probability. In a frequent situation the absorption of two photons leads to the formation of two semiquinones in the same reaction centre and energy is not stored in a stable way. The way out of this dilemma clearly is to switch the two quinones in series and to allow protonation and release only to the final quinone, which is then Q_B in the electron transfer chain, as it is seen in the reaction centres of purple bacteria and photosystem II. A considerable increase in the efficiency of light energy conversion, especially under low light conditions, must result.

5. ASPECTS OF MEMBRANE PROTEIN STRUCTURE

5.1 *The membrane anchor of the cytochrome subunit*

The X-ray structure analysis established that the L and M subunits are firmly integrated into the membrane, both possessing five transmembrane helices, whereas the H subunit is anchored to the membrane by one transmembrane helix. The X-ray work showed no indication of any intramembraneous part of the cytochrome subunit. Nevertheless, in the hands of the biochemists it behaved like a membrane protein and aggregated easily. A strange observation during the protein sequencing was that upon Edman-degradation of the isolated cytochrome subunit no N-terminal amino acid could be identified after the first degradation step, but a normal sequence could be obtained starting with the second amino acid from the N-terminus. K. A. Weyer was then able to isolate a modified amino-terminal amino acid with the help of F. Lottspeich, and to elucidate the structure of this modified amino-terminal amino acid together with W. Schäfer using mass spectrometry (Weyer et al., 1987a, 1987c). The result is shown in *Figure 19*. The N-

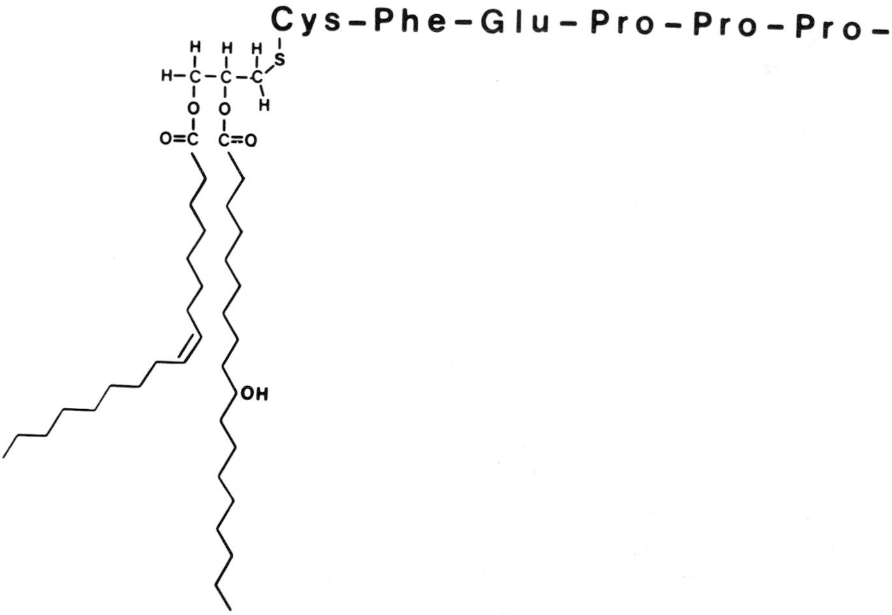

Figure 19: The N-terminus of the cytochrome subunit. Two fatty acids are esterified to the N-terminal S-glycero-cysteine. The fatty acids are a mixture of 18:OH (2 isomers) and 18:1 (3 isomers) acids roughly in a 1:1 ratio, which are represented by oleic-acid and 11-hydroxy-stearic acid in the figure. (Taken from Weyer et al., 1987c).

Chemistry 1988

terminal amino acid is a cysteine linked to a glycerol residue via a thioether bridge. Two fatty acids are then esterified to the two OH-groups of the glycerol. The fatty acids are a statistical mixture of singly unsaturated and singly hydroxylated C_{18} fatty acids. These experiments firmly established that the cytochrome subunit also possesses a membrane anchor, but this is now of a lipid type and not of a peptide type. The membrane anchor is very similar to that of the bacterial lipoproteins (see e. g. Pugsley, et al., 1986, Yu et al., 1986). The reaction centre cytochrome subunit is the first cytochrome molecule known to contain such a membrane anchor.

5.2 *Protein lipid contacts*

The contact between lipids and protein occurs at the surface of the proteins. Therefore a look on the surface of the protein complex might be very informative. For this purpose a space filling model of the reaction centre is shown as *Figure 20*. Carbon atoms approaching the surface of the reaction centre are shown as white spheres. A central section perpendicular to the approximate twofold rotation axis can be seen where carbon atoms form the surface of the protein almost exclusively. They are mainly side chain atoms of the amino acids leucine, isoleucine and phenylalanine. This central zone must correspond to the hydrophobic part of the protein surface which in the membrane is in contact with the alkane chains of lipids. Approaching the cytoplasmic rim of that central zone a row of nitrogen atoms is seen at the protein surface. These nitrogen atoms are side chain atoms of the basic amino acids arginine and histidine. The role of these basic residues might be

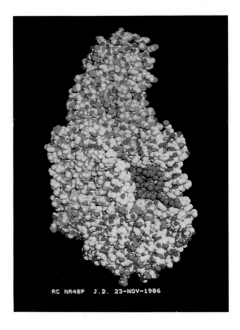

Figure 20: Space filling model of the photosynthetic reaction centre from *Rps. viridis.* Carbon atoms are shown in white nitrogen atoms in blue, oxygen atoms in red and sulphur in yellow. The visible atoms of a bacteriopheophytins approaching the surface are represented in brown.

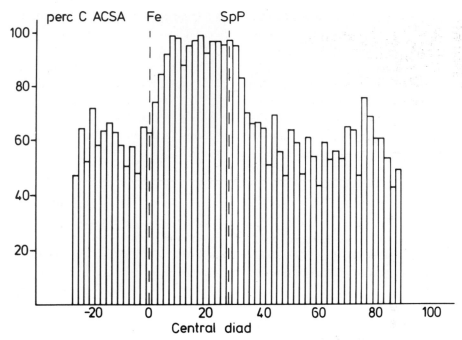

Figure 21: Percentage (perc) of the accessible surface area (ACSA) occupied by carbon atoms shown for 3 Å thick layers perpendicular to the noncrystallographic twofold rotation axis, which runs through the ferrous non-heme-iron atom (Fe) and the special pair (SpP).

to determine the position of the reaction centre perpendicular to the membrane via specific interaction between negatively charged amino acid side chains of the reaction centre protein subunits.

Figure 21 presents the percentage of the "accessible surface area" which is covered by carbon atoms, shown in layers perpendicular to the central twofold rotation axis. The twofold rotation axis runs through the non-heme iron atom near the cytoplasmic side and relates the special pair bacteriochlorophylls near the periplasmic side of the membrane. Two important conclusions can be drawn from *Figure 21*: (i) The primary electron donor (special pair), is located in the hydrophobic non-polar part of the membrane, whereas the non-heme iron atom is already in that zone where the protein surface is polar and most likely interacts with the polar head groups of the lipids. (ii) The thickness of the hydrophobic zone perpendicular to the membrane is 30 to 31 Å only. This value is smaller than expected for a lipid bilayer composed of lipids with C_{18} fatty acids.

5.3 *Distribution of amino acids and bound water molecule*

Figure 22a shows the distribution of the strongly basic amino acids arginine and lysine, and of the strongly acidic amino acids, glutamic acid and aspartic acid, which at neutral pH possess electric charges at the ends of their side chains. A central zone, where none of these amino acids are found, has a thickness of about 25 Å and is thus slightly thinner than the hydrophobic surface zone shown in *Figures 20* and *21*. The slight discrepancy is due to

A

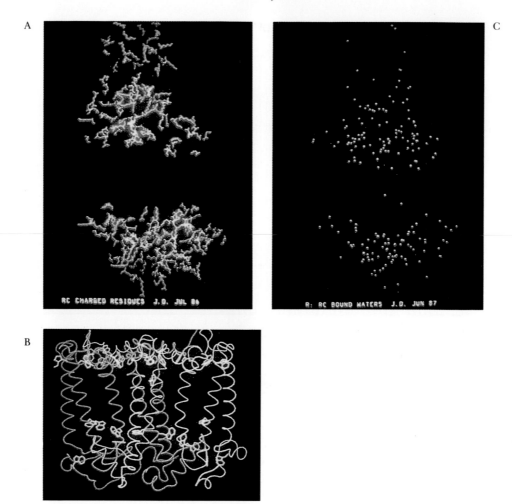

B

C

Figure 22: (A) Distribution of the "charged" amino acids in the photosynthetic reaction centre from *Rps. viridis.* The negatively charged amino acids (aspartate, glutamate) are shown in red, the positively charged amino acids (arginine, lysine) in blue. (C) Distribution of tryptophan residues (green) in the L (brownish) and M-subunits (blue). (B) Distribution of bound water molecules in the reaction centre. The reaction centres and the L- and M subunits subunits are always shown from a view parallel to the membrane.

two arginine residues and one glutamic acid residue, which are apparently in a hydrophobic environment without counter charges. The role of the positive charges of the arginine side chains seems to be structural. They possibly cancel the partial negative charge at the carboxy-terminal ends of the short helices in the connections of the long D and E transmembrane helices. These short helices partly intrude into the hydrophobic zone of the membrane and a positive charge of the arginine side chains seems to be necessary for the change of the direction of the peptide chain. The glutamic acid (L104) seems to be protonated, thus neutral, and to form a hydrogen bond with one of the bacteriopheophytins (Michel et al., 1986b).

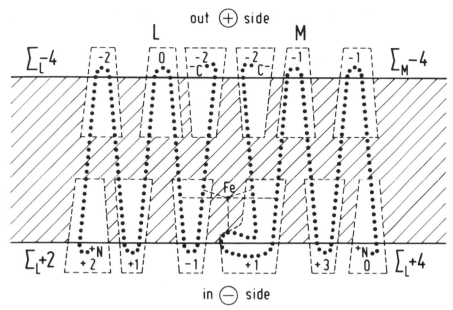

Figure 23: Schematic drawing of the transmembrane helices and the helix connections of the L and M subunits from the *Rps. viridis* reaction centre in the membrane to show the net charges at the ends of the helices and the helix connections. The negatively charged interior of the cell is indicated by the minus sign at the bottom, the positively charged extracellular medium by the plus sign at the top. (Taken from Michel & Deisenhofer, 1987).

Within the L and M subunits the glutamates and aspartates, the lysine and arginine residues show an interesting asymmetric distribution with respect to cytoplasmic and periplasmic sides. If one calculates "net charges" of the peptide chains on the periplasmic side of the membrane and compares them with the net charges of the cytoplasmic side (assuming that all glutamic acid residues, aspartic acid residues and the carboxy-termini are negatively charged, whereas all the arginine and lysine residues and the amino-termini are positively charged) one finds that the cytoplasmic ends of the transmembrane helices and their respective connections are nearly always less negatively charged than their counterparts on the periplasmic side. This phenomenon is illustrated schematically in *Figure 23:* As a result the cytoplasmic part of the M-subunit carries four positive net charges and the periplasmic part four negative charges, the cytoplasmic part of the L-subunit two positive charges and the periplasmic part four negative charges. The charge asymmetry becomes even more pronounced if one considers the existence of the firmly bound non-heme iron atom on the cytoplasmic side and the presumed protonation of glutamic acid L104. Thus these membrane proteins are strong electric dipoles. This result can be correlated with the fact that the interior of bacteria is negatively charged, due to the action of electrogenic ion pumps. This means that the L- and M-subunits are oriented in the membrane in the energetically more favorable manner. *Vice versa,* the combination of the electric field across the membrane, established by the ion pumps, and the anisotropic distribution of negatively and positively charged amino acids in the protein may be one of the factors, which determine the orientation of membrane proteins with respect to the inside and outside of the cell.

In the L- and M-subunits the remarkably uneven distribution of the amino acid tryptophan as shown in *Figure 22b* was quite unexpected. About two thirds of the tryptophans are found at the ends of transmembrane helices or in the helix connections on the periplasmic site. Only a few tryptophan residues are seen in the hydrophobic zone, where they are in contact with pigments. The residual tryptophans are located in the hydrophobic to polar transition zone or the polar part of the L- and M-subunits near the cytoplasmic surface. The indole rings of the tryptophans are oriented preferentially towards the hydrophobic zone of the membrane.

Figure 22c shows the distribution of the bound water molecules which have been tentatively identified by the X-ray crystallographic analysis. Only five of them are found in the hydrophobic intramembranous zone. A closer inspection shows that they may perform an important structural role. Fig. 24 shows one of these water molecules and its probable hydrogen-bonding pattern. It apparently crosslinks two transmembrane helices, one of the L-subunit, the other of the M-subunit by donating hydrogen bonds to two peptide oxygen atoms. Another hydrogen bond with an asparagine side chain is possible. How much these water molecules contribute to the stability of the reaction centre structure has to be determined in the future.

5.4 *Crystal packing and detergent binding*

As outlined in paragraph 1, the most promising strategy was to crystallize the reaction centres within the detergent micelles. According to this concept the crystal lattice should be formed by polar interactions between polar surface domains of the reaction centre. This expectation was confirmed by the results of the structural analysis. Mainly the polar surfaces of the cytochrome subunit and the H-subunit are involved in the crystal packing, to a minor extent also the polar surface part of the M-subunit.

As expected for detergents in a micelle most of the detergent is crystallographically not ordered and cannot be seen in the electron density map with one exception: The single transmembrane helix of the H-subunit, two transmembrane helices of the M-subunit, and part of the pigments seem to form a pocket where one detergent molecule is bound. Its polar head-group apparently undergoes specific interactions with the protein near the cytoplasmic end of the hydrophobic surface zone. Specific binding of this particular detergent molecule might explain, why crystals of the photosynthetic reaction centre from *Rhodopseudomonas viridis* could be grown only with N,N-dimethyldodecylamine-N-oxide as detergent, but not when octylglucopyranoside or similar detergents were used.

In collaboration with M. Roth and A. Bentley-Lewit from the Institut Laue-Langevin in Grenoble the detergent micelle could be visualized by neutron crystallography and H_2O/D_2O contrast variation. A rather flat, monolayer-like ring of detergent molecules surrounding the hydrophobic

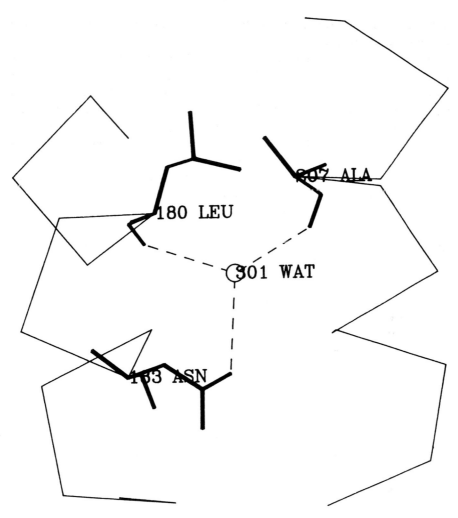

Figure 24: A firmly bound water-molecule (301 WAT) in the hydrophobic part of the membrane crosslinking two transmembrane helices by forming hydrogen bonds with the peptide oxygen atoms of leucine L180 alanine M207. Another hydrogen bond with the side chain of asparagine L183 is possible.

surface zone of the reaction centre became visible. Regions, where the detergent micelles are in contact, can also be seen. Therefore, attractive interactions between detergent micelles may also contribute to the stability of the protein's crystal lattice. In general, the strategy to crystallize membrane proteins within their detergent micelles (Michel, 1983; Garavito et al., 1986) now seems to be proven. However, the progress made in crystallizing membrane proteins other than bacterial photosynthetic reaction centres and bacterial porins has been unexpectedly slow: well diffracting crystals of membrane proteins have only been obtained in these cases. The necessary fine tuning with respect to the size of the detergent micelle and the size of the polar head group of the detergent is still a formidable task which has to be solved empirically for each individual membrane protein.

Acknowledgements: We wish to express our sincere gratitude to persons and organizations who helped to make our work possible and successful: Our colleagues Otto Epp, Heidi Grünberg, Friedrich Lottspeich, Kunio Miki, Wolfram Schäfer, Irmgard Sinning, Karl-Alois Weyer and Wolfgang Zinth, made important direct contributions as mentioned in the text. Dieter Oesterhelt and Robert Huber, their departments and the whole Max-Planck-Institut für Biochemie provided facilities and a stimulating and supportive atmosphere. Dieter Oesterhelt's backing, especially during the initial, frustrating phases, and the stable environment in the Max-Planck-Gesellschaft allowed us to start a project of unknown duration and outcome. Financially, the project was supported by the Deutsche Forschungsgemeinschaft through Sonderforschungsbereich 143 (Teilprojekt A3 to H.M., Teilprojekt A6 to J.D. and H.M.) and the Max-Planck-Gesellschaft.

REFERENCES

Allen, J. P. & Feher, G. (1984). Proc. Natl. Acad. Sci. USA *81*, 4795−4799.

Allen, J. P., Feher, G., Yeates, T. O., Rees, D. C., Deisenhofer, J., Michel, H. & Huber, R. (1986). Proc. Natl. Acad. Sci. USA *83*, 8589−8593.

Alt, J., Morris, J., Westhoff, I. & Herrmann, R. G. (1984) Curr. Genet. *8*, 597−606.

Baumeister, W. & Vogell, W., eds. (1980) Electron Microscopy at Molecular Dimensions, Springer, Berlin.

Bjorkmann, P. J., Saper, M. A., Samraoui, B., Bennett, W. S., Strominger, J. L. & Wiley, D. C. (1987) Nature (London) *329*, 506−512.

Breton, J., Martin, J. L., Migus, A., Antonetti, A. & Orszag, A. (1986) Proc. Natl. Acad. Sci. USA *83*, 5121−5125.

Chang, C. II., Schiffer, M., Tiede, D., Smith, U. & Norris, J. (1985) J. Mol. Biol. *186*, 201−203.

Chang, C. H., Tiede, D., Tang, J., Smith, U., Norris, J. & Schiffer, M. (1986) FEBS Lett. *205*, 82−86.

Clayton, R. K. & Clayton, B. J. (1978) Biochim. Biophys. Acta *501*, 478−487.

Debus, R. J., Feher, G. & Okamura, M. Y. (1986) Biochemistry *25*, 2276−2287.

Debus, R. J., Barry, B. A., Babcock, G. T. & McIntosh, L. (1988) Proc. Natl. Acad. Sci. USA *85*, 427−430.

Deisenhofer, J., Epp, O., Miki, K., Huber, R. & Michel, H. (1984) J. Mol. Biol. *180*, 385−398.

Deisenhofer, J, Epp, O., Miki, K., Huber, R. & Michel, H. (1985a) Nature *318*, 618−624.

Deisenhofer, J., Remington, S. J. & Steigemann, W. (1985b) Meth. Enzymol. *115*, 303−323.

Deisenhofer, J. & Michel, H. (1988) in: The Photosynthetic Bacterial Reaction Center Structure and Dynamics (Breton, J. & Vermeglio, A., eds.), pp 1−3, Plenum Press, New York.

Dracheva, S. M., Drachev, L. A., Konstantinov, A. A., Semenov, A. Y., Skulachev, V. P., Arutjunjan, A. M., Shuvalov, V. A. & Zaberezhnaya, S. M. (1988) Eur. J. Biochem. *171*, 253−264.

Feher, G. & Okamura, M. Y. (1978) in: The Photosynthetic Bacteria. Clayton, R. K. & Sistrom, W. R. (eds.) Plenum Press, New York, London, pp. 349−386.

Fleming, G. R., Martin, J. L. & Breton, J. (1988) Nature *333*, 190−192.

Garavito, R. M. & J. P. Rosenbusch (1980) J. Cell Biol. *86*, 327−329.

Garavito, R. M., Markovic-Housley, Z. & Jenkins, J. A. (1986) J. Crystal Growth *76*, 701−709.

Gast, P., Michalski, T. J., Hunt, J. E. & Norris, J. R. (1985) FEBS Lett. *179*, 325−328.

Henderson, R. & Unwin, P. N. T. (1975) Nature (London) *257*, 28−32.

Hoganson, C. W. & Babcock, G. T. (1988) Biochemistry *27*, 5848−5855.

Happe, M. & Overath, P. (1976) Biochem. Biophys. Res. Commun. *72*, 1504−1511.

Holschuh, K., Bottomley, W. & Whitfield, P. R. (1984) Nucleic Acid Res. *12*, 8819−8834.

Huber, R., Schneider, M., Epp, O., Mayr, I., Messerschmidt, A., Pflugrath, J. & Kayser, H. (1987) J. Mol. Biol. *195*, 423−434.

Jack, A. & Levitt, M. (1978) Acta Crystallogr. *A34*, 931−935.

Jones, T. A., Bartels, K. & Schwager, P. (1977) in: The Rotation Method in Crystallography (Arndt, U. W. & Wonacott, A. J., eds.) pp 105−117, North Holland, Amsterdam.

Jones, T. A., (1978) J. Appl. Crystallogr. *11*, 268−272.

Kirmaier, C., Holten, D. & Parson, W. W. (1985a) Biochim. Biophys. Acta *810*, 33−48.

Kirmaier, C., Holten, D. & Parson, W. W. (1985b) Biochim. Biophys. Acta *810*, 49−61.

Kirmaier, C., Holten, D., Debus, R. J., Feher, G. & Okamura, M. Y. (1986). Proc. Natl. Acad. Sci. USA *83*, 6407−6411.

Kirmaier, C. & Holten, D. (1987) Photosynthesis Research *13*, 225−260.

Knapp, E. W., Fischer, S. F., Zinth, W., Sander, M., Kaiser, W., Deisenhofer, J. & Michel, H. (1985) Proc. Natl. Acad. Sci. USA *82*, 8463−8467.

Lee, B. & Richards, F. M. (1971) J. Mol. Biol. *55*, 379−400.

Luzzati, P. V. (1952). Acta Crystallogr. *5*, 802−810.

Mathews, F. S., Argos, P. & Levine, M. (1972) Cold Spring Harbor Symp. Quant. Biol. *36*, 387−397.

Michel, H., (1982a) EMBO J. *1*, 1267−1271.

Michel, H. (1982b) J. Mol. Biol. *158*, 567−572.

Michel, H., (1983) Trends Biochem. Sci. *8*, 56−59.

Michel, H., Osterhelt, D. & Henderson, R. (1980) Proc. Natl. Acad. Sci. USA 77, 338−342.

Michel, H. & Oesterhelt, D. (1980) Proc. Natl. Acad. Sci. USA 77, 1283−1285.

Michel, H., Deisenhofer, J., Miki, K., Weyer, K. A. & Lottspeich, F. (1983) in: Structure and Function of Membrane Proteins (Quagliariello, E. & Palmieri, F., Eds.) pp 191−197, Elsevier, Amsterdam.

Michel, H., Weyer, K. A., Gruenberg, H. & Lottspeich, F. (1985) EMBO J. *4*, 1667−1672.

Michel, H., Weyer K. A., Gruenberg, H., Dunger, I., Oesterhelt, D. & Lottspeich, F. (1986a) J. *5*, 1144−1158.

Michel, H., Epp, O. & Deisenhofer, J. (1986b) EMBO-J. *5*, 2445−2451.

Michel, H. & Deisenhofer, J. (1987) Chem. Scripta *27B*, 173−180.

Michel, H. & Deisenhofer, J. (1988) Biochemistry *27*, 1−7.

Michel-Beyerle, M. E., Plato, M., Deisenhofer, J., Michel, H., Bixon, M. & Jortner, J. (1988) Biochim. Biophys. Acta *932*, 52−70.

Nakatani, H. Y., Ke, B., Dolan, E. & Arntzen, C. J. (1984) Biochim. Biophys. Acta *765*, 347−352.

Nanba, O. & Satoh, K. (1987) Proc. Natl. Acad. Sci. USA *84*, 109−112.

Newcomer, M. E., Jones, T. A., Aqvist, J., Sundelin, J., Eriksson, U., Rask, I. & Peterson, P. A. (1984) EMBO J. *3*, 1451−1454.

Norris, J. R., Uphaus, R. A., Crespi, H. L. & Katz, J. J. (1971) Proc. Natl. Acad. Sci. USA *68*, 625−628.

Oesterhelt, D. & Stoeckenius, W. (1971) Nature New Biol. *233*, 149−152.

Oesterhelt, D. (1972) Hoppe−Seyler's Z. Physiol. Chem. *353*, 1554−1555.

Okamura, M. Y., Feher, G. & Nelson, W. (1982) in: Photosynthesis: Energy Conversion by Plants and Bacteria, Vol. 1 (Govindjee, Ed.) pp 195−272, Academic Press, New York.

Omata, T., Murata, M. & Satoh, K. (1984) Biochim. Biophys. Acta *765*, 403−405.

Parson, W. W. (1987) in: New Comprehensive Biochemistry: Photosynthesis (Amesz, J., Ed.) pp 43−61, Elsevier.

Parson, W. W. & Ke, B. (1982) in: Photosynthesis: Energy Conversion by Plants and Bacteria, Vol. 1 (Govindjee, Ed.), pp 331−384, Academic Press, New York.

Pfister, K., Steinback, K. E., Gardner, G. & Arntzen, C. J. (1981) Proc. Natl. Acad. Sci. USA *78*, 981−985.

Pugsley, A. P., Chapon, C. & Schwartz, M. (1986) J. Biol. Chem. *256*, 2194−2198.

Rasmussen, O. F., Bookjans, G., Stummann, B. M. & Henningen, K. W. (1984) Plant Mol. Biol. *3*, 191−199.

Rochaix, J. D., Dron, M., Rahire, M. & Maloe, P. (1984) Plant Mol. Biol. *3*, 363−370.

Rossmann, M. G. *(1979)* J. Appl. Crystallogr. *12*, 225−238.

Schmid, M. F., Weaver, L. H., Holmes, M. A., Gruetter, M. G., Ohlendorf, D. H., Reynolds, R. A., Remington, S. J. & Matthews, B. W. (1981) Acta Crystallogr. *A37*, 701−710.

Schwager, P., Bartels, K. & Jones, A. (1975) J. Appl. Crystallogr. *8*, 275−280.

Shuvalov, V. H. & Klimov, U. V. (1976) Biochim. Biophys. Acta *440*, 587−599.

Sinning, I. & Michel, H. (1987a) in Progress in Photosynthesis Research, Vol 3 (Biggins, J., Ed.) pp III.11.771−III.11.773, Martinus Nijhoff, Dordrecht.

Sinning. I. & Michel, H. (1987b) Z. Naturforsch. *42*, 751−754.

Tiede, D. M., Prince, R. C., Reed, G. H. & Dutton, P. L. (1976) FEBS Lett. *65*, 301−304.

Tronrud, D. E., Ten Eyck, L. F. & Matthews, B. W. (1987) Acta Crystallogr. *A43*, 489−501.

Varghese, J. N., Laver, W. G. & Colman, P. M. (1983) Nature (London) *303*, 35−40.

Vermaas, W. F., Rutherford, A. W. & Hansson, Ö. (1988) Proc. Natl. Acad. Sci. USA *85*, 8477−8481.

de Vitry, C. & Diner, B. (1984) FEBS lett. *167*, 327−331.

Wang, B. C. (1985) Meth. Enzymol. *115*, 90−112.

Weyer, K. A., Schäfer, W., Lottspeich, F. & Michel, H. (1987a) Biochemistry *26*, 2909−2914.

Weyer, K. A., Lottspeich, F., Gruenberg, H., Lang, F., Oesterhelt, D. & Michel, H. (1987b) EMBO J. *6*, 2197−2202.

Weyer, K. A., Schäfer, W. Lottspeich, F. & Michel, H. (1987c) in: Cytochrome Systems, Papa, S., Chanee, B. & Ernster, L. (eds.) Plenum Publishing Corporation pp. 325−331.

Williams, J. C., Steiner, L. A., Ogden, R. C., Simon, M. I. & Feher, G. (1983) Proc.. Natl. Acad. Sci. USA *80*, 6505−6509.

Williams, J. C., Steiner, L. A., Feher, G. & Simon, M. I. (1984) Proc. Natl. Acad. Sci. USA *81*, 7303−7307.

Williams, J. C., Steiner, L. A. & Feher, G. (1986) PROTEINS: Structure, Function and Genetic *1*, 312−325.

Wilson, I. A., Skehel, J. J. & Wiley, D. C. (1981) Nature (London) *289*, 366−373.

Youvan, D. C., Bylina, E. J., Alberti, M., Begusch, H. & Hearst, J. E. (1984) Cell (Cambridge, Mass.) *37*, 949−957.

Yu, F. Inouye, S. & Inouye, M. (1986) J. Biol. Chem. *261*, 2284−2288.

Zinth, W., Kaiser, W. & Michel, H. (1983). Biochim. Biophys. Acta *723*, 128−131.

Zinth, W., Knapp, E. W., Fischer, S. F., Kaiser, W., Deisenhofer, J. & Michel, H. (1985) Chem. Phys. Lett. *119*, 1−4.

Zulauf, M., Weckstrom, K., Hayter, J. B., Degiorgio, V. & Corti, M. (1985) J. Phys. Chem. *89*, 3411−3417.

Zurawski, G., Bohnert, H. J., Whitfeld, P. R. & Bottomley, W. (1982) Proc. Natl. Acad. Sci. USA *79*, 7699−7703.

Robert Huber

ROBERT HUBER

I was born February 20, 1937 in München as the first child of Sebastian and Helene Huber. My father was cashier at a bank and my mother kept the house and brought up the children, me and my younger sister, a difficult task during the war, a continuous struggle for some milk and bread and search for air-raid shelters. There was no Grammar school in 1945 and 1946 and I entered the Humanistische Karls-Gymnasium in München 1947 with intense teaching of Latin and Greek, some natural science and a few optional monthly hours of chemistry. I learned easily and had time to follow my inclination for sports (light athletics and skiing) and chemistry, which I taught myself by reading all textbooks I could get.

I left the Gymnasium with the Abitur in 1956 and began to study chemistry at the Technische Hochschule (later Technische Universität) in München, where I also made the Diploma in Chemistry in 1960. A stipend of the Bayerisches Ministerium für Erziehung und Kultur and later of the Studienstiftung des Deutschen Volkes helped to relieve financial problems of my family and allowed me to study without delay. The most impressive teachers I remember were W. Hieber and the logical flow and impressive diction of his lectures in inorganic chemistry; E.O. Fischer, the young star in metalloorganic chemistry; F. Weygand and his deep knowledge of organic chemistry; and G. Joos and G. Scheibe, the physicist and physicochemist, respectively. I joined the crystallographer W. Hoppe's laboratory for my diploma work on crystallographic studies of the insect metamorphosis hormone ecdysone. Part of these studies were made in Karlson's laboratory at the Physiologisch-Chemisches Institut der Universität München, where I found by a simple crystallograpic experiment the molecular weight and probable steroid nature of ecdysone which Hoppe and I later elucidated in atomic detail after my thesis work which was on the crystal structure of a diazo compound (1963). This discovery convinced me of the power of crystallography and led me to continue in this field.

After a number of structure determinations of organic compounds and methodical development of Patterson search techniques I began in 1967, with Hoppe's and Braunitzer's support, crystallographic work on the insect protein erythrocruorin (with Formanek). The elucidation of this structure and its resemblance to the mammalian globins as determined by Perutz and Kendrew in their classical studies suggested for the first time a universal globin fold. In 1971 the University of Basel offered me a chair of structural biology at the Biozentrum and the Max-Planck-Gesellschaft the position of a director at the Max-Planck-Institut für Biochemie, which I accepted. I remained associated with the Technische Universität München, where I became Professor in 1976.

In 1970, I had begun work on the basic pancreatic trypsin inhibitor which has later become the model compound for the development of protein NMR, molecular dynamics, and experimental folding studies in other laboratories. Work in the field of proteolytic enzymes and their natural inhibitors has been continued and extended to many different inhibitor classes, proteases, their proenzymes, and complexes between them (with Bode, Bartels, Chen, Fehlhammer, Deisenhofer, Loebermann, Kukla, Papamokos, Rühlmann, Steigemann, Tokuoka, Wang, Walter, Weber, Wei) including recently inhibitors of cysteine proteases (with Musil, Bode, Engh) and other hydrolytic enzymes like α-amylase (whith Pflugrath, Wiegand) and creatine hydrolase (with Hoeffken). The potential of these systems for drug and protein design has spurred our interest until today.

Early in the seventies I initiated work on immunoglobulins and their fragments, which culminated in the elucidation of several fragments, an intact antibody and its Fc fragment, the first glycoprotein to be analysed in atomic detail (with Colman, Deisenhofer, Epp, Marquart, Matsushima). Work was extended to proteins interacting with immunoglobulins and to complement proteins (with Paques, Jones, Deisenhofer). We also studied a variety of enzymes leading to the elucidation of the structure and the chemical nature of the selenium moiety in glutathione peroxidase (with Ladenstein, Epp). We determined the structures of citrate synthase in different states of ligation (with Remington, Wiegand) and recently of a very large multienzyme complex, heavy riboflavin synthase (with Ladenstein).

Early in the 1980s we began with studies of proteins involved in excitation energy and electron transfer, light-harvesting proteins (with Schirmer, Bode), later bilin-binding protein, the reaction centre (with Deisenhofer, Epp, Miki in collaboration with Michel) and ascorbate oxidase (with Messerschmidt, Ladenstein) which are described in my lecture.

Most of these structural studies were collaborative undertakings with other laboratories, many of them from foreign countries.

We had discovered that some of the proteins analysed showed large-scale flexibility which was functionally significant. The trypsinogen system was investigated (with Bode) in great detail by low temperature crystallography, γ-ray spectroscopy, chemical modification, and molecular dynamics calculations. However, it required some years before the scientific community in general accepted that flexibility and disorder are very relevant molecular properties also in other systems.

The development of methods of protein crystallography has been in the focus of my laboratory's work from the beginning and led to the development of refinement in protein crystallography (with Steigemann, Deisenhofer, Remington), to the development of Patterson search methods (with Bartels and Fehlhammer), to methods and suites of computer programmes for intensity data evaluation and absorption correction (FILME, with Bartels, Bennett, Schwager), for protein crystallographic computing (PROTEIN, with Steigemann), for computer graphics and electron density interpretation and refinement (FRODO, Jones), and for area detector data collec-

tion (MADNES, Pflugrath, Messerschmidt). These methods and programmes are in use in many laboratories in the world today.

I married Christa Essig in 1960. We have four children. The eldest
daughter (1961) and the two sons (1963, 1966) have been or are studying
economics. The youngest daughter (1976) shows some interest in biology, a
last hope.

Affiliations

member of the Deutsche Chemische Gesellschaft
member of the Gesellschaft für Biologische Chemie
honorary member of the American Society of Biological Chemists
honorary member of the Swedish Society for Biophysics
member of EMBO and EMBO council

Honors

1972 E.K.-Frey Medaille (Gesellschaft für Chirurgie)
1977 Otto-Warburg Medaille (Gesellschaft für Biologische Chemie)
1982 Emil von Behring Medaille (Universität Marburg)
1987 Keilin Medal (Biochemical Society, London)
1987 Richard-Kuhn Medaille (Gesellschaft Deutscher Chemiker)
1987 Dr. h.c., Université Catholique de Louvain
1988 Member of the Bayerische Akademie der Wissenschaften
1988 Nobel Prize for Chemistry
1989 E. K. Frey-E. Werle Gedächtnismedaille
1989 Dr. h. c., University of Ljubljana
1990 Kone Award, Association of Clinical Biochemists, United Kingdom
1991 Dr. h. c. for Medicine and Surgery, Università 'Tor Vergata', Rome

Key references

Citations in the Nobel Lecture concern works on proteins involved in light
energy and electron transfer. The following references are therefore key
references to studies of other systems. There are 270 publications from my
laboratory.

— R. Huber and W. Hoppe
    "Zur Chemie des Ecdysons VII. Die Kristall- und Molekülstrukturanalyse des
    Insektenverpuppungshormons Ecdyson mit der automatisierten Faltmolekül-
    methode", Chem.Ber. 7 (1965) 2403.
— E. Huber, O. Epp and H. Formanek
    "The Environment of the Haem Group in Erythrocruorin *(Chironomus thummi)*",
    J.Mol.Biol. 42 (1969) 591.
— R. Huber, D. Kukla, A. Rühlmann, O. Epp and H. Formanek
    "The basic Trypsin Inhibitor of Bovine Pancreas. I. Structure Analysis and
    Conformation of the Polypeptide Chain", Naturwissenschaften 57 (1970) 389.
— R. Huber, O. Epp, H. Formanek and W. Steigmann
    "The Atomic Structure of Erythrocruorin in the light of the chemical sequence
    and its comparison with Myoglobin", Eur.J.Biochem. 19 (1971) 42−50.

— A. Rühlmann, D. Kukla, P. Schwager, K. Bartels and R. Huber
"Structure of the Complex formed by Bovine Trypsin and Bovine Pancreatic Trypsin Inhibitor. Crystal Structure Determination and Stereochemistry of the Contact Region", J.Mol. Biol. *77* (1973) 417–436.

— O. Epp, P. Colman, H. Fehlhammer, W. Bode, W. Palm, M. Schiffer and R. Huber
"Crystal and Molecular Structure of a Dimer composed of the variable Portions of the Bence-Jones Protein REI", Eur.J. Biochem. *45* (1974) 513–524.

— R. Huber, D. Kukla, W. Bode, P. Schwager, K. Bartels, J. Deisenhofer and W. Steigemann
"Structure of the Complex formed by Bovine Trypsin and Bovine Pancreatic Trypsin Inhibitor. II. Crystallographic Refinement at 1.9 Å Resolution", J. Mol.Biol. *89* (1974) 73–101.

— P.M. Colman, J. Deisenhofer, R. Huber and W. Palm
"Structure of the Human Antibody Molecule Kol (Immungloubuline G1) An Electron Density Map at 5 Å Resolution", J.Mol.Biol. *100* (1976) 257–282.

— R. Huber
"Antibody Structure", Trends in Biochemical Sciences *1* (1976) 174–178.

— R. Huber, J. Deisenhofer, P.M. Colman, M. Matsushima and W. Palm
"Crystallographic structure studies of an IgG molecule and an Fc fragment", Nature *264* (1976) 415–420.

— H. Fehlhammer, W. Bode and R. Huber
"Crystal structure of Bovine Trypsinogen at 1.8 Å Resolution. II. Crystallographic Refinement, Refined Crystal Structure and Comparison with Bovine Trypsin. J.Mol.Biol. *111* (1977) 415–438.

— W. Bode, P. Schwager and R. Huber
"The Transition of Bovine Trypsinogen to a Trypsin-like State upon Strong Ligand Binding. The Refined Crystal Structures of the Bovine Trypsinogen-Pancreatic Trypsin Inhibitor Complex and of its Ternary Complex with Ile-Val at 1.9 Å Resolution", J.Mol.Biol. *118* (1978) 99–112.

— E. Weber, W. Steigemann, T.A. Jones and R. Huber
"The Structure of Oxy-Erythrocruorin at 1.4 Å Resolution", J.Mol.Biol. *120* (1978) 327–336.

— R. Huber and W. Bode
"Structural Basis of the Activation and Action of Trypin", In: Accounts of Chemical Research *11* (1978) 114–122.

— M. Matsushima, M. Marquart, T.A. Jones, P.M. Colman, K. Bartels, R. Huber and W. Palm
"Crystal Structure of the Human Fab Fragment Kol and its Comparison with the intact Kol Molecule", J.Mol.Biol. *121* (1978) 441–459.

— J. Deisenhofer, T.A. Jones, R. Huber, J. Sjödahl and J. Sjöquist
"Crystallization, Crystal Structure Analysis and Atomic Model of the Complex Formed by a Human Fc Fragment and Fragment B of Protein A from Staphylococcus aureus", Hoppe-Seyler's Z. Physiol.Chem *359* (1978) 975–985.

— R. Ladenstein, O. Epp, K. Bartels, A. Jones, R. Huber and A. Wendel
"Structure Analysis and Molecular Model of the Selenoenzyme Glutathione Peroxidase at 2.8 Å-Resolution", J.Mol.Biol. *134* (1979) 199–218.

— M. Marquart, J. Deisenhofer, R. Huber and W. Palm
"Crystallographic Refinement and Atomic Models of the Intact Immunoglobulin Molecule Kol and its Antigen-binding Fragment at 3.0 Å and 1.9 Å Resolution", J.Mol.Biol. *141* (1980) 369–391.

— R. Huber, H. Scholze. E.P. Paques and J. Deisenhofer
"Crystal Structure Analysis and Molecular Model of Human C3a Anaphylatoxin", Hoppe-Seyler's Z. Physiol.Chem. *361* (1980) 1389–1399.

— R. Huber
"Spatial Structure of Immunoglobulin Molecules", Klin. Wochenschr. *58* (1980) 1217—1231.

— E. Weber, E. Papamokos, W. Bode, R. Huber, I. Kato and M. Laskowski, Jr..
"Crystallization, Crystal Structural Analysis and Molecular Model of the Third Domain of Japanese Quail Ovomucoid, a Kazal Type Inhibitor", J.Mol.Biol. *149* (1981) 109—123.

— S. Remington, G. Wiegand and R. Huber
"Crystallographic Refinement and Atomic Models of Two Different Forms of Citrate Synthase at 2.7 and 1.7 Å Resolution", J.Mol.Biol. *158* (1982) 111—152.

— E. Papamokos, E. Weber, W. Bode, R. Huber, M.W. Empie, I. Kato and M. Laskowski, Jr.
"Crystallographic Refinement of Japanese Quail Ovomucoid, a Kazal-type Inhibitor, and Model Building Studies of Complexes with Serine Proteases", J.Mol.Biol. *158* (1982) 515—537.

— M., Bolognesi, G. Gatti, E. Menegatti, M. Guarneri, M. Marquart, E. Papamokos and R. Huber
"Three-dimensional Structure of the Complex Between Pancreatic Secretory Trypsin Inhibitor (Kazal Type) and Trypsinogen at 1.8 Å Resolution. Structure Solution, Crystallographic Refinement and Preliminary Structural Interpretation", J. Mol. Biol. *162* (1982) 839—868.

— M. Marquart, J. Walter, J. Deisenhofer, W. Bode and R. Huber
"The Geometry of the Reactive Site and of the Peptide Groups in Trypsin, Trypsinogen and its Complexes with Inhibitors", Acta Cryst. *B39* (1983) 480—490.

— G. Wiegand, S. Remington, J. Deisenhofer and R. Huber
"Crystal Structure Analysis and Molecular Model of a Complex of Citrate Synthase with Oxaloacetate and S-Acetonyl-coenzyme A"., J.Mol.Biol. *174* (1984) 205—219.

— W.S. Bennett and R. Huber
"Structural and Functional Aspects of Domain Motions in Proteins", CRC Crit. Rev. Biochem. *15* (1984) 291—384.

— H. Löbermann, R. Tokuoka, J. Deisenhofer and R. Huber
"Human α_1-Proteinase Inhibitor. Crystal Structure Analysis of Two Crystal Modifications, Molecular Model and Preliminary Analysis of the Implications for Function", J.Mol.Biol. *177* (1984) 531—556.

— R. Huber
"Three-Dimensional Structure of Antibodies". In: Behring Institute Mitteilungen (P. Gronski, F.R. Seiler, eds.), Die Medizinische Verlagsgesellschaft mbH *76* (1984) 1—14.

— J. Deisenhofer, O. Epp, K. Miki, R. Huber and H. Michel
"X-ray Structure Analysis of a Membrane Protein Complex. Electron Density Map at 3 Å Resolution and a Model of the Chromophores of the Photosynthetic Reaction Center from *Rhodopseudomonas viridis*", J.Mol.Biol. *180* (1984) 385—398.

— A. Wlodawer, J. Walter, R. Huber and L. Sjölin
"Structure of Bovine Pancreatic Trypsin Inhibitor, Results of Joint Neutron and X-ray Refinement of Crystal Form II", J.Mol.Biol *180* (1984) 301—320.

— J. Deisenhofer. H. Michel and R. Huber
"The structural basis of photosynthetic light reactions in bacteria", Trends Biochem. Sci. *10* (1985) 243—248.

— T. Schirmer, W. Bode, R. Huber, W. Sidler and H. Zuber
"X-ray Crystallographic Structure of the Light-harvesting Biliprotein C-Phycocyanin from the Thermophilic Cyanobacterium *Mastigocladus laminosus* and its resemblance to Globin Structure", J.Mol.Biol. *184* (1985) 257—277.

— D. Wang, W. Bode and R. Huber
"Bovine Chymotrypsinogen A X-ray Crystal Structure Analysis and Refinement of a New Crystal Form at 1.8 Å Resolution", J.Mol.Biol. *185* (1985) 595−624.

— J. Deisenhofer, O. Epp, K. Miki, R. Huber and H. Michel
"Structure of the protein subunits in the photosynthetic reaction centre of *Rhodopseudomonas viridis* at 3 Å resolution", Nature *318* (1985) 618−624.

— T. Schirmer, R. Huber, M. Schneider, W. Bode, M. Miller and M.L. Hackert
"Crystal Structure Analysis and Refinement at 2.5 Å of Hexameric C-phycocyanin from the Cyanobacterium *Agmenellum quadruplicatum*. The Molecular Model and its Implications for Light-harvesting", J.Mol.Biol. *188* (1986) 651−676.

— J.W. Pflugrath, G. Wiegand, R. Huber and L. Vertesy
"Crystal Structure Determination, Refinement and the Molecular Model of the α-Amylase Inhibitor", J.Mol.Biol. *189*, 383−386.

— W. Bode, A.-Z. Wei, R. Huber, E. Meyer, J. Travis and S. Neumann
"X-ray crystal structure of the complex of human leukocyte elastase (PMN elastase) and the third domain of the turkey ovomucoid", The EMBO J. *5* (1986) 2453−2458.

— J.P. Allen, G. Feher, T.O. Yeates, D.C. Rees, J. Deisenhofer, H. Michel and R. Huber
"Structural homology of reaction centers from *Rhodopseudomonas sphaeroides* and *Rhodopseudomonas viridis* as determined by X-ray diffraction", Proc. Natl. Acad. Sci. USA *83* (1986) 8589−8593.

— R. Ladenstein, A. Bacher, and R. Huber
"Some Observations of a Correlation between the Symmetry of large Heavy-atom Complexes and their Binding Sites on Proteins", J.Mol.Biol. *195* (1987) 751−753.

— T. Schirmer, W. Bode and R. Huber
"Refined Three-dimensional Structures of Two Cyanobacterial C-phycocyanins at 2.1 and 2.5 Å Resolution", J.Mol.Biol. *196*, 677−695.

— A. Brünger, R. Huber and M. Karplus
"Trypsinogen-Trypsin Transition: A Molecular Dynamics Study of Induced Conformational Change in the Activation Domain", Biochem. *26* (1987) 5153−5162.

— R. Huber, M. Schneider, I. Mayr, R. Müller, R. Deutzmann, F. Suter, H. Zuber, H. Falk and H. Kayser
"Molecular Structure of the Bilin Binding Protein (BBP) from *Pieris brassicae* after Refinement at 2.0 Å Resolution", J.Mol.Biol. *198* (1987) 499−513.

— R. Huber
"Flexibility and Rigidity of Proteins and Protein-Pigment Complexes", Angew. Chem. Int. Ed. Engl. *27* (1988) 79−88.

— M.G. Grütter, G. Fendrich, R. Huber and W. Bode
"The 2.5 Å X-ray crystal structure of the acid-stable proteinase inhibitor from human mucous secretions analysed in its complex with bovine α-chymotrypsin", The EMBO Journal *7* (1988) 345−351.

— J-P. Declerc, B. Tinant, J. Parello, G. Etienne and R. Huber
"Crystal Structure Determination and Refinement of Pike 4.10 Parvalbumin (Minor Component from *Esox lucius*)", J.Mol.Biol. *202*, (1988) 349−353.

— W. Bode, R. Engh, D. Musil, U. Thiele, R. Huber, A. Karshikov, J. Brzin, J. Kos and V. Turk
"The 2.0 Å X-ray crystal structure of chicken egg white cystatin and its possible mode of interaction with cysteine proteinases", The EMBO J. *7* (1988) 2593−2599.

— R. Ladenstein, M. Schneider, R. Huber, H.-D. Bartunik, K. Wilson, K. Schott and A. Bacher
"Heavy Riboflavin Synthase from *Bacillus subtilis*. Crystal Structure Analysis of the Icosahedral β_{60} Capsid at 3.3 Å Resolution", J.Mol.Biol. *203* (1988) 1045−1070.

— H.W. Hoeffken, S.H. Knof, P.A. Bartlett, R. Huber, H. Moellering, G. Schumacher
"Crystal Structure Determination, Refinement and Molecular Model of Creatine Amidinohydrolase from *Pseudomonas putida*", J.Mol.Biol. *204* (1988) 417 — 433.

A STRUCTURAL BASIS OF LIGHT ENERGY AND ELECTRON TRANSFER IN BIOLOGY

Nobel Lecture, December 8, 1988

by
ROBERT HUBER

Max-Planck-Institut für Biochemie, 8033 Martinsried

Dedicated to Christa

ABBREVIATIONS

PBS, phycobilisomes; light harvesting organelles peripheral to the thylakoid membrane in cyanobacteria, which carry out oxygenic photosynthesis and have photosystems I and II;

PE, PEC, PC, APC, phycoerythrin, phycoerythrocyanin, phycocyanin, allophycocyanin; biliprotein components in PBS with covalently attached tetrapyrrole (bilin) pigments;

PS I, II; photosynthetic reaction centers in chloroplasts and cyanobacteria;

RBP; retinol binding protein;

BBP, bilin (biliverdin IX); binding protein in *Pieris brassicae;*

A *Rps. viridis,* bacteriochlorophyll-b containing purple bacterium carrying out anoxygenic photosynthesis;

RC, reaction centre;

C, H, L, M; the four subunits of the reaction centre from *Rps. viridis:* the cytochrome c subunit (C), with 4 haems displaying two redox potentials (c_{553}, c_{558}) is located on the periplasmic side of the membrane; the L- and M-subunits are integrated in the membrane and their polypeptide chains span the membrane with 5 α-helices each, labelled A, B, C, D, E; they bind the bacteriochlorophyll-b (BChl-b or BC), bacteriopheophytin-b (BPh-b or BP), menaquinone-9 (Q_A), ubiquinone-9 (UQ, Q_B) and Fe^{2+} cofactors; the subscripts $_{P, A, M, L}$ indicate pair, accessory, M-, L-subunit association, respectively; the H-subunit is located on the cytoplasmic side and its N-terminal α-helical segment (H) spans the membrane;

P680, P960; primary electron donors in PS II and the RC of *Rps. viridis,* respectively, indicating the long wavelength absorption maxima;

P^*, D^*; electronically excited states of P and D A;

LHC, light harvesting complexes;

$LH_{a,b}$; light harvesting protein pigment complexes in BChl-a,b containing bacteria;

Car, carotenoids;

Sor; Soretbands of Chl and BChl;

PCY, plastocyanin; electron carrier in the photosynthetic apparatus of plants;

LAC, laccase; oxidase in plants and fungi;

AO, ascorbate oxidase; oxidase in plants;

CP, ceruloplasmin; oxidase in mammalian plasma.

SUMMARY

Aspects of intramolecular light energy and electron transfer will be discussed for three protein cofactor complexes, whose three-dimensional structures have been elucidated by X-ray crystallography: Components of light harvesting cyanobacterial phycobilisomes, the purple bacterial reaction centre, and the blue multi-copper oxidases. A wealth of functional data is available for these systems which allow specific correlations between structure and function and general conclusions about light energy and electron transfer in biological materials to be made.

*

INTRODUCTION

All life on Earth depends ultimately on the sun, whose radiant energy is captured by plants and other organisms capable of growing by photosynthesis. They use sunlight to synthesize organic substances which serve as building materials or stores of energy. This was clearly formulated by L. Boltzmann, who stated that 'there exist between the sun and the earth a colossal difference in temperature .... The equalization of temperature between these two bodies, a process which must occur because it is based on the law of probability will, because of the enormous distance and magnitude involved, last millions of years. The energy of the sun may, before reaching the temperature of the earth, assume improbable transition forms. It thus becomes possible to utilize the temperature drop between the sun and the earth to perform work as is the case with the temperature drop between steam and water .... To make the most use of this transition, green plants spread the enormous surface of their leaves and, in a still unknown way, force the energy of the sun to carry out chemical syntheses before it cools down to the temperature level of the Earth's surface. These chemical syntheses are to us in our laboratories complete mysteries .... (Boltzmann, 1886).

Today many of these 'mysteries' have been resolved by biochemical research and the protein components and their basic catalytic functions have been defined (Calvin & Bassham, 1962).

I will focus in my lecture on Boltzmann's 'improbable transition forms', namely, excited electronic states and charge transfer states in modern terminology, the structures of biological materials involved and the interplay of cofactors (pigments and metals) and proteins. I will discuss some aspects of the photosynthetic centre of *Rps. viridis* (see the original publications cited later and short reviews (Deisenhofer et al., 1985 a, 1986, 1989)) and of functionally related systems, whose structures have been studied in my laboratory: Light harvesting cyanobacterial phycobilisomes and blue oxidases. A wealth of structural and functional data is available for these three systems, which make them uniquely appropriate examples from which to derive general principles of light energy and electron transfer in biological materials. Indeed, there are very few systems known in sufficient detail for such purposes.*

We strive to understand the underlying physical principles of light and electron conduction in biological materials with considerable hope for success as these processes appear to be more tractable than other biological reactions, which involve diffusive motions of substrates and products and intramolecular motions. Large-scale motions have been identified in many proteins and shown to be essential for many functions (Bennett & Huber, 1983; Huber, 1988). Theoretical treatments of these reactions have to take flexibility and solvent into account and become theoretically tractable only by applying the rather severe approximations of molecular dynamics (Karplus & McCammon, 1981; Burkert & Allinger, 1982) or by limiting the system to a few active site residues, which can then be treated by quantum mechanical methods.

Light and electron transfer processes seem to be amenable to a more quantitative theoretical treatment. The substrates are immaterial or very small and the transfer processes on which I focus are intramolecular and far removed from solvent. Molecular motions seem to be unimportant, as shown by generally small temperature dependences. The components active in energy and electron transport are cofactors, which, in a first approximation suffice for a theoretical analysis, simplifying calculations considerably.

1. Models for energy and electron transfer

To test theories developed for energy and electron transfer appropriate model compounds are essential. Although it would be desirable these models need not be mimics of the biological structures.

Förster's theory of inductive resonance (Förster, 1948, 1967) treats the cases of strong and very weak coupling in energy transfer. Strong interactions lead to optical spectra which are very different from the component spectra. Examples include concentrated solutions of some dyes, crystalline arrays, and the BC_p discussed in Section 3.1. The electronic excitation is in this case delocalized over a molecular assembly. Very weak coupling produces little or no alteration of the absorption spectra but the luminescence properties may be quite different. Structurally defined models for this case

---

* The structure of the *Rb. sphaeroides* RC is closely related to the *Rps. viridis* RC (Allen et al., 1986, 1987; Chang et al., 1986). A green bacterial bacteriochlorophyll-a containing light harvesting protein is well defined in structure (Tronrud et al., 1986) but not in function. In the multiheme cytochromes (Pierrot et al., 1982; Higuchi et al., 1984) the existence or significance of intramolecular electron transfer is unclear.

are scarce. The controlled deposited dye layers of Kuhn and Frommherz (Kuhn, 1970; Frommherz & Reinbold, 1988) may serve this purpose and have demonstrated the general validity of Förster's theory, but with deviations.

Synthetic models with electron transfer are abundant and have recently been supplemented by appropriately chemically modified proteins (e.g. Mayo et al., 1986; Gray, 1986; McGoutry et al., 1987). They are covered in reviews (see, e.g. Taube & Gould, 1969; Hopfield, 1974; Cramer & Crofts, 1982; Eberson, 1982; Marcus & Sutin, 1985; Mikkelsen et al., 1987; Kebarle & Chowdhury, 1987; McLendon, 1988). *Figure 1* shows essential elements of such models: Donor D (of electrons) and acceptor A may be connected by a bridging ligand (B) with a pendant group (P) embedded in a matrix M.

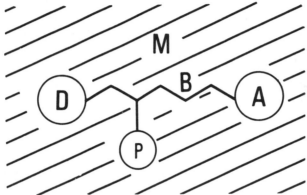

Figure 1. Determinants of electron transfer models. D: Donor, A: Acceptor, B: Bridge, P: Pendant group, M: Matrix.

Models with porphyrins as donors and quinones as acceptors are mimics of the RC (Gust et al., 1987; Schmidt et al., 1988). Models with peptide bridging ligands (Isied et al., 1985) merit interest especially in relation to the blue oxidases. The effect of pendant groups (P), which are not in the direct line of electron transfer (Taube and Gould, 1969) is noteworthy in relation to the unused electron transfer branches in the RC and the blue oxidases. It is clear, however, that the biological systems are substantially more complex than synthetic models. The protein matrix is inhomogeneous and unique in each case. Despite these shortcomings, theory and models provide the framework within which the factors controlling the transfer of excitation energy and electrons and competing processes are to be evaluated.

1.1. *Determinants of energy and electron transfer*

The important factors are summarized in *Table 1*. They may be derived from Förster's theory and forms of Marcus' theory (Marcus & Sutin, 1985) for excitation and electron transfer, respectively. These theoretical treatments may in turn be derived from classical considerations or from Fermi's Golden Rule with suitable approximations (see e.g. Barltrop & Coyle, 1978). Excitation and electron transfer depend on the geometric relation between

Table 1

	factors controlling rates
Excitation energy transfer $D^* + A \rightarrow D + A^*$ (very weak coupling)	distance and orientation (coupling of excited states); spectral overlap of emission and absorpation of D and A; refractive index of medium;
electron transfer from excited state $D^* + A \rightarrow D^+ + A^-$ and from ground state $D^- + A \rightarrow D + A^-$	distance and orientation (electronic coupling, orbital overlap); free energy change ("driving force"); reorganization in D and A; orientation polarization of medium;

donors and acceptors. Excitation energy transfer may occur over wide distances when the transition dipole moments are favourably aligned. Fast electron transfer requires sufficient electronic orbital overlap. Fast electron transfer over wide distances must therefore involve a series of closely spaced intermediate carriers with low lying unoccupied molecular orbitals or suitable ligands bridging donor and acceptor. Bridging ligands may actively participate in the transfer process and form ligand radical intermediates (chemical mechanism) or the electron may at no time be in a bound state of the ligands (resonance mechanism) (Hains, 1975). The spectral overlap and the 'driving force', for energy and electron transfer, respectively, have obvious effects on the transfer rates and are largely determined by the chemical nature and geometry of donors and acceptors. Nuclear reorganization of donor, acceptor, and the surrounding medium accompanying electron transfer is an important factor but difficult to evaluate in a complex protein system even qualitatively; we observe that the protein typically binds donors and acceptors firmly and rigidly, keeping reactant reorganization effects small. Surrounding polar groups may slow rapid electron transfer due to their reorientation. However, a polar environment also contributes to the energetics by stabilizing ion pairs ($D^+ A^-$) or lowering activation and tunneling barriers and may increase 'driving force' and rate. Energy transfer also depends on the medium and is disfavoured in media with a high refractive index.

Table 2

	competing processes
Excitation energy transfer $D^* + A \rightarrow D + A^*$ (very weak coupling)	non-radiative relaxation of D^* by photoisomerization and other conformational changes; excited state proton transfer ; intersystem crossing; chemical reactions of D^*, A^*, D^+, A^- with the matrix; fluorescence radiation of D^*;
electron transfer from excited state $D^* + A \rightarrow D^+ + A^-$	energy transfer; as above; back reaction to ground state D, A;
from ground state $D^- + A \rightarrow D + A^-$	———

Processes competing with productive energy and electron transfer from excited states 'lurk' everywere (*Table 2*). Quite generally, they are minimized by high transfer rates and conformational rigidity of the cofactors imposed by the protein.

I will discuss these factors in relation to the biological structures later on.

2. The role of cofactors

The naturally occurring amino acids are transparent to visible light and seem also to be unsuitable as single electron carriers with the exception of tyrosine. Tyrosyl radicals have been identified in PS II as Z^* and D^* intermediates, which are involved in electron transfer from the water splitting manganese protein complex to the photooxidized $P680^+$ (for reviews, see Barber, 1987; Prince, 1988). Their identification has been assisted by the observation that Tyr L162 lies in the electron transfer path from the

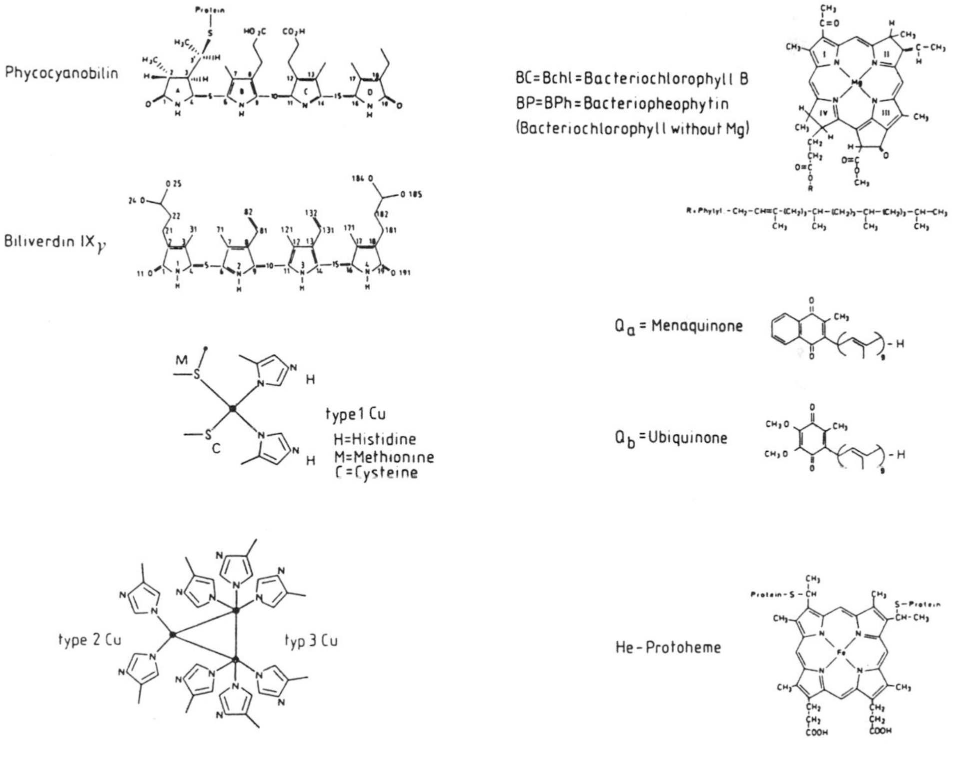

Figure 2. Cofactors in PC, BBP, AO, RC. Phycocyanobilins are covalently bound by thioether linkages to the protein. Biliverdin IXγ is non-covalently bound to BBP. Type-1, type-2, type-3 copper ions are linked to AO by coordination to the amino acid residues indicated. 4 BChl-b and 2 BPh-b are bound to the RC. A pair of BChl-b serves as the primary electron donor, a menaquinone-9 is the primary electron acceptor (Q_A) and an ubiquione-9 the secondary acceptor (Q_B). The 4 haem groups are bound by thioether linkages to the cytochrome c.

cytochrome to BC_{LP} in the bacterial RC (Deisenhofer et al., 1985) (see 3.2.2.2 and Figure 10c). A tyrosyl radical is not generated in the bacterial system, because the redox potential of $P960^+$ is insufficient.

Generally therefore cofactors, pigments and metal ions, serve as light energy acceptors and redox active elements in biological materials.

Figure 2 is a gallery of the pigments and metals clusters which will be discussed further on, namely the bile pigments, phycocyanobilin and biliverdin IXγ in the light harvesting complexes, the BChl-b, BP-b, and quinones in the purple bacterial RC and the copper centres in the blue oxidases.

The physical chemical properties of these cofactors determine the coarse features of the protein pigment complexes, but the protein part exerts a decisive influence on the spectral and redox properties.

3. The role of the protein

The role of the protein follows a hierarchy in determining the properties of the functional protein cofactor complexes shown in *Table 3*. These interactions are different for the various systems and shall be described separately, except point 1, as there are common features in the action of the protein as a polydentate ligand ascribed to a 'rack mechanism'.

Table 3. Hierarchy of protein cofactor interactions

1. Influence on configuration and conformation of the cofactors by the nature and geometry of ligands (the protein as a *polydentate ligand*).
2. Determation of the spatial arrangements of arrays of cofactors (the protein as a *scaffold*).
3. The protein as the *medium*.
4. Mediation of the interaction with other components in the supramolecular biological system.

3.1. *The protein as a polydentate ligand*

The 'rack mechanism' was introduced by Lumry and Eyring (Lumry & Eyring, 1954) and Gray and Malmström (Gray & Malmström, 1983) to explain unusual reactivities, spectral and redox properties of amino acids and cofactors by the distortion enforced by the protein.

A comparison of isolated and protein-bound bile pigments gives a clear demonstration of this effect. Isolated bile pigments in solution and in the crystalline state prefer a macrocyclic helical geometry with configuration ZZZ and conformation syn, syn, syn and show weak absorption in the visible range and low fluorescence quantum yield (Scharnagl et al., 1983; Huber et al., 1987a,b). When bound as cofactors to light harvesting phycocyanins they have strong absorption in the visible range and high fluorescence yield *(Figure 3)*. The auxochromic shift, essential for the light harvesting functions, is due to a strained conformation of the chromophore, which has configuration ZZZ and conformation anti, syn, anti stabilized by tight polar interactions with the protein (Schirmer et al., 1985, 1986, 1987) *(Figure 4)*. Particularly noteworthy is an aspartate residue (A87 here) bound to the central pyrrole nitrogens and conserved in all pigment sites. It influences protonation, charge, and spectral properties of the tetrapyrrole systems.

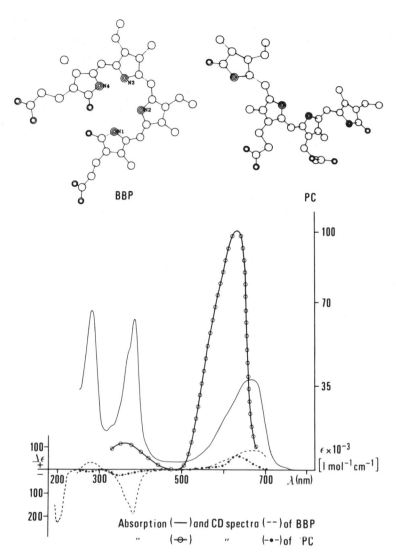

Figure 3. Tetrapyrrole structures in PC and BBP and the associated optical and circular dichroism spectra (Schirmer et al. 1987; Huber et al., 1987 b).

Tight binding is also effective against deexcitation by conformational changes. The structure shown in Figure 3 as representative of the free pigment is in fact observed in a bilin binding protein from insects (Huber et al., 1987a,b). This protein serves a different function and prefers the low energy conformer. The open chain tetrapyrrole bilins are conformationally adaptable, a property, which makes them appropriate cofactors for different purposes.

The cyclic BChl in the RC is conformationally restrained but responds to the environment by twisting and bending of the macrocycle. This may be one cause for the different electron transfer properties of the two pigment branches in the RC as will be discussed later. A more profound influence of

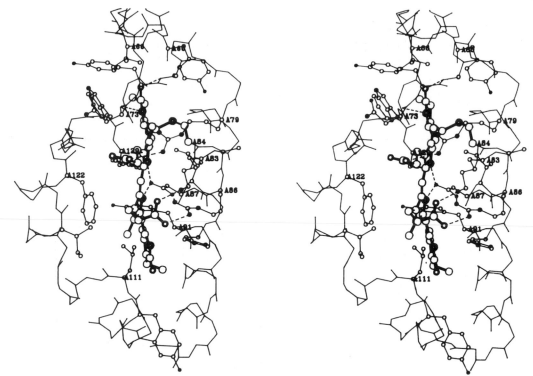

Figure 4. Stereo drawing of phycocyanobilin A84 (thick bonds) and its protein environment (thin bonds). All polar groups of the bilin except those of the terminal D pyrrole ring are bound by hydrogen bonds and salt links to protein groups (Schirmer et al., 1987).

the protein on the RC pigment system is seen in the absorption spectra, which differ from the composite spectra of the individual components *(Figure 5)*. The protein binds a pair of BChl-b (BC$_P$) so that the two BChl-b interact strongly between their pyrrole rings I including the acetyl substituents and the central magnesium ions (Deisenhofer et al., 1984). Alignment of the transition dipole moments and close approach cause excitonic coupling which partially explains the long wavelength absorption band P960 (Knapp et al., 1985).

The optical spectra are even more perturbed in blue copper proteins compared with cupric ions in normal tetragonal coordination *(Figure 6)*. The redox potential is also raised to about 300−500 mV vs 150 mV for Cu^{2+} (aq) (Gray & Solomon, 1981). These effects are caused by the distorted tetraedral coordination of the type-1 copper (a strained conformation stabilizing the cuprous state) and a charge transfer transition from a ligand cysteine S$^-$→ Cu^{2+} (Blair et al., 1985; Gray & Malmström, 1983).

The examples presented demonstrate the influence of the protein on the cofactors by various mechanisms, stabilization of unstable conformers and strained ligand geometries and the generation of contacts between pigments leading to strong electronic interaction.

The fixation of the relative arrangements of systems of cofactors is the basis of the energy and charge transfer properties in each system.

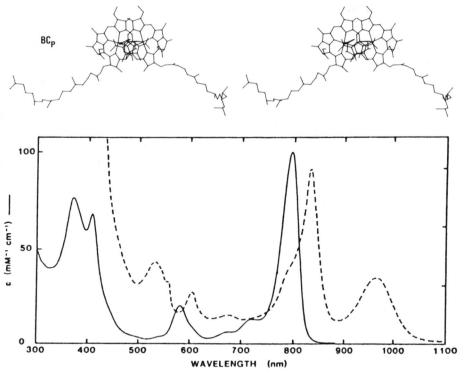

Figure 5. Stereodrawing of the special pair BC$_P$ in the RC (Deisenhofer el al. 1984) mainly responsible for the spectral alterations and the long wavelength absorption of the RC of *Rps. viridis* (– – –) compared with the spectra of BChl-b in ether solution (–) (spectra from Parson et al. . . . 1985).

Figure 6. The type-1 copper and its ligands in AO in stereo. The coordination of the copper is to His A446, His A513, Met A518, Cys A508 (Messerschmidt et al., 1989). The optical absorption spectra of "blue" copper in copper proteins (–) are compared with normal tetragonal copper (– – –) (spectrum from Gray & Solomon, 1981).

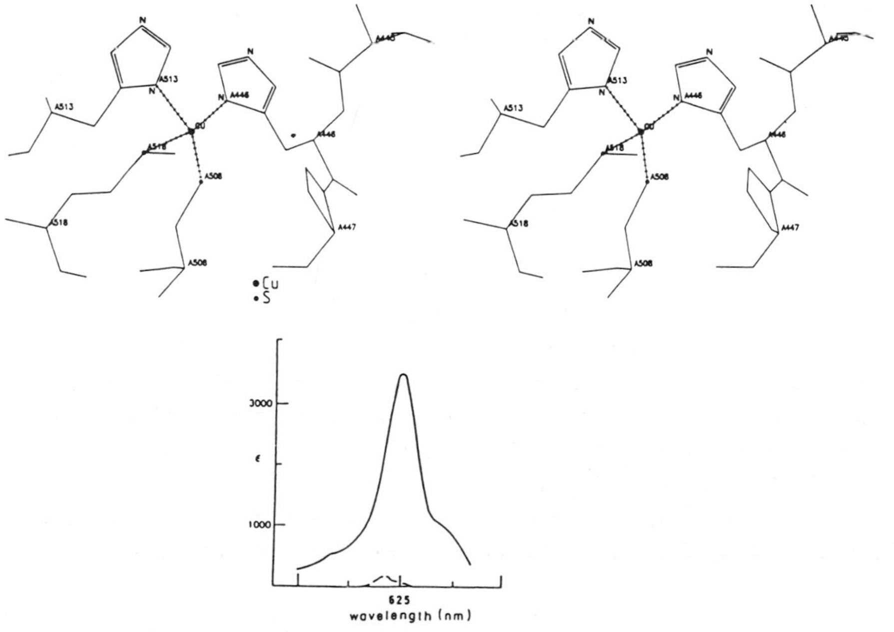

3.2. *Protein as a scaffold*

3.2.1. *Light harvesting by phycobilisomes*

The limited number of pigment molecules associated with RC would absorb only a small portion of incident sunlight. The RC are therefore associated with LHC, which may be located within the photosynthetic membrane, or form layers or antenna-like organelles in association with the photosynthetic membrane. Cyanobacteria have particularly intricate light harvesting systems, the PBS organelles peripheral to the thylakoid membrane. They absorb light of shorter wavelengths than do PS I and II, so that a wide spectral range of sunlight is used (*Figure 7*). The PBS are assembled from components with finely tuned spectral properties such that the light energy is channeled along an energy gradient to PS II.

3.2.1.1. Morphology

PBS consist of biliproteins and linker polypeptides. Biochemical and electron microscopy studies (Gantt et al., 1976; Mörschel et al., 1977; Bryant et al., 1979; Nies & Wehrmeyer, 1981) lead to the model representative of a hemidiscoidal PBS in *Figure 7*. Accordingly PBS rods are assembled in a polar way from PE or PEC and PC, which is attached to a central core of APC. APC is next to the photosynthetic membrane and close to PS II (for a review see, MacColl & Guard-Friar, 1987). The PC component consists of α- and β-protein subunits, which are arranged as $(\alpha\beta)_6$ disc-like aggregates with dimensions 120 Å x 60 Å (for reviews, see Scheer, 1982; Cohen-Bazire & Briant, 1982; Glazer, 1985; Zilinskas & Greenwald, 1986; Zuber, 1985, 1986).

From crystallographic analyses, a detailed picture of PC and PEC components has emerged (Schirmer et al. 1985, 1986, 1987; Duerring, 1988; Duerring et al., 1989). Amino acid sequence homology suggests that all components have similar structures.

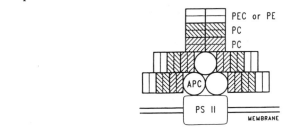

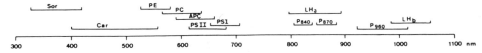

Figure 7. Scheme of a typical PBS with the arrangements of the components and the putative spatial relationship to the thylakoid and PS II (for reviews see, MacColl & Guard-Friar, 1987; Nies & Wehrmeyer, 1981). The component labelled PS II is thought to represent PS II and the phycobilisome attachment sites. The main absorption bands of photosynthetic protein cofactor complexes in photosynthetic organisms are also shown. The PBS components absorb differently to cover a wide spectral range and permit energy flow from PEC/PE via PC and APC to PS II.

3.2.1.2. Structure of phycocyanin

The PC α-subunit and β-subunits have 162 and 172 amino acid residues, respectively (in *Mastigocladus laminosus*). Phycocyanobilin chromophores are linked via thioether bonds to cysteine residues at position 84 of both chains (A84, B84) and at position 155 of the β-subunit (B155) (Frank et al., 1978). Both subunits have similar structures and are folded into eight α-helices (X, Y, A, B, E, F, G, H: see Figure 13). A84 and B84 are attached to helix E, B155 to the G-H loop. α-helices X and Y form a protruding anti-parallel pair essential for formation of the (αβ) unit.

The isolated protein forms (αβ)₃-trimers with C3 symmetry and hexamers (αβ)₆ as head to head associated trimers with D3 symmetry *(Figure 8)*. The inter-trimer contact is exclusively mediated by the α-subunits, which are linked by an intricate network of polar bonds. The inter-hexamer contacts within the crystal (and in the native PBS rods) are made by the β-subunits (Schirmer et al., 1987).

3.2.1.3. Oligomeric aggregates: spectral properties and energy transfer

The spectral properties, absorption strength and quantum yield of fluorescence of biliproteins depend on the state of aggregation. The absorption spectrum of the (αβ) unit resembles the sum of the spectra of the constituent subunits, but the fluorescence quantum yield is somewhat higher. Upon trimer formation, the absorption is red-shifted and its strength and the quantum yield of fluorescence increased (Glazer et al., 1973; Mimuro et al.,

Figure 8. Stereodrawing of the polypeptide chain fold of a (αβ)₆ hexamer of PC seen along the disk axis (upper panel). The scheme (lower panel) indicates the packing of subunits in the hexamer seen from the side.

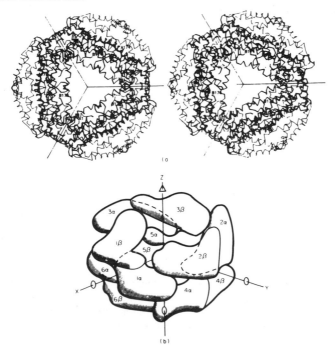

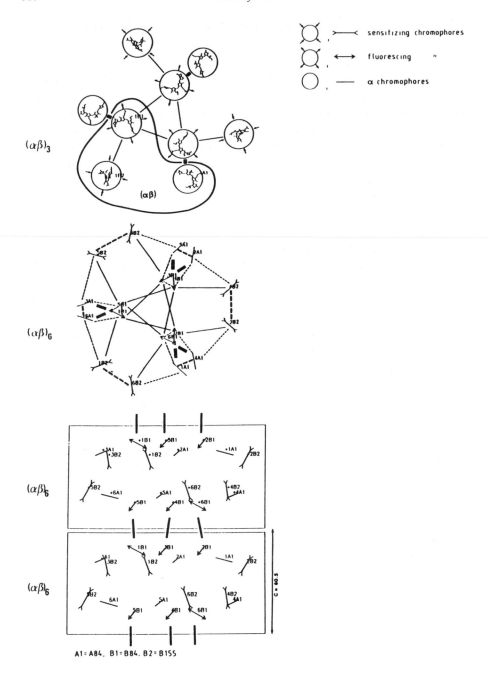

Figure 9a. Arrangement of chromophores and preferred energy transfer pathways in $(\alpha\beta)_3$ trimers, $(\alpha\beta)_6$ hexamers and stacked hexamers based on Table 10 in Schirmer el al., (1987). For the trimer the detailed structures of the chromophores are drawn, otherwise their approximate transition dipole directions are indicated. For the trimer and hexamer the view is along the disk axis; for the stacked hexamers it is perpendicular. In the stacked hexamers only the inter-hexamer transfers are indicated. The strength of coupling is indicated by the thickness of the connecting lines. Transfer paths within and between the trimers are represented by full and broken lines, respectively.

1986; for a review, see Glazer, 1985). In the $(\alpha\beta)_6$-linker complexes, the fluorescence is further increased and the absorption spectrum further altered (Lundell et al., 1981).

These observations can be rationalized by the structure of the aggregates. Formation of $(\alpha\beta)$ units causes little change in the environment of the chromophores. They remain quite separated with distances > 36 Å *(Figure 9a)*. Upon trimer formation, the environment of chromophore A84 changes profoundly by approach of chromophore B84 of a related unit (Figure 9a, upper panel). In the hexamer (Figure 9a, middle panel) the A84 and B155 chromophores interact pairwise strongly across the trimer interface. Also the molecular structures become more rigid with increasing size of the aggregates as seen in the crystals of the trimeric and hexameric aggregates (Schirmer et al., 1986, 1987). Rigidity hinders excitation relaxation by isomerization and thus increases the fluorescence quantum yield.

The chromophores can be divided into subsets of s (sensitizing) and f (fluorescing) chromophores (Teale & Dale, 1970; Zickendraht-Wendelstadt et al., 1980). The s-chromophores absorb at the blue edge of the absorption band and transfer the excitation energy rapidly to the f-chromophores. This transfer is accompanied by depolarization (Hefferle et al., 1983). Excitation at the red absorption edge (f-chromophores), however, results in little depolarization, suggesting that the energy is transferred along stacks of similarly oriented f-chromophores (Gillbro et al., 1985). The assignment of the chromophores to s and f was made by steady-state spectroscopy on different aggregates (Mimuro et al., 1986), by chemical modification guided by the spatial structure (Siebzehnrübl et al., 1987) and conclusively by measurement of linear dichroism and polarized fluorescence in single crystals (Schirmer & Vincent, 1987). Accordingly B155 is the s-, B84 the f-, and A84 the intermediate chromophore.

Light energy is transferred rapidly within 50 to 100 psec from the tips of the PBS to the core (for a review, see e.g. Glazer, 1985; Porter et al., 1978; Searle et al., 1978; Wendler et al., 1984; Yamazaki et al., 1984; Gillbro et al., 1985; Holzwarth, 1986). The transfer times from the periphery to the base are several orders of magnitude faster than the intrinsic fluorescence life-times of the isolated components (Porter et al., 1978; Hefferle et al., 1983). The distances between the chromophores within and between the hexamers are too large for strong (excitonic) coupling, but efficient energy transfer by inductive resonance occurs. A Förster radius of about 50 Å has been suggested by Grabowski & Gantt (1978). The relative orientations and distances of the chromophores as obtained by Schirmer et al. (1987) were the basis for the calculation of the energy transfer rates in Figure 9a. It shows the preferred energy transfer pathways in $(\alpha\beta)$ units, $(\alpha\beta)_3$ trimers, $(\alpha\beta)_6$ hexamers and stacked disks as models for native antenna rods. There is very weak coupling of the chromophores in the $(\alpha\beta)$ units. Some energy transfer takes place, however, as indicated by steady-state polarization measurements (Switalski & Sauer, 1984; Mimuro et al., 1986) probably between B155 and B84. Trimer formation generates strong coupling be-

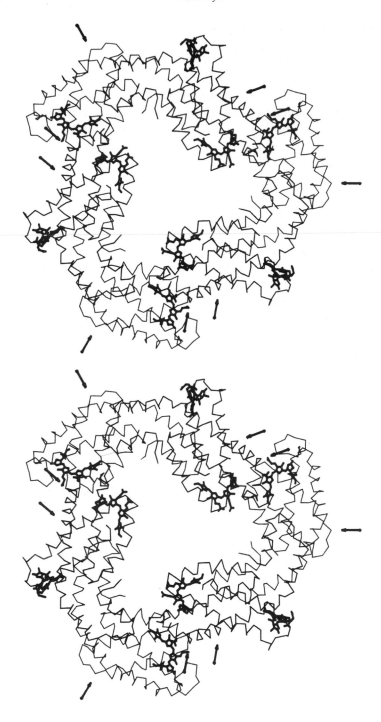

Figure 9b. Model of PE (αβ)₃ on the basis of PC with the locations of the additional phycoery-throbilins indicated by arrows.

tween A84 and B84, but B155 is integrated only weakly. In the hexamer many additional transfer pathways are opened and B155 is efficiently coupled. Hexamers are obviously the functional units, as the energy can be distributed and concentrated on the central f-chromophores, which couple the stacks of hexamers. Kinetic studies (Glazer et al., 1985; Gillbro et al., 1985; Holzwarth, 1985; Mimuro et al., 1986) have confirmed the picture of energy transfer along the rods as a random walk (trap or diffusion limited) along a one-dimensional array of f-chromophores. Sauer et al. (1987) have successfully simulated the observed energy transfer kinetics in PC aggregates on the basis of the structures using Förster's mechanism. The PEC component at the tips of PBS rods is extremely similar to PC (Duerring, 1989; Duerring et al., 1989). Its short wavelength absorbing chromophore A84 is located at the periphery *(Figure 10a)* as are the additional chromophores in PE which is also a tip component *(Figure 9b)*.

The phycobilisome rods act as light collectors and energy concentrators from the peripheral onto the central chromophores, that is, as excitation energy funnels from the periphery to centre and from the tip to the bottom.

We may expect functional modulations by the linker polypeptides. Some of them are believed to be located in the central channel of the hexamers, where they may interact with B84.

3.2.2. *Electron transfer in the reaction centre*[*]

3.2.2.1. Reaction centre, composition[**]

The RC of *Rps. viridis* is a complex of four protein subunits, C, L, M, H and cofactors arranged as in *Figure 10a*. As shown by the amino acid sequence they consist of 336, 273, 323, and 258 residues, respectively (Michel et al., 1985; Michel et al., 1986a; Weyer et al., 1987). The c-type cytochrome has four haem groups covalently bound via thioether linkages. The cofactors are four BChl-b (BC_{MP}, BC_{LP}, BC_{LA}, BC_{MA}), two BPh-b (BP_M, BP_L), one menaquinone-9 (Q_A), and a ferrous iron involved in electron transfer. A second quinone (ubiquinone-9) (Q_B), which is a component of the functional complex, is partially lost during preparation and crystallization of the RC.

3.2.2.2. Chromophore arrangement and electron transfer

The chromophores are arranged in L- and M-branches related by an axis of approximately two-fold symmetry which meet at BC_P (Deisenhofer et al., 1984). This axis is normal to the plane of the membrane.

While many of the optical properties of the pigment system are rather well understood on the basis of the spatial structure (Knapp et al., 1985), electron transfer is less well understood. The excited BC_P is quenched by electron transfer to BP_L in 3 psec and further on to the primary acceptor Q_A in about 200 psec, driven by the redox potential gradient between

---

[*] A historical background of the development of concepts and key features of the purple bacterial reaction centre is given by Parson (1978).

[**] The arrangement of the reaction centre in the thylakoid membranes of *Rps. viridis* as obtained by electron microscopy is described by Stark et al. (1984).

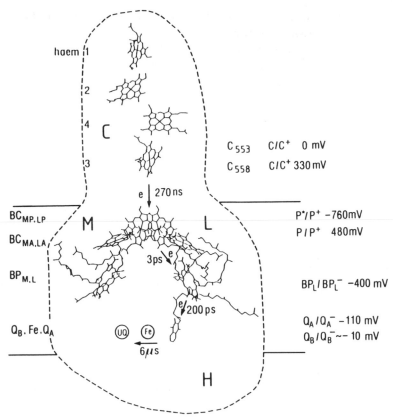

haem 1

2

4

C

3

C_{553} C/C^+ 0 mV
C_{558} C/C^+ 330 mV

e ↓ 270 ns

$BC_{MP.LP}$ M   L
$BC_{MA.LA}$

$BP_{M.L}$

$Q_B . Fe . Q_A$

3 ps ↘ e

e ↓ 200 ps

(UQ)   (Fe)

6 μs

H

P^*/P^+ −760 mV
P/P^+ 480 mV

BP_L/BP_L^- −400 mV

Q_A/Q_A^- −110 mV
Q_B/Q_B^- ∼− 10 mV

Figure 10a. Scheme of the structure of the RC cofactor system, the outline of the protein subunits (C, M, L, H), the electron transfer $t_{1/2}$ times, and the redox potentials of defined intermediates (for references see text section 3.2.2.2.).

P^*/P^+ (about -760 mV) and Q_A/Q_A^- (about -110 mV). The redox potential of BP^- is intermediate with about -400 mV (Cogdell and Crofts, 1972; Carithers and Parson, 1975; Prince et al., 1976; Netzel et al., 1977; Bolton, 1978; Holten et al., 1978; Woodbury et al., 1985; Breton et al., 1986). These functional data are summarized in Figure 10a. General factors controlling the transfer rates have been summarized in Table 1 and are detailed for the RC here:

Fast electron transfer requires effective overlap of the molecular orbitals. The orbital interaction decreases exponentially with the edge to edge distance of donor and acceptor and is insignificant at distances larger than about 10 Å (Kavarnos and Turro, 1986; McLendon, 1988). In the RC the distance between BC_P and Q_A is far too large to allow fast direct electron transfer; instead the electron migrates *via* BP_L. BP_L^- is a spectroscopically and kinetically well defined intermediate. Although located between BC_P and BP_L, BC_{LA}^- is not an intermediate but is probably involved in electron transfer by a "superexchange" mechanism mediating a strong quantum mechanical coupling (Fleming et al., 1988; for a review see Barber, 1988). The distance between BP_L and Q_A seems large for a fast transfer. Indeed the

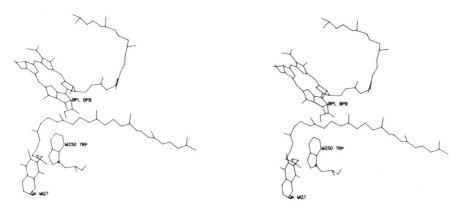

Figure 10b. Stereo drawing of the arrangement of BP_L, Trp M250 and Q_A in the L-branch of the RC pigment system.

gap is bridged by the aromatic side chain of Trp M250 in the L branch of the pigment system (*Figur 10b*) (Deisenhofer et al., 1985; Michel et al., 1986b), which might mediate coupling *via* appropriate orbitals. In addition, the isoprenoid side chain of Q_A is close to BP_L. Electron transfer via long connecting chains by through-bond coupling of donor and acceptor orbitals has been observed (Pasman et al., 1982; Moore et al., 1984; Kavarnos & Turro, 1986) but there is only Van der Waals contact here.

A second important factor for electron transfer is the free energy change (ΔG), which is governed by the chemical nature of the components, by geometrical factors, and by the environment (solvent polarity). It depends on the ionization potential of the donor in its excited state, the electron affinity of the acceptor, and on the coulombic interaction of the radical ion pair, which is probably small as donor and acceptor (BC_P and Q_A in the RC) are far apart. The effect of the environment may be substantial by stabilizing the radical ion pair by ionic interactions and hydrogen bonds. ΔG is *a* determinant of the activation energy of electron transfer. Another is nuclear rearrangements of the reactants and the environment. As the charge on donor and acceptor develops the nuclear configurations change. These changes are likely to be small in the RC as the BChl-b macrocycles are relatively rigid and tightly packed in the protein and the charge is distributed over the extended aromatic electron systems. Reorientable dipolar groups (peptide groups and side chains) may contribute strongly to the energy barrier of electron transfer. A matrix with high electronic polarizability on the other hand stabilizes the developing charge in the transition state of the reaction and reduces the activation energy. An alternative picture is that the potential energy barrier to electron tunnelling is decreased. Aromatic compounds which are concentrated in the vicinity of the electron carriers in the RC have these characteristics (see Trp M250).

The electron transfer from P* to Q_A occurs with very low activation energy (Arnold and Clayton, 1960; Parson, 1974; Parson and Cogdell, 1975; Carithers and Parson, 1975; Bolton, 1978; Kirmaier et al., 1985; Woodbury et al., 1985) and proceeds readily at 1°K. Thermally activated processes, nuclear motions, and collisions are therefore not important for the initial very fast charge separation steps. There is even a slight increase in rate with temperature decrease either due to a closer approach of the pigments at low temperature, or to changes of the vibrational levels wich may lead.to a more favourable Franck-Condon factor.

The electron transfer between primary and secondary quinone acceptors, Q_A and Q_B is rather different from the previous processes, because it is much slower (about 6 µs at pH7, derived from Carithers and Parson, 1975) and has a substantial activation energy of about 8 kcal mol^{-1}. In *Rps. viridis* Q_A is a menaquinone-9 and Q_B a ubiquinone-9, which differ in their redox potentials in solution by about 100 mV. In other purple bacteria both Q_A and Q_B are ubiquinones. The redox potential difference required for efficient electron transfer in these cases is generated by the asymmetric protein matrix. The protein matrix is also responsible for the quite different functional properties of Q_A and Q_B. Q_A accepts only one electron (leading to a semiquinone anion), which is transferred to Q_B before the next electron transfer can occur. Q_B, however, accepts two electrons and is protonated to form a hydroquinone, which diffuses from the RC (two-electron gate (Wraight, 1982)). Q_B is close to Glu L212, which opens a path to the H-subunit and may protonate Q_B. The path between Q_A and Q_B is very different from the environment of the primary electron transfer compo-nents. The line connecting Q_A and Q_B (the Q_B binding site has been inferred from the binding mode of competitive inhibitors and ubiquinone-1 in *Rps. viridis* crystals (Deisenhofer et al., 1985)) is occupied by the iron and its five coordinating ligands, four histidine (M217, M264, L190, L230) and a glutamic acid (M232) residue. His M 217 forms a hydrogen bond to Q_A. His L 190 is close to Q_B. Q_A and Q_B have an edge to edge distance of about 15 Å which might explain the slow transfer. If electron transfer and protonation are coupled, the observed pH dependence of the electron transfer rate of Q_A to Q_B (Kleinfeld et al., 1985) could be explained and nuclear motions required for proton transfer may generate the observed activation energy barrier. The role of the charged Fe-His$_4$-Glu complex in the Q_A to Q_B electron transfer is poorly understood at present, as it also occurs in the absence of the iron (Debus et al., 1986). Its role seems to be predominantly structural.

The cycle of electron transfer is closed by rereduction of the BC_p^* from the cytochrome bridging a distance of about 11Å between pyrrole ring I of haem 3 and pyrrole ring II of BC_{LP}. The transfer time is 270 ns (Holten et al., 1978; elder measurements in Case et al., 1970), considerably slower than the initial processes. Tyr L162, which is located midway (*Figure 10c*) may facilitate electron transfer by mediating electronic coupling between the widely spaced donor and acceptor. The biphasic temperature depen-dance indicates a complex mechanism in which at high temperatures nucle-ar motions play a role (for a review see, Dutton & Prince, 1978; DeVault & Chance, 1966).

The favourable rate controlling factors discussed are a necessary, but not sufficient condition for electron transfer, which competes with other

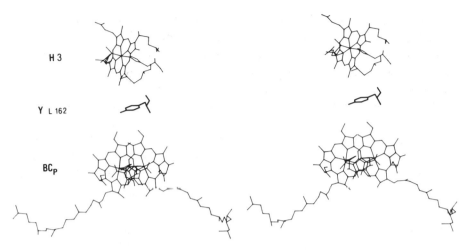

H 3

Y L 162

BC$_p$

Figure 10c. Stereo drawing of haem 3 (H3) of the cytochrome c, the special pair BC$_p$ and the intercalating Tyr L162 (Y L162) of the L-subunit (Deisenhofer et al. 1985). The His and Met ligands to the iron of H3 and the His ligands to the magnesium ions of the BC$_p$ are also shown.

quenching processes summarized in Table 2 and detailed for the RC:

Energy transfer from P* back to the LHC or to other pigments may be favourable from orientation and proximity considerations but is disfavoured for energetic reasons. The special pair absorbs usually (but not in *Rps. viridis* where the maximal absorption of the RC and the LHC are at 960 and 1020 nm respectively, see *Figure 7*) at longer wavelengths than other pigments of the photosynthetic apparatus and represents the light energy sink. The natural radiative lifetime of the excited singlet state P* is around 20 ns (Slooten, 1972; Parson and Cogdell, 1975) and may serve as an estimate of the times involved in the other wasteful quenching processes. Clearly electron transfer is much faster. Non-radiative relaxation of BC$_p$* by isomerizations and conformational changes is unlikely for the cyclic pigment systems tightly packed in the protein matrix.

The back reaction $P^+ Q_A^-$ to $P Q_A$ has a favourable driving force (*Figure 10a*) and proceeds independent of temperature, but is slow and insignificant under physiological conditions (for a review see, Bolton, 1978). The physical basis for this has yet to be explained. It may be related to a gating function of BP$_L$ by its negative redox potential compared to Q_A, to electronic properties of P^+ disfavouring charge transfer, and to conformational changes induced by electron transfer.

The profound influence of the protein matrix on electron transfer in the RC is obvious in the observed asymmetry of electron transfer in the two branches of the BChl-b and BPh-b pigments. Only the branch more closely associated with the L-subunit is active. An explanation is offered by the fact that the protein environment of both branches, although provided by homologous proteins (L and M), is rather different in particular by the Trp M250 located between BP$_L$ and Q_A and the numerous differences in the Q_A and Q_B binding sites (Deisenhofer et al., 1985; Michel et al., 1986b).

Asymmetry is observed in the BC_p due to different distortions and hydrogen bonding of the macrocycles and in the slightly different spatial arrangements of the BC_A and BP. It is suggested to facilitate electron release into the L-branch (Michel-Beyerle et al., 1988). The M-branch may have influence as a pendant group though.

The protein matrix also serves to dissipate the excess energy of about 650 mV (Prince et al., 1976) of the excited special pair (P* Q_A) over the radical ion pair $P^+Q_A^-$. These processes are probably very fast.

In summary, the very fast electron transfer from BC_p^* to Q_A occurs between closely spaced aromatic macrocycles with matched redox potentials. The protein matrix in which the pigments are tightly held is lined predominantly with apolar amino acid side chains with a high proportion of aromatic residues. The electron path is removed from bulk water.

3.2.3. *The blue oxidases*

Oxidases catalyse the reduction of dioxygen in single electron transfers from substrates. Dioxygen requires 4 electrons and 4 protons to be reduced to two water molecules. Oxidases must provide recognition sites for the two substrates, a storage site for electrons and/or means to stabilize reactive partially reduced oxygen intermediates (Malmström, 1978, 1982; Farver & Pecht, 1984)

The 'blue' oxidases are classified corresponding to distinct spectroscopic properties of the three types of copper which they contain: Type-1 Cu^{++} is responsible for the deep blue color of these proteins; type-2 or normal Cu^{++} has undetectable optical absorption; type-1 and type-2 cupric ions are paramagnetic; type-3 copper has a strong absorption around 330 nm and is antiferromagnetic, indicating coupling of a pair of cupric ions. The characteristic optical and electron paramagnetic resonance spectra disappear upon reduction.

Studies of the catalytic and redox properties of the 'blue' oxidases are well documented in several recent reviews (e.g. for laccase, Reinhammar, 1984; for ascorbate oxidase, Mondovì & Avigliano, 1984; for ceruloplasmin, Rydén, 1984). Basically type-1 Cu^{++} is reduced by electron transfer from the substrate. The electron is transferred on to the type-3 and type-2 copper ions. The second substrate, dioxygen, is associated with the type-3 and/or type-2 copper ions.

3.2.3.1. Ascorbate oxidase, composition and copper arrangement

Ascorbate oxidase is a polypeptide of 553 amino acid residues folded into three tightly associated domains (Messerschmidt et al., 1989). It is a dimer in solution, but the functional unit is the monomer. It belongs to the group of 'blue' oxidases together with laccase and ceruloplasmin (Malkin & Malmström, 1970).

Structures of copper proteins containing only one of the different copper types are known: Plastocyanin has a 'blue' type-1 copper, which is coordinated to two histidine residues and the sulfur atoms of cysteine and methionine as a distorted tetrahedron (Guss & Freeman, 1983). Cu-Zn-superoxide dismutase contains a type-2 copper, which has 4 histidine ligands with slightly distorted quadratic coordination (Richardson et al., 1975). Hemocyanin of *Panulirus interruptus* has type-3 copper, a pair of copper ions 3.4 Å apart with 6 histidine ligands (Gaykema et al., 1984).

In domain 3 of ascorbate oxidase (see section 4.4.) a copper ion is found in a strongly distorted tetrahedral (approaching trigonal pyramidal geometry) coordination by the ligands His, Cys, His, Met as had been shown in *Figure 6*. It resembles the blue type-1 copper in plastocyanin. Between domain 1 and domain 3, a trinuclear copper site is enclosed and shown in *Figure 11a*. Four (-His-X-His-) amino acid sequences provide the eight histidine ligands. The trinuclear copper site is subdivided in a pair of coppers (Cu31, Cu32) with 2×3 histidyl (A108, A451, A507; A64, A106, A509) ligands forming a trigonal prism. It represents the type-3 copper pair, as a comparable arrangement is observed in hemocyanin. The remaining copper (Cu 2) has two histidyl ligands (A62, A449). It is type-2 copper. The trinuclear copper cluster is the site where dioxygen binds, but the structural details including the presence of additional non-protein ligands require clarification. The close spatial association of the three copper ions in the cluster suggests facile electron exchange. It may function as an electron storage site and cooperative three-electron donor to dioxygen, to irreversibly break the $O-O$ bond.

3.2.3.2. Intramolecular electron transfer in ascorbate oxidase

Electrons are transferred from the type-1 copper to the trinuclear site. The shortest pathway is *via* Cys-A508 and His-A507 or His-A509. The (His-X-His-) segment links electron donor and acceptor bridging a distance of 12 Å

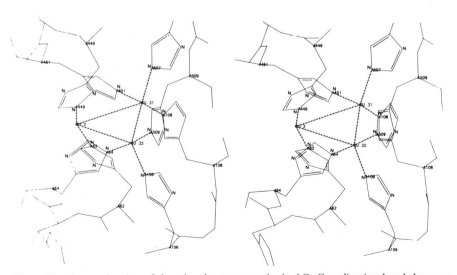

Figure 11a. Stereo drawing of the trinuclear copper site in AO. Coordination bonds between the copper ions and the protein residues are marked ($-\ -\ -\ -$) (Messerschmidt et al., 1989).

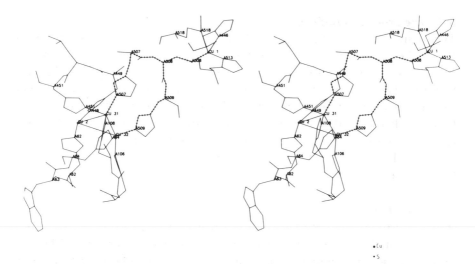

Figure 11b. Stereo drawing of the tridentate peptide ligand (-His 507-Cys 508-His 509-) bridging type-1 copper (Cu 1) and the trinuclear cluster (Cu 31, Cu 32, Cu 2) (Messerschmidt et al. 1989).

(*Figure 11b*). The cysteine sulfur and the imidazole components of the bridging ligand have low lying unoccupied molecular orbitals and may favour a chemical mechanism of electron transfer but the intervening aliphatic and peptide chains are unlikely to form transient radicals and may participate by resonance. The optical absorption of the blue copper assigned to a cysteine $S^- \rightarrow Cu^{2+}$ charge transfer transition supports the suggested electron pathway.

The putative electron path branches at the C^α atom of Cys A508. Model compounds have shown inequivalence and faster transfer in the N-C direction of amide linking groups (Schmidt et al., 1988). This may apply also to the blue oxidases and cause preferred transfer to A507.

The redox potential differences between the type-1 copper and the type-3 copper are -40 mV in ascorbate oxidase. Unfortunately, there are no direct measurements of the intramolecular electron transfer rates available. The turnover number serves as a lower limit and is 7.5×10^3 sec^{-1} in AO (Dawson, 1966; Gerwin et al., 1974) indicating a quite rapid transfer despite the long distance and small driving force. The electron pathway is intramolecular and removed from bulk water.

The characteristic distribution of redox centres as mono- and trinuclear sites in the blue oxidases may be found also in the most complex oxidase, cytochrome oxidase (see the hypothetical model of Holm et al., 1987) and in the water-splitting manganese protein complex of PS II, which carries out the reverse reaction of the oxidases. For its (Mn)$_4$ cofactor either two binuclear or a tetranuclear metal centre is favoured (Babcock, 1987), but mono- and trinuclear arrangements can not be excluded.

3.3. *The protein as medium*

The boundary between the action of the protein as ligand and as medium is fluid. The protein medium is microscopically extremely complex in structure, polarity and polarisability, which may influence energy and electron transfer. There is no obvious common structural scheme in the protein systems discussed except a high proportion of aromatic residues (particularly tryptophans) bordering the electron transfer paths in RC and AO and their wide separation from bulk water by internal location within the protein and the hydrocarbon bilayer (in RC). These effects have been mentioned in sections 1.3 and 3.2.2.2.

4. **Structural relationships and internal repeats**

All four protein systems mentioned show internal repetition of structural motifs or similarities to other proteins of known folding patterns. This is a quite common phenomenon and not confined to energy and electron transfer proteins. It is also not uncommon that these relationships often remained undetected on the basis of the amino acid sequences, ultimately a reflection of our ignorance about the sequence structure relationships. An analysis of structural relationships will shed light on evolution and function of the protein systems and is thus appropriate here.

4.1. *Retinol and bilin binding proteins*

The simplest case is shown in *Figure 12*, where BBP (Huber et al., 1987) is compared to RBP (Newcomer et al., 1984). The structural similarity is obvious for the bottom of the β-barrel structure, while the upper part which is involved in binding of the pigments, biliverdin and retinol, differs greatly. The molecule is apparently divided in framework and hypervariable segments which determine binding specificity in analogy to the immunoglobulins (Huber, 1984). The relationship suggests carrier functions for BBP as for RBP, although it serves also for pigmentation in butterflys.

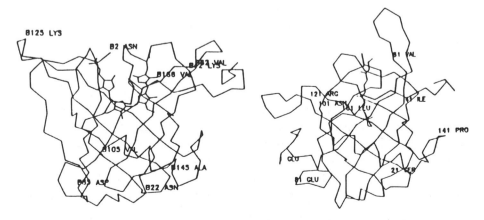

Figure 12. Comparison of the polypeptide chain folds of BBP and RBP with bound cofactors.

4.2. *Phycocyanin*

The PC consist of two polypeptide chains α and β which are clearly related in structure (*Figure 13*) and originate probably from a common precursor.

The α-subunit is shorter in the GH turn and lacks the s-chromophore B155 (see section 3.2.1.3.). The loss or acquisition of chromophores during evolution may be less important than differentiation of the α and β subunits, which occupy non-equivalent positions in the (αβ)₃ trimer, so that the homologous chromophores A84 and B84 are non-equivalent with B84 lying on the inner wall of the disk. In addition the α and β subunits play very different roles in the formation of the (αβ)₆ hexamer as had been shown in *Figure 8*. Symmetrical precursor hexamers might have existed and could

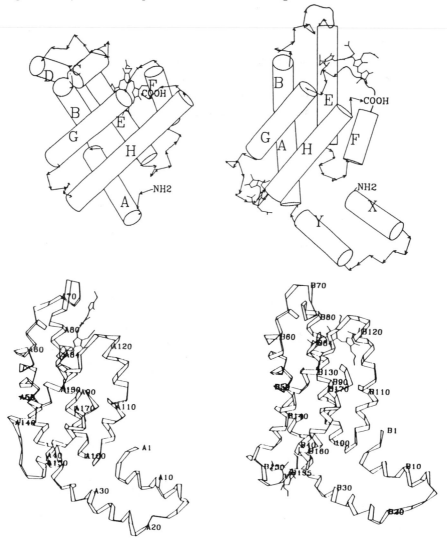

Figure 13. Polypeptide chain folds of the α- and β-subunits of phycocyanin (Schirmer el al. 1987) (lower part, left and right) and comparison of the arrangements of α-helices in myoglobin and phycocyanin (upper part, left and right).

have formed stacks, but would lack the differentiation of the chromophores, in particular the inequivalence and close interaction of A84 and
B84 in the trimer. Functional improvement has probably driven divergent
evolution of the α- and β-subunits.

A most surprising similarity was discovered between the PC subunits and
the globins shown in *Figure 13*. The globular helical assemblies A to H show
similar topology. The N-terminal X,Y α-helices forming a U-shaped extension in PC is essential for formation of the αβ substructure. The amino acid
sequence comparison after structural superposition reveals some homology
suggesting divergent evolution of phycobiliproteins and globins (Schirmer
et al., 1987), however, what function a precursor of light harvesting and
oxygen binding proteins might have had remains mysterious.

4.3. *Reaction centre*

The RC lacks symmetry across the membrane plane, not surprising for a
complex, which catalyses a vectorial process across the membrane. However, there is quasi-symmetry relating the L- and M subunits and the
pigment system. Structural similarity and amino acid sequence homology

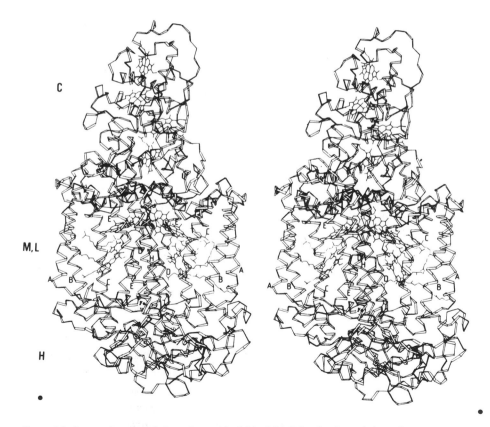

Figure 14. Stereo drawing of the polypeptide fold of the RC subunits and the cofactor system.
The membrane-spanning α-helices of the L- and M-subunits (A, B, C, D, E in sequential and A,
B, C, E, D in spatial order) and the H-subunit (H) are labelled (Deisenhofer el al., 1985).

between the L- and M-subunits suggest a common evolutionary origin. This relationship is extended to the PS II components D1 and D2 on the basis of sequence homology and conservation of residues involved in cofactor binding (for reviews, see Trebst, 1986; Michel & Deisenhofer, 1988). The putative precursor was a symmetrical dimer with identical electron transfer pathways. The interaction with the H subunit introduces asymmetry, particularly noteworthy at the N-terminal transmembrane α-helix of the H-subunit (H), which is close to the E transmembrane α-helix of the M-subunit and the L-branch of the pigment system and Q_A (*Figure 14*). The improvement of the interaction with the H-subunit, which appears to play a role in the electron transfer from Q_A to Q_B and in protonation of Q_B might have driven divergent evolution of the L- and M-subunits at the expense of the inactivation of the M pigment branch. However, the electron transfer from BC_p to Q_A is extremely fast and not rate-limiting for the overall reaction. The evolutionary conservation of the M branch of pigments may be of functional significance in light harvesting and electron transfer as a pendant group. There are also structural reasons, as its deletion would generate void space.

The cytochrome subunit adds to the asymmetry of the L-M complex and shows itself an internal duplication (Deisenhofer et al., 1985). All four heme groups are associated with a helix-turn-helix motif, but the turns are short for haem groups 1 and 3 and long for 2 and 4.

4.4. *Blue oxidases*

Gene multiplication and divergent evolution is most evident in the blue oxidase, ascorbate oxidase. *Figure 15* shows the polypeptide chain of 553 amino acid residues folded into 3 closely associated domains of similar topology (Messerschmidt et al., 1989). Although nearly twice as large, they resemble the simple, small copper protein plastocyanin (Guss & Freeman, 1983) *(Figure 16)*. In the blue oxidase domains I and III enclose the trinuclear copper cluster in a quasi-symmetrical fashion, but only domain III contains the type 1 copper, the electron donor to the trinuclear site. A potential electron transfer pathway in domain I is not realized, reminiscent of the M-branch of pigments in the RC. Similar to the H subunit in the RC, the linking domain II introduces asymmetry in AO, which might have driven evolutionary divergence of domains I and III.

The proteins plastocyanin, ascorbate oxidase, laccase, and ceruloplasmin are members of a family of copper proteins as indicated by structural relations and sequence homology (Messerschmidt et al., 1989; Ohkawa et al., 1988; Germann et al., 1988; Takahashi et al., 1984). They provide a record from which an evolutionary tree may be proposed *(Figure 17)*. The simplest molecule is plastocyanin containing only a type-1 copper. A dimer of plastocyanin-like molecules could provide the 2x4 histidyl ligands for the trinuclear copper cluster, representing a symmetrical oxidase. From this hypothetical precursor the modern blue oxidases and ceruloplasmin might have evolved following different paths of gene (domain) insertion and loss

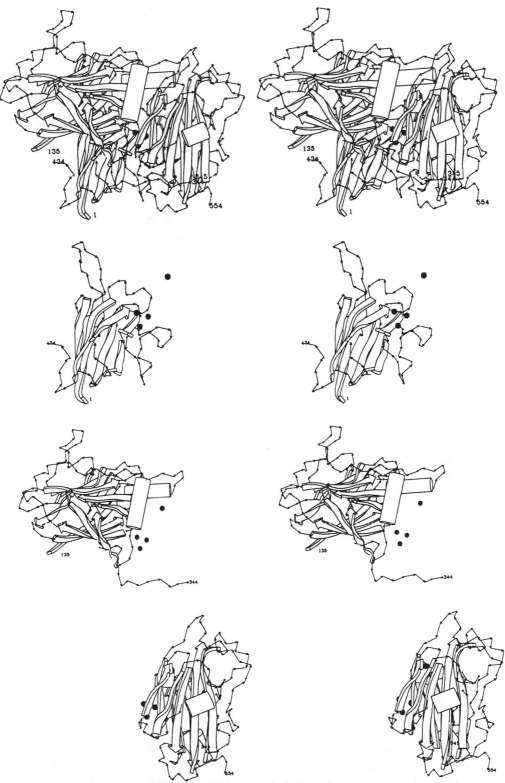

Figure 15. Stereo drawing of the polypeptide chain folds of AO and explosion view of its three domains from top to bottom (Messerschmidt et al. 1989). β-strands are indicated as arrows and α-helices as cylinders (produced by the plot program of Lesk & Hardman, 1982).

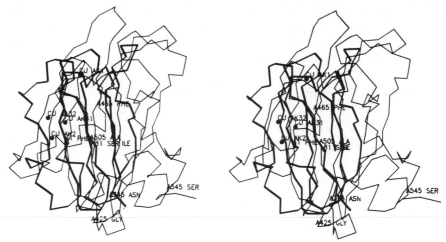

Figure 16. Stereo drawing and superposition of domain III of AO (thin lines) and PCY (thick lines). The trinuclear copper site in AO is buried between domain I (not shown) and domain III.

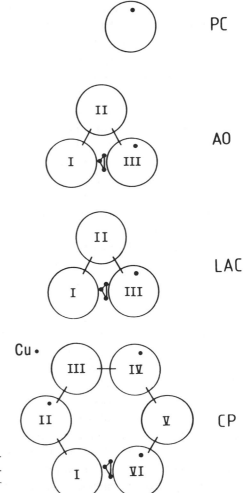

Figure 17. Homologous domains in plastocyanin, ascorbate oxidase, laccase and ceruloplasmin. The mono- and trinuclear copper sites are indicated.

or aquisition of coppers. In both the arrangement of the N- and C-terminal domains, which contain the functional copper cluster has been preserved. Recombinant DNA technology has the tools to reconstruct the hypothetical precursor oxidase. This is under investigation.

5. Implications from the structure of the reaction centre for membrane proteins in general

The structures of water soluble proteins show a seemingly unlimited diversity, although they are built from only a few defined secondary structural elements as helices, β-sheets and turns and despite their construction from domains and recurring structural motifs. The proteins discussed provide ample evidence. That there seems to be a limited set of basic folds may be related to the evolution of proteins from a basic set of structures and/or to constraints by protein stability and rates of folding. These basic folding motifs do not represent rigid building blocks, however, but adapt to sequence changes and respond to the environment and association with other structural elements. Adaptability and plasticity (which is not to be confused with flexibility) is related to the fact that the entire protein and solvent system must attain the global minimum, not its individual components. Water is a good hydrogen bond donor and acceptor and is thus able to saturate polar surface exposed peptide groups nearly as well as intraprotein hydrogen bonds do (except for entropic effects).

Membrane proteins face the inert hydrocarbon part of the phospholipid bilayer and must satisfy their hydrogen bonds intramolecularly. Only two secondary structures form closed hydrogen bonding arrangements of their main chains, which satisfy this condition, namely the helix and the β-barrel. For assemblies of α-helices packing rules have been derived which predict certain preferred angles between the helix axes although with a broad distribution. Similarly, the arrangement of strands in β-sheets and β-barrels follows defined rules (Chothia, 1984).

5.1. *Structure of the membrane associated parts of the RC*

The structure of the RC may support some conclusions about membrane proteins in general, of which the RC structure was the first to be determined at atomic resolution after the low resolution structure of bacteriorhodopsin which has some common features (Henderson & Unwin, 1975). The RC has 11 transmembrane α-helices, which consist of 26 residues (H-subunit) or $24-30$ residues (L- and M-subunits) appropriate lengths to span the membrane. The amino acid sequences of these segments are devoid of charged residues *(Figure 18)*. Few charged residues occur close to the ends of the α-helices. Glycine residues initiate and terminate almost all α-helical segments, both the transmembrane and the connecting α-helices. It is well known from soluble proteins that glycine residues are abundant in turns and often associated with flexible regions of proteins (Bennett and Huber 1984). They may be important for the insertion into the membrane by allowing rearrangements. The angles between the axes of the contacting

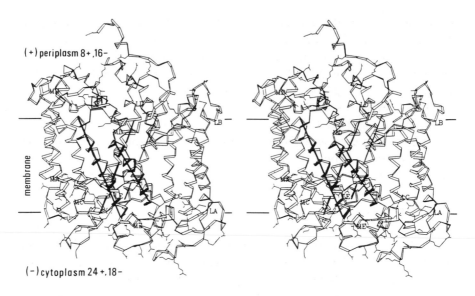

(+) periplasm 8+ ,16–

membrane

(–) cytoplasm 24 +,18–

Figure 18. Stereo drawing of the polypeptide chains of the L- and M-subunits of the RC in ribbon representation. The N-terminal residues of the membrane-spanning α-helices are labelled (including the prefix M and L) and the tetra-helical motif of the D and E α-helices is marked by shading and lines. The side chains of charged residues are drawn. Asp, Glu, and carboxy-termini as negatively charged, and Lys, Arg and amino-termini are counted as positively charged and added for the cytoplasmic and periplasmic sides (Deisenhofer el al., 1985, 1988).

α-helices of the L- and M-complex are inclined by 20° to 30°, a preferred angular range for the packing of the α-helices in soluble proteins. They have features in common with buried α-helices in large globular proteins, which are also characterized by the absence of charged residues and the preference of glycines and prolines at the termini (Loebermann et al., 1984; Remington et al., 1982). In addition, the D E α-helices of the L- and M-subunits (*Figure 18*) find counterparts in soluble proteins. They are associated around the local diad axis and form the centre of the LM module, which binds the iron and the BC_p. The four D and E α-helices of the L- and M-subunits are arranged as a bundle tied together by the iron ion and splay out towards the cytoplasmic side to accommodate the large special pair. This motif is quite common in soluble electron transfer proteins (Weber and Salemme, 1980). I will resume this discussion later and suggest appropriate substructures of soluble proteins as models for pore forming membrane proteins.

5.2. *Membrane insertion*

The structure of the RC is similarly important for our views of the mechanism of integration of membrane proteins into the phospholipid bilayer. The RC is composed of components which are quite differently arranged with respect to the membrane: The C-subunit is located on the periplasmic side. The H-subunit is folded into two parts: a globular part located on the cytoplasmic side and a transmembrane α-helix. The L- and M-subunits are incorporated into the phospholipid bilayer. Consequently, C has to be completely translocated across the membrane from its intracellular site of synthesis. In the H-, L-, and M-subunits the transmembrane α-helices are embedded in the bilayer. Only the N-terminal segment of H and the C-termini and connecting segments of the α-helices located at the periplasmic side of L and M (A − B, C − D) require transfer.

It is interesting to note that only the cytochrome gene possesses a prokaryotic signal sequence, as indicated by the sequence of the gene (Weyer et al., 1987). Transfer of the large hydrophilic C-subunit may require a complex translocation system and a signal sequence, while H, L, and M may spontaneously insert into the bilayer due to the affinity of the contiguous hydrophobic segments with the phospholipids (for a review of this and related problems see Rapoport, 1986). A "simple" dissolution still requires transfer across the membrane of those charged residues which are located at the periplasmic side (Deisenhofer et al., 1985; Michel et al., 1986b). The increasingly favourable protein lipid interaction which develops with insertion may assist in this process. M and L have considerably more charged residues at the cytoplasmic side (41) than at the periplasmic side (24), providing a lower activation energy barrier for correct insertion. The net charge distribution of the LM complex is asymmetric with 6 positive charges at the cytoplasmic side and 8 negative charges at the periplasmic side. As the intracellular membrane potential is negative, the observed orientation of the LM complex is energetically favoured (Figure 18).

The H-subunit has a very polar amino acid sequence at the C-terminus of the transmembrane α-helix with a stretch of 7 consecutive charged residues (H33 − H39) (Deisenhofer et al., 1985; Michel et al., 1985) which may efficiently stop membrane insertion. Similarly, there are 3 to 11 charged residues in each of the connecting segments of the α-helices at the cytoplasmic side of the L- and M-subunits, which might stop the transfer of α-helices or α-helical pairs (Engelman et al., 1986). As an alternative to sequential insertion the L-, M-subunits may be inserted into the membrane as assembled protein pigment complexes, because they cohere tightly by protein protein and protein cofactor interactions.

5.3. *Models of pore forming proteins*

It is not obvious whether the structural principles observed in the RC apply also to 'pore' or 'channel' forming α-helical proteins. These could, in principle, elaborate quite complex structures within the aqueous channel (Lodish, 1988), but available evidence at low resolution for gap junction proteins (Milks et al., 1988) indicates in this specific case a simple hexameric arrangement of membrane spanning amphiphilic α-helices, whose polar sides face the aqueous channel.

Guided by the observation that rules for structure and packing of α-helices derived for soluble proteins apply also to the RC, we may derive models for membrane pore forming proteins from appropriate soluble protein substructures. The penta-helical pore seen at high resolution in the icosaedral multi-enzyme complex riboflavin synthase seems to be a suitable model (Ladenstein et al., 1988) *(Figure 19)*. 5 amphiphilic α-helices of 23 residues each are nearly perpendicular to the capsid surface. The coiled coil of α-helices has a right-handed twist and forms a pore for the putative import of substrates and export of products. They pack with their apolar sides against the central 4-stranded β-sheet of the protein, which mimics the hydrocarbon part of a phospholipid bilayer and project charged residues into the aqueous channel.

Similar modelling of membrane protein structures may be extended to another class of membrane proteins which have β-structures spanning the outer membrane, the bacterial porins (Kleffel et al., 1985). In soluble proteins β-barrels observed have 4 to 8 or more strands. The lower limit is determined by the distortion of regular hydrogen bonds. An upper limit may be given by the possible sizes of stable protein domains. A four-stranded β-barrel with 4 parallel strands duplicated head to head with symmetry D4 is seen in the ovomucoid octamer (Weber et al., 1981). The β-strands lean against the hydrophobic core of the molecule and project their (short) polar residues into the channel (which is extremely narrow here).

6. Some thoughts on the future of protein crystallography

Thirty years after the elucidation of the first protein crystal structures by Perutz and Kendrew and after steady development, protein crystallography is undergoing a revolution. Recent technical and methodical developments enable us to analyse large functional protein complexes like the RC (Deisenhofer et al., 1985; Allen et al., 1987), large virus structures (to mention only Harrison et al., 1978; Rossmann et al., 1985; Hogle et al., 1985), protein DNA complexes (to mention only Ollis et al., 1985), and multi-enzyme complexes like riboflavin synthase (Ladenstein et al., 1988).

The significance of these studies for understanding biological functions is obvious and has excited the interest of the scientific community in general.

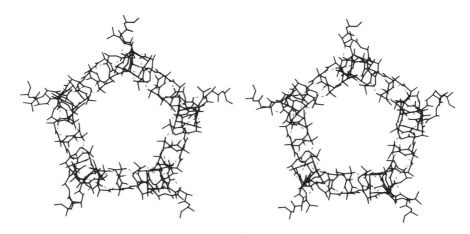

Figure 19. Penta-helical pore in heavy riboflavin synthase in stereo (Ladenstein et al., 1988).

In addition, it was recognized that detailed structural information is a prerequisite for rational design of drugs and proteins. For an illustration I chose human leucocyte elastase which is an important pathogenic agent. On the basis of its three-dimensional structure *(Figure 20)* (Bode et al., 1986) and the criteria of optimal stereochemical fit potent inhibitors are now being synthesized or natural inhibitors modified by use of recombinant DNA technology in many scientific and commercial institutions. Other, equally important proteins are similarly studied. This field especially benefits from the facile molecular modelling software (e.g. FRODO, Jones, 1978) and a standard and depository of structural data, the Protein Data Bank (Bernstein et al., 1977).

Success *and* the new technical and methodical developments spur protein crystallography's progress. These new developments are indeed remarkable: Area detectors for automatic recording of diffraction intensities have been designed. Brilliant X-ray sources (synchrotrons) are available for very fast measurements and now permit use of very small crystals or radiation sensitive materials. Their polychromatic radiation is used to obtain diffraction data sets within milliseconds by Laue techniques (Hajdu et al., 1988) and their tunability allows the optimal use of anomalous dispersion effects (Hendrickson et al., 1988; Guss et al., 1988).

Refinement methods including crystallographic and conformational energy terms provide improved protein models. Methods which allow the analysis of large protein complexes with internal symmetry averaging procedures were developed (Bricogne, 1976) leading from blurred to remarkably clear pictures. A priori information of a relationship to known proteins can be used to great advantage as it is possible to solve an unknown crystal structure using a known model of a variant structure by a method discovered and named the 'Faltmolekül' method by my teacher, W. Hoppe. It has become a very powerful tool in protein crystallography.

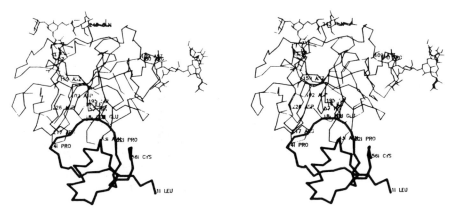

Figure 20. Stereo drawing of the complex between human leucocyte elastase (thin lines) and turkey ovomucoid inhibitor (thick lines) (Bode el al., 1986).

With this last paragraph I wish to pay tribute to W. Hoppe who in 1957 laid the foundation to Patterson search methods by discovering that the Patterson function (the Fourier transform of the diffraction intensities) of molecular crystals can be decomposed into sums of intra- and intermolecular vector sets (Hoppe, 1957) from which orientation and translation of the molecules can be derived when their approximate structure is known *(Figure 21)*. Hoppe's method was profoundly elaborated, computerized, and reformulated (Rossmann & Blow, 1962; Huber, 1965; Crowther & Blow, 1967). It provided a short-cut to the crystal structure of the RC of *Rb. sphaeroides* which was solved on the basis of the molecular structure of the RC of *Rps. viridis* and subsequently refined (Allen et al., 1986, 1987). The molecular architectures are very similar although the *Rb. sphaeroides* RC lacks the permanently bound cytochrome. The structure solution was independently confirmed using similar methods by Chang et al., (1986). With the Faltmolekül method the orientation and location of a molecule in a crystal cell can be determined. The detailed molecular structure and its deviations from the parent model have to be worked out by crystallographic refinement, to which W. Steigemann and J. Deisenhofer (in his thesis work) laid a foundation in my laboratory (Huber et al., 1974; Deisenhofer & Steigemann, 1975).

Recently, NMR techniques (nuclear magnetic resonance) have demonstrated their capability to determine three-dimensional structure of small proteins in solution. In one case, a detailed comparison between crystal and solution structure has shown very good correspondance (Kline et al., 1986; Pflugrath et al., 1986) but future developments will be needed to extend the power of the method to larger protein structures.

Protein crystallography is the only tool to unravel in detail the architecture of the large protein complexes described here and will continue in the foreseeable future to be the only experimental method that provides atomic resolution data on atom-atom and molecule-molecule interactions. It is the successful analytic method E. Fischer addressed in his 9th Faraday Lecture by pointing out that 'the precise nature of the assimilation process... will only be accomplished when biological research, aided by improved analytical methods, has succeeded in following the changes which take place in the actual chlorophyll granules' (Fischer, 1907). Yet an ultimate goal for which we all struggle is the solution of the folding problem. The growing number of known protein structures and the design of single residue variants by recombinant DNA technology and their analysis by protein crystallography has brought us nearer to this goal. We are able to study contributions of individual residue to rates of folding, structure, stability, and function. Also theoretical analysis of protein structures has progressed (to mention only Levitt & Sharon, 1988) but a clue to the code relating sequence and structure is not in sight (Jaenicke, 1988). Like Carl von Linné who 250 years ago created a system of plants on the basis of morphology (Genera plantarum, Leiden 1737), we classify proteins by their shapes and structures. Whether this may lead to a solution of the folding problem is unclear, but it

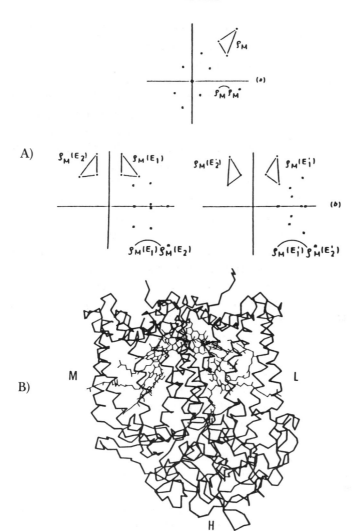

Figure 21. Faltmolekül construct A) $\rho_M\rho_M^*$ (a) and ρ_M (E1) ρ_M^* (E2) (b)
are the intra- and intermolecular vector sets of a triangular structure ρ_M, respectively. Their sum
represents the Patterson function. The intramolecular vector set can be constructed from the
molecular structure, is located at the origin, and permits the determination of the orientation.
From the intermolecular vector set the translation component relative to the mirror line can be
derived. In (b) the intermolecular vector sets corresponding to two different orientations of ρ_M
are shown (Huber, 1985).

B)
Drawing of the main chain of the M-, L-, H-subunits and the cofactors which served as search
model to solve the phase problem for the *Rb. sphaeroides* RC crystal structure. For the calculation all homologous main chain and side chain atoms were included (Allen et al., 1986).

is certain that the end of protein crystallography will only come through protein crystallography.

ACKNOWLEDGEMENT

J. Deisenhofer's and my interest in structural studies of the photosynthetic reaction centre of *Rps. viridis* was raised by the establishment of D. Oesterhelt's department in Martinsried in 1980; he brought with him H. Michel, with whom a fruitful collaboration on the analysis of the crystal structure of this large protein complex began. Later other members of my group, O. Epp and K. Miki, became involved. We had been studying enzymes, proteases and their natural inhibitors, immunoglobulins and had developed methods to improve data collection, electron density map interpretation and crystallographic refinement. The tools were available to attack a problem which was and still is the largest asymmetric protein analysed at atomic resolution today.

The "heureka" moment of protein crystallography is at the very end when one sees for the first time a new macromolecule with the eyes of a discoverer of unknown territories. To reach this moment much, sometimes tedious, work has to be done with the ever present possibility of failure. I am deeply grateful to my collaborators, those who are with me and those who had left, for their dedicated and patient work over many years. I mention by name those involved in the studies of the light harvesting cyanobacterial proteins and the blue oxidases: W. Bode, M. Duerring, R. Ladenstein, A. Messerschmidt, T. Schirmer. These projects were collaborative undertakings with biochemists in Switzerland (H. Zuber, W. Sidler), USA (M. L. Hackert) and Italy (M. Bolognesi, A. Marchesini, A. Finazzi-Agro).

Scientific work needs a stimulating environment which was provided at the Max-Planck-Institut für Biochemie and it needs steady financial support, which was provided by the Max-Planck-Gesellschaft and the Deutsche Forschungsgemeinschaft.

I thank R. Engh, S. Knof, R. Ladenstein, M. Duerring, E. Meyer for their helpful comments on this manuscript.

LITERATURE

Allen, J. P., Feher, G., Yeates, T. O., Rees, D. C., Deisenhofer, J., Michel, H., Huber, R. (1986) Proc. Natl. Acad. Sci. USA *83*, 8589–8593.

Allen, J. P., Feher, G., Yeates, T. O., Kemiya, H., Rees, D. C. (1987) Proc. Natl. Acad. Sci. USA *84*, 6162–6166.

Arnold, W., and Clayton, R. D. (1960), Proc. Natl. Acad. Sci. USA *46*, 769–776.

Babcock, G. T. (1987) Oxygen-Evolving Process in Photosynthesis (J. Amesz, ed.) Elsevier Science Publ.

Barber, J. (1987) Trends Biochem. Sci. *12*, 321–326.

Barber, J. (1988) Nature *333*, 114.

Barltrop, J. A. and Coyle, J. D. (1978) Principles of Photochemistry, John Wiley and Sons, Chichester.

Bernstein, F. C., Koetzle, T. F., Williams, G. J. B., Meyer, Jr., E. F., Brice, M. D.,

Rodgers, J. R., Kennard, O., Shimonouchi, T., Tasumi, M (1977) J. Mol. Biol. *112*, 535 – 542.

Bennett, W. S. and Huber, R. (1984), CRC Critical Reviews in Biochemistry *15*, 291 – 384.

Blair, D. F., Campbell, G. W., Schoonover, J. R., Chan, S. I., Gray, H. B., Malmström, B. G., Pecht, I., Swanson, B. F., Woodneff, W. H., Cho, W. K., English, A. R., Fry, A. H., Lum, V., and Norton, K. A. (1985) J. Amer. Chem. Soc. *10*, 5755 – 5766.

Bode, W., Wei, A., Huber, R., Meyer, E., Travis, P., Neumann, S. (1986) EMBO J. *5*, 2453 – 2458.

Bolton, J. R. (1978) in: The Photosynthetic Bacteria, (R. K. Clayton and W. R. Sistrom, eds.), Plenum Press, New York and London, pp. 419 – 442.

Boltzmann, L. (1886) Der Zweite Hauptsatz der mechanischen Wärmetheorie (Essay in Populäre Schriften) L. Boltzmann-Gesamtausgabe Bd. 7 (1919) Akad. Druck- und Verlagsanstalt Vieweg, Wiesbaden, pp 25 – 46 (English translation from Arnon, D. L. (1961) in: Light and Life (W. D. McElroy, B. Glass, eds.) The Johns Hopkins Press, Baltimore, pp. 489 – 569.

Breton, J. (1985) Biochim. Biophys. Acta *810*, 235 – 245.

Breton, J., Farkas, D. L., and Parson, W. W. (1985) Biochim. Biophys. Acta *808*, 421 – 427.

Breton, J., Martin, J.-L., Migus, A., Antonetti, A., and Orszag, A. (1986) Proc. Natl. Acad. Sci. USA *83*, 5121 – 5125.

Bricogne, G. (1976) Acta Crystallogr. A *32*, 832 – 847.

Bryant, D. A., Guglielmi, G., Tandeau de Marsac, N., Castets, A.-M. & Cohen-Bazire, G. (1979) Arch. Microbiol. *123*, 113 – 127.

Burkert, U., Allinger, N. L. (1982) Molecular Mechanics, American Chemical Society.

Calvin, M. and Bassham, J. A. (1962) in: The Photosynthesis of Carbon Compounds, Benjamin, New York, pp. 1 – 127.

Carithers, R. P., and Parson, W. W. (1975) Biochim. Biophys. Acta *387*, 194 – 211.

Case, G. D., Parson, W. W., and Thornber, J. P. (1970) Biochim. Biophys. Acta *223*, 122 – 128.

Chang, C.-H., Tiede, D., Tang, J., Smith, U., Norris, J., Schiffer, M. (1986) FEBS Lett. *205*, 82 – 86.

Chothia, C. (1984) Ann. Rev. Biochem. *53*, 537 – 572.

Cogdell, R. J. and Crofts, A. R. (1972) FEBS Lett. *27*,176 – 178.

Cohen-Bazire, G. & Briant, D. A. (1982) in: The Biology of Cyanobacteria (Carr, N. G. & Whitton, B. eds.), Blackwell, London, pp. 143 – 189.

Crowther, R. A. & Blow, D. M. (1967) Acta Cryst. *23*, 544 – 548.

Cramer, W. A., and Crofts, A. R. (1982) in: Electron and Proton Transport in Photosynthesis: Energy Conversion by Plants and Bacteria Vol 1, Academic Press Inc., pp. 387.

Dawson, C. R. (1966) in: The Biochemistry of Copper (Peisach, J., Aison, P. and Blumberg, W. E., eds.), Academic Press, New York, pp. 305 – 337.

Debus, R. J., Feher, G., and Okamura, M. Y. (1986) Biochemistry *25*, 2276 – 2287.

Deisenhofer, J., Huber, R., Michel, H. (1986) Nachr. Chem. Tech. Lab. *34*, 416 – 422.

Deisenhofer, J., Steigemann, W. (1975) Acta Cryst. B *31*, 238 – 280.

Deisenhofer, J., Epp, O., Miki, K., Huber, R., and Michel, H., (1984) J. Mol. Biol. *180*, 385 – 398.

Deisenhofer, J., Epp, O., Miki, K., Huber, R., and Michel, H., (1985) Nature *318*, 618 – 624.

Deisenhofer, J., Michel, H., and Huber, R. (1985a) Trends Biochem. Sci. *10*, 243 – 248.

Deisenhofer, J., Huber, R., Michel, H. (1989) in: Prediction of Protein Structure and the Principles of Protein Conformation (G. D. Fasman, ed.) Plenum Pub. Corp., New York, (in press.).

DeVault D., and Chance, B. (1966) Biophys. J. *6*, 825 – 847.

Duerring, M. (1988) Thesis, Technical University, München.

Duerring, M., Bode, W., Huber, R., Ruembeli, R., Zuber, H. (1989) to be submitted.

Duerring, M., Huber, R. and Bode W. (1988) FEBS Letters, *236*, 167 – 170.

Dutton, P. L., and Prince, R. C. (1978) in: The Photosynthetic Bacteria (R. K. Clayton and W. R. Sistrom, eds.), Plenum Press, New York and London, pp. 525 – 565.

Eberson, L. (1982) Adv. Phys. Org. Chem. *18*, 79 – 185.

Engelman, D. M., Steitz, T. A., and Goldman, A. (1986) Ann. Rev. Biophys. Chem. *15*, 321 – 353.

Farver, O., Pecht, I. (1984) in: Copper Proteins and Copper Enzymes (R. Lontie, ed.) *1*, CRC Press, Inc., Boca Raton, Florida, pp. 183 – 214.

Fischer, E. (1907) J. Chem. Soc. *91*, 1749 – 1765.

Fleming, G. R., Marti, J. L. and Breton, J. (1988) Nature *333*, 190 – 192.

Förster, T. (1948) Ann. Physik *2*, 55 – 75.

Förster, T. (1967) in: Comprehensive Biochemistry (M. Florkin, E. H. Stotz, eds.), Vol *22*, pp. 61 – 80, Elsevier, Amsterdam.

Frank, G., Sidler, W., Widmer, H. and Zuber, H. (1978) Hoppe-Seyler's Z. Physiol. Chem. *359*, 1491 – 1507.

Frommherz, P. and Reinbold, G. (1988) Thin Solid Films *160*, 347 – 353.

Gaykema, W. P. J., Hol, W. G. J., Verijken, J. M., Soeter, N. M., Bak, H. J., and Beintema, J. J. (1984) Nature *309*, 23 – 29.

Gantt, E., Lipschultz, C. A., Zilinskas, B. (1976) Biochim. Biophys. Acta *430*, 375 – 388.

Germann, U. A., Müller, G., Hunziker, P. E., Lerch, K. (1988) J. Biol. Chem. *263*, 885 – 896.

Gillbro, T., Sandström, Å., Sundström, V., Wendler, J. and Holzwarth, A. R. (1985) Biochem. Biophys. Acta *808*, 52 – 65.

Gerwin, B., Burstein, S. R., and Westley, J. (1974) J. Biol. Chem. *249*, 2005 – 2008.

Glazer, A. N., Fang, S. and Brown, D. M. (1973) J. Biol. Chem. *16*, 5679 – 5685.

Glazer, A. N. (1985) Ann. Rev. Biophys. Chem. *19*, 47 – 77.

Grabowski, J. and Gantt, E. (1978) Photochem. Photobiol. *28*, 39 – 45.

Gray, H. B., and Solomon, E. I. (1981) in: Copper proteins (T. G. Spiro, ed.) J. Wiley & Sons, New York, pp. 1 – 39.

Gray, H. B., and Malmström, B. G. (1983) Comments Inorg. Chem. *2*, 203 – 209.

Gray, H. B. (1986) Chem. Soc. Rev. *15*, 17 – 30.

Gust, D., Moore, T. A., Lidell, P. A., Nemeth, G. A., Makings, L. R., Moore, A. L., Barrett, D., Pessiki, P. J., Bensasson, R. V., Rougée, M., Chachaty, C., De Schryver, F. C., Van der Anweraer, M., Holzwarth, A. R., and Connolly, J. S. (1987) J. Amer. Chem. Soc. *109*, 846 – 856.

Guss, J. M., Freeman, H. C. (1983) J. Mol. Biol. *169*, 521 – 563.

Guss, J. M., Merritt, E. A., Phizackerly, R. P., Hedman, B., Murata M., Hodgson, K. O., Freeman, H. C. (1988) Science *241*, 806 – 811.

Hajdu, J., Acharya, K. R., Stuart, D. A., Barford, D., and Johnson, L. (1988) Trends Biochem. Sci. *13*, 104 – 109.

Hains, A. (1975) Acc. Chem. Res. *8*, 264 – 272.

Harrison, S. C., Olsson, A. J., Schutt, C. E., Winkler, F. K., Bricogne, G. (1978) Nature (London) *276*, 368 – 373.

Hendrickson, W. A., Smith, J. L., Phizackerly, R. P., Merritt, E. A. (1988) Proteins *4*, 77 – 88.

Hefferle, P., Nies, M., Wehrmeyer, W. and Schneider, S. (1983) Photobiochem. Photobiophys. *5*, 41 – 51.

Henderson, R. and Unwin, P. N. T. (1975) *Nature* 257, 28 – 32.

Higuchi, Y., Kusunoki, M., Matsuura, Y., Yasuoka, N., Kakudo, M. (1984) J. Mol. Biol. *172*, 109 – 139.

Hogle, J. M., Chow, M., Filman, D. J. (1985) Science *229*, 1358 – 1365.

Holm, L., Saraste, M., and Wikström, M. (1987) EMBO J. *6*, 2819 – 2823.

Holten, D., Windsor, M. W., Parson, W. W. and Thornber, J. P. (1978) Biochem. Biophys. Acta *501*, 112 – 126.

Holzwarth, A. R. (1985) in: Antennas and Reaction Centers of Photosynthetic Bacteria (Michel-Beyerle, M. E., ed.), Springer-Verlag, Berlin, pp. 45 – 52.

Holzwarth, A. R. (1986) Photochem. Photobiol. *43*, 707 – 725.

Hopfield, J. J. (1974) Proc. Natl. Acad. Sci. USA *71*, 3640 – 3644.

Hoppe, W. (1957) Acta Cryst. *10*, 750 – 751.

Huber, R. (1965) Acta Cryst. *19*, 353 – 356.

Huber, R. (1985) in: Molecular Replacement. Proceedings of the Daresbury Study Weekend (P. Machin, ed.) Daresbury Laboratory, pp. 58 – 61.

Huber, R., Kukla, D., Bode, W., Schwager, P., Bartels, K., Deisenhofer, J., Steigemann, W. (1974) J. Mol. Biol. *89*, 70 – 101.

Huber, R. (1984) in: Behring Institute Mitteilungen *76*, 1 – 14 (Gronski, P., Seiler, F. R., eds.), Die Medizinische Verlagsgesellschaft mbH, Marburg.

Huber, R., Schneider, M., Epp, O., Mayr, I., Messerschmidt, A., Pflugrath, J. and Kayser, H. (1987a) J. Mol. Biol. *195*, 423 – 434.

Huber, R., Schneider, M., Mayr, I., Müller, R., Deutzmann, R., Suter, F., Zuber, H., Falk, H. and Kayser, H. (1987b) J. Mol. Biol. *198*, 499 – 513.

Huber, R. (1988) Angew. Chem. Int. Ed. Engl. *27*, 79 – 88.

Isied, S., Vassilian, A., Magnuson, R., and Schwarz, H. (1985) J. Am. Chem. Soc. 107, 7432 – 7438.

Jaenicke, R. (1988) in: 39. Mosbacher Kolloquium (E.-L. Winnacker and R. Huber, eds.) pp. 16 – 36, Springer-Verlag, Berlin Heidelberg.

Jones, A. T. (1978) J. Appl. Cryst. *11*, 268 – 272.

Karplus, M., McCammon, J. A. (1981) CRC Crit. Rev. Biochem. *9*, 293 – 349.

Kavarnos, G. J. and Turro, N. J. (1986) Chem. Rev. *86*, 401 – 449.

Kebarle, P., and Chowdhury, S. (1987) Chem. Rev. *87*, 513 – 534.

Kirmaier, Ch., Holten, D., and Parson, W. W. (1985) Biochim. Biophys. Acta *810*, 33 – 48.

Kirmaier, Ch., Holton, D., and Parson, W. W. (1985). Biochem. Biophys. Acta *810*, 49 – 61.

Kleffel, B., Garavito, R. M., Baumeister, N., Rosenbusch, J. P. (1985) EMBO J. *4*, 1589 – 1592.

Kleinfeld, D., Okamura, M. Y., Feher, G. (1985) Biochem. Biophys. Acta 809, 291 – 310.

Kline, A. D., Braun, W., Wüthrich, K. (1986) J. Mol. Biol. *189*, 377 – 382.

Knapp, E. W., Fischer, S. F., Zinth, W., Sander, M., Kaiser, W., Deisenhofer, J., and Michel, H. (1985) Proc. Natl. Acad. Sci. USA *82*, 8463 – 8467.

Kuhn, H. (1970) J. Chem. Phys. *53*, 101 – 108.

Ladenstein, R., Schneider, M., Huber, R., Bartunik, H.-D., Wilson, K., Schott, K., Bacher, A. (1988) J. Mol. Biol. *203*, 1045 – 1070.

Levitt, M., Sharon, R. (1988) Proc. Natl. Acad. Sci. USA *85*, 7557 – 7561.

Lesk, A. M. and Hardman, K. D. (1982) Science *216*, 539 – 540.

Lodish, H. F. (1988) Trends Biochem. Sci. *13*, 332 – 334.

Loebermann, H., Tokuoka, R., Deisenhofer, J., Huber, R. (1984) J. Mol. Biol. *177*, 531 – 556.

Lumry, R., Eyring, H. (1954) J. Phys. Chem. US *58*, 110 – 2.

Lundell, D. J., Williams, R. C. and Glazer, A. N. (1981) J. Biol. Chem. *256*, 3580 – 3952.

MacColl, R., and Guard-Friar, D. (1987) Phycobiliproteins, CRC Press, Inc., Boca Raton, pp. 157 – 173.

Malkin, R. and Malmström, B. G. (1970) Adv. Enzymol. *33*, 177 – 243.

Malmström, B. G. (1978) New Trends, Bio-inorganic Chemistry, pp. 59 – 77, Academic Press.

Malmström, B. G. (1982) Ann. Rev. Biochem. *51*, 21 – 59.

Marcus, R. A. and Sutin, N. (1985) Biochim. Biophys. Acta *811*, 265 – 322.

Mayo, S. L., Ellis, W. R., Crutchley, R. J. and Gray H. B. (1986) Science *233*, 948 – 952.

McGoutry, J. L., Peterson-Kennedy, S. E., Ruo, W. Y. and Hoffman, B. M. (1987) Biochemistry *26*, 8302 – 8312.

McLendon, G. (1988) Acc. Chem. Res. *21*, 160 – 167.

Messerschmidt, A., Rossi, A., Ladenstein, R., Huber, R., Bolognesi, M., Gatti, G., Marchesini, A., Petruzzelli, T., Finazzi-Agrò, A. (1989) J. Mol. Biol. *206*, 513 – 530.

Michel, H., Weyer, K. A., Gruenberg, H., and Lottspeich, F. (1985) EMBO J. *4*, 1667 – 1672.

Michel, H., Epp, O., Deisenhofer, J. (1986b) EMBO J. *5*, 2445 – 2451.

Michel, H., Weyer, K. A., Gruenberg, H., Dunger, I., Oesterhelt, D. and Lottspeich, F. (1986a) EMBO J. *5*, 1149 – 1158.

Michel, H., Deisenhofer, J., (1988) Biochemistry *27*, 1 – 7.

Michel-Beyerle, M. E., Plato, M., Deisenhofer, J., Michel, H., Bixon, M., Jortner, J. (1988) Biochem. Biophys. Acta *932*, 52 – 70.

Mikkelsen, K. V., and Ratner, M. A. (1987) Chem. Rev. 87, 113 – 153.

Milks, L. C., Kumar, N. M., Houghten, R., Unwin, N., Gilula, N. B. (1988) EMBO J. *7*, 2967 – 2975.

Mimuro, M., Flüglistaller, P., Rümbeli, R. and Zuber, H. (1986) Biochim. Biophys. Acta *848*, 155 – 166.

Moore, T. A., Gust. D., Mathis, P., Bialoiq, J.-C., Chachaty, C., Bensasson, R. V., Land, E. J., Doizi, D., Liddell, P. A., Lehman, W. R., Nemeth, G. A., Moore, A. L. (1984) Nature *307*, 630 – 632.

Mörschel, E., Koller, K.-P., Wehrmeyer, W. & Schneider, H. (1977) Cytobiologie *16*, 118 – 129.

Mondovì, B. and Avigliano, L. (1984) in: Copper Proteins and Copper Enzymes (Lontie, L. ed.) *3*, CRC Press Inc., Boca Raton, Florida, pp. 101 – 118.

Netzel, T. L., Rentzepis, P. M., Tiede, D. M., Prince, R. C., and Dutton, P. L. (1977) Biochim. Biophys. Acta, *460*, 467 – 479.

Newcomer, M. E., Jones, T. A., Åqvist, J., Sundelin, J., Eriksson, U., Rask, I. & Peterson, P. A. (1984) EMBO J. *3*, 1451 – 1454.

Nies, M. and Wehrmeyer, W. (1981) Arch. Microbiol. *129*, 374 – 379.

Ohkawa, J., Okada, N., Shinmyo, A. and Takano, M. (1988) Proc. Nat. Acad. Sci. USA, submitted.

Ollis, D., Brick, P., Hamlin, R., Xuong, N. G., Steitz, T. A. (1985) Nature *313*, 762 – 766.

Parson, W. W. (1974) Ann. Rev. Microbiol. *28*, 41 – 59.

Parson, W. W., and Cogdell, R. J. (1975) Biochem. Biophys. Acta *416*, 105 – 149.

Parson, W. W. (1978) in: The Photosynthetic Bacteria (R. K. Clayton, W. R. Sistrom, eds.) Plenum Press, New York, London, pp. 317 – 322.

Parson, W. W., Scherz, A., Warshel, A. (1985) in: Antennas and Reaction Centers of Photosynthetic Bacteria (M. E. Michel-Beyerle, ed.) Springer Verlag, Berlin, pp. 122 – 133,

Pasman, P., Rob, F., Verhoeven, J. W. (1982) J. Am. Chem. Soc. *104*, 5127 – 5133.

Pflugrath, J. W., Wiegand, W., Huber, R. & Vertesy, L. (1986) J. Mol. Biol. *189*, 383 – 386.

Pierrot, M., Haser, R., Frey, M., Payan, F. & Astier, J.P. (1982) J. Biol. Chem. *257*, 14341—14348.

Porter, G., Tredwell, C.J., Searle, G. F. W. and Barber, J. (1978) Biochim. Biophys. Acta *501*, 232—245.

Prince, R.C., Leigh, J.S., and Dutton, P.L. (1976) Biochim. Biophys. Acta *440*, 622—636.

Prince, R.C. (1988) Trends Biochem. Sci. *13*, 286—288.

Rapoport, T.A. (1986) CRC Critical Rev. Biochem. *20*, 73—137.

Reinhammar, B. (1984) in: Copper Proteins and Copper Enzymes (Lontie, L. ed.) Vol. 3, pp. 1—35, CRC Press Inc., Boca Raton, Florida.

Remington, S., Wiegand, G., Huber, R. (1982) J. Mol. Biol. *158*, 111—152.

Richardson, J.S., Thomas, K.A., Rubin, B.H., and Richardson, D.C. (1975) Proc. Natl. Acad. Sci. USA *72*, 1349—1353.

Rossmann, M.G. and Blow, D.M. (1962) Acta Cryst. *15*, 24—31.

Rossmann, M.G., Arnold, E., Erickson, J.W., Frankenberger, E.A., Griffith, J.P., Hecht, H.-J., Johnson, J.E., Kamer, G., Luo, M., Mosser, A. G., Rueckert, R., Sherry, B., and Vriand, G. Nature (London) *317*, 145—153.

Rydén, L. (1984) in: Copper Proteins and Copper Enzymes (Lontie, L. ed.) *3*, CRC Press Inc., Boca Raton, Florida, pp. 34—100.

Sauer, K., Scheer, H., and Sauer, P. (1987) Photochem. Photobiol. *46*, 427—440.

Scharnagl, C., Köst-Reyes, E., Schneider, S., Köst, H.-P., Scheer, H. (1983) Z. Naturforsch. *38c*, 951—959.

Scheer, H. (1982) in: Light Reaction Path of Photosynthesis (Fong, F. K., ed.) Springer, Berlin, pp. 7—45.

Schirmer, T., Bode, W., Huber, R., Sidler, W. and Zuber, H. (1985) J. Mol. Biol. *184*, 257—277.

Schirmer, T., Huber, R., Schneider, M., Bode, W., Miller, M. and Hackert, M. L. (1986) J. Mol. Biol. *188*, 651—676.

Schirmer, T., Bode, W., and Huber, R. (1987) J. Mol. Biol. *196*, 677—695.

Schirmer, T. & Vincent M. G. (1987) Biochem. Biophys. Acta *893*, 379—385.

Schmidt, J. A., McIntosh, A. R., Weedon, A. C., Bolton, J. R., Connolly, J. S., Hurley, J. K., and Wasielewski, M. R. (1988) J. Amer. Chem. Soc. *110*, 1733—1740.

Searle, G. F. W., Barber, J., Porter, G., Tredwell, C. J. (1978) Biochem. Biophys. Acta *501*, 246—256.

Siebzehnrübl, S., Fischer, R., and Scheer, H. (1987) Z. Naturforsch. *42i*, 258—262.

Slooten, L. (1972) Biochim. Biophys. Acta *256*, 452—466.

Stark, N., Kuhlbrandt, W., Wildhaber, I., Wehrli, E. & Mühlethaler, K. (1984) EMBO J. *3*, 777—783.

Switalski, S. C., and Sauer, J. (1984) Photochem. Photobiol. *40*, 423—427.

Takahashi, N., Ortel, T. L., Putnam, F. W. (1984) Proc. Natl. Acad. Sci. *81*, 390—394.

Taube, H., and Gould, E. S. (1969) Acc. Chem. Res. *2*, 321—329.

Teale, F. W. J. and Dale, R. E. (1970) Biochem. J. *116*, 161—169.

Trebst, A. (1986) Z. Naturforsch. *41c*, 240—245.

Tronrud, D. E., Schmid, M. F., and Matthews, B. W. (1986) J. Mol. Biol. *188*, 443—454.

Weber, P. C., and Salemme, F. R. (1980) Nature (London) *287*, 82—84.

Weber, E., Papamokos, E., Bode, W., Huber, R., Kato, F., Laskowski, M. (1981) J. Mol. Biol. *149*, 109—123.

Wendler, J., Holzwarth, A. R., Wehrmeyer, W. (1984) Biochem. Biophys. Acta *765*, 58—67.

Weyer, K. A., Lottspeich, F., Gruenberg, H., Lang, F., Oesterhelt, D., and Michel, H. (1987) EMBO J. *6*, 2197—2202.

Chemistry 1988

Woodbury, N. W., Becker, M., Middendorf, D., and Parson, W. W. (1985) Biochemistry *24*, 7516—7521.

Wraight, C. A. (1982) in: Function of Quinones in Energy Conserving Systems (B. L. Trumpower, ed.) Academic Press, London, pp, 181—197.

Yamazaki, I., Mimuro, M., Murao, T., Yamazaki, T., Yoshihara, K., Fujita, Y. (1984) Photochem. Photobiol. *39*, 233—240

Zickendraht-Wendelstadt, B., Friedrich, J., and Rüdiger, W. (1980) Photochem. Photobiol. *31*, 367—376.

Zilinskas, B. A. and Greenwald, L. S. (1986) Photosynth. Res. *10*, 7—35.

Zuber, H. (1985) Photochem. Photobiol. *42*, 821—844.

Zuber, H. (1986) Trends Biochem. Sci. *11*, 414—419.

Chemistry 1989

SIDNEY ALTMAN and THOMAS R. CECH

for their discovery of catalytic properties of RNA

THE NOBEL PRIZE IN CHEMISTRY

Speech by Professor Bertil Andersson of the Royal Swedish Academy of
Sciences.
Translation from the Swedish text.

Your Majesties, Your Royal Highnesses, Ladies and Gentlemen,

The cells making up such living organisms as bacteria, plants, animals and
human beings can be looked upon as chemical miracles. Simultaneously
occurring in each and every one of these units of life, invisible to the naked
eye, are thousands of different chemical reactions, necessary to the main-
tenance of biological processes. Among the large number of components
responsible for cell functions, two groups of molecules are outstandingly
important. They are the nucleic acids — carriers of genetic information —
and the proteins, which catalyze the metabolism of cells through their ability
to act as enzymes.

Genetic information is programmed like a chemical code in deoxyribonu-
cleic acid, better known by its abbreviated name of DNA. The cell, however,
cannot decipher the genetic code of the DNA molecule directly. Only when
the code has been transferred, with the aid of enzymes, to another type of
nucleic acid, ribonucleic acid or RNA, can it be interpreted by the cell and
used as a template for producing protein. Genetic information, in other
words, flows from the genetic code of DNA to RNA and finally to the
proteins, which in turn build up cells and organisms having various functions. This is the molecular reason for a frog looking different from a
chaffinch and a hare being able to run faster than a hedgehog.

Life would be impossible without enzymes, the task of which is to catalyze
the diversity of chemical reactions which take place in biological cells. What
is a catalyst and what makes catalysis such a pivotal concept in chemistry?
The actual concept is not new. It was minted as early as 1835 by the famous
Swedish scientist Jöns Jacob Berzelius, who described a catalyst as a molecule capable of putting life into dormant chemical reactions. Berzelius had
observed that chemical processes, in addition to the reagents, often needed
an auxiliary substance — a catalyst — to occur. Let us consider ordinary
water, which consists of oxygen and hydrogen. These two substances do not
react very easily with one another. Instead, small quantities of the metal
platinum are needed to accelerate or catalyze the formation of water.
Today, perhaps, the term catalyst is most often heard in connection with
purification of vehicle exhausts, a process in which the metals platinum and
rhodium catalyze the degradation of the contaminant nitrous oxides.

As I said earlier, living cells also require catalysis. A certain enzyme, for
example, is needed to catalyze the breakdown of starch into glucose and
then other enzymes are needed to burn the glucose and supply the cell with
necessary energy. In green plants, enzymes are needed which can convert

atmospheric carbon dioxide into complicated carbon compounds such as starch and cellulose.

As recently as the early 1980s, the generally accepted view among scientists was that enzymes were proteins. The idea of proteins having a monopole of biocatalytic capacity has been deeply rooted, and created a fundamental dogma of biochemistry. This is the very basic perspective in which we have to regard the discovery today being rewarded with the Nobel Prize for Chemistry. When Sidney Altman showed that the enzyme denoted RNaseP only needed RNA in order to function, and when Thomas Cech discovered self-catalytic splicing of a nucleic acid fragment from an immature RNA molecule, this dogma was well and truly holed below the waterline. They had shown that RNA can have catalytic capacity and can function as an enzyme. The discovery of catalytic RNA came as a great surprise and was indeed met with a certain amount of scepticism. Who could ever have suspected that scientists, as recently as in our own decade, were missing such a fundamental component in their understanding of the molecular prerequisites of life? Altman's and Cech's discoveries not only mean that the introductory chapters of our chemistry and biology textbooks will have to be rewritten, they also herald a new way of thinking and are a call to new biochemical research.

The discovery of catalytic properties in RNA also gives us a new insight into the way in which biological processes once began on this earth, billions of years ago. Researchers have wondered which were the first biological molecules. How could life begin if the DNA molecules of the genetic code can only be reproduced and deciphered with the aid of protein enzymes, and proteins can only be produced by means of genetic information from DNA? Which came first, the chicken or the egg? Altman and Cech have now found the missing link. Probably it was the RNA molecule that came first. This molecule has the properties needed by an original biomolecule, because it is capable of being both genetic code and enzyme at one and the same time.

Professor Altman, Professor Cech, you have made the unexpected discovery that RNA is not only a molecule of heredity in living cells, but also can serve as a biocatalyst. This finding, which went against the most basic dogma in biochemistry, was initially met with scepticism by the scientific community. However, your personal determination and experimental skills have overcome all resistance, and today your discovery of catalytic RNA opens up new and exciting possibilities for future basic and applied chemical research.

In recognition of your important contributions to chemistry, the Royal Swedish Academy of Sciences has decided to confer upon you this year's Nobel Prize for Chemistry. It is a privilege and pleasure for me to convey to you the warmest congratulations of the Academy and to ask you to receive your prizes from the hands of His Majesty the King.

Sidney Alt

SIDNEY ALTMAN

I was born in Montreal in 1939, the second son of poor immigrants. My mother worked in a textile mill and my father in a grocery store before they met and married. It was from them that I learned that hard work in stable surroundings could yield rewards, even if only in infinitesimally small increments.

For our immediate family and relatives, Canada was a land of opportunity. However, it was made clear to the first generation of Canadian-born children that the path to opportunity was through education. No sacrifice was too great to forward our education and, fortunately, books and the tradition of study were not unknown in our family.

I am conscious of two events that sparked my early interest in science, the first being the appearance of the A-bomb. The mystique associated with the bomb, the role that scientists played in it, and its general importance could not fail to impress even a six-year old. About seven years later I was given a book about the periodic table of the elements. For the first time I saw the elegance of scientific theory and its predictive power. I should mention that while I was growing up, Einstein was presented as a worthy role model for a young boy who was good at his studies. I added various writers of fiction and stars of ice hockey and baseball to my pantheon.

By the time I reached high school my father's grocery store had made our life adequately comfortable and I was able to choose, without any practical encumbrances, the subjects that I wanted to pursue in college. My intention was to enroll at McGill University but an unexpected series of events led me to study physics at the Massachusetts Institute of Technology. There I experienced four years of over-stimulation among brilliant, arrogant and zany peers and outstanding teachers. Lee Grodzins supervised my senior thesis in nuclear physics and provided me with a wonderful research experience and with his friendship. During my final semester at MIT, I took a short introductory course in molecular biology to find out what all the excitement was about. That course, taught by Cyrus Levinthal, familiarized me with nucleic acids and molecular genetics and prepared me for future encounters with these topics.

I spent eighteen months as a graduate student in physics at Columbia University, waiting unhappily for an opportunity to work in a laboratory and wondering if I should continue in physics. Eight months later, having left Columbia, I was studying physics in a summer program and working in Colorado when I decided to enroll as a graduate student in biophysics. George Gamow, the physicist, had steered me to Leonard Lerman, then working on the intercalation of acridines into DNA at the University of Colorado Medical Center. In the excellent department chaired by Theodore T. Puck, Lerman provided the guidance, friendship and critical analysis that enabled me to enjoy molecular biology in a productive manner. After working on the effects of acridines on the replication of bacteriophage T4 DNA, I joined Mathew Meselson's laboratory at Harvard University to study a DNA endonuclease involved in the replication and recombination of T4 DNA. Two years later I was privileged to become a member of the group led by Sydney Brenner and Francis Crick at the Medical Research Council Laboratory of Molecular Biology in Cambridge, England. As an ex-physicist, I felt as if I was joining the equivalent of Bohr's group in Copenhagen in the 1920's. It turned out to be scientific heaven.

At the MRC laboratory I started the work that led to the discovery of RNase P and the enzymatic properties of the RNA subunit of that enzyme. John D. Smith, as well as several post-doctoral colleagues, provided me with much good advice that enabled me to test my ideas. The discovery of the first radiochemically pure precursor to a tRNA molecule enabled me to get a job as an assistant professor at Yale University in 1971, a difficult time to get any job at all.

My career at Yale followed a standard academic pattern with promotion through the ranks until I became Professor in 1980. I was Chairman of my department from 1983 – 1985 and in 1985 became the Dean of Yale College for four years, an experience that not only provided me with the opportunity to make many new friends, mostly outside the sciences, but also revealed to me the full panorama of human and academic problems that exist in a university community. On July 1, 1989 I returned to the post of Professor on a fulltime basis.

I have been blessed with outstanding mentors, people who became personal friends and who have illuminated so many aspects of human creativity for me with their intellectual power, expertise as scientists and qualities as human beings. In particular, they are Leonard Lerman, Mathew Meselson, Sydney Brenner and Lee Grodzins. There are, of course, many others whose names I cannot list here. My life has been enormously enriched by my marriage to Ann Körner in 1972. My wife is my colleague, mentor and friend in every respect. She and our two wonderful children, Daniel, born in 1974 and Leah, born in 1977, have contributed immeasurably to whatever success I have achieved.

Selected bibliography:
Altman, S. (1971) Isolation of tyrosine tRNA precursor molecules. *Nature New Biology* 229: 19 – 21.
Altman, S. and Smith, J. D. (1971) Tyrosine tRNA precursor molecule polynucleotide sequence. *Nature New Biology* 233: 35 – 39.
Robertson, H. D., Altman, S. and Smith, J. D. (1972) Purification and properties of a specific *Escherichia coli* ribonuclease which cleaves a tyrosine transfer ribonucleic acid precursor. *J. Biol. Chem.* 247: 5243 – 5251.
Stark, B. C., Kole, R., Bowman, E. J. and Altman, S. (1978) Ribonuclease P: an

enzyme with an essential RNA component. *Proc. Nat. Acad. Sci. USA.* 75: 3717 – 3721.

Kole, R. and Altman, S. (1979) Reconstitution of RNase P activity from inactive RNA and protein. *Proc. Nat. Acad. Sci. USA.* 76: 3795 – 3799.

Kole, R., Baer, M., Stark, B. and Altman, S. (1980) *E. coli* RNase P has a required RNA component *in vivo. Cell* 19: 881 – 887.

Reed, R., Baer, M., Guerrier-Takada, C., Donis-Keller, H. and Altman, S. (1982) Nucleotide sequence of the gene encoding the RNA subunit (M1 RNA) of Ribonuclease P from *Escherichia coli. Cell* 30: 627 – 636.

Guerrier-Takada, C., Gardiner, K., Marsh, T., Pace, N. and Altman, S. (1983) The RNA moiety of Ribonuclease P is the catalytic subunit of the enzyme. *Cell* 35: 849 – 857.

Guerrier-Takada, C. and Altman, S. (1984) Catalytic Activity of an RNA molecule prepared by transcription *in vitro. Science* 223: 285 – 286.

Altman, S. (1984) Aspects of biochemical catalysis. *Cell* 36: 237 – 239.

Guerrier-Takada, C., Haydock, K., Allen, L. and Altman, S. (1986) Metal ion requirements and other aspects of the reaction catalyzed by M1 RNA, the RNA subunit of Ribonuclease P from *Escherichia coli. Biochemistry* 25: 1509 – 1515.

Guerrier-Takada, C. and Altman, S. (1986) M1 RNA with large terminal deletions still retains its catalytic activity. *Cell* 45: 177 – 183.

McClain, W. H., Guerrier-Takada, C. and Altman, S. (1987) Model substrates for an RNA enzyme. *Science* 238: 527 – 530.

Bartkiewicz, M., Gold, H. and Altman, S. (1989) Identification and characterization of an RNA molecule that copurifies with RNase P activity from HeLa cells. *Genes and Development* 3: 488 – 499.

ENZYMATIC CLEAVAGE OF RNA BY RNA

Nobel Lecture, December 8, 1989

SIDNEY ALTMAN

Department of Biology, Yale University, New Haven, CT 06520, USA

Introduction

The transfer of genetic information from nucleic acid to protein inside cells can be represented as shown in Fig. 1. This simple scheme reflects accurately the fact that the information contained in the linear arrangement of the subunits of DNA is copied accurately into the linear arrangement of subunits of RNA which, in turn, is translated by machinery inside the cell into proteins, the macromolecules responsible for governing many of the important biochemical processes *in vivo*. The function of the straightforward transfer of information is carried out by a class of molecules called messenger RNAs (mRNAs). The diagram shown does not elaborate on the properties of other RNA molecules that are transcribed from DNA, namely transfer RNA (tRNA) and ribosomal RNA (rRNA) and many other minor species of RNA found *in vivo* that had no identifiable function prior to 1976, nor does it indicate that the information in DNA and RNA can be replicated as daughter DNA and RNA molecules, respectively (see Crick, 1970, for further discussion).

Ribosomes are complexes, which in *Escherichia coli,* are made of about 50 proteins and three RNA molecules. It is on these particles that mRNA directs the synthesis of protein from free amino acids. tRNA molecules (Fig. 2) perform an adaptor function in the sense that they match particular amino acids to a group of three specific nucleotides in the mRNA to be translated and ensure that the growing polypeptide (protein) chain contains the right linear sequence of amino acid subunits. Thus, rRNA and tRNA participate in the process of information transfer inside cells but they clearly do so in a comparatively complex manner. mRNA, understandably,

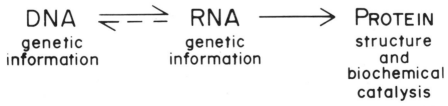

Figure 1. A representation of the flow of information inside cells from DNA to protein. This diagram is not a complete representation of the central dogma (see Crick, 1970).

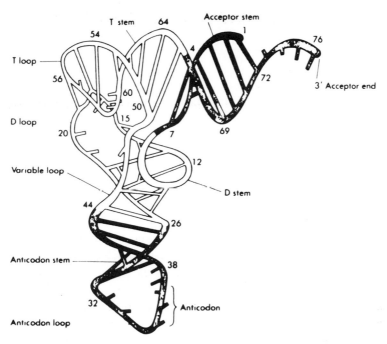

Figure 2. A diagram illustrating the folding of the yeast tRNA^{Phe} molecule. The ribose-phosphate backbone is drawn as a continuous ribbon and internal hydrogen-bonding is indicated by crossbars. Positions of single bases are indicated by bars that are intentionally shortened. The anticodon and acceptor arms are shaded. (Reprinted with permission from Watson, J.D., Hopkins, N.H., Roberts, J.W., Steitz, J.A. and Weiner, A.M., Molecular Biology of the Gene, 4th ed., Benjamin/Cummings, Menlo Park, 1987).

has a very short half-life inside cells but rRNA and tRNA are relatively stable molecules since they are part of the translational machinery that must be used over and over.

My work on RNA began as a study of certain mutants that disrupted the ability of tRNA molecules to function normally during translation (Altman, 1971). This research, in turn, led to the identification of another stable RNA molecule that had, unexpectedly, all the properties of an enzyme (Guerrier-Takada et al., 1983). Aside from its intrinsic interests to students of catalysis and enzymology, our finding of enzymatic activity associated with RNA has stimulated reconsideration of the role of RNA in biochemical systems today (see Cech, 1987, and Altman, 1989, for reviews) as illustrated in Fig. 1, and of the nature of complex biochemical systems eons ago (Darnell and Doolittle, 1986; Westheimer, 1986; Weiner and Maizels, 1987; Joyce, 1989).

As was first pointed out over twenty years ago by Woese (1967), Crick (1968) and Orgel (1968), if RNA can act as a catalyst the origin of the genetic code plays a much less critical role in the early stages of evolution of the first biochemical systems that were capable of replicating themselves. Indeed, the variety of biochemical reactions now known to be governed by RNA, as outlined in Table 1 (Altman, 1989), allows one to consider the

Table 1. *Some Properties of Catalytic RNAs*

RNA	End Groups[a]	Cofactor[b]	Mechanism
1. Group I introns	5'-P, 3'-OH	Yes	Transesterification
2. Group II introns	5'-P, 3'-OH	No	Transesterification
3. M1 RNA	5'-P, 3'-OH	No	Hydrolysis
4. Viroid/satellite	5'-OH, 2',3'-cyclic phosphate	No	Transesterification
5. Lead ion/tRNA	5'-OH, 2'3'-cyclic phosphate	No	Similar to RNAse A

[a] The end groups are those produced during the initial cleavage step of self-splicing reactions or during the usual cleavage reactions of other RNA species.
[b] This column refers to the use of a nucleotide cofactor.

possibility that a large number of diverse enzymatic reactions took place in the absence of protein. To add further substance to these ideas about life on earth over a billion years ago, it is important to understand exactly how catalytic RNA, as we know it, works and what role it plays *in vivo* today. This discussion deals primarily with the catalytic RNA subunit of the enzyme ribonuclease P from *Escherichia coli*.

A BRIEF ACCOUNT OF STUDIES OF RIBONUCLEASE P

Finding the substrate
In October, 1969 I arrived at the MRC laboratory of Molecular Biology in Cambridge, England ostensibly to study the three-dimensional structure of tRNA through the use of physical-chemical methods. On my arrival, Sydney Brenner and Francis Crick informed me that the crystal structure of yeast tRNA[Phe] had recently been solved (Kim et al., 1974; Robertus et al., 1974) and that there was no further need to engage in the studies originally outlined for me. I was further instructed to get settled, to think about a new problem for a week or two, and then to return for another discussion. Although some of my colleagues remember me as being upset by that conversation with Brenner and Crick, the feeling must have passed quickly because I only recall being presented with a marvelous opportunity to follow my own ideas.

I proposed to make acridine-induced mutants of tRNA[Tyr] from *E. coli* to determine if altering spatial relationships in tRNA, by deleting or adding a nucleotide to its sequence, would drastically alter the function of the molecule. Since Brenner and John D. Smith and their colleagues (Abelson et al., 1970; Russell et al., 1970; Smith et al., 1971) had just completed a classic series of studies of base-substitution mutants of tRNA[Tyr], they were not overly excited by the prospect of someone simply producing more mutants. Nevertheless, Brenner and Crick did not prevent me from pushing ahead and John Smith, in time, provided valuable advice about the genetics of the system in use in the laboratory.

The mutants I made lacked the usual function of suppressor tRNAs and

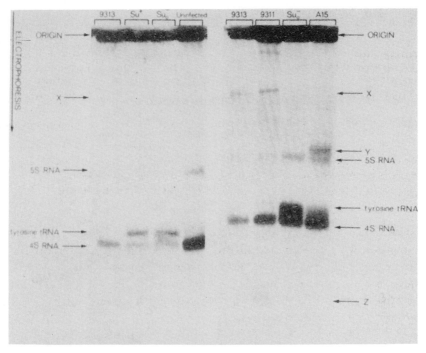

Figure 3. Separation by electrophoresis of labeled RNA from *E. coli* infected with derivatives of bacteriophage φ80 carrying various genes for tRNA^{Tyr}. The figure shows an autoradiogram of polyacrylamide gels. Experimental details are given in Altman (1971). Each column in the gel patterns is titled according to the tRNA^{Tyr} gene carried by the infecting phage. 9313 and 9311 are acridine-induced mutants of the suppressor tRNA^{Tyr}su$_3$$^+$. A15 is a mutant derivative carrying the G15−A15 mutation and su$_0$- is the wild type tRNA^{Tyr} gene. (Reprinted with permission).

made no mature tRNA *in vivo,* but they reverted at a very high rate (about 1%) to wild type. These properties indicated that there might be an unstable duplication or partial duplication of the gene for tRNA in the DNA that contained the information for the tRNA. Furthermore, it seemed likely that RNA would be transcribed from this mutant gene. I reasoned that if I could isolate the RNA transcript, which had to be unstable since no mature tRNA was made, I might be able to understand the nature of the duplication event.

The simple expedient of quickly pouring an equal volume of phenol into a growing culture of *E. coli* labeled with $^{32}PO_4^{3-}$ enabled me to isolate and characterize not only the transcript of the gene for tRNA^{Tyr} mutated by acridines, but also transcripts of the gene for tRNA^{Tyr} (Fig. 3; Altman, 1971) which had been previously mutated by other means by Brenner, Smith and their colleagues. The ability to isolate these gene transcripts, which contained sequences in addition to the mature tRNA sequences at both ends of the molecules (Fig. 4; Altman and Smith, 1971) and were, therefore, tRNA precursor molecules, depended on the rapid phenol extraction technique and the fact that the mutated molecules were less susceptible to attack by

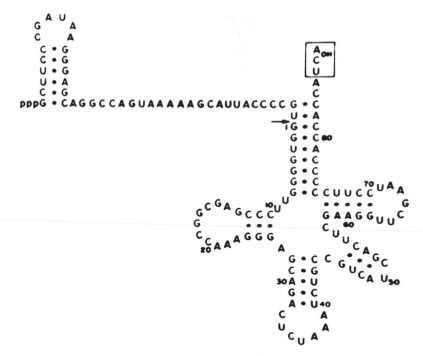

Figure 4. Nucleotide sequence of the precursor to tRNA^Tyr^su_3^+. The arrow pointing toward the sequence indicates the site of cleavage by RNase P on the 5' side of nucleotide 1 of the mature tRNA sequence. The boxed nucleotides are extra nucleotides at the 3' terminus (after Altman and Smith, 1971).

intra-cellular ribonucleases than the transcripts of the wild-type gene. The "extra" sequences, themselves, though of interest because no such segments of gene transcripts had been characterized at that time, proved not to be particularly revealing *per se.*

Although the earlier work of Darnell (Bernhardt and Darnell, 1967), Burdon (Burdon, 1971; 1974) and their coworkers had shown that tRNAs were probably made from precursor molecules in eucaryotic cells, further characterization of the enzymes involved in the biosynthesis of tRNA, or tRNA processing events, could not proceed without a radiochemically pure, homogeneous substrate of the kind that I had isolated.

When the the precursor to tRNA^Tyr^ was mixed with an extract of *E. coli,* it was immediately apparent that enzymatic activities were present in the cell extract that could remove the "extra" nucleotides from both the 5' and 3' ends of the mature tRNA sequence (Fig. 5; Altman and Smith, 1971; Robertson et al., 1972). The activity that processed the 5' end of the tRNA precursor, which we named Ribonuclease P, did so by one endonucleolytic cleavage event in contrast to what appeared to be non-specific exonucleolytic degradation at the 3' end of the molecule. In fact, no ribonucleases with such limited specificity with respect to the site of cleavage as that exhibited by RNase P were known at that time, so the novelty of this reaction assured our continuing interest in it. Some characterization of the reaction was

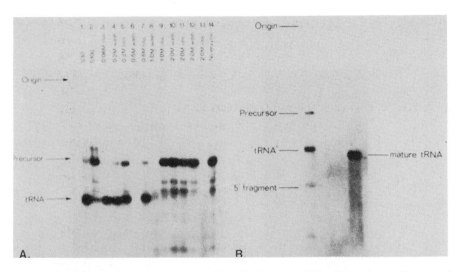

Figure 5. A. RNase P activity in extracts of *E. coli*. Extracts of *E. coli*, partial purification of
RNase P, and cleavage reactions were carried out as generally described in Robertson et al.,
(1972). The substrate used was the precursor to *E. coli* tRNA^Tyr. B. Separation by gel electrophoresis of the products of cleavage in vitro of the precursor to tRNA^TyrA25 by a partially
purified preparation of RNase P. The 3' end fragment ('tRNA') includes the additional nucleotides of the precursor. (Reprinted with permission from Robertson et al., 1972).

immediately carried out in collaboration with Hugh Robertson and John
Smith (Robertson et al., 1972).

Characterization of Ribonuclease P from **Escherichia coli**

At the MRC laboratory, we showed that RNase P produced 5' phosphate
and 3' hydroxyl groups at its site of cleavage (Robertson et al., 1972), unlike
most non-specific nucleases which produce 5' hydroxyl and 3' phosphate
groups. This observation fitted with the fact that mature tRNAs have a 5'
phosphate at their 5' termini. While some progress was made in terms of
chromatographic purification of the enzyme, in retrospect the most striking
observation made in the early studies was that "it is possible that the active
form of RNase P, which must have a strong negative charge, could be
associated with some nucleic acid." The next important step was taken a few
years later by Benjamin Stark, a graduate student in my laboratory, who
showed that an RNA of high molecular weight copurified with the enzymatic activity and, in a classic experiment, he demonstrated that this RNA
molecule was essential for enzymatic activity (Stark et al., 1978). (The RNA
was named M1 RNA and was later shown to be similar to a stable RNA
species [band IX] of unknown function that had been described by Ikemura
and Dahlberg (1973) as one of a series of minor RNA species found in *E.
coli*; the protein subunit of RNase P from *E. coli* was named C5 protein.)

The essential role of the RNA component was established by first treating
RNase P with micrococcal nuclease, an enzyme that destroys RNA, and
subsequently assaying the treated enzyme for RNase P activity: there was
none after treatment with micrococcal nuclease (Fig. 6) or, for that matter,

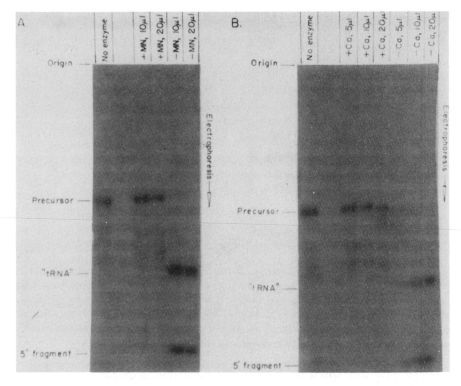

Figure 6. Inactivation of RNase P with RNase pretreatment. Control reactions were performed without micrococcal nuclease (MN) in the pretreatment mixture (A) or without $CaCl_2$ (B). RNase P pretreated with MN had less that 5% the activity of control reaction RNase P. The extent of inactivation can be varied by changing the reaction conditions as shown. (Reprinted with permission from Stark et al., 1978).

after treatment with various proteinases. Thus, under the conditions we were then using (that is, buffers that contained 10 mM $MgCl_2$), both protein and RNA components were shown to be essential for enzymatic activity. Concurrently we showed that the enzyme had a buoyant density in CsCl of 1.72 g/ml (Stark et al., 1978), characteristic of an RNA-protein complex that consists predominantly of RNA. Velocity sedimentation experiments had previously determined the sedimentation coefficient to be 12.5 S (Robertson et al., 1972).

While the biochemical purification was proceeding, the study of mutants of *E. coli* made by Schedl and Primakoff (1973; Schedl et al., 1974), Shimura, Ozeki and their coworkers (Ozeki et al., 1974; Sakano et al., 1974) showed that RNase P was an essential enzyme in *E. coli* for the biosynthesis of all tRNAs and that both RNA and protein subunits were required *in vivo*. Furthermore, additional work from the laboratories of William McClain (reviewed in McClain, 1977) and John Carbon (Carbon et al., 1974) added to the evidence that RNase P was responsible for the processing of many different tRNA precursor molecules. Although appropriate genetic analyses could not be performed, we also showed that RNase P-like activities existed in the extracts of cells from many other organisms, including humans (Altman and Robertson, 1973; Garber et al., 1978). These early studies showed that RNase P was capable of cleaving many different tRNA precursor molecules and that there was no identifiable similarity in terms of nucleotide sequence around the sites of cleavage. The manner in which the enzyme recognized its sites of cleavage in different substrates with such selectivity seemed worthy of study, and recognition of some feature of the structure in solution, common to all tRNA precursor molecules, was suspected.

When Stark's experiments were published we did not have the temerity to suggest, nor did we suspect, that the RNA component alone of RNase P could be responsible for its catalytic activity. The fact that an enzyme had an essential RNA subunit, in itself, seemed heretical enough. Shortly thereafter, however, when Ryszard Kole demonstrated that the enzyme consisted of an RNA (M1 RNA) and a protein subunit (C5 protein; Mr $\sim$ 14,000), which were not covalently linked and which could be separated into inactive subunits and then reconstituted to form an active enzyme (Kole and Altman, 1979), the similarities in chemical composition and properties of assembly of this system to that of the ribosome were sufficiently striking that we could not escape thinking about the possibility that the RNA, at the very least, participated in the formation of the active site of the enzyme. Indeed, making the comparison with ribosomes proved to be important in overcoming some resistance to the idea that an enzyme could have an RNA subunit. From a purely chemical point of view, there was no reason why RNA could not participate in formation of the active site (Kole and Altman, 1981) or even in catalysis itself.

The advent of recombinant DNA technology and powerful systems for the transcription *in vitro* of isolated pieces of DNA enabled us to characterize in some detail the RNA subunit of RNase P (377 nucleotides in length; Reed et al., 1982) and to prepare large quantities for biochemical experiments. Concurrent progress in our purification of the protein subunit prepared us for a series of experiments, conducted in collaboration with Norman Pace's group, in which we made hybrid enzymes with subunits from *E. coli* (prepared in our laboratory) and from *B. subtilis* (prepared in Pace's laboratory). As an offshoot of these experiments, Cecilia Guerrier-Takada in my laboratory was testing reconstituted enzymes under ionic conditions optimal for the activity of the holoenzyme from *B. subtilis* and different from the ones we had previously usually employed. She found, in control experiments, that the RNA subunit from *E. coli*, exhibited catalytic activity of its own in buffers that contained 60 mM $MgCl_2$ (An example of such reactions is shown in Fig. 7; in fact the catalytic activity of M1 RNA is evident when the concentration of Mg^{2+} is greater than 20 mM; Guerrier-Takada et al., 1983). The protein subunit of the enzyme increased the k_{cat} by a factor of 10-20 but had little effect on the K_m. These observations were possible because of the purity of our preparations of M1 RNA and the use of a natural substrate, the precursor to tRNA^{Tyr} from *E. coli*.

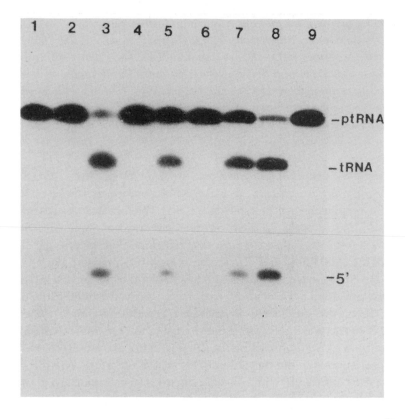

Figure 7. Dependence of the catalytic activity of M1 RNA on the concentration of Mg^{2+}. The precursor to $tRNA^{Tyr}$, abbreviated as pTyr, is the substrate. *Lane 1:* pTyr alone; 10 mM $MgCl_2$. *Lane 2:* M1 RNA added; 10 mM $MgCl_2$. *Lane 3:* M1 RNA and C5 protein added: 10 mM $MgCl_2$. *Lane 4:* pTyr alone; 100 mM $MgCl_2$. *Lane 5:* M1 RNA added; 100 mM $MgCl_2$. *Lane 6:* M1 RNA and C5 protein added, 100 mM $MgCl_2$. *Lane 7:* M1 RNA added; 100 mM $MgCl_2$/4% polyethylene glycol. *Lane 8:* M1 RNA and C5 protein (20-fold excess) added; 100 mM $MgCl_2$. *Lane 9:* C5 protein added; 10 mM Mg Cl_2. Reactions were carrried out as described by Guerrier-Takada et al., (1983).

We quickly determined that M1 RNA had all the properties of a true enzyme as defined in biochemistry textbooks (Fruton and Simmonds, 1958; p. 211): it was unchanged (in size) during the course of the reaction; it had a true turnover number as measured by Michaelis-Menten analysis of the kinetics (Fig. 8) and, therefore, it was a catalyst; it was needed in only small amounts and it was stable. Soon thereafter we proposed a model of the secondary structure of M1 RNA based on its susceptibility to nucleases in solution and some simple notions of the stability of RNA structures. We also rapidly outlined the general ionic requirements of the reaction (Table 2; Guerrier-Takada et al., 1986)). The curve of the dependence of the rate of the reaction on the pH is flat between 5 and 9 and is suggestive of the involvement of more than one group with a pK_a not characteristic of those found on nucleotides alone in solution. It is reasonable to expect, therefore, that the active site of M1 RNA is embedded in a folded structure and that the local environment of the active site will not be precisely identical to that of the aqueous buffer in which the whole molecule is dissolved.

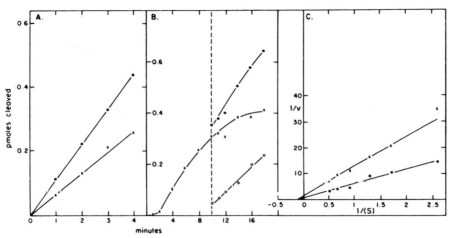

Figure 8. Kinetic analysis of the M1 RNA and RNase P reactions with the precursor to tRNA[Tyr] (pTyr) as substrate. A. Comparison of the kinetics of reconstituted (by dialysis) *E. coli* RNase P in buffer that contained 5 mM $MgCl_2$ and M1 RNA in buffer that contained 60 mM $MgCl_z$ that had been treated in the same way. (●) RNase P activity; (x) M1 RNA activity. B. Kinetics of M1 RNA action in buffer that contained 60 mM $MgCl_2$. M1 RNA was incubated with a five-fold excess of pTyr. 10 min after the start of incubation a further three-fold excess of pTyr, or buffer alone, were added to the reaction mixture. (●) pTyr added after 10 min; (x) buffer alone added after 10 min; (o) net added pTyr cleaved after 10 min (all in pmoles). C. Determination of K_m and V_{max} for the reactions shown in A. A Lineweaver-Burk double reciprocal plot was constructed from the appropriate kinetic data. (●) RNase P in buffer that contained 5 mM $MgCl_2$; :(x) M1 RNA in buffer that contained 60 mM $MgCl_2$. Units: 1/S (pmoles x $5x10^{-4})^{-1}$; 1/v [(pmoles substrate cleaved/min)-1]. (Reprinted from Guerrier-Takada et al., 1983).

Table 2. *Catalytic Activity of M1 RNA.*

M1 RNA active[a]:
   $\geq$ 20 mM $MgCl_2$
     10 mM $MgCl_2$ plus C5 protein
     10 mM $MgCl_2$ plus 5 mM polyamine

M1 RNA not active:
     10 mM $MgCl_2$

---

[a] The table summarizes data presented in Guerrier-Takada et al., (1983). The complete composition of reaction mixtures is given in the reference.

These findings complemented those of Cech's group (Cech et al., 1981; Cech and Bass, 1986, for review) on self-splicing RNA and started intense speculation about the role RNA may have played in the origin of life. However, our immediate interest was in determining precisely how the enzyme works, what its role is *in vivo,* and how it manages to recognize 60 or so different substrates in *E. coli* with no apparent sequence specificity around the site of cleavage.

RECENT WORK

Structure

The original model of the secondary structure of M1 RNA has been extensively refined by phylogenetic analysis (Fig. 9) carried out primarily by Pace and coworkers (James et al., 1988). However, this analysis has not yet yielded a satisfactory correlation between the phenotypes of mutants (Lumelsky and Altman, 1988) and features of the secondary structure of M1 RNA or its analogs from other bacteria (see below). It does provide the basis for hypotheses about the regions of M1 RNA that are essential for function (Waugh et al., 1989), as indicated by evolutionary conservation, and it

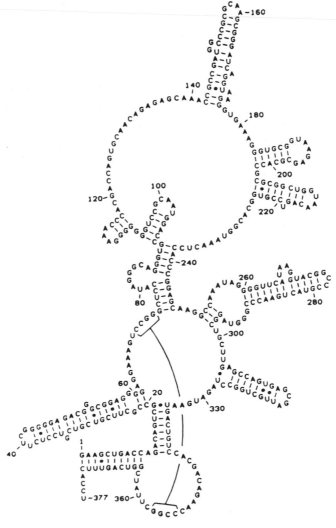

Figure 9. A model for the secondary structure of M1 RNA based on extensive phylogenetic analysis of the nucleotide sequence of the RNA subunit of RNase P from several eubacteria. (Reprinted with permission from James et al., 1988).

highlights the necessity of determining the three-dimensional features of the structure. To this end, additional phylogenetic comparisons, utilizing the data concerning the homolog of M1 RNA from several eucaryotic species (Miller and Martin, 1983; Krupp et al., 1986; Lee and Engelke, 1989), and crystallographic studies are in progress. One observation of continuing interest from these studies is that the evolutionary clock for both the RNA and protein subunits of RNase P seems to be a very fast one in comparison with that for rRNAs (Lawrence et al., 1987; Gold, 1988). Although the function of RNase P, as judged by the antigenic properties of the protein (Gold et al., 1988; Mamula et al., 1989), its ability to cleave various substrates and to reconstitute active enzyme with subunits from different organisms (Guerrier-Takada et al., 1983; Gold and Altman, 1986; Lawrence et al., 1987) has been highly conserved, the nucleotide sequences of the genes for the subunits of the enzyme have drifted extremely rapidly (Gold, 1988; Bartkiewicz et al., 1989).

Mechanism

The detailed mechanism of the reaction catalyzed by RNase P is not known but two proposals have been made. In one case (Guerrier-Takada et al., 1986), a variation of the S_N2 in-line displacement mechanism has been suggested in which a complex between one Mg ion and six water molecules facilitates the nucleophilic action of a water molecule in solution (Fig. 10). Investigations of the rRNA self-splicing reaction in *Tetrahymena* in Tom Cech's laboratory indicate that the original proposal for the mechanism of the RNase P reaction may also be relevant in the self-splicing reaction (Cech, 1987; McSwiggen and Cech, 1989). In the other proposal for the mechanism of the RNase P reaction (Reich et al., 1988), the nucleophile is derived from groups on the surface of the enzyme and the role of the magnesium ion is not as clearly specified. Attempts are underway in our laboratory to test the first model, by the insertion of a phosphothioate bond at the cleavage site and analysis of the stereochemistry of the cleaved product.

While many aspects of structure-function relationships and clues to the mechanism of the reaction may be revealed if the crystal structure of the enzyme becomes available, the determination of a crystal structure may prove to be elusive. We have, therefore, embarked on an attempt to identify regions of M1 RNA that are critical for the reaction by cross-linking the substrate to the enzyme by irradiation with ultraviolet light. Such experiments have revealed that a cross-link is formed between a nucleotide close to the site of cleavage in the substrate (C -3) and residue C92 in M1 RNA (Guerrier-Takada et al., 1989). If C92 is deleted from M1 RNA, the kinetics of the enzymatic reaction, and its site of cleavage with particular substrates, are significantly altered. Furthermore, the region of secondary structure around C92 in M1 RNA resembles that of the tRNA E site in 23S rRNA (Fig. 11). Additional studies have shown that this site is important in the binding of the aminoacyl stem of a tRNA precursor to the enzyme and that,

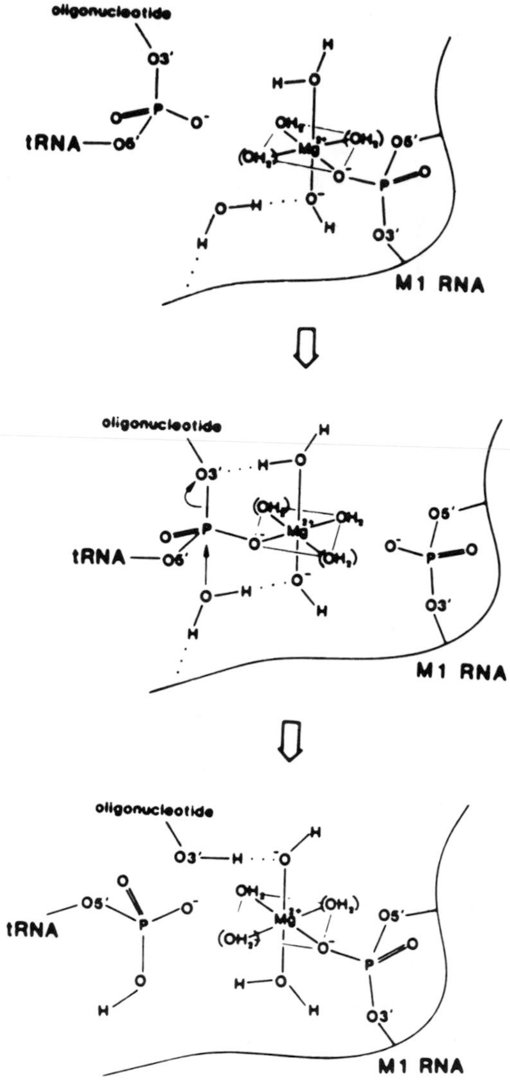

Figure 10. Hypothetical electronic mechanism of tRNA precursor hydrolysis by M1 RNA of RNase P. The reaction is catalyzed by an Mg-H$_2$0 complex that is initially bound to a phosphate of M1 RNA. Mg^{2+} is formally shown as hexacoordinated, but it may well be tetracoordinated as indicated by the parentheses around the two equatorial water ligands. In the top panel, a water molecule from the solvent that will participate in hydrolysis is positioned by a hydrogen bond to an O or N atom in M1 RNA. In the middle and bottom panels, the tRNA precursor substrate is bound by the water molecule attached to M1 RNA and passes through a transition state prior to cleavage of the "extra" oligonucleotide and prior to the addition of OH to its O5' terminal phosphate. After the reaction steps shown here, a solvent water chain between the axial ligands of Mg^{2+} recocks the enzyme for the next cycle. (Reprinted with permission from Guerrier-Takada et al., 1986. Copyright 1986 American Chemical Society).

as in the binding of tRNA to the E site of 23S rRNA (Moazed and Noller, 1989), the 3' terminal CCA sequence plays a critical role in the interaction of the enzyme with the substrate. These results, in addition to allowing the

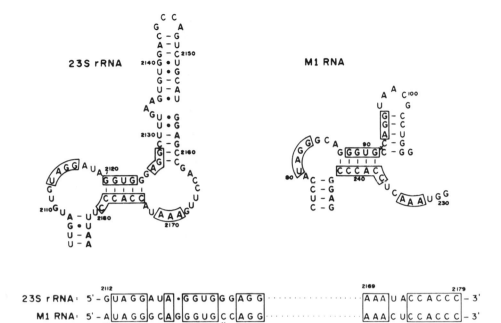

Figure 11. Comparison of part of the E site of 23S rRNA with a region in M1 RNA that surrounds the crosslink with the substrate. The secondary structures are taken from Moazed and Noller (1989) and James et al. (1988). The "x" marks C92, the nucleotide in M1 RNA that is crosslinked to the substrate. Nucleotides shown in boxes are found in approximately the same relative positions in the structures shown (see Guerrier-Takada et al., 1989).

identification of domains with similar structural and functional properties in RNA molecules with very different cellular functions, delineate a region of importance for function in M1 RNA and suggest further experiments for a more detailed definition of the interactions between enzyme and substrate and of the particular steps in the enzymatic reaction that involve this region.

Recognition of the substrate
Early notions of the features important for the recognition by RNase P of its substrate focussed either on the possibility of Watson-Crick pairing of nucleotide sequences common to all tRNAs (e.g. CCA and UUCG) with M1 RNA (Guerrier-Takada and Altman, 1986) or some other, incompletely specified, measuring mechanism that recognizes the three-dimensional structure of the tRNA moiety of the precursor (Bothwell et al., 1976). Results of several experiments indicated that extensive pairing between enzyme and substrate was not essential for the enzymatic reaction (Guerrier-Takada and Altman, 1986; Baer et al., 1987) and attention was focussed on the conformation in solution of the substrate. It had been demonstrated early on that RNase P from any one source can cleave tRNA precursors from any other source. Thus, when an unusual tRNA which lacked the D stem and loop was found, namely, tRNA[Ser] from bovine mitochondria (de

Bruijn and Klug, 1983), we examined whether M1 RNA could cleave an analog of a precursor to that tRNA.

In collaboration with Bill McClain (McClain et al., 1987), we showed not only that the D stem and loop were not essential for recognition, but also that the anticodon stem and loop were also dispensable (Fig. 12). A substrate that consisted merely of a single-stranded region at the 5' end of a mini-tRNA, containing only the T stem and loop stacked on the aminoacyl acceptor stem, was cleaved almost as efficiently as the parent tRNA precursor from which it was derived. This minimal substrate contains an RNA helix that is analogous to one part of the intact three-dimensional structure of a normal tRNA (Fig. 2; Kim et al., 1974; Robertus et al., 1974)). Recently, we have also shown that neither the loop segment of the structure, nor more than six base pairs in the helical region, and, in separate experiments, that no more than one "extra" nucleotide at the 5' terminus are needed for cleavage by M1 RNA. While substrates with these features are not cleaved as efficiently as either a normal precursor tRNA or pAT1 (the substrate shown in Fig. 12 that resembles a hairpin), they must, nevertheless, contain sufficient recognition elements to allow the reaction to proceed. All these model substrates have the 3' terminal CCA sequence and none are cleaved at detectable levels when the CCA sequence is altered. Therefore, it appears that the CCA sequence and at least one half turn of an RNA helix play an essential role in substrate recognition.

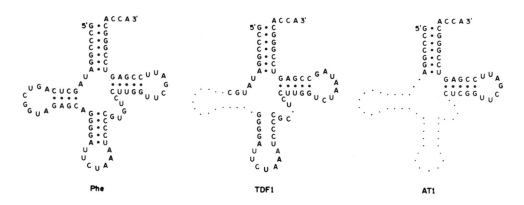

Figure 12. A. Structure of mature tRNA[Phe], and derivatives of it, encoded by synthetic genes. The sequences are shown without the modified nucleotides characteristic of mature tRNA[Phe] because they are not present in the transcripts made *in vitro*. The transcripts contain "extra" nucleotides at both ends of the molecules (after McClain et al., 1987). B. Structure of the AT-1 precursor drawn in a stem and loop structure similar to the corresponding region in tRNA[Phe]. The arrows mark the site of cleavage by RNase P or M1 RNA. The sequence of mature AT-1 is shown in A. (Reprinted with permission from McClain et al., 1987).

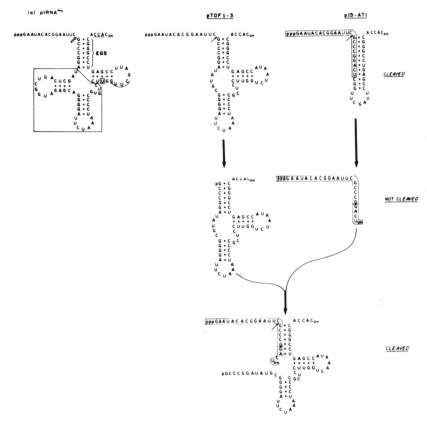

Figure 13. Scheme for formation of substrates for RNase P by hybridization of two oligoribonucleotides. TDF-1 (see Fig. 3) was prepared by RNAse P cleavage *in vitro* of its precursor molecule
that had been transcribed *in vitro* (see McClain et al., 1987). A portion (boxed sequence) of the
precursor to AT-1 (Fig. 3) was prepared by transcription *in vitro* of a restriction fragment of the
DNA encoding the AT-1 synthetic gene (A.C. Forster, personal communication).

Conclusions from experiments with model substrates have to be tempered by the knowledge that some recognition elements, which appear to
play a prominent role in these examples, may not play as important a role
and may be supplemented by other elements in the normal tRNA precursors found in cells. It is certainly the case that a change in the D or
anticodon stems of a normal tRNA precursor can have a dramatic effect on
the rate of cleavage by RNase P even though these entire regions of the
substrate are absent in the model substrates.

Through the hybridization of two oligoribonucleotides as shown in Fig.
13, we can create and manipulate novel substrates. An "external guide
sequence" which can guide RNase P to its target, can be hybridized to any
other RNA of known sequence and will form the downstream, or 3' part, of
the substrate. RNase P should then cleave the hybrid target at the junction
between the single- and double-stranded region at the 5' side of the double-
stranded region. This new method presents opportunities to investigate
more precisely the details of the recognition mechanism and it also pro-

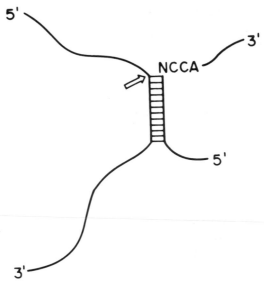

Figure 14. Targeting of RNA for cleavage by RNase P. An external guide sequence (EGS) is shown by the shorter line ending in NCCA. N is most frequently found to be A in tRNA molecules. The region of the EGS shown as hydrogen-bonded is designed to be complementary to a region of known sequence in the RNA to be targeted.

vides, in principle, a means to inactivate any mRNA of known sequence *in vivo*. Aside from the problem of the expression of the external guide sequence *in vivo*, the method does have the advantage that RNase P is already present in cells of all types. Providing that the hybrid can be designed to be compatible with the cleavage-site specificity of the enzyme in the particular host organism of choice, the target RNA should be inactivated.

In this example of the use of one oligoribonucleotide to target another RNA that is to be cleaved by RNase P, substrate recognition by the enzyme resembles, in a formal sense, selection of the site of cleavage by some of the other known RNA catalysts. Group I introns and the satellite and similar RNAs use guide sequences (Cech, 1987; Altman, 1989, for reviews) in the selection of cleavage sites or to form structures in which a cleavage site becomes defined. In virtually all other respects, these reactions are quite distinct from that carried out by RNase P.

The Past, Present and Future of RNase P

The discovery of RNA catalysis has led to new hypotheses about the origin of the earliest self-replicating biochemical systems from which the question of the origin of the genetic code can be excluded. Models of these early systems rely entirely on RNA as the genetic material and as the source of catalytic activity (Fig. 15; Darnell and Doolittle, 1986; Weiner and Maizels, 1987; Joyce, 1989). All this speculation clearly presupposes that what we see in present-day systems reflects, in some manner the properties of RNA over

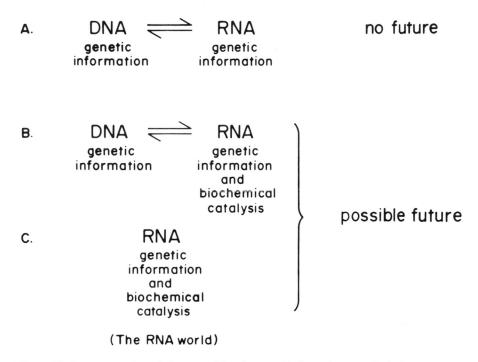

A. DNA ⇌ RNA   no future
    genetic   genetic
    information   information

B. DNA ⇌ RNA
    genetic   genetic
    information   information
    and
    biochemical
    catalysis   possible future
C.   RNA
    genetic
    information
    and
    biochemical
    catalysis

(The RNA world)

Figure 15. A representation of three possible schemes of information transfer before proteins were part of the scheme.

a billion years ago. Should that indeed be the case, the richness of biochemical mechanisms exhibited by RNA (Table 1) is impressive and can allow for rather complex systems to develop in the absence of protein and DNA. In this limited context, we shall consider some aspects of the reactions governed by RNase P *in vitro*.

Although M1 RNA can cleave very simple substrates, it is apparent that these particular cleavage reactions cannot occur *in vivo* today because such cleavages would occur too frequently in the population of RNA in any cell: that is, the entire population of RNA molecules would be too susceptible to degradation by RNase P. However, one can imagine that in an RNA world, there was considerable advantage to having an RNA molecule that could identify many sites in very long molecules generated by enzymatic or non-enzymatic mechanisms. The proliferation of many smaller molecules from larger ones would give rise to the possibility of a great variety of conformations of RNA in solution, some of which may have endowed RNA with catalytic activity or other useful functions very long transcripts did not have.

Setting aside for the moment the details of the origin of the genetic code and the appearance of proteins, one can ask, however, why a protein subunit became associated with M1 RNA. We recently showed that the protein subunit of RNase P can alter the site of cleavage and affect the rate of the reaction in a manner sensitive to the nature of the particular substrate being used (Guerrier-Takada et al., 1988; 1989). Thus, it is possible that proteins may have fine-tuned the site specificity of RNA enzymes by enhancing the rates of reaction at particular sites and with particular substrates. What we see today as the "normal" cleavage sites of RNA enzymes may have been selected for over the eons, in conjunction with the appearance of protein cofactors, as physiological conditions changed during evolutionary time. The "unselected" reactions, for example those with very small hairpin substrates, became in consequence second- or lower-order reactions and are no longer relevant to events *in vivo*.

Finally, why do RNA enzymes only cleave phosphodiester bonds? There are three answers that come readily to mind. First and most trivially, RNA enzymes may cleave other classes of bonds and we just have not yet made the critical observations or found the right reaction conditions (the last part of the answer is a generic response to questions about the lack of success in performing certain reactions *in vitro*). Second, it is possible that, in the RNA world, perhaps RNA molecules could only cleave phosphodiester bonds: it was a primitive world and no other reactions were governed by enzymes. Once proteins appeared on the scene there was no further need to diversify RNA enzymes. Lastly, and most important, the chemistry of RNA enzymes, when sufficiently well-understood, may indicate to us that there is a compelling reason why RNA molecules cleave only phosphoester bonds. The validity, or lack of it, of this last answer can be tested by direct experimentation, and therein lies the work of the next several years.

ACKNOWLEDGEMENTS

My indebtedness to so many people makes it impractical to list them all here. Nevertheless, I wish to express my gratitude to my parents, my family, my teachers (especially Leonard Lerman, Mathew Meselson, Sydney Brenner and Lee Grodzins), my professional colleagues and collaborators (especially, Hugh Robertson and Bill McClain) and my students and coworkers in my laboratory (especially Cecilia Guerrier-Takada, Ben Stark, Ryszard Kole, Robin Reed and Madeline Baer). They have all tolerated my bouts of obsessiveness and have shared the moments of discovery and pleasure. The taxpayers of the United States, through the agencies of the National Institutes of Health and the National Science Foundation, have generously supported my work.

REFERENCES

Abelson, J. N., Gefter, M. L., Barnett, L., Landy, A. and Russell, R. L. (1970) J. Mol. Biol. *47* 15 – 28.

Altman, S. (1971) Nature New Biology *229* 19 – 21.

Altman, S. (1989) Adv. Enzymol., ed A. Meister (J. Wiley, NY) Vol. 62, pp. 1 – 36.

Altman, S. and J. D. Smith (1971) Nature New Biology *233* 35 – 39.

Altman, S. and Robertson, H. D. (1973) Molec. Cell. Biochem. *1* 83 – 93.

Baer, M. F., Reilly, R. M., McCorkle, G. M., Hai, T-Y, Altman, S. and RajBhandary, U. L. (1989) J. Biol. Chem. *263* 2344 – 2351.

Bartkiewicz, M., Gold, H. and Altman, S. (1989) Genes and Development *3* 488 – 499.

Bernhardt, D. and Darnell, J. E. (1967) J. Mol. Biol. *42* 43 – 56

Bothwell, A. L. M., Stark, B. C. and Altman, S. (1976) Proc. Nat. Acad. Sci. USA *73* 1912 – 1916.

Burdon, R. H. (1971) Prog. Nucl. Acids Res. Mol. Biol. *11* 33 – 79.

Burdon, R. H. (1974) Brookhaven Symp. Biol. *26* 138 – 153.

Carbon, J., Chang, S. and Kirk, L. L. (1974) Brookhaven Symp. Biol. *26* 26 – 36.

Cech, T. R. (1987) Science *236* 1532 – 1539.

Cech, T. R. and Bass, B.L. (1986) Annu. Rev. Biochem. *55* 599 – 629.

Cech, T. R., Zaug, A. J. and Grabowski, P. J. (1981) Cell *27* 487 – 496.

Chang, D.D. and Clayton, D.A. (1987) Science *235* 1178 – 1184.

Crick, F. (1968) J. Mol. Biol. *38* 367 – 379.

Crick, F. (1970) Nature *277* 561 – 563.

Darnell, J. E. and Doolittle, W. F. (1986) Proc. Natl. Acad. Sci. USA *83* 1271 – 1275.

de Bruijn, M. H. L. and Klug, A. (1983) EMBO J. *2* 1309 – 1321.

Fruton, J., and Simmonds, S. (1958) General Biochemistry, 2nd ed. (New York: J. Wiley and Sons).

Garber, R. L., Siddiqui, M. A. Q. and Altman, S. (1978) Proc. Nat. Acad. Sci. USA. *75* 635 – 639.

Gilbert, W. (1986) Nature *319* 618.

Gold, H. A. (1988) Ph.D. Thesis, Yale University, New Haven, CT USA.

Gold, H. A. and Altman, S. (1986) Cell *44* 243 – 249.

Gold, H. A., Craft, J., Hardin, J. A., Bartkiewicz, M. and Altman, S. (1988) Proc. Nat. Acad. Sci. USA *85* 5483 – 5487.

Guerrier-Takada, C. and Altman, S. (1984) Biochemistry *23* 6327 – 6334.

Guerrier-Takada, C. and Altman, S. (1986) Cell *45* 177 – 183.

Guerrier-Takada, C. Gardiner, K., Marsh, T., Pace, N., and Altman, S. (1983) Cell *35* 849 – 857.

Guerrier-Takada, C., Haydock, K., Allen, L. and Altman, S. (1986) Biochemistry *25* 1509 – 1515.

Guerrier-Takada, C., Knap, A. K., Lumelsky, N. and Altman, S. (1989) Science, *246* 1578 – 1584.

Guerrier-Takada, C., van Belkum, A., Pleij, C. W. A. and Altman, S. (1988) Cell *53* 267 – 272.

Guthrie, C., Seidman, J. G., Altman, S., Barrell, B. G., Smith, J. D. and McClain, W. H. (1973) Nature New Biol. *246* 6 – 11.

Ikemura, T. and Dahlberg, J. E. (1973) J. Biol. Chem. *248* 5024 – 5032.

James, B., Olsen, G. J., Lin, J. and Pace, N. (1988) Cell *52* 19 – 26.

Joyce, G. F. (1989) Nature *338* 217 – 223.

Kim, S. H., Suddath, F. L., Quigley, G. J., McPherson, A., Sussman, J. L., Wang, A. H. J., Seeman, N. C. and Rich, A. (1974) Science *185* 435 – 440.

Kole, R. and Altman, S. (1979) Proc. Natl. Acad. Sci. USA *76* 3795 – 3799.

Kole, R. and Altman, S. (1981) Biochem. *20* 1902 – 1906.

Krupp, G., Cherayil, B., Frendeway, D., Nishikawa, S. and Soll, D. (1986) EMBO J. *5* 1697−1703.

Lawrence, N. P., Richman, A., Amini, R. and Altman, S. (1987) Proc. Natl. Acad. Sci. USA *84* 6825−6829.

Lee, J-Y., and Engelke, D. R. (1989) Molec. Cell. Biol. *9* 2536−2543.

Lumelsky, N. and Altman, S. (1988) J. Mol. Biol. *202* 443−454.

Mamula, M.J., Baer, M., Craft, J. and Altman, S. (1989) Proc. Nat. Acad. Sci. USA, *86* 8717−8721.

McClain, W. H. (1977) Accts. Chem. Res. *10* 418−425.

McClain, W. H., Guerrier-Takada, C. and Altman, S. (1987) Science *238* 527−530.

McSwiggen, J. A. and Cech, T. R. (1989) Science *244* 679−683.

Miller, D. L. and Martin, N. C. (1983) Cell *34* 911−917.

Moazed, D. and Noller, H. F. (1989) Cell *57* 585−597.

Orgel, L. (1968) J. Mol. Biol. *38* 381−393.

Ozeki, H., Sakano, H., Yamada, S., Ikemura, T. and Shimura, Y. (1974) Brookhaven Symp. Biol. *26* 89−105.

Reed, R., Baer, M., Guerrier-Takada, C., Donis-Keller, H. and Altman, S. (1982) Cell *30* 627−636.

Reich, C. I., Olsen, G. J., Pace, B. and Pace, N. R. (1988) Science *239* 178−181.

Robertson, H. D. (1986) Nature *322* 16−17.

Robertson, H. D., Altman, S. and Smith, J. D. (1972) J. Biol. Chem. *247* 5243−5251.

Robertus, J. D., Ladner, J. E., Finch, J. T., Rhodes, D., Brown, R. S., Clark, B. F. C. and Klug, A. (1974) Nature *250* 546−551.

Russell, R. L., Abelson, J. N., Landy, A., Gefter, M. L., Brenner, S. and Smith, J. D. (1970) J. Mol. Biol. *47* 1−13.

Sakano, H., Yamada, S., Ikemura, T., Shimura, Y. and Ozeki, H. (1974) Nucleic Acids Res. *1* 355−371.

Schedl, P., and Primakoff, P. (1973) Proc. Natl. Acad. Sci. USA *70* 2091−2095.

Schedl, P., Primakoff, P. and Roberts, J. (1974) Brookhaven Symp. Biol. *26* 53−76.

Smith, J. D., Barnett, L., Brenner, S. and Russell, R.L. (1971) J. Mol. Biol. *54* 1−14.

Stark, B. C., Kole, R., Bowman, E. J. and Altman, S. (1978) Proc. Natl. Acad. Sci. USA *75* 3717−3721.

Waugh, D. S., Green, C. J. and Pace, N. R. (1989) Science *244* 1569−1571.

Weiner, A. M. and Maizels, N. (1987) Proc. Nat. Acad. Sci. USA *84* 7383−7387.

Westheimer, F. H. (1986) Nature *319* 534−535.

Woese, C. R. (1967) The Origins of the Genetic Code (Harper and Row, New York)

Thomas R. Cech

THOMAS R. CECH

Grandfather Josef, a shoemaker, immigrated to the U. S. from Bohemia in 1913. My other grandparents, also of Czech origin, were first-generation Americans. My father was and is a physician, my mother the homemaker. I was born in Chicago on December 8, 1947.

The safe streets and good schools of Iowa City, Iowa provided the backdrop for the childhood years of my sister Barbara, my brother Richard and myself. My father, who loved physics as much as medicine, interjected a scientific approach and point of view into most every family discussion. I discovered science for myself in fourth grade, collecting rocks and minerals and worrying about how they were formed. By the time I was in junior high school, I would knock on Geology professors' doors at the University of Iowa, asking to see models of crystal structures and to discuss meteorites and fossils.

In 1966 I entered Grinnell College, where I was to derive as much enjoyment studying Homer's *Odyssey*, Dante's *Inferno*, and Constitutional History as Chemistry. I met Carol over the melting point apparatus in a make-up Organic Chemistry lab, starting the partnership of our lives that is now more than 20 years old.

The Chemistry I appreciated the most from textbooks was physical chemistry. However, undergraduate research experiences at Argonne National Laboratory and at Lawrence Berkeley Laboratory taught me that I didn't have a long enough attention span for the elaborate plumbing and electronics of gas-phase chemical physics. I was later attracted to biological chemistry because of the almost daily interplay of experimental design, observation, and interpretation.

Berkeley, 1970. Carol and I chose the University of California as much for the excitement of life there as for the excellence of its Chemistry Department. My thesis advisor, John Hearst, had an enthusiasm for chromosome structure and function that proved infectious; I have not yet recovered, nor do I wish to. Long days in the laboratory were punctuated by occasional backpacking trips in the alpine splendor of the Sierra Nevada.

In 1975 we obtained our Ph.D.'s and moved to postdoctoral positions in Cambridge, Massachusetts, Carol at Harvard and I at M.I.T. I strengthened my knowledge of biology in Mary Lou Pardue's laboratory, and enjoyed being part of the interactive scientific scene at M.I.T.

We began our first faculty positions at the University of Colorado, Boulder in 1978. I was initially attracted by the enthusiasm and energy of the faculty; I have stayed because in my field the intellectual environment here would be very hard to equal. We have benefitted from very fine colleagues, with whom we have shared many great dinners and ski trips to the nearby Rocky Mountains. More recently, life has been transformed by the addition to our family of two energetic daughters, Allison (born 1982) and Jennifer (1986). It promises to return to normal sometime in the next century.

Because of my research group's discoveries, more than a dozen national and international awards preceded the Nobel Prize for Chemistry in 1989. Among them were the Pfizer Award in Enzyme Chemistry (American Chemical Society), the Award in Molecular Biology (U. S. National Academy of Sciences), the Heineken Prize (Royal Netherlands Academy of Arts and Sciences), and the Lasker Award. I received an honorary D. Sc. degree from Grinnell College in 1987 and from the University of Chicago in 1991. I have been elected to the U. S. National Academy of Sciences (1987) and to the American Academy of Arts and Sciences (1988). In 1987 I was awarded a lifetime Professorship by the American Cancer Society, and in 1988 became Investigator of the Howard Hughes Medical Institute.

SELF-SPLICING AND ENZYMATIC ACTIVITY OF AN INTERVENING SEQUENCE RNA FROM *TETRAHYMENA*

Nobel Lecture, December 8, 1989

by

THOMAS R. CECH

Howard Hughes Medical Institute, Department of Chemistry and Biochemistry, University of Colorado, Boulder, CO 80309-0215, USA

A living cell requires thousands of different chemical reactions to utilize energy, move, grow, respond to external stimuli and reproduce itself. While these reactions take place spontaneously, they rarely proceed at a rate fast enough for life. Enzymes, biological catalysts found in all cells, greatly accelerate the rates of these chemical reactions and impart on them extraordinary specificity.

In 1926, James B. Sumner crystallized the enzyme urease and found that it was a protein. Skeptics argued that the enzymatic activity might reside in a trace component of the preparation rather than in the protein (Haldane, 1930), and it took another decade for the generality of Sumner's finding to be established. As more and more examples of protein enzymes were found, it began to appear that biological catalysis would be exclusively the realm of proteins. In 1981 and 1982, my research group and I found a case in which RNA, a form of genetic material, was able to cleave and rejoin its own nucleotide linkages. This self-splicing RNA provided the first example of a catalytic active site formed of ribonucleic acid.

This lecture gives a personal view of the events that led to our realization of RNA self-splicing and the catalytic potential of RNA. It provides yet another illustration of the circuitous path by which scientific inquiry often proceeds. The decision to expend so many words describing the early experiments means that much of our current knowledge about the system will not be mentioned. For a more comprehensive view of the mechanism and structure of the *Tetrahymena* self-splicing RNA and RNA catalysis in general, the reader is directed to a number of recent reviews (Cech & Bass, 1986; Cech, 1987, 1988a, 1990; Burke, 1988; Altman, 1989). Possible medical and pharmaceutical implications of RNA catalysis have also been described recently (Cech, 1988b).

Why *Tetrahymena?*
In the pre-recombinant DNA era of the early 1970's, much of the research on the structure and function of eukaryotic chromosomes utilized entire

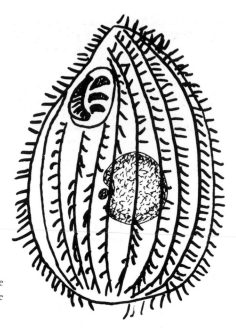

Figure 1. Tetrahymena thermophila, showing the transcriptionallyactive macronucleus and the germ-line micronucleus.

genomes as experimental systems. My own research with John Hearst in Berkeley and with Mary Lou Pardue at M.I.T. concerned the organization of DNA sequences and chromosomal proteins in the mouse genome. During my stay at M.I.T., I began to be dissatisfied with this global approach and became interested in the prospect of being able to dissect the structure and expression of some particular gene. Thus, when I set up my own laboratory in Boulder in 1978, I turned my attention entirely to the rDNA (gene for the large ribosomal RNAs) of the ciliated protozoan, *Tetrahymena* (Figure 1).

Unlike most nuclear genes, which are embedded in giant chromosomes, the genes for rRNA in *Tetrahymena* are located on small DNA molecules in the nucleoli; they are extrachromosomal (Engberg et al., 1974; Gall, 1974). Furthermore, in the transcriptionally active macronucleus the gene is amplified to a level of $\approx 10,000$ copies (Yao et al., 1974). These properties made it possible to purify a significant amount of the rDNA. The ability to purify the gene was not in itself a major attraction, because by this time the availability of recombinant DNA techniques ensured that no gene would long escape isolation and sequence analysis. Rather the attractive feature was the prospect of being able to isolate the gene complete with its associated structural proteins and proteins that regulated transcription (the synthesis of an RNA copy by RNA polymerase).

One feature of the *Tetrahymena* rDNA that was of only peripheral interest to me at that time was the presence of an intervening sequence (IVS) or intron, which interrupted the rRNA-coding sequences of the rDNA of some strains of *Tetrahymena pigmentosa* (Wild & Gall, 1979). In the course of mapping the RNA-coding regions of the rDNA of *Tetrahymena thermophila*,

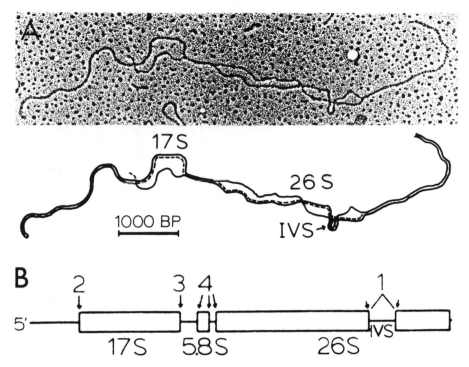

Figure 2. (A) Visualization of the RNA-coding portions of the *T. thermophila* rDNA by electron microscopy. The mature, processed rRNA was hybridized to the DNA under R-loop conditions. In the interpretation, each solid line indicates a single strand of DNA and each dashed line a strand of RNA. (Reproduced from Cech & Rio, 1979) (B) The pre-rRNA, thin lines representing portions that are removed during processing and open boxes representing the mature rRNA sequences. Numbered arrows indicate the usual order of RNA processing events: 1, splicing; 2 − 4, endonucleolytic cleavages.

Don Rio and I found that this species also harbored an IVS in its rDNA (Figure 2; Cech & Rio, 1979; independently described by Din et al., 1979). Although intervening sequences had been discovered only two years before, by Phil Sharp's lab at M.I.T. and a group from Cold Spring Harbor Laboratory, there were already a large number of examples. Thus, the finding of another IVS was hardly cause for us to be distracted from our plan to investigate the proteins that regulated rDNA transcription.

RNA Splicing *in Vitro*
The first step towards biochemical dissection of the transcriptional process was to see if rRNA synthesis would proceed in a crude cell-free system. We isolated nuclei from *T. thermophila* (the nuclei provided both the RNA polymerase and the ribosomal chromatin templates) and incubated them with the nucleoside triphosphates and salts necessary for transcription. We also included α-amanitin, a mushroom toxin known to inhibit the polymerases that transcribe mRNA, tRNA and other small RNAs, which enabled us to focus on synthesis of the large rRNA.

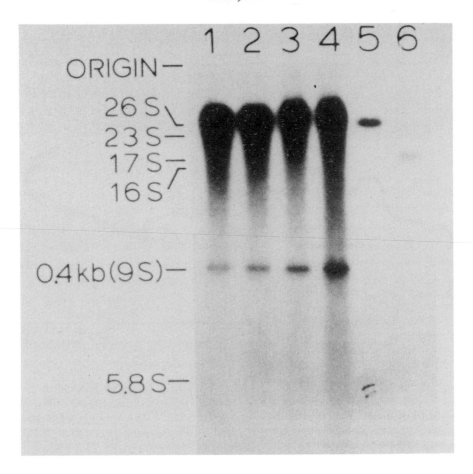

Figure 3. Transcription and splicing of pre-rRNA *in vitro* in isolated *T. thermophila* nuclei. (Lanes 1−4) RNA produced by incubation of nuclei at 30°C for times ranging from 5−60 min. The 0.4 kilobase species is the excised IVS RNA. (Lanes 5 and 6) Purified *Tetrahymena* 26S and 17S rRNAs, serving as molecular weight standards. RNA was analyzed by polyacrylamide gel electrophoresis. (Reproduced from Zaug & Cech, 1980; copyright by Cell Press)

When the products of these *in vitro* transcription reactions were separated by gel electrophoresis, they were found to consist of a somewhat heterogeneous distribution of high molecular weight RNA ≥26 S, the size expected for full-length pre-rRNA (Figure 3). In addition, there was a discrete low molecular weight product (≈9 S). The small RNA accumulated post-transcriptionally (Zaug & Cech, 1980). Thus, it seemed likely to be one of several short regions of the pre-rRNA that was cut out and ultimately discarded during the maturation process. These candidate regions included an external transcribed spacer at the 5' end, the internal transcribed spacers flanking the 5.8 S rRNA, and the IVS (Figure 2B).

Driven more by curiosity than by any conviction that the results would be of central importance to our research goals, I encouraged Art Zaug to identify the sequences encoding the small RNA. He confirmed that the small RNA was encoded by the rRNA gene, and then mapped it to the intervening sequence (Zaug & Cech, 1980; see also Carin et al., 1980).

This was a finding of considerable excitement: the intervening sequence, synthesized as part of the pre-rRNA in our *in vitro* transcription reactions, was also being cleanly excised from the pre-rRNA *in vitro*. Despite a great deal of interest in the mechanism of RNA splicing (Darnell, 1978; Abelson, 1979; Crick, 1979; Lerner et al., 1980), in only one other case − pre-tRNA in yeast − had RNA splicing been confirmed to occur *in vitro* (Knapp et al., 1979; Peebles et al., 1979). It seemed reasonable that rRNA splicing in *Tetrahymena* might follow a quite different path than tRNA splicing in yeast, so that detailed study of both systems would be justified.

Furthermore, in each *Tetrahymena* cell there were 10,000 identical genes each pumping out unspliced pre-rRNA at the rate of one copy per gene per sec. I reasoned that the nuclei might contain an unusually high concentration of the splicing enzyme to accomplish so much reaction, which could facilitate isolation of the first splicing enzyme. Little did we guess that the splicing "enzyme" would not exist in the traditional sense, but that something much more interesting lay in wait for us.

Self-splicing Unrecognized
Our strategy for purifying the splicing enzyme was conventional. We would find conditions in which pre-rRNA transcription occurred but splicing was inhibited, and purify the accumulated pre-rRNA to use as a substrate. We would then treat the pre-rRNA with extracts of *Tetrahymena* nuclei, which we already knew contained splicing activity, and would use gel electrophoresis to monitor the splicing reaction. Finally, we would obtain ever purer subfractions of the nuclear extract that retained activity, eventually isolating the splicing enzyme.

Isolation of unspliced pre-rRNA substrate proved to be straightforward (Cech et al., 1981). Art Zaug purified this RNA by standard SDS-phenol extraction and used it in a series of RNA splicing reactions. One set of test tubes contained substrate RNA and nuclear extract dissolved in the same solution of simple salts and nucleotides that had been conducive to RNA transcription and splicing in our earlier experiments with intact nuclei. Another tube, containing the same components except with the nuclear extract omitted, was to serve as the "splicing minus" control.

The very first attempt was successful. The RNA in the tubes containing the nuclear extract gave rise to the small band of RNA characteristic of the IVS. Surprisingly, however, the control RNA incubated only in salts and nucleotides produced the same amount of IVS. "Well, Art, this looks very encouraging, except you must have made some mistake making up the control sample." Yet several careful repetitions of the experiment gave the same result: release of the IVS occurred independent of the addition of nuclear extract, and therefore apparently independent of any enzyme. I became concerned that we weren't observing RNA splicing at all; perhaps the RNA we had been calling "precursor" had already been spliced *in vivo*, and what we were observing *in vitro* was some disaggregation or release of the IVS from already spliced RNA. Clearly the reaction would be worthy of further pursuit only if we could show that a chemical transformation was occurring *in vitro*, and that it was the same cutting and rejoining that occurred in the living cell. We would have to teach ourselves some RNA chemistry.

The Mystery of the Extra G

To verify the accuracy of RNA splicing in nuclei, we decided to determine the nucleotide sequence of the IVS RNA product. (It was not obvious that this would be particularly illuminating, since we had already shown that the IVS product came from the IVS region of the rDNA and that within the error limits of our measurements it was the correct size to account for the entire IVS.) When Art Zaug labeled the RNA on its 5' end with ^{32}P and subjected it to sequencing reactions, he determined a sequence 5'-GAAAUAGNAA... (where N represents an unidentified nucleotide). The DNA sequence across the exon-IVS junction had been reported earlier for *T. pigmentosa* by Wild and Sommer (1980), and was being determined for *T. thermophila* by Nancy Kan in Joe Gall's laboratory. The *T. thermophila* DNA sequence predicted that the IVS RNA would begin with 5'-AAAUAG-CAA....

Thus, our RNA sequence was a perfect match to the DNA sequence except for the extra G residue on its 5' end. Art Zaug meticulously checked and rechecked the identity of this terminal nucleotide using a variety of enzyme treatments and chromatography systems until there was no doubt: the IVS RNA began with an ordinary guanosine residue, linked to the next nucleotide by a standard 3' — 5' phosphodiester bond such as that produced by RNA polymerase. Clearly the Gall lab, known for the high quality of their science, must have made an error. We telephoned them, advising them that they had determined most of the sequence correctly but had apparently missed one G right at the 5' end of the IVS. Much to our surprise, they defended every nucleotide of their sequence: no ambiguity in the DNA sequence, at least in that region, and no chance of a G at the 5' splice site.

At about the same time, I was working to define the minimum components necessary for the release of the IVS from pre-rRNA. The original experiments had been done in a "transcription cocktail" that included the four nucleoside triphosphates, building blocks for RNA synthesis. I found that removal of three of the NTPs had no effect on the reaction, but the fourth, GTP, was required in micromolar concentration. In addition, IVS release required $MgCl_2$ and was stimulated by certain salts such as $(NH_4)_2SO_4$.

Was it a coincidence that GTP, the nucleotide required for IVS release in our simple *in vitro* system, was also the nucleotide that was found unexpectedly at the 5' end of the excised IVS? Or might there be a causal relationship between the two observations? The obvious hypothesis was that GTP was required so that it could be added to the 5' end of the IVS during splicing.

The test was simple: mix ^{32}P-labeled GTP with unlabeled pre-rRNA, and look for labeling of the IVS RNA concomitant with its excision. The experiment was the strangest I had ever performed. On the one hand, its success was a straightforward prediction from our existing knowledge of the system. On the other hand, it seemed incredibly naive and unrealistic to expect that simple mixing of a nucleotide with phenol-extracted, protein-ase-treated RNA could possibly result in formation of a covalent bond. I certainly didn't want to be embarrassed in front of my graduate students and colleagues by the failure of such an experiment, so I did it very quietly.

The next day I ran the gel, exposed it for autoradiography, and developed the X-ray film. In the sample containing ^{32}P-GTP plus $MgCl_2$ there was a bright signal of radioactivity at the position of the IVS RNA. In a sample containing ^{32}P-GTP but no $MgCl_2$, and in a sample in which ^{32}P-ATP had been substituted for the ^{32}P-GTP, there was no labeled IVS.

Over the next weeks, we confirmed several major features of the reaction. GTP addition was stoichiometric, one GTP per IVS RNA. The GTP was added precisely to the 5' end of the IVS by a normal 3' — 5' phosphodiester bond. Finally, the triphosphate was unnecessary; GMP and even guanosine were active. The last of these observations eliminated one otherwise very reasonable hypothesis, that the GTP was providing an energy source for ligation much as ATP is used by phage T4 RNA ligase. Had that been the case, forms such as guanosine and GMP which are missing the phosphoan-hydride moiety would have been inactive.

It took only a few moments of thought to devise a simple splicing pathway that integrated our new information about the reaction. Addition of guano-sine to the phosphorus atom at the 5' splice site must be occurring by a transfer of phosphate esters, or transesterification reaction (Figure 4A). Such a reaction would free the 5' exon, leaving a new 3' hydroxyl group at its 3' end. The simplest way to proceed from such proposed intermediates to the final observed products was to invoke a second transesterification reaction: attack of the 3' hydroxyl of the 5' exon at the 3' splice site. Thus, a single active site capable of promoting transesterification could be responsi-ble for the entire splicing reaction.

The model shown in Figure 4A has undergone little change since we first described it in 1981 and drew a more explicit version in 1982 (Cech et al., 1981; Zaug & Cech, 1982). It has been a good predictor of a great many other IVS-catalyzed reactions since then (e.g., Zaug et al., 1983; Tabak et al., 1987). Furthermore, the model has been strongly bolstered by the isolation and characterization of the proposed intermediates (Inoue et al., 1986), by the demonstration of the reversibility of the reactions (Sullivan & Cech, 1985; Woodson & Cech, 1989), and by determination of the stereo-chemical course (McSwiggen & Cech, 1989; Rajagopol et al., 1989). An atomic-level model of transesterification occurring by an $S_N2(P)$ mechanism is presented in Figure 4B.

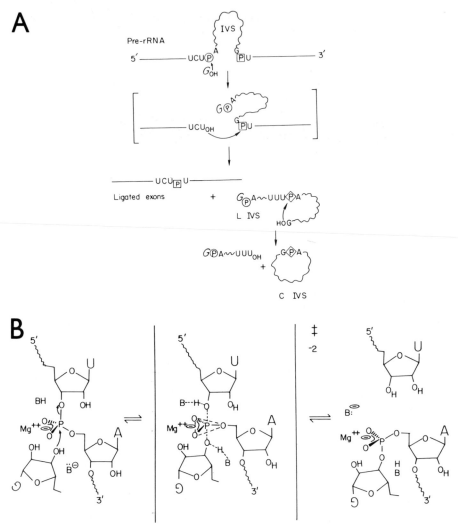

Figure 4. (A) Self-splicing of the *Tetrahymena* pre-rRNA by consecutive transesterification reactions. Straight lines, exons (mature rRNA sequences); wavy line, IVS; circle, 5' splice-site phosphate; square, 3' splice-site phosphate; diamond, cyclization site phosphate. (B) Model for the initial step involving nucleophilic attack by the 3'-hydroxyl group of guanosine on the phosphorus atom at the 5' splice site. The hypothesis of an in-line, S_N2 (P) reaction with inversion of configuration around phosphorus was subsequently confirmed (McSwiggen & Cech, 1989). The hypotheses of acid-base catalysis and coordination of Mg^{2+} to the phosphate, enhancing the electrophilicity of the phosphorus atom and stabilizing the trigonal bipyramid transition state, are untested. (Reproduced from Cech, 1987; copyright by the AAAS)

Self-splicing Recognized

The guanosine-addition reaction provided us with the proof we needed: specific RNA bond breakage and formation were occurring in our simple *in vitro* splicing reaction. Such a chemically difficult reaction between very unreactive molecules certainly had to be catalyzed. But what was the catalyst?

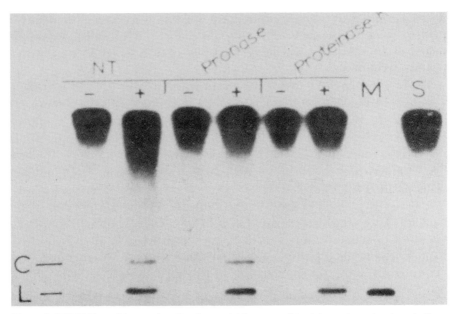

Figure 5. IVS RNA excision and cyclization activities are resistant to protease treatment. (Lane S) Pre-rRNA' purified by SDS-phenol extraction. This RNA was then treated exhaustively with proteases or (NT) not treated. Incubation with GTP was performed, (lanes -) in the absence of MgCl₂ or (lanes +) with 10 mM MgCl₂. Excision of the IVS RNA as a linear molecule (L) and subsequent conversion to a circular form (C) were undiminished by protease treatment. (Lane M) isolated linear and circular IVS RNAs. (Reproduced from Grabowski et al., 1983, by permission of Alan R. Liss, Inc.)

Our first hypothesis was that the splicing activity was a protein tightly bound (perhaps even covalently bonded) to the pre-rRNA isolated from *Tetrahymena* nuclei. This would have to be a very unusual protein-RNA complex to survive the multiple forms of abuse to which we had subjected it: boiling in the presence of the detergent SDS, SDS-phenol extraction at temperatures as high as 65°C, and extensive treatment with several nonspecific proteases (Figure 5). That we took this hypothesis seriously provides an indication of how deeply we were steeped in the prevailing wisdom that only proteins were capable of highly efficient and specific biological catalysis.

In the same paper, we described an alternative hypothesis:

The resistance of the splicing activity to phenol extraction, SDS and proteases can also be interpreted in a more straightforward manner. The rRNA precursor might be able to undergo splicing without the participation of a protein enzyme. A portion of the RNA chain could be folded in such a way that it formed an active site or sites that bound the guanosine cofactor and catalyzed the various bond-cleavage and ligation events. If one of the RNA molecules produced in the reaction (for example, the free IVS) retained is activity and catalyzed additional splicing events, then it would be an example of an RNA enzyme. (Cech et al., 1981)

As we accumulated negative result upon negative result trying to identify a protein stuck to the pre-rRNA, the alternative "RNA only" hypothesis began to appear more and more attractive. But how could we obtain a positive result to prove it?

The best strategy we could devise was to synthesize the RNA in as artificial a manner as possible, so that it was never in contact with the *Tetrahymena* cells that up to now had been our sole source of the RNA. Complete chemical synthesis of an RNA the size of the pre-rRNA (or even the size of the IVS) was and still is beyond the scope of available technology. The next best approach was to synthesize RNA from a recombinant DNA template using purified RNA polymerase.

A bacterial plasmid (Figure 6) encoding the IVS and a portion of the flanking rRNA sequences, situated so as to allow transcription by *Escherichia coli* RNA polymerase, was already being constructed in the lab by Kelly Kruger. The original purpose was to facilitate synthesis of large quantities of pre-rRNA substrate for isolation of the splicing enzyme. The cloning

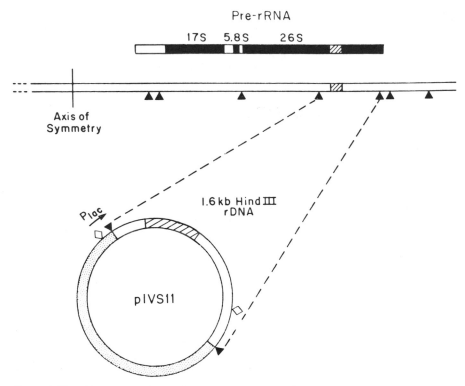

Figure 6. Plasmid constructed to enable synthesis of an artificial, shortened version of the pre-rRNA. (Top) Diagram of the natural pre-rRNA, with transcribed spacers shown as open boxes, mature rRNA sequences as solid boxes, and the IVS as a hatched box. (Middle) The *T. thermophila* rDNA. One half of a palindromic rDNA molecule is shown. (Bottom) Plasmid pIVSll containing the 1.6 kilobase *Hind* III fragment of the rDNA inserted adjacent to P_{lac}, a promoter for transcription *by E. coli* RNA polymerase. Restriction endonuclease sites are *Hind* III (▲) and *Eco* RI (◇). (Adapted from Kruger et al., 1982; copyright by Cell Press)

took longer than any of us had anticipated, such that by the time it was accomplished early in 1982 we were desperate to have the plasmid for a different purpose: to produce synthetic pre-rRNA for the self-splicing test.

The plasmid was grown up in *E. coli*, carefully deproteinized, and incubated with purified *E. coli* RNA polymerase under conditions that we already knew were inhibitory for splicing. The polymerase was then destroyed and the RNA purified by gel electrophoresis under denaturing conditions. Upon addition of GTP, MgCl$_2$ and salt, the IVS RNA was released from this artificial, shortened pre-rRNA. The site at which GTP broke the RNA chain was exactly the position that served as the 5' splice site *in vivo*, providing some confidence that self-splicing was relevant to splicing as it occurred in the living cell.

We held a relatively subdued celebration in the lab. Between sips of champagne we compiled a list of possible general names for RNA molecules able to lower the activation energy for specific biochemical reactions. It was then that we coined the term "ribozyme," for a ribonucleic acid with enzyme-like properties.

RNA in Circles

Concurrent with our studies of pre-rRNA splicing in isolated nuclei, we were pursuing a post-splicing phenomenon: the conversion of the excised intervening sequence RNA into a circular RNA molecule in isolated *Tetrahymena* nuclei (Grabowski et al., 1981). The circular form survived treatment with protease and various denaturants, which suggested that it was a covalently closed circle of RNA. As Paula Grabowski, then a graduate student in the laboratory, characterized the cyclization reaction, she found that it occurred with extensively deproteinized RNA in a simple buffered MgCl$_2$ solution. I dismissed her original observation of protein-free cyclization as being an artifact of incomplete denaturation of the linear RNA. Yet, as the experiments proved reproducible, I began to derive some solace in the knowledge that two researchers, Zaug and Grabowski, studying two different reactions, splicing and cyclization, were both finding activity in the absence of added protein. Having two strange results somehow made me more comfortable than just a single strange result. The two sets of observations came together when we found that the plasmid transcripts that underwent self-splicing produced IVS RNA that underwent self-cyclization (Kruger et al., 1982).

The circular IVS RNA was not formed by end-to-end joining of the linear form. Instead, the 3' end of the linear IVS attacked an internal phosphorus atom near the 5' end of the molecule, clipping off a short oligonucleotide in the process (Figure 4A). Thus cyclization, like RNA splicing, occurred by transesterification (Zaug et al., 1983).

How Does the RNA Do It?

At first glance, RNA seemed ill suited to be a catalyst. With its four rather similar bases, RNA would appear greatly limited in its ability to form a specific substrate-binding pocket. In contrast, the 20 amino acids found in proteins explore a wide range of sizes and shapes, hydrophilicity and hydrophobicity. In terms of promoting chemistry, RNA has a dearth of functional groups that are ionizable near neutral pH, whereas proteins have histidine and cysteine (pK$_a$'s in the range of 6–8). How then, was the *Tetrahymena* IVS able to catalyze transesterification?

The first glimpse of the catalytic mechanism came from detailed studies of the guanosine requirement. Brenda Bass tested every available guanosine analog and found great variation in their activity (Bass & Cech, 1984, 1986). Derivatives carrying bulky substituents on the 7 or 8 position of the guanine base or on the 5' position of the ribose sugar were as active as guanosine, indicating that these positions did not interact with the IVS RNA. Other derivatives were fully active but only at high concentration, or were inactive as substrates but acted as competitive inhibitors of the reaction of guanosine. Based on the K$_m$ or K$_i$ of these guanosine analogs, we could assign free energy contributions to individual functional groups of guanosine. All the data pointed to the IVS containing a well-behaved binding site for guanosine. The site has recently been located within the IVS in elegant work by Michel et al. (1989).

The existence of a G-binding site explained the high specificity for guanosine. In addition, by orienting the nucleophile with respect to the 5'-splice site phosphate, the G-site would contribute substantially to rate acceleration. RNA catalysis was suddenly in a familiar context; the loss of entropy and orientation of reacting groups achieved by formation of a specific enzyme-substrate complex is central to catalysis by protein enzymes (Jencks, 1969; Fersht, 1985).

The other reactant in the first transesterification step, the phosphorus atom at the 5' splice-site, is also held in place by a binding interaction. As first proposed by Davies et al. (1982) and also apparent in models of Michel et al. (1982), a 5' exon-binding site within the IVS base-pairs to the last few nucleotides of the 5' exon. This pairing interaction specifies the site of guanosine addition and also holds the 5' exon into place for the second step of splicing, exon ligation (Waring et al., 1986; Been & Cech, 1986; Price et al., 1987; Barfod & Cech, 1989).

The IVS does much more than simply hold the reacting groups in place. In the absence of guanosine, hydrolysis occurs specifically at the splice-site phosphodiester bonds, producing 5' phosphate/3' hydroxyl termini (Zaug et al., 1984; Inoue et al., 1986). A 3' hydroxyl is the same as the product of self-splicing but opposite to the product of random alkaline hydrolysis of RNA. This site-specific hydrolysis reaction reflects the ability of the catalytic center of the IVS to activate the splice-site phosphates (or perhaps activate the nucleophile, in which case it must be able to activate OH⁻ as well as the hydroxyl of guanosine). While the structural basis of this activation is unknown, reasonable hypotheses include stabilization of the pentacoordinate transition-state structure of the phosphate and specific coordination of a Mg^{2+} ion (Zaug et al., 1984; Guerrier-Takada et al., 1986; Cech, 1987;

Grosshans & Cech, 1989; Sugimoto et al., 1989). Once again, in a general sense the RNA catalyst is recapitulating a major catalytic strategy of protein enzymes, or vice versa.

Tetrahymena is Not Alone

While we were intently characterizing splicing of the *Tetrahymena* IVS, we were also wondering when (and perhaps if) related intervening sequences would be found in other organisms. The differences in nucleotide sequences near the splice sites made it seem unlikely that *Tetrahymena* rRNA splicing would be related to nuclear tRNA or mRNA splicing. The related intervening sequences came from an unexpected direction: yeast mitochondria. This was unanticipated because mitochondria, thought to have arisen from symbiotic prokaryotes, do not usually have genes or modes of gene expression similar to those of eukaryotes. Furthermore, yeast is evolutionarily extremely distant from *Tetrahymena*.

In 1982, several groups identified short sequence elements that were conserved among a group of fungal mitochondrial intervening sequences (Burke & RajBhandary, 1982; Davies et al., 1982; Michel et al., 1982). Furthermore both Michel and the Davies group had proposed that the

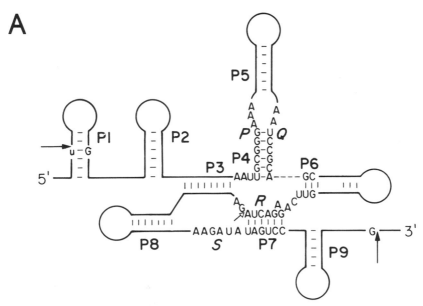

Figure 7. Secondary structure of group I intervening sequences determined by Michel et al. (1982) and Davies et al. (1982) by the method of comparative sequence analysis (Fox & Woese, 1975; Noller & Woese, 1981). (A) Generalized structure indicating conserved base-paired elements Pl — P9. Conserved sequence elements *P, Q, R* and *S* are represented by their most common nucleotide sequences. Filled arrows, 5' and 3' splice sites. Open arrow, site of insertion of extra stem-loop(s) in group IA structures. (B) Secondary structure of the *T. thermophila* rRNA IVS, with the 5' exon-binding site and the core structure most conserved in group I IVSs shaded. Guanosine-binding site (G-site) was located by Michel et al. (1989). UV-induced crosslink (X-link) identifies a tertiary structure interaction (Downs & Cech, 1990). Lowercase letters, exons. Uppercase letters, IVS. (Adapted from Cech, 1988b, by permission of Elsevier)

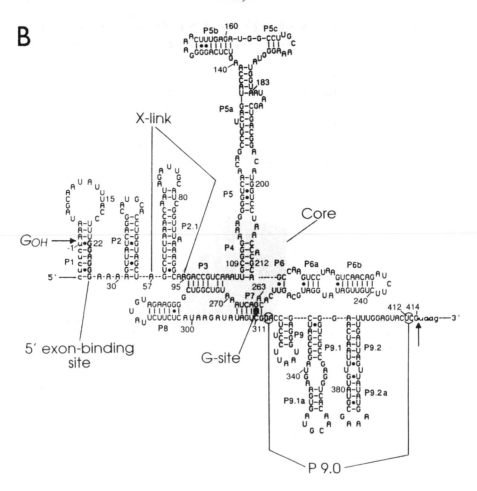

conserved sequence blocks interacted to form a common set of short base-paired regions, serving to fold the intervening sequences into the same fundamental secondary structure (Figure 7A). Michel called these the group I introns, to distinguish them from a second group of mitochondrial introns that shared a different structure (Michel & Dujon, 1983).

The *Tetrahymena* rRNA IVS contained the conserved sequence elements and secondary structures characteristic of the mitochondrial group I (Michel & Dujon, 1983; Waring et al., 1983). Furthermore, a very similar structure model of the *Tetrahymena* IVS was independently derived by another method, free energy minimization as constrained by experimentally determined sites of cleavage of the folded RNA by various nucleases (Cech et al., 1983). A current version of the secondary structure is shown in Figure 7B.

This convergence of two previously noninteracting sets of ideas was important in several respects. In terms of splicing mechanisms, it was now unlikely that the *Tetrahymena* intron would be unique; we could extend our knowledge of its mechanism by comparing and contrasting splicing of different members of the group. Second, a believable model of the secondary structure of the *Tetrahymena* IVS was now in hand, and one could begin to formulate structure-function relationships. Finally, the similarity between the *Tetrahymena* rRNA and fungal mitochondrial introns might be revealing their origin; perhaps they were transposable elements able to enter both nuclear and mitochondrial compartments (Cech et al., 1983).

Waiting for Number Two

If RNA catalysis were of any general significance to biology, there would be additional examples. Throughout 1982 and most of 1983, none came forth. Yet there seemed to be some reasonable candidates. In the fall of 1983, we wrote an article in which we speculated:

> Several enzymes, such as RNase P, 1,4-α-glucan branching enzyme and potato *o*-diphenol oxidase, have RNA components essential for their catalytic activities. The peptidyl transferase activity of ribosomes also requires an RNA-protein complex. It remains to be seen whether the RNA is directly involved in the active site of any of these ribonucleoprotein enzymes. (Bass and Cech, 1984)

The speculation about RNase P was already outdated when our paper was published in 1984, because by that time Guerrier-Takada et al. (1983) had announced that the RNA component was the catalytic subunit of that ribonucleoprotein enzyme. This was followed by the report early in 1984 that a synthetic *E. coli* RNase P RNA, transcribed *in vitro* from a recombinant DNA template, also had intrinsic catalytic activity; the possibility that catalysis was due to protein contamination was thereby eliminated (Guerrier-Takada and Altman, 1984). Similarly, the RNA subunit of *Bacillus subtilis* RNase P acted as an enzyme *in vitro* (Guerrier-Takada et al., 1983; Marsh & Pace, 1985). The RNase P discovery was very exciting to us. Not only did it provide a second example of an RNA molecule that lowered the activation energy for a specific biochemical reaction, but it was the first proven case of an RNA molecule that catalyzed a reaction without itself undergoing any net change. RNA catalysis was not restricted to the realm of intramolecular catalysis.

Within the next year, self-splicing of additional group I intervening sequences was reported. Garriga and Lambowitz (1984) found that the first IVS of the cytochrome *b* pre-mRNA from *Neurospora* mitochondria underwent self-splicing *in vitro*. Several self-splicing group I RNAs from yeast mitochondria were characterized by van der Horst and Tabak (1985). Most unexpectedly, self-splicing group I IVSs were found in three bacteriophage T4 mRNAs (Belfort et al., 1985; Ehrenman et al., 1986; Gott et al., 1986); RNA splicing took place in a prokaryote. In all cases, splicing occurred by the same G-addition pathway as splicing of the *Tetrahymena* nuclear rRNA IVS.

The mitochondrial group II intervening sequences have conserved sequences and secondary structures distinct from those of group I (Michel and Dujon, 1983). Peebles et al. (1986) and Van der Veen et al. (1986)

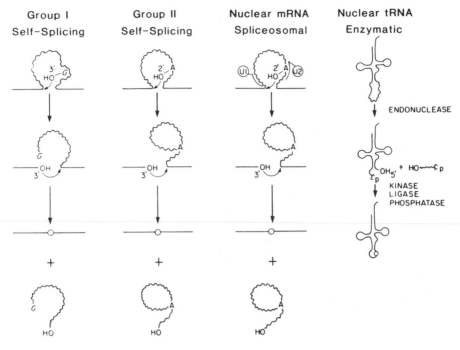

Figure 8. Mechanism of splicing of the four major groups of precursor RNAs. Wavy lines indicate IVSs, smooth lines indicate flanking exons. For nuclear mRNA splicing, many components assemble with the pre-mRNA to form the spliceosome; only the Ul and U2 small nuclear ribonucleoproteins are shown here. (Adapted from Cech, 1986a; copyright by Cell Press)

discovered that pre-mRNA containing a group II intervening sequence was self-splicing *in vitro*. The reaction did not require guanosine, and occurred by formation of a branched "lariat" RNA. The proposed mechanism is shown in Figure 8.

The fundamental chemistry of group II RNA splicing appears to be the same as that of nuclear pre-mRNAs, which do not self-splice. Instead, nuclear mRNA splicing requires assembly of the substrate with a large complex of proteins and small nuclear ribonucleoproteins to form the spliceosome (Brody & Abelson, 1985; Frendewey & Keller, 1985; Grabowski et al., 1985). The mechanistic similarities have led to the speculation that nuclear mRNA splicing may also be RNA catalyzed, with much of the catalysis being provided in the form of the small nuclear RNAs (Kruger et al., 1982; Maniatis and Reed, 1987).

Enzymologists Outraged
Although our description of RNA self-splicing was shocking to many, it was quickly accepted by the scientific community. In contrast, our use of the words "catalysis" and "enzyme-like" to describe the phenomenon provoked some much more heated reactions.

Our reasons for emphasizing the relationship between RNA self-splicing and biological catalysis might be better appreciated in the context of the

Table 1. *Defining characteristics of biological catalysis.*

Characteristic	Self-splicing	L − 19 IVS RNA
1. Rate acceleration	√	√
2. Specificity	√	√
3. Catalyst regenerated	*	√

* No, although the active site is preserved through the reactions. (From Cech, 1988c, by permission of Alan R. Liss Inc.)

definition given in Table 1. First, biological catalysts achieve rate accelerations of the order of 10^6- to 10^{13}-fold, bringing the reactions into a time scale that is useful for living systems. Self-splicing clearly meets this criterion; although the quantitation of rate acceleration can be done only for the site-specific hydrolysis reaction promoted by the RNA active site, even this relatively slow side-reaction occurs 10^{10}-fold faster than the estimated uncatalyzed rate (Figure 9). Second, the extraordinary specificity of biological catalysts is evident in the self-splicing reaction; the molecule selects GTP as the attacking group for the first step of the reaction and is able to choose 2 of the 6000 nucleotides in the pre-rRNA as splice sites. On the other hand, the IVS RNA is clearly not regenerated in exactly the same form as it entered the reaction; after all, the purpose of self-splicing is to convert pre-rRNA to ligated exons plus excised IVS. Nevertheless, enzymologists do speak of intramolecular catalysis (Jencks, 1969; Bender et al., 1984; Fersht, 1985), and we thought that such a descriptor was particularly appropriate for self-splicing.

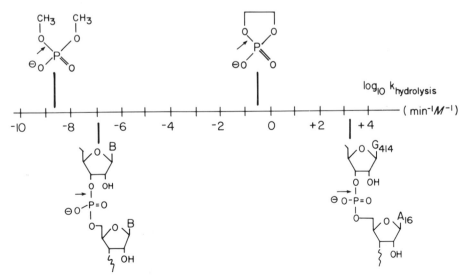

Figure 9. The IVS RNA has an extremely efficient catalytic center. Second-order rate constants (42°C) for the alkaline hydrolysis of phosphate diesters are displayed on a logarithmic scale. Arrows designate the P-O bond that undergoes cleavage. Above line, data for dimethyl phosphate and ethylene phosphate from Kumamoto et al. (1956) and Haake and Westheimer (1961). Below line, hydrolysis of RNA (left) uncatalyzed and (right) catalyzed by the catalytic center of the *Tetrahymena* IVS. (Reproduced from Cech, 1987; copyright by the AAAS)

The enzymologists who reviewed Bass & Cech (1984) were far from convinced. All three referees wrote thoughtful reviews, expressing considerable interest in the data but chastising us for our naivete about enzymes. For example:

> Enzymes are true catalysts: they speed the rate of a reaction, but are themselves unchanged by the reaction. In the present instance, the ribozyme acts in a "one-shot" reaction and is permanently changed so that it can no longer cycle as does a true catalyst. The authors are well aware of this, but appear to ignore this key feature in making their comparisons.

How fundamental was this distinction between self-splicing and catalysis? Opinions would probably vary as much today as they did six years ago. Instead of engaging in a protracted argument, we decided to test experimentally whether the self-splicing RNA could be converted into a multiple-turnover catalyst. Arthur Zaug made a slight alteration of the self-splicing IVS, removing its first 19 nucleotides. The resulting L − 19 IVS RNA met all three criteria of a biological catalyst (Table 1 and Zaug & Cech, 1986a).

Shortened forms of the *Tetrahymena* IVS have multiple enzymatic activities. Depending on the substrates with which they are presented, these RNA enzymes can catalyze nucleotidyltransfer, phosphotransfer and hydrolysis reactions; nucleotidyltransfer can result in either endonucleolytic cleavage or polymerization of RNA (Zaug & Cech, 1986a,b; Zaug et al., 1986; Kay & Inoue, 1987; Been & Cech, 1986, 1988; Doudna & Szostak, 1989).

As an example, consider the reaction diagrammed in Figure 10. A form of the IVS RNA missing both its splice sites catalyzes the cleavage of other RNA molecules after sequences resembling that of the normal 5' exon, CUCU. The reaction is an intermolecular version of the first step of RNA self-splicing, in which RNA cleavage is accompanied by covalent joining of guanosine to the 5' end of the 3' cleavage product. Because the IVS RNA is

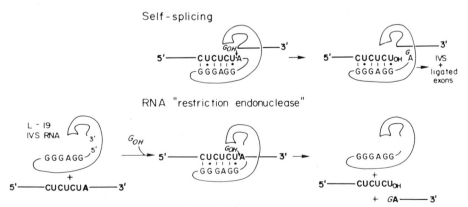

Figure 10. A shortened version of the *Tetrahymena* IVS RNA has enzymatic activity as an endonuclease. The mechanism is an intermolecular version of the first step of pre-rRNA self-splicing. Thin letters and lines, IVS sequences; bold letters and thick lines, exon sequences (top) or substrate RNA sequences (bottom); G in italics, free guanosine or GTP. (Reproduced by permission from Zaug et al., 1986; copyright Macmillan Journals Limited)

unaltered by the reaction, it can sequentially bind and process a large number of substrate RNA molecules. The k_{cat}/K_m for cleavage of an oligoribonucleotide substrate containing a CCCUCU recognition sequence is 10^8 M^{-1} min^{-1} (Herschlag & Cech, 1990), well within the range of protein enzymes.

Site-directed mutagenesis of the 5' exon-binding site of the IVS redirects substrate specificity as predicted by the rules of Watson-Crick base-pairing (Zaug et al., 1986; Murphy & Cech, 1989). Thus, it has been possible to create a whole set of "RNA restriction endonucleases" that may be of use for the sequence-specific cleavage of RNA.

Converting a self-processing RNA into an RNA enzyme by physically separating its internal substrate from its catalytic center has been generally successful. The "hammerhead" and "hairpin" ribozymes, both found in plant infectious agents (viroids or viral satellite RNAs), undergo self-cleavage leaving 2',3'-cyclic phosphate termini (Prody et al., 1986; Forster & Symons, 1987). Both have been converted into RNA enzymes (Uhlenbeck, 1987; Haseloff & Gerlach, 1988; Hampel & Tritz, 1989). A group II IVS can cleave RNA containing a 5' splice-site or ligated exon RNA in an intermolecular reaction (Jacquier & Rosbash, 1986; Jarrell et al., 1988), although in these cases multiple turnover was not demonstrated. Even the intramolecular lead-cleavage reaction of tRNA can be converted into an enzymatic system (Sampson et al., 1987). The ability of pieces of a structured RNA molecule to self-assemble, recreating an active unit, therefore appears to be a general property of the RNA biopolymer. In addition, if the fragment which undergoes reaction is secured to the rest of the molecule by a relatively weak interaction such as a few base-pairs, it will dissociate quickly enough to permit multiple turnovers.

Origin of Life Fantasies

The discoveries of RNA self-splicing and the enzymatic activity of RNase P RNA rekindled earlier speculation concerning the possible role of RNA in the origin of life (Woese, 1967; Crick, 1968; Orgel, 1968). Contemporary cells depend on a complex interplay of nucleic acids and proteins, the former serving as informational molecules and the latter as the catalysts that replicate and express the information. Certainly the first self-reproducing biochemical system also had an absolute need for both informational and catalytic molecules. The dilemma was therefore *Which came first, the nucleic acid or the protein, the information or the function?* One solution would be the co-evolution of nucleic acids and proteins (Eigen, 1971). The finding that RNA can be a catalyst as well as an information-carrier lent plausibility to an alternative scenario: the first self-reproducing system could have consisted of RNA alone (Sharp, 1985; Pace & Marsh, 1985; Orgel, 1986).

Perhaps coincidentally or perhaps because of its ancestry, one of the reactions catalyzed by the *Tetrahymena* IVS RNA enzyme is a nucleotidyl transfer reaction with fundamental similarity to the reaction catalyzed by RNA replicases (Zaug & Cech, 1986a; Been & Cech, 1988). A specific model

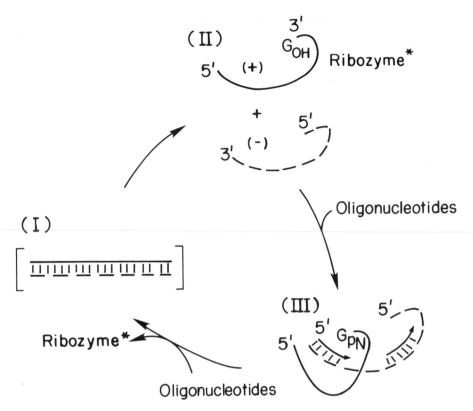

Figure 11. Hypothetical model for RNA self-replication involving an RNA catalyst (ribozyme*).
Double-stranded RNA (I) undergoes strand separation to give ribozyme* ((+) strand) and the
complementary (−) strand (II). The ribozyme* catalyzes synthesis of a new (+) strand, using the
(−) strand as a template (III). The detailed mechanism is described by Cech (1986b). Completion of synthesis reforms the double-stranded RNA (I). A second cycle is needed to achieve
replication of the starting material.

for RNA self-replication based on the properties of the catalytic center of
the *Tetrahymena* IVS RNA is given in Figure 11. The RNA enzyme, ribozyme*, differs from the L − 19 IVS RNA in that it utilizes an external
rather than an internal template (Cech, 1986b). Separation of the template
region from the rest of the catalytic center of the RNA with retention of
activity has recently been achieved by Doudna and Szostak (1989).

Now that we have examples of catalytic RNAs, it has been entertaining to
look back at earlier speculations about the catalytic potential of RNA. As a
representative example, consider the following:

... in the evolutionary scheme, folded nucleic acid structures were abandoned by nature in favour of proteins. Although nucleic acids may have
performed many enzymatic tasks in primitive cells, this is much more
efficiently done by proteins. ... With nucleic acids the four bases are all of
one structural type, though the existence of many modified bases points
to an evolutionary proliferation giving much greater possibilities in form-

ing structures. Nevertheless, the use of bases cannot match the enormous flexibility provided by having twenty amino acids which fall into three or four different structural types. (Klug et al., 1974)

One implication is that nucleic acids, if they could have any such activity, would by nature be inferior catalysts. At one time I had a similar bias, but it was gradually dispelled as we quantitated the rate acceleration and specificity inherent to the catalytic center of the *Tetrahymena* ribozyme. The second conclusion, regarding the limited versatility of RNA catalysts, still strikes me as being correct. In all well established examples, the substrate for an RNA catalyst or ribonucleoprotein enzyme is RNA or DNA or, in the case of protein synthesis on ribosomes, the closely related aminoacyl-tRNA. To its credit, RNA can form a specific binding site for at least one amino acid (Yarus, 1988), and there is evidence that covalent linkage of the terminal protein to poliovirus RNA is at least in part RNA-catalyzed (Tobin et al., 1989). Nevertheless, it seems unlikely that RNA can match the enormous variety of binding sites that can be formed from amino acid side chains. The list of RNA catalysts is still growing quite rapidly. Yet it seems likely that if an entire list of biological catalysts is ever complete, it will include more proteins than RNA molecules.

ACKNOWLEDGEMENTS

I dedicate this lecture to my wife Carol and to my parents, Robert and Annette. I am indebted to my coworkers over the past decade, who performed the experiments and contributed many of the ideas about RNA catalysis described here. In addition to those whose work I have described, there are many who made important contributions that have gone unmentioned here but not at all unappreciated. Funding for our research has been provided principally by the National Institutes of Health and the American Cancer Society, and more recently by the Howard Hughes Medical Institute; I thank them for their uninterrupted and enthusiastic support. Finally, I have enjoyed the comradeship and helpful criticism of the RNA research community, both in Boulder and worldwide.

REFERENCES

Abelson J. (1979) *Ann. Rev. Biochem.* **48**, 1035–1069.
Altman S. (1989) *Adv. Enzym.* **62**, 1–36.
Barfod, E. T. & Cech, T. R. (1989) *Molec. Cell. Biol.* **9**, 3657–3666.
Bass, B. L. & Cech, T. R. (1984) *Nature* **308**, 820–826.
Bass, B. L. & Cech, T. R. (1986) *Biochemistry* **25**, 4473–4477.
Been, M. D. & Cech, T. R. (1986) *Cell* **47**, 207–216.
Been, M. D. & Cech, T. R. (1988) *Science* **239**, 1412–1416.
Belfort, M., Pedersen-Lane, J., West, D., Ehrenman, K., Maley, G., Chu, F. & Maley, F. (1985) *Cell* **41**, 375–382.
Bender, M. L., Bergeron, R. J. & Komiyama, M. (1984) *The Bioorganic Chemistry of Enzymatic Catalysis* (John Wiley, New York).

Brody, E. & Abelson, J. (1985) *Science* **228,** 963 – 966.

Burke, J.M. (1988) *Gene* **73,** 273 – 294.

Burke, J. & RajBhandary, U. L. (1982) *Cell* **31,** 509 – 520.

Carin, M., Jensen, B. F., Jentsch, K. D., Leer, J. C., Nielson, O. F. & Westergaard, O. (1980) *Nucleic Acids Res.* **8,** 5551 – 5566.

Cech, T. R. (1986a) *Cell* **44,** 207 – 210.

Cech, T. R. (1986b) *Proc. Natl. Acad. Sci. USA* **83,** 4360 – 4363.

Cech, T. R. (1987) *Science* **236,** 1532 – 1539.

Cech, T. R. (1988a) *Gene* **73,** 259 – 271.

Cech, T. R. (1988b) *J. Am. Med. Assn.* **260,** 3030 – 3034.

Cech, T. R. (1988c) *Harvey Lec.* **82,** 123 – 144.

Cech, T. R. (1990) *Ann. Rev. Biochem.* **59,** in press.

Cech, T. R. & Bass, B. L. (1986) *Ann. Rev. Biochem.* **55,** 599 – 629.

Cech, T. R. & Rio, D. C. (1979) *Proc. Natl. Acad. Sci. USA* **76,** 5051 – 5055.

Cech, T. R., Zaug, A. J. & Grabowski, P. J. (1981) *Cell* **27,** 487 – 496.

Cech, T. R., Tanner, N. K., Tinoco, I., Jr., Weir, B. R., Zuker, M. & Perlman, P. S. (1983) *Proc. Natl. Acad. Sci. USA* **80,** 3903 – 3907.

Crick, F. H. C. (1968) *J. Mol. Biol.* **38,** 372.

Crick, F. (1979) *Science* **204,** 264 – 271.

Darnell, J. E., Jr. (1978) *Science* **202,** 1257 – 1260.

Davies, R. W., Waring, R. B., Ray, J. A., Brown, T. A. & Scazzocchio, C. (1982) *Nature* **300,** 719 – 724.

Din, N., Engberg, J., Kaffenberger, W. & Eckert, W. (1979) *Cell* **18,** 525 – 532.

Doudna, J. A. & Szostak, J. W. (1989) *Nature* **339,** 519 – 522.

Downs, W. D. & Cech, T. R. (1990) *Biochemistry*, in press.

Ehrenman, K., Pedersen-Lane, J., West, D., Herman, R., Maley, F. & Belfort, M. (1986) *Proc. Natl. Acad. Sci. USA* **83,** 5875 – 5879.

Eigen, M. (1971) *Naturwissenschaften* **58,** 465 – 523.

Engberg, J., Nilsson, J. R., Pearlman, R. E. & Leick, V. (1974) *Proc. Natl. Acad. Sci. USA* **71,** 894 – 898.

Fersht, A. (1985) *Enzyme Structure and Mechanism*, ed. **2,** (Freeman, New York).

Forster, A. C. & Symons, R. H. (1987) *Cell* **49,** 211 – 220.

Fox, G. & Woese, C. R. (1975) *Nature* **256,** 505 – 507.

Frendewey, D. & Keller, W. (1985) *Cell* **42,** 355 – 367.

Gall, J. G. (1974) *Proc. Natl. Acad. Sci. USA* **71,** 3078 – 3081.

Garriga, G. & Lambowitz, A. M. (1984) *Cell* **39,** 631 – 641.

Gott, J. M., Shub, D. A. & Belfort, M. (1986) *Cell* **47,** 81 – 87.

Grabowski, P. J., Zaug, A. J. & Cech, T. R. (1981) *Cell* **23,** 467 – 476.

Grabowski, P. J., Brehm, S. L., Zaug, A. J., Kruger, K. & Cech, T. R. (1983) In *GENE EXPRESSION*, D. H. Hamer and M. J. Rosenberg, eds. (Alan R. Liss, New York) Vol. **8,** 327 – 342.

Grabowski, P. J., Seiler, S. R. & Sharp, P. A. (1985) *Cell* **42,** 345 – 353.

Grosshans, C. A. & Cech, T. R. (1989) *Biochemistry* **28,** 6888 – 6894.

Guerrier-Takada C., & Altman, S. (1984) *Science* **223,** 285 – 286.

Guerrier-Takada, C., Gardiner, K., Marsh, T., Pace, N. & Altman, S. (1983) *Cell* **35,** 849 – 857.

Guerrier-Takada, C., Haydock, K., Allen, L. & Altman, S. (1986) *Biochemistry* **25,** 1509 – 1515.

Haake, P. C. & Westheimer, F. H. (1961) *J. Am. Chem. Soc.* **83,** 1102 – 1109.

Haldane, J. B. S. (1930) *Enzymes* (Longmans, Green & Co., Great Britain). (1965, M. I. T. Press, Cambridge, Mass.).

Hampel, A. & Tritz, R. (1989) *Biochemistry* **28,** 4929 – 4933.

Haseloff, J. & Gerlach, W. L. (1988) *Nature* **334,** 585 – 591.

Herschlag, D. & Cech, T. R. (1990) *Nature* **344,** 405 – 409.

Inoue, T., Sullivan, F. X. & Cech, T. R. (1986) *J. Mol. Biol.* **189,** 143 – 165.

Jacquier, A. & Rosbash, M. (1986) *Science* **234,** 1099 – 1104.

Jarrell, K. A., Peebles, C. L., Dietrich, R. C., Romiti, S. L. & Perlman, P. S. (1988) *J. Biol. Chem.* **263,** 3432 – 3439.

Jencks, W. P. (1969) *Catalysis in Chemistry and Enzymology,* (McGraw-Hill, New York).

Kay, P. S. & Inoue, T. (1987) *Nature* **327,** 343 – 346.

Klug, A., Ladner, J. & Robertus, J. D. (1974) *J. Mol. Biol.* **89,** 511 – 516.

Knapp, G., Ogden, R. C., Peebles, C. L. & Abelson, J. (1979) *Cell* **18,** 37 – 45.

Kruger, K., Grabowski, P. J., Zaug, A. J., Sands, J., Gottschling, D. E. & Cech, T. R (1982) *Cell* **31,** 147 – 157.

Kumamoto, J., Cox, J. R., Jr. & Westheimer, F. H. (1956) *J. Am. Chem. Soc.* **78,** 4858 – 4860.

Lerner, M. R., Boyle, J. A., Mount, S. M., Wolin, S. L. & Steitz, J. A. (1980) *Nature* **283,** 220 – 224.

McSwiggen, J. A., & Cech, T. R. (1989) *Science* **244,** 679 – 683.

Maniatis, T. & Reed, R. (1987) *Nature* **325,** 673 – 678.

Marsh, T. L. & Pace, N. R. (1985) *Science* **229,** 79 – 81.

Michel, F. & Dujon, B. (1983) *EMBO J.* **2,** 33 – 38.

Michel, F., Jacquier, A. & Dujon, B. (1982) *Biochimie* **64,** 867 – 881.

Michel, F., Hanna, M., Green, R., Bartel, D. P. & Szostak, J. W. (1989) *Nature* **342,** 391 – 395.

Murphy, F. & Cech, T. R. (1989) *Proc. Natl. Acad. Sci. USA* **86,** 9218 – 9222.

Noller, H. F. & Woese, C. R. (1981) *Science* **212,** 403 – 410.

Orgel, L. E. (1986) *J. Theor. Biol.* **123,** 127 – 149.

Pace, N. R. & Marsh, T. L. (1985) *Orig. Life* **16,** 97 – 116.

Peebles, C. L., Ogden, R. C., Knapp, G. & Abelson, J. (1979) *Cell* **18,** 27 – 35.

Peebles, C. L., Perlman, P. S., Mecklenburg, K. L., Petrillo, M. L., Tabor, J. H., Jarrell, K. A. & Cheng, H.-L. (1986) *Cell* **44,** 213 – 223.

Price, J. V., Engberg, J. & Cech, T. R. (1987) *J. Mol. Biol.* **196,** 49 – 60.

Prody, G. A., Bakos, J. T., Buzayan, J. M., Schneider, I. R. & Bruening, G. (1986) *Science* **231,** 1577 – 1580.

Rajagopal, J., Doudna, J. A. & Szostak, J. W. (1989) *Science* **244,** 692 – 694.

Sampson, J. R., Sullivan, F. X., Behlen, L. S., DiRenzo, A. B. & Uhlenbeck, O. C. (1987) *Cold Spring Harbor Symp. Quant. Biol.,* Vol. LII, (Cold Spring Harbor, New York), pp. 267 – 277.

Sharp, P.A. (1985) *Cell* **42,** 397 – 400.

Sugimoto, N., Tomka, M., Kierzek, R., Bevilacqua, P. C. & Turner, D. H. (1989) *Nucleic Acids Res.* **17,** 355 – 371.

Sullivan, F. X. & Cech, T. R. (1985) *Cell* **42,** 639 – 648.

Tabak, H. F., Van der Horst, G., Kamps, A. M. J. E. & Arnberg, A. C. (1987) *Cell* **48,** 101 – 110.

Tobin, G. J., Young, D. C. & Flanegan, J. B. (1989) *Cell* **59,** 511 – 519.

Uhlenbeck, O. C. (1987) *Nature* **328,** 596 – 600.

Van der Horst, G. & Tabak, H. F. (1985) *Cell* **40,** 759 – 766.

Van der Veen, R., Arnberg, A. C., Van der Horst, G., Bonen, L., Tabak, H. F. & Grivell, L. A. (1986) *Cell* **44,** 225 – 234.

Waring, R. B., Scazzocchio, C., Brown, T. A. & Davies, R. W. (1983) *J. Mol. Biol.* **167,** 595 – 605.

Waring, R. B., Towner, P., Minter, S. J. & Davies, R. W. (1986) *Nature* **321,** 133 – 139.

Wild, M. A. & Gall, J. G. (1979) *Cell* **16,** 565 – 573.

Wild, M. A. & Sommer, R. (1980) *Nature* **283,** 693 – 694.

Woese, C. (1967) In *The Genetic Code: The Molecular Basis for Genetic Expression* (Harper and Row, New York), p. 186.

Woodson, S. A. & Cech, T. R. (1989) *Cell* **57,** 335 – 345.

Yao, M.-C., Kimmel, A. R. & Gorovsky, M. A. (1974) *Proc. Natl. Acad. Sci. USA* **71,** 3082 – 3086.

Yarus, M. (1988) *Science* **240,** 1751 – 1758.

Zaug, A. J. & Cech, T. R. (1980) *Cell* **19,** 331 – 338.

Zaug, A. J. & Cech, T. R. (1982) *Nucleic Acids Res.* **10,** 2823 – 2838.

Zaug, A. J. & Cech, T. R. (1986a) *Science* **231,** 470 – 475.

Zaug, A. J. & Cech, T. R. (1986b) *Biochemistry* **25,** 4478 – 4482.

Zaug, A. J., Grabowski, P. J. & Cech, T. R. (1983) *Nature* **301,** 578 – 583.

Zaug, A. J., Kent, J. R. & Cech, T. R. (1984) *Science* **224,** 574 – 578.

Zaug, A. J., Been, M. D. & Cech, T. R. (1986) *Nature* **324,** 429 – 433.

Chemistry 1990

ELIAS JAMES COREY

for his development of the theory and methodology of organic synthesis

THE NOBEL PRIZE IN CHEMISTRY

Speech by Professor Salo Gronowitz of the Royal Swedish Academy of Sciences.
Translation from the Swedish text.

Your Majesties, Your Royal Highnesses, Ladies and Gentlemen,

Organic synthesis — the preparation of complicated organic compounds, using simple and cheap starting materials — is one of the prerequisites of our civilization, the chemical age in which we live. As late as the 1820s it was still believed that organic natural products such as sugar, camphor or morphine were endowed with a special vital force and therefore could not be prepared in the laboratory. In 1828, the German chemist Friedrich Wöhler crushed this dogma by preparing the organic compound urea from inorganic ammonium cyanate. Today, the most complicated natural products can be prepared, which in their three-dimensional structure are completely identical with those isolated from nature. This year's Nobel Laureate has made extremely important contributions in this area.

In spite of these developments, the delusion still thrives in many quarters that there are some special advantages in the natural product over the synthetically prepared one. However, Vitamin C is Vitamin C, regardless of whether it is isolated from citrus fruits or synthesized in chemical factories.

The development of organic synthesis during a period of a little over one hundred years has afforded efficient industrial methods for the preparation of paints and dyes, pharmaceuticals and vitamins, insecticides and herbicides, which increase our harvests, plastics and textile fibers, which clothe humanity. Organic synthesis has contributed to the high standards of living and health and the longevity enjoyed at least in the Western world. It is understandable that contributions to organic synthesis have often been rewarded with the Nobel Prize in Chemistry.

The synthesis of complicated organic compounds often shows elements of artistic creation. Many earlier syntheses were performed more or less intuitively, so that their planning was difficult to perceive. Asking a chemist why he chose precisely the starting materials and reactions that so elegantly led to the desired result would probably be as meaningless as asking Picasso why he painted as he did.

The process of synthetic planning has been compared to a game of three-dimensional chess using 40 pieces on each side. But the problem of synthesis may be even harder than this. Over 35,000 usable methods of synthesis are described in the chemical literature, each with its possibilities and its limitations. During synthesis, moreover, new methods appear which can modify the strategy. It is like allowing new moves during a game of chess.

Beginning in the 1960s Professor E.J. Corey coined the term and developed the concept of retrosynthetic analysis. Starting from the structure of

the molecule he was to produce, the target molecule, he established rules
for how it should be dissected into smaller parts, and what strategic bonds
should be broken. In this way, less complicated building blocks were obtained, which could later be assembled in the process of synthesis. These
building blocks were then analyzed in the same way until simple compounds
had been reached, whose synthesis was already described in the literature or
which were commercially available. Corey showed that strict logical retrosynthetic analysis was amenable to computer programming. He is the leader
in this rapidly developing field.

Through his brilliant analysis of the theory of organic synthesis, Corey has
been able to carry out total syntheses of around a hundred naturally
occurring biologically active compounds, according to simple logical principles, which previously were very difficult to achieve. Only a few of his
achievements in organic synthesis can be mentioned here. In 1978, he
prepared gibberellic acid, which belongs to a class of very important plant
hormones of complicated structure. Corey has furthermore synthesized
gingkolid B, which is the active substance in an extract from the ginkgo tree
and is used as a folk medicine in China.

Corey's most important syntheses are concerned with prostaglandins and
related compounds. These often very instable compounds are responsible
for multifarious and vital regulatory functions of significance in reproduc-
tion, blood coagulation and normal and pathological processess in the
immune system. Their importance is witnessed by the awarding of the 1982
Nobel Prize in Physiology or Medicine to Professors Sune Bergström, Bengt
Samuelsson and Sir John Vane for their discovery of prostaglandins and
closely related biologically active compounds.

With enormous skill Corey has carried out the total syntheses of a large
number of such compounds. It is thanks to Corey's contributions that many
of these important pharmaceuticals are commercially available.

To perform these total syntheses successfully, Corey was also obliged to
develop some fifty entirely new or considerably improved synthesis reactions. His systematic use of different types of organometallic reagents has
revolutionized recent techniques of synthesis in many respects. In recent
years, he has also introduced a number of very effective enzyme-like catalysts, which yield only one mirror isomer of the target product in certain
types of synthetically important reactions. No other chemist has developed
such a comprehensive and varied assortment of methods, often showing the
simplicity of genius, which have become commonplace in organic synthesis
laboratories.

Corey has thus been rewarded with the Prize for three intimately connected contributions, which form a whole. It can be summarized in the following way. Through retrosynthetic analysis and introduction of new synthetic
reactions, he has succeeded in preparing biologically important natural
products, previously thought impossible to achieve. Corey's contributions
have turned the art of synthesis into a science.

Professor Corey,

In these few minutes I have tried to explain your immense impact on the theory and methodology of organic synthesis. In recognition of your important contribution to chemistry, the Royal Swedish Academy of Sciences has decided to confer upon you this years' Nobel Prize for Chemistry. It is an honour and pleasure for me to extend to you the congratulations of the Royal Swedish Academy of Sciences and to ask you to receive your Prize from the hands of His Majesty the King.

Elias J. Corey

ELIAS JAMES COREY

My birth in July 1928 in Methuen, Massachusetts was followed just eighteen months later by the death of my father, Elias, a successful business man in that community 30 miles north of Boston. My mother, Fatina (née Hasham), changed my name from William to Elias shortly after my father's passing. I do not remember my father, but all his friends and associates made it clear that he was a remarkably gifted and much admired person. I have always been guided by a desire to be a worthy son to the father I cannot remember and to the loving, courageous mother who raised me, my brother, and two sisters through the trials of the Depression and World War II. My grandparents on both sides, who emigrated from Lebanon to the United States, also knew how to cope with adversity, as Christians in a tragically torn country, under the grip of the Ottoman empire.

In 1931, our family grew to include my mother's sister, Naciby, and her husband, John Saba, who had no children of their own. We all lived together in a spacious house in Methuen, still a gathering place for family reunions. My uncle and aunt were like second parents to us. As a youngster I was rather independent, preferring such sports as football, baseball and hiking to work. However, when my aunt, who was much stricter than my mother, assigned a household chore, it had to be taken seriously. From her I learned to be efficient and to take pleasure in a job well done, no matter how mundane. We were a very close, happy and hardworking family with everything that we needed, despite the loss of my father and the hard economic times. Uncle John died in 1957, and too soon afterwards, in 1960, my aunt passed away. My mother died in 1970 at the age of seventy. They all lived to see each of the four children attain a measure of success.

From the ages of five to twelve I attended the Saint Laurence O'Toole elementary school in Lawrence, a city next to Methuen, and was taught by sisters of the Catholic order of Notre Dame de Namour. I enjoyed all my subjects there. I do not remember ever learning any science, except for mathematics. I graduated from Lawrence Public High School at the age of sixteen and entered the Massachusetts Institute of Technology, just a few weeks later, in July, 1945, with excellent preparation, since most of my high school teachers had been dedicated and able. Although my favorite subject was mathematics, I had no plan for a career, except the notion that electronic engineering might be attractive, since it utilized mathematics at an interesting technological frontier. My first courses at M.I.T. were in the basic sciences: mathematics, physics and chemistry, all of which were wonderful. I became a convert to chemistry before even taking an engineering course because of the excellence and enthusiasm of my teachers, the central position of chemistry in the sciences and the joy of solving problems in the laboratory. Organic chemistry was especially fascinating with its intrinsic beauty and its great relevance to human health. I had many superb teachers at M.I.T., including Arthur C. Cope, John C. Sheehan, John D. Roberts and Charles Gardner Swain. I graduated from M.I.T. after three years and, at the suggestion of Professor Sheehan, continued there as a graduate member of his pioneering program on synthetic penicillins. My doctoral work was completed by the end of 1950 and, at the age of twenty-two, I joined the University of Illinois at Urbana-Champaign as an Instructor in Chemistry under the distinguished chemists Roger Adams and Carl S Marvel. I am forever grateful to them for giving me such a splendid opportunity, as well as for their help and friendship over many years.

Because my interests in chemistry ranged from the theoretical and quantitative side to the biological end of the spectrum, I decided to maintain a broad program of teaching and research and to approach chemistry as a discipline without internal boundaries. My research in the first three years, which had to be done with my own hands and a few undergraduate students, was in physical organic chemistry. It had to do with the application of molecular orbital theory to the understanding of the transition states for various reactions in three dimensional (i.e. stereochemical) detail. The stereoelectronic ideas which emerged from this work are still widely used in chemistry and mechanistic enzymology. By 1954, as an Assistant Professor with a group of three graduate students, I was able to initiate more complex experimental projects, dealing with the structure, stereochemistry and synthesis of natural products. As a result of the success of this research, I was appointed in 1956, at age twenty-seven, as Professor of Chemistry. My research group grew and the scope of our work broadened to include other topics: enantioselective synthesis, metal complexes, new reactions for synthesis and enzyme chemistry. The pace of discovery accelerated.

In the fall of 1957, I received a Guggenheim fellowship and my first sabbatical leave. It was divided between Harvard, to which I had been invited by the late Prof. Robert B. Woodward, and Europe. The last four months of 1957 would prove eventful. In September, shortly after the beginning of my stay at Harvard, my uncle John passed away. At least I had been lucky enough to have seen him just two days before. I was deeply affected by the loss of this fine and generous man whom I loved as a real father. In solitude and sadness I returned to my work and a very deep immersion in studies which proved to be pivotal to my future research. In early October several of the key ideas for a logical and general way of thinking about chemical synthesis came to me. The application of these insights led to rapid and unusual solutions to several specific synthetic problems of interest to me at the time. I showed one such plan (for the molecule longifolene) to R. B. Woodward and was pleased by his enthusiastic response. Later in 1957 I visited Switzerland, London and Lund, the last as a guest of Prof. Karl Sune Bergström. It was at Lund, in Bergström's

Department, that I became intrigued by the prostaglandins. Our research in the mid 1960's led to the first chemical syntheses of prostaglandins and to involvement in the burgeoning field of eicosanoids ever since.

In the spring of 1959 I received an offer of a Professorship at Harvard, which I accepted with alacrity since I wanted to be near my family and since the Chemistry Department at Harvard was unsurpassed. The Harvard faculty in 1959 included Paul D. Bartlett, Konrad Bloch, Louis F. Fieser, George B. Kistiakowski, E. G. Rochow, Frank H. Westheimer, E. B. Wilson and R. B. Woodward, all giants in the field of Chemistry. Roger Adams, who was always very kind and encouraging to me, gave his blessing even though years before he had declined a professorial appointment at Harvard. I have always regarded the offer of a Professorship at Harvard as the most gratifying of my professional honors.

At Harvard my research group grew in size and quality, and developed a spirit and dynamism which has been a continuing delight to me. I was able to start many new scientific projects and to teach an advanced graduate course on chemical synthesis. Using the concepts of retrosynthetic analysis under guidance of broad strategies, first-year graduate students could be taught in just three months to design sophisticated chemical syntheses. My research interests soon evolved to include the following areas: synthesis of complex, bioactive molecules; the logic of chemical synthesis; new methods of synthesis; molecular catalysts and robots; theoretical organic chemistry and reaction mechanisms; organometallic chemistry; bioorganic and enzyme chemistry; prostaglandins and other eicosanoids and their relevance to medicine; application of computers to organic chemical problems, especially to retrosynthetic analysis. My personal scientific aspirations can be similarly summarized: to be creative over a broad range of the chemical sciences; to sustain that creativity over many years; to raise the power of research in chemistry to a qualitatively higher level; and to develop new generations of outstanding chemists.

In September, 1961, I married Claire Higham, a graduate of the University of Illinois. We have three children. David Reid is a graduate of Harvard (A.B. 1985) and the University of California, Berkeley (Ph.D., 1990), who is currently a Postdoctoral Fellow in Chemistry/Molecular Biology at the University of California Medical School at San Francisco. Our second son, John, graduated from Harvard (A.B. 1987) and the Paris Conservatory of Music (1990) and is now carrying out advanced studies in classical music composition at the latter institution. Our daughter, Susan, graduated from Harvard with a major in anthropology (A.B. 1990) and plans graduate work in Education. Claire and I live near the Harvard Campus in Cambridge, as we have for nearly thirty years. My leisure interests include outdoor activities and music.

I am very proud of the many graduate students and postdoctoral fellows from all over the world who have worked in my research group. Their discoveries in my laboratory and their subsequent achievements in science have been a source of enormous satisfaction. The Corey research family now includes about one hundred fifty university professors and an even larger number of research scientists in the pharmaceutical and chemical industry. It has been my good fortune to have been involved in the education of scholars and leaders in every area of chemical research, and especially, to have contributed to the scientific development of many different countries. My research family has been an extraordinarily important part of my life. Much of the credit for what I have achieved belongs to that professional family, my wonderful teachers and faculty colleagues, and not least, to my own dear personal family.

Honorary Degrees:

A.M.:	Harvard University, 1959	D.Sc.:	University of Liege, 1985
D.Sc.:	University of Chicago, 1968	D.Sc.:	University of Illinois (Urbana), 1985
D.Sc.:	Hofstra University, 1974	D.Sc.:	Kenyon College, 1989
D.Sc.:	Colby College, 1977	Ph.D.:	University of Helsinki, 1990
D.Sc.:	Oxford University (U.K.), 1982	D.Sc.:	Merrimack College, 1990
		LL.D.:	Hokkaido University, 1990

Awards:

Award in Pure Chemistry, American Chemical Society, 1960

Chevreul Medal, Chemistry Society of France, 1964

Fritzsche Award, American Chemical Society, 1967

Intra-Science Foundation Award, 1967

Harrison Howe Award (Rochester Section, American Chemical Society), 1970

Award for Creative Work in Synthetic Organic Chemistry, American Chemical Society, 1971

Centenary Medal, Chemical Society of London, 1971

Ciba Foundation Medal, 1972

Evans Award (Ohio State University), 1972

Linus Pauling Award (Puget Sound and Washington Sections, American Chemical Society), 1973

The Dickson Prize in Science (Carnegie Mellon University), 1973

The George Ledlie Prize in Science (Harvard University), 1973

The Remsen Award (Maryland Section, American Chemical Society), 1974

Arthur C. Cope Award (American Chemical Society), 1976

Nichols Medal (New York Section, American Chemical Society), 1977

Buchman Memorial Award (California Institute of Technology), 1978

The Franklin Medal (The Franklin Institute), 1978

Scientific Achievement Award Medal (The City College of the City University of New York), 1979

J. G. Kirkwood Award (Yale University), 1980

C. S. Hamilton Award (University of Nebraska), 1980

Chemical Pioneer Award, American Institute of Chemists, 1981

Lewis S. Rosenstiel Award for Distinguished Work in Basic Medical Research (Brandeis University), 1981

Paul Karrer Award (University of Zurich), 1982

Medal of Excellence (University of Helsinki), 1982

Tetrahedron Prize for Creativity in Organic Chemistry, 1983

Willard Gibbs Award (Chicago Section, American Chemical Society), 1984

Paracelsus Award (Swiss Chemical Society), 1984

Madison Marshall Award (American Chemical Society, Alabama Section), 1985

V. D. Mattia Award (Roche Institute of Molecular Biology), 1985

Wolf Prize in Chemistry (Wolf Foundation), 1986

Silliman Award (Yale University), 1986

Robert Robinson Medal, Royal Society of Chemistry, 1988

National Medal of Science, President of the United States, 1988

Japan Prize in Science (Science and Technology Foundation of Japan), 1989

Gold Medal Award (American Institute of Chemists), 1989

Order of the Rising Sun, Gold and Silver Star, Government of Japan, 1989

Janot Medal, University of Paris, 1990

Fellowships:

Alfred P. Sloan Foundation, 1955 — 57

Guggenheim, 1957 — 58 and 1968 — 69

American Academy of Arts and Sciences (elected 1960)

National Academy of Sciences (elected 1966)

Honorary Life Membership, The Franklin Institute, 1978

Honorary Member, Royal Society of Chemistry, U.K., 1981

Honorary Foreign Member, Chemical Society of Finland, 1982

Honorary Member, Pharmaceutical Society of Japan, 1989

Honorary Member, Chemical Society of Japan, 1990

Publications:

Over 700 publications in scientific journals.

THE LOGIC OF CHEMICAL SYNTHESIS: MULTISTEP SYNTHESIS OF COMPLEX CARBOGENIC MOLECULES

Nobel Lecture, December 8, 1990

by

ELIAS JAMES COREY

Department of Chemistry, Harvard University, Cambridge, Massachusetts, USA

Carbogens, members of the family of carbon-containing compounds, can exist in an infinite variety of compositions, forms and sizes. The naturally occurring carbogens, or organic substances as they are known more traditionally, constitute the matter of all life on earth, and their science at the molecular level defines a fundamental language of that life. The chemical synthesis of these naturally occurring carbogens and many millions of unnatural carbogenic substances has been one of the major enterprises of science in this century. That fact is affirmed by the award of the Nobel Prize in Chemistry for 1990 for the "development of the theory and methodology of organic synthesis". Chemical synthesis is uniquely positioned at the heart of chemistry, the central science, and its impact on our lives and society is all pervasive. For instance, many of today's medicines are synthetic and many of tomorrow's will be conceived and produced by synthetic chemists. To the field of synthetic chemistry belongs an array of responsibilities which are crucial for the future of mankind, not only with regard to the health, material and economic needs of our society, but also for the attainment of an understanding of matter, chemical change and life at the highest level of which the human mind is capable.

The post World War II period encompassed remarkable achievement in chemical synthesis. In the first two decades of this period chemical syntheses were developed which could not have been anticipated in the earlier part of this century. For the first time, several very complex molecules were assembled by elaborately conceived multistep processes, for example vitamin A (O. Isler, 1949), cortisone (R. B. Woodward, R. Robinson, 1951), morphine (M. Gates, 1956), penicillin (J. C. Sheehan, 1957), and chlorophyll (R. B. Woodward, 1960).[1] This striking leap forward, which was recognized by the award of the Nobel Prize in Chemistry to R. B. Woodward in 1965,[2] was followed by an equally dramatic scientific advance during the past three decades, in which chemical synthesis has been raised to a qualitatively higher level of sophistication. Today, in many laboratories around the world chemists are synthesizing at an astonishing rate complex carbogenic structures which could not have been made effectively in the 1950's or early

1960's. This advance has been propelled by the availability of more power-ful conceptual processes for the planning of chemical syntheses, the use of new chemical methods, in the form of reactions and reagents, and the advent of improved methods for analysis, separation and determination of structure. Many talented investigators all over the world have contributed to the latest surge of chemical synthesis. Their efforts constitute a collective undertaking of vast dimensions, even though made independently, and their ideas and discoveries interact synergistically to the benefit of all. I am happy to have been selected by the Nobel Committee for contributions to the science of chemical synthesis, but I am even more pleased that this important field of science has again received high recognition.

Genesis

In the fall of 1947, as an undergraduate at the Massachusetts Institute, I took a course in Advanced Synthetic Organic Chemistry, taught by the distinguished chemist A. C. Cope, in which the major reactions of synthesis were surveyed. It was explained that very few new synthetic methods re-mained to be found, since only five important reactions had been discov-ered in the preceding fifty years; and we students were advised to learn how to devise chemical syntheses using the available portfolio of known con-structions. We were given numerous molecular structures as synthetic prob-lems. After doing a few of the problem sets, I had developed sufficient skill and experience to handle all of the remaining assignments with ease, much as I had learned to use the English language, to prove mathematical theo-rems, or to play chess. My new found competence in chemical problem solving seemed to result from an automatic "know how" rather than from the conscious application of well-defined procedures. Nonetheless, even though I had mastered the classical reactions, designing syntheses of mole-cules beyond the modest level of complexity of these instructional problems still eluded me. Molecules such as morphine, cholesterol, penicillin, or sucrose were so forbidding that they defined the frontiers of 1947; each seemed to be unique and to require a very high level of creativity and invention. Much of my research over the years has been devoted to probing those frontiers and advancing the level of synthetic science by an approach consisting of three integral components: the development of more general and powerful ways of thinking about synthetic problems, the invention of new general reactions and reagents for synthesis, and the design and execution of efficient multistep syntheses of complex molecules at the limits of contemporary synthetic science.

Retrosynthetic Analysis

During the first half of this century most syntheses were developed by selecting an appropriate starting material, after a trial and error search for commercially available compounds having a structural resemblance to the target of synthesis. Suitable reactions were then sought for elaboration of the chosen starting material to the desired product. Synthetic planning in most instances was strongly dependent on an assumed starting point. In the fall of 1957 I came upon a simple idea which led to an entirely different way of designing a chemical synthesis. In this approach the target structure is subjected to a deconstruction process which corresponds to the reverse of a synthetic reaction, *so as to convert that target structure to simpler precursor structures, without any assumptions with regard to starting materials.* Each of the precursors so generated is then examined in the same way, and the process is repeated until simple or commercially available structures result. This "retrosynthetic" or "antithetic" procedure constitutes the basis of a general logic of synthetic planning which was developed and demonstrated in practice over the ensuing decade.[3, 4, 5] In an early example, retrosynthetic planning for the tricyclic sesquiterpene longifolene (1) (Chart I) produced several attractive pathways for synthesis, one of which was selected and validated by experimental execution.[6] The basic ideas of retrosynthetic analysis were used to design many other syntheses and to develop a computer program for generating possible synthetic routes to a complex target structure without any input of potential starting materials or intermediates for the synthesis.[4, 5] The principles of retrosynthetic analysis have been summarized most recently in the textbook, "The Logic of Chemical Synthesis"[7] which was written for advanced undergraduate and graduate students of chemistry. The retrosynthetic way of thinking about chemical synthesis also provided a logical and efficient way to teach synthetic planning to intermediate and advanced students, a good example of the intimate link between teaching and research in an academic setting. A brief synopsis of the retrosynthetic planning of syntheses will now be given.

Retrosynthetic[8] (or *antithetic*) analysis is a problem-solving technique for transforming the structure of a *synthetic target* (TGT) molecule to a sequence of progressively simpler structures along a pathway which ultimately leads to simple or commercially available starting materials for a chemical synthesis. The transformation of a molecule to a synthetic precursor is accomplished by the application of a *transform*, the exact reverse of a *synthetic reaction*, to a target structure. Each structure derived antithetically from a TGT then itself becomes a TGT for further analysis. Repetition of this process eventually produces a tree of intermediates having chemical structures as nodes and pathways from bottom to top corresponding to possible synthetic routes to the TGT. Such trees, called EXTGT trees since they grow out from the TGT, can be quite complex since a high degree of branching is possible at each node and since the vertical pathways can include many steps. This central fact implies the need for strategies which control or guide the generation of EXTGT trees so as to avoid explosive branching and the proliferation of useless pathways.

Each retrosynthetic step requires the presence of a target structure of a keying structural subunit or *retron* which allows the application of a particular transform. For example, the retron for the aldol transform consists of the subunit HO-C-C-C=O, and it is the presence of this subunit which permits transform function, e.g. as follows:

Transforms vary in terms of their power to simplify a target structure. The most powerful of simplifying transforms, which reduce molecular complexity in the retrosynthetic direction, occupy a special position in the hierarchy of all transforms. Their application, even when the appropriate retron is absent, may justify the use of a number of non-simplifying transforms to generate that retron. In general, simplifying transforms function to modify structural elements which contribute to molecular complexity: molecular size, cyclic connectivity (topology), stereocenter content, element and functional group content, chemical reactivity, structural instability, and density of complicating elements.

Molecular complexity is important to strategy selection. For each type of molecular complexity there is a collection of general strategies for dealing with that complexity. For instance, in the case of a complex polycyclic structure, strategies for the simplification of the molecular network, i.e. topological strategies, must play an important part in transform selection. However, the most efficient mode of retrosynthetic analysis lies not in the separate application of individual strategies, but in the concurrent application of as many different independent strategies as possible.

The major types of strategies[7] which are of value in retrosynthetic analysis may be summarized briefly as follows.[8]

1. Transform-based strategies — long range search or look-ahead to apply a powerfully simplifying transform (or a tactical combination of simplifying transforms) to a TGT with certain appropriate keying features. The retron required for application of a powerful transform may not be present in a complex TGT and a number of antithetic steps (subgoals) may be needed to establish it.

2. Structure-goal strategies — directed at the structure of a potential intermediate or potential starting material. Such a goal greatly narrows a retrosynthetic search and allows the application of bidirectional search techniques.

3. Topological strategies — the identification of one or more individual bond disconnections or correlated bond-pair disconnections as strategic. Topological strategies may also lead to the recognition of a key substructure for disassembly or to the use of rearrangement transforms.

4. Stereochemical strategies — general strategies which clear, *i.e.* remove, stereocenters and stereorelationships under stereocontrol. Such stereocontrol can arise from transform-mechanism control or substrate-structure control. In the case of the former the retron for a particular transform contains critical stereochemical information (ab-

solute or relative) on one or more stereocenters. Stereochemical strategies may also dictate the retention of certain stereocenter(s) during retrosynthetic processing or the joining of atoms in three-dimensional proximity. A major function of stereochemical strategies is the achievement of an experimentally valid clearance of stereo-centers, including clearance of molecular chirality.

5. Functional group-based strategies. The retrosynthetic reduction of molecular complexity involving functional groups (FG's) takes various forms. Single FG's or pairs of FG's (and the interconnecting atom path) can key directly the disconnection of a TGT skeleton to form simpler molecules or signal the application of transforms which re-place functional groups by hydrogen. Functional group interchange (FGI) is a commonly used tactic for generating the retrons of simplify-ing transforms from a TGT. FG's may key transforms which stereo-selectively remove stereocenters, break topologically strategic bonds or join proximate atoms to form rings.

6. "Other" types of strategies. The recognition of substructural units within a TGT which represent major obstacles to synthesis often provides major strategic input. Certain other strategies result from the requirements of a particular problem, for example a requirement that several related target structures be synthesized from a common intermediate. A TGT which resists retrosynthetic simplification may require the invention of new chemical methodology. The recognition of obstacles to synthesis provides a stimulus for the discovery of such novel processes. The application of a chain of hypotheses to guide the search for an effective line of retrosynthetic analysis is important.

Other strategies deal with optimization of a synthetic design *after* a set of pathways has been generated antithetically, specifically for the ordering of synthetic steps, the use of protection or activation steps, or the determina-tion of alternate paths.

Systematic and rigorous retrosynthetic analysis is the *broad principle* of synthetic problem solving under which the individual strategies take their place. Another overarching idea is the use *concurrently* of as many independent strategies as possible to guide the search for retrosynthetic pathways. *The greater the number of strategies which are used in parallel to develop a line of analysis, the easier the analysis and the simpler the emerging synthetic plan is likely to be.*[10]

An abbreviated form of the 1957 retrosynthetic plan for the synthesis of longifolene (1) is shown in Chart I. Changes in the retrosynthetic direction are indicated by a double arrow ($\Rightarrow$) to distinguish them from the synthetic direction of chemical reactions ($\rightarrow$) and the number below indicates the number of transforms required for the retrosynthetic change if greater than one. The selection of transforms was initially guided by a topological strategy (disconnection of bond *a* in 1). The Michael transform, which simplifies structure 2 to precursor 3, can be found by general transform selection procedures. [5, 9] The starting materials for the synthesis which emerge from retrosynthetic analysis, **4** and **5**, have little resemblance to the target structure **1**.

A detailed explanation of this example of retrosynthetic analysis has been given. [6b, 10] During the past 20 years systematic retrosynthetic thinking has permeated all areas of carbogenic synthesis. It is no longer possible to teach the subject of carbogenic synthesis effectively without the extensive use of retrosynthetic concepts and thinking.

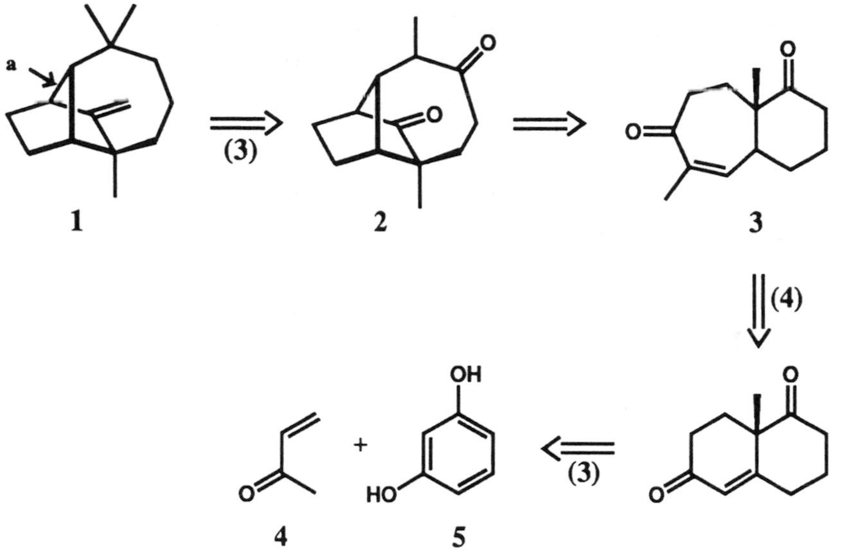

Retrosynthetic Analysis For Longifolene (1957)

Chart I

Computer-Assisted Retrosynthetic Analysis

The use of computers to generate possible pathways for chemical synthesis, which was first demonstrated in the 1960's, [4, 5, 9, 11] was made possible by the development of the retrosynthetic methods outlined above and the re-quired computer methodology. Graphical input of structures by hand draw-ing using an electrostatic tablet and stylus, in the natural manner of a chemist, and output to a video terminal [4, 12] provided an extraordinarily simple and effective interface between chemist and machine. Chemical structures were represented in the machine by means of atom and bond tables, and manipulated by appropriate instructions. Algorithms were de-vised for perception by machine of structural features, patterns, and sub-units which are needed for synthetic analysis. Techniques were developed for storage and retrieval of information on chemical transforms (including retron recognition and keying) using a higher level "chemical English" language. The program (LHASA) was designed to be interactive, with any level of control or input desired by the user, and to emulate the problem-

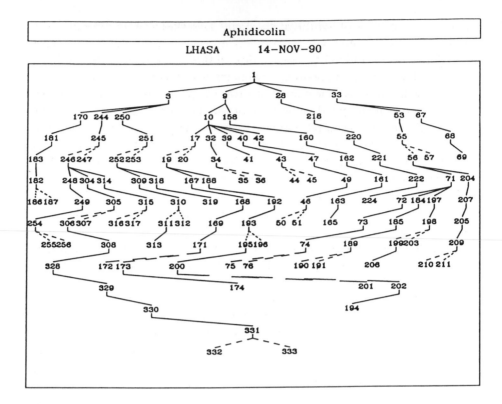

Chart II

Chart III

solving approaches of synthetic chemists. The chemical knowledge base, written so as to be intelligible to a practicing chemist, contains all the types of information required for generation and evaluation of retrosynthetic changes, for example, data on individual transforms and their mechanisms, scope, and limitations. The performance of the program is below that which should eventually be possible because of the immensity of the task and the modest size of our research effort. However, despite present limitations, including a modest knowledge base of about 2000 transforms, LHASA is capable of providing interesting suggestions of synthetic pathways for challenging targets. The present level of capability of LHASA can best be appreciated by its performance on specific problems. Shown in Chart II is the EXTGT tree generated by LHASA for the antiviral agent aphidicolin, using just one particular line of analysis. The pathway in this tree from intermediates **332** and **333** to aphidicolin consists of the structures shown in Chart III.[13] The suggestion by LHASA of such non-obvious pathways is both stimulating and valuable to a chemist.

The field of computer assisted synthetic analysis is fascinating in its own right, and surely one of the most interesting problems in the area of machine intelligence. Because of the enormous memory and speed of modern machines and the probability of continuing advances, it seems clear that computers can play an important role in synthetic design. However, before that potential can be realized, many difficult computing problems must be solved. Multistep retrosynthetic look-ahead, even under strategic guidance, requires complex and powerful software. Vast amounts of information — structural, stereochemical, and chemical — must be generated and analyzed using all available chemical knowledge. A massive undertaking will be required.

New Synthetic Methodology and Multistep Synthesis of Complex Molecules

The invention of new reactions and reagents has revolutionized the field of carbogenic synthesis, literally placing an extraordinarily powerful new chemistry alongside the classical reactions of the pre-1950 period. Without this methodology, the achievements of modern chemical synthesis would not have been possible. Two early landmarks in this advance, the discovery of the Wittig synthesis of olefins and the hydroboration of olefins, were highlighted by the award of the Nobel Prize in Chemistry for 1979 to G. Wittig and H. C. Brown. More recent developments have provided many methods which are noteworthy for their great chemical selectivity and stereochemical control and for their suitability in the construction of complex molecules. Indeed, many new synthetic processes have been discovered as a result of a perceived need in connection with specific problems involving novel or complicated structures and a deliberate search for suitable methodology. The rational design of such methods depends on the use of mechanistic reaction theory, stereochemical principles, and a wide range of chemistry involving many elements and ephemeral reactive intermediates.

A key to the success of many of the multistep chemical syntheses which

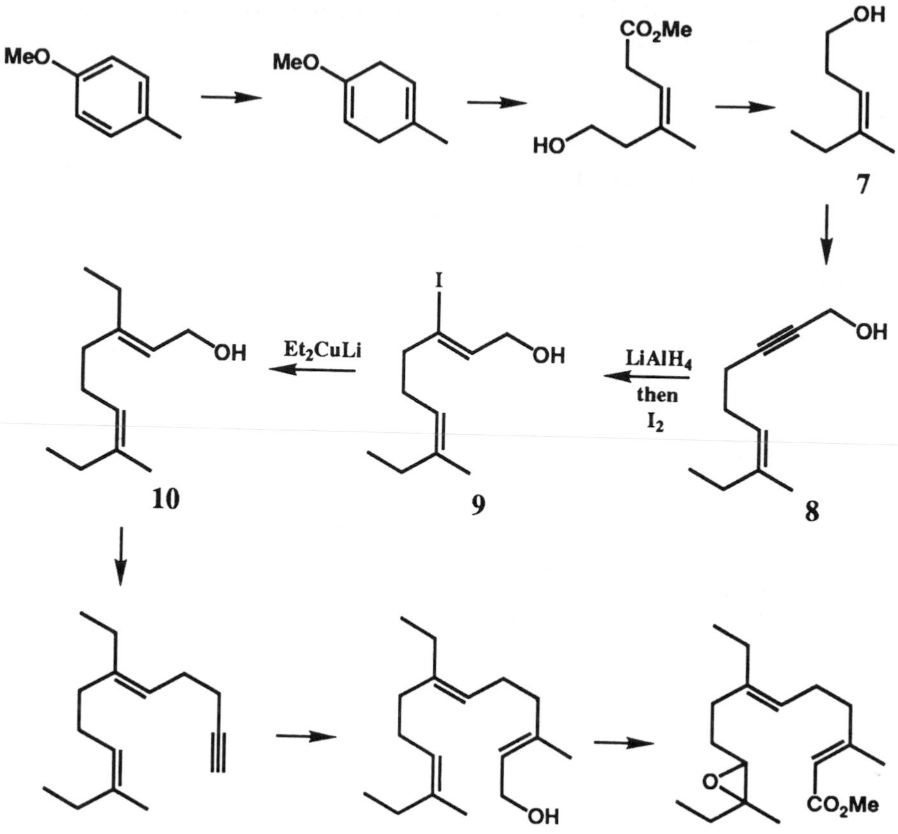

Chart IV

have been demonstrated in our laboratories over the years has been the invention of new methodology. Since more than fifty such methods have been developed in our laboratory, it is impossible to summarize this aspect of synthetic research in a brief article. However, a few examples may serve to illustrate the effectiveness of these new methods in providing access to rare and valuable carbogenic compounds and their impact on the whole field of chemical synthesis.

The discovery and identification of the insect juvenile hormone of *Cecropia* (**6**, now known as JH-I)[14] in 1967 generated immense interest because of the potential of such nontoxic compounds for insect control.[15] Chemical synthesis was essential because of the extreme paucity of material from natural sources. Despite the apparent simplicity of structure **6**, a stereospecific route for the synthesis was not obvious, because no general methods existed in 1967 for the stereocontrolled generation of the trisubstituted olefinic units which it contains. The first stereospecific synthesis of **6**[16] was possible using new methodology which was specifically devised for this application. An abbreviated version of the synthesis is shown in Chart IV. The first olefinic intermediate (7) was synthesized from *p*-methoxytoluene in a way which guarantees the Z-configuration of the stereocenter. Reaction of 8 with LiAlH₄ (*trans* hydroalumination) followed by iodine (replacement of Al by I) produced 9 stereospecifically by a novel sequence.[16] The replacement of iodine in 9 by ethyl was effected by another new process, cross coupling of a vinylic iodide with an organocopper reagent, which provided 10 stereospecifically. A completely analogous series of reactions converted 10 to triene 11, from which JH-I (6) was obtained by a novel selective oxidation sequence. Thus, the synthesis outlined in Chart IV depended on no less than four new synthetic methods. Three of these methods have come into very general use. The coupling of carbon groups using organocopper chemistry is now a major method of chemical synthesis.[17] The related carbometallation of acetylenes, also developed in connection with the synthesis of 6,[16b,18] has been extended in many directions, and this approach has become commonplace for the stereospecific construction of trisubstituted double bonds, a frequently occurring type of structural unit in biologically active natural substances.

The ready availability of the insect juvenile hormone 6 permitted a wide range of biological studies and an understanding of the best ways of using such "third generation" agents for insect control. Inexpensive synthetic mimics of insect juvenile hormone are now produced commercially as environmentally safe insect control products.

Numerous naturally occurring microbial substances, especially antibiotics, are members of the "macrolide" structural family and contain a lactone functional group as part of a many membered ring. The key to the successful synthesis of complex macrolides such as erythronolide B (12), the precursor of the erythromycin antibiotics, was the development of new methodology for macrolactone ring formation (Chart V). Our group developed a very mild, effective and general method for this synthetic operation, the double-activation method,[19] which subsequently has been widely used. Thio-ester 13, produced by total synthesis,[20] was converted to macrolactone 15 simply by heating in toluene solution. Internal proton transfer in 13 generates the internal ion pair 14, which is doubly activated for the internal carbonyl addition required for ring closure to 15. Erythronolide B was obtained from 15 by a simple reaction sequence.[20] The double-activation method has been used effectively for the synthesis of a number of other remarkable natural macrocyclic lactones (Chart V) including brefeldin A (16), an inhibitor of protein transport and processing in mammalian cells,[21] the microbial iron transporter enterobactin (17),[22] and the marine eicosanoid, hybridalactone (18).[23]

Bilobalide (19) is a complex and unusual molecule produced by the ginkgo tree, *Ginkgo biloba*. An effective synthesis of 19 was made possible by the development of a remarkable reaction for the formation of five-membered rings (Chart VI).[24] The readily available chiral diester 20 was converted by Claisen acylation to the acetylenic keto diester 21. Treatment of 21

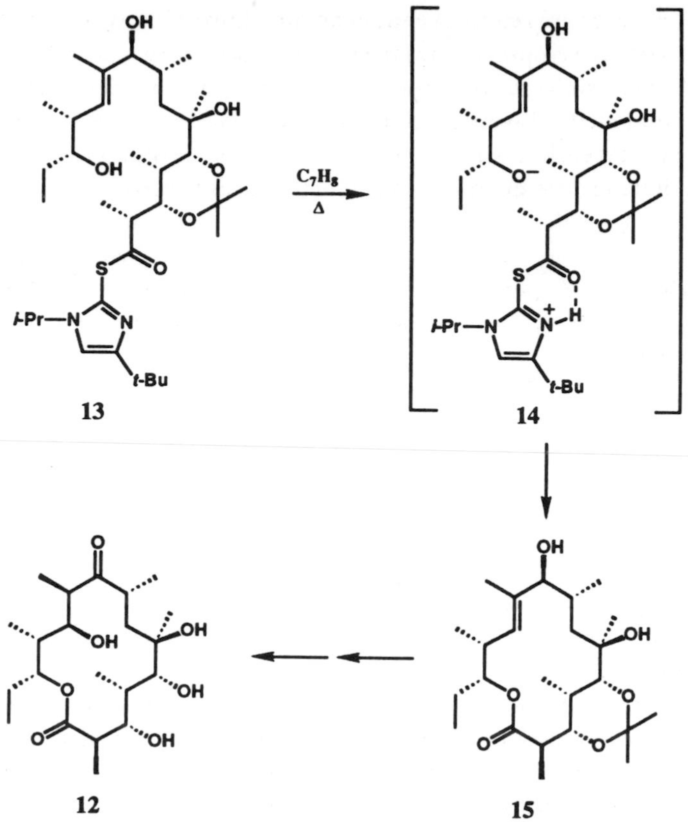

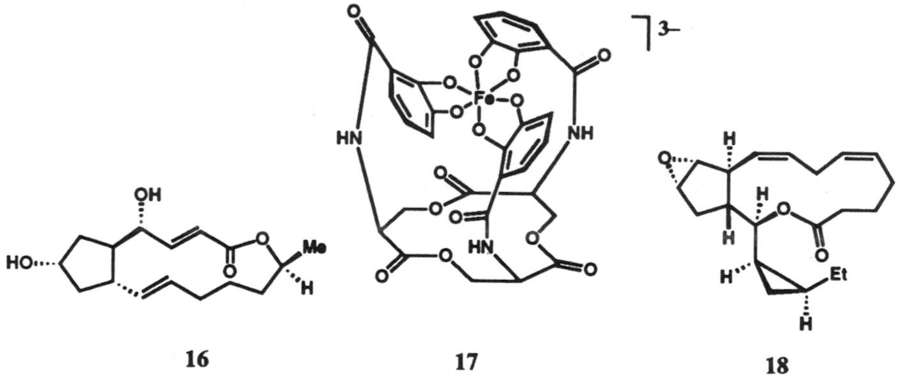

Chart V

with base effected a novel ring closure to give the bicyclic ketone **22**, which was then transformed into bilobalide (**19**) by a multistep sequence. There are several variants on this cyclization methodology which demonstrate considerable scope.[25]

A wide range of reagents and reactants have played a role in the new methodology developed in our group: transition metals; metal-ion complexes; and silico, sulfo, boro, alumino, phospho, and stanno carbogens. The new general methods which have resulted include processes for ring formation, chain extension, oxidation, reduction, functional group transformation, activation, protection, and stereochemical control.

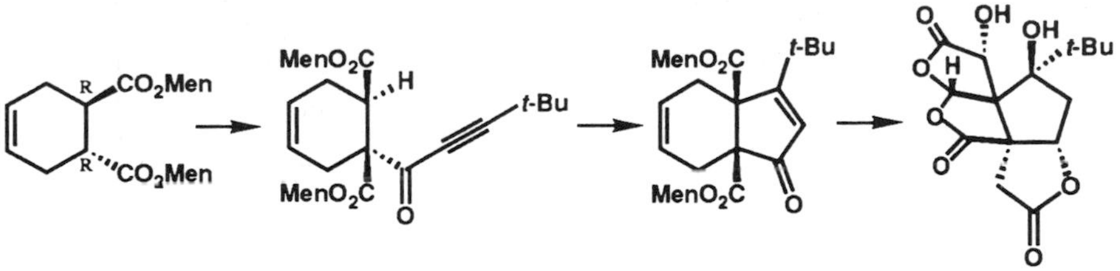

Chart VI

Multistep Synthesis — General

To a synthetic chemist, the complex molecules of nature are as beautiful as any of her other creations. The perception of that beauty depends on the understanding of chemical structures and their transformations, and, as with a treasured work of art, deepens as the subject is studied, perhaps even to a level approaching romance. It is no wonder that the synthetic chemist of today is filled with joy by the discovery of a new naturally occurring structure and the appearance of yet another challenge to synthesis. It makes no difference that the realization of a difficult synthesis entails long hours of study, thought and physical effort, since a complex chemical synthesis is an exciting adventure which leads to a beautiful creation. I believe that the case for molecular synthesis, as a high intellectual endeavor and as a scientific art form, can stand on these merits. The chemist who designs and completes an original and esthetically pleasing multistep synthesis is like the composer, artist or poet who, with great individuality, fashions new forms of beauty from the interplay of mind and spirit.

It is fortunate that molecular synthesis also serves the utilitarian function of producing quantities of rare or novel substances which satisfy human needs, especially with regard to health, and the scientific function of stimulating research and education throughout the whole discipline of chemistry.

Our research group has been responsible for the creation of more than one hundred new multistep syntheses of interesting molecules, our sonatas and string quartets. The step-by-step construction of most of these targets of synthesis is outlined in the *"Logic of Chemical Synthesis"*[7] and is discussed in detail in the original research papers referred to therein. The structures of a small, and somewhat random, selection of these synthetic targets are shown in Charts VIIa and VIIb. A few comments will be given here on the syntheses of each of these to provide an overview of this aspect of our research.

Maytansine[26a] and aplasmomycin,[26b] each scarce and therapeutically interesting, were synthesized enantioselectively and with control of stereochemistry using novel methodology for assembling the molecular skeleton and
for forming the macrocyclic unit. These syntheses and that of erythronolide
B[19, 26c] provided early demonstrations that such complex, macrocyclic molecules can be made efficiently by multistep total synthesis.

Gibberellic acid resisted total synthesis, despite studies in several leading
laboratories, for more than two decades because of an unusually forbidding
arrangement of structural subunits. The first successful synthesis, and
subsequent improved versions, [26d] required a deep and complex retrosynthetic analysis[26e] and a number of new concepts and methods. An entirely
different strategic approach was utilized for the first synthesis of the biosyn-
thetically related plant regulator antheridic acid[26f] which confirmed the
proposed gross structure and clarified the stereochemistry. The availability
of synthetic antheridic acid is essential to the further study of this rare and
potent plant hormone.

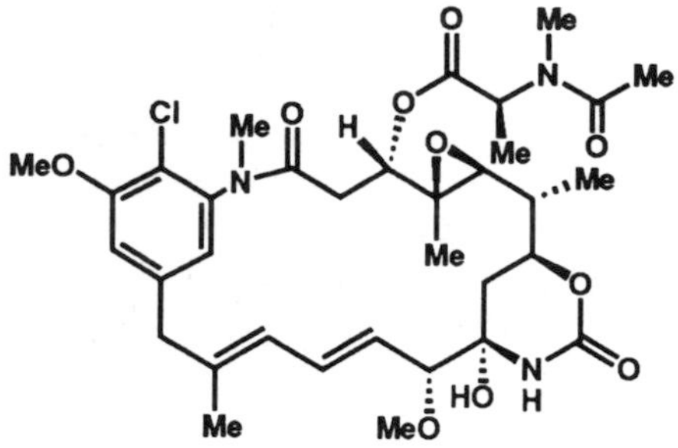

Maytansine 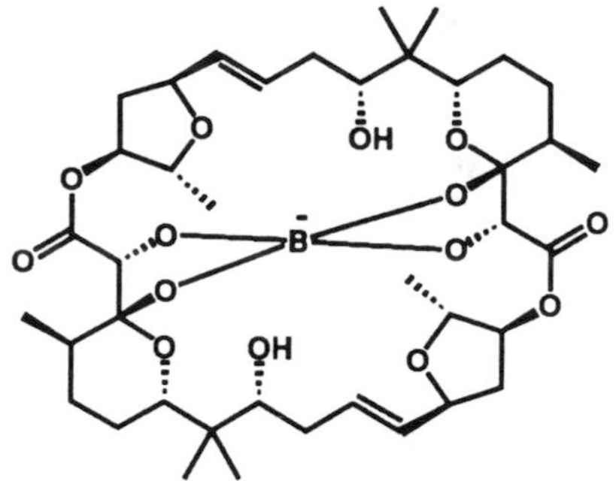   Aplasmomycin

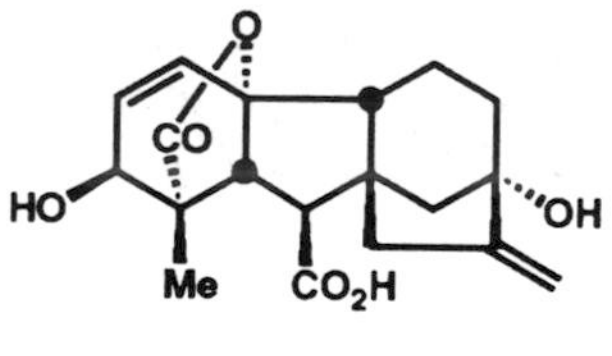

Gibberellic Acid 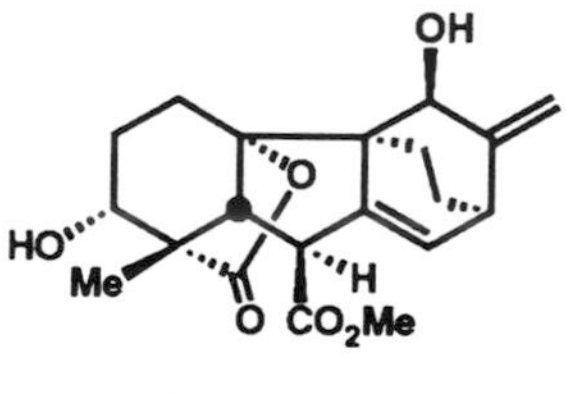   Antheridic Acid 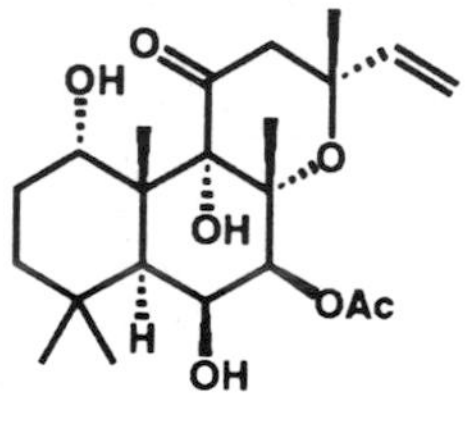   Forskolin

Chart VII a

Forskolin, the first known activator of the enzyme adenylate cyclase, is a
promising therapeutic agent which is available only in limited quantity from
plant sources. An efficient multistep synthesis of forskolin, which is both
enantio- and stereocontrolled, has been developed[26g] based on several new
synthetic methods. The synthesis of picrotoxinin, known since 1811 and a
potent inhibitor of the neurotransmitter γ-aminobutyric acid, would not
have been possible without retrosynthetic analysis and new methodology.[26h]
Perhydrohistrionicotoxin, a rare and highly bioactive alkaloid from poison-

ous frogs, is a useful tool in neuroscience which was easily produced by total synthesis.[26i]

Pseudopterosin E,[26j] a powerful antiinflammatory agent, and venustatriol,[26k] an antiviral agent, are biosynthesized by marine organisms in only trace amounts. Both are available in chiral form by efficient enantiocontrolled multistep syntheses.

Last, but no means least, in this brief summary of our studies on the total synthesis of complex molecules, is the case of ginkgolide B, an unusual substance for several reasons. Ginkgolide B is biosynthesized in the roots of the unique and ancient ginkgo tree, *Ginkgo biloba*, by an extraordinarily complex biosynthetic process, for reasons that are unknown. It is an active

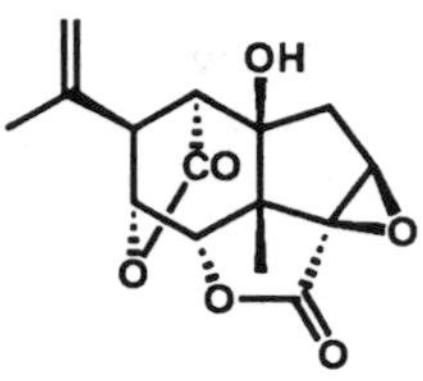

Picrotoxinin

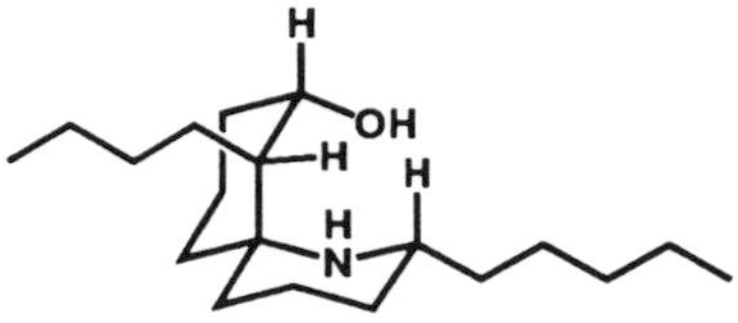

Perhydrohistrionicotoxin

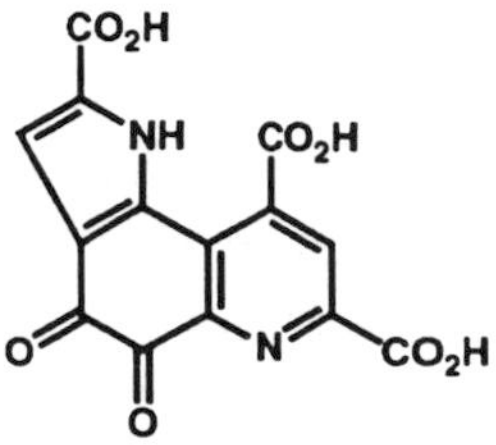

Coenzyme PQQ

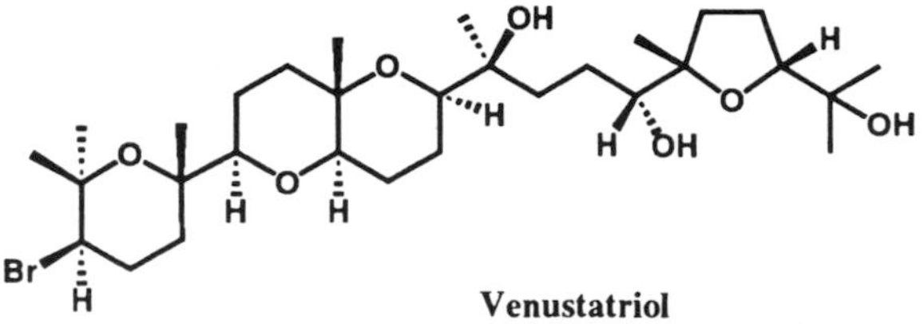

Venustatriol

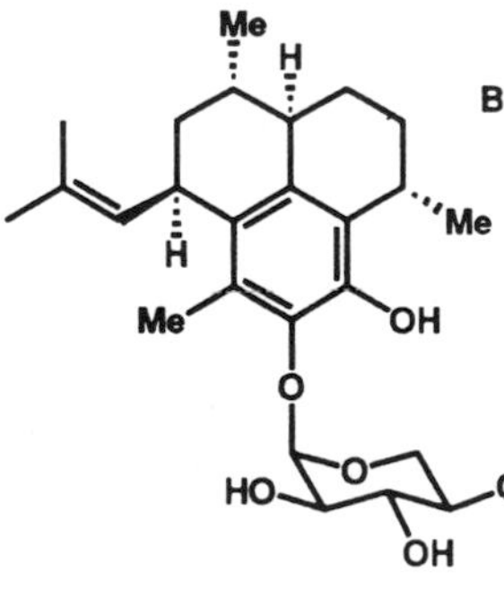

Pseudopterosin E

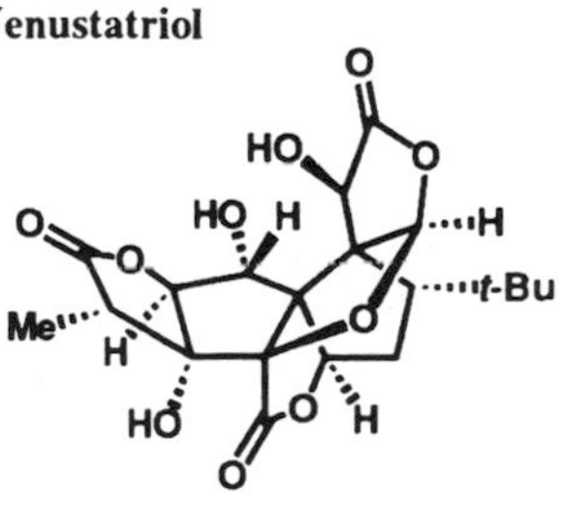

Ginkgolide B

Chart VII b

ingredient in the medicinal extract of ginkgo which is now widely used in oriental and western medicine. The total synthesis of ginkgolide B posed a challenge for synthesis in the 1980's which was comparable to the most difficult problems of earlier eras, for example, steroids in the 1950's or vitamin B-12 and gibberellic acid in the 1970's. That challenge was met in just three years of research, thanks again to the power of modern retrosynthetic planning and to the invention of new tools for this particular synthesis.[26l,27]

Multistep Synthesis Exemplified — The Prostaglandins

The prostaglandins, the first of the known eicosanoids, were detected as bioactive substances more than fifty years ago. However, it was not until the pioneering work of K. Sune Bergström and his group in Sweden in the 1950's and 1960's that the structures of the various members of the prostaglandin family were determined.[28] For that research Bergström and Bengt Samuelsson received (together with John Vane) the Nobel Prize in Medicine for 1982, and deservedly so since, as has been written of the prostaglandins: "Their actions and the pharmacologic agents that influence their formation affect almost every aspect of medical practice."[29] The occurrence of only trace amounts of prostaglandins (PG's) in mammalian sources and the potent effect of these twenty-carbon carboxylic compounds on muscle and blood vessels indicated the need for an effective synthesis that would make available all of the PG's in ample amounts for the study of their physiologic effects and therapeutic uses. The problem of chemical synthesis was complicated by uncertainties regarding the stability and chemistry of PG's, and the existence of three different families (PG_1's, PG_2's and PG_3's) each consisting of several members. The first total synthesis of the principal PG's, demonstrated by 1967,[30,31] made available the important members of the first family (PGA_1, PGE_1 and $PGF_{1\alpha}$) and allowed an evaluation of their chemical properties which facilitated the design of a general synthetic route to all of the prostaglandins. This general synthesis of prostaglandins provided access to all PG's from a single intermediate, commonly known as the Corey lactone aldehyde.[32,33] In various forms this flexible synthesis has been used by laboratories all over the world to prepare not only naturally occurring PG's but also countless structural analogs on any scale.[34]

The original version of the 1969 general synthesis of PG's is summarized briefly in Chart VIII. The bicycloheptenone 23 was synthesized stereospecifically by a novel Cu(II) catalyzed Diels-Alder reaction followed by alkaline hydrolysis of the resulting adduct. Alkaline peroxide converted 23 to the hydroxy acid 24 which was readily resolved using (+)-ephedrine. Lactonization and functional group interchange operations transformed 24 into the Corey lactone aldehyde 25, a versatile precursor of all of the PG's and analogs thereof. Enone 26 (Am = C_5H_{11}), produced stereospecifically from 25 by Horner-Emmons coupling, upon reduction with zinc borohydride generated the required 15-(S)-alcohol along with the 15-(R)-diastereomer which was separated and recycled via 26 to the 15-(S)-alcohol. Protection of the hydroxyl groups at C(11) and C(15) afforded the corresponding bistetrahydropyranyl (bis THP) ether, 27. Reduction of the lactone function of 27 to lactol (R_2AlH) and Wittig coupling produced 28 stereospecifically. Acidic hydrolysis of 28 afforded $PGF_{2\alpha}$ (30), whereas oxidation of 28 followed by hydrolysis gave PGE_2(29). Hydrogenation of the 5,6-double bond in 28 followed by these same final steps produced $PGF_{1\alpha}$ and PGE_1. A parallel series of transformations was used to convert 25 to $PGE_{3\alpha}$ and PGE_3.

Although the 1969 bicycloheptenone route to PG's was highly effective

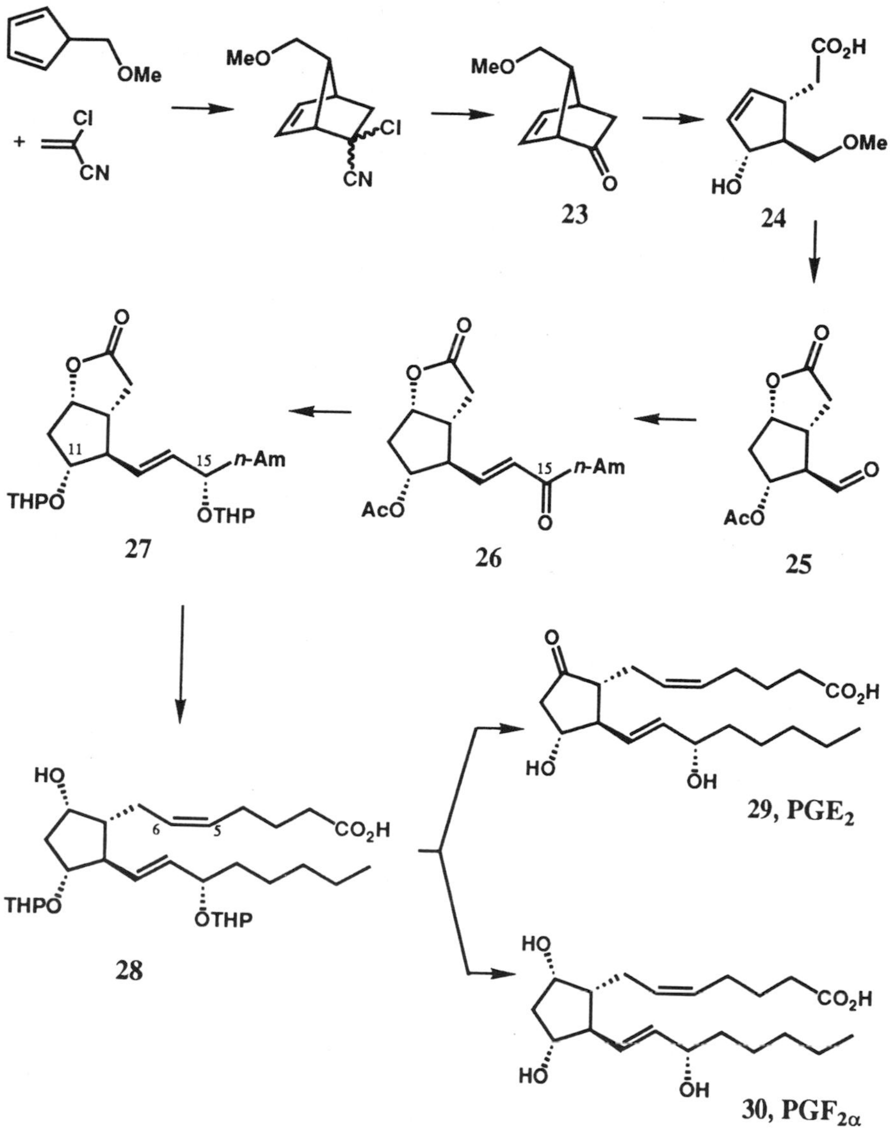

Chart VIII

for the synthesis on a large scale, it was not the ultimate. The Diels-Alder route to **23** produced racemic material which in turn necessitated the resolution of hydroxy acid **24**. Another problem was the lack of stereospecificity in the reduction of the C(15) keto group of **26**. Both of these limitations were overcome by the invention of novel methodology which has simultaneously opened up large new areas of synthetic endeavor.

The problem of controlling the stereochemistry of reduction of the 15-keto group in **26** was solved in a number of different ways (Chart IX). First, the use of a bulky trialkylborohydride reagent with a suitably chosen "controller" group at the C(11) oxygen, for example phenylcarbamoyl, resulted

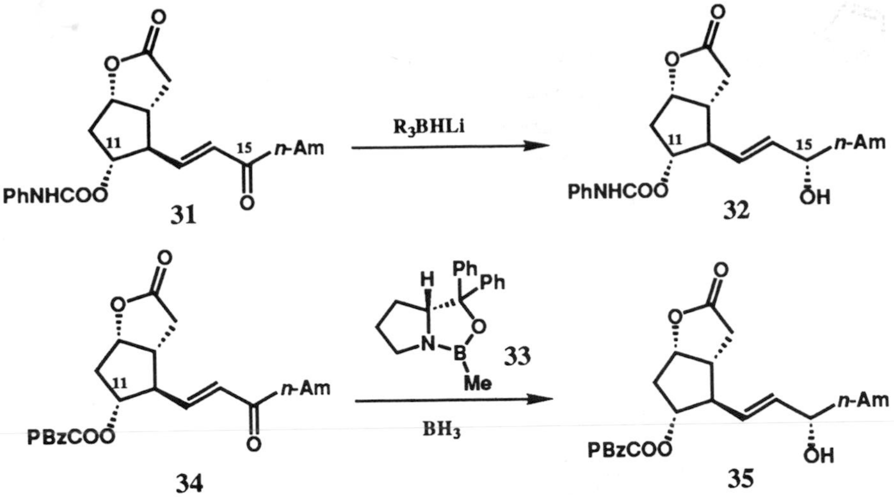

Chart IX

in reduction of C(15) to give the required 15-(S) product with greater than 10:1 diastereoselectivity.[33,35] Further, the small amount of 15-(R) by-product was easily separated for recycling. Second, using the chiral catalyst **33** (10 mole %) and borane (0.6 mole equivalent) in tetrahydrofuran as solvent at ambient temperature, the 15-ketone **34** was reduced to the 15-(S) alcohol **35** with 9:1 diastereoselectivity.[33,36]

Oxazaborolidine **33** is a remarkable catalytic reagent. It controls the absolute stereochemistry of reduction of a large variety of ketones in addition to accelerating the rate of reduction by borane. The absolute stereochemistry of the reduction product of an achiral ketone R_SCOR_L, where R_S is smaller than R_L, is predictable.[37,38] The observed catalysis and enantioselectivity are in accord with the mechanism outlined in Chart X. The catalyst **33** has been shown to complex with borane stereospecifically to give a species which is activated for binding to the ketonic substrate. Complexation at the sterically more available lone pair of the carbonyl oxygen and internal hydride transfer then leads to the observed enantiomeric secondary alcohol. In this mechanism the reagent **33** literally acts like a *molecular robot:* It first picks up and holds one of the reactants, BH_3, and becomes activated toward the other. It attaches to the ketone in a precise three dimensional assembly that facilitates a transfer of hydrogen between the two reactants to form a *specific* enantiomer of the reduction product. The molecular robot finally discharges these products and repeats the reaction cycle. Such action by the small molecule **33** resembles catalysis by enzymes, which can also be regarded as molecular robots. Of course, **33** lacks the substrate-shape and size discrimination of enzymes because it is too small to possess a binding pocket or distal multicontact recognition sites.

The action of the molecular robot **33** represents a new direction for chemical synthesis. In the past, most synthetic constructions have depended

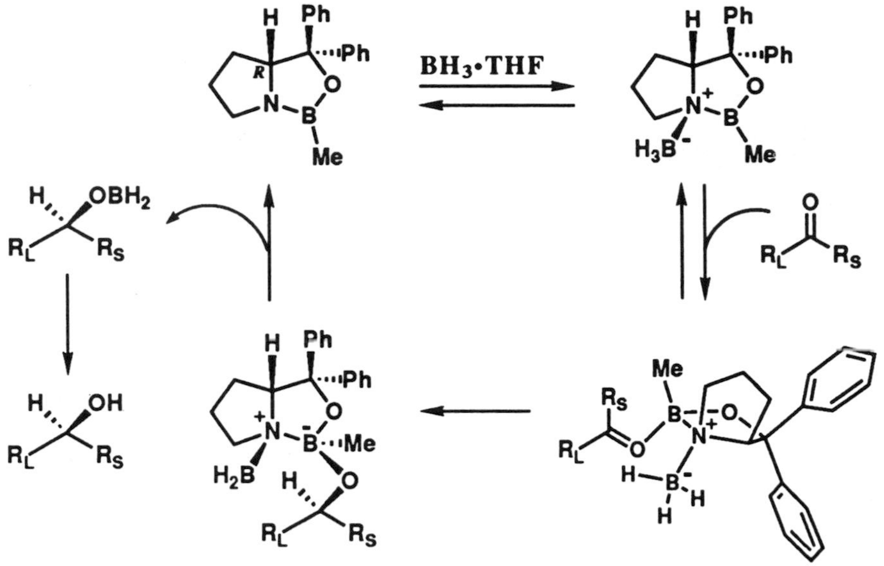

Chart X

on pairwise collisions between reactants without the help of a robot-like assembler. It is likely that synthetic chemistry will produce many new molecular robots in the future as a part of its advance to greater heights.

Two solutions were developed for the enantioselective synthesis of bicycloheptenone **23** (Chart VIII) of the correct chirality for the production of natural prostaglandins. In the first of these, a stereocontrolled aluminium chloride-catalyzed Diels-Alder reaction between benzyloxymethylcyclopentadiene (**37**) (Chart XI) and the acrylate ester of 8-phenylmenthol (**36**) was used with the result that the required adduct **38** was formed with very high (32:1) enantioselectivity.[39] Adduct **38** was converted via ketone **23** to iodo lactone **39** which was obtained in enantiomerically pure form in high yield by a single recrystallization and converted to the standard PG intermediate **40**. In addition, 8-phenylmenthol was recovered efficiently. The enantioselective formation of **38** can be understood from the geometry shown for the complex **36●AlCl₃** and the steric screening by phenyl of the *si* (rear) face of the acrylate α,β-double bond. The efficiency of the 8-phenylmenthol controller stimulated the development of other controllers (chiral auxiliaries) for use in enantioselective synthesis.[40]

More recently, this achievement has been surpassed by the development of a *molecular robot* which assembles the achiral components, as shown in Chart XII, to give the required Diels-Alder adduct in 94% yield and almost 50:1 enantioselectivity.[38,41] Conversion to enantiomerically pure iodolactone **39** was accomplished using standard procedures.[42]

The general synthesis of prostaglandins by the 1969 pathway can now be carried out efficiently and with total stereochemical control, in a way that could not have been foreseen twenty years ago. Such progress augurs well

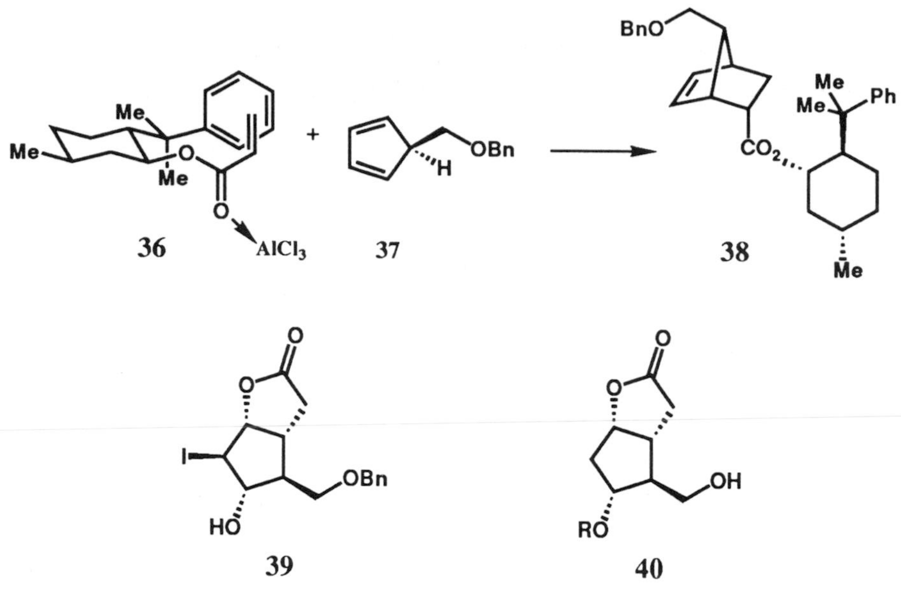

Chart XI

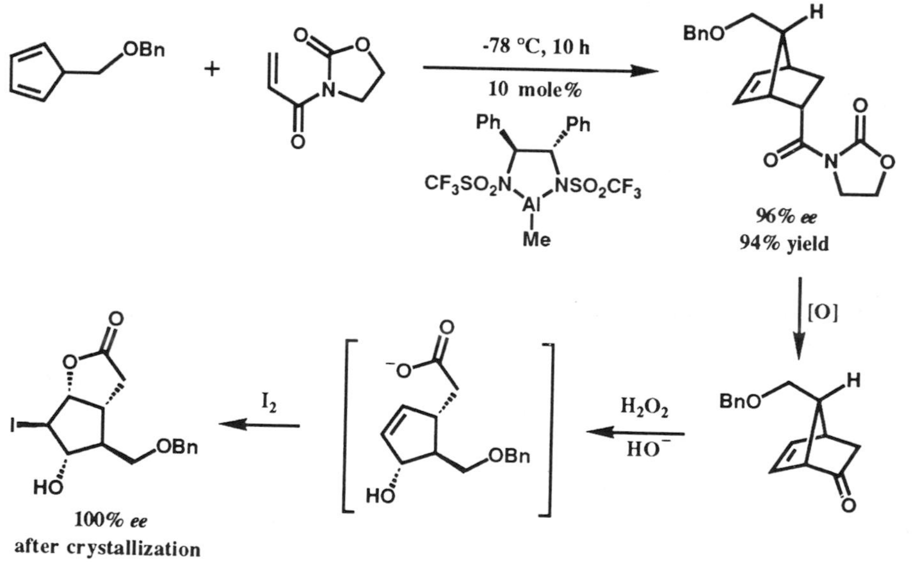

Chart XII

for the future of chemical synthesis. It is not unlikely that today's chemical synthesis, magnificent as it may now appear, will prove to be rudimentary as compared to that of the next century.

The trail of research which originated with the synthesis of prostaglandins was followed for more than two decades, eventually including the synthesis

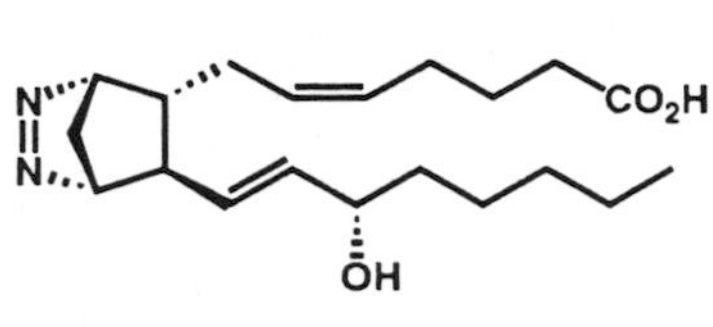

41

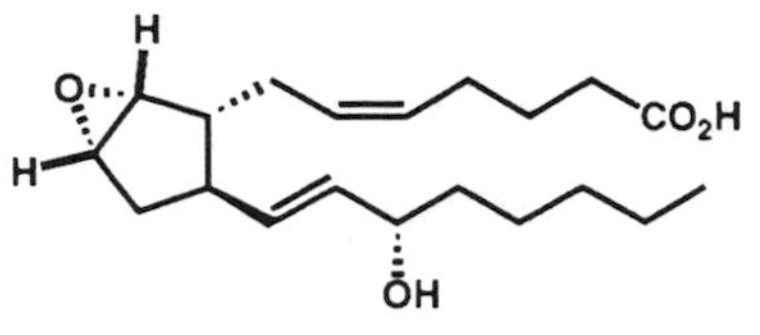

42

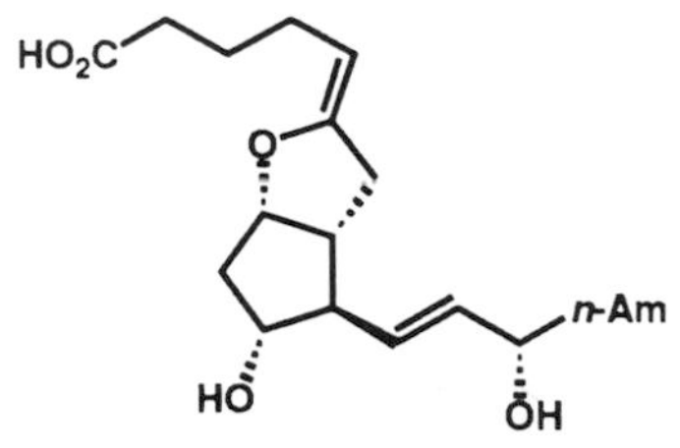

43, PGI$_2$

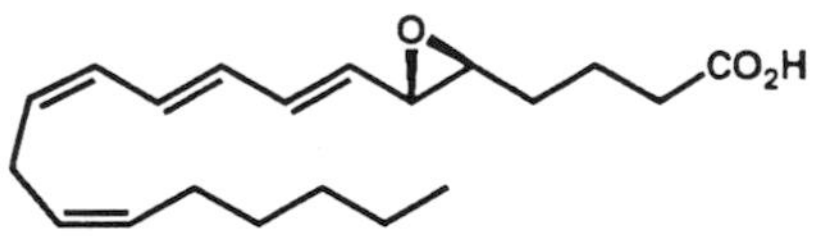

44, LTA$_4$

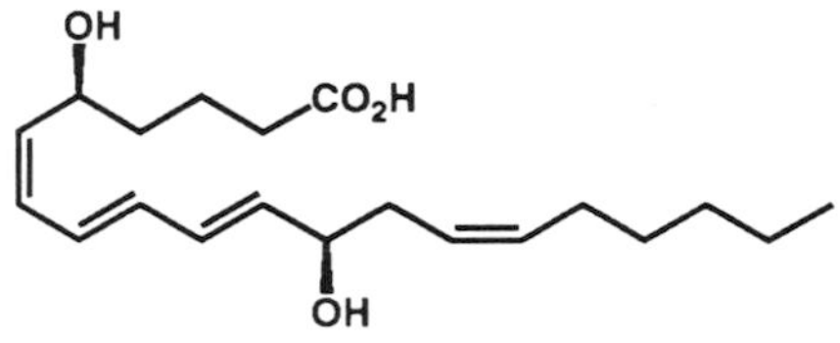

46, LTB$_4$

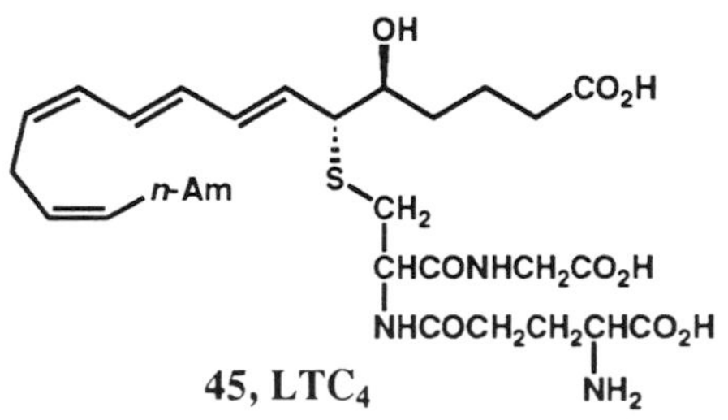

45, LTC$_4$

Chart XIII

of many other members of the eicosanoid (twenty-carbon) class of mammalian cell regulators. A few of the highlights of this program deserve mention. The biosynthesis of prostaglandins occurs by the oxidative conversion of the C$_{20}$ unsaturated acid arachidonic acid to the bicyclic endoperoxide PGH$_2$, which serves as a precursor not only of PGE$_2$ and PGF$_{2\alpha}$, but also of PGI$_2$ and thromboxane A$_2$.[43] We were able to synthesize a stable, active azo analog (**41**) of the unstable PGH$_2$[44] (Chart XIII) and stable, active analogs (e.g. **42**) of the unstable thromboxane A$_2$,[45] as well as PGI$_2$ itself (**43**).[46] In 1977 we suggested the structure of the unstable eicosanoid which is known as leukotriene A$_4$ (LTA$_4$) (**44**) and, by early 1979, had synthesized that structure in advance of isolation from natural sources.[47] Collaborative research between our group and the Karolinska team headed by Bengt Samuelsson established that LTA$_4$ combines with glutathione to form the primary "slow reacting substance," now known as LTC$_4$ (**45**).[47] The detailed stereochemistry of the chemotactic leukotriene LTB$_4$ (**46**) was first established by synthesis.[47, 48] The chemical syntheses of these leukotrienes made these compounds available in quantity for the many hundreds of biological

studies which ensued. Useful new compounds which are active as antago-
nists of $PGF_{2\alpha}$,[49] thromboxane A_2,[50] and LTB_4,[51] and as inhibitors of leukotriene biosynthesis[52] have also emerged from our synthetic program.

It is my hope that our studies in the eicosanoid field[34] will prove to be a
harbinger of future programs in academic synthetic research, since there is
an unparalleled opportunity for the application of chemical synthesis to
biological and medical problems at a fundamental level.

Acknowledgement
It gives me great pleasure to express my indebtedness to many individuals
who deserve credit for the accomplishments summarized herein: my teachers at the Lawrence High School and the Massachusetts Institute of Technology, my colleagues at the University of Illinois and Harvard University,
my wonderful collaborators in research, and my very dear family. Research
grants from the National Science Foundation, the National Institutes of
Health, Pfizer Inc. and numerous other donors have provided the necessary
financial support.

REFERENCES

 1. N. Anand, J. S. Bindra, and S. Ranganathan, *Art in Organic Synthesis* (Holden-
    Day, Inc., San Francisco, first edition, 1970). This book contains a summary of
    these and other noteworthy syntheses of the period 1940 – 1970.
 2. R. B. Woodward, *Les Prix Nobel en 1965,* (Almquist and Wiksell, Intl., Stock-
    holm, 1966) p. 192.
 3. E. J. Corey, *Pure and Applied Chem.,* **14,** 19 (1967).
 4. E. J. Corey and W. T. Wipke, *Science,* **166,** 178 (1969).
 5. E. J. Corey, *Quart. Rev. Chem. Soc.,* **25,** 455 (1971).
 6. (a) E. J. Corey, M. Ohno, P. A. Vatakencherry, and R. B. Mitra, *J. Am. Chem.
    Soc.,* **83,** 1251 (1961); (b) *idem. J. Am. Chem. Soc.,* **86,** 478 (1964).
 7. E. J. Corey and X.-M. Cheng, *The Logic of Chemical Synthesis* (John Wiley and
    Sons, Inc., New York, 1989).
 8. This section is essentially taken from ref. 7.
 9. E. J. Corey, R. D. Cramer, III, and W. J. Howe, *J. Am. Chem. Soc.,* **94,** 440
    (1972).
10. Ref. 7, Chapter 6.
11. E. J. Corey, A. K. Long, and S. D. Rubenstein, *Science,* **228,** 408 (1985).
12. E. J. Corey, W. T. Wipke, R. D. Cramer, III, and W. J. Howe, *J. Am. Chem. Soc.,*
    94, 421 (1972).
13. This computer analysis was performed by Mr. John Kappos of our LHASA
    group.
14. H. Röller, K.-H. Dahm, C. C. Sweeley, and B. M. Trost, *Angew. Chem. Int. Ed.
    Engl.,* **6,** 179 (1967).
15. S. S. Tobe and B. Stay, *Adv. Insect. Physiol.,* **18,** 305 (1985).
16. (a) E. J. Corey, J. A. Katzenellenbogen, N. W. Gilman, S. A. Roman, and B. W.
    Erickson, *J. Am. Chem. Soc.,* **90,** 5618 (1968); (b) E. J. Corey, *Bull. Soc. Ent.
    Suisse,* **44,** 87 (1971); (c) Ref. 7, p. 146; (d) E. J. Corey and G. H. Posner, *J. Am.
    Chem. Soc.,* **89,** 3911 (1967).
17. G. H. Posner, *An Introduction to Organic Synthesis Using Organocopper Reagents*
    (John Wiley and Sons, Inc., New York, 1980).
18. E. J. Corey and J. A. Katzenellenbogen, *J. Am. Chem. Soc.,* **91,** 1851 (1969).

19. (a) E. J. Corey and K. C. Nicolaou, *J. Am. Chem. Soc.*, **96**, 5614 (1974); (b) E. J. Corey, K. C. Nicolaou, and L. S. Melvin, Jr., *J. Am. Chem. Soc.*, **97**, 654 (1975); (c) E. J. Corey, D. J. Brunelle, and P. J. Stork, *Tetrahedron Lett.*, 3405 (1976); (d) E. J. Corey and D. J. Brunelle, *Tetrahedron Lett.*, 3409 (1976).

20. E. J. Corey, S. Kim, S.-e. Yoo, K. C. Nicolaou, L. S. Melvin, Jr., D.J. Brunelle, J. R. Falck, E. J. Trybulski, R. Lett, and P. W. Sheldrake, *J. Am. Chem. Soc.*, **100**, 4620 (1978).

21. (a) E. J.Corey and R. H. Wollenberg, *Tetrahedron Lett.*, 4705 (1976); (b) E. J. Corey, R. H. Wollenberg, and D. R. Williams, *Tetrahedron Lett.*, 2243 (1977); E. J. Corey and P. Carpino, *Tetrahedron Lett.*, **31**, 7555, 1990.

22. E. J. Corey and S. Bhattacharyya, *Tetrahedron Lett.*, 3919 (1977).

23. E. J. Corey and B. De, *J. Am. Chem. Soc.*, **106**, 2735 (1984).

24. (a) E. J. Corey and W.-g. Su, *J. Am. Chem. Soc.*, **109**, 7534 (1987); (b) E. J. Corey and W.-g. Su, *Tetrahedron Lett.*, **29**, 3423 (1988).

25. (a) E. J. Corey, W.-g. Su, and I. N. Houpis, *Tetrahedron Lett.*, **27**, 5951 (1986); (b) E. J. Corey and W.-g. Su, *Tetrahedron Lett.*, **28**, 5241 (1987).

26. (a) Ref. 7, pp. 116–123; (b) ref. 7, pp. 128–133; (c) ref. 7, pp. 104–107; (d) ref. 7, pp. 205–211; (e) ref. 7, pp. 84–85; (f) ref. 7, pp. 212–214; (g) ref. 7, pp. 230–233; (h) ref. 7, pp. 86–87, 178–179; (i) ref. 7, pp. 83–84, 136–137; (j) ref. 7, pp. 237–238; (k) ref. 7, pp. 234–236; (l) ref. 7, pp. 89–91; 221–226.

27. E. J. Corey, *Chem. Soc. Rev.*, **17**, 111 (1988).

28. S. Bergström, *Science*, **157**, 382 (1967).

29. J. A. Oats, G. A. Fitzgerald, R. A. Branch, E. K. Jackson, H. R. Knapp, and L. J. Roberts, II, *New Eng. J. Med.*, **319**, 689 (1988).

30. (a) E. J. Corey, N. H. Andersen, R. M. Carlson, J. Paust, E. Vedejs, I. Vlattas, and R. E. K. Winter, *J. Am. Chem. Soc.*, **90**, 3245 (1968); (b) E. J. Corey, I. Vlattas, N. H. Andersen, and K. Harding, *J. Am. Chem. Soc.*, **90**, 3247 (1968); (c) E. J. Corey, I. Vlattas, and K. Harding, *J. Am. Chem. Soc.*, **91**, 235 (1969); (d) E. J. Corey, *Ann. New York Acad. Sci.*, **180**, 24 (1971).

31. Ref. 7, pp. 250–254.

32. (a) E. J. Corey, N. M. Weinshenker, T. K. Schaaf, and W. Huber, *J. Am. Chem. Soc.*, **91**, 5675 (1969); (b) E. J. Corey, T. K. Schaaf, W. Huber, U. Koelliker, and N. M. Weinshenker, *J. Am. Chem. Soc.*, **92**, 397 (1970); (c) E. J. Corey, R. Noyori, and T. K. Schaaf, *J. Am. Chem. Soc.*, **92**, 2586 (1970).

33. Ref. 7, pp. 255–296.

34. For another general account of this project and subsequent studies on eicosanoids, see E. J. Corey, Japan Prize in Science for 1989, Annual Report of the Science and Technology Foundation of Japan, pp. 95–109, 1989.

35. E. J. Corey, S. M. Albonico, U. Koelliker, T. K. Schaaf, and R. K. Varma, *J. Am. Chem. Soc.*, **93**, 1491 (1971).

36. E. J. Corey, R. K. Bakshi, S. Shibata, C.-P. Chen, and V. K. Singh, *J. Am. Chem. Soc.*, **109**, 7925 (1987).

37. E. J. Corey, R. K. Bakshi, and S. Shibata, *J. Am. Chem. Soc.*, **109**, 555, (1987).

38. E. J. Corey, *Pure and Appl. Chem.*, **62**, 1209 (1990).

39. E. J. Corey and H. E. Ensley, *J. Am. Chem. Soc.*, **97**, 6908 (1975).

40. (a) W. Oppolzer, *Angew. Chem. Int. Ed. Engl.*, **23**, 876 (1984); (b) G. Helmchen, R. Karge, and J. Weetman, *Modern Synthetic Methods* Vol. 4 (Springer-Verlag, Berlin, 1986), R. Scheffold, ed., p. 261.

41. E. J. Corey, R. Imwinkelried, S. Pikul, and Y. B. Xiang, *J. Am. Chem. Soc.*, **111**, 5493 (1989).

42. E. J. Corey and N. Imai, *Tetrahedron Lett,* in press (1991).

43. N. A. Nelson, R. C. Kelly, and R. A. Johnson, *Chem. Eng. News*, **60**, 30 (1982).

44. (a) E. J. Corey, K. C. Nicolaou, Y. Machida, C. L. Malmsten, and B. Samuelsson,

Proc. Nat. Acad. Sci. USA, **72**, 3355 (1975); (b) E. J. Corey, K. Narasaka, and M. Shibasaki, *J. Am. Chem. Soc.*, **98**, 6417 (1976).

45. (a) T. K. Schaaf, D. L. Bussolotti, M. J. Parry, and E. J. Corey, *J. Am. Chem. Soc.*, **103**, 6502 (1981); (b) E. J. Corey and W.-g. Su, *Tetrahedron Lett.*, **31**, 2677 (1990).

46. E. J. Corey, G. E. Keck, and I. Székely, *J. Am. Chem. Soc.*, **99**, 2006 (1977); (b) E. J. Corey, H. L. Pearce, I. Székely, and M. Ishiguro, *Tetrahedron Lett.*, 1023 (1978).

47. (a) E. J. Corey, *Experientia*, **38**, 1259 (1982); (b) Ref. 7, pp. 312−317; (c) E. J. Corey, Y. Arai, and C. Mioskowski, *J. Am. Chem. Soc.*, **101**, 6748 (1979); (d) E. J. Corey, D. A. Clark, G. Goto, A. Marfat, C. Mioskowski, B. Samuelsson, S. Hammarström, *J. Am. Chem. Soc.*, **102**, 1436, 3663 (1980).

48. (a) E. J. Corey, A. Marfat, G. Goto, and F. Brion, *J. Am. Chem. Soc.*, **102**, 7984 (1980); (b) E. J. Corey, A. Marfat, J. E. Munroe, K. S. Kim, P.B. Hopkins, and F. Brion, *Tetrahedron Lett.*, **22**, 1077 (1981).

49. R. B. Stinger, T. M. Fitzpatrick, E. J. Corey, P. W. Ramwell, J. C. Rose, and P. A. Kot, *J. Pharm. Exp. Ther.*, **220**, 521 (1982).

50. E. J. Corey and W.-g. Su, *Tetrahedron Lett.*, **31**, 3833 (1990).

51. H. J. Showell, I. G. Otterness, A. Marfat, and E. J. Corey, *Biochem. Biophys. Res. Commun.*, **106**, 741 (1982).

52. Ref. 7, pp. 345−352.